# PERSPECTIVES ON SUPERSYMMETRY II

## ADVANCED SERIES ON DIRECTIONS IN HIGH ENERGY PHYSICS

*Published*

Vol. 1 – High Energy Electron–Positron Physics (*eds. A. Ali and P. Söding*)
Vol. 2 – Hadronic Multiparticle Production (*ed. P. Carruthers*)
Vol. 3 – CP Violation (*ed. C. Jarlskog*)
Vol. 4 – Proton–Antiproton Collider Physics (*eds. G. Altarelli and L. Di Lella*)
Vol. 5 – Perturbative QCD (*ed. A. Mueller*)
Vol. 6 – Quark–Gluon Plasma (*ed. R. C. Hwa*)
Vol. 7 – Quantum Electrodynamics (*ed. T. Kinoshita*)
Vol. 9 – Instrumentation in High Energy Physics (*ed. F. Sauli*)
Vol. 10 – Heavy Flavours (*eds. A. J. Buras and M. Lindner*)
Vol. 11 – Quantum Fields on the Computer (*ed. M. Creutz*)
Vol. 12 – Advances of Accelerator Physics and Technologies (*ed. H. Schopper*)
Vol. 13 – Perspectives on Higgs Physics (*ed. G. L. Kane*)
Vol. 14 – Precision Tests of the Standard Electroweak Model (*ed. P. Langacker*)
Vol. 15 – Heavy Flavours II (*eds. A. J. Buras and M. Lindner*)
Vol. 16 – Electroweak Symmetry Breaking and New Physics at the TeV Scale (*eds. T. L. Barklow, S. Dawson, H. E. Haber and J. L. Siegrist*)
Vol. 17 – Perspectives on Higgs Physics II (*ed. G. L. Kane*)
Vol. 18 – Perspectives on Supersymmetry (*ed. G. L. Kane*)
Vol. 19 – Linear Collider Physics in the New Millennium (*eds. K. Fujii, D. J. Miller and A. Soni*)
Vol. 20 – Lepton Dipole Moments (*eds. B. Lee Roberts and William J. Marciano*)

*Forthcoming*

Vol. 8 – Standard Model, Hadron Phenomenology and Weak Decays on the Lattice (*ed. G. Martinelli*)

Advanced Series on
Directions in High Energy Physics — Vol. 21

# PERSPECTIVES ON SUPERSYMMETRY II


Editor

**Gordon L Kane**

University of Michigan

NEW JERSEY • LONDON • SINGAPORE • BEIJING • SHANGHAI • HONG KONG • TAIPEI • CHENNAI

*Published by*

World Scientific Publishing Co. Pte. Ltd.
5 Toh Tuck Link, Singapore 596224
*USA office:* 27 Warren Street, Suite 401-402, Hackensack, NJ 07601
*UK office:* 57 Shelton Street, Covent Garden, London WC2H 9HE

**British Library Cataloguing-in-Publication Data**
A catalogue record for this book is available from the British Library.

**PERSPECTIVES ON SUPERSYMMETRY II**
**Advanced Series on Directions in High Energy Physics — Vol. 21**



ISBN-13 978-981-4307-48-2
ISBN-10 981-4307-48-3
ISBN-13 978-981-4307-49-9 (pbk)
ISBN-10 981-4307-49-1 (pbk)

# Introduction

Supersymmetry is a hypothetical symmetry of the basic laws of nature. It proposes the surprising idea that the basic laws are symmetric under interchanging bosons and fermions in the appropriate manner. We do not yet know for sure that nature is actually supersymmetric at the scale of weak interactions, but there is considerable indirect evidence that it is.

One of the remarkable things about supersymmetry is that it was not invented for typical reasons. Most physical theories are invented in response to a puzzle, or to explain data, or to improve the consistency of previous theories. Yang-Mills theories are one exception – they were an extension of Abelian gauge theories to non-Abelian ones, and have turned out to be of great importance for understanding nature. Supersymmetry was also an extension, found in the early 1970's both by examining certain field theories and also by trying to understand if a boson-fermion symmetry could be consistent with relativistic quantum field theory. The key point for our purposes is that supersymmetry was not invented to solve any puzzle or explain any observation.

During the decade after the discovery of supersymmetry, it was slowly realized by a number of people that it had the potential to solve an astonishing number of major physical problems. Even though it was not invented to do so, supersymmetry could provide a solution to the hierarchy problem (this and other terminology are clearly explained in Steve Martin's fine pedagogical Chapter); provided a derivation of the Higgs mechanism (which was technically satisfactory for the Standard Model (SM) but lacked a physical basis there) and in the process successfully predict a heavy top quark; allowed the unification of the SM forces (that supersymmetry was needed to achieve this was confirmed in the early 1990's, a decade after it was predicted); provided a connection between the SM forces and gravity; and provided a candidate for the cold dark matter of the universe (the lightest superpartner (LSP)) before astronomy demonstrated the need for non-baryonic cold dark matter. All of this was in place by the early 1980's.

More recently, it has been recognized that the scalar potential of supersymmetry, the potential energy from the scalars and their interactions, could provide the potential energy that determines the course of inflation(s), and the scalars themselves can be the inflaton(s) (see Randall's chapter). The parameters in that scalar potential also determine how scalars will behave at colliders, and can affect neutrino masses, baryogenesis and other phenomena. The study of fermion masses becomes a serious research problem in a supersymmetric framework since the Yukawa couplings are contained in the superpotential, and supersymmetry affects our approach to proton decay, baryogenesis, and CP violation. It also suggests approaches to understanding the cosmological constant. Given all of these successes and opportunities, it is clear why so many physicists expect the world to be supersymmetric.

Is nature actually supersymmetric on the electroweak scale? The way people will finally be convinced is by the explicit detection of superpartners of the SM particles. Contrary to what is sometimes stated, it is not surprising that superpartners have not yet been detected. First, theoretically all particles can get masses from two sources, the breaking of the electroweak symmetry and the breaking of supersymmetry. One set of particles only gets mass from the electroweak breaking, and all of those except one (the Higgs boson) have been observed. The masses of all those particles would vanish if the electroweak symmetry were restored. It is entirely reasonable that the particles that get mass from both sources should be somewhat heavier and not yet observed. The one exception, the Higgs boson, is very difficult to produce and to detect at hadron colliders — it should be observed at FNAL in 2011 if the machine and detectors function well, and also after several $\text{fb}^{-1}$ at the LHC, where it is a problem of statistics in several channels. Second, phenomenologically there are no general bounds on superpartners masses that make the non-colored ones heavier than about $M_Z$, or colored ones heavier than about $M_{\text{top}}$, so searches have not yet reached levels where one might have expected to find the superpartners. (All published limits depend on guesses for parameters and on models.) The Chapter by Feng, Grivaz, and Nachtman summarizes recent limits.

It would be very nice if clean, unambiguous experimental signals could appear one day. But a little thought tells us that is unlikely — probably impossible. Consider colliders. What will probably happen is that as energy and/or luminosity increases at LHC, a few events of superpartner production will occur. Perhaps such events have already occurred at the Tevatron collider. Each event has two escaping LSP's, so it is never possible to find a dramatic $Z$-like two body peak, or even a $W$-like peak with one escaping particle. Even worse, often several channels look similar to detectors so simple features can be obscured. And usually there are SM processes that can fake any particular signature, as well as ways to fake signatures because detectors are imperfect.

Thus to make progress it is essential to proceed with limited amounts of incomplete information. Without theory input it is entirely possible that signals would not be noticed, hidden under backgrounds since there was no guide to what cuts to use. Further, a particular signal might be encouraging but not convincing — only when combined with other signals that were related by the theory but not directly experimentally could a strong case be made.

The first goal, of course, is to establish that superpartners indeed exist at the EW scale. Given that, then the goal becomes the mesurement of the parameters of the Lagrangian, so that theories implying a particular Lagrangian can be tested, and the form the Lagrangian takes will suggest how SUSY is broken. Measuring the basic parameters may be feasible at LHC but will be very difficult if possible.

The way superpartners will behave — particularly their experimental signatures — is largely determined by the quantum numbers of the LSP. Theoretical and phenomenological arguments probably combine to imply the LSP is either the lightest neutralino (the superpartners of $\gamma$, $Z$, and the neutral Higgs fields of supersymmetry can quantum mechanically mix — the mass eigenstates are "neutralinos") or the gravitino. Almost all of the studies of supersymmetric models can be classified according to what is the LSP. One category is "unstable LSP"; see Dreiner's chapter. There could be a light gravitino LSP ($\widetilde{\text{G}}$LSP), or an LSP that is mainly bino ($\widetilde{\text{B}}$LSP), or mainly higgsino ($\tilde{h}$LSP), or mainly wino ($\tilde{\text{w}}$LSP). The implications for SUSY-breaking and experimental signature are very different for the various cases — it will probably be possible to recognize which is being observed once superpartners are detected. ($\widetilde{\text{G}}$LSP corresponds to gauge-mediated SUSY-breaking, and the others gravity-mediated SUSY-breaking.) A wino-LSP world has been theoretically well motivated for over a decade, and phenomenologically encouraging in recent years (see the chapter by Feldman and Kane).

Once superpartners are detected it will be a delightful challenge to extract physics information from the data. As discussed above, often signals will come from several channels. The relationship of what is observed by experimenters to the masses and couplings of the superpartners will be complicated and nonlinear, and the relationship to the parameters of the effective Lagrangian in the electroweak symmetry basis is even more difficult to extract. To connect with theoretical work it is necessary to measure $\tan\beta$, $\mu$, and the soft-breaking masses, which are only indirectly related to the particle masses.

Actual measurements of effects of superpartners will produce cross sections and distributions, excesses of events with some set of particles and perhaps with missing energy. They do not produce direct measurements of the masses or couplings of superpartners, and determination of the soft-breaking parameters or $\mu$ or $\tan\beta$ is even less likely. Most distributions get contributions from several processes as well. How can we proceed to extract the physics parameters of interest from data in such a nonlinear situation?

Most analyses have proceeded by assuming a model, and perhaps also assuming values for some parameters, until the problem is reduced to a simpler one. That is of use since there is no guarantee a solution can be obtained, so if a consistent solution can be found it may be a physical one.

As data about superpartners is increasingly available, more and more of the parameters of the effective Lagrangians $\mathcal{L}_{EFF}$ of the theory at the electroweak scale will be measured. Constraints from rare decays, CP violation, baryogenesis etc. will be included. Then, assuming the theory to be perturbative to the scale where the gauge coupling unify, the effective Lagrangian at that scale will be deduced by using renormalization group equations. It is not necessary (nor expected) that there be a desert in between, but only that the theory be perturbative. Intermediate matter and scales are expected. There will be consistency checks that allow the perturbativity to be confirmed. Since constraints occur at both ends it is not just an extrapolation. Unification may or may not involve a unified gauge group.

String theory, on the other hand, is naturally around or somewhat above the unification scale. If the way to select the vacuum was known, and also how SUSY was broken (perhaps once the vacuum is known the latter will be determined), then the $\mathcal{L}_{EFF}$ could be predicted, and compared with the $\mathcal{L}_{EFF}$ deduced from data. In practice I expect it to be the other way, as it has been throughout the history of physics — once we know the experimental $\mathcal{L}_{EFF}$ deduced from data the patterns of parameters will be recognizable and will tell someone how SUSY is broken and how the vacuum is selected. After that it will be possible to derive it from string theory. String theorists and sphenomenologists will meet at the unification scale.

The present book is designed to fill several gaps in a coherent way. Anyone with a basic grasp of the Standard Model of particle physics can learn to understand the achievements and goals and issues of supersymmetry from the initial fine chapters of S. Martin, and of K. Dienes and C. Kolda. Supersymmetry is an active, ongoing area of research. The chapter from Dienes and Kolda is a masterful introduction to most of the open theoretical questions in supersymmetry — it can be viewed as a list of opportunities for theoretical researchers.

A number of chapters focus on perhaps the most important question at present — how do we detect the superpartners if they are there. Others look in more detail at theoretical questions. Some focus on the connections of experiment and theory, or aspects of the connection of unification or Planck scale theory with low energy data. All of the chapters have introductions that are useful to non-experts, and references that point to places to learn more, as well as summaries of the current status of their subject.

Some topics are not included in this book because they are well covered recently in what can be thought of as a companion volume, "Perspectives on Higgs Physics II", edited for World Scientific. These include the supersymmetric Higgs sector properties and ways to detect the supersymmetric Higgs bosons, the general upper limit of about 180 GeV on the mass of the lightest Higgs boson that is the only quantitative upper limit that is a test of supersymmetry, the role of supersymmetry in electroweak baryogenesis, and more.

Supersymmetry is an adolescent field even though there are thousands of papers. The situation with supersymmetry today is actually quite analogous to the situation of the Standard Model in 1973. The theory was attractive and convincing, and all experimental evidence was indirect. Then within a short time the Standard Model was in place and broadly accepted. The field of supersymmetry is similarly poised to move rapidly once data on superpartners begins to come. The initial challenge will be to go from the raw data to measure the basic parameters of the theory.

Then the central problem of the field is to understand how supersymmetry is broken. The result of the breaking is a set of soft-breaking masses and interaction parameters. Measuring these soft parameters (which are not the masses of the physical particles) will be the central problem for experimenters. Soon after these are measured perhaps their pattern will be recognized as one generated by a particular physical mechanism. Then the cosmological implications of supersymmetry and its connections to formulating a fundamental Planck scale theory will become the central topics of a mature field.

If supersymmetry is indeed discovered, it may be extraordinarily important for our understanding of the underlying laws of nature. The Standard Model describes the world we see, and is the elegant and powerful synthesis of four centuries of physical science. Supersymmetry takes us beyond the Standard Model, to new opportunities to grasp deeper aspects of nature. It lets us join the collider or electroweak scale, where experiments are done, with the Planck scale where the laws of nature are naturally formulated. The discovery of supersymmetry will open a window to let us probe the underlying theory and test it experimentally. In particular, it will allow us to test string theories, and to formulate and test a theory of our string vacuum.

It is a pleasure for me to acknowledge many students, postdocs, and colleagues (overlapping categories) for collaboration on supersymmetry research, and for shaping my views. I am particularly grateful to Daniel Feldman for encouraging the publication of this second edition, for discussions, and for help with editing the manuscript.

*Gordon Kane*

## Contents

# A Supersymmetry Primer

Stephen P. Martin

*Department of Physics, Northern Illinois University, DeKalb IL 60115*
*and*
*Fermi National Accelerator Laboratory, P.O. Box 500, Batavia IL 60510*

I provide a pedagogical introduction to supersymmetry. The level of discussion is aimed at readers who are familiar with the Standard Model and quantum field theory, but who have had little or no prior exposure to supersymmetry. Topics covered include: motivations for supersymmetry, the construction of supersymmetric Lagrangians, supersymmetry-breaking interactions, the Minimal Supersymmetric Standard Model (MSSM), $R$-parity and its consequences, the origins of supersymmetry breaking, the mass spectrum of the MSSM, decays of supersymmetric particles, experimental signals for supersymmetry, and some extensions of the minimal framework.

## 1.1. Introduction

The Standard Model of high-energy physics, augmented by neutrino masses, provides a remarkably successful description of presently known phenomena. The experimental frontier has advanced into the TeV range with no unambiguous hints of additional structure. Still, it seems clear that the Standard Model is a work in progress and will have to be extended to describe physics at higher energies. Certainly, a new framework will be required at the reduced Planck scale $M_{\rm P} = (8\pi G_{\rm Newton})^{-1/2} = 2.4 \times 10^{18}$ GeV, where quantum gravitational effects become important. Based only on a proper respect for the power of Nature to surprise us, it seems nearly as obvious that new physics exists in the 16 orders of magnitude in energy between the presently explored territory near the electroweak scale, $M_W$, and the Planck scale.

The mere fact that the ratio $M_P/M_W$ is so huge is already a powerful clue to the character of physics beyond the Standard Model, because of the infamous "hierarchy problem".[1] This is not really a difficulty with the Standard Model itself, but rather a disturbing sensitivity of the Higgs potential to new physics in almost any imaginable extension of the Standard Model. The electrically neutral part of the Standard Model Higgs field is a complex scalar $H$ with a classical potential

$$V = m_H^2|H|^2 + \lambda|H|^4 . \tag{1.1}$$

The Standard Model requires a non-vanishing vacuum expectation value (VEV) for $H$ at the minimum of the potential. This will occur if $\lambda > 0$ and $m_H^2 < 0$, resulting in $\langle H \rangle = \sqrt{-m_H^2/2\lambda}$. Since we know experimentally that $\langle H \rangle$ is approximately 174 GeV, from measurements of the properties of the weak interactions, it must be that $m_H^2$ is very roughly of order $-(100$ GeV$)^2$. The problem is that $m_H^2$ receives enormous quantum corrections from the virtual effects of every particle that couples, directly or indirectly, to the Higgs field.

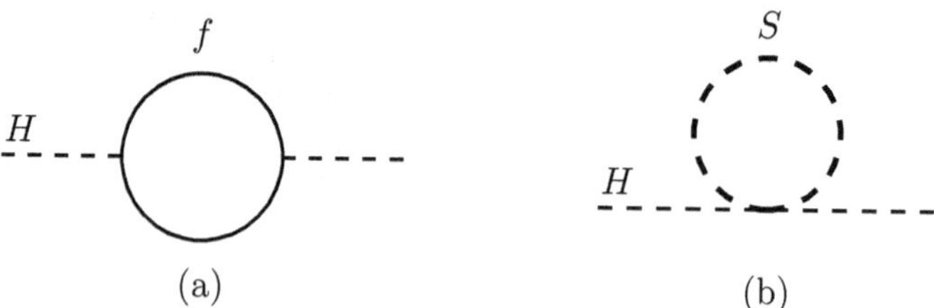


Fig. 1.1. One-loop quantum corrections to the Higgs squared mass parameter $m_H^2$, due to (a) a Dirac fermion $f$, and (b) a scalar $S$.

For example, in Figure 1.1a we have a correction to $m_H^2$ from a loop containing a Dirac fermion $f$ with mass $m_f$. If the Higgs field couples to $f$ with a term in the Lagrangian $-\lambda_f H\overline{f}f$, then the Feynman diagram in Figure 1.1a yields a correction

$$\Delta m_H^2 = -\frac{|\lambda_f|^2}{8\pi^2}\Lambda_{UV}^2 + \dots . \tag{1.2}$$

Here $\Lambda_{UV}$ is an ultraviolet momentum cutoff used to regulate the loop integral; it should be interpreted as at least the energy scale at which new physics enters to alter the high-energy behavior of the theory. The ellipses represent terms proportional to $m_f^2$, which grow at most logarithmically with $\Lambda_{UV}$ (and actually differ for the real and imaginary parts of $H$). Each of the leptons and quarks of the Standard Model can play the role of $f$;

for quarks, eq. (1.2) should be multiplied by 3 to account for color. The largest correction comes when $f$ is the top quark with $\lambda_f \approx 1$. The problem is that if $\Lambda_{\rm UV}$ is of order $M_{\rm P}$, say, then this quantum correction to $m_H^2$ is some 30 orders of magnitude larger than the required value of $m_H^2 \sim -(100\ {\rm GeV})^2$. This is only directly a problem for corrections to the Higgs scalar boson squared mass, because quantum corrections to fermion and gauge boson masses do not have the direct quadratic sensitivity to $\Lambda_{\rm UV}$ found in eq. (1.2). However, the quarks and leptons and the electroweak gauge bosons $Z^0$, $W^\pm$ of the Standard Model all obtain masses from $\langle H \rangle$, so that the entire mass spectrum of the Standard Model is directly or indirectly sensitive to the cutoff $\Lambda_{\rm UV}$.

One could imagine that the solution is to simply pick a $\Lambda_{\rm UV}$ that is not too large. But then one still must concoct some new physics at the scale $\Lambda_{\rm UV}$ that not only alters the propagators in the loop, but actually cuts off the loop integral. This is not easy to do in a theory whose Lagrangian does not contain more than two derivatives, and higher-derivative theories generally suffer from a failure of either unitarity or causality.[2] In string theories, loop integrals are nevertheless cut off at high Euclidean momentum $p$ by factors $e^{-p^2/\Lambda_{\rm UV}^2}$. However, then $\Lambda_{\rm UV}$ is a string scale that is usually[a] thought to be not very far below $M_{\rm P}$. Furthermore, there are contributions similar to eq. (1.2) from the virtual effects of any arbitrarily heavy particles that might exist, and these involve the masses of the heavy particles, not just the cutoff.

For example, suppose there exists a heavy complex scalar particle $S$ with mass $m_S$ that couples to the Higgs with a Lagrangian term $-\lambda_S |H|^2 |S|^2$. Then the Feynman diagram in Figure 1.1b gives a correction

$$\Delta m_H^2 = \frac{\lambda_S}{16\pi^2}\left[\Lambda_{\rm UV}^2 - 2m_S^2 \ln(\Lambda_{\rm UV}/m_S) + \ldots\right]. \qquad (1.3)$$

If one rejects the possibility of a physical interpretation of $\Lambda_{\rm UV}$ and uses dimensional regularization on the loop integral instead of a momentum cutoff, then there will be no $\Lambda_{\rm UV}^2$ piece. However, even then the term proportional to $m_S^2$ cannot be eliminated without the physically unjustifiable tuning of a counter-term specifically for that purpose. So $m_H^2$ is sensitive to the masses of the *heaviest* particles that $H$ couples to; if $m_S$ is very large, its effects on the Standard Model do not decouple, but instead make it difficult to understand why $m_H^2$ is so small.

[a]Some recent attacks on the hierarchy problem, not reviewed here, are based on the proposition that the ultimate cutoff scale is actually close to the electroweak scale, rather than the apparent Planck scale.

This problem arises even if there is no direct coupling between the Standard Model Higgs boson and the unknown heavy particles. For example, suppose there exists a heavy fermion $F$ that, unlike the quarks and leptons of the Standard Model, has vector-like quantum numbers and therefore gets a large mass $m_F$ without coupling to the Higgs field. [In other words, an arbitrarily large mass term of the form $m_F\overline{F}F$ is not forbidden by any symmetry, including weak isospin $SU(2)_L$.] In that case, no diagram like Figure 1.1a exists for $F$. Nevertheless there will be a correction to $m_H^2$ as long as $F$ shares some gauge interactions with the Standard Model Higgs field; these may be the familiar electroweak interactions, or some unknown gauge forces that are broken at a very high energy scale inaccessible to experiment. In either case, the two-loop Feynman diagrams in Figure 1.2 yield a correction

$$\Delta m_H^2 = C_H T_F \left(\frac{g^2}{16\pi^2}\right)^2 \left[a\Lambda_{\rm UV}^2 + 24 m_F^2 \ln(\Lambda_{\rm UV}/m_F) + \ldots\right], \quad (1.4)$$

where $C_H$ and $T_F$ are group theory factors[b] of order 1, and $g$ is the appropriate gauge coupling. The coefficient $a$ depends on the method used to cut off the momentum integrals. It does not arise at all if one uses dimensional regularization, but the $m_F^2$ contribution is always present with the given coefficient. The numerical factor $(g^2/16\pi^2)^2$ may be quite small (of order $10^{-5}$ for electroweak interactions), but the important point is that these contributions to $\Delta m_H^2$ are sensitive both to the largest masses and to the ultraviolet cutoff in the theory, presumably of order $M_{\rm P}$. The "natural" squared mass of a fundamental Higgs scalar, including quantum corrections, therefore seems to be more like $M_{\rm P}^2$ than the experimentally favored value! Even very indirect contributions from Feynman diagrams with three

Fig. 1.2. Two-loop corrections to the Higgs squared mass parameter involving a heavy fermion $F$ that couples only indirectly to the Standard Model Higgs through gauge interactions.

[b]Specifically, $C_H$ is the quadratic Casimir invariant of $H$, and $T_F$ is the Dynkin index of $F$ in a normalization such that $T_F = 1$ for a Dirac fermion (or two Weyl fermions) in a fundamental representation of $SU(n)$.

or more loops can give unacceptably large contributions to $\Delta m_H^2$. The argument above applies not just for heavy particles, but for arbitrary high-scale physical phenomena such as condensates or additional compactified dimensions.

It could be that there is no fundamental Higgs boson, as in technicolor models, top-quark condensate models, and models in which the Higgs boson is composite. Or it could be that the ultimate ultraviolet cutoff scale is much lower than the Planck scale. These ideas are certainly worth exploring, although they often present difficulties in their simplest forms. But, if the Higgs boson is a fundamental particle, and there really is physics far above the electroweak scale, then we have two remaining options: either we must make the rather bizarre assumption that there do not exist *any* high-mass particles or effects that couple (even indirectly or extremely weakly) to the Higgs scalar field, or else some striking cancellation is needed between the various contributions to $\Delta m_H^2$.

The systematic cancellation of the dangerous contributions to $\Delta m_H^2$ can only be brought about by the type of conspiracy that is better known to physicists as a symmetry. Comparing eqs. (1.2) and (1.3) strongly suggests that the new symmetry ought to relate fermions and bosons, because of the relative minus sign between fermion loop and boson loop contributions to $\Delta m_H^2$. (Note that $\lambda_S$ must be positive if the scalar potential is to be bounded from below.) If each of the quarks and leptons of the Standard Model is accompanied by two complex scalars with $\lambda_S = |\lambda_f|^2$, then the $\Lambda_{\rm UV}^2$ contributions of Figures 1.1a and 1.1b will neatly cancel.[3] Clearly, more restrictions on the theory will be necessary to ensure that this success persists to higher orders, so that, for example, the contributions in Figure 1.2 and eq. (1.4) from a very heavy fermion are canceled by the two-loop effects of some very heavy bosons. Fortunately, the cancellation of all such contributions to scalar masses is not only possible, but is actually unavoidable, once we merely assume that there exists a symmetry relating fermions and bosons, called a *supersymmetry.*

A supersymmetry transformation turns a bosonic state into a fermionic state, and vice versa. The operator $Q$ that generates such transformations must be an anticommuting spinor, with

$$Q|\text{Boson}\rangle = |\text{Fermion}\rangle, \qquad Q|\text{Fermion}\rangle = |\text{Boson}\rangle. \tag{1.5}$$

Spinors are intrinsically complex objects, so $Q^\dagger$ (the hermitian conjugate of $Q$) is also a symmetry generator. Because $Q$ and $Q^\dagger$ are fermionic operators, they carry spin angular momentum 1/2, so it is clear that supersymmetry must be a spacetime symmetry. The possible forms for such symmetries in an interacting quantum field theory are highly restricted by the Haag-Lopuszanski-Sohnius extension of the Coleman-Mandula theorem.[4] For realistic theories that, like the Standard Model, have chiral fermions (i.e., fermions whose left- and right-handed pieces transform differently under the gauge group) and thus the possibility of parity-violating interactions, this theorem implies that the generators $Q$ and $Q^\dagger$ must satisfy an algebra of anticommutation and commutation relations with the schematic form

$$\{Q, Q^\dagger\} = P^\mu, \tag{1.6}$$

$$\{Q, Q\} = \{Q^\dagger, Q^\dagger\} = 0, \tag{1.7}$$

$$[P^\mu, Q] = [P^\mu, Q^\dagger] = 0, \tag{1.8}$$

where $P^\mu$ is the four-momentum generator of spacetime translations. Here we have ruthlessly suppressed the spinor indices on $Q$ and $Q^\dagger$; after developing some notation we will, in section 1.3.1, derive the precise version of eqs. (1.6)-(1.8) with indices restored. In the meantime, we simply note that the appearance of $P^\mu$ on the right-hand side of eq. (1.6) is unsurprising, since it transforms under Lorentz boosts and rotations as a spin-1 object while $Q$ and $Q^\dagger$ on the left-hand side each transform as spin-1/2 objects.

The single-particle states of a supersymmetric theory fall into irreducible representations of the supersymmetry algebra, called *supermultiplets*. Each supermultiplet contains both fermion and boson states, which are commonly known as *superpartners* of each other. By definition, if $|\Omega\rangle$ and $|\Omega'\rangle$ are members of the same supermultiplet, then $|\Omega'\rangle$ is proportional to some combination of $Q$ and $Q^\dagger$ operators acting on $|\Omega\rangle$, up to a spacetime translation or rotation. The squared-mass operator $-P^2$ commutes with the operators $Q$, $Q^\dagger$, and with all spacetime rotation and translation operators, so it follows immediately that particles inhabiting the same irreducible supermultiplet must have equal eigenvalues of $-P^2$, and therefore equal masses.

The supersymmetry generators $Q, Q^\dagger$ also commute with the generators of gauge transformations. Therefore particles in the same supermultiplet must also be in the same representation of the gauge group, and so must have the same electric charges, weak isospin, and color degrees of freedom.

Each supermultiplet contains an equal number of fermion and boson degrees of freedom. To prove this, consider the operator $(-1)^{2s}$ where $s$ is the spin angular momentum. By the spin-statistics theorem, this operator has eigenvalue $+1$ acting on a bosonic state and eigenvalue $-1$ acting on a fermionic state. Any fermionic operator will turn a bosonic state into a fermionic state and vice versa. Therefore $(-1)^{2s}$ must anticommute with every fermionic operator in the theory, and in particular with $Q$ and $Q^\dagger$. Now, within a given supermultiplet, consider the subspace of states $|i\rangle$ with the same eigenvalue $p^\mu$ of the four-momentum operator $P^\mu$. In view of eq. (1.8), any combination of $Q$ or $Q^\dagger$ acting on $|i\rangle$ must give another state $|i'\rangle$ with the same four-momentum eigenvalue. Therefore one has a completeness relation $\sum_i |i\rangle\langle i| = 1$ within this subspace of states. Now one can take a trace over all such states of the operator $(-1)^{2s}P^\mu$ (including each spin helicity state separately):

$$\begin{aligned}\sum_i \langle i|(-1)^{2s}P^\mu|i\rangle &= \sum_i \langle i|(-1)^{2s}QQ^\dagger|i\rangle + \sum_i \langle i|(-1)^{2s}Q^\dagger Q|i\rangle \\ &= \sum_i \langle i|(-1)^{2s}QQ^\dagger|i\rangle + \sum_i\sum_j \langle i|(-1)^{2s}Q^\dagger|j\rangle\langle j|Q|i\rangle \\ &= \sum_i \langle i|(-1)^{2s}QQ^\dagger|i\rangle + \sum_j \langle j|Q(-1)^{2s}Q^\dagger|j\rangle \\ &= \sum_i \langle i|(-1)^{2s}QQ^\dagger|i\rangle - \sum_j \langle j|(-1)^{2s}QQ^\dagger|j\rangle \\ &= 0. \end{aligned} \tag{1.9}$$

The first equality follows from the supersymmetry algebra relation eq. (1.6); the second and third from use of the completeness relation; and the fourth from the fact that $(-1)^{2s}$ must anticommute with $Q$. Now $\sum_i\langle i|(-1)^{2s}P^\mu|i\rangle = p^\mu\ \mathrm{Tr}[(-1)^{2s}]$ is just proportional to the number of bosonic degrees of freedom $n_B$ minus the number of fermionic degrees of freedom $n_F$ in the trace, so that

$$n_B = n_F \tag{1.10}$$

must hold for a given $p^\mu \neq 0$ in each supermultiplet.

The simplest possibility for a supermultiplet consistent with eq. (1.10) has a single Weyl fermion (with two spin helicity states, so $n_F = 2$) and two real scalars (each with $n_B = 1$). It is natural to assemble the two real scalar degrees of freedom into a complex scalar field; as we will see below this provides for convenient formulations of the supersymmetry algebra, Feynman rules, supersymmetry-violating effects, etc. This combination of a two-component Weyl fermion and a complex scalar field is called a *chiral* or *matter* or *scalar* supermultiplet.

The next-simplest possibility for a supermultiplet contains a spin-1 vector boson. If the theory is to be renormalizable, this must be a gauge boson that is massless, at least before the gauge symmetry is spontaneously broken. A massless spin-1 boson has two helicity states, so the number of bosonic degrees of freedom is $n_B = 2$. Its superpartner is therefore a massless spin-1/2 Weyl fermion, again with two helicity states, so $n_F = 2$. (If one tried to use a massless spin-3/2 fermion instead, the theory would not be renormalizable.) Gauge bosons must transform as the adjoint representation of the gauge group, so their fermionic partners, called *gauginos*, must also. Since the adjoint representation of a gauge group is always its own conjugate, the gaugino fermions must have the same gauge transformation properties for left-handed and for right-handed components. Such a combination of spin-1/2 gauginos and spin-1 gauge bosons is called a *gauge* or *vector* supermultiplet.

If we include gravity, then the spin-2 graviton (with 2 helicity states, so $n_B = 2$) has a spin-3/2 superpartner called the gravitino. The gravitino would be massless if supersymmetry were unbroken, and so it has $n_F = 2$ helicity states.

There are other possible combinations of particles with spins that can satisfy eq. (1.10). However, these are always reducible to combinations[c] of chiral and gauge supermultiplets if they have renormalizable interactions, except in certain theories with "extended" supersymmetry. Theories with extended supersymmetry have more than one distinct copy of the supersymmetry generators $Q, Q^\dagger$. Such models are mathematically amusing, but evidently do not have any phenomenological prospects. The reason is that extended supersymmetry in four-dimensional field theories cannot allow for chiral fermions or parity violation as observed in the Standard Model. So we will not discuss such possibilities further, although extended supersymmetry in higher-dimensional field theories might describe the real world if the extra dimensions are compactified in an appropriate way, and extended supersymmetry in four dimensions provides interesting toy models. The ordinary, non-extended, phenomenologically viable type of supersymmetric model is sometimes called $N = 1$ supersymmetry, with $N$ referring to the number of supersymmetries (the number of distinct copies of $Q, Q^\dagger$).

In a supersymmetric extension of the Standard Model,[5–7] each of the known fundamental particles is therefore in either a chiral or gauge super-

[c]For example, if a gauge symmetry were to spontaneously break without breaking supersymmetry, then a massless vector supermultiplet would "eat" a chiral supermultiplet, resulting in a massive vector supermultiplet with physical degrees of freedom consisting of a massive vector ($n_B = 3$), a massive Dirac fermion formed from the gaugino and the chiral fermion ($n_F = 4$), and a real scalar ($n_B = 1$).

multiplet, and must have a superpartner with spin differing by 1/2 unit. The first step in understanding the exciting phenomenological consequences of this prediction is to decide exactly how the known particles fit into supermultiplets, and to give them appropriate names. A crucial observation here is that only chiral supermultiplets can contain fermions whose left-handed parts transform differently under the gauge group than their right-handed parts. All of the Standard Model fermions (the known quarks and leptons) have this property, so they must be members of chiral supermultiplets.[d] The names for the spin-0 partners of the quarks and leptons are constructed by prepending an "s", for scalar. So, generically they are called *squarks* and *sleptons* (short for "scalar quark" and "scalar lepton"), or sometimes *sfermions*. The left-handed and right-handed pieces of the quarks and leptons are separate two-component Weyl fermions with different gauge transformation properties in the Standard Model, so each must have its own complex scalar partner. The symbols for the squarks and sleptons are the same as for the corresponding fermion, but with a tilde ( $\widetilde{}$ ) used to denote the superpartner of a Standard Model particle. For example, the superpartners of the left-handed and right-handed parts of the electron Dirac field are called left- and right-handed selectrons, and are denoted $\widetilde{e}_L$ and $\widetilde{e}_R$. It is important to keep in mind that the "handedness" here does not refer to the helicity of the selectrons (they are spin-0 particles) but to that of their superpartners. A similar nomenclature applies for smuons and staus: $\widetilde{\mu}_L$, $\widetilde{\mu}_R$, $\widetilde{\tau}_L$, $\widetilde{\tau}_R$. The Standard Model neutrinos (neglecting their very small masses) are always left-handed, so the sneutrinos are denoted generically by $\widetilde{\nu}$, with a possible subscript indicating which lepton flavor they carry: $\widetilde{\nu}_e$, $\widetilde{\nu}_\mu$, $\widetilde{\nu}_\tau$. Finally, a complete list of the squarks is $\widetilde{q}_L$, $\widetilde{q}_R$ with $q = u, d, s, c, b, t$. The gauge interactions of each of these squark and slepton fields are the same as for the corresponding Standard Model fermions; for instance, the left-handed squarks $\widetilde{u}_L$ and $\widetilde{d}_L$ couple to the $W$ boson, while $\widetilde{u}_R$ and $\widetilde{d}_R$ do not.

It seems clear that the Higgs scalar boson must reside in a chiral supermultiplet, since it has spin 0. Actually, it turns out that just one chiral supermultiplet is not enough. One reason for this is that if there were only one Higgs chiral supermultiplet, the electroweak gauge symmetry would suffer a gauge anomaly, and would be inconsistent as a quantum theory. This is because the conditions for cancellation of gauge anomalies include

[d] In particular, one cannot attempt to make a spin-1/2 neutrino be the superpartner of the spin-1 photon; the neutrino is in a doublet, and the photon is neutral, under weak isospin.

$\text{Tr}[T_3^2 Y] = \text{Tr}[Y^3] = 0$, where $T_3$ and $Y$ are the third component of weak isospin and the weak hypercharge, respectively, in a normalization where the ordinary electric charge is $Q_{\text{EM}} = T_3 + Y$. The traces run over all of the left-handed Weyl fermionic degrees of freedom in the theory. In the Standard Model, these conditions are already satisfied, somewhat miraculously, by the known quarks and leptons. Now, a fermionic partner of a Higgs chiral supermultiplet must be a weak isodoublet with weak hypercharge $Y = 1/2$ or $Y = -1/2$. In either case alone, such a fermion will make a non-zero contribution to the traces and spoil the anomaly cancellation. This can be avoided if there are two Higgs supermultiplets, one with each of $Y = \pm 1/2$, so that the total contribution to the anomaly traces from the two fermionic members of the Higgs chiral supermultiplets vanishes by cancellation. As we will see in section 1.5.1, both of these are also necessary for another completely different reason: because of the structure of supersymmetric theories, only a $Y = 1/2$ Higgs chiral supermultiplet can have the Yukawa couplings necessary to give masses to charge $+2/3$ up-type quarks (up, charm, top), and only a $Y = -1/2$ Higgs can have the Yukawa couplings necessary to give masses to charge $-1/3$ down-type quarks (down, strange, bottom) and to the charged leptons. We will call the $SU(2)_L$-doublet complex scalar fields with $Y = 1/2$ and $Y = -1/2$ by the names $H_u$ and $H_d$, respectively.[e] The weak isospin components of $H_u$ with $T_3 = (1/2, -1/2)$ have electric charges 1, 0 respectively, and are denoted $(H_u^+, H_u^0)$. Similarly, the $SU(2)_L$-doublet complex scalar $H_d$ has $T_3 = (1/2, -1/2)$ components $(H_d^0, H_d^-)$. The neutral scalar that corresponds to the physical Standard Model Higgs boson is in a linear combination of $H_u^0$ and $H_d^0$; we will discuss this further in section 1.7.1. The generic nomenclature for a spin-1/2 superpartner is to append "-ino" to the name of the Standard Model particle, so the fermionic partners of the Higgs scalars are called higgsinos. They are denoted by $\widetilde{H}_u$, $\widetilde{H}_d$ for the $SU(2)_L$-doublet left-handed Weyl spinor fields, with weak isospin components $\widetilde{H}_u^+$, $\widetilde{H}_u^0$ and $\widetilde{H}_d^0$, $\widetilde{H}_d^-$.

We have now found all of the chiral supermultiplets of a minimal phenomenologically viable extension of the Standard Model. They are summarized in Table 1.1, classified according to their transformation properties under the Standard Model gauge group $SU(3)_C \times SU(2)_L \times U(1)_Y$, which combines $u_L, d_L$ and $\nu, e_L$ degrees of freedom into $SU(2)_L$ doublets. Here we follow a standard convention, that all chiral supermultiplets are defined

[e]Other notations in the literature have $H_1, H_2$ or $H, \overline{H}$ instead of $H_u, H_d$. The notation used here has the virtue of making it easy to remember which Higgs VEVs gives masses to which type of quarks.

Table 1.1. Chiral supermultiplets in the Minimal Supersymmetric Standard Model. The spin-0 fields are complex scalars, and the spin-1/2 fields are left-handed two-component Weyl fermions.

| Names | | spin 0 | spin 1/2 | $SU(3)_C,\, SU(2)_L,\, U(1)_Y$ |
|---|---|---|---|---|
| squarks, quarks | $Q$ | $(\widetilde{u}_L\ \ \widetilde{d}_L)$ | $(u_L\ \ d_L)$ | $(\,\mathbf{3},\ \mathbf{2}\,,\ \frac{1}{6})$ |
| (×3 families) | $\overline{u}$ | $\widetilde{u}^*_R$ | $u^\dagger_R$ | $(\,\overline{\mathbf{3}},\ \mathbf{1},\ -\frac{2}{3})$ |
| | $\overline{d}$ | $\widetilde{d}^*_R$ | $d^\dagger_R$ | $(\,\overline{\mathbf{3}},\ \mathbf{1},\ \frac{1}{3})$ |
| sleptons, leptons | $L$ | $(\widetilde{\nu}\ \widetilde{e}_L)$ | $(\nu\ \ e_L)$ | $(\,\mathbf{1},\ \mathbf{2}\,,\ -\frac{1}{2})$ |
| (×3 families) | $\overline{e}$ | $\widetilde{e}^*_R$ | $e^\dagger_R$ | $(\,\mathbf{1},\ \mathbf{1},\ 1)$ |
| Higgs, higgsinos | $H_u$ | $(H_u^+\ \ H_u^0)$ | $(\widetilde{H}_u^+\ \ \widetilde{H}_u^0)$ | $(\,\mathbf{1},\ \mathbf{2}\,,\ +\frac{1}{2})$ |
| | $H_d$ | $(H_d^0\ \ H_d^-)$ | $(\widetilde{H}_d^0\ \ \widetilde{H}_d^-)$ | $(\,\mathbf{1},\ \mathbf{2}\,,\ -\frac{1}{2})$ |

in terms of left-handed Weyl spinors, so that the *conjugates* of the right-handed quarks and leptons (and their superpartners) appear in Table 1.1. This protocol for defining chiral supermultiplets turns out to be very useful for constructing supersymmetric Lagrangians, as we will see in section 1.3. It is also useful to have a symbol for each of the chiral supermultiplets as a whole; these are indicated in the second column of Table 1.1. Thus, for example, $Q$ stands for the $SU(2)_L$-doublet chiral supermultiplet containing $\widetilde{u}_L, u_L$ (with weak isospin component $T_3 = 1/2$), and $\widetilde{d}_L, d_L$ (with $T_3 = -1/2$), while $\overline{u}$ stands for the $SU(2)_L$-singlet supermultiplet containing $\widetilde{u}^*_R, u^\dagger_R$. There are three families for each of the quark and lepton supermultiplets, Table 1.1 lists the first-family representatives. A family index $i = 1, 2, 3$ can be affixed to the chiral supermultiplet names ($Q_i$, $\overline{u}_i, \ldots$) when needed, for example $(\overline{e}_1, \overline{e}_2, \overline{e}_3) = (\overline{e}, \overline{\mu}, \overline{\tau})$. The bar on $\overline{u}$, $\overline{d}$, $\overline{e}$ fields is part of the name, and does not denote any kind of conjugation.

The Higgs chiral supermultiplet $H_d$ (containing $H_d^0$, $H_d^-$, $\widetilde{H}_d^0$, $\widetilde{H}_d^-$) has exactly the same Standard Model gauge quantum numbers as the left-handed sleptons and leptons $L_i$, for example ($\widetilde{\nu}$, $\widetilde{e}_L$, $\nu$, $e_L$). Naively, one might therefore suppose that we could have been more economical in our assignment by taking a neutrino and a Higgs scalar to be superpartners, instead of putting them in separate supermultiplets. This would amount to the proposal that the Higgs boson and a sneutrino should be the same particle. This attempt played a key role in some of the first attempts to connect supersymmetry to phenomenology,[5] but it is now known to not work. Even ignoring the anomaly cancellation problem mentioned above, many insoluble phenomenological problems would result, including lepton-number non-conservation and a mass for at least one of the neutrinos in

Table 1.2. Gauge supermultiplets in the Minimal Supersymmetric Standard Model.

| Names | spin 1/2 | spin 1 | $SU(3)_C$, $SU(2)_L$, $U(1)_Y$ |
|---|---|---|---|
| gluino, gluon | $\widetilde{g}$ | $g$ | ( $\mathbf{8}$, $\mathbf{1}$ , 0) |
| winos, W bosons | $\widetilde{W}^{\pm}$ $\widetilde{W}^0$ | $W^{\pm}$ $W^0$ | ( $\mathbf{1}$, $\mathbf{3}$ , 0) |
| bino, B boson | $\widetilde{B}^0$ | $B^0$ | ( $\mathbf{1}$, $\mathbf{1}$ , 0) |

gross violation of experimental bounds. Therefore, all of the superpartners of Standard Model particles are really new particles, and cannot be identified with some other Standard Model state.

The vector bosons of the Standard Model clearly must reside in gauge supermultiplets. Their fermionic superpartners are generically referred to as gauginos. The $SU(3)_C$ color gauge interactions of QCD are mediated by the gluon, whose spin-1/2 color-octet supersymmetric partner is the gluino. As usual, a tilde is used to denote the supersymmetric partner of a Standard Model state, so the symbols for the gluon and gluino are $g$ and $\widetilde{g}$ respectively. The electroweak gauge symmetry $SU(2)_L \times U(1)_Y$ is associated with spin-1 gauge bosons $W^+, W^0, W^-$ and $B^0$, with spin-1/2 superpartners $\widetilde{W}^+, \widetilde{W}^0, \widetilde{W}^-$ and $\widetilde{B}^0$, called *winos* and *bino*. After electroweak symmetry breaking, the $W^0$, $B^0$ gauge eigenstates mix to give mass eigenstates $Z^0$ and $\gamma$. The corresponding gaugino mixtures of $\widetilde{W}^0$ and $\widetilde{B}^0$ are called zino ($\widetilde{Z}^0$) and photino ($\widetilde{\gamma}$); if supersymmetry were unbroken, they would be mass eigenstates with masses $m_Z$ and 0. Table 1.2 summarizes the gauge supermultiplets of a minimal supersymmetric extension of the Standard Model.

The chiral and gauge supermultiplets in Tables 1.1 and 1.2 make up the particle content of the Minimal Supersymmetric Standard Model (MSSM). The most obvious and interesting feature of this theory is that none of the superpartners of the Standard Model particles has been discovered as of this writing. If supersymmetry were unbroken, then there would have to be selectrons $\widetilde{e}_L$ and $\widetilde{e}_R$ with masses exactly equal to $m_e = 0.511...$ MeV. A similar statement applies to each of the other sleptons and squarks, and there would also have to be a massless gluino and photino. These particles would have been extraordinarily easy to detect long ago. Clearly, therefore, *supersymmetry is a broken symmetry* in the vacuum state chosen by Nature.

An important clue as to the nature of supersymmetry breaking can be obtained by returning to the motivation provided by the hierarchy problem.

Supersymmetry forced us to introduce two complex scalar fields for each Standard Model Dirac fermion, which is just what is needed to enable a cancellation of the quadratically divergent ($\Lambda_{\rm UV}^2$) pieces of eqs. (1.2) and (1.3). This sort of cancellation also requires that the associated dimensionless couplings should be related (for example $\lambda_S = |\lambda_f|^2$). The necessary relationships between couplings indeed occur in unbroken supersymmetry, as we will see in section 1.3. In fact, unbroken supersymmetry guarantees that the quadratic divergences in scalar squared masses must vanish to all orders in perturbation theory.[f] Now, if broken supersymmetry is still to provide a solution to the hierarchy problem even in the presence of supersymmetry breaking, then the relationships between dimensionless couplings that hold in an unbroken supersymmetric theory must be maintained. Otherwise, there would be quadratically divergent radiative corrections to the Higgs scalar masses of the form

$$\Delta m_H^2 = \frac{1}{8\pi^2}(\lambda_S - |\lambda_f|^2)\Lambda_{\rm UV}^2 + \ldots. \tag{1.11}$$

We are therefore led to consider "soft" supersymmetry breaking. This means that the effective Lagrangian of the MSSM can be written in the form

$$\mathcal{L} = \mathcal{L}_{\rm SUSY} + \mathcal{L}_{\rm soft}, \tag{1.12}$$

where $\mathcal{L}_{\rm SUSY}$ contains all of the gauge and Yukawa interactions and preserves supersymmetry invariance, and $\mathcal{L}_{\rm soft}$ violates supersymmetry but contains only mass terms and coupling parameters with *positive* mass dimension. Without further justification, soft supersymmetry breaking might seem like a rather arbitrary requirement. Fortunately, we will see in section 1.6 that theoretical models for supersymmetry breaking do indeed yield effective Lagrangians with just such terms for $\mathcal{L}_{\rm soft}$. If the largest mass scale associated with the soft terms is denoted $m_{\rm soft}$, then the additional nonsupersymmetric corrections to the Higgs scalar squared mass must vanish in the $m_{\rm soft} \to 0$ limit, so by dimensional analysis they cannot be proportional to $\Lambda_{\rm UV}^2$. More generally, these models maintain the cancellation of quadratically divergent terms in the radiative corrections of all scalar masses, to all orders in perturbation theory. The corrections also cannot

[f] A simple way to understand this is to recall that unbroken supersymmetry requires the degeneracy of scalar and fermion masses. Radiative corrections to fermion masses are known to diverge at most logarithmically in any renormalizable field theory, so the same must be true for scalar masses in unbroken supersymmetry.

go like $\Delta m_H^2 \sim m_{\rm soft}\Lambda_{\rm UV}$, because in general the loop momentum integrals always diverge either quadratically or logarithmically, not linearly, as $\Lambda_{\rm UV} \to \infty$. So they must be of the form

$$\Delta m_H^2 = m_{\rm soft}^2 \left[\frac{\lambda}{16\pi^2} \ln(\Lambda_{\rm UV}/m_{\rm soft}) + \ldots\right]. \tag{1.13}$$

Here $\lambda$ is schematic for various dimensionless couplings, and the ellipses stand both for terms that are independent of $\Lambda_{\rm UV}$ and for higher loop corrections (which depend on $\Lambda_{\rm UV}$ through powers of logarithms).

Because the mass splittings between the known Standard Model particles and their superpartners are just determined by the parameters $m_{\rm soft}$ appearing in $\mathcal{L}_{\rm soft}$, eq. (1.13) tells us that the superpartner masses cannot be too huge. Otherwise, we would lose our successful cure for the hierarchy problem, since the $m_{\rm soft}^2$ corrections to the Higgs scalar squared mass parameter would be unnaturally large compared to the square of the electroweak breaking scale of 174 GeV. The top and bottom squarks and the winos and bino give especially large contributions to $\Delta m_{H_u}^2$ and $\Delta m_{H_d}^2$, but the gluino mass and all the other squark and slepton masses also feed in indirectly, through radiative corrections to the top and bottom squark masses. Furthermore, in most viable models of supersymmetry breaking that are not unduly contrived, the superpartner masses do not differ from each other by more than about an order of magnitude. Using $\Lambda_{\rm UV} \sim M_{\rm P}$ and $\lambda \sim 1$ in eq. (1.13), one finds that $m_{\rm soft}$, and therefore the masses of at least the lightest few superpartners, should be at the most about 1 TeV or so, in order for the MSSM scalar potential to provide a Higgs VEV resulting in $m_W, m_Z = 80.4, 91.2$ GeV without miraculous cancellations. This is the best reason for the optimism among many theorists that supersymmetry will be discovered at the Fermilab Tevatron or the CERN Large Hadron Collider, and can be studied at a future $e^+e^-$ linear collider.

However, it should be noted that the hierarchy problem was *not* the historical motivation for the development of supersymmetry in the early 1970's. The supersymmetry algebra and supersymmetric field theories were originally concocted independently in various disguises[8–11] bearing little resemblance to the MSSM. It is quite impressive that a theory developed for quite different reasons, including purely aesthetic ones, can later be found to provide a solution for the hierarchy problem.

One might also wonder whether there is any good reason why all of the superpartners of the Standard Model particles should be heavy enough to have avoided discovery so far. There is. All of the particles in the

MSSM that have been found so far have something in common; they would necessarily be massless in the absence of electroweak symmetry breaking. In particular, the masses of the $W^{\pm}$, $Z^0$ bosons and all quarks and leptons are equal to dimensionless coupling constants times the Higgs VEV $\sim$ 174 GeV, while the photon and gluon are required to be massless by electromagnetic and QCD gauge invariance. Conversely, all of the undiscovered particles in the MSSM have exactly the opposite property; each of them can have a Lagrangian mass term in the absence of electroweak symmetry breaking. For the squarks, sleptons, and Higgs scalars this follows from a general property of complex scalar fields that a mass term $m^2|\phi|^2$ is always allowed by all gauge symmetries. For the higgsinos and gauginos, it follows from the fact that they are fermions in a real representation of the gauge group. So, from the point of view of the MSSM, the discovery of the top quark in 1995 marked a quite natural milestone; the already-discovered particles are precisely those that had to be light, based on the principle of electroweak gauge symmetry. There is a single exception: one neutral Higgs scalar boson should be lighter than about 135 GeV if the minimal version of supersymmetry is correct, for reasons to be discussed in section 1.7.1. In non-minimal models that do not have extreme fine tuning of parameters, and that remain perturbative up to the scale of apparent gauge coupling unification, the lightest Higgs scalar boson can have a mass up to about 150 GeV.

An important feature of the MSSM is that the superpartners listed in Tables 1.1 and 1.2 are not necessarily the mass eigenstates of the theory. This is because after electroweak symmetry breaking and supersymmetry breaking effects are included, there can be mixing between the electroweak gauginos and the higgsinos, and within the various sets of squarks and sleptons and Higgs scalars that have the same electric charge. The lone exception is the gluino, which is a color octet fermion and therefore does not have the appropriate quantum numbers to mix with any other particle. The masses and mixings of the superpartners are obviously of paramount importance to experimentalists. It is perhaps slightly less obvious that these phenomenological issues are all quite directly related to one central question that is also the focus of much of the theoretical work in supersymmetry: "How is supersymmetry broken?" The reason for this is that most of what we do not already know about the MSSM has to do with $\mathcal{L}_{\rm soft}$. The structure of supersymmetric Lagrangians allows little arbitrariness, as we will see in section 1.3. In fact, all of the dimensionless couplings and all but one mass term in the supersymmetric part of the MSSM Lagrangian correspond directly to parameters in the ordinary Standard Model that have already been measured by experiment. For example, we will find out that the supersymmetric coupling of a gluino to a squark and a quark is determined by the QCD coupling constant $\alpha_S$. In contrast, the supersymmetry-breaking part of the Lagrangian contains many unknown parameters and, apparently, a considerable amount of arbitrariness. Each of the mass splittings between Standard Model particles and their superpartners correspond to terms in the MSSM Lagrangian that are purely supersymmetry-breaking in their origin and effect. These soft supersymmetry-breaking terms can also introduce a large number of mixing angles and CP-violating phases not found in the Standard Model. Fortunately, as we will see in section 1.5.4, there is already strong evidence that the supersymmetry-breaking terms in the MSSM are actually not arbitrary at all. Furthermore, the additional parameters will be measured and constrained as the superpartners are detected. From a theoretical perspective, the challenge is to explain all of these parameters with a predictive model for supersymmetry breaking.

The rest of the discussion is organized as follows. Section 1.2 provides a list of important notations. In section 1.3, we will learn how to construct Lagrangians for supersymmetric field theories. Soft supersymmetry-breaking couplings are described in section 1.4. In section 1.5, we will apply the preceding general results to the special case of the MSSM, introduce the concept of $R$-parity, and emphasize the importance of the structure of the soft terms. Section 1.6 outlines some considerations for understanding the origin of supersymmetry breaking, and the consequences of various proposals. In section 1.7, we will study the mass and mixing angle patterns of the new particles predicted by the MSSM. Their decay modes are considered in section 1.8. The discussion will be lacking in historical accuracy or perspective; the reader is encouraged to consult the many outstanding books,[12–24] review articles[25–48] and the reprint volume,[49] which contain a much more consistent guide to the original literature.

## 1.2. Interlude: Notations and Conventions

This section specifies my notations and conventions. Four-vector indices are represented by letters from the middle of the Greek alphabet $\mu, \nu, \rho, \ldots = 0, 1, 2, 3$. The contravariant four-vector position and momentum of a particle are

$$x^\mu = (t, \vec{x}), \qquad\qquad p^\mu = (E, \vec{p}), \tag{1.14}$$

while the four-vector derivative is

$$\partial_\mu = (\partial/\partial t, \vec{\nabla}). \tag{1.15}$$

The spacetime metric is

$$\eta_{\mu\nu} = \text{diag}(-1, +1, +1, +1), \tag{1.16}$$

so that $p^2 = -m^2$ for an on-shell particle of mass $m$.

It is overwhelmingly convenient to employ two-component Weyl spinor notation for fermions, rather than four-component Dirac or Majorana spinors. The Lagrangian of the Standard Model (and any supersymmetric extension of it) violates parity; each Dirac fermion has left-handed and right-handed parts with completely different electroweak gauge interactions. If one used four-component spinor notation instead, then there would be clumsy left- and right-handed projection operators

$$P_L = (1 - \gamma_5)/2, \qquad P_R = (1 + \gamma_5)/2 \tag{1.17}$$

all over the place. The two-component Weyl fermion notation has the advantage of treating fermionic degrees of freedom with different gauge quantum numbers separately from the start, as Nature intended for us to do. But an even better reason for using two-component notation here is that in supersymmetric models the minimal building blocks of matter are chiral supermultiplets, each of which contains a single two-component Weyl fermion. Since two-component fermion notation may be unfamiliar to some readers, I now specify my conventions by showing how they correspond to the four-component spinor language. A four-component Dirac fermion $\Psi_D$ with mass $M$ is described by the Lagrangian

$$\mathcal{L}_{\text{Dirac}} = i\overline{\Psi}_D \gamma^\mu \partial_\mu \Psi_D - M\overline{\Psi}_D \Psi_D\,. \tag{1.18}$$

For our purposes it is convenient to use the specific representation of the 4×4 gamma matrices given in 2×2 blocks by

$$\gamma^\mu = \begin{pmatrix} 0 & \sigma^\mu \\ \overline{\sigma}^\mu & 0 \end{pmatrix}, \qquad \gamma_5 = \begin{pmatrix} -1 & 0 \\ 0 & 1 \end{pmatrix}, \tag{1.19}$$

where

$$\sigma^0 = \overline{\sigma}^0 = \begin{pmatrix} 1 & 0 \\ 0 & 1 \end{pmatrix}, \qquad \sigma^1 = -\overline{\sigma}^1 = \begin{pmatrix} 0 & 1 \\ 1 & 0 \end{pmatrix},$$

$$\sigma^2 = -\overline{\sigma}^2 = \begin{pmatrix} 0 & -i \\ i & 0 \end{pmatrix}, \qquad \sigma^3 = -\overline{\sigma}^3 = \begin{pmatrix} 1 & 0 \\ 0 & -1 \end{pmatrix}. \tag{1.20}$$

In this representation, a four-component Dirac spinor is written in terms of 2 two-component, complex, anticommuting objects $\xi_\alpha$ and $(\chi^\dagger)^{\dot{\alpha}} \equiv \chi^{\dagger\dot{\alpha}}$ with two distinct types of spinor indices $\alpha = 1,2$ and $\dot{\alpha} = 1,2$:

$$\Psi_D = \begin{pmatrix} \xi_\alpha \\ \chi^{\dagger\dot{\alpha}} \end{pmatrix}. \tag{1.21}$$

It follows that

$$\overline{\Psi}_D = \Psi_D^\dagger \begin{pmatrix} 0 & 1 \\ 1 & 0 \end{pmatrix} = (\, \chi^\alpha \quad \xi^\dagger_{\dot{\alpha}} \,) \,. \tag{1.22}$$

Undotted (dotted) indices from the beginning of the Greek alphabet are used for the first (last) two components of a Dirac spinor. The field $\xi$ is called a "left-handed Weyl spinor" and $\chi^\dagger$ is a "right-handed Weyl spinor". The names fit, because

$$P_L \Psi_D = \begin{pmatrix} \xi_\alpha \\ 0 \end{pmatrix}, \qquad P_R \Psi_D = \begin{pmatrix} 0 \\ \chi^{\dagger\dot{\alpha}} \end{pmatrix}. \tag{1.23}$$

The Hermitian conjugate of any left-handed Weyl spinor is a right-handed Weyl spinor:

$$\psi^\dagger_{\dot{\alpha}} \equiv (\psi_\alpha)^\dagger = (\psi^\dagger)_{\dot{\alpha}}\,, \tag{1.24}$$

and vice versa:

$$(\psi^{\dagger\dot{\alpha}})^\dagger = \psi^\alpha. \tag{1.25}$$

Therefore, any particular fermionic degrees of freedom can be described equally well using a left-handed Weyl spinor (with an undotted index) or by a right-handed one (with a dotted index). By convention, all names of fermion fields are chosen so that left-handed Weyl spinors do not carry daggers and right-handed Weyl spinors do carry daggers, as in eq. (1.21).

The heights of the dotted and undotted spinor indices are important; for example, comparing eqs. (1.18)-(1.22), we observe that the matrices $(\sigma^\mu)_{\alpha\dot{\alpha}}$ and $(\overline{\sigma}^\mu)^{\dot{\alpha}\alpha}$ defined by eq. (1.20) carry indices with the heights as indicated. The spinor indices are raised and lowered using the antisymmetric symbol $\epsilon^{12} = -\epsilon^{21} = \epsilon_{21} = -\epsilon_{12} = 1$, $\epsilon_{11} = \epsilon_{22} = \epsilon^{11} = \epsilon^{22} = 0$, according to

$$\xi_\alpha = \epsilon_{\alpha\beta}\xi^\beta, \qquad \xi^\alpha = \epsilon^{\alpha\beta}\xi_\beta, \qquad \chi^\dagger_{\dot{\alpha}} = \epsilon_{\dot{\alpha}\dot{\beta}}\chi^{\dagger\dot{\beta}}, \qquad \chi^{\dagger\dot{\alpha}} = \epsilon^{\dot{\alpha}\dot{\beta}}\chi^\dagger_{\dot{\beta}}. \tag{1.26}$$

This is consistent since $\epsilon_{\alpha\beta}\epsilon^{\beta\gamma} = \epsilon^{\gamma\beta}\epsilon_{\beta\alpha} = \delta_\alpha^\gamma$ and $\epsilon_{\dot{\alpha}\dot{\beta}}\epsilon^{\dot{\beta}\dot{\gamma}} = \epsilon^{\dot{\gamma}\dot{\beta}}\epsilon_{\dot{\beta}\dot{\alpha}} = \delta_{\dot{\alpha}}^{\dot{\gamma}}$.

As a convention, repeated spinor indices contracted like

$$^\alpha{}_\alpha \qquad \text{or} \qquad {}_{\dot{\alpha}}{}^{\dot{\alpha}} \tag{1.27}$$

can be suppressed. In particular,

$$\xi\chi \equiv \xi^\alpha\chi_\alpha = \xi^\alpha\epsilon_{\alpha\beta}\chi^\beta = -\chi^\beta\epsilon_{\alpha\beta}\xi^\alpha = \chi^\beta\epsilon_{\beta\alpha}\xi^\alpha = \chi^\beta\xi_\beta \equiv \chi\xi \quad (1.28)$$

with, conveniently, no minus sign in the end. [A minus sign appeared in eq. (1.28) from exchanging the order of anticommuting spinors, but it disappeared due to the antisymmetry of the $\epsilon$ symbol.] Likewise, $\xi^\dagger\chi^\dagger$ and $\chi^\dagger\xi^\dagger$ are equivalent abbreviations for $\chi^\dagger_{\dot\alpha}\xi^{\dagger\dot\alpha} = \xi^\dagger_{\dot\alpha}\chi^{\dagger\dot\alpha}$, and in fact this is the complex conjugate of $\xi\chi$:

$$\xi^\dagger\chi^\dagger = \chi^\dagger\xi^\dagger = (\xi\chi)^*. \quad (1.29)$$

In a similar way, one can check that

$$\xi^\dagger\overline{\sigma}^\mu\chi = -\chi\sigma^\mu\xi^\dagger = (\chi^\dagger\overline{\sigma}^\mu\xi)^* = -(\xi\sigma^\mu\chi^\dagger)^* \quad (1.30)$$

stands for $\xi^\dagger_{\dot\alpha}(\overline{\sigma}^\mu)^{\dot\alpha\alpha}\chi_\alpha$, etc. The anti-commuting spinors here are taken to be classical fields; for quantum fields the complex conjugation in the last two equations would be replaced by Hermitian conjugation in the Hilbert space operator sense.

Some other identities that will be useful below include:

$$\xi\sigma^\mu\overline{\sigma}^\nu\chi = \chi\sigma^\nu\overline{\sigma}^\mu\xi = (\chi^\dagger\overline{\sigma}^\nu\sigma^\mu\xi^\dagger)^* = (\xi^\dagger\overline{\sigma}^\mu\sigma^\nu\chi^\dagger)^*, \quad (1.31)$$

and the Fierz rearrangement identity:

$$\chi_\alpha\,(\xi\eta) = -\xi_\alpha\,(\eta\chi) - \eta_\alpha\,(\chi\xi), \quad (1.32)$$

and the reduction identities

$$\sigma^\mu_{\alpha\dot\alpha}\,\overline{\sigma}^{\dot\beta\beta}_\mu = -2\delta_\alpha^\beta\delta_{\dot\alpha}^{\dot\beta}, \quad (1.33)$$

$$\sigma^\mu_{\alpha\dot\alpha}\,\sigma_{\mu\beta\dot\beta} = -2\epsilon_{\alpha\beta}\epsilon_{\dot\alpha\dot\beta}, \quad (1.34)$$

$$\overline{\sigma}^{\mu\dot\alpha\alpha}\,\overline{\sigma}^{\dot\beta\beta}_\mu = -2\epsilon^{\alpha\beta}\epsilon^{\dot\alpha\dot\beta}, \quad (1.35)$$

$$\left[\sigma^\mu\overline{\sigma}^\nu + \sigma^\nu\overline{\sigma}^\mu\right]_\alpha{}^\beta = -2\eta^{\mu\nu}\delta_\alpha^\beta, \quad (1.36)$$

$$\left[\overline{\sigma}^\mu\sigma^\nu + \overline{\sigma}^\nu\sigma^\mu\right]^{\dot\beta}{}_{\dot\alpha} = -2\eta^{\mu\nu}\delta_{\dot\alpha}^{\dot\beta}, \quad (1.37)$$

$$\overline{\sigma}^\mu\sigma^\nu\overline{\sigma}^\rho = -\eta^{\mu\nu}\overline{\sigma}^\rho - \eta^{\nu\rho}\overline{\sigma}^\mu + \eta^{\mu\rho}\overline{\sigma}^\nu + i\epsilon^{\mu\nu\rho\kappa}\overline{\sigma}_\kappa, \quad (1.38)$$

$$\sigma^\mu\overline{\sigma}^\nu\sigma^\rho = -\eta^{\mu\nu}\sigma^\rho - \eta^{\nu\rho}\sigma^\mu + \eta^{\mu\rho}\sigma^\nu - i\epsilon^{\mu\nu\rho\kappa}\sigma_\kappa, \quad (1.39)$$

where $\epsilon^{\mu\nu\rho\kappa}$ is the totally antisymmetric tensor with $\epsilon^{0123} = +1$.

With these conventions, the Dirac Lagrangian eq. (1.18) can now be rewritten:

$$\mathcal{L}_{\text{Dirac}} = i\xi^\dagger\overline{\sigma}^\mu\partial_\mu\xi + i\chi^\dagger\overline{\sigma}^\mu\partial_\mu\chi - M(\xi\chi + \xi^\dagger\chi^\dagger) \quad (1.40)$$

where we have dropped a total derivative piece $-i\partial_\mu(\chi^\dagger\overline{\sigma}^\mu\chi)$, which does not affect the action.

A four-component Majorana spinor can be obtained from the Dirac spinor of eq. (1.22) by imposing the constraint $\chi = \xi$, so that

$$\Psi_{\rm M} = \begin{pmatrix} \xi_\alpha \\ \xi^{\dagger\dot{\alpha}} \end{pmatrix}, \qquad\qquad \overline{\Psi}_{\rm M} = (\, \xi^\alpha \quad \xi^\dagger_{\dot{\alpha}}\,). \tag{1.41}$$

The four-component spinor form of the Lagrangian for a Majorana fermion with mass $M$,

$$\mathcal{L}_{\rm Majorana} = \frac{i}{2}\overline{\Psi}_{\rm M}\gamma^\mu\partial_\mu\Psi_{\rm M} - \frac{1}{2}M\overline{\Psi}_{\rm M}\Psi_{\rm M} \tag{1.42}$$

can therefore be rewritten as

$$\mathcal{L}_{\rm Majorana} = i\xi^\dagger\overline{\sigma}^\mu\partial_\mu\xi - \frac{1}{2}M(\xi\xi + \xi^\dagger\xi^\dagger) \tag{1.43}$$

in the more economical two-component Weyl spinor representation. Note that even though $\xi_\alpha$ is anticommuting, $\xi\xi$ and its complex conjugate $\xi^\dagger\xi^\dagger$ do not vanish, because of the suppressed $\epsilon$ symbol, see eq. (1.28). Explicitly, $\xi\xi = \epsilon^{\alpha\beta}\xi_\beta\xi_\alpha = \xi_2\xi_1 - \xi_1\xi_2 = 2\xi_2\xi_1$.

More generally, any theory involving spin-1/2 fermions can always be written in terms of a collection of left-handed Weyl spinors $\psi_i$ with

$$\mathcal{L} = i\psi^{\dagger i}\overline{\sigma}^\mu\partial_\mu\psi_i + \ldots \tag{1.44}$$

where the ellipses represent possible mass terms, gauge interactions, and Yukawa interactions with scalar fields. Here the index $i$ runs over the appropriate gauge and flavor indices of the fermions; it is raised or lowered by Hermitian conjugation. Gauge interactions are obtained by promoting the ordinary derivative to a gauge-covariant derivative:

$$\mathcal{L} = i\psi^{\dagger i}\overline{\sigma}^\mu D_\mu\psi_i + \ldots \tag{1.45}$$

with

$$D_\mu\psi_i = \partial_\mu\psi_i - ig_a A^a_\mu T^{aj}_i\psi_j, \tag{1.46}$$

where $g_a$ is the gauge coupling corresponding to the Hermitian Lie algebra generator matrix $T^a$ with vector field $A^a_\mu$.

There is a different $\psi_i$ for the left-handed piece and for the hermitian conjugate of the right-handed piece of a Dirac fermion. Given any expression involving bilinears of four-component spinors

$$\Psi_i = \begin{pmatrix} \xi_i \\ \chi^\dagger_i \end{pmatrix}, \tag{1.47}$$

labeled by a flavor or gauge-representation index $i$, one can translate into two-component Weyl spinor language (or vice versa) using the dictionary:

$$\overline{\Psi}_i P_L \Psi_j = \chi_i \xi_j, \qquad \overline{\Psi}_i P_R \Psi_j = \xi_i^\dagger \chi_j^\dagger, \tag{1.48}$$

$$\overline{\Psi}_i \gamma^\mu P_L \Psi_j = \xi_i^\dagger \overline{\sigma}^\mu \xi_j, \qquad \overline{\Psi}_i \gamma^\mu P_R \Psi_j = \chi_i \sigma^\mu \chi_j^\dagger \tag{1.49}$$

etc.

Let us now see how the Standard Model quarks and leptons are described in this notation. The complete list of left-handed Weyl spinors can be given names corresponding to the chiral supermultiplets in Table 1.1:

$$Q_i = \begin{pmatrix} u \\ d \end{pmatrix}, \begin{pmatrix} c \\ s \end{pmatrix}, \begin{pmatrix} t \\ b \end{pmatrix}, \tag{1.50}$$

$$\overline{u}_i = \overline{u},\ \overline{c},\ \overline{t}, \tag{1.51}$$

$$\overline{d}_i = \overline{d},\ \overline{s},\ \overline{b} \tag{1.52}$$

$$L_i = \begin{pmatrix} \nu_e \\ e \end{pmatrix}, \begin{pmatrix} \nu_\mu \\ \mu \end{pmatrix}, \begin{pmatrix} \nu_\tau \\ \tau \end{pmatrix}, \tag{1.53}$$

$$\overline{e}_i = \overline{e},\ \overline{\mu},\ \overline{\tau}. \tag{1.54}$$

Here $i = 1, 2, 3$ is a family index. The bars on these fields are part of the names of the fields, and do *not* denote any kind of conjugation. Rather, the unbarred fields are the left-handed pieces of a Dirac spinor, while the barred fields are the names given to the conjugates of the right-handed piece of a Dirac spinor. For example, $e$ is the same thing as $e_L$ in Table 1.1, and $\overline{e}$ is the same as $e_R^\dagger$. Together they form a Dirac spinor:

$$\begin{pmatrix} e \\ \overline{e}^\dagger \end{pmatrix} \equiv \begin{pmatrix} e_L \\ e_R \end{pmatrix}, \tag{1.55}$$

and similarly for all of the other quark and charged lepton Dirac spinors. (The neutrinos of the Standard Model are not part of a Dirac spinor, at least in the approximation that they are massless.) The fields $Q_i$ and $L_i$ are weak isodoublets, which always go together when one is constructing interactions invariant under the full Standard Model gauge group $SU(3)_C \times SU(2)_L \times U(1)_Y$. Suppressing all color and weak isospin indices, the kinetic and gauge part of the Standard Model fermion Lagrangian density is then

$$\mathcal{L} = iQ^{\dagger i} \overline{\sigma}^\mu D_\mu Q_i + i\overline{u}_i^\dagger \overline{\sigma}^\mu D_\mu \overline{u}^i + i\overline{d}_i^\dagger \overline{\sigma}^\mu D_\mu \overline{d}^i \tag{1.56}$$

$$+ iL^{\dagger i} \overline{\sigma}^\mu D_\mu L_i + i\overline{e}_i^\dagger \overline{\sigma}^\mu D_\mu \overline{e}^i$$

with the family index $i$ summed over, and $D_\mu$ the appropriate Standard Model covariant derivative. For example,

$$D_\mu \begin{pmatrix} \nu_e \\ e \end{pmatrix} = \left[\partial_\mu - igW^a_\mu(\tau^a/2) - ig'Y_L B_\mu\right] \begin{pmatrix} \nu_e \\ e \end{pmatrix} \tag{1.57}$$

$$D_\mu \overline{e} = \left[\partial_\mu - ig'Y_{\overline{e}} B_\mu\right] \overline{e} \tag{1.58}$$

with $\tau^a$ $(a = 1, 2, 3)$ equal to the Pauli matrices, $Y_L = -1/2$ and $Y_{\overline{e}} = +1$. The gauge eigenstate weak bosons are related to the mass eigenstates by

$$W^\pm_\mu = (W^1_\mu \mp iW^2_\mu)/\sqrt{2}, \tag{1.59}$$

$$\begin{pmatrix} Z_\mu \\ A_\mu \end{pmatrix} = \begin{pmatrix} \cos\theta_W & -\sin\theta_W \\ \sin\theta_W & \cos\theta_W \end{pmatrix} \begin{pmatrix} W^3_\mu \\ B_\mu \end{pmatrix}. \tag{1.60}$$

Similar expressions hold for the other quark and lepton gauge eigenstates, with $Y_Q = 1/6$, $Y_{\overline{u}} = -2/3$, and $Y_{\overline{d}} = 1/3$. The quarks also have a term in the covariant derivative corresponding to gluon interactions proportional to $g_3$ (with $\alpha_S = g_3^2/4\pi$) with generators $T^a = \lambda^a/2$ for $Q$, and in the complex conjugate representation $T^a = -(\lambda^a)^*/2$ for $\overline{u}$ and $\overline{d}$, where $\lambda^a$ are the Gell-Mann matrices.

## 1.3. Supersymmetric Lagrangians

In this section we will describe the construction of supersymmetric Lagrangians. Our aim is to arrive at a recipe that will allow us to write down the allowed interactions and mass terms of a general supersymmetric theory, so that later we can apply the results to the special case of the MSSM. We will not use the superfield[50] language, which is often more elegant and efficient for those who know it, but might seem rather cabalistic to some. Our approach is therefore intended to be complementary to the superspace and superfield derivations given in other works. We begin by considering the simplest example of a supersymmetric theory in four dimensions.

### 1.3.1. *The simplest supersymmetric model: A free chiral supermultiplet*

The minimum fermion content of a field theory in four dimensions consists of a single left-handed two-component Weyl fermion $\psi$. Since this is an intrinsically complex object, it seems sensible to choose as its superpartner a complex scalar field $\phi$. The simplest action we can write down for these fields just consists of kinetic energy terms for each:

$$S = \int d^4x \, (\mathcal{L}_{\text{scalar}} + \mathcal{L}_{\text{fermion}}) \, , \tag{1.61}$$

$$\mathcal{L}_{\text{scalar}} = -\partial^\mu \phi^* \partial_\mu \phi, \qquad \mathcal{L}_{\text{fermion}} = i\psi^\dagger \overline{\sigma}^\mu \partial_\mu \psi. \tag{1.62}$$

This is called the massless, non-interacting *Wess-Zumino model*,[10] and it corresponds to a single chiral supermultiplet as discussed in the Introduction.

A supersymmetry transformation should turn the scalar boson field $\phi$ into something involving the fermion field $\psi_\alpha$. The simplest possibility for the transformation of the scalar field is

$$\delta\phi = \epsilon\psi, \qquad \delta\phi^* = \epsilon^\dagger \psi^\dagger, \tag{1.63}$$

where $\epsilon^\alpha$ is an infinitesimal, anticommuting, two-component Weyl fermion object parameterizing the supersymmetry transformation. Until section 1.6.5, we will be discussing global supersymmetry, which means that $\epsilon^\alpha$ is a constant, satisfying $\partial_\mu \epsilon^\alpha = 0$. Since $\psi$ has dimensions of $[\text{mass}]^{3/2}$ and $\phi$ has dimensions of [mass], it must be that $\epsilon$ has dimensions of $[\text{mass}]^{-1/2}$. Using eq. (1.63), we find that the scalar part of the Lagrangian transforms as

$$\delta\mathcal{L}_{\text{scalar}} = -\epsilon\partial^\mu\psi \, \partial_\mu\phi^* - \epsilon^\dagger \partial^\mu \psi^\dagger \, \partial_\mu \phi. \tag{1.64}$$

We would like for this to be canceled by $\delta\mathcal{L}_{\text{fermion}}$, at least up to a total derivative, so that the action will be invariant under the supersymmetry transformation. Comparing eq. (1.64) with $\mathcal{L}_{\text{fermion}}$, we see that for this to have any chance of happening, $\delta\psi$ should be linear in $\epsilon^\dagger$ and in $\phi$, and should contain one spacetime derivative. Up to a multiplicative constant, there is only one possibility to try:

$$\delta\psi_\alpha = -i(\sigma^\mu \epsilon^\dagger)_\alpha \, \partial_\mu\phi, \qquad \delta\psi^\dagger_{\dot\alpha} = i(\epsilon\sigma^\mu)_{\dot\alpha} \, \partial_\mu\phi^*. \tag{1.65}$$

With this guess, one immediately obtains

$$\delta\mathcal{L}_{\text{fermion}} = -\epsilon\sigma^\mu\overline{\sigma}^\nu \partial_\nu\psi \, \partial_\mu\phi^* + \psi^\dagger\overline{\sigma}^\nu\sigma^\mu\epsilon^\dagger \, \partial_\mu\partial_\nu\phi \, . \tag{1.66}$$

This can be put in a slightly more useful form by employing the Pauli matrix identities eqs. (1.36), (1.37) and using the fact that partial derivatives commute ($\partial_\mu\partial_\nu = \partial_\nu\partial_\mu$). Equation (1.66) then becomes

$$\begin{aligned} \delta\mathcal{L}_{\text{fermion}} &= \epsilon\partial^\mu\psi \, \partial_\mu\phi^* + \epsilon^\dagger\partial^\mu\psi^\dagger \, \partial_\mu\phi \\ &\quad -\partial_\mu \left( \epsilon\sigma^\nu\overline{\sigma}^\mu\psi \, \partial_\nu\phi^* + \epsilon\psi \, \partial^\mu\phi^* + \epsilon^\dagger\psi^\dagger \, \partial^\mu\phi \right) . \end{aligned} \tag{1.67}$$

The first two terms here just cancel against $\delta\mathcal{L}_{\text{scalar}}$, while the remaining contribution is a total derivative. So we arrive at

$$\delta S = \int d^4x \;\; (\delta\mathcal{L}_{\text{scalar}} + \delta\mathcal{L}_{\text{fermion}}) = 0, \tag{1.68}$$

justifying our guess of the numerical multiplicative factor made in eq. (1.65).

We are not quite finished in showing that the theory described by eq. (1.61) is supersymmetric. We must also show that the supersymmetry algebra closes; in other words, that the commutator of two supersymmetry transformations parameterized by two different spinors $\epsilon_1$ and $\epsilon_2$ is another symmetry of the theory. Using eq. (1.65) in eq. (1.63), one finds

$$(\delta_{\epsilon_2}\delta_{\epsilon_1} - \delta_{\epsilon_1}\delta_{\epsilon_2})\phi \equiv \delta_{\epsilon_2}(\delta_{\epsilon_1}\phi) - \delta_{\epsilon_1}(\delta_{\epsilon_2}\phi) = i(-\epsilon_1\sigma^\mu\epsilon_2^\dagger + \epsilon_2\sigma^\mu\epsilon_1^\dagger)\,\partial_\mu\phi. \tag{1.69}$$

This is a remarkable result; in words, we have found that the commutator of two supersymmetry transformations gives us back the derivative of the original field. Since $\partial_\mu$ corresponds to the generator of spacetime translations $P_\mu$, eq. (1.69) implies the form of the supersymmetry algebra that was foreshadowed in eq. (1.6) of the Introduction. (We will make this statement more explicit before the end of this section.)

All of this will be for nothing if we do not find the same result for the fermion $\psi$. Using eq. (1.63) in eq. (1.65), we get

$$(\delta_{\epsilon_2}\delta_{\epsilon_1} - \delta_{\epsilon_1}\delta_{\epsilon_2})\psi_\alpha = -i(\sigma^\mu\epsilon_1^\dagger)_\alpha\,\epsilon_2\partial_\mu\psi + i(\sigma^\mu\epsilon_2^\dagger)_\alpha\,\epsilon_1\partial_\mu\psi. \tag{1.70}$$

This can be put into a more useful form by applying the Fierz identity eq. (1.32) with $\chi = \sigma^\mu\epsilon_1^\dagger$, $\xi = \epsilon_2$, $\eta = \partial_\mu\psi$, and again with $\chi = \sigma^\mu\epsilon_2^\dagger$, $\xi = \epsilon_1$, $\eta = \partial_\mu\psi$, followed in each case by an application of the identity eq. (1.30). The result is

$$\begin{aligned}(\delta_{\epsilon_2}\delta_{\epsilon_1} - \delta_{\epsilon_1}\delta_{\epsilon_2})\psi_\alpha &= i(-\epsilon_1\sigma^\mu\epsilon_2^\dagger + \epsilon_2\sigma^\mu\epsilon_1^\dagger)\,\partial_\mu\psi_\alpha \\ &\quad + i\epsilon_{1\alpha}\,\epsilon_2^\dagger\overline{\sigma}^\mu\partial_\mu\psi - i\epsilon_{2\alpha}\,\epsilon_1^\dagger\overline{\sigma}^\mu\partial_\mu\psi.\end{aligned} \tag{1.71}$$

The last two terms in (1.71) vanish on-shell; that is, if the equation of motion $\overline{\sigma}^\mu\partial_\mu\psi = 0$ following from the action is enforced. The remaining piece is exactly the same spacetime translation that we found for the scalar field.

The fact that the supersymmetry algebra only closes on-shell (when the classical equations of motion are satisfied) might be somewhat worrisome, since we would like the symmetry to hold even quantum mechanically. This can be fixed by a trick. We invent a new complex scalar field $F$, which does not have a kinetic term. Such fields are called *auxiliary*, and they are really just book-keeping devices that allow the symmetry algebra to close off-shell. The Lagrangian density for $F$ and its complex conjugate is simply

$$\mathcal{L}_{\text{auxiliary}} = F^* F \, . \tag{1.72}$$

The dimensions of $F$ are [mass]$^2$, unlike an ordinary scalar field, which has dimensions of [mass]. Equation (1.72) implies the not-very-exciting equations of motion $F = F^* = 0$. However, we can use the auxiliary fields to our advantage by including them in the supersymmetry transformation rules. In view of eq. (1.71), a plausible thing to do is to make $F$ transform into a multiple of the equation of motion for $\psi$:

$$\delta F = -i\epsilon^\dagger \overline{\sigma}^\mu \partial_\mu \psi, \qquad \delta F^* = i\partial_\mu \psi^\dagger \overline{\sigma}^\mu \epsilon. \tag{1.73}$$

Once again we have chosen the overall factor on the right-hand sides by virtue of foresight. Now the auxiliary part of the Lagrangian density transforms as

$$\delta\mathcal{L}_{\text{auxiliary}} = -i\epsilon^\dagger \overline{\sigma}^\mu \partial_\mu \psi \, F^* + i\partial_\mu \psi^\dagger \overline{\sigma}^\mu \epsilon \, F, \tag{1.74}$$

which vanishes on-shell, but not for arbitrary off-shell field configurations. Now, by adding an extra term to the transformation law for $\psi$ and $\psi^\dagger$:

$$\delta\psi_\alpha = -i(\sigma^\mu \epsilon^\dagger)_\alpha \, \partial_\mu \phi + \epsilon_\alpha F, \qquad \delta\psi^\dagger_{\dot\alpha} = i(\epsilon\sigma^\mu)_{\dot\alpha} \, \partial_\mu \phi^* + \epsilon^\dagger_{\dot\alpha} F^*, \tag{1.75}$$

one obtains an additional contribution to $\delta\mathcal{L}_{\text{fermion}}$, which just cancels with $\delta\mathcal{L}_{\text{auxiliary}}$, up to a total derivative term. So our "modified" theory with $\mathcal{L} = \mathcal{L}_{\text{scalar}} + \mathcal{L}_{\text{fermion}} + \mathcal{L}_{\text{auxiliary}}$ is still invariant under supersymmetry transformations. Proceeding as before, one now obtains for each of the fields $X = \phi, \phi^*, \psi, \psi^\dagger, F, F^*$,

$$(\delta_{\epsilon_2}\delta_{\epsilon_1} - \delta_{\epsilon_1}\delta_{\epsilon_2})X = i(-\epsilon_1\sigma^\mu\epsilon_2^\dagger + \epsilon_2\sigma^\mu\epsilon_1^\dagger) \, \partial_\mu X \tag{1.76}$$

using eqs. (1.63), (1.73), and (1.75), but now without resorting to any of the equations of motion. So we have succeeded in showing that supersymmetry is a valid symmetry of the Lagrangian off-shell.

In retrospect, one can see why we needed to introduce the auxiliary field $F$ in order to get the supersymmetry algebra to work off-shell. On-shell, the complex scalar field $\phi$ has two real propagating degrees of freedom, matching the two spin polarization states of $\psi$. Off-shell, however, the Weyl fermion $\psi$ is a complex two-component object, so it has four real degrees of freedom. (Going on-shell eliminates half of the propagating degrees of freedom for $\psi$, because the Lagrangian is linear in time derivatives, so that

Table 1.3. Counting of real degrees of freedom in the Wess-Zumino model.

| | $\phi$ | $\psi$ | $F$ |
|---|---|---|---|
| on-shell ($n_B = n_F = 2$) | 2 | 2 | 0 |
| off-shell ($n_B = n_F = 4$) | 2 | 4 | 2 |

the canonical momenta can be reexpressed in terms of the configuration variables without time derivatives and are not independent phase space coordinates.) To make the numbers of bosonic and fermionic degrees of freedom match off-shell as well as on-shell, we had to introduce two more real scalar degrees of freedom in the complex field $F$, which are eliminated when one goes on-shell. This counting is summarized in Table 1.3. The auxiliary field formulation is especially useful when discussing spontaneous supersymmetry breaking, as we will see in section 1.6.

Invariance of the action under a symmetry transformation always implies the existence of a conserved current, and supersymmetry is no exception. The *supercurrent* $J^\mu_\alpha$ is an anticommuting four-vector. It also carries a spinor index, as befits the current associated with a symmetry with fermionic generators.[51] By the usual Noether procedure, one finds for the supercurrent (and its hermitian conjugate) in terms of the variations of the fields $X = \phi, \phi^*, \psi, \psi^\dagger, F, F^*$:

$$\epsilon J^\mu + \epsilon^\dagger J^{\dagger\mu} \equiv \sum_X \delta X \frac{\delta \mathcal{L}}{\delta(\partial_\mu X)} - K^\mu, \tag{1.77}$$

where $K^\mu$ is an object whose divergence is the variation of the Lagrangian density under the supersymmetry transformation, $\delta\mathcal{L} = \partial_\mu K^\mu$. Note that $K^\mu$ is not unique; one can always replace $K^\mu$ by $K^\mu + k^\mu$, where $k^\mu$ is any vector satisfying $\partial_\mu k^\mu = 0$, for example $k^\mu = \partial^\mu \partial_\nu a^\nu - \partial_\nu \partial^\nu a^\mu$. A little work reveals that, up to the ambiguity just mentioned,

$$J^\mu_\alpha = (\sigma^\nu \overline{\sigma}^\mu \psi)_\alpha \, \partial_\nu \phi^*, \qquad J^{\dagger\mu}_{\dot\alpha} = (\psi^\dagger \overline{\sigma}^\mu \sigma^\nu)_{\dot\alpha} \, \partial_\nu \phi. \tag{1.78}$$

The supercurrent and its hermitian conjugate are separately conserved:

$$\partial_\mu J^\mu_\alpha = 0, \qquad \partial_\mu J^{\dagger\mu}_{\dot\alpha} = 0, \tag{1.79}$$

as can be verified by use of the equations of motion. From these currents one constructs the conserved charges

$$Q_\alpha = \sqrt{2} \int d^3\vec{x}\, J^0_\alpha, \qquad Q^\dagger_{\dot\alpha} = \sqrt{2} \int d^3\vec{x}\, J^{\dagger 0}_{\dot\alpha}, \tag{1.80}$$

which are the generators of supersymmetry transformations. (The factor of $\sqrt{2}$ normalization is included to agree with an arbitrary historical convention.) As quantum mechanical operators, they satisfy

$$\left[\epsilon Q+\epsilon^\dagger Q^\dagger, X\right]=-i\sqrt{2}\,\delta X \tag{1.81}$$

for any field $X$, up to terms that vanish on-shell. This can be verified explicitly by using the canonical equal-time commutation and anticommutation relations

$$[\phi(\vec{x}),\pi(\vec{y})] = [\phi^*(\vec{x}),\pi^*(\vec{y})] = i\delta^{(3)}(\vec{x}-\vec{y}), \tag{1.82}$$

$$\{\psi_\alpha(\vec{x}),\psi^\dagger_{\dot\alpha}(\vec{y})\} = (\sigma^0)_{\alpha\dot\alpha}\,\delta^{(3)}(\vec{x}-\vec{y}) \tag{1.83}$$

derived from the free field theory Lagrangian eq. (1.61). Here $\pi=\partial_0\phi^*$ and $\pi^*=\partial_0\phi$ are the momenta conjugate to $\phi$ and $\phi^*$ respectively.

Using eq. (1.81), the content of eq. (1.76) can be expressed in terms of canonical commutators as

$$\left[\epsilon_2 Q+\epsilon_2^\dagger Q^\dagger,\,\left[\epsilon_1 Q+\epsilon_1^\dagger Q^\dagger,\,X\right]\right]-\left[\epsilon_1 Q+\epsilon_1^\dagger Q^\dagger,\,\left[\epsilon_2 Q+\epsilon_2^\dagger Q^\dagger,\,X\right]\right]$$
$$=2(\epsilon_1\sigma^\mu\epsilon_2^\dagger-\epsilon_2\sigma^\mu\epsilon_1^\dagger)\,i\partial_\mu X, \tag{1.84}$$

up to terms that vanish on-shell. The spacetime momentum operator is $P^\mu=(H,\vec{P})$, where $H$ is the Hamiltonian and $\vec{P}$ is the three-momentum operator, given in terms of the canonical fields by

$$H=\int d^3\vec{x}\left[\pi^*\pi+(\vec\nabla\phi^*)\cdot(\vec\nabla\phi)+i\psi^\dagger\vec\sigma\cdot\vec\nabla\psi\right], \tag{1.85}$$

$$\vec{P}=-\int d^3\vec{x}\left(\pi\vec\nabla\phi+\pi^*\vec\nabla\phi^*+i\psi^\dagger\overline{\sigma}^0\vec\nabla\psi\right). \tag{1.86}$$

It generates spacetime translations on the fields $X$ according to

$$[P^\mu,X]=i\partial^\mu X. \tag{1.87}$$

Rearranging the terms in eq. (1.84) using the Jacobi identity, we therefore have

$$\left[\left[\epsilon_2 Q+\epsilon_2^\dagger Q^\dagger,\,\epsilon_1 Q+\epsilon_1^\dagger Q^\dagger\right],\,X\right]=2(\epsilon_1\sigma_\mu\epsilon_2^\dagger-\epsilon_2\sigma_\mu\epsilon_1^\dagger)\,[P^\mu,X], \tag{1.88}$$

for any $X$, up to terms that vanish on-shell, so it must be that

$$\left[\epsilon_2 Q+\epsilon_2^\dagger Q^\dagger,\,\epsilon_1 Q+\epsilon_1^\dagger Q^\dagger\right]=2(\epsilon_1\sigma_\mu\epsilon_2^\dagger-\epsilon_2\sigma_\mu\epsilon_1^\dagger)\,P^\mu. \tag{1.89}$$

Now by expanding out eq. (1.89), one obtains the precise form of the supersymmetry algebra relations

$$\{Q_\alpha,Q^\dagger_{\dot\alpha}\}=-2\sigma^\mu_{\alpha\dot\alpha}P_\mu, \tag{1.90}$$

$$\{Q_\alpha,Q_\beta\}=0,\qquad\qquad \{Q^\dagger_{\dot\alpha},Q^\dagger_{\dot\beta}\}=0, \tag{1.91}$$

as promised in the Introduction. [The commutator in eq. (1.89) turns into anticommutators in eqs. (1.90) and (1.91) in the process of extracting the anticommuting spinors $\epsilon_1$ and $\epsilon_2$.] The results

$$[Q_\alpha, P^\mu] = 0, \qquad [Q^\dagger_{\dot\alpha}, P^\mu] = 0 \tag{1.92}$$

follow immediately from eq. (1.87) and the fact that the supersymmetry transformations are global (independent of position in spacetime). This demonstration of the supersymmetry algebra in terms of the canonical generators $Q$ and $Q^\dagger$ requires the use of the Hamiltonian equations of motion, but the symmetry itself is valid off-shell at the level of the Lagrangian, as we have already shown.

### 1.3.2. *Interactions of chiral supermultiplets*

In a realistic theory like the MSSM, there are many chiral supermultiplets, with both gauge and non-gauge interactions. In this subsection, our task is to construct the most general possible theory of masses and non-gauge interactions for particles that live in chiral supermultiplets. In the MSSM these are the quarks, squarks, leptons, sleptons, Higgs scalars and higgsino fermions. We will find that the form of the non-gauge couplings, including mass terms, is highly restricted by the requirement that the action is invariant under supersymmetry transformations. (Gauge interactions will be dealt with in the following subsections.)

Our starting point is the Lagrangian density for a collection of free chiral supermultiplets labeled by an index $i$, which runs over all gauge and flavor degrees of freedom. Since we will want to construct an interacting theory with supersymmetry closing off-shell, each supermultiplet contains a complex scalar $\phi_i$ and a left-handed Weyl fermion $\psi_i$ as physical degrees of freedom, plus a complex auxiliary field $F_i$, which does not propagate. The results of the previous subsection tell us that the free part of the Lagrangian is

$$\mathcal{L}_{\text{free}} = -\partial^\mu \phi^{*i} \partial_\mu \phi_i + i\psi^{\dagger i}\overline{\sigma}^\mu \partial_\mu \psi_i + F^{*i} F_i, \tag{1.93}$$

where we sum over repeated indices $i$ (not to be confused with the suppressed spinor indices), with the convention that fields $\phi_i$ and $\psi_i$ always carry lowered indices, while their conjugates always carry raised indices. It is invariant under the supersymmetry transformation

$$\delta\phi_i = \epsilon\psi_i, \qquad \delta\phi^{*i} = \epsilon^\dagger\psi^{\dagger i}, \tag{1.94}$$

$$\delta(\psi_i)_\alpha = -i(\sigma^\mu\epsilon^\dagger)_\alpha\, \partial_\mu\phi_i + \epsilon_\alpha F_i, \quad \delta(\psi^{\dagger i})_{\dot\alpha} = i(\epsilon\sigma^\mu)_{\dot\alpha}\, \partial_\mu\phi^{*i} + \epsilon^\dagger_{\dot\alpha} F^{*i}, \tag{1.95}$$

$$\delta F_i = -i\epsilon^\dagger\overline{\sigma}^\mu\partial_\mu\psi_i, \qquad \delta F^{*i} = i\partial_\mu\psi^{\dagger i}\overline{\sigma}^\mu\epsilon\,. \tag{1.96}$$

We will now find the most general set of renormalizable interactions for these fields that is consistent with supersymmetry. We do this working in the field theory before integrating out the auxiliary fields. To begin, note that in order to be renormalizable by power counting, each term must have field content with total mass dimension $\leq 4$. So, the only candidate terms are:

$$\mathcal{L}_{\text{int}} = \left( -\frac{1}{2} W^{ij} \psi_i \psi_j + W^i F_i + x^{ij} F_i F_j \right) + \text{c.c.} - U, \tag{1.97}$$

where $W^{ij}$, $W^i$, $x^{ij}$, and $U$ are polynomials in the scalar fields $\phi_i, \phi^{*i}$, with degrees 1, 2, 0, and 4, respectively. [Terms of the form $F^{*i} F_j$ are already included in eq. (1.93), with the coefficient fixed by the transformation rules (1.94)-(1.96).]

We must now require that $\mathcal{L}_{\text{int}}$ is invariant under the supersymmetry transformations, since $\mathcal{L}_{\text{free}}$ was already invariant by itself. This immediately requires that the candidate term $U(\phi_i, \phi^{*i})$ must vanish. If there were such a term, then under a supersymmetry transformation eq. (1.94) it would transform into another function of the scalar fields only, multiplied by $\epsilon \psi_i$ or $\epsilon^\dagger \psi^{\dagger i}$, and with no spacetime derivatives or $F_i$, $F^{*i}$ fields. It is easy to see from eqs. (1.94)-(1.97) that nothing of this form can possibly be canceled by the supersymmetry transformation of any other term in the Lagrangian. Similarly, the dimensionless coupling $x^{ij}$ must be zero, because its supersymmetry transformation likewise cannot possibly be canceled by any other term. So, we are left with

$$\mathcal{L}_{\text{int}} = \left( -\frac{1}{2} W^{ij} \psi_i \psi_j + W^i F_i \right) + \text{c.c.} \tag{1.98}$$

as the only possibilities. At this point, we are not assuming that $W^{ij}$ and $W^i$ are related to each other in any way. However, soon we will find out that they *are* related, which is why we have chosen to use the same letter for them. Notice that eq. (1.28) tells us that $W^{ij}$ is symmetric under $i \leftrightarrow j$.

It is easiest to divide the variation of $\mathcal{L}_{\text{int}}$ into several parts, which must cancel separately. First, we consider the part that contains four spinors:

$$\delta \mathcal{L}_{\text{int}}|_{4-\text{spinor}} = \left[ -\frac{1}{2} \frac{\delta W^{ij}}{\delta \phi_k} (\epsilon \psi_k)(\psi_i \psi_j) - \frac{1}{2} \frac{\delta W^{ij}}{\delta \phi^{*k}} (\epsilon^\dagger \psi^{\dagger k})(\psi_i \psi_j) \right] + \text{c.c.} \tag{1.99}$$

The term proportional to $(\epsilon \psi_k)(\psi_i \psi_j)$ cannot cancel against any other term. Fortunately, however, the Fierz identity eq. (1.32) implies

$$(\epsilon \psi_i)(\psi_j \psi_k) + (\epsilon \psi_j)(\psi_k \psi_i) + (\epsilon \psi_k)(\psi_i \psi_j) = 0, \tag{1.100}$$

so this contribution to $\delta\mathcal{L}_{\text{int}}$ vanishes identically if and only if $\delta W^{ij}/\delta\phi_k$ is totally symmetric under interchange of $i, j, k$. There is no such identity available for the term proportional to $(\epsilon^\dagger\psi^{\dagger k})(\psi_i\psi_j)$. Since that term cannot cancel with any other, requiring it to be absent just tells us that $W^{ij}$ cannot contain $\phi^{*k}$. In other words, $W^{ij}$ is analytic (or holomorphic) in the complex fields $\phi_k$.

Combining what we have learned so far, we can write

$$W^{ij} = M^{ij} + y^{ijk}\phi_k \tag{1.101}$$

where $M^{ij}$ is a symmetric mass matrix for the fermion fields, and $y^{ijk}$ is a Yukawa coupling of a scalar $\phi_k$ and two fermions $\psi_i\psi_j$ that must be totally symmetric under interchange of $i, j, k$. It is therefore possible, and it turns out to be convenient, to write

$$W^{ij} = \frac{\delta^2}{\delta\phi_i\delta\phi_j}W \tag{1.102}$$

where we have introduced a useful object

$$W = \frac{1}{2}M^{ij}\phi_i\phi_j + \frac{1}{6}y^{ijk}\phi_i\phi_j\phi_k, \tag{1.103}$$

called the *superpotential*. This is not a scalar potential in the ordinary sense; in fact, it is not even real. It is instead an analytic function of the scalar fields $\phi_i$ treated as complex variables.

Continuing on our vaunted quest, we next consider the parts of $\delta\mathcal{L}_{\text{int}}$ that contain a spacetime derivative:

$$\delta\mathcal{L}_{\text{int}}|_\partial = \left(iW^{ij}\partial_\mu\phi_j\,\psi_i\sigma^\mu\epsilon^\dagger + iW^i\,\partial_\mu\psi_i\sigma^\mu\epsilon^\dagger\right) + \text{c.c.} \tag{1.104}$$

Here we have used the identity eq. (1.30) on the second term, which came from $(\delta F_i)W^i$. Now we can use eq. (1.102) to observe that

$$W^{ij}\partial_\mu\phi_j = \partial_\mu\left(\frac{\delta W}{\delta\phi_i}\right). \tag{1.105}$$

Therefore, eq. (1.104) will be a total derivative if

$$W^i = \frac{\delta W}{\delta\phi_i} = M^{ij}\phi_j + \frac{1}{2}y^{ijk}\phi_j\phi_k\,, \tag{1.106}$$

which explains why we chose its name as we did. The remaining terms in $\delta\mathcal{L}_{\text{int}}$ are all linear in $F_i$ or $F^{*i}$, and it is easy to show that they cancel, given the results for $W^i$ and $W^{ij}$ that we have already found.

Actually, we can include a linear term in the superpotential without disturbing the validity of the previous discussion at all:

$$W = L^i\phi_i + \frac{1}{2}M^{ij}\phi_i\phi_j + \frac{1}{6}y^{ijk}\phi_i\phi_j\phi_k. \tag{1.107}$$

Here $L^i$ are parameters with dimensions of [mass]$^2$, which affect only the scalar potential part of the Lagrangian. Such linear terms are only allowed when $\phi_i$ is a gauge singlet, and there are no such gauge singlet chiral supermultiplets in the MSSM with minimal field content. I will therefore omit this term from the remaining discussion of this section. However, this type of term does play an important role in the discussion of spontaneous supersymmetry breaking, as we will see in section 1.6.1.

To recap, we have found that the most general non-gauge interactions for chiral supermultiplets are determined by a single analytic function of the complex scalar fields, the superpotential $W$. The auxiliary fields $F_i$ and $F^{*i}$ can be eliminated using their classical equations of motion. The part of $\mathcal{L}_{\rm free} + \mathcal{L}_{\rm int}$ that contains the auxiliary fields is $F_iF^{*i} + W^iF_i + W_i^*F^{*i}$, leading to the equations of motion

$$F_i = -W_i^*, \qquad\qquad F^{*i} = -W^i\,. \tag{1.108}$$

Thus the auxiliary fields are expressible algebraically (without any derivatives) in terms of the scalar fields.

After making the replacement[g] eq. (1.108) in $\mathcal{L}_{\rm free} + \mathcal{L}_{\rm int}$, we obtain the Lagrangian density

$$\mathcal{L} = -\partial^\mu\phi^{*i}\partial_\mu\phi_i + i\psi^{\dagger i}\overline{\sigma}^\mu\partial_\mu\psi_i - \frac{1}{2}\left(W^{ij}\psi_i\psi_j + W_{ij}^*\psi^{\dagger i}\psi^{\dagger j}\right) - W^iW_i^*. \tag{1.109}$$

Now that the non-propagating fields $F_i, F^{*i}$ have been eliminated, it follows from eq. (1.109) that the scalar potential for the theory is just given in terms of the superpotential by

$$\begin{aligned} V(\phi,\phi^*) = W^kW_k^* = F^{*k}F_k = M_{ik}^*M^{kj}\phi^{*i}\phi_j \qquad\qquad (1.110)\\ + \frac{1}{2}M^{in}y_{jkn}^*\phi_i\phi^{*j}\phi^{*k} + \frac{1}{2}M_{in}^*y^{jkn}\phi^{*i}\phi_j\phi_k + \frac{1}{4}y^{ijn}y_{kln}^*\phi_i\phi_j\phi^{*k}\phi^{*l}\,. \end{aligned}$$

This scalar potential is automatically bounded from below; in fact, since it is a sum of squares of absolute values (of the $W^k$), it is always non-negative. If we substitute the general form for the superpotential eq. (1.103) into

[g]Since $F_i$ and $F^{*i}$ appear only quadratically in the action, the result of instead doing a functional integral over them at the quantum level has precisely the same effect.

eq. (1.109), we obtain for the full Lagrangian density

$$\mathcal{L} = -\partial^\mu \phi^{*i} \partial_\mu \phi_i - V(\phi, \phi^*) + i\psi^{\dagger i} \overline{\sigma}^\mu \partial_\mu \psi_i - \frac{1}{2} M^{ij} \psi_i \psi_j - \frac{1}{2} M^*_{ij} \psi^{\dagger i} \psi^{\dagger j}$$
$$- \frac{1}{2} y^{ijk} \phi_i \psi_j \psi_k - \frac{1}{2} y^*_{ijk} \phi^{*i} \psi^{\dagger j} \psi^{\dagger k}. \tag{1.111}$$

Now we can compare the masses of the fermions and scalars by looking at the linearized equations of motion:

$$\partial^\mu \partial_\mu \phi_i = M^*_{ik} M^{kj} \phi_j + \ldots, \tag{1.112}$$
$$i\overline{\sigma}^\mu \partial_\mu \psi_i = M^*_{ij} \psi^{\dagger j} + \ldots, \qquad i\sigma^\mu \partial_\mu \psi^{\dagger i} = M^{ij} \psi_j + \ldots. \tag{1.113}$$

One can eliminate $\psi$ in terms of $\psi^\dagger$ and vice versa in eq. (1.113), obtaining [after use of the identities eqs. (1.36) and (1.37)]:

$$\partial^\mu \partial_\mu \psi_i = M^*_{ik} M^{kj} \psi_j + \ldots, \qquad \partial^\mu \partial_\mu \psi^{\dagger j} = \psi^{\dagger i} M^*_{ik} M^{kj} + \ldots. \tag{1.114}$$

Therefore, the fermions and the bosons satisfy the same wave equation with exactly the same squared-mass matrix with real non-negative eigenvalues, namely $(M^2)_i{}^j = M^*_{ik} M^{kj}$. It follows that diagonalizing this matrix by redefining the fields with a unitary matrix gives a collection of chiral supermultiplets, each of which contains a mass-degenerate complex scalar and Weyl fermion, in agreement with the general argument in the Introduction.

### 1.3.3. *Lagrangians for gauge supermultiplets*

The propagating degrees of freedom in a gauge supermultiplet are a massless gauge boson field $A^a_\mu$ and a two-component Weyl fermion gaugino $\lambda^a$. The index $a$ here runs over the adjoint representation of the gauge group ($a = 1, \ldots, 8$ for $SU(3)_C$ color gluons and gluinos; $a = 1, 2, 3$ for $SU(2)_L$ weak isospin; $a = 1$ for $U(1)_Y$ weak hypercharge). The gauge transformations of the vector supermultiplet fields are

$$\delta_{\text{gauge}} A^a_\mu = \partial_\mu \Lambda^a + g f^{abc} A^b_\mu \Lambda^c, \tag{1.115}$$
$$\delta_{\text{gauge}} \lambda^a = g f^{abc} \lambda^b \Lambda^c, \tag{1.116}$$

where $\Lambda^a$ is an infinitesimal gauge transformation parameter, $g$ is the gauge coupling, and $f^{abc}$ are the totally antisymmetric structure constants that define the gauge group. The special case of an Abelian group is obtained by just setting $f^{abc} = 0$; the corresponding gaugino is a gauge singlet in that case. The conventions are such that for QED, $A^\mu = (V, \vec{A})$ where $V$ and $\vec{A}$ are the usual electric potential and vector potential, with electric and magnetic fields given by $\vec{E} = -\vec{\nabla} V - \partial_0 \vec{A}$ and $\vec{B} = \vec{\nabla} \times \vec{A}$.

Table 1.4. Counting of real degrees of freedom for each gauge supermultiplet.

| | $A_\mu$ | $\lambda$ | $D$ |
|---|---|---|---|
| on-shell ($n_B = n_F = 2$) | 2 | 2 | 0 |
| off-shell ($n_B = n_F = 4$) | 3 | 4 | 1 |

The on-shell degrees of freedom for $A^a_\mu$ and $\lambda^a_\alpha$ amount to two bosonic and two fermionic helicity states (for each $a$), as required by supersymmetry. However, off-shell $\lambda^a_\alpha$ consists of two complex, or four real, fermionic degrees of freedom, while $A^a_\mu$ only has three real bosonic degrees of freedom; one degree of freedom is removed by the inhomogeneous gauge transformation eq. (1.115). So, we will need one real bosonic auxiliary field, traditionally called $D^a$, in order for supersymmetry to be consistent off-shell. This field also transforms as an adjoint of the gauge group [i.e., like eq. (1.116) with $\lambda^a$ replaced by $D^a$] and satisfies $(D^a)^* = D^a$. Like the chiral auxiliary fields $F_i$, the gauge auxiliary field $D^a$ has dimensions of [mass]$^2$ and no kinetic term, so it can be eliminated on-shell using its algebraic equation of motion. The counting of degrees of freedom is summarized in Table 1.4.

Therefore, the Lagrangian density for a gauge supermultiplet ought to be

$$\mathcal{L}_{\text{gauge}} = -\frac{1}{4} F^a_{\mu\nu} F^{\mu\nu a} + i\lambda^{\dagger a} \overline{\sigma}^\mu D_\mu \lambda^a + \frac{1}{2} D^a D^a, \tag{1.117}$$

where

$$F^a_{\mu\nu} = \partial_\mu A^a_\nu - \partial_\nu A^a_\mu + g f^{abc} A^b_\mu A^c_\nu \tag{1.118}$$

is the usual Yang-Mills field strength, and

$$D_\mu \lambda^a = \partial_\mu \lambda^a + g f^{abc} A^b_\mu \lambda^c \tag{1.119}$$

is the covariant derivative of the gaugino field. To check that eq. (1.117) is really supersymmetric, one must specify the supersymmetry transformations of the fields. The forms of these follow from the requirements that they should be linear in the infinitesimal parameters $\epsilon, \epsilon^\dagger$ with dimensions of [mass]$^{-1/2}$, that $\delta A^a_\mu$ is real, and that $\delta D^a$ should be real and proportional to the field equations for the gaugino, in analogy with the role of the auxiliary field $F$ in the chiral supermultiplet case. Thus one can guess, up to multiplicative factors, that[h]

$$\delta A_\mu^a = -\frac{1}{\sqrt{2}}\left(\epsilon^\dagger\overline{\sigma}_\mu\lambda^a + \lambda^{\dagger a}\overline{\sigma}_\mu\epsilon\right), \tag{1.120}$$

$$\delta\lambda_\alpha^a = \frac{i}{2\sqrt{2}}(\sigma^\mu\overline{\sigma}^\nu\epsilon)_\alpha\, F_{\mu\nu}^a + \frac{1}{\sqrt{2}}\epsilon_\alpha\, D^a, \tag{1.121}$$

$$\delta D^a = \frac{i}{\sqrt{2}}\left(-\epsilon^\dagger\overline{\sigma}^\mu D_\mu\lambda^a + D_\mu\lambda^{\dagger a}\overline{\sigma}^\mu\epsilon\right). \tag{1.122}$$

The factors of $\sqrt{2}$ are chosen so that the action obtained by integrating $\mathcal{L}_{\text{gauge}}$ is indeed invariant, and the phase of $\lambda^a$ is chosen for future convenience in treating the MSSM.

It is now a little bit tedious, but straightforward, to also check that

$$(\delta_{\epsilon_2}\delta_{\epsilon_1} - \delta_{\epsilon_1}\delta_{\epsilon_2})X = i(-\epsilon_1\sigma^\mu\epsilon_2^\dagger + \epsilon_2\sigma^\mu\epsilon_1^\dagger)D_\mu X \tag{1.123}$$

for $X$ equal to any of the gauge-covariant fields $F_{\mu\nu}^a$, $\lambda^a$, $\lambda^{\dagger a}$, $D^a$, as well as for arbitrary covariant derivatives acting on them. This ensures that the supersymmetry algebra eqs. (1.90)-(1.91) is realized on gauge-invariant combinations of fields in gauge supermultiplets, as they were on the chiral supermultiplets [compare eq. (1.76)]. This check requires the use of identities eqs. (1.31), (1.33) and (1.38). If we had not included the auxiliary field $D^a$, then the supersymmetry algebra eq. (1.123) would hold only after using the equations of motion for $\lambda^a$ and $\lambda^{\dagger a}$. The auxiliary fields satisfies a trivial equation of motion $D^a = 0$, but this is modified if one couples the gauge supermultiplets to chiral supermultiplets, as we now do.

### 1.3.4. *Supersymmetric gauge interactions*

Now we are ready to consider a general Lagrangian density for a supersymmetric theory with both chiral and gauge supermultiplets. Suppose that the chiral supermultiplets transform under the gauge group in a representation with hermitian matrices $(T^a)_i{}^j$ satisfying $[T^a, T^b] = if^{abc}T^c$. [For example, if the gauge group is $SU(2)$, then $f^{abc} = \epsilon^{abc}$, and the $T^a$ are 1/2 times the Pauli matrices for a chiral supermultiplet transforming in the

[h]The supersymmetry transformations eqs. (1.120)-(1.122) are non-linear for non-Abelian gauge symmetries, since there are gauge fields in the covariant derivatives acting on the gaugino fields and in the field strength $F_{\mu\nu}^a$. By adding even more auxiliary fields besides $D^a$, one can make the supersymmetry transformations linear in the fields. The version here, in which those extra auxiliary fields have been removed by gauge transformations, is called "Wess-Zumino gauge".[52]

fundamental representation.] Since supersymmetry and gauge transformations commute, the scalar, fermion, and auxiliary fields must be in the same representation of the gauge group, so

$$\delta_{\rm gauge} X_i = ig\Lambda^a (T^a X)_i \tag{1.124}$$

for $X_i = \phi_i, \psi_i, F_i$. To have a gauge-invariant Lagrangian, we now need to replace the ordinary derivatives in eq. (1.93) with covariant derivatives:

$$\partial_\mu \phi_i \to D_\mu \phi_i = \partial_\mu \phi_i - igA^a_\mu (T^a \phi)_i \tag{1.125}$$

$$\partial_\mu \phi^{*i} \to D_\mu \phi^{*i} = \partial_\mu \phi^{*i} + igA^a_\mu (\phi^* T^a)^i \tag{1.126}$$

$$\partial_\mu \psi_i \to D_\mu \psi_i = \partial_\mu \psi_i - igA^a_\mu (T^a \psi)_i . \tag{1.127}$$

Naively, this simple procedure achieves the goal of coupling the vector bosons in the gauge supermultiplet to the scalars and fermions in the chiral supermultiplets. However, we also have to consider whether there are any other interactions allowed by gauge invariance and involving the gaugino and $D^a$ fields, which might have to be included to make a supersymmetric Lagrangian. Since $A^a_\mu$ couples to $\phi_i$ and $\psi_i$, it makes sense that $\lambda^a$ and $D^a$ should as well.

In fact, there are three such possible interaction terms that are renormalizable (of field mass dimension $\leq 4$), namely

$$(\phi^* T^a \psi)\lambda^a, \qquad \lambda^{\dagger a}(\psi^\dagger T^a \phi), \qquad \text{and} \qquad (\phi^* T^a \phi) D^a. \tag{1.128}$$

Now one can add them, with unknown dimensionless coupling coefficients, to the Lagrangians for the chiral and gauge supermultiplets, and demand that the whole mess be real and invariant under supersymmetry, up to a total derivative. Not surprisingly, this is possible only if the supersymmetry transformation laws for the matter fields are modified to include gauge-covariant rather than ordinary derivatives. Also, it is necessary to include one strategically chosen extra term in $\delta F_i$, so:

$$\delta \phi_i = \epsilon \psi_i \tag{1.129}$$

$$\delta \psi_{i\alpha} = -i(\sigma^\mu \epsilon^\dagger)_\alpha \, D_\mu \phi_i + \epsilon_\alpha F_i \tag{1.130}$$

$$\delta F_i = -i\epsilon^\dagger \overline{\sigma}^\mu D_\mu \psi_i + \sqrt{2} g (T^a \phi)_i \, \epsilon^\dagger \lambda^{\dagger a}. \tag{1.131}$$

After some algebra one can now fix the coefficients for the terms in eq. (1.128), with the result that the full Lagrangian density for a renormalizable supersymmetric theory is

$$\begin{aligned} \mathcal{L} = {} & \mathcal{L}_{\rm chiral} + \mathcal{L}_{\rm gauge} \\ & - \sqrt{2} g (\phi^* T^a \psi)\lambda^a - \sqrt{2} g \lambda^{\dagger a} (\psi^\dagger T^a \phi) + g(\phi^* T^a \phi) D^a . \end{aligned} \tag{1.132}$$

Here $\mathcal{L}_{\rm chiral}$ means the chiral supermultiplet Lagrangian found in section 1.3.2 [e.g., eq. (1.109) or (1.111)], but with ordinary derivatives replaced everywhere by gauge-covariant derivatives, and $\mathcal{L}_{\rm gauge}$ was given in eq. (1.117). To prove that eq. (1.132) is invariant under the supersymmetry transformations, one must use the identity

$$W^i(T^a\phi)_i = 0. \qquad (1.133)$$

This is precisely the condition that must be satisfied anyway in order for the superpotential, and thus $\mathcal{L}_{\rm chiral}$, to be gauge invariant, since the left side is proportional to $\delta_{\rm gauge} W$.

The second line in eq. (1.132) consists of interactions whose strengths are fixed to be gauge couplings by the requirements of supersymmetry, even though they are not gauge interactions from the point of view of an ordinary field theory. The first two terms are a direct coupling of gauginos to matter fields; this can be thought of as the "supersymmetrization" of the usual gauge boson couplings to matter fields. The last term combines with the $D^aD^a/2$ term in $\mathcal{L}_{\rm gauge}$ to provide an equation of motion

$$D^a = -g(\phi^* T^a \phi). \qquad (1.134)$$

Thus, like the auxiliary fields $F_i$ and $F^{*i}$, the $D^a$ are expressible purely algebraically in terms of the scalar fields. Replacing the auxiliary fields in eq. (1.132) using eq. (1.134), one finds that the complete scalar potential is (recall that $\mathcal{L}$ contains $-V$):

$$V(\phi,\phi^*) = F^{*i}F_i + \frac{1}{2}\sum_a D^aD^a = W_i^*W^i + \frac{1}{2}\sum_a g_a^2(\phi^*T^a\phi)^2. \qquad (1.135)$$

The two types of terms in this expression are called "$F$-term" and "$D$-term" contributions, respectively. In the second term in eq. (1.135), we have now written an explicit sum $\sum_a$ to cover the case that the gauge group has several distinct factors with different gauge couplings $g_a$. [For instance, in the MSSM the three factors $SU(3)_C$, $SU(2)_L$ and $U(1)_Y$ have different gauge couplings $g_3$, $g$ and $g'$.] Since $V(\phi,\phi^*)$ is a sum of squares, it is always greater than or equal to zero for every field configuration. It is an interesting and unique feature of supersymmetric theories that the scalar potential is completely determined by the *other* interactions in the theory. The $F$-terms are fixed by Yukawa couplings and fermion mass terms, and the $D$-terms are fixed by the gauge interactions.

By using Noether's procedure [see eq. (1.77)], one finds the conserved supercurrent

$$J^\mu_\alpha = (\sigma^\nu\overline{\sigma}^\mu\psi_i)_\alpha\, D_\nu\phi^{*i} + i(\sigma^\mu\psi^{\dagger i})_\alpha\, W_i^* \\ -\frac{1}{2\sqrt{2}}(\sigma^\nu\overline{\sigma}^\rho\sigma^\mu\lambda^{\dagger a})_\alpha\, F^a_{\nu\rho} + \frac{i}{\sqrt{2}} g_a\phi^* T^a\phi\,(\sigma^\mu\lambda^{\dagger a})_\alpha, \qquad (1.136)$$

generalizing the expression given in eq. (1.78) for the Wess-Zumino model. This result will be useful when we discuss certain aspects of spontaneous supersymmetry breaking in section 1.6.5.

### 1.3.5. *Summary: How to build a supersymmetric model*

In a renormalizable supersymmetric field theory, the interactions and masses of all particles are determined just by their gauge transformation properties and by the superpotential $W$. By construction, we found that $W$ had to be an analytic function of the complex scalar fields $\phi_i$, which are always defined to transform under supersymmetry into left-handed Weyl fermions. We should mention that in an equivalent language, $W$ is said to be a function of chiral *superfields*.[50] A superfield is a single object that contains as components all of the bosonic, fermionic, and auxiliary fields within the corresponding supermultiplet, for example $\Phi_i \supset (\phi_i, \psi_i, F_i)$. (This is analogous to the way in which one often describes a weak isospin doublet or color triplet by a multicomponent field.) The gauge quantum numbers and the mass dimension of a chiral superfield are the same as that of its scalar component. In the superfield formulation, one writes instead of eq. (1.107)

$$W = L^i\Phi_i + \frac{1}{2}M^{ij}\Phi_i\Phi_j + \frac{1}{6}y^{ijk}\Phi_i\Phi_j\Phi_k, \qquad (1.137)$$

which implies exactly the same physics. The derivation of all of our preceding results can be obtained somewhat more elegantly using superfield methods, which have the advantage of making invariance under supersymmetry transformations manifest by defining the Lagrangian in terms of integrals over a "superspace" with fermionic as well as ordinary commuting coordinates. We have avoided this extra layer of notation on purpose, in favor of the more pedestrian, but more familiar and accessible, component field approach. The latter is at least more appropriate for making contact with phenomenology in a universe with supersymmetry breaking. The only (occasional) use we will make of superfield notation is the purely cosmetic one of following the common practice of specifying superpotentials like eq. (1.137) rather than (1.107). The specification of the superpotential is really a code for the terms that it implies in the Lagrangian, so the reader may feel free to think of the superpotential either as a function of the scalar fields $\phi_i$ or as the same function of the superfields $\Phi_i$.

Given the supermultiplet content of the theory, the form of the superpotential is restricted by the requirement of gauge invariance [see eq. (1.133)]. In any given theory, only a subset of the parameters $L^i$, $M^{ij}$, and $y^{ijk}$ are allowed to be non-zero. The parameter $L^i$ is only allowed if $\Phi_i$ is a gauge singlet. (There are no such chiral supermultiplets in the MSSM with the minimal field content.) The entries of the mass matrix $M^{ij}$ can only be non-zero for $i$ and $j$ such that the supermultiplets $\Phi_i$ and $\Phi_j$ transform under the gauge group in representations that are conjugates of each other. (In the MSSM there is only one such term, as we will see.) Likewise, the Yukawa couplings $y^{ijk}$ can only be non-zero when $\Phi_i$, $\Phi_j$, and $\Phi_k$ transform in representations that can combine to form a singlet.

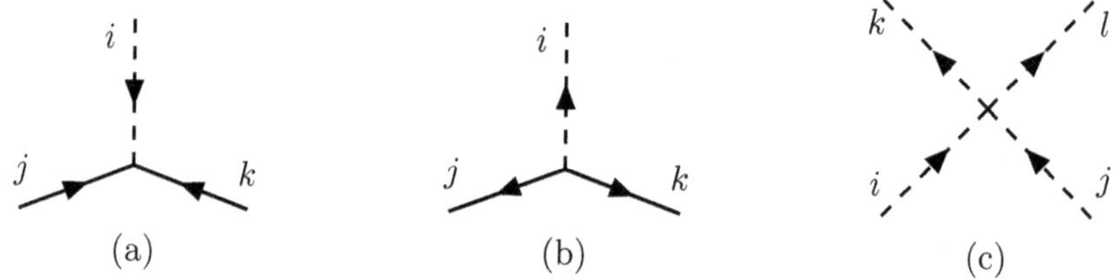


Fig. 1.3. The dimensionless non-gauge interaction vertices in a supersymmetric theory: (a) scalar-fermion-fermion Yukawa interaction $y^{ijk}$, (b) the complex conjugate interaction interaction $y_{ijk}$, and (c) quartic scalar interaction $y^{ijn}y^*_{kln}$.

The interactions implied by the superpotential eq. (1.137) (with $L^i = 0$) were listed in eqs. (1.111), (1.111), and are shown[i] in Figures 1.3 and 1.4. Those in Figure 1.3 are all determined by the dimensionless parameters $y^{ijk}$. The Yukawa interaction in Figure 1.3a corresponds to the next-to-last term in eq. (1.111). For each particular Yukawa coupling of $\phi_i\psi_j\psi_k$ with strength $y^{ijk}$, there must be equal couplings of $\phi_j\psi_i\psi_k$ and $\phi_k\psi_i\psi_j$, since $y^{ijk}$ is completely symmetric under interchange of any two of its indices as shown in section 1.3.2. The arrows on the fermion and scalar lines point in the direction for propagation of $\phi$ and $\psi$ and opposite the direction of propagation of $\phi^*$ and $\psi^\dagger$. Thus there is also a vertex corresponding to

[i]Here, the auxiliary fields have been eliminated using their equations of motion ("integrated out"). One could instead give Feynman rules that include the auxiliary fields, or directly in terms of superfields on superspace, although this is usually less useful in practical phenomenological applications.

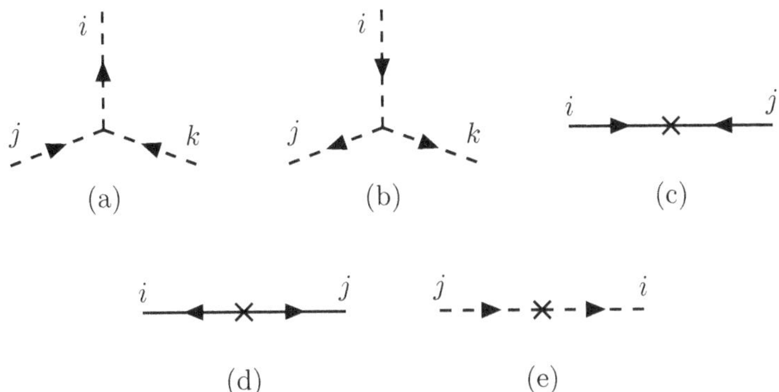


Fig. 1.4. Supersymmetric dimensionful couplings: (a) $(\text{scalar})^3$ interaction vertex $M^*_{in}y^{jkn}$ and (b) the conjugate interaction $M^{in}y^*_{jkn}$, (c) fermion mass term $M^{ij}$ and (d) conjugate fermion mass term $M^*_{ij}$, and (e) scalar squared-mass term $M^*_{ik}M^{kj}$.

the one in Figure 1.3a but with all arrows reversed, corresponding to the complex conjugate [the last term in eq. (1.111)]. It is shown in Figure 1.3b. There is also a dimensionless coupling for $\phi_i\phi_j\phi^{*k}\phi^{*l}$, with strength $y^{ijn}y^*_{kln}$, as required by supersymmetry [see the last term in eq. (1.111)]. The relationship between the Yukawa interactions in Figures 1.3a,b and the scalar interaction of Figure 1.3c is exactly of the special type needed to cancel the quadratic divergences in quantum corrections to scalar masses, as discussed in the Introduction [compare Figure 1.1, and eq. (1.11)].

Figure 1.4 shows the only interactions corresponding to renormalizable and supersymmetric vertices with coupling dimensions of [mass] and $[\text{mass}]^2$. First, there are $(\text{scalar})^3$ couplings in Figure 1.4a,b, which are entirely determined by the superpotential mass parameters $M^{ij}$ and Yukawa couplings $y^{ijk}$, as indicated by the second and third terms in eq. (1.111). The propagators of the fermions and scalars in the theory are constructed in the usual way using the fermion mass $M^{ij}$ and scalar squared mass $M^*_{ik}M^{kj}$. The fermion mass terms $M^{ij}$ and $M_{ij}$ each lead to a chirality-changing insertion in the fermion propagator; note the directions of the arrows in Figure 1.4c,d. There is no such arrow-reversal for a scalar propagator in a theory with exact supersymmetry; as depicted in Figure 1.4e, if one treats the scalar squared-mass term as an insertion in the propagator, the arrow direction is preserved.

Figure 1.5 shows the gauge interactions in a supersymmetric theory. Figures 1.5a, b, c occur only when the gauge group is non-Abelian, for example for $SU(3)_C$ color and $SU(2)_L$ weak isospin in the MSSM. Figures 1.5a

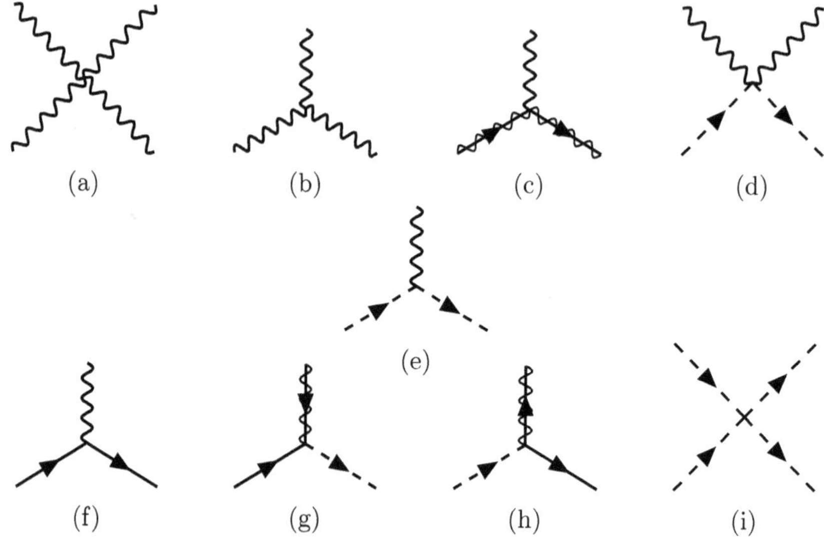


Fig. 1.5. Supersymmetric gauge interaction vertices.

and 1.5b are the interactions of gauge bosons, which derive from the first term in eq. (1.117). In the MSSM these are exactly the same as the well-known QCD gluon and electroweak gauge boson vertices of the Standard Model. (We do not show the interactions of ghost fields, which are necessary only for consistent loop amplitudes.) Figures 1.5c,d,e,f are just the standard interactions between gauge bosons and fermion and scalar fields that must occur in any gauge theory because of the form of the covariant derivative; they come from eqs. (1.119) and (1.125)-(1.127) inserted in the kinetic part of the Lagrangian. Figure 1.5c shows the coupling of a gaugino to a gauge boson; the gaugino line in a Feynman diagram is traditionally drawn as a solid fermion line superimposed on a wavy line. In Figure 1.5g we have the coupling of a gaugino to a chiral fermion and a complex scalar [the first term in the second line of eq. (1.132)]. One can think of this as the "supersymmetrization" of Figure 1.5e or 1.5f; any of these three vertices may be obtained from any other (up to a factor of $\sqrt{2}$) by replacing two of the particles by their supersymmetric partners. There is also an interaction in Figure 1.5h which is just like Figure 1.5g but with all arrows reversed, corresponding to the complex conjugate term in the Lagrangian [the second term in the second line in eq. (1.132)]. Finally in Figure 1.5i we have a scalar quartic interaction vertex [the last term in eq. (1.135)], which is also determined by the gauge coupling.

The results of this section can be used as a recipe for constructing the supersymmetric interactions for any model. In the case of the MSSM, we already know the gauge group, particle content and the gauge transformation properties, so it only remains to decide on the superpotential. This we will do in section 1.5.1.

## 1.4. Soft Supersymmetry Breaking Interactions

A realistic phenomenological model must contain supersymmetry breaking. From a theoretical perspective, we expect that supersymmetry, if it exists at all, should be an exact symmetry that is broken spontaneously. In other words, the underlying model should have a Lagrangian density that is invariant under supersymmetry, but a vacuum state that is not. In this way, supersymmetry is hidden at low energies in a manner analogous to the fate of the electroweak symmetry in the ordinary Standard Model.

Many models of spontaneous symmetry breaking have indeed been proposed and we will mention the basic ideas of some of them in section 1.6. These always involve extending the MSSM to include new particles and interactions at very high mass scales, and there is no consensus on exactly how this should be done. However, from a practical point of view, it is extremely useful to simply parameterize our ignorance of these issues by just introducing extra terms that break supersymmetry explicitly in the effective MSSM Lagrangian. As was argued in the Introduction, the supersymmetry-breaking couplings should be soft (of positive mass dimension) in order to be able to naturally maintain a hierarchy between the electroweak scale and the Planck (or any other very large) mass scale. This means in particular that dimensionless supersymmetry-breaking couplings should be absent.

The possible soft supersymmetry-breaking terms in the Lagrangian of a general theory are

$$\begin{aligned}\mathcal{L}_{\text{soft}} &= -\left(\frac{1}{2}M_a\,\lambda^a\lambda^a + \frac{1}{6}a^{ijk}\phi_i\phi_j\phi_k + \frac{1}{2}b^{ij}\phi_i\phi_j + t^i\phi_i\right) + \text{c.c.} \\ &\quad -(m^2)^i_j\phi^{j*}\phi_i,\end{aligned} \tag{1.138}$$

$$\mathcal{L}_{\text{maybe soft}} = -\frac{1}{2}c_i^{jk}\phi^{*i}\phi_j\phi_k + \text{c.c.} \tag{1.139}$$

They consist of gaugino masses $M_a$ for each gauge group, scalar squared-mass terms $(m^2)_i^j$ and $b^{ij}$, and (scalar)$^3$ couplings $a^{ijk}$ and $c_i^{jk}$, and "tadpole" couplings $t^i$. The last of these can only occur if $\phi_i$ is a gauge singlet, and so is absent from the MSSM. One might wonder why we have not included possible soft mass terms for the chiral supermultiplet fermions, like $\mathcal{L} = -\frac{1}{2}m^{ij}\psi_i\psi_j + \text{c.c.}$ Including such terms would be redundant; they can always be absorbed into a redefinition of the superpotential and the terms $(m^2)^i_j$ and $c_i^{jk}$.

It has been shown rigorously that a softly broken supersymmetric theory with $\mathcal{L}_{\text{soft}}$ as given by eq. (1.138) is indeed free of quadratic divergences in quantum corrections to scalar masses, to all orders in perturbation theory.[53] The situation is slightly more subtle if one tries to include the non-analytic (scalar)$^3$ couplings in $\mathcal{L}_{\text{maybe soft}}$. If any of the chiral supermultiplets in the theory are singlets under all gauge symmetries, then non-zero $c_i^{jk}$ terms can lead to quadratic divergences, despite the fact that they are formally soft. Now, this constraint need not apply to the MSSM, which does not have any gauge-singlet chiral supermultiplets. Nevertheless, the possibility of $c_i^{jk}$ terms is nearly always neglected. The real reason for this is that it is difficult to construct models of spontaneous supersymmetry breaking in which the $c_i^{jk}$ are not negligibly small. In the special case of a theory that has chiral supermultiplets that are singlets or in the adjoint representation of a simple factor of the gauge group, then there are also possible soft supersymmetry-breaking Dirac mass terms between the corresponding fermions $\psi_a$ and the gauginos:[54–59]

$$\mathcal{L} = -M^a_{\text{Dirac}}\lambda^a\psi_a + \text{c.c.} \tag{1.140}$$

This is not relevant for the MSSM with minimal field content, which does not have adjoint representation chiral supermultiplets. Therefore, equation (1.138) is usually taken to be the general form of the soft supersymmetry-breaking Lagrangian. For some interesting exceptions, see Refs. 54–65.

The terms in $\mathcal{L}_{\text{soft}}$ clearly do break supersymmetry, because they involve only scalars and gauginos and not their respective superpartners. In fact, the soft terms in $\mathcal{L}_{\text{soft}}$ are capable of giving masses to all of the scalars and gauginos in a theory, even if the gauge bosons and fermions in chiral supermultiplets are massless (or relatively light). The gaugino masses $M_a$ are always allowed by gauge symmetry. The $(m^2)^i_j$ terms are allowed for $i, j$ such that $\phi_i$, $\phi^{j*}$ transform in complex conjugate representations of each other under all gauge symmetries; in particular this is true of course when $i = j$, so every scalar is eligible to get a mass in this way if supersymmetry is broken. The remaining soft terms may or may not be allowed by the symmetries. The $a^{ijk}$, $b^{ij}$, and $t^i$ terms have the same form as the $y^{ijk}$, $M^{ij}$, and $L^i$ terms in the superpotential [compare eq. (1.138) to eq. (1.107)

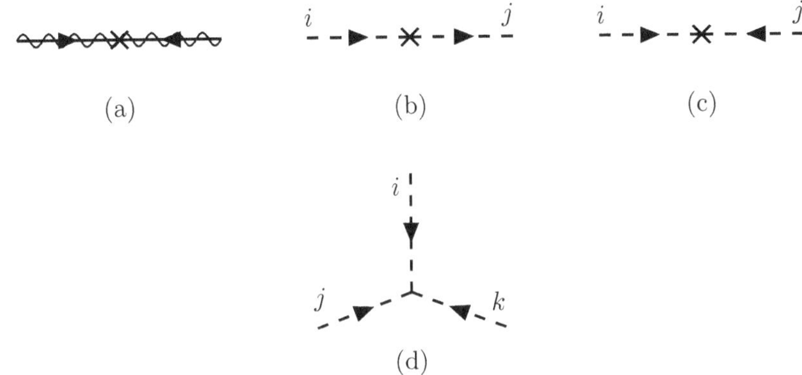


Fig. 1.6. Soft supersymmetry-breaking terms: (a) Gaugino mass $M_a$; (b) non-analytic scalar squared mass $(m^2)^i_j$; (c) analytic scalar squared mass $b^{ij}$; and (d) scalar cubic coupling $a^{ijk}$.

or eq. (1.137)], so they will each be allowed by gauge invariance if and only if a corresponding superpotential term is allowed.

The Feynman diagram interactions corresponding to the allowed soft terms in eq. (1.138) are shown in Figure 1.6. For each of the interactions in Figures 1.6a,c,d there is another with all arrows reversed, corresponding to the complex conjugate term in the Lagrangian. We will apply these general results to the specific case of the MSSM in the next section.

## 1.5. The Minimal Supersymmetric Standard Model

In sections 1.3 and 1.4, we have found a general recipe for constructing Lagrangians for softly broken supersymmetric theories. We are now ready to apply these general results to the MSSM. The particle content for the MSSM was described in the Introduction. In this section we will complete the model by specifying the superpotential and the soft supersymmetry-breaking terms.

### 1.5.1. *The superpotential and supersymmetric interactions*

The superpotential for the MSSM is

$$W_{\rm MSSM} = \overline{u}\mathbf{y_u}QH_u - \overline{d}\mathbf{y_d}QH_d - \overline{e}\mathbf{y_e}LH_d + \mu H_u H_d \,. \tag{1.141}$$

The objects $H_u$, $H_d$, $Q$, $L$, $\overline{u}$, $\overline{d}$, $\overline{e}$ appearing here are chiral superfields corresponding to the chiral supermultiplets in Table 1.1. (Alternatively, they can be just thought of as the corresponding scalar fields, as was done in section 1.3, but we prefer not to put the tildes on $Q$, $L$, $\overline{u}$, $\overline{d}$, $\overline{e}$ in order to reduce clutter.) The dimensionless Yukawa coupling parameters $\mathbf{y_u}, \mathbf{y_d}, \mathbf{y_e}$ are 3×3 matrices in family space. All of the gauge [$SU(3)_C$ color and $SU(2)_L$ weak isospin] and family indices in eq. (1.141) are suppressed. The "$\mu$ term", as it is traditionally called, can be written out as $\mu(H_u)_\alpha(H_d)_\beta\epsilon^{\alpha\beta}$, where $\epsilon^{\alpha\beta}$ is used to tie together $SU(2)_L$ weak isospin indices $\alpha, \beta = 1, 2$ in a gauge-invariant way. Likewise, the term $\overline{u}\mathbf{y_u}QH_u$ can be written out as $\overline{u}^{ia}\,(\mathbf{y_u})_i{}^j\,Q_{j\alpha a}\,(H_u)_\beta\epsilon^{\alpha\beta}$, where $i = 1, 2, 3$ is a family index, and $a = 1, 2, 3$ is a color index which is lowered (raised) in the $\mathbf{3}$ ($\overline{\mathbf{3}}$) representation of $SU(3)_C$.

The $\mu$ term in eq. (1.141) is the supersymmetric version of the Higgs boson mass in the Standard Model. It is unique, because terms $H_u^* H_u$ or $H_d^* H_d$ are forbidden in the superpotential, which must be analytic in the chiral superfields (or equivalently in the scalar fields) treated as complex variables, as shown in section 1.3.2. We can also see from the form of eq. (1.141) why both $H_u$ and $H_d$ are needed in order to give Yukawa couplings, and thus masses, to all of the quarks and leptons. Since the superpotential must be analytic, the $\overline{u}QH_u$ Yukawa terms cannot be replaced by something like $\overline{u}QH_d^*$. Similarly, the $\overline{d}QH_d$ and $\overline{e}LH_d$ terms cannot be replaced by something like $\overline{d}QH_u^*$ and $\overline{e}LH_u^*$. The analogous Yukawa couplings would be allowed in a general non-supersymmetric two Higgs doublet model, but are forbidden by the structure of supersymmetry. So we need both $H_u$ and $H_d$, even without invoking the argument based on anomaly cancellation mentioned in the Introduction.

The Yukawa matrices determine the current masses and CKM mixing angles of the ordinary quarks and leptons, after the neutral scalar components of $H_u$ and $H_d$ get VEVs. Since the top quark, bottom quark and tau lepton are the heaviest fermions in the Standard Model, it is often useful to make an approximation that only the $(3, 3)$ family components of each of $\mathbf{y_u}$, $\mathbf{y_d}$ and $\mathbf{y_e}$ are important:

$$\mathbf{y_u} \approx \begin{pmatrix} 0 & 0 & 0 \\ 0 & 0 & 0 \\ 0 & 0 & y_t \end{pmatrix}, \quad \mathbf{y_d} \approx \begin{pmatrix} 0 & 0 & 0 \\ 0 & 0 & 0 \\ 0 & 0 & y_b \end{pmatrix}, \quad \mathbf{y_e} \approx \begin{pmatrix} 0 & 0 & 0 \\ 0 & 0 & 0 \\ 0 & 0 & y_\tau \end{pmatrix}. \tag{1.142}$$

In this limit, only the third family and Higgs fields contribute to the MSSM superpotential. It is instructive to write the superpotential in terms of the separate $SU(2)_L$ weak isospin components $[Q_3 = (t\,b)$, $L_3 = (\nu_\tau\,\tau)$, $H_u = (H_u^+\,H_u^0)$, $H_d = (H_d^0\,H_d^-)$, $\overline{u}_3 = \overline{t}$, $\overline{d}_3 = \overline{b}$, $\overline{e}_3 = \overline{\tau}]$, so:

$$W_{\rm MSSM} \approx y_t(\overline{t}tH_u^0 - \overline{t}bH_u^+) - y_b(\overline{b}tH_d^- - \overline{b}bH_d^0) - y_\tau(\overline{\tau}\nu_\tau H_d^- - \overline{\tau}\tau H_d^0)$$
$$+\mu(H_u^+H_d^- - H_u^0H_d^0). \qquad (1.143)$$

The minus signs inside the parentheses appear because of the antisymmetry of the $\epsilon^{\alpha\beta}$ symbol used to tie up the $SU(2)_L$ indices. The other minus signs in eq. (1.141) were chosen so that the terms $y_t\overline{t}tH_u^0$, $y_b\overline{b}bH_d^0$, and $y_\tau\overline{\tau}\tau H_d^0$, which will become the top, bottom and tau masses when $H_u^0$ and $H_d^0$ get VEVs, each have overall positive signs in eq. (1.143).

Since the Yukawa interactions $y^{ijk}$ in a general supersymmetric theory must be completely symmetric under interchange of $i, j, k$, we know that $\mathbf{y_u}$, $\mathbf{y_d}$ and $\mathbf{y_e}$ imply not only Higgs-quark-quark and Higgs-lepton-lepton couplings as in the Standard Model, but also squark-Higgsino-quark and slepton-Higgsino-lepton interactions. To illustrate this, Figures 1.7a,b,c show some of the interactions involving the top-quark Yukawa coupling $y_t$.

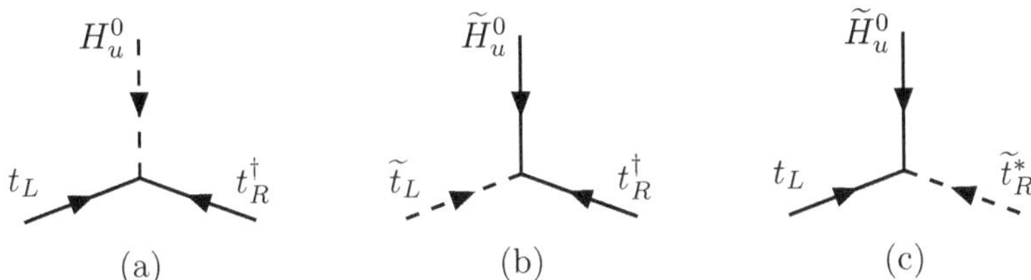


Fig. 1.7. The top-quark Yukawa coupling (a) and its "supersymmetrizations" (b), (c), all of strength $y_t$.

Figure 1.7a is the Standard Model-like coupling of the top quark to the neutral complex scalar Higgs boson, which follows from the first term in eq. (1.143). For variety, we have used $t_L$ and $t_R^\dagger$ in place of their synonyms $t$ and $\overline{t}$ (see the discussion near the end of section 1.2). In Figure 1.7b, we have the coupling of the left-handed top squark $\widetilde{t}_L$ to the neutral higgsino field $\widetilde{H}_u^0$ and right-handed top quark, while in Figure 1.7c the right-handed top anti-squark field (known either as $\widetilde{\overline{t}}$ or $\widetilde{t}_R^*$ depending on taste) couples to $\widetilde{H}_u^0$ and $t_L$. For each of the three interactions, there is another with $H_u^0 \to H_u^+$ and $t_L \to -b_L$ (with tildes where appropriate), corresponding to the second part of the first term in eq. (1.143). All of these interactions are required by supersymmetry to have the same strength $y_t$. These couplings are dimensionless and can be modified by the introduction of soft supersymmetry breaking only through finite (and small) radiative correc-

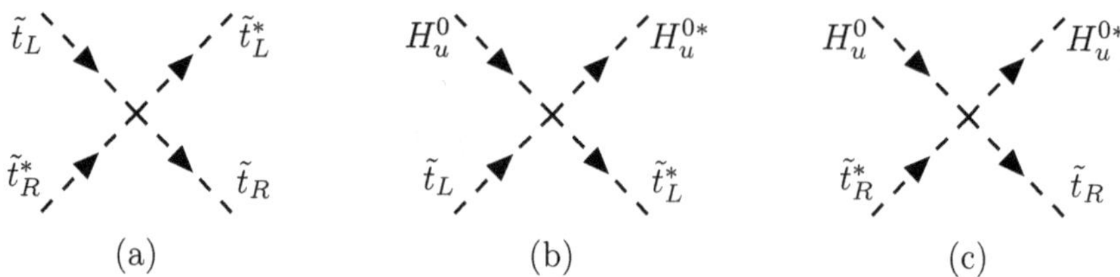


Fig. 1.8. Some of the (scalar)$^4$ interactions with strength proportional to $y_t^2$.

tions, so this equality of interaction strengths is also a prediction of softly broken supersymmetry. A useful mnemonic is that each of Figures 1.7a,b,c can be obtained from any of the others by changing two of the particles into their superpartners.

There are also scalar quartic interactions with strength proportional to $y_t^2$, as can be seen from Figure 1.3c or the last term in eq. (1.111). Three of them are shown in Figure 1.8. Using eq. (1.111) and eq. (1.143), one can see that there are five more, which can be obtained by replacing $\widetilde{t}_L \to \widetilde{b}_L$ and/or $H_u^0 \to H_u^+$ in each vertex. This illustrates the remarkable economy of supersymmetry; there are many interactions determined by only a single parameter. In a similar way, the existence of all the other quark and lepton Yukawa couplings in the superpotential eq. (1.141) leads not only to Higgs-quark-quark and Higgs-lepton-lepton Lagrangian terms as in the ordinary Standard Model, but also to squark-higgsino-quark and slepton-higgsino-lepton terms, and scalar quartic couplings [(squark)$^4$, (slepton)$^4$, (squark)$^2$(slepton)$^2$, (squark)$^2$(Higgs)$^2$, and (slepton)$^2$(Higgs)$^2$]. If needed, these can all be obtained in terms of the Yukawa matrices $\mathbf{y_u}$, $\mathbf{y_d}$, and $\mathbf{y_e}$ as outlined above.

However, the dimensionless interactions determined by the superpotential are usually not the most important ones of direct interest for phenomenology. This is because the Yukawa couplings are already known to be very small, except for those of the third family (top, bottom, tau). Instead, production and decay processes for superpartners in the MSSM are typically dominated by the supersymmetric interactions of gauge-coupling strength, as we will explore in more detail in sections 1.8. The couplings of the Standard Model gauge bosons (photon, $W^\pm$, $Z^0$ and gluons) to the MSSM particles are determined completely by the gauge invariance of the kinetic terms in the Lagrangian. The gauginos also couple to (squark, quark) and (slepton, lepton) and (Higgs, higgsino) pairs as illustrated in the general case in Figure 1.5g,h and the first two terms in the second line in eq. (1.132). For instance, each of the squark-quark-gluino couplings is

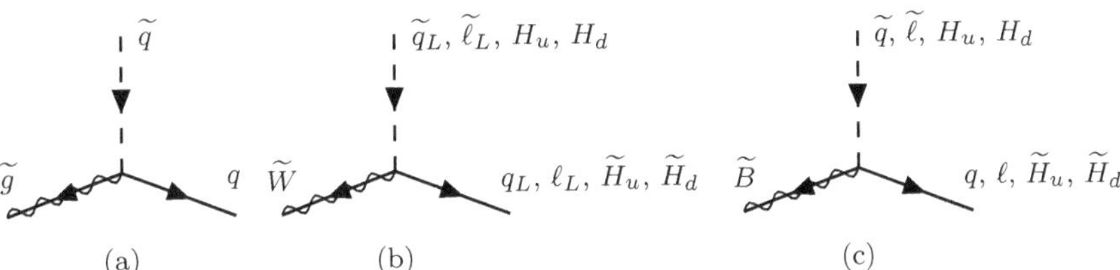


Fig. 1.9. Couplings of the gluino, wino, and bino to MSSM (scalar, fermion) pairs.

given by $\sqrt{2}g_3(\widetilde{q}T^a q\widetilde{g} + \text{c.c.})$ where $T^a = \lambda^a/2$ $(a = 1\ldots 8)$ are the matrix generators for $SU(3)_C$. The Feynman diagram for this interaction is shown in Figure 1.9a. In Figures 1.9b,c we show in a similar way the couplings of (squark, quark), (lepton, slepton) and (Higgs, higgsino) pairs to the winos and bino, with strengths proportional to the electroweak gauge couplings $g$ and $g'$ respectively. For each of these diagrams, there is another with all arrows reversed. Note that the winos only couple to the left-handed squarks and sleptons, and the (lepton, slepton) and (Higgs, higgsino) pairs of course do not couple to the gluino. The bino coupling to each (scalar, fermion) pair is also proportional to the weak hypercharge $Y$ as given in Table 1.1. The interactions shown in Figure 1.9 provide, for example, for decays $\widetilde{q} \to q\widetilde{g}$ and $\widetilde{q} \to \widetilde{W}q'$ and $\widetilde{q} \to \widetilde{B}q$ when the final states are kinematically allowed to be on-shell. However, a complication is that the $\widetilde{W}$ and $\widetilde{B}$ states are not mass eigenstates, because of splitting and mixing due to electroweak symmetry breaking, as we will see in section 1.7.2.

There are also various scalar quartic interactions in the MSSM that are uniquely determined by gauge invariance and supersymmetry, according to the last term in eq. (1.135), as illustrated in Figure 1.5i. Among them are (Higgs)$^4$ terms proportional to $g^2$ and $g'^2$ in the scalar potential. These are the direct generalization of the last term in the Standard Model Higgs potential, eq. (1.1), to the case of the MSSM. We will have occasion to identify them explicitly when we discuss the minimization of the MSSM Higgs potential in section 1.7.1.

The dimensionful couplings in the supersymmetric part of the MSSM Lagrangian are all dependent on $\mu$. Using the general result of eq. (1.111), $\mu$ provides for higgsino fermion mass terms

$$-\mathcal{L}_{\text{higgsino mass}} = \mu(\widetilde{H}_u^+\widetilde{H}_d^- - \widetilde{H}_u^0\widetilde{H}_d^0) + \text{c.c.}, \tag{1.144}$$

as well as Higgs squared-mass terms in the scalar potential

$$-\mathcal{L}_{\text{supersymmetric Higgs mass}} = |\mu|^2\big(|H_u^0|^2 + |H_u^+|^2 + |H_d^0|^2 + |H_d^-|^2\big). \tag{1.145}$$

Since eq. (1.145) is non-negative with a minimum at $H_u^0 = H_d^0 = 0$, we cannot understand electroweak symmetry breaking without including a negative supersymmetry-breaking squared-mass soft term for the Higgs scalars. An explicit treatment of the Higgs scalar potential will therefore have to wait until we have introduced the soft terms for the MSSM. However, we can already see a puzzle: we expect that $\mu$ should be roughly of order $10^2$ or $10^3$ GeV, in order to allow a Higgs VEV of order 174 GeV without too much miraculous cancellation between $|\mu|^2$ and the negative soft squared-mass terms that we have not written down yet. But why should $|\mu|^2$ be so small compared to, say, $M_{\rm P}^2$, and in particular why should it be roughly of the same order as $m_{\rm soft}^2$? The scalar potential of the MSSM seems to depend on two types of dimensionful parameters that are conceptually quite distinct, namely the supersymmetry-respecting mass $\mu$ and the supersymmetry-breaking soft mass terms. Yet the observed value for the electroweak breaking scale suggests that without miraculous cancellations, both of these apparently unrelated mass scales should be within an order of magnitude or so of 100 GeV. This puzzle is called "the $\mu$ problem". Several different solutions to the $\mu$ problem have been proposed, involving extensions of the MSSM of varying intricacy. They all work in roughly the same way; the $\mu$ term is required or assumed to be absent at tree-level before symmetry breaking, and then it arises from the VEV(s) of some new field(s). These VEVs are in turn determined by minimizing a potential that depends on soft supersymmetry-breaking terms. In this way, the value of the effective parameter $\mu$ is no longer conceptually distinct from the mechanism of supersymmetry breaking; if we can explain why $m_{\rm soft} \ll M_{\rm P}$, we will also be able to understand why $\mu$ is of the same order. Some other attractive solutions for the $\mu$ problem are proposed in Refs. 66–68. From the point of view of the MSSM, however, we can just treat $\mu$ as an independent parameter.

The $\mu$-term and the Yukawa couplings in the superpotential eq. (1.141) combine to yield (scalar)$^3$ couplings [see the second and third terms on the right-hand side of eq. (1.111)] of the form

$$\mathcal{L}_{\text{supersymmetric (scalar)}^3} = \mu^* (\widetilde{\overline{u}} \mathbf{y_u} \widetilde{u} H_d^{0*} + \widetilde{\overline{d}} \mathbf{y_d} \widetilde{d} H_u^{0*} + \widetilde{\overline{e}} \mathbf{y_e} \widetilde{e} H_u^{0*} + \widetilde{\overline{u}} \mathbf{y_u} \widetilde{d} H_d^{-*} + \widetilde{\overline{d}} \mathbf{y_d} \widetilde{u} H_u^{+*} + \widetilde{\overline{e}} \mathbf{y_e} \widetilde{\nu} H_u^{+*}) + \text{c.c.} \tag{1.146}$$

Figure 1.10 shows some of these couplings, proportional to $\mu^* y_t$, $\mu^* y_b$, and $\mu^* y_\tau$ respectively. These play an important role in determining the mixing of top squarks, bottom squarks, and tau sleptons, as we will see in section 1.7.4.

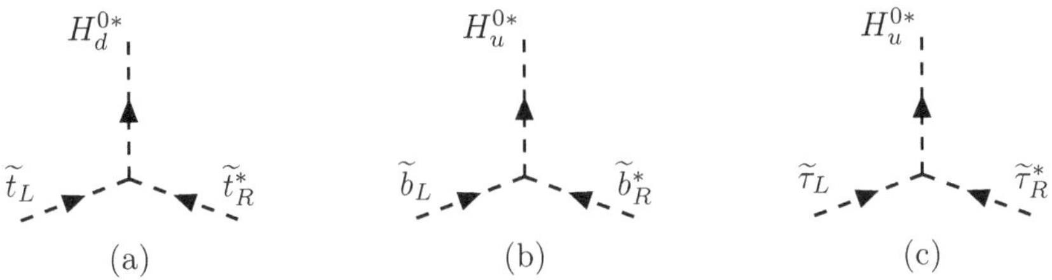


Fig. 1.10. Some of the supersymmetric (scalar)$^3$ couplings proportional to $\mu^* y_t$, $\mu^* y_b$, and $\mu^* y_\tau$. When $H_u^0$ and $H_d^0$ get VEVs, these contribute to (a) $\widetilde{t}_L, \widetilde{t}_R$ mixing, (b) $\widetilde{b}_L, \widetilde{b}_R$ mixing, and (c) $\widetilde{\tau}_L, \widetilde{\tau}_R$ mixing.

### 1.5.2. *R-parity (also known as matter parity) and its consequences*

The superpotential eq. (1.141) is minimal in the sense that it is sufficient to produce a phenomenologically viable model. However, there are other terms that one can write that are gauge-invariant and analytic in the chiral superfields, but are not included in the MSSM because they violate either baryon number (B) or total lepton number (L). The most general gauge-invariant and renormalizable superpotential would include not only eq. (1.141), but also the terms

$$W_{\Delta \mathrm{L}=1} = \frac{1}{2}\lambda^{ijk} L_i L_j \overline{e}_k + \lambda'^{ijk} L_i Q_j \overline{d}_k + \mu'^i L_i H_u \tag{1.147}$$

$$W_{\Delta \mathrm{B}=1} = \frac{1}{2}\lambda''^{ijk} \overline{u}_i \overline{d}_j \overline{d}_k \tag{1.148}$$

where family indices $i = 1, 2, 3$ have been restored. The chiral supermultiplets carry baryon number assignments B $= +1/3$ for $Q_i$; B $= -1/3$ for $\overline{u}_i, \overline{d}_i$; and B $= 0$ for all others. The total lepton number assignments are L $= +1$ for $L_i$, L $= -1$ for $\overline{e}_i$, and L $= 0$ for all others. Therefore, the terms in eq. (1.147) violate total lepton number by 1 unit (as well as the individual lepton flavors) and those in eq. (1.148) violate baryon number by 1 unit.

The possible existence of such terms might seem rather disturbing, since corresponding B- and L-violating processes have not been seen experimentally. The most obvious experimental constraint comes from the non-observation of proton decay, which would violate both B and L by 1 unit. If both $\lambda'$ and $\lambda''$ couplings were present and unsuppressed, then the lifetime of the proton would be extremely short. For example, Feynman

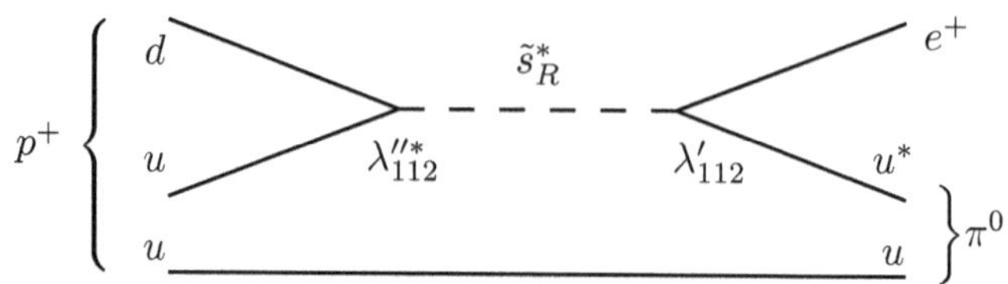


Fig. 1.11. Squarks would mediate disastrously rapid proton decay if $R$-parity were violated by both $\Delta$B $=1$ and $\Delta$L $=1$ interactions. This example shows $p \to e^+\pi^0$ mediated by a strange (or bottom) squark.

diagrams like the one in Figure 1.11[j] would lead to $p^+ \to e^+\pi^0$ (shown) or $e^+K^0$ or $\mu^+\pi^0$ or $\mu^+K^0$ or $\nu\pi^+$ or $\nu K^+$ etc. depending on which components of $\lambda'$ and $\lambda''$ are largest.[k] As a rough estimate based on dimensional analysis, for example,

$$\Gamma_{p\to e^+\pi^0} \sim m^5_{\rm proton} \sum_{i=2,3} |\lambda'^{11i}\lambda''^{11i}|^2/m^4_{\tilde{d}_i}, \tag{1.149}$$

which would be a tiny fraction of a second if the couplings were of order unity and the squarks have masses of order 1 TeV. In contrast, the decay time of the proton into lepton+meson final states is known experimentally to be in excess of $10^{32}$ years. Therefore, at least one of $\lambda'^{ijk}$ or $\lambda''^{11k}$ for each of $i = 1, 2$; $j = 1, 2$; $k = 2, 3$ must be extremely small. Many other processes also give strong constraints on the violation of lepton and baryon numbers.[69,70]

One could simply try to take B and L conservation as a postulate in the MSSM. However, this is clearly a step backward from the situation in the Standard Model, where the conservation of these quantum numbers is *not* assumed, but is rather a pleasantly "accidental" consequence of the fact that there are no possible renormalizable Lagrangian terms that violate B or L. Furthermore, there is a quite general obstacle to treating B and L as fundamental symmetries of Nature, since they are known to be necessarily violated by non-perturbative electroweak effects[71] (even though those effects are calculably negligible for experiments at ordinary energies). Therefore, in the MSSM one adds a new symmetry, which has the effect

[j]In this diagram and others below, the arrows on propagators are often omitted for simplicity, and external fermion label refer to physical particle states rather than 2-component fermion fields.

[k]The coupling $\lambda''$ must be antisymmetric in its last two flavor indices, since the color indices are combined antisymmetrically. That is why the squark in Figure 1.11 can be $\widetilde{s}$ or $\widetilde{b}$, but not $\widetilde{d}$, for $u, d$ quarks in the proton.

of eliminating the possibility of B and L violating terms in the renormalizable superpotential, while allowing the good terms in eq. (1.141). This new symmetry is called "$R$-parity"[7] or equivalently "matter parity".[72]

Matter parity is a multiplicatively conserved quantum number defined as

$$P_M = (-1)^{3(\mathrm{B}-\mathrm{L})} \tag{1.150}$$

for each particle in the theory. It is easy to check that the quark and lepton supermultiplets all have $P_M = -1$, while the Higgs supermultiplets $H_u$ and $H_d$ have $P_M = +1$. The gauge bosons and gauginos of course do not carry baryon number or lepton number, so they are assigned matter parity $P_M = +1$. The symmetry principle to be enforced is that a candidate term in the Lagrangian (or in the superpotential) is allowed only if the product of $P_M$ for all of the fields in it is $+1$. It is easy to see that each of the terms in eqs. (1.147) and (1.148) is thus forbidden, while the good and necessary terms in eq. (1.141) are allowed. This discrete symmetry commutes with supersymmetry, as all members of a given supermultiplet have the same matter parity. The advantage of matter parity is that it can in principle be an *exact* and fundamental symmetry, which B and L themselves cannot, since they are known to be violated by non-perturbative electroweak effects. So even with exact matter parity conservation in the MSSM, one expects that baryon number and total lepton number violation can occur in tiny amounts, due to non-renormalizable terms in the Lagrangian. However, the MSSM does not have renormalizable interactions that violate B or L, with the standard assumption of matter parity conservation.

It is often useful to recast matter parity in terms of $R$-parity, defined for each particle as

$$P_R = (-1)^{3(\mathrm{B}-\mathrm{L})+2s} \tag{1.151}$$

where $s$ is the spin of the particle. Now, matter parity conservation and $R$-parity conservation are precisely equivalent, since the product of $(-1)^{2s}$ for the particles involved in any interaction vertex in a theory that conserves angular momentum is always equal to $+1$. However, particles within the same supermultiplet do not have the same $R$-parity. In general, symmetries with the property that fields within the same supermultiplet have different transformations are called $R$ symmetries; they do not commute with supersymmetry. Continuous $U(1)$ $R$ symmetries are often encountered in the model-building literature; they should not be confused with $R$-parity, which is a discrete $Z_2$ symmetry. In fact, the matter parity version of $R$-parity makes clear that there is really nothing intrinsically "$R$" about it; in other words it secretly does commute with supersymmetry, so its name is somewhat suboptimal. Nevertheless, the $R$-parity assignment is very useful for phenomenology because all of the Standard Model particles and the Higgs bosons have even $R$-parity ($P_R = +1$), while all of the squarks, sleptons, gauginos, and higgsinos have odd $R$-parity ($P_R = -1$).

The $R$-parity odd particles are known as "supersymmetric particles" or "sparticles" for short, and they are distinguished by a tilde (see Tables 1.1 and 1.2). If $R$-parity is exactly conserved, then there can be no mixing between the sparticles and the $P_R = +1$ particles. Furthermore, every interaction vertex in the theory contains an even number of $P_R = -1$ sparticles. This has three extremely important phenomenological consequences:

- The lightest sparticle with $P_R = -1$, called the "lightest supersymmetric particle" or LSP, must be absolutely stable. If the LSP is electrically neutral, it interacts only weakly with ordinary matter, and so can make an attractive candidate[73] for the non-baryonic dark matter that seems to be required by cosmology.
- Each sparticle other than the LSP must eventually decay into a state that contains an odd number of LSPs (usually just one).
- In collider experiments, sparticles can only be produced in even numbers (usually two-at-a-time).

We *define* the MSSM to conserve $R$-parity or equivalently matter parity. While this decision seems to be well-motivated phenomenologically by proton decay constraints and the hope that the LSP will provide a good dark matter candidate, it might appear somewhat artificial from a theoretical point of view. After all, the MSSM would not suffer any internal inconsistency if we did not impose matter parity conservation. Furthermore, it is fair to ask why matter parity should be exactly conserved, given that the discrete symmetries in the Standard Model (ordinary parity $P$, charge conjugation $C$, time reversal $T$, etc.) are all known to be inexact symmetries. Fortunately, it *is* sensible to formulate matter parity as a discrete symmetry that is exactly conserved. In general, exactly conserved, or "gauged" discrete symmetries[74] can exist provided that they satisfy certain anomaly cancellation conditions[75] (much like continuous gauged symmetries). One particularly attractive way this could occur is if B$-$L is a continuous gauge symmetry that is spontaneously broken at some very high energy scale. A continuous $U(1)_{\rm B-L}$ forbids the renormalizable terms that violate B and L,[76,77] but this gauge symmetry must be spontaneously broken, since there is no corresponding massless vector boson. However, if gauged $U(1)_{\rm B-L}$ is only broken by scalar VEVs (or other order parameters) that carry even integer values of 3(B−L), then $P_M$ will automatically survive as an exactly conserved discrete remnant subgroup.[77] A variety of extensions of the MSSM in which exact $R$-parity conservation is guaranteed in just this way have been proposed (see for example[77,78]).

It may also be possible to have gauged discrete symmetries that do not owe their exact conservation to an underlying continuous gauged symmetry, but rather to some other structure such as can occur in string theory. It is also possible that $R$-parity is broken, or is replaced by some alternative discrete symmetry.

### 1.5.3. *Soft supersymmetry breaking in the MSSM*

To complete the description of the MSSM, we need to specify the soft supersymmetry breaking terms. In section 1.4, we learned how to write down the most general set of such terms in any supersymmetric theory. Applying this recipe to the MSSM, we have:

$$\begin{aligned}
\mathcal{L}_{\rm soft}^{\rm MSSM} = & -\frac{1}{2}\left(M_3\widetilde{g}\widetilde{g} + M_2\widetilde{W}\widetilde{W} + M_1\widetilde{B}\widetilde{B} + {\rm c.c.}\right) \\
& -\left(\widetilde{\overline{u}}\,\mathbf{a_u}\,\widetilde{Q}H_u - \widetilde{\overline{d}}\,\mathbf{a_d}\,\widetilde{Q}H_d - \widetilde{\overline{e}}\,\mathbf{a_e}\,\widetilde{L}H_d + {\rm c.c.}\right) \\
& -\widetilde{Q}^\dagger\,\mathbf{m^2_Q}\,\widetilde{Q} - \widetilde{L}^\dagger\,\mathbf{m^2_L}\,\widetilde{L} - \widetilde{\overline{u}}\,\mathbf{m^2_{\overline{u}}}\,\widetilde{\overline{u}}^\dagger - \widetilde{\overline{d}}\,\mathbf{m^2_{\overline{d}}}\,\widetilde{\overline{d}}^\dagger - \widetilde{\overline{e}}\,\mathbf{m^2_{\overline{e}}}\,\widetilde{\overline{e}}^\dagger \\
& - m^2_{H_u}H^*_uH_u - m^2_{H_d}H^*_dH_d - (bH_uH_d + {\rm c.c.})\,. \qquad (1.152)
\end{aligned}$$

In eq. (1.152), $M_3$, $M_2$, and $M_1$ are the gluino, wino, and bino mass terms. Here, and from now on, we suppress the adjoint representation gauge indices on the wino and gluino fields, and the gauge indices on all of the chiral supermultiplet fields. The second line in eq. (1.152) contains the (scalar)$^3$ couplings [of the type $a^{ijk}$ in eq. (1.138)]. Each of $\mathbf{a_u}$, $\mathbf{a_d}$, $\mathbf{a_e}$ is a complex $3 \times 3$ matrix in family space, with dimensions of [mass]. They are in one-to-one correspondence with the Yukawa couplings of the superpotential. The third line of eq. (1.152) consists of squark and slepton mass terms of the $(m^2)_i^j$ type in eq. (1.138). Each of $\mathbf{m^2_Q}$, $\mathbf{m^2_{\overline{u}}}$, $\mathbf{m^2_{\overline{d}}}$, $\mathbf{m^2_L}$, $\mathbf{m^2_{\overline{e}}}$ is a $3 \times 3$ matrix in family space that can have complex entries, but they must be hermitian so that the Lagrangian is real. (To avoid clutter, we do not put tildes on the $\mathbf{Q}$ in $\mathbf{m^2_Q}$, etc.) Finally, in the last line of eq. (1.152) we have supersymmetry-breaking contributions to the Higgs potential; $m^2_{H_u}$ and $m^2_{H_d}$ are squared-mass terms of the $(m^2)_i^j$ type, while $b$ is the only squared-mass term of the type $b^{ij}$ in eq. (1.138) that can occur in the MSSM.[1] As argued in the Introduction, we expect

$$M_1,\ M_2,\ M_3,\ \mathbf{a_u},\ \mathbf{a_d},\ \mathbf{a_e} \sim m_{\rm soft}, \tag{1.153}$$

$$\mathbf{m^2_Q},\ \mathbf{m^2_L},\ \mathbf{m^2_{\overline{u}}},\ \mathbf{m^2_{\overline{d}}},\ \mathbf{m^2_{\overline{e}}},\ m^2_{H_u},\ m^2_{H_d},\ b \sim m^2_{\rm soft}, \tag{1.154}$$

with a characteristic mass scale $m_{\rm soft}$ that is not much larger than 1000 GeV. The expression eq. (1.152) is the most general soft supersymmetry-breaking Lagrangian of the form eq. (1.138) that is compatible with gauge invariance and matter parity conservation in the MSSM.

Unlike the supersymmetry-preserving part of the Lagrangian, the above $\mathcal{L}^{\rm MSSM}_{\rm soft}$ introduces many new parameters that were not present in the ordinary Standard Model. A careful count[79] reveals that there are 105 masses, phases and mixing angles in the MSSM Lagrangian that cannot be rotated away by redefining the phases and flavor basis for the quark and lepton supermultiplets, and that have no counterpart in the ordinary Standard Model. Thus, in principle, supersymmetry *breaking* (as opposed to supersymmetry itself) appears to introduce a tremendous arbitrariness in the Lagrangian.

### 1.5.4. *Hints of an organizing principle*

Fortunately, there is already good experimental evidence that some powerful organizing principle must govern the soft supersymmetry breaking Lagrangian. This is because most of the new parameters in eq. (1.152) imply flavor mixing or CP violating processes of the types that are severely restricted by experiment.[80–105]

For example, suppose that $\mathbf{m^2_{\overline{e}}}$ is not diagonal in the basis $(\widetilde{e}_R, \widetilde{\mu}_R, \widetilde{\tau}_R)$ of sleptons whose superpartners are the right-handed parts of the Standard Model mass eigenstates $e, \mu, \tau$. In that case, slepton mixing occurs, so the individual lepton numbers will not be conserved, even for processes that only involve the sleptons as virtual particles. A particularly strong limit on this possibility comes from the experimental bound on the process $\mu \rightarrow e\gamma$, which could arise from the one-loop diagram shown in Figure 1.12a. The symbol "×" on the slepton line represents an insertion coming from $-(\mathbf{m^2_{\overline{e}}})_{21}\widetilde{\mu}^*_R\widetilde{e}_R$ in $\mathcal{L}^{\rm MSSM}_{\rm soft}$, and the slepton-bino vertices are determined by the weak hypercharge gauge coupling [see Figures 1.5g,h and eq. (1.132)].

[1]The parameter called $b$ here is often seen elsewhere as $B\mu$ or $m^2_{12}$ or $m^2_3$.

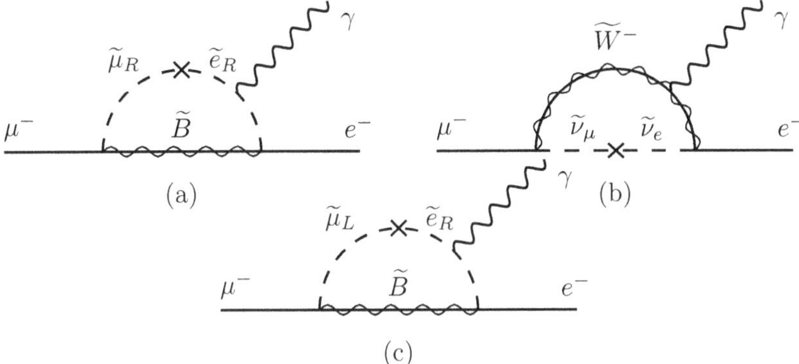


Fig. 1.12. Some of the diagrams that contribute to the process $\mu^- \to e^-\gamma$ in models with lepton flavor-violating soft supersymmetry breaking parameters (indicated by $\times$). Diagrams (a), (b), and (c) contribute to constraints on the off-diagonal elements of $\mathbf{m^2_{\bar{e}}}$, $\mathbf{m^2_{L}}$, and $\mathbf{a_e}$, respectively.

The result of calculating this diagram gives,[82,85] approximately,

$$\mathrm{Br}(\mu \to e\gamma) = \left(\frac{|m^2_{\tilde{\mu}^*_R \tilde{e}_R}|}{m^2_{\tilde{\ell}_R}}\right)^2 \left(\frac{100\,\mathrm{GeV}}{m_{\tilde{\ell}_R}}\right)^4 10^{-6} \times \begin{cases} 15 & \text{for } m_{\tilde{B}} \ll m_{\tilde{\ell}_R}, \\ 5.6 & \text{for } m_{\tilde{B}} = 0.5 m_{\tilde{\ell}_R}, \\ 1.4 & \text{for } m_{\tilde{B}} = m_{\tilde{\ell}_R}, \\ 0.13 & \text{for } m_{\tilde{B}} = 2 m_{\tilde{\ell}_R}, \end{cases} \tag{1.155}$$

where it is assumed for simplicity that both $\tilde{e}_R$ and $\tilde{\mu}_R$ are nearly mass eigenstates with almost degenerate squared masses $m^2_{\tilde{\ell}_R}$, that $m^2_{\tilde{\mu}^*_R \tilde{e}_R} \equiv (\mathbf{m^2_{\bar{e}}})_{21} = [(\mathbf{m^2_{\bar{e}}})_{12}]^*$ can be treated as a perturbation, and that the bino $\widetilde{B}$ is nearly a mass eigenstate. This result is to be compared to the present experimental upper limit $\mathrm{Br}(\mu \to e\gamma)_{\mathrm{exp}} < 1.2 \times 10^{-11}$ from.[106] So, if the right-handed slepton squared-mass matrix $\mathbf{m^2_{\bar{e}}}$ were "random", with all entries of comparable size, then the prediction for $\mathrm{Br}(\mu \to e\gamma)$ would be too large even if the sleptons and bino masses were at 1 TeV. For lighter superpartners, the constraint on $\tilde{\mu}_R, \tilde{e}_R$ squared-mass mixing becomes correspondingly more severe. There are also contributions to $\mu \to e\gamma$ that depend on the off-diagonal elements of the left-handed slepton squared-mass matrix $\mathbf{m^2_{L}}$, coming from the diagram shown in Figure 1.12b involving the charged wino and the sneutrinos, as well as diagrams just like Figure 1.12a but with left-handed sleptons and either $\widetilde{B}$ or $\widetilde{W}^0$ exchanged. Therefore, the slepton squared-mass matrices must not have significant mixings for $\widetilde{e}_L, \widetilde{\mu}_L$ either.

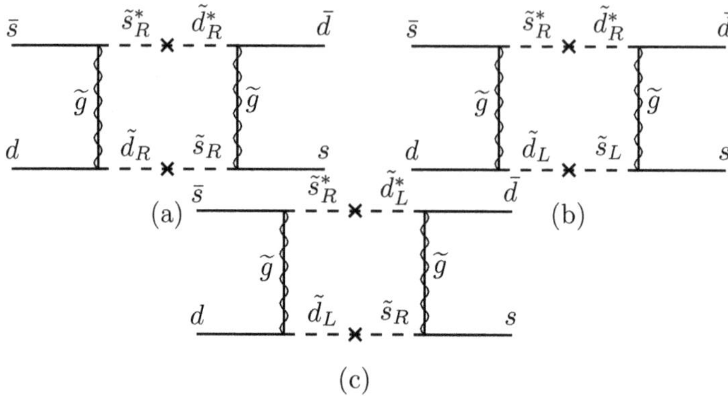


Fig. 1.13. Some of the diagrams that contribute to $K^0 \leftrightarrow \overline{K}^0$ mixing in models with strangeness-violating soft supersymmetry breaking parameters (indicated by $\times$). These diagrams contribute to constraints on the off-diagonal elements of (a) $\mathbf{m^2_{\overline{d}}}$, (b) the combination of $\mathbf{m^2_{\overline{d}}}$ and $\mathbf{m^2_{Q}}$, and (c) $\mathbf{a_d}$.

Furthermore, after the Higgs scalars get VEVs, the $\mathbf{a_e}$ matrix could imply squared-mass terms that mix left-handed and right-handed sleptons with different lepton flavors. For example, $\mathcal{L}_{\rm soft}^{\rm MSSM}$ contains $\widetilde{\bar{e}}\mathbf{a_e}\widetilde{L}H_d + \text{c.c.}$ which implies terms $-\langle H_d^0\rangle(\mathbf{a_e})_{12}\widetilde{e}_R^*\widetilde{\mu}_L - \langle H_d^0\rangle(\mathbf{a_e})_{21}\widetilde{\mu}_R^*\widetilde{e}_L + \text{c.c.}$ These also contribute to $\mu \to e\gamma$, as illustrated in Figure 1.12c. So the magnitudes of $(\mathbf{a_e})_{12}$ and $(\mathbf{a_e})_{21}$ are also constrained by experiment to be small, but in a way that is more strongly dependent on other model parameters.[85] Similarly, $(\mathbf{a_e})_{13}, (\mathbf{a_e})_{31}$ and $(\mathbf{a_e})_{23}, (\mathbf{a_e})_{32}$ are constrained, although more weakly,[86] by the experimental limits on $\text{Br}(\tau \to e\gamma)$ and $\text{Br}(\tau \to \mu\gamma)$.

There are also important experimental constraints on the squark squared-mass matrices. The strongest of these come from the neutral kaon system. The effective Hamiltonian for $K^0 \leftrightarrow \overline{K}^0$ mixing gets contributions from the diagrams in Figure 1.13, among others, if $\mathcal{L}_{\rm soft}^{\rm MSSM}$ contains terms that mix down squarks and strange squarks. The gluino-squark-quark vertices in Figure 1.13 are all fixed by supersymmetry to be of QCD interaction strength. (There are similar diagrams in which the bino and winos are exchanged, which can be important depending on the relative sizes of the gaugino masses.) For example, suppose that there is a non-zero right-handed down-squark squared-mass mixing $(\mathbf{m^2_{\overline{d}}})_{21}$ in the basis corresponding to the quark mass eigenstates. Assuming that the supersymmetric correction to $\Delta m_K \equiv m_{K_L} - m_{K_S}$ following from Figure 1.13a and others does not exceed, in absolute value, the experimental value $3.5 \times 10^{-12}$ MeV,

Ref. 95 obtains:

$$\frac{|\mathrm{Re}[(m^2_{\tilde{s}^*_R \tilde{d}_R})^2]|^{1/2}}{m^2_{\tilde{q}}} < \left(\frac{m_{\tilde{q}}}{500\,\mathrm{GeV}}\right) \times \begin{cases} 0.02 \text{ for } m_{\tilde{g}} = 0.5 m_{\tilde{q}}, \\ 0.05 \text{ for } m_{\tilde{g}} = m_{\tilde{q}}, \\ 0.11 \text{ for } m_{\tilde{g}} = 2 m_{\tilde{q}}. \end{cases} \tag{1.156}$$

Here nearly degenerate squarks with mass $m_{\tilde{q}}$ are assumed for simplicity, with $m^2_{\tilde{s}^*_R \tilde{d}_R} = (\mathbf{m^2_{\overline{d}}})_{21}$ treated as a perturbation. The same limit applies when $m^2_{\tilde{s}^*_R \tilde{d}_R}$ is replaced by $m^2_{\tilde{s}^*_L \tilde{d}_L} = (\mathbf{m^2_Q})_{21}$, in a basis corresponding to the down-type quark mass eigenstates. An even more striking limit applies to the combination of both types of flavor mixing when they are comparable in size, from diagrams including Figure 1.13b. The numerical constraint is:[95]

$$\frac{|\mathrm{Re}[m^2_{\tilde{s}^*_R \tilde{d}_R} m^2_{\tilde{s}^*_L \tilde{d}_L}]|^{1/2}}{m^2_{\tilde{q}}} < \left(\frac{m_{\tilde{q}}}{500\,\mathrm{GeV}}\right) \times \begin{cases} 0.0008 \text{ for } m_{\tilde{g}} = 0.5 m_{\tilde{q}}, \\ 0.0010 \text{ for } m_{\tilde{g}} = m_{\tilde{q}}, \\ 0.0013 \text{ for } m_{\tilde{g}} = 2 m_{\tilde{q}}. \end{cases} \tag{1.157}$$

An off-diagonal contribution from $\mathbf{a_d}$ would cause flavor mixing between left-handed and right-handed squarks, just as discussed above for sleptons, resulting in a strong constraint from diagrams like Figure 1.13c. More generally, limits on $\Delta m_K$ and $\epsilon$ and $\epsilon'/\epsilon$ appearing in the neutral kaon effective Hamiltonian severely restrict the amounts of $\widetilde{d}_{L,R}$, $\widetilde{s}_{L,R}$ squark mixings (separately and in various combinations), and associated CP-violating complex phases, that one can tolerate in the soft squared masses.

Weaker, but still interesting, constraints come from the $D^0, \overline{D}^0$ system, which limits the amounts of $\widetilde{u}, \widetilde{c}$ mixings from $\mathbf{m^2_{\overline{u}}}$, $\mathbf{m^2_Q}$ and $\mathbf{a_u}$. The $B^0_d, \overline{B}^0_d$ and $B^0_s, \overline{B}^0_s$ systems similarly limit the amounts of $\widetilde{d}, \widetilde{b}$ and $\widetilde{s}, \widetilde{b}$ squark mixings from soft supersymmetry-breaking sources. More constraints follow from rare $\Delta F = 1$ meson decays, notably those involving the parton-level processes $b \to s\gamma$ and $b \to s\ell^+\ell^-$ and $c \to u\ell^+\ell^-$ and $s \to de^+e^-$ and $s \to d\nu\bar{\nu}$, all of which can be mediated by flavor mixing in soft supersymmetry breaking. There are also strict constraints on CP-violating phases in the gaugino masses and (scalar)$^3$ soft couplings following from limits on the electric dipole moments of the neutron and electron.[83] Detailed limits can be found in the literature,[80–105] but the essential lesson from experiment is that the soft supersymmetry-breaking Lagrangian cannot be arbitrary or random.

All of these potentially dangerous flavor-changing and CP-violating effects in the MSSM can be evaded if one assumes (or can explain!) that supersymmetry breaking is suitably "universal". Consider an idealized limit in which the squark and slepton squared-mass matrices are flavor-blind, each proportional to the $3 \times 3$ identity matrix in family space:

$$\mathbf{m^2_Q} = m^2_Q \mathbf{1}, \quad \mathbf{m^2_{\overline{u}}} = m^2_{\overline{u}} \mathbf{1}, \quad \mathbf{m^2_{\overline{d}}} = m^2_{\overline{d}} \mathbf{1}, \quad \mathbf{m^2_L} = m^2_L \mathbf{1}, \quad \mathbf{m^2_{\overline{e}}} = m^2_{\overline{e}} \mathbf{1}. \tag{1.158}$$

Then all squark and slepton mixing angles are rendered trivial, because squarks and sleptons with the same electroweak quantum numbers will be degenerate in mass and can be rotated into each other at will. Supersymmetric contributions to flavor-changing neutral current processes will therefore be very small in such an idealized limit, up to mixing induced by $\mathbf{a_u}$, $\mathbf{a_d}$, $\mathbf{a_e}$. Making the further assumption that the (scalar)$^3$ couplings are each proportional to the corresponding Yukawa coupling matrix,

$$\mathbf{a_u} = A_{u0}\, \mathbf{y_u}, \qquad \mathbf{a_d} = A_{d0}\, \mathbf{y_d}, \qquad \mathbf{a_e} = A_{e0}\, \mathbf{y_e}, \tag{1.159}$$

will ensure that only the squarks and sleptons of the third family can have large (scalar)$^3$ couplings. Finally, one can avoid disastrously large CP-violating effects by assuming that the soft parameters do not introduce new complex phases. This is automatic for $m^2_{H_u}$ and $m^2_{H_d}$, and for $m^2_Q$, $m^2_{\overline{u}}$, etc. if eq. (1.158) is assumed; if they were not real numbers, the Lagrangian would not be real. One can also fix $\mu$ in the superpotential and $b$ in eq. (1.152) to be real, by appropriate phase rotations of fermion and scalar components of the $H_u$ and $H_d$ supermultiplets. If one then assumes that

$$\arg(M_1),\ \arg(M_2),\ \arg(M_3),\ \arg(A_{u0}),\ \arg(A_{d0}),\ \arg(A_{e0}) = 0 \text{ or } \pi, \tag{1.160}$$

then the only CP-violating phase in the theory will be the usual CKM phase found in the ordinary Yukawa couplings. Together, the conditions eqs. (1.158)-(1.160) make up a rather weak version of what is often called the hypothesis of *soft supersymmetry-breaking universality.* The MSSM with these flavor- and CP-preserving relations imposed has far fewer parameters than the most general case. Besides the usual Standard Model gauge and Yukawa coupling parameters, there are 3 independent real gaugino masses, only 5 real squark and slepton squared mass parameters, 3 real scalar cubic coupling parameters, and 4 Higgs mass parameters (one of which can be traded for the known electroweak breaking scale).

It must be mentioned in passing that there are at least three other possible types of explanations for the suppression of flavor violation in the MSSM that could replace the universality hypothesis of eqs. (1.158)-(1.160). They can be referred to as the "irrelevancy", "alignment", and "$R$-symmetry" hypotheses for the soft masses. The "irrelevancy" idea is that the sparticles masses are *extremely* heavy, so that their contributions to flavor-changing and CP-violating diagrams like Figures 1.13a,b are suppressed, as can be seen for example in eqs. (1.155)-(1.157). In practice, however, the degree of suppression needed typically requires $m_{\rm soft}$ much larger than 1 TeV for at least some of the scalar masses; this seems to go directly against the motivation for supersymmetry as a cure for the hierarchy problem as discussed in the Introduction. Nevertheless, it has been argued that this is a sensible possibility.[107,108] The "alignment" idea is that the squark squared-mass matrices do not have the flavor-blindness indicated in eq. (1.158), but are arranged in flavor space to be aligned with the relevant Yukawa matrices in just such a way as to avoid large flavor-changing effects.[56,109] The alignment models typically require rather special flavor symmetries. The third possibility is that the theory is (approximately) invariant under a continuous $U(1)_R$ symmetry.[62] This requires that the MSSM is supplemented, as in,[59] by additional chiral supermultiplets in the adjoint representations of $SU(3)_c$, $SU(2)_L$, and $U(1)_Y$, as well as an additional pair of Higgs chiral supermultiplets. The gaugino masses in this theory are purely Dirac, of the type in eq. (1.140), and the couplings $\mathbf{a_u}$, $\mathbf{a_d}$, and $\mathbf{a_e}$ are absent. This implies a very efficient suppression of flavor-changing effects,[62,63] even if the squark and slepton mass eigenstates are light, non-degenerate, and have large mixings in the basis determined by the Standard Model quark and lepton mass eigenstates. This can lead to unique and intriguing collider signatures.[62,65] However, we will not consider these possibilities further here.

The soft-breaking universality relations eqs. (1.158)-(1.160), or stronger (more special) versions of them, can be presumed to be the result of some specific model for the origin of supersymmetry breaking, although there is considerable disagreement among theorists as to what the specific model should actually be. In any case, they are indicative of an assumed underlying simplicity or symmetry of the Lagrangian at some very high energy scale $Q_0$. If we used this Lagrangian to compute masses and cross-sections and decay rates for experiments at ordinary energies near the electroweak scale, the results would involve large logarithms of order $\ln(Q_0/m_Z)$ coming from loop diagrams. As is usual in quantum field theory, the large logarithms can be conveniently resummed using renormalization group (RG) equations, by treating the couplings and masses appearing in the Lagrangian as running parameters. Therefore, eqs. (1.158)-(1.160) should be interpreted as boundary conditions on the running soft parameters at the scale $Q_0$, which is likely very far removed from direct experimental probes. We must then RG-evolve all of the soft parameters, the superpotential parameters, and the gauge couplings down to the electroweak scale or comparable scales where humans perform experiments.

At the electroweak scale, eqs. (1.158) and (1.159) will no longer hold, even if they were exactly true at the input scale $Q_0$. However, to a good approximation, key flavor- and CP-conserving properties remain. This is because, as we will see in section 1.5.5 below, RG corrections due to gauge interactions will respect the form of eqs. (1.158) and (1.159), while RG corrections due to Yukawa interactions are quite small except for couplings involving the top, bottom, and tau flavors. Therefore, the (scalar)$^3$ couplings and scalar squared-mass mixings should be quite negligible for the squarks and sleptons of the first two families. Furthermore, RG evolution does not introduce new CP-violating phases. Therefore, if universality can be arranged to hold at the input scale, supersymmetric contributions to flavor-changing and CP-violating observables can be acceptably small in comparison to present limits (although quite possibly measurable in future experiments).

One good reason to be optimistic that such a program can succeed is the celebrated apparent unification of gauge couplings in the MSSM.[110] The 1-loop RG equations for the Standard Model gauge couplings $g_1, g_2, g_3$ are

$$\beta_{g_a} \equiv \frac{d}{dt} g_a = \frac{1}{16\pi^2} b_a g_a^3,$$

$$(b_1, b_2, b_3) = \begin{cases} (41/10,\ -19/6,\ -7) & \text{Standard Model} \\ (33/5,\ 1,\ -3) & \text{MSSM} \end{cases} \tag{1.161}$$

where $t = \ln(Q/Q_0)$, with $Q$ the RG scale. The MSSM coefficients are larger because of the extra MSSM particles in loops. The normalization for $g_1$ here is chosen to agree with the canonical covariant derivative for grand unification of the gauge group $SU(3)_C \times SU(2)_L \times U(1)_Y$ into $SU(5)$ or $SO(10)$. Thus in terms of the conventional electroweak gauge couplings $g$ and $g'$ with $e = g\sin\theta_W = g'\cos\theta_W$, one has $g_2 = g$ and $g_1 = \sqrt{5/3}g'$. The quantities $\alpha_a = g_a^2/4\pi$ have the nice property that their reciprocals run linearly with RG scale at one-loop order:

$$\frac{d}{dt}\alpha_a^{-1} = -\frac{b_a}{2\pi} \qquad (a = 1, 2, 3). \tag{1.162}$$

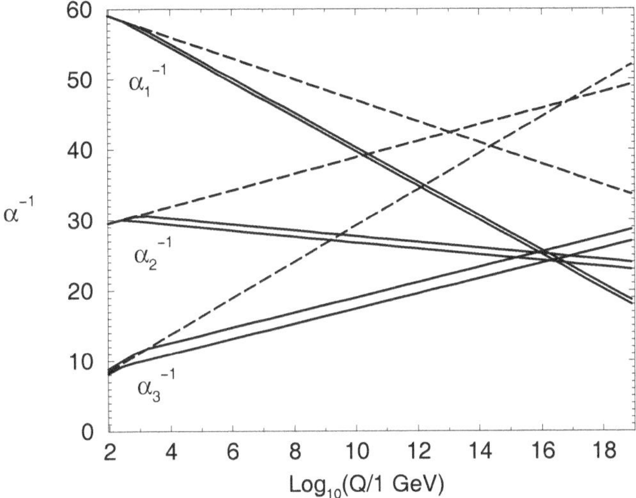


Fig. 1.14. RG evolution of the inverse gauge couplings $\alpha_a^{-1}(Q)$ in the Standard Model (dashed lines) and the MSSM (solid lines). In the MSSM case, the sparticle mass thresholds are varied between 250 GeV and 1 TeV, and $\alpha_3(m_Z)$ between 0.113 and 0.123. Two-loop effects are included.

Figure 1.14 compares the RG evolution of the $\alpha_a^{-1}$, including two-loop effects, in the Standard Model (dashed lines) and the MSSM (solid lines). Unlike the Standard Model, the MSSM includes just the right particle content to ensure that the gauge couplings can unify, at a scale $M_U \sim 2 \times 10^{16}$ GeV. While the apparent unification of gauge couplings at $M_U$ might be just an accident, it may also be taken as a strong hint in favor of a grand unified theory (GUT) or superstring models, both of which can naturally accommodate gauge coupling unification below $M_{\rm P}$. Furthermore, if this hint is taken seriously, then we can reasonably expect to be able to apply a similar RG analysis to the other MSSM couplings and soft masses as well. The next section discusses the form of the necessary RG equations.

### 1.5.5. *Renormalization group equations for the MSSM*

In order to translate a set of predictions at an input scale into physically meaningful quantities that describe physics near the electroweak scale, it is necessary to evolve the gauge couplings, superpotential parameters, and soft terms using their renormalization group (RG) equations. This ensures that the loop expansions for calculations of observables will not suffer from very large logarithms.

As a technical aside, some care is required in choosing regularization and renormalization procedures in supersymmetry. The most popular regularization method for computations of radiative corrections within the Standard Model is dimensional regularization (DREG), in which the number of spacetime dimensions is continued to $d = 4 - 2\epsilon$. Unfortunately, DREG introduces a spurious violation of supersymmetry, because it has a mismatch between the numbers of gauge boson degrees of freedom and the gaugino degrees of freedom off-shell. This mismatch is only $2\epsilon$, but can be multiplied by factors up to $1/\epsilon^n$ in an $n$-loop calculation. In DREG, supersymmetric relations between dimensionless coupling constants ("supersymmetric Ward identities") are therefore not explicitly respected by radiative corrections involving the finite parts of one-loop graphs and by the divergent parts of two-loop graphs. Instead, one may use the slightly different scheme known as regularization by dimensional reduction, or DRED, which does respect supersymmetry.[111] In the DRED method, all momentum integrals are still performed in $d = 4 - 2\epsilon$ dimensions, but the vector index $\mu$ on the gauge boson fields $A^a_\mu$ now runs over all 4 dimensions to maintain the match with the gaugino degrees of freedom. Running couplings are then renormalized using DRED with modified minimal subtraction ($\overline{\text{DR}}$) rather than the usual DREG with modified minimal subtraction ($\overline{\text{MS}}$). In particular, the boundary conditions at the input scale should presumably be applied in a supersymmetry-preserving scheme like $\overline{\text{DR}}$. One loop $\beta$-functions are always the same in these two schemes, but it is important to realize that the $\overline{\text{MS}}$ scheme does violate supersymmetry, so that $\overline{\text{DR}}$ is preferred[m] from that point of view. (The NSVZ scheme[116] also respects supersymmetry and has some very useful properties, but with a less obvious connection to calculations of physical observables. It is also possible, but not always very practical, to work consistently within the $\overline{\text{MS}}$ scheme, as long as one translates all $\overline{\text{DR}}$ couplings and masses into their $\overline{\text{MS}}$ counterparts.[117–119])

A general and powerful result known as the *supersymmetric non-renormalization theorem*[120] governs the form of the renormalization group equations for supersymmetric theories. This theorem implies that the logarithmically divergent contributions to a particular process can always be written in terms of wave-function renormalizations, without any coupling

---

[m]Even the DRED scheme may not provide a supersymmetric regulator, because of either ambiguities or inconsistencies (depending on the precise method) appearing at five-loop order at the latest.[112] Fortunately, this does not seem to cause practical difficulties.[113,114] See also Ref. 115 for an interesting proposal that avoids doing violence to the number of spacetime dimensions.

vertex renormalization.[n] It can be proved most easily using superfield techniques. For the parameters appearing in the superpotential eq. (1.107), the implication is that

$$\beta_{y^{ijk}} \equiv \frac{d}{dt} y^{ijk} = \gamma^i_n y^{njk} + \gamma^j_n y^{ink} + \gamma^k_n y^{ijn}, \tag{1.163}$$

$$\beta_{M^{ij}} \equiv \frac{d}{dt} M^{ij} = \gamma^i_n M^{nj} + \gamma^j_n M^{in}, \tag{1.164}$$

$$\beta_{L^i} \equiv \frac{d}{dt} L^i = \gamma^i_n L^n, \tag{1.165}$$

where the $\gamma^i_j$ are anomalous dimension matrices associated with the superfields, which generally have to be calculated in a perturbative loop expansion. [Recall $t = \ln(Q/Q_0)$, where $Q$ is the renormalization scale, and $Q_0$ is a reference scale.] The anomalous dimensions and RG equations for softly broken supersymmetry are now known up to 3-loop order, with some partial 4-loop results; they have been given in Refs. 121–126. There are also relations, good to all orders in perturbation theory, that give the RG equations for soft supersymmetry couplings in terms of those for the supersymmetric couplings.[116,127] Here we will only use the 1-loop approximation, for simplicity.

In general, at 1-loop order,

$$\gamma^i_j = \frac{1}{16\pi^2}\left[\frac{1}{2} y^{imn} y^*_{jmn} - 2g_a^2 C_a(i)\delta^i_j\right], \tag{1.166}$$

where $C_a(i)$ are the quadratic Casimir group theory invariants for the superfield $\Phi_i$, defined in terms of the Lie algebra generators $T^a$ by

$$(T^a T^a)_i{}^j = C_a(i)\delta^j_i \tag{1.167}$$

with gauge couplings $g_a$. Explicitly, for the MSSM supermultiplets:

$$C_3(i) = \begin{cases} 4/3 \text{ for } \Phi_i = Q, \overline{u}, \overline{d}, \\ 0 \quad\; \text{for } \Phi_i = L, \overline{e}, H_u, H_d, \end{cases} \tag{1.168}$$

$$C_2(i) = \begin{cases} 3/4 \text{ for } \Phi_i = Q, L, H_u, H_d, \\ 0 \quad\; \text{for } \Phi_i = \overline{u}, \overline{d}, \overline{e}, \end{cases} \tag{1.169}$$

$$C_1(i) = 3Y_i^2/5 \text{ for each } \Phi_i \text{ with weak hypercharge } Y_i. \tag{1.170}$$

[n] Actually, there *is* vertex renormalization working in a supersymmetric gauge theory in which auxiliary fields have been integrated out, but the sum of divergent contributions for a process always has the form of wave-function renormalization. This is related to the fact that the anomalous dimensions of the superfields differ, by gauge-fixing dependent terms, from the anomalous dimensions of the fermion and boson component fields.[31]

For the one-loop renormalization of gauge couplings, one has in general

$$\beta_{g_a} = \frac{d}{dt} g_a = \frac{1}{16\pi^2} g_a^3 \Big[\sum_i I_a(i) - 3C_a(G)\Big], \tag{1.171}$$

where $C_a(G)$ is the quadratic Casimir invariant of the group [0 for $U(1)$, and $N$ for $SU(N)$], and $I_a(i)$ is the Dynkin index of the chiral supermultiplet $\phi_i$ [normalized to 1/2 for each fundamental representation of $SU(N)$ and to $3Y_i^2/5$ for $U(1)_Y$]. Equation (1.161) is a special case of this.

The 1-loop renormalization group equations for the general soft supersymmetry breaking Lagrangian parameters appearing in eq. (1.138) are:

$$\beta_{M_a} \equiv \frac{d}{dt} M_a = \frac{1}{16\pi^2} g_a^2 \Big[2\sum_n I_a(n) - 6C_a(G)\Big] M_a, \tag{1.172}$$

$$\beta_{a^{ijk}} \equiv \frac{d}{dt} a^{ijk} = \frac{1}{16\pi^2}\left[\frac{1}{2} a^{ijp} y^*_{pmn} y^{kmn} + y^{ijp} y^*_{pmn} a^{mnp} \right.$$
$$\left. + g_a^2 C_a(i)(4M_a y^{ijk} - 2a^{ijk})\right] + (i \leftrightarrow k) + (j \leftrightarrow k), \tag{1.173}$$

$$\beta_{b^{ij}} \equiv \frac{d}{dt} b^{ij} = \frac{1}{16\pi^2}\left[\frac{1}{2} b^{ip} y^*_{pmn} y^{jmn} + \frac{1}{2} y^{ijp} y^*_{pmn} b^{mn} + M^{ip} y^*_{pmn} a^{mnj} \right.$$
$$\left. + g_a^2 C_a(i)(4M_a M^{ij} - 2b^{ij})\right] + (i \leftrightarrow j), \tag{1.174}$$

$$\beta_{t^i} \equiv \frac{d}{dt} t^i = \frac{1}{16\pi^2}\left[\frac{1}{2} y^{imn} y^*_{mnp} t^p + a^{imn} y^*_{mnp} L^p \right.$$
$$\left. + M^{ip} y^*_{pmn} b^{mn}\right], \tag{1.175}$$

$$\beta_{(m^2)_i^j} \equiv \frac{d}{dt} (m^2)_i^j = \frac{1}{16\pi^2}\left[\frac{1}{2} y^*_{ipq} y^{pqn} (m^2)_n^j + \frac{1}{2} y^{jpq} y^*_{pqn} (m^2)_i^n \right.$$
$$+ 2 y^*_{ipq} y^{jpr} (m^2)_r^q + a^*_{ipq} a^{jpq}$$
$$\left. - 8 g_a^2 C_a(i) |M_a|^2 \delta_i^j + 2 g_a^2 (T^a)_i{}^j \mathrm{Tr}(T^a m^2)\right]. \tag{1.176}$$

Applying the above results to the special case of the MSSM, we will use the approximation that only the third-family Yukawa couplings are significant, as in eq. (1.142). Then the Higgs and third-family superfield anomalous dimensions are diagonal matrices, and from eq. (1.166) they are, at 1-loop order:

$$\gamma_{H_u} = \frac{1}{16\pi^2}\left[3 y_t^* y_t - \frac{3}{2} g_2^2 - \frac{3}{10} g_1^2\right], \tag{1.177}$$

$$\gamma_{H_d} = \frac{1}{16\pi^2}\left[3y_b^* y_b + y_\tau^* y_\tau - \frac{3}{2}g_2^2 - \frac{3}{10}g_1^2\right], \tag{1.178}$$

$$\gamma_{Q_3} = \frac{1}{16\pi^2}\left[y_t^* y_t + y_b^* y_b - \frac{8}{3}g_3^2 - \frac{3}{2}g_2^2 - \frac{1}{30}g_1^2\right], \tag{1.179}$$

$$\gamma_{\overline{u}_3} = \frac{1}{16\pi^2}\left[2y_t^* y_t - \frac{8}{3}g_3^2 - \frac{8}{15}g_1^2\right], \tag{1.180}$$

$$\gamma_{\overline{d}_3} = \frac{1}{16\pi^2}\left[2y_b^* y_b - \frac{8}{3}g_3^2 - \frac{2}{15}g_1^2\right], \tag{1.181}$$

$$\gamma_{L_3} = \frac{1}{16\pi^2}\left[y_\tau^* y_\tau - \frac{3}{2}g_2^2 - \frac{3}{10}g_1^2\right], \tag{1.182}$$

$$\gamma_{\overline{e}_3} = \frac{1}{16\pi^2}\left[2y_\tau^* y_\tau - \frac{6}{5}g_1^2\right]. \tag{1.183}$$

[The first and second family anomalous dimensions in the approximation of eq. (1.142) follow by setting $y_t$, $y_b$, and $y_\tau$ to 0 in the above.] Putting these into eqs. (1.163), (1.164) gives the running of the superpotential parameters with renormalization scale:

$$\beta_{y_t} \equiv \frac{d}{dt}y_t = \frac{y_t}{16\pi^2}\left[6y_t^* y_t + y_b^* y_b - \frac{16}{3}g_3^2 - 3g_2^2 - \frac{13}{15}g_1^2\right], \tag{1.184}$$

$$\beta_{y_b} \equiv \frac{d}{dt}y_b = \frac{y_b}{16\pi^2}\left[6y_b^* y_b + y_t^* y_t + y_\tau^* y_\tau - \frac{16}{3}g_3^2 - 3g_2^2 - \frac{7}{15}g_1^2\right], \tag{1.185}$$

$$\beta_{y_\tau} \equiv \frac{d}{dt}y_\tau = \frac{y_\tau}{16\pi^2}\left[4y_\tau^* y_\tau + 3y_b^* y_b - 3g_2^2 - \frac{9}{5}g_1^2\right], \tag{1.186}$$

$$\beta_{\mu} \equiv \frac{d}{dt}\mu = \frac{\mu}{16\pi^2}\left[3y_t^* y_t + 3y_b^* y_b + y_\tau^* y_\tau - 3g_2^2 - \frac{3}{5}g_1^2\right]. \tag{1.187}$$

The one-loop RG equations for the gauge couplings $g_1$, $g_2$, and $g_3$ were already listed in eq. (1.161). The presence of soft supersymmetry breaking does not affect eqs. (1.161) and (1.184)-(1.187). As a result of the supersymmetric non-renormalization theorem, the $\beta$-functions for each supersymmetric parameter are proportional to the parameter itself. One consequence of this is that once we have a theory that can explain why $\mu$ is of order $10^2$ or $10^3$ GeV at tree-level, we do not have to worry about $\mu$ being made very large by radiative corrections involving the masses of some very heavy unknown particles; all such RG corrections to $\mu$ will be directly proportional to $\mu$ itself and to some combinations of dimensionless couplings.

The one-loop RG equations for the three gaugino mass parameters in the MSSM are determined by the same quantities $b_a^{\rm MSSM}$ that appear in the gauge coupling RG eqs. (1.161):

$$\beta_{M_a} \equiv \frac{d}{dt} M_a = \frac{1}{8\pi^2} b_a g_a^2 M_a \qquad (b_a = 33/5,\ 1,\ -3) \tag{1.188}$$

for $a = 1,2,3$. It follows that the three ratios $M_a/g_a^2$ are each constant (RG scale independent) up to small two-loop corrections. Since the gauge couplings are observed to unify at $Q = M_U = 2 \times 10^{16}$ GeV, it is a popular assumption that the gaugino masses also unify[o] near that scale, with a value called $m_{1/2}$. If so, then it follows that

$$\frac{M_1}{g_1^2} = \frac{M_2}{g_2^2} = \frac{M_3}{g_3^2} = \frac{m_{1/2}}{g_U^2} \tag{1.189}$$

at any RG scale, up to small (and known) two-loop effects and possibly much larger (and not so known) threshold effects near $M_U$. Here $g_U$ is the unified gauge coupling at $Q = M_U$. The hypothesis of eq. (1.189) is particularly powerful because the gaugino mass parameters feed strongly into the RG equations for all of the other soft terms, as we are about to see.

Next we consider the 1-loop RG equations for the analytic soft parameters $\mathbf{a_u}$, $\mathbf{a_d}$, $\mathbf{a_e}$. In models obeying eq. (1.159), these matrices start off proportional to the corresponding Yukawa couplings at the input scale. The RG evolution respects this property. With the approximation of eq. (1.142), one can therefore also write, at any RG scale,

$$\mathbf{a_u} \approx \begin{pmatrix} 0 & 0 & 0 \\ 0 & 0 & 0 \\ 0 & 0 & a_t \end{pmatrix}, \quad \mathbf{a_d} \approx \begin{pmatrix} 0 & 0 & 0 \\ 0 & 0 & 0 \\ 0 & 0 & a_b \end{pmatrix}, \quad \mathbf{a_e} \approx \begin{pmatrix} 0 & 0 & 0 \\ 0 & 0 & 0 \\ 0 & 0 & a_\tau \end{pmatrix}, \tag{1.190}$$

which defines[p] running parameters $a_t$, $a_b$, and $a_\tau$. In this approximation, the RG equations for these parameters and $b$ are

$$16\pi^2 \frac{d}{dt} a_t = a_t \left[ 18 y_t^* y_t + y_b^* y_b - \frac{16}{3} g_3^2 - 3 g_2^2 - \frac{13}{15} g_1^2 \right] + 2 a_b y_b^* y_t$$
$$+ y_t \left[ \frac{32}{3} g_3^2 M_3 + 6 g_2^2 M_2 + \frac{26}{15} g_1^2 M_1 \right], \tag{1.191}$$

[o]In GUT models, it is automatic that the gauge couplings and gaugino masses are unified at all scales $Q \geq M_U$, because in the unified theory the gauginos all live in the same representation of the unified gauge group. In many superstring models, this can also be a good approximation.

[p]Rescaled soft parameters $A_t = a_t/y_t$, $A_b = a_b/y_b$, and $A_\tau = a_\tau/y_\tau$ are commonly used in the literature. We do not follow this notation, because it cannot be generalized beyond the approximation of eqs. (1.142), (1.190) without introducing horrible complications such as non-polynomial RG equations, and because $a_t$, $a_b$ and $a_\tau$ are the couplings that actually appear in the Lagrangian anyway.

$$16\pi^2 \frac{d}{dt} a_b = a_b \Big[ 18 y_b^* y_b + y_t^* y_t + y_\tau^* y_\tau - \frac{16}{3} g_3^2 - 3 g_2^2 - \frac{7}{15} g_1^2 \Big] + 2 a_t y_t^* y_b$$
$$+ 2 a_\tau y_\tau^* y_b + y_b \Big[ \frac{32}{3} g_3^2 M_3 + 6 g_2^2 M_2 + \frac{14}{15} g_1^2 M_1 \Big], \qquad (1.192)$$

$$16\pi^2 \frac{d}{dt} a_\tau = a_\tau \Big[ 12 y_\tau^* y_\tau + 3 y_b^* y_b - 3 g_2^2 - \frac{9}{5} g_1^2 \Big] + 6 a_b y_b^* y_\tau$$
$$+ y_\tau \Big[ 6 g_2^2 M_2 + \frac{18}{5} g_1^2 M_1 \Big], \qquad (1.193)$$

$$16\pi^2 \frac{d}{dt} b = b \Big[ 3 y_t^* y_t + 3 y_b^* y_b + y_\tau^* y_\tau - 3 g_2^2 - \frac{3}{5} g_1^2 \Big]$$
$$+ \mu \Big[ 6 a_t y_t^* + 6 a_b y_b^* + 2 a_\tau y_\tau^* + 6 g_2^2 M_2 + \frac{6}{5} g_1^2 M_1 \Big]. \qquad (1.194)$$

The $\beta$-function for each of these soft parameters is *not* proportional to the parameter itself, because couplings that violate supersymmetry are not protected by the supersymmetric non-renormalization theorem. So, even if $a_t$, $a_b$, $a_\tau$ and $b$ vanish at the input scale, the RG corrections proportional to gaugino masses appearing in eqs. (1.191)-(1.194) ensure that they will not vanish at the electroweak scale.

Next let us consider the RG equations for the scalar squared masses in the MSSM. In the approximation of eqs. (1.142) and (1.190), the squarks and sleptons of the first two families have only gauge interactions. This means that if the scalar squared masses satisfy a boundary condition like eq. (1.158) at an input RG scale, then when renormalized to any other RG scale, they will still be almost diagonal, with the approximate form

$$\mathbf{m^2_Q} \approx \begin{pmatrix} m^2_{Q_1} & 0 & 0 \\ 0 & m^2_{Q_1} & 0 \\ 0 & 0 & m^2_{Q_3} \end{pmatrix}, \quad \mathbf{m^2_{\overline{u}}} \approx \begin{pmatrix} m^2_{\overline{u}_1} & 0 & 0 \\ 0 & m^2_{\overline{u}_1} & 0 \\ 0 & 0 & m^2_{\overline{u}_3} \end{pmatrix}, \qquad (1.195)$$

etc. The first and second family squarks and sleptons with given gauge quantum numbers remain very nearly degenerate, but the third-family squarks and sleptons feel the effects of the larger Yukawa couplings and so their squared masses get renormalized differently. The one-loop RG equations for the first and second family squark and slepton squared masses are

$$16\pi^2 \frac{d}{dt} m^2_{\phi_i} = - \sum_{a=1,2,3} 8 C_a(i) g_a^2 |M_a|^2 + \frac{6}{5} Y_i g_1^2 S \qquad (1.196)$$

for each scalar $\phi_i$, where the $\sum_a$ is over the three gauge groups $U(1)_Y$, $SU(2)_L$ and $SU(3)_C$, with Casimir invariants $C_a(i)$ as in eqs. (1.168)-(1.170), and $M_a$ are the corresponding running gaugino mass parameters.

Also,

$$S \equiv \mathrm{Tr}[Y_j m^2_{\phi_j}] = m^2_{H_u} - m^2_{H_d} + \mathrm{Tr}[\mathbf{m^2_Q} - \mathbf{m^2_L} - 2\mathbf{m^2_{\overline{u}}} + \mathbf{m^2_{\overline{d}}} + \mathbf{m^2_{\overline{e}}}]. \quad (1.197)$$

An important feature of eq. (1.196) is that the terms on the right-hand sides proportional to gaugino squared masses are negative, so[q] the scalar squared-mass parameters grow as they are RG-evolved from the input scale down to the electroweak scale. Even if the scalars have zero or very small masses at the input scale, they can obtain large positive squared masses at the electroweak scale, thanks to the effects of the gaugino masses.

The RG equations for the squared-mass parameters of the Higgs scalars and third-family squarks and sleptons get the same gauge contributions as in eq. (1.196), but they also have contributions due to the large Yukawa ($y_{t,b,\tau}$) and soft ($a_{t,b,\tau}$) couplings. At one-loop order, these only appear in three combinations:

$$X_t = 2|y_t|^2(m^2_{H_u} + m^2_{Q_3} + m^2_{\overline{u}_3}) + 2|a_t|^2, \quad (1.198)$$
$$X_b = 2|y_b|^2(m^2_{H_d} + m^2_{Q_3} + m^2_{\overline{d}_3}) + 2|a_b|^2, \quad (1.199)$$
$$X_\tau = 2|y_\tau|^2(m^2_{H_d} + m^2_{L_3} + m^2_{\overline{e}_3}) + 2|a_\tau|^2. \quad (1.200)$$

In terms of these quantities, the RG equations for the soft Higgs squared-mass parameters $m^2_{H_u}$ and $m^2_{H_d}$ are

$$16\pi^2 \frac{d}{dt} m^2_{H_u} = 3X_t - 6g_2^2|M_2|^2 - \frac{6}{5}g_1^2|M_1|^2 + \frac{3}{5}g_1^2 S, \quad (1.201)$$
$$16\pi^2 \frac{d}{dt} m^2_{H_d} = 3X_b + X_\tau - 6g_2^2|M_2|^2 - \frac{6}{5}g_1^2|M_1|^2 - \frac{3}{5}g_1^2 S. \quad (1.202)$$

Note that $X_t$, $X_b$, and $X_\tau$ are generally positive, so their effect is to decrease the Higgs masses as one evolves the RG equations down from the input scale to the electroweak scale. If $y_t$ is the largest of the Yukawa couplings, as suggested by the experimental fact that the top quark is heavy, then $X_t$ will typically be much larger than $X_b$ and $X_\tau$. This can cause the RG-evolved $m^2_{H_u}$ to run negative near the electroweak scale, helping to destabilize the point $H_u = H_d = 0$ and so provoking a Higgs VEV (for a linear combination of $H_u$ and $H_d$, as we will see in section 1.7.1), which is just what we want.[r] Thus a large top Yukawa coupling favors the breakdown of the electroweak

[q]The contributions proportional to $S$ are relatively small in most known realistic models.
[r]One should think of "$m^2_{H_u}$" as a parameter unto itself, and not as the square of some mythical real number $m_{H_u}$. So there is nothing strange about having $m^2_{H_u} < 0$. However, strictly speaking $m^2_{H_u} < 0$ is neither necessary nor sufficient for electroweak symmetry breaking; see section 1.7.1.

symmetry breaking because it induces negative radiative corrections to the Higgs squared mass.

The third-family squark and slepton squared-mass parameters also get contributions that depend on $X_t$, $X_b$ and $X_\tau$. Their RG equations are given by

$$16\pi^2 \frac{d}{dt} m^2_{Q_3} = X_t + X_b - \frac{32}{3} g_3^2 |M_3|^2 - 6 g_2^2 |M_2|^2 - \frac{2}{15} g_1^2 |M_1|^2 + \frac{1}{5} g_1^2 S, \tag{1.203}$$

$$16\pi^2 \frac{d}{dt} m^2_{\overline{u}_3} = 2X_t - \frac{32}{3} g_3^2 |M_3|^2 - \frac{32}{15} g_1^2 |M_1|^2 - \frac{4}{5} g_1^2 S, \tag{1.204}$$

$$16\pi^2 \frac{d}{dt} m^2_{\overline{d}_3} = 2X_b - \frac{32}{3} g_3^2 |M_3|^2 - \frac{8}{15} g_1^2 |M_1|^2 + \frac{2}{5} g_1^2 S, \tag{1.205}$$

$$16\pi^2 \frac{d}{dt} m^2_{L_3} = X_\tau - 6 g_2^2 |M_2|^2 - \frac{6}{5} g_1^2 |M_1|^2 - \frac{3}{5} g_1^2 S, \tag{1.206}$$

$$16\pi^2 \frac{d}{dt} m^2_{\overline{e}_3} = 2X_\tau - \frac{24}{5} g_1^2 |M_1|^2 + \frac{6}{5} g_1^2 S. \tag{1.207}$$

In eqs. (1.201)-(1.207), the terms proportional to $|M_3|^2$, $|M_2|^2$, $|M_1|^2$, and $S$ are just the same ones as in eq. (1.196). Note that the terms proportional to $X_t$ and $X_b$ appear with smaller numerical coefficients in the $m^2_{Q_3}$, $m^2_{\overline{u}_3}$, $m^2_{\overline{d}_3}$ RG equations than they did for the Higgs scalars, and they do not appear at all in the $m^2_{L_3}$ and $m^2_{\overline{e}_3}$ RG equations. Furthermore, the third-family squark squared masses get a large positive contribution proportional to $|M_3|^2$ from the RG evolution, which the Higgs scalars do not get. These facts make it plausible that the Higgs scalars in the MSSM get VEVs, while the squarks and sleptons, having large positive squared mass, do not.

An examination of the RG equations (1.191)-(1.194), (1.196), and (1.201)-(1.207) reveals that if the gaugino mass parameters $M_1$, $M_2$, and $M_3$ are non-zero at the input scale, then all of the other soft terms will be generated too. This implies that models in which gaugino masses dominate over all other effects in the soft supersymmetry breaking Lagrangian at the input scale can be viable. On the other hand, if the gaugino masses were to vanish at tree-level, then they would not get any contributions to their masses at one-loop order; in that case the gauginos would be extremely light and the model would not be phenomenologically acceptable.

Viable models for the origin of supersymmetry breaking typically make predictions for the MSSM soft terms that are refinements of eqs. (1.158)-(1.160). These predictions can then be used as boundary conditions for the RG equations listed above. In the next section we will study the ideas that go into making such predictions, before turning to their implications for the MSSM spectrum in section 1.7.

## 1.6. Origins of Supersymmetry Breaking

### 1.6.1. *General considerations for spontaneous supersymmetry breaking*

In the MSSM, supersymmetry breaking is simply introduced explicitly. However, we have seen that the soft parameters cannot be arbitrary. In order to understand how patterns like eqs. (1.158), (1.159) and (1.160) can emerge, it is necessary to consider models in which supersymmetry is spontaneously broken. By definition, this means that the vacuum state $|0\rangle$ is not invariant under supersymmetry transformations, so $Q_\alpha|0\rangle \neq 0$ and $Q^\dagger_{\dot\alpha}|0\rangle \neq 0$. Now, in global supersymmetry, the Hamiltonian operator $H$ is related to the supersymmetry generators through the algebra eq. (1.90):

$$H = P^0 = \frac{1}{4}(Q_1 Q_1^\dagger + Q_1^\dagger Q_1 + Q_2 Q_2^\dagger + Q_2^\dagger Q_2). \tag{1.208}$$

If supersymmetry is unbroken in the vacuum state, it follows that $H|0\rangle = 0$ and the vacuum has zero energy. Conversely, if supersymmetry is spontaneously broken in the vacuum state, then the vacuum must have positive energy, since

$$\langle 0|H|0\rangle = \frac{1}{4}\Big(\|Q_1^\dagger|0\rangle\|^2 + \|Q_1|0\rangle\|^2 + \|Q_2^\dagger|0\rangle\|^2 + \|Q_2|0\rangle\|^2\Big) > 0 \tag{1.209}$$

if the Hilbert space is to have positive norm. If spacetime-dependent effects and fermion condensates can be neglected, then $\langle 0|H|0\rangle = \langle 0|V|0\rangle$, where $V$ is the scalar potential in eq. (1.135). Therefore, supersymmetry will be spontaneously broken if the expectation value of $F_i$ and/or $D^a$ does not vanish in the vacuum state.

If any state exists in which all $F_i$ and $D^a$ vanish, then it will have zero energy, implying that supersymmetry is not spontaneously broken in the true ground state. Conversely, one way to guarantee spontaneous supersymmetry breaking is to look for models in which the equations $F_i = 0$ and $D^a = 0$ cannot all be simultaneously satisfied for *any* values of the fields. Then the true ground state necessarily has broken supersymmetry, as does the vacuum state we live in (if it is different).

However, another possibility is that the vacuum state in which we live is not the true ground state (which may preserve supersymmetry), but is instead a higher energy metastable supersymmetry-breaking state with lifetime at least of order the present age of the universe.[128–130] Finite temperature effects can indeed cause the early universe to prefer the metastable supersymmetry-breaking local minimum of the potential over the supersymmetry-breaking global minimum.[131]

Regardless of whether the vacuum state is stable or metastable, the spontaneous breaking of a global symmetry always implies a massless Nambu-Goldstone mode with the same quantum numbers as the broken symmetry generator. In the case of global supersymmetry, the broken generator is the fermionic charge $Q_\alpha$, so the Nambu-Goldstone particle ought to be a massless neutral Weyl fermion, called the *goldstino*. To prove it, consider a general supersymmetric model with both gauge and chiral supermultiplets as in section 1.3. The fermionic degrees of freedom consist of gauginos ($\lambda^a$) and chiral fermions ($\psi_i$). After some of the scalar fields in the theory obtain VEVs, the fermion mass matrix has the form:

$$\mathbf{m}_{\rm F} = \begin{pmatrix} 0 & \sqrt{2} g_b (\langle\phi^*\rangle T^b)^i \\ \sqrt{2} g_a (\langle\phi^*\rangle T^a)^j & \langle W^{ji}\rangle \end{pmatrix} \tag{1.210}$$

in the ($\lambda^a$, $\psi_i$) basis. [The off-diagonal entries in this matrix come from the first term in the second line of eq. (1.132), and the lower right entry can be seen in eq. (1.109).] Now observe that $\mathbf{m}_{\rm F}$ annihilates the vector

$$\widetilde{G} = \begin{pmatrix} \langle D^a\rangle/\sqrt{2} \\ \langle F_i\rangle \end{pmatrix}. \tag{1.211}$$

The first row of $\mathbf{m}_{\rm F}$ annihilates $\widetilde{G}$ by virtue of the requirement eq. (1.133) that the superpotential is gauge invariant, and the second row does so because of the condition $\langle\partial V/\partial\phi_i\rangle = 0$, which must be satisfied at a local minimum of the scalar potential. Equation (1.211) is therefore proportional to the goldstino wavefunction; it is non-trivial if and only if at least one of the auxiliary fields has a VEV, breaking supersymmetry. So we have proved that if global supersymmetry is spontaneously broken, then there must be a massless goldstino, and that its components among the various fermions in the theory are just proportional to the corresponding auxiliary field VEVs.

There is also a useful sum rule that governs the tree-level squared masses of particles in theories with spontaneously broken supersymmetry. For a general theory of the type discussed in section 1.3, the squared masses of the real scalar degrees of freedom are the eigenvalues of the matrix

$$\mathbf{m}^2_{\rm S} = \begin{pmatrix} W^*_{jk} W^{ik} + g_a^2[(T^a\phi)_j(\phi^* T^a)^i + T^{ai}_j D^a] & W^*_{ijk} W^k + g_a^2 (T^a\phi)_i (T^a\phi)_j \\ W^{ijk} W^*_k + g_a^2 (\phi^* T^a)^i (\phi^* T^a)^j & W^*_{ik} W^{jk} + g_a^2[(T^a\phi)_i(\phi^* T^a)^j + T^{aj}_i D^a] \end{pmatrix}, \tag{1.212}$$

since the quadratic part of the tree-level potential is

$$V = (\,\phi^{*j} \quad \phi_j\,)\,\mathbf{m}_{\rm S}^2 \begin{pmatrix} \phi_i \\ \phi^{*i} \end{pmatrix}. \tag{1.213}$$

Here $W^{ijk} = \delta^3 W/\delta\phi_i\delta\phi_j\delta\phi_k$, and the scalar fields on the right-hand side of eq. (1.212) are understood to be replaced by their VEVs. It follows that the sum of the real scalar squared-mass eigenvalues is

$$\mathrm{Tr}(\mathbf{m}_{\rm S}^2) \;=\; 2W^*_{ik}W^{ik} + 2g_a^2[C_a(i)\phi^{*i}\phi_i + \mathrm{Tr}(T^a)D^a], \tag{1.214}$$

with the Casimir invariants $C_a(i)$ defined by eq. (1.167). Meanwhile, the squared masses of the two-component fermions are given by the eigenvalues of

$$\mathbf{m}_{\rm F}^\dagger\mathbf{m}_{\rm F} \;=\; \begin{pmatrix} 2g_a g_b(\phi^* T^a T^b \phi) & \sqrt{2}g_b(T^b\phi)_k W^{ik} \\ \sqrt{2}g_a(\phi^* T^a)^k W^*_{jk} & W^*_{jk}W^{ik} + 2g_a^2(T^a\phi)_j(\phi^* T^a)^i \end{pmatrix}, \tag{1.215}$$

so the sum of the two-component fermion squared masses is

$$\mathrm{Tr}(\mathbf{m}_{\rm F}^\dagger\mathbf{m}_{\rm F}) \;=\; W^*_{ik}W^{ik} + 4g_a^2 C_a(i)\phi^{*i}\phi_i. \tag{1.216}$$

Finally, the vector squared masses are:

$$\mathbf{m}_{\rm V}^2 \;=\; g_a^2(\phi^*\{T^a, T^b\}\phi), \tag{1.217}$$

so

$$\mathrm{Tr}(\mathbf{m}_{\rm V}^2) \;=\; 2g_a^2 C_a(i)\phi^{*i}\phi_i. \tag{1.218}$$

It follows that the *supertrace* of the tree-level squared-mass eigenvalues, defined in general by a weighted sum over all particles with spin $j$:

$$\mathrm{STr}(m^2) \;\equiv\; \sum_j (-1)^{2j}(2j+1)\mathrm{Tr}(m_j^2), \tag{1.219}$$

satisfies the sum rule

$$\mathrm{STr}(m^2) = \mathrm{Tr}(\mathbf{m}_{\rm S}^2) - 2\mathrm{Tr}(\mathbf{m}_{\rm F}^\dagger\mathbf{m}_{\rm F}) + 3\mathrm{Tr}(\mathbf{m}_{\rm V}^2) = 2g_a^2\mathrm{Tr}(T^a)D^a = 0. \tag{1.220}$$

The last equality assumes that the traces of the $U(1)$ charges over the chiral superfields are 0. This holds for $U(1)_Y$ in the MSSM, and more generally for any non-anomalous gauge symmetry. The sum rule eq. (1.220) is often a useful check on models of spontaneous supersymmetry breaking.

### 1.6.2. *Fayet-Iliopoulos (D-term) supersymmetry breaking*

Supersymmetry breaking with a non-zero $D$-term VEV can occur through the Fayet-Iliopoulos mechanism.[132] If the gauge symmetry includes a $U(1)$ factor, then one can introduce a term linear in the corresponding auxiliary field of the gauge supermultiplet:

$$\mathcal{L}_{\text{Fayet-Iliopoulos}} = -\kappa D \tag{1.221}$$

where $\kappa$ is a constant with dimensions of $[\text{mass}]^2$. This term is gauge-invariant and supersymmetric by itself. [Note that for a $U(1)$ gauge symmetry, the supersymmetry transformation $\delta D$ in eq. (1.122) is a total derivative.] If we include it in the Lagrangian, then $D$ may be forced to get a non-zero VEV. To see this, consider the relevant part of the scalar potential from eqs. (1.117) and (1.132):

$$V = \kappa D - \frac{1}{2}D^2 - gD\sum_i q_i|\phi_i|^2. \tag{1.222}$$

Here the $q_i$ are the charges of the scalar fields $\phi_i$ under the $U(1)$ gauge group in question. The presence of the Fayet-Iliopoulos term modifies the equation of motion eq. (1.134) to

$$D = \kappa - g\sum_i q_i|\phi_i|^2. \tag{1.223}$$

Now suppose that the scalar fields $\phi_i$ that are charged under the $U(1)$ all have non-zero superpotential masses $m_i$. (Gauge invariance then requires that they come in pairs with opposite charges.) Then the potential will have the form

$$V = \sum_i |m_i|^2|\phi_i|^2 + \frac{1}{2}(\kappa - g\sum_i q_i|\phi_i|^2)^2. \tag{1.224}$$

Since this cannot vanish, supersymmetry must be broken; one can check that the minimum always occurs for non-zero $D$. For the simplest case in which $|m_i|^2 > gq_i\kappa$ for each $i$, the minimum is realized for all $\phi_i = 0$ and $D = \kappa$, with the $U(1)$ gauge symmetry unbroken. As further evidence that supersymmetry has indeed been spontaneously broken, note that the scalars then have squared masses $|m_i|^2 - gq_i\kappa$, while their fermion partners have squared masses $|m_i|^2$. The gaugino remains massless, as can be understood from the fact that it is the goldstino, as argued on general grounds in section 1.6.1.

For non-Abelian gauge groups, the analog of eq. (1.221) would not be gauge-invariant and is therefore not allowed, so only $U(1)$ $D$-terms can drive spontaneous symmetry breaking. In the MSSM, one might imagine that the $D$ term for $U(1)_Y$ has a Fayet-Iliopoulos term as the principal source of supersymmetry breaking. Unfortunately, this cannot work, because the squarks and sleptons do not have superpotential mass terms. So, at least some of them would just get non-zero VEVs in order to make eq. (1.223) vanish. That would break color and/or electromagnetism, but not supersymmetry. Therefore, a Fayet-Iliopoulos term for $U(1)_Y$ must be subdominant compared to other sources of supersymmetry breaking in the MSSM, if not absent altogether. One could instead attempt to trigger supersymmetry breaking with a Fayet-Iliopoulos term for some other $U(1)$ gauge symmetry, which is as yet unknown because it is spontaneously broken at a very high mass scale or because it does not couple to the Standard Model particles. However, if this is the dominant source for supersymmetry breaking, it proves difficult to give appropriate masses to all of the MSSM particles, especially the gauginos. In any case, we will not discuss $D$-term breaking as the ultimate origin of supersymmetry violation any further (although it may not be ruled out[133]).

### 1.6.3. *O'Raifeartaigh (F-term) supersymmetry breaking*

Models where spontaneous supersymmetry breaking is ultimately due to a non-zero $F$-term VEV, called O'Raifeartaigh models,[134] have brighter phenomenological prospects. The idea is to pick a set of chiral supermultiplets $\Phi_i \supset (\phi_i, \psi_i, F_i)$ and a superpotential $W$ in such a way that the equations $F_i = -\delta W^*/\delta\phi^{*i} = 0$ have no simultaneous solution. Then $V = \sum_i |F_i|^2$ will have to be positive at its minimum, ensuring that supersymmetry is broken.

The simplest example that does this has three chiral supermultiplets with

$$W = -k\Phi_1 + m\Phi_2\Phi_3 + \frac{y}{2}\Phi_1\Phi_3^2. \tag{1.225}$$

Note that $W$ contains a linear term, with $k$ having dimensions of $[\text{mass}]^2$. Such a term is allowed if the corresponding chiral supermultiplet is a gauge singlet. In fact, a linear term is necessary to achieve $F$-term breaking at tree-level in renormalizable theories,[s] since otherwise setting all $\phi_i = 0$ will always give a supersymmetric global minimum with all $F_i = 0$. Without loss of generality, we can choose $k$, $m$, and $y$ to be real and positive (by a

[s]Non-polynomial superpotential terms, for example arising from non-perturbative effects, can avoid this requirement.

phase rotation of the fields). The scalar potential following from eq. (1.225) is

$$V = |F_1|^2 + |F_2|^2 + |F_3|^2, \tag{1.226}$$

$$F_1 = k - \frac{y}{2}\phi_3^{*2}, \qquad F_2 = -m\phi_3^*, \qquad F_3 = -m\phi_2^* - y\phi_1^*\phi_3^*. \tag{1.227}$$

Clearly, $F_1 = 0$ and $F_2 = 0$ are not compatible, so supersymmetry must indeed be broken. If $m^2 > yk$ (which we assume from now on), then the absolute minimum of the potential is at $\phi_2 = \phi_3 = 0$ with $\phi_1$ undetermined, so $F_1 = k$ and $V = k^2$ at the minimum. The fact that $\phi_1$ is undetermined is an example of a "flat direction" in the scalar potential; this is a common feature of supersymmetric models.[t]

If we presciently choose to expand $V$ around $\phi_1 = 0$, the mass spectrum of the theory consists of 6 real scalars with tree-level squared masses

$$0,\ 0,\ m^2,\ m^2,\ m^2 - yk,\ m^2 + yk. \tag{1.228}$$

Meanwhile, there are 3 Weyl fermions with squared masses

$$0,\ m^2,\ m^2. \tag{1.229}$$

The non-degeneracy of scalars and fermions is a clear sign that supersymmetry has been spontaneously broken. [Note that the sum rule eq. (1.220) is indeed satisfied by these squared masses.] The 0 eigenvalues in eqs. (1.228) and (1.229) correspond to the complex scalar $\phi_1$ and its fermionic partner $\psi_1$. However, $\phi_1$ and $\psi_1$ have different reasons for being massless. The masslessness of $\phi_1$ corresponds to the existence of the flat direction, since any value of $\phi_1$ gives the same energy at tree-level. This flat direction is an accidental feature of the classical scalar potential, and in this case it is removed ("lifted") by quantum corrections. This can be seen by computing the Coleman-Weinberg one-loop effective potential.[136] A little calculation reveals that the global minimum is indeed fixed at $\phi_1 = \phi_2 = \phi_3 = 0$, with the complex scalar $\phi_1$ receiving a small positive-definite squared mass equal to

$$m^2_{\phi_1} = \frac{1}{16\pi^2} y^2 m^2 \left[\ln(1 - r^2) - 1 + \frac{1}{2}(r + 1/r)\ln\left(\frac{1+r}{1-r}\right)\right], \tag{1.230}$$

where $r = yk/m^2$. [Equation (1.230) reduces to $m^2_{\phi_1} = y^4k^2/48\pi^2 m^2$ in the limit $yk \ll m^2$.] In contrast, the Weyl fermion $\psi_1$ remains exactly massless, because it is the goldstino, as predicted in section 1.6.1.

[t] More generally, "flat directions" are non-compact lines and surfaces in the space of scalar fields along which the scalar potential vanishes. The classical scalar potential of the MSSM would have many flat directions if supersymmetry were not broken.[135]

The O'Raifeartaigh superpotential determines the mass scale of supersymmetry breaking $\sqrt{F_1}$ in terms of a dimensionful parameter $k$ put in by hand. This appears somewhat artificial, since $k$ will have to be tiny compared to $M_{\rm P}^2$ in order to give the right order of magnitude for the MSSM soft terms. We would like to have a mechanism that can instead generate such scales naturally. This can be done in models of dynamical supersymmetry breaking, in which the small (compared to $M_{\rm P}$) mass scales associated with supersymmetry breaking arise by dimensional transmutation. In other words, they generally feature a new asymptotically free non-Abelian gauge symmetry with a gauge coupling $g$ that is perturbative at $M_{\rm P}$ and gets strong in the infrared at some smaller scale $\Lambda \sim e^{-8\pi^2/|b|g_0^2} M_{\rm P}$, where $g_0$ is the running gauge coupling at $M_{\rm P}$ with negative beta function $-|b|g^3/16\pi^2$. Just as in QCD, it is perfectly natural for $\Lambda$ to be many orders of magnitude below the Planck scale. Supersymmetry breaking may then be best described in terms of the effective dynamics of the strongly coupled theory. Supersymmetry is still broken by the VEV of an $F$ field, but it may be the auxiliary field of a composite chiral supermultiplet built out of fields that are charged under the new strongly coupled gauge group.

Constructing non-perturbative models that actually break supersymmetry in an acceptable way is not a simple business. It is particularly difficult if one requires that the supersymmetry-breaking vacuum state is the true ground state (classically, the global minimum of the potential). One can prove using the Witten index[137,138] that any strongly coupled gauge theory with only vector-like, massive matter cannot spontaneously break supersymmetry in its ground state. Furthermore, a theory that has a generic superpotential and spontaneously breaks supersymmetry in its ground state must[139] have a continuous $U(1)$ $R$-symmetry, a quite non-trivial requirement. (However, effective superpotentials generated by non-perturbative dynamics are often not generic, so this requirement can be evaded.[139]) Many models that spontaneously break supersymmetry in their ground states have been found; for reviews see Ref. 140.

However, as noted in section 1.6.1, the supersymmetry-breaking vacuum state in which we live may instead correspond to only a local minimum of the potential. It has recently been shown by Intriligator, Seiberg, and Shih[130] that even supersymmetric Yang-Mills theories with vector-like matter can have metastable vacuum states with non-vanishing $F$-terms that break supersymmetry, and lifetimes that can be arbitrarily long. (The simplest model that does this is just supersymmetric $SU(N_c)$ gauge theory, with $N_f$ massive flavors of quark and antiquark supermultiplets, with

$N_c + 1 \leq N_f < 3N_c/2$.) The possibility of a metastable vacuum state simplifies model building and opens up many new possibilities.[130,141]

Finding the ultimate cause of supersymmetry breaking is one of the most important goals for the future. However, for many purposes, one can simply assume that an $F$-term has obtained a VEV, without worrying about the specific dynamics that caused it. For understanding collider phenomenology, the most immediate concern is usually the nature of the couplings of the $F$-term VEV to the MSSM fields. This is the subject we turn to next.

### 1.6.4. *The need for a separate supersymmetry-breaking sector*

It is now clear that spontaneous supersymmetry breaking (dynamical or not) requires us to extend the MSSM. The ultimate supersymmetry-breaking order parameter cannot belong to any of the MSSM supermultiplets; a $D$-term VEV for $U(1)_Y$ does not lead to an acceptable spectrum, and there is no candidate gauge singlet whose $F$-term could develop a VEV. Therefore one must ask what effects *are* responsible for spontaneous supersymmetry breaking, and how supersymmetry breakdown is "communicated" to the MSSM particles. It is very difficult to achieve the latter in a phenomenologically viable way working only with renormalizable interactions at tree-level, even if the model is extended to involve new supermultiplets including gauge singlets. First, on general grounds it would be problematic to give masses to the MSSM gauginos, because the results of section 1.3 inform us that renormalizable supersymmetry never has any (scalar)-(gaugino)-(gaugino) couplings that could turn into gaugino mass terms when the scalar gets a VEV. Second, at least some of the MSSM squarks and sleptons would have to be unacceptably light, and should have been discovered already. This can be understood from the existence of sum rules that can be obtained in the same way as eq. (1.220) when the restrictions imposed by flavor symmetries are taken into account. For example, in the limit in which lepton flavors are conserved, the selectron mass eigenstates $\tilde{e}_1$ and $\tilde{e}_2$ could in general be mixtures of $\tilde{e}_L$ and $\tilde{e}_R$. But if they do not mix with other scalars, then part of the sum rule decouples from the rest, and one obtains:

$$m^2_{\tilde{e}_1} + m^2_{\tilde{e}_2} = 2m^2_e, \tag{1.231}$$

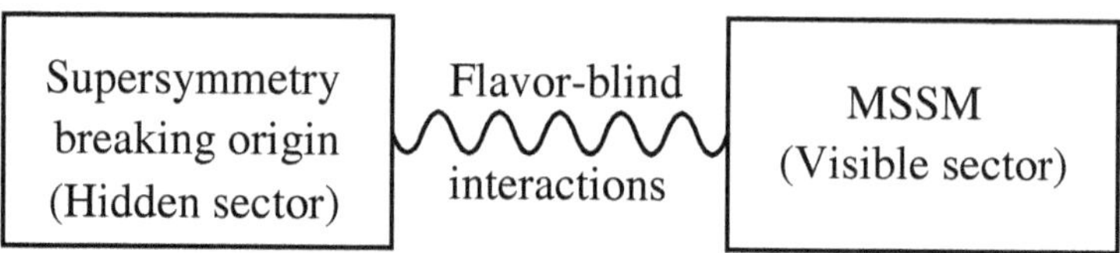


Fig. 1.15. The presumed schematic structure for supersymmetry breaking.

which is of course ruled out by experiment. Similar sum rules follow for each of the fermions of the Standard Model, at tree-level and in the limits in which the corresponding flavors are conserved. In principle, the sum rules can be evaded by introducing flavor-violating mixings, but it is very difficult to see how to make a viable model in this way. Even ignoring these problems, there is no obvious reason why the resulting MSSM soft supersymmetry-breaking terms in this type of model should satisfy flavor-blindness conditions like eqs. (1.158) or (1.159).

For these reasons, we expect that the MSSM soft terms arise indirectly or radiatively, rather than from tree-level renormalizable couplings to the supersymmetry-breaking order parameters. Supersymmetry breaking evidently occurs in a "hidden sector" of particles that have no (or only very small) direct couplings to the "visible sector" chiral supermultiplets of the MSSM. However, the two sectors do share some interactions that are responsible for mediating supersymmetry breaking from the hidden sector to the visible sector, resulting in the MSSM soft terms. (See Figure 1.15.) In this scenario, the tree-level squared mass sum rules need not hold, even approximately, for the physical masses of the visible sector fields, so that a phenomenologically viable superpartner mass spectrum is, in principle, achievable. As a bonus, if the mediating interactions are flavor-blind, then the soft terms appearing in the MSSM will automatically obey conditions like eqs. (1.158), (1.159) and (1.160).

There have been two main competing proposals for what the mediating interactions might be. The first (and historically the more popular) is that they are gravitational. More precisely, they are associated with the new physics, including gravity, that enters near the Planck scale. In this "gravity-mediated", or *Planck-scale-mediated supersymmetry breaking* (PMSB) scenario, if supersymmetry is broken in the hidden sector by a VEV $\langle F \rangle$, then the soft terms in the visible sector should be roughly

$$m_{\text{soft}} \sim \langle F \rangle / M_{\text{P}}, \tag{1.232}$$

by dimensional analysis. This is because we know that $m_{\rm soft}$ must vanish in the limit $\langle F\rangle \to 0$ where supersymmetry is unbroken, and also in the limit $M_{\rm P} \to \infty$ (corresponding to $G_{\rm Newton} \to 0$) in which gravity becomes irrelevant. For $m_{\rm soft}$ of order a few hundred GeV, one would therefore expect that the scale associated with the origin of supersymmetry breaking in the hidden sector should be roughly $\sqrt{\langle F\rangle} \sim 10^{10}$ or $10^{11}$ GeV. Another possibility is that the supersymmetry breaking order parameter is a gaugino condensate $\langle 0|\lambda^a\lambda^b|0\rangle = \delta^{ab}\Lambda^3 \neq 0$. If the composite field $\lambda^a\lambda^b$ is part of an auxiliary field $F$ for some (perhaps composite) chiral superfield, then by dimensional analysis we expect supersymmetry breaking soft terms of order

$$m_{\rm soft} \sim \Lambda^3/M_{\rm P}^2, \tag{1.233}$$

with, effectively, $\langle F\rangle \sim \Lambda^3/M_{\rm P}$. In that case, the scale associated with dynamical supersymmetry breaking should be more like $\Lambda \sim 10^{13}$ GeV.

A second possibility is that the flavor-blind mediating interactions for supersymmetry breaking are the ordinary electroweak and QCD gauge interactions. In this *gauge-mediated supersymmetry breaking* (GMSB) scenario, the MSSM soft terms come from loop diagrams involving some *messenger* particles. The messengers are new chiral supermultiplets that couple to a supersymmetry-breaking VEV $\langle F\rangle$, and also have $SU(3)_C \times SU(2)_L \times U(1)_Y$ interactions, which provide the necessary connection to the MSSM. Then, using dimensional analysis, one estimates for the MSSM soft terms

$$m_{\rm soft} \sim \frac{\alpha_a}{4\pi}\frac{\langle F\rangle}{M_{\rm mess}} \tag{1.234}$$

where the $\alpha_a/4\pi$ is a loop factor for Feynman diagrams involving gauge interactions, and $M_{\rm mess}$ is a characteristic scale of the masses of the messenger fields. So if $M_{\rm mess}$ and $\sqrt{\langle F\rangle}$ are roughly comparable, then the scale of supersymmetry breaking can be as low as about $\sqrt{\langle F\rangle} \sim 10^4$ GeV (much lower than in the gravity-mediated case!) to give $m_{\rm soft}$ of the right order of magnitude.

### 1.6.5. *The goldstino and the gravitino*

As shown in section 1.6.1, the spontaneous breaking of global supersymmetry implies the existence of a massless Weyl fermion, the goldstino. The goldstino is the fermionic component of the supermultiplet whose auxiliary field obtains a VEV.

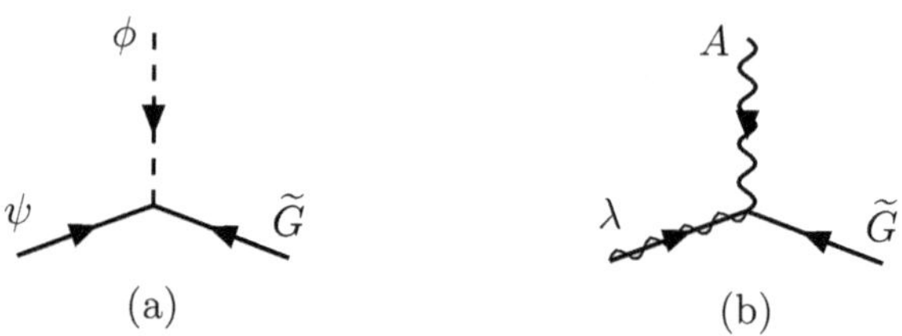


Fig. 1.16. Goldstino/gravitino $\tilde{G}$ interactions with superpartner pairs $(\phi, \psi)$ and $(\lambda, A)$.

We can derive an important property of the goldstino by considering the form of the conserved supercurrent eq. (1.136). Suppose for simplicity[u] that the only non-vanishing auxiliary field VEV is $\langle F\rangle$ with goldstino superpartner $\widetilde{G}$. Then the supercurrent conservation equation tells us that

$$0 = \partial_\mu J^\mu_\alpha = -i\langle F\rangle(\sigma^\mu \partial_\mu \widetilde{G}^\dagger)_\alpha + \partial_\mu j^\mu_\alpha + \dots \tag{1.235}$$

where $j^\mu_\alpha$ is the part of the supercurrent that involves all of the other supermultiplets, and the ellipses represent other contributions of the goldstino supermultiplet to $\partial_\mu J^\mu_\alpha$, which we can ignore. [The first term in eq. (1.235) comes from the second term in eq. (1.136), using the equation of motion $F_i = -W_i^*$ for the goldstino's auxiliary field.] This equation of motion for the goldstino field allows us to write an effective Lagrangian

$$\mathcal{L}_{\text{goldstino}} = i\widetilde{G}^\dagger \overline{\sigma}^\mu \partial_\mu \widetilde{G} - \frac{1}{\langle F\rangle}(\widetilde{G}\partial_\mu j^\mu + \text{c.c.}), \tag{1.236}$$

which describes the interactions of the goldstino with all of the other fermion-boson pairs.[142] In particular, since $j^\mu_\alpha = (\sigma^\nu \overline{\sigma}^\mu \psi_i)_\alpha \partial_\nu \phi^{*i} - \sigma^\nu \overline{\sigma}^\rho \sigma^\mu \lambda^{\dagger a} F^a_{\nu\rho}/2\sqrt{2} + \dots$, there are goldstino-scalar-chiral fermion and goldstino-gaugino-gauge boson vertices as shown in Figure 1.16. Since this derivation depends only on supercurrent conservation, eq. (1.236) holds independently of the details of how supersymmetry breaking is communicated from $\langle F\rangle$ to the MSSM sector fields $(\phi_i, \psi_i)$ and $(\lambda^a, A^a)$. It may appear strange at first that the interaction couplings in eq. (1.236) get larger in the limit $\langle F\rangle$ goes to zero. However, the interaction term $\widetilde{G}\partial_\mu j^\mu$ contains two derivatives, which turn out to always give a kinematic factor proportional to the squared-mass difference of the superpartners when they are on-shell, i.e. $m^2_\phi - m^2_\psi$ and $m^2_\lambda - m^2_A$ for Figures 1.16a and 1.16b respectively. These can be non-zero only by virtue of supersymmetry breaking, so they must

[u] More generally, if supersymmetry is spontaneously broken by VEVs for several auxiliary fields $F_i$ and $D^a$, then one should make the replacement $\langle F\rangle \to (\sum_i |\langle F_i\rangle|^2 + \frac{1}{2}\sum_a \langle D^a\rangle^2)^{1/2}$ everywhere in the following.

also vanish as $\langle F\rangle \to 0$, and the interaction is well-defined in that limit. Nevertheless, for fixed values of $m_\phi^2 - m_\psi^2$ and $m_\lambda^2 - m_A^2$, the interaction term in eq. (1.236) can be phenomenologically important if $\langle F\rangle$ is not too large.[142–145]

The above remarks apply to the breaking of global supersymmetry. However, taking into account gravity, supersymmetry must be promoted to a local symmetry. This means that the spinor parameter $\epsilon^\alpha$, which first appeared in section 1.3.1, is no longer a constant, but can vary from point to point in spacetime. The resulting locally supersymmetric theory is called *supergravity*.[146,147] It necessarily unifies the spacetime symmetries of ordinary general relativity with local supersymmetry transformations. In supergravity, the spin-2 graviton has a spin-3/2 fermion superpartner called the gravitino, which we will denote $\widetilde{\Psi}_\mu^\alpha$. The gravitino has odd $R$-parity ($P_R = -1$), as can be seen from the definition eq. (1.151). It carries both a vector index ($\mu$) and a spinor index ($\alpha$), and transforms inhomogeneously under local supersymmetry transformations:

$$\delta\widetilde{\Psi}_\mu^\alpha = \partial_\mu \epsilon^\alpha + \ldots . \tag{1.237}$$

Thus the gravitino should be thought of as the "gauge" field of local supersymmetry transformations [compare eq. (1.115)]. As long as supersymmetry is unbroken, the graviton and the gravitino are both massless, each with two spin helicity states. Once supersymmetry is spontaneously broken, the gravitino acquires a mass by absorbing ("eating") the goldstino, which becomes its longitudinal (helicity $\pm 1/2$) components. This is called the *super-Higgs* mechanism, and it is analogous to the ordinary Higgs mechanism for gauge theories, by which the $W^\pm$ and $Z^0$ gauge bosons in the Standard Model gain mass by absorbing the Nambu-Goldstone bosons associated with the spontaneously broken electroweak gauge invariance. The massive spin-3/2 gravitino now has four helicity states, of which two were originally assigned to the would-be goldstino. The gravitino mass is traditionally called $m_{3/2}$, and in the case of $F$-term breaking it can be estimated as[148]

$$m_{3/2} \sim \langle F\rangle / M_{\mathrm{P}}. \tag{1.238}$$

This follows simply from dimensional analysis, since $m_{3/2}$ must vanish in the limits that supersymmetry is restored ($\langle F\rangle \to 0$) and that gravity is turned off ($M_P \to \infty$). Equation (1.238) implies very different expectations for the mass of the gravitino in gravity-mediated and in gauge-mediated models, because they usually make very different predictions for $\langle F\rangle$.

In the Planck-scale-mediated supersymmetry breaking case, the gravitino mass is comparable to the masses of the MSSM sparticles [compare eqs. (1.232) and (1.238)]. Therefore $m_{3/2}$ is expected to be at least of order 100 GeV or so. Its interactions will be of gravitational strength, so the gravitino will not play any role in collider physics, but it can be important in cosmology.[149] If it is the LSP, then it is stable and its primordial density could easily exceed the critical density, causing the universe to become matter-dominated too early. Even if it is not the LSP, the gravitino can cause problems unless its density is diluted by inflation at late times, or it decays sufficiently rapidly.

In contrast, gauge-mediated supersymmetry breaking models predict that the gravitino is much lighter than the MSSM sparticles as long as $M_{\rm mess} \ll M_{\rm P}$. This can be seen by comparing eqs. (1.234) and (1.238). The gravitino is almost certainly the LSP in this case, and all of the MSSM sparticles will eventually decay into final states that include it. Naively, one might expect that these decays are extremely slow. However, this is not necessarily true, because the gravitino inherits the non-gravitational interactions of the goldstino it has absorbed. This means that the gravitino, or more precisely its longitudinal (goldstino) components, can play an important role in collider physics experiments. The mass of the gravitino can generally be ignored for kinematic purposes, as can its transverse (helicity $\pm 3/2$) components, which really do have only gravitational interactions. Therefore in collider phenomenology discussions one may interchangeably use the same symbol $\widetilde{G}$ for the goldstino and for the gravitino of which it is the longitudinal (helicity $\pm 1/2$) part. By using the effective Lagrangian eq. (1.236), one can compute that the decay rate of any sparticle $\widetilde{X}$ into its Standard Model partner $X$ plus a goldstino/gravitino $\widetilde{G}$ is

$$\Gamma(\widetilde{X} \to X\widetilde{G}) = \frac{m_{\widetilde{X}}^5}{16\pi \langle F \rangle^2} \left(1 - m_X^2/m_{\widetilde{X}}^2\right)^4 . \tag{1.239}$$

This corresponds to either Figure 1.16a or 1.16b, with $(\widetilde{X}, X) = (\phi, \psi)$ or $(\lambda, A)$ respectively. One factor $(1 - m_X^2/m_{\widetilde{X}}^2)^2$ came from the derivatives in the interaction term in eq. (1.236) evaluated for on-shell final states, and another such factor comes from the kinematic phase space integral with $m_{3/2} \ll m_{\widetilde{X}}, m_X$.

If the supermultiplet containing the goldstino and $\langle F \rangle$ has canonically normalized kinetic terms, and the tree-level vacuum energy is required to vanish, then the estimate eq. (1.238) is sharpened to

$$m_{3/2} = \langle F \rangle / \sqrt{3} M_{\rm P}. \tag{1.240}$$

In that case, one can rewrite eq. (1.239) as

$$\Gamma(\widetilde{X} \to X\widetilde{G}) = \frac{m_{\widetilde{X}}^5}{48\pi M_{\rm P}^2 m_{3/2}^2}\left(1 - m_X^2/m_{\widetilde{X}}^2\right)^4, \tag{1.241}$$

and this is how the formula is sometimes presented, although it is less general since it assumes eq. (1.240). The decay width is larger for smaller $\langle F\rangle$, or equivalently for smaller $m_{3/2}$, if the other masses are fixed. If $\widetilde{X}$ is a mixture of superpartners of different Standard Model particles $X$, then each partial width in eq. (1.239) should be multiplied by a suppression factor equal to the square of the cosine of the appropriate mixing angle. If $m_{\widetilde{X}}$ is of order 100 GeV or more, and $\sqrt{\langle F\rangle} \lesssim$ few $\times 10^6$ GeV [corresponding to $m_{3/2}$ less than roughly 1 keV according to eq. (1.240)], then the decay $\widetilde{X} \to X\widetilde{G}$ can occur quickly enough to be observed in a modern collider detector. This implies some interesting phenomenological signatures, which we will discuss further in sections 1.8.5.

We now turn to a more systematic analysis of the way in which the MSSM soft terms arise.

### 1.6.6. *Planck-scale-mediated supersymmetry breaking models*

Consider the class of models defined by the feature that the spontaneous supersymmetry-breaking sector connects with our MSSM only (or dominantly) through gravitational-strength interactions.[150,151] This means that the supergravity effective Lagrangian contains non-renormalizable terms that communicate between the two sectors and are suppressed by powers of the Planck mass $M_{\rm P}$. These will include

$$\begin{aligned}\mathcal{L}_{\rm NR} = &-\frac{1}{M_{\rm P}}F\left(\frac{1}{2}f_a\lambda^a\lambda^a + \frac{1}{6}y'^{ijk}\phi_i\phi_j\phi_k + \frac{1}{2}\mu'^{ij}\phi_i\phi_j\right) + \text{c.c.}\\ &-\frac{1}{M_{\rm P}^2}FF^*\, k^i_j\phi_i\phi^{*j}\end{aligned} \tag{1.242}$$

where $F$ is the auxiliary field for a chiral supermultiplet in the hidden sector, and $\phi_i$ and $\lambda^a$ are the scalar and gaugino fields in the MSSM, and $f^a$, $y'^{ijk}$, and $k^i_j$ are dimensionless constants. By themselves, the terms in eq. (1.242) are not supersymmetric, but it is possible to show that they are part of a non-renormalizable supersymmetric Lagrangian (see Appendix) that contains other terms that we may ignore. Now if one assumes that $\sqrt{\langle F\rangle} \sim 10^{10}$ or $10^{11}$ GeV, then $\mathcal{L}_{\rm NR}$ will give us nothing other than a

Lagrangian of the form $\mathcal{L}_{\rm soft}$ in eq. (1.138), with MSSM soft terms of order $m_{\rm soft} \sim \langle F \rangle / M_{\rm P}$ = a few hundred GeV.

Note that couplings of the form $\mathcal{L}_{\rm maybe\ soft}$ in eq. (1.139) do not arise from eq. (1.242). They actually are expected to occur, but the largest term from which they could come is:

$$\mathcal{L} = -\frac{1}{M_{\rm P}^3} F F^* x_i^{jk} \phi^{*i} \phi_j \phi_k + {\rm c.c.}, \tag{1.243}$$

so in this model framework they are of order $\langle F \rangle^2 / M_{\rm P}^3 \sim m_{\rm soft}^2 / M_{\rm P}$, and therefore negligible.

The parameters $f_a$, $k^i_j$, $y'^{ijk}$ and $\mu'^{ij}$ in $\mathcal{L}_{\rm NR}$ are to be determined by the underlying theory. This is a difficult enterprise in general, but a dramatic simplification occurs if one assumes a "minimal" form for the normalization of kinetic terms and gauge interactions in the full, non-renormalizable supergravity Lagrangian (see Appendix). In that case, there is a common $f_a = f$ for the three gauginos; $k^i_j = k\delta^i_j$ is the same for all scalars; and the other couplings are proportional to the corresponding superpotential parameters, so that $y'^{ijk} = \alpha y^{ijk}$ and $\mu'^{ij} = \beta \mu^{ij}$ with universal dimensionless constants $\alpha$ and $\beta$. Then the soft terms in $\mathcal{L}_{\rm soft}^{\rm MSSM}$ are all determined by just four parameters:

$$m_{1/2} = f \frac{\langle F \rangle}{M_{\rm P}}, \quad m_0^2 = k \frac{|\langle F \rangle|^2}{M_{\rm P}^2}, \quad A_0 = \alpha \frac{\langle F \rangle}{M_{\rm P}}, \quad B_0 = \beta \frac{\langle F \rangle}{M_{\rm P}}. \tag{1.244}$$

In terms of these, the parameters appearing in eq. (1.152) are:

$$M_3 = M_2 = M_1 = m_{1/2}, \tag{1.245}$$

$$\mathbf{m^2_Q} = \mathbf{m^2_{\overline{u}}} = \mathbf{m^2_{\overline{d}}} = \mathbf{m^2_L} = \mathbf{m^2_{\overline{e}}} = m_0^2\, \mathbf{1}, \qquad m^2_{H_u} = m^2_{H_d} = m_0^2, \tag{1.246}$$

$$\mathbf{a_u} = A_0 \mathbf{y_u}, \qquad \mathbf{a_d} = A_0 \mathbf{y_d}, \qquad \mathbf{a_e} = A_0 \mathbf{y_e}, \tag{1.247}$$

$$b = B_0 \mu, \tag{1.248}$$

at a renormalization scale $Q \approx M_{\rm P}$. It is a matter of some controversy whether the assumptions going into this parameterization are well-motivated on purely theoretical grounds,[v] but from a phenomenological perspective they are clearly very nice. This framework successfully evades the most dangerous types of flavor changing and CP violation as discussed in section 1.5.4. In particular, eqs. (1.246) and (1.247) are just stronger versions of eqs. (1.158) and (1.159), respectively. If $m_{1/2}$, $A_0$ and $B_0$ all have the same complex phase, then eq. (1.160) will also be satisfied.

[v] The familiar flavor blindness of gravity expressed in Einstein's equivalence principle does not, by itself, tell us anything about the form of eq. (1.242), and in particular need not imply eqs. (1.245)-(1.247). (See Appendix.)

Equations (1.245)-(1.248) also have the virtue of being highly predictive. [Of course, eq. (1.248) is content-free unless one can relate $B_0$ to the other parameters in some non-trivial way.] As discussed in sections and 1.5.4 and 1.5.5, they should be applied as RG boundary conditions at the scale $M_{\rm P}$. The RG evolution of the soft parameters down to the electroweak scale will then allow us to predict the entire MSSM spectrum in terms of just five parameters $m_{1/2}$, $m_0^2$, $A_0$, $B_0$, and $\mu$ (plus the already-measured gauge and Yukawa couplings of the MSSM). A popular approximation is to start this RG running from the unification scale $M_U \approx 2 \times 10^{16}$ GeV instead of $M_{\rm P}$. The reason for this is more practical than principled; the apparent unification of gauge couplings gives us a strong hint that we know something about how the RG equations behave up to $M_U$, but unfortunately gives us little guidance about what to expect at scales between $M_U$ and $M_{\rm P}$. The errors made in neglecting these effects are proportional to a loop suppression factor times $\ln(M_{\rm P}/M_U)$. These corrections hopefully can be partly absorbed into a redefinition of $m_0^2$, $m_{1/2}$, $A_0$ and $B_0$ at $M_U$, but in many cases can lead to other important effects.[152] The framework described in the above few paragraphs has been the subject of the bulk of phenomenological studies of supersymmetry. It is sometimes referred to as the *minimal supergravity* (MSUGRA) or *supergravity-inspired* scenario for the soft terms. A few examples of the many useful numerical RG studies of the MSSM spectrum that have been performed in this framework can be found in Ref. 153.

Particular models of gravity-mediated supersymmetry breaking can be even more predictive, relating some of the parameters $m_{1/2}$, $m_0^2$, $A_0$ and $B_0$ to each other and to the mass of the gravitino $m_{3/2}$. For example, three popular kinds of models for the soft terms are:

- Dilaton-dominated:[154] $m_0^2 = m_{3/2}^2, \quad m_{1/2} = -A_0 = \sqrt{3} m_{3/2}$.
- Polonyi:[155] $m_0^2 = m_{3/2}^2, \quad A_0 = (3 - \sqrt{3}) m_{3/2}, \quad m_{1/2} = \mathcal{O}(m_{3/2})$.
- "No-scale":[156] $m_{1/2} \gg m_0, A_0, m_{3/2}$.

Dilaton domination arises in a particular limit of superstring theory. While it appears to be highly predictive, it can easily be generalized in other limits.[157] The Polonyi model has the advantage of being the simplest possible model for supersymmetry breaking in the hidden sector, but it is rather *ad hoc* and does not seem to have a special place in grander schemes like superstrings. The "no-scale" limit may appear in a low-energy limit of superstrings in which the gravitino mass scale is undetermined at tree-level (hence the name). It implies that the gaugino masses dominate over other sources of supersymmetry breaking near $M_P$. As we saw in section 1.5.5, RG evolution feeds $m_{1/2}$ into the squark, slepton, and Higgs squared-mass parameters with sufficient magnitude to give acceptable phenomenology at the electroweak scale. More recent versions of the no-scale scenario, however, also can give significant $A_0$ and $m_0^2$ at $M_P$. In many cases $B_0$ can also be predicted in terms of the other parameters, but this is quite sensitive to model assumptions. For phenomenological studies, $m_{1/2}$, $m_0^2$, $A_0$ and $B_0$ are usually just taken to be imperfect but convenient independent parameters of our ignorance of the supersymmetry breaking mechanism.

### 1.6.7. *Gauge-mediated supersymmetry breaking models*

In gauge-mediated supersymmetry breaking (GMSB) models,[158,159] the ordinary gauge interactions, rather than gravity, are responsible for the appearance of soft supersymmetry breaking in the MSSM. The basic idea is to introduce some new chiral supermultiplets, called messengers, that couple to the ultimate source of supersymmetry breaking, and also couple indirectly to the (s)quarks and (s)leptons and Higgs(inos) of the MSSM through the ordinary $SU(3)_C \times SU(2)_L \times U(1)_Y$ gauge boson and gaugino interactions. There is still gravitational communication between the MSSM and the source of supersymmetry breaking, of course, but that effect is now relatively unimportant compared to the gauge interaction effects.

In contrast to Planck-scale mediation, GMSB can be understood entirely in terms of loop effects in a renormalizable framework. In the simplest such model, the messenger fields are a set of left-handed chiral supermultiplets $q$, $\overline{q}$, $\ell$, $\overline{\ell}$ transforming under $SU(3)_C \times SU(2)_L \times U(1)_Y$ as

$$q \sim (\mathbf{3}, \mathbf{1}, -\frac{1}{3}), \quad \overline{q} \sim (\overline{\mathbf{3}}, \mathbf{1}, \frac{1}{3}), \quad \ell \sim (\mathbf{1}, \mathbf{2}, \frac{1}{2}), \quad \overline{\ell} \sim (\mathbf{1}, \mathbf{2}, -\frac{1}{2}). \quad (1.249)$$

These supermultiplets contain messenger quarks $\psi_q, \psi_{\overline{q}}$ and scalar quarks $q, \overline{q}$ and messenger leptons $\psi_\ell, \psi_{\overline{\ell}}$ and scalar leptons $\ell, \overline{\ell}$. All of these particles must get very large masses so as not to have been discovered already. Assume they do so by coupling to a gauge-singlet chiral supermultiplet $S$ through a superpotential:

$$W_{\text{mess}} = y_2 S\ell\overline{\ell} + y_3 Sq\overline{q}. \quad (1.250)$$

The scalar component of $S$ and its auxiliary ($F$-term) component are each supposed to acquire VEVs, denoted $\langle S \rangle$ and $\langle F_S \rangle$ respectively. This can be accomplished either by putting $S$ into an O'Raifeartaigh-type model,[158] or by a dynamical mechanism.[159] Exactly how this happens is an interesting and important question, without a clear answer at present. Here, we will simply parameterize our ignorance of the precise mechanism of supersymmetry breaking by asserting that $S$ participates in another part of the superpotential, call it $W_{\rm breaking}$, which provides for the necessary spontaneous breaking of supersymmetry.

Let us now consider the mass spectrum of the messenger fermions and bosons. The fermionic messenger fields pair up to get mass terms:

$$\mathcal{L} = -y_2\langle S\rangle\psi_\ell\psi_{\overline{\ell}} - y_3\langle S\rangle\psi_q\psi_{\overline{q}} + \text{c.c.} \tag{1.251}$$

as in eq. (1.111). Meanwhile, their scalar messenger partners $\ell, \overline{\ell}$ and $q, \overline{q}$ have a scalar potential given by (neglecting $D$-term contributions, which do not affect the following discussion):

$$V = \left|\frac{\delta W_{\rm mess}}{\delta \ell}\right|^2 + \left|\frac{\delta W_{\rm mess}}{\delta \overline{\ell}}\right|^2 + \left|\frac{\delta W_{\rm mess}}{\delta q}\right|^2 + \left|\frac{\delta W_{\rm mess}}{\delta \overline{q}}\right|^2 + \left|\frac{\delta}{\delta S}(W_{\rm mess} + W_{\rm breaking})\right|^2 \tag{1.252}$$

as in eq. (1.111). Now, suppose that, at the minimum of the potential,

$$\langle S\rangle \neq 0, \tag{1.253}$$

$$\langle \delta W_{\rm breaking}/\delta S\rangle = -\langle F_S^*\rangle \neq 0, \tag{1.254}$$

$$\langle \delta W_{\rm mess}/\delta S\rangle = 0. \tag{1.255}$$

Replacing $S$ and $F_S$ by their VEVs, one finds quadratic mass terms in the potential for the messenger scalar leptons:

$$\begin{aligned} V = {} & |y_2\langle S\rangle|^2\big(|\ell|^2 + |\overline{\ell}|^2\big) + |y_3\langle S\rangle|^2\big(|q|^2 + |\overline{q}|^2\big) \\ & - \big(y_2\langle F_S\rangle \ell\overline{\ell} + y_3\langle F_S\rangle q\overline{q} + \text{c.c.}\big) \\ & + \text{quartic terms.} \end{aligned} \tag{1.256}$$

The first line in eq. (1.256) represents supersymmetric mass terms that go along with eq. (1.251), while the second line consists of soft supersymmetry-breaking masses. The complex scalar messengers $\ell, \overline{\ell}$ thus obtain a squared-mass matrix equal to:

$$\begin{pmatrix} |y_2\langle S\rangle|^2 & -y_2^*\langle F_S^*\rangle \\ -y_2\langle F_S\rangle & |y_2\langle S\rangle|^2 \end{pmatrix} \tag{1.257}$$

with squared mass eigenvalues $|y_2\langle S\rangle|^2 \pm |y_2\langle F_S\rangle|$. In just the same way, the scalars $q, \overline{q}$ get squared masses $|y_3\langle S\rangle|^2 \pm |y_3\langle F_S\rangle|$.

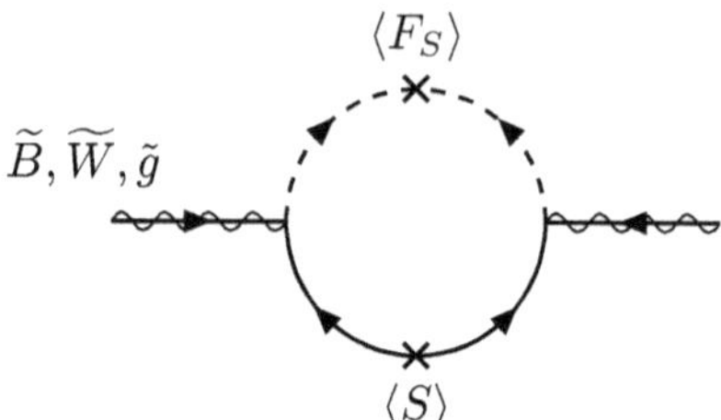


Fig. 1.17. Contributions to the MSSM gaugino masses in gauge-mediated supersymmetry breaking models come from one-loop graphs involving virtual messenger particles.

So far, we have found that the effect of supersymmetry breaking is to split each messenger supermultiplet pair apart:

$$\ell, \overline{\ell}: \quad m^2_{\rm fermions} = |y_2 \langle S \rangle|^2 , \quad m^2_{\rm scalars} = |y_2 \langle S \rangle|^2 \pm |y_2 \langle F_S \rangle| , \quad (1.258)$$

$$q, \overline{q}: \quad m^2_{\rm fermions} = |y_3 \langle S \rangle|^2 , \quad m^2_{\rm scalars} = |y_3 \langle S \rangle|^2 \pm |y_3 \langle F_S \rangle| . \quad (1.259)$$

The supersymmetry violation apparent in this messenger spectrum for $\langle F_S \rangle \neq 0$ is communicated to the MSSM sparticles through radiative corrections. The MSSM gauginos obtain masses from the 1-loop Feynman diagram shown in Figure 1.17. The scalar and fermion lines in the loop are messenger fields. Recall that the interaction vertices in Figure 1.17 are of gauge coupling strength even though they do not involve gauge bosons; compare Figure 1.5g. In this way, gauge-mediation provides that $q, \overline{q}$ messenger loops give masses to the gluino and the bino, and $\ell, \overline{\ell}$ messenger loops give masses to the wino and bino fields. Computing the 1-loop diagrams, one finds[159] that the resulting MSSM gaugino masses are given by

$$M_a = \frac{\alpha_a}{4\pi} \Lambda, \qquad (a = 1, 2, 3), \tag{1.260}$$

in the normalization for $\alpha_a$ discussed in section 1.5.4, where we have introduced a mass parameter

$$\Lambda \equiv \langle F_S \rangle / \langle S \rangle . \tag{1.261}$$

(Note that if $\langle F_S \rangle$ were 0, then $\Lambda = 0$ and the messenger scalars would be degenerate with their fermionic superpartners and there would be no contribution to the MSSM gaugino masses.) In contrast, the corresponding MSSM gauge bosons cannot get a corresponding mass shift, since they are protected by gauge invariance. So supersymmetry breaking has been successfully communicated to the MSSM ("visible sector"). To a good approximation, eq. (1.260) holds for the running gaugino masses at an RG scale $Q_0$ corresponding to the average characteristic mass of the heavy

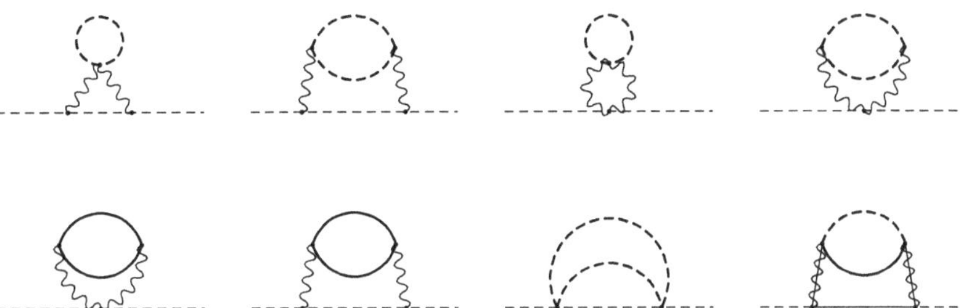

Fig. 1.18. MSSM scalar squared masses in gauge-mediated supersymmetry breaking models arise in leading order from these two-loop Feynman graphs. The heavy dashed lines are messenger scalars, the solid lines are messenger fermions, the wavy lines are ordinary Standard Model gauge bosons, and the solid lines with wavy lines superimposed are the MSSM gauginos.

messenger particles, roughly of order $M_{\rm mess} \sim y_I\langle S\rangle$ for $I = 2, 3$. The running mass parameters can then be RG-evolved down to the electroweak scale to predict the physical masses to be measured by future experiments.

The scalars of the MSSM do not get any radiative corrections to their masses at one-loop order. The leading contribution to their masses comes from the two-loop graphs shown in Figure 1.18, with the messenger fermions (heavy solid lines) and messenger scalars (heavy dashed lines) and ordinary gauge bosons and gauginos running around the loops. By computing these graphs, one finds that each MSSM scalar $\phi_i$ gets a squared mass given by:

$$m^2_{\phi_i} = 2\Lambda^2 \left[ \left(\frac{\alpha_3}{4\pi}\right)^2 C_3(i) + \left(\frac{\alpha_2}{4\pi}\right)^2 C_2(i) + \left(\frac{\alpha_1}{4\pi}\right)^2 C_1(i) \right], \quad (1.262)$$

with the quadratic Casimir invariants $C_a(i)$ as in eqs. (1.167)-(1.170). The squared masses in eq. (1.262) are positive (fortunately!).

The terms $\mathbf{a_u}$, $\mathbf{a_d}$, $\mathbf{a_e}$ arise first at two-loop order, and are suppressed by an extra factor of $\alpha_a/4\pi$ compared to the gaugino masses. So, to a very good approximation one has, at the messenger scale,

$$\mathbf{a_u} = \mathbf{a_d} = \mathbf{a_e} = 0, \quad (1.263)$$

a significantly stronger condition than eq. (1.159). Again, eqs. (1.262) and (1.263) should be applied at an RG scale equal to the average mass of the messenger fields running in the loops. However, evolving the RG equations down to the electroweak scale generates non-zero $\mathbf{a_u}$, $\mathbf{a_d}$, and $\mathbf{a_e}$ proportional to the corresponding Yukawa matrices and the non-zero gaugino masses, as indicated in section 1.5.5. These will only be large for the third-family squarks and sleptons, in the approximation of eq. (1.142). The parameter $b$ may also be taken to vanish near the messenger scale, but this is quite model-dependent, and in any case $b$ will be non-zero when it is RG-evolved to the electroweak scale. In practice, $b$ can be fixed in terms of the other parameters by the requirement of correct electroweak symmetry breaking, as discussed below in section 1.7.1.

Because the gaugino masses arise at *one*-loop order and the scalar squared-mass contributions appear at *two*-loop order, both eq. (1.260) and (1.262) correspond to the estimate eq. (1.234) for $m_{\rm soft}$, with $M_{\rm mess} \sim y_I \langle S \rangle$. Equations (1.260) and (1.262) hold in the limit of small $\langle F_S \rangle / y_I \langle S \rangle^2$, corresponding to mass splittings within each messenger supermultiplet that are small compared to the overall messenger mass scale. The sub-leading corrections in an expansion in $\langle F_S \rangle / y_I \langle S \rangle^2$ turn out[160] to be quite small unless there are very large messenger mass splittings.

The model we have described so far is often called the minimal model of gauge-mediated supersymmetry breaking. Let us now generalize it to a more complicated messenger sector. Suppose that $q, \overline{q}$ and $\ell, \overline{\ell}$ are replaced by a collection of messengers $\Phi_I, \overline{\Phi}_I$ with a superpotential

$$W_{\rm mess} = \sum_I y_I S \Phi_I \overline{\Phi}_I . \tag{1.264}$$

The bar is used to indicate that the left-handed chiral superfields $\overline{\Phi}_I$ transform as the complex conjugate representations of the left-handed chiral superfields $\Phi_I$. Together they are said to form a "vector-like" (real) representation of the Standard Model gauge group. As before, the fermionic components of each pair $\Phi_I$ and $\overline{\Phi}_I$ pair up to get squared masses $|y_I \langle S \rangle|^2$ and their scalar partners mix to get squared masses $|y_I \langle S \rangle|^2 \pm |y_I \langle F_S \rangle|$. The MSSM gaugino mass parameters induced are now

$$M_a = \frac{\alpha_a}{4\pi} \Lambda \sum_I n_a(I) \qquad (a = 1, 2, 3) \tag{1.265}$$

where $n_a(I)$ is the Dynkin index for each $\Phi_I + \overline{\Phi}_I$, in a normalization where $n_3 = 1$ for a $\mathbf{3} + \overline{\mathbf{3}}$ of $SU(3)_C$ and $n_2 = 1$ for a pair of doublets of $SU(2)_L$. For $U(1)_Y$, one has $n_1 = 6Y^2/5$ for each messenger pair with weak hypercharges $\pm Y$. In computing $n_1$ one must remember to add up the contributions for each component of an $SU(3)_C$ or $SU(2)_L$ multiplet. So, for example, $(n_1, n_2, n_3) = (2/5, 0, 1)$ for $q + \overline{q}$ and $(n_1, n_2, n_3) = (3/5, 1, 0)$ for $\ell + \overline{\ell}$. Thus the total is $\sum_I (n_1, n_2, n_3) = (1, 1, 1)$ for the minimal model, so that eq. (1.265) is in agreement with eq. (1.260). On general group-theoretic grounds, $n_2$ and $n_3$ must be integers, and $n_1$ is always an integer multiple of $1/5$ if fractional electric charges are confined.

The MSSM scalar masses in this generalized gauge mediation framework are now:

$$m_{\phi_i}^2 = 2\Lambda^2\Big[\Big(\frac{\alpha_3}{4\pi}\Big)^2 C_3(i)\sum_I n_3(I) + \Big(\frac{\alpha_2}{4\pi}\Big)^2 C_2(i)\sum_I n_2(I) + \Big(\frac{\alpha_1}{4\pi}\Big)^2 C_1(i)\sum_I n_1(I)\Big]. \tag{1.266}$$

In writing eqs. (1.265) and (1.266) as simple sums, we have implicitly assumed that the messengers are all approximately equal in mass, with

$$M_{\text{mess}} \approx y_I\langle S\rangle. \tag{1.267}$$

Equation (1.266) is still not a bad approximation if the $y_I$ are not very different from each other, because the dependence of the MSSM mass spectrum on the $y_I$ is only logarithmic (due to RG running) for fixed $\Lambda$. However, if large hierarchies in the messenger masses are present, then the additive contributions to the gaugino masses and scalar squared masses from each individual messenger multiplet $I$ should really instead be incorporated at the mass scale of that messenger multiplet. Then RG evolution is used to run these various contributions down to the electroweak or TeV scale; the individual messenger contributions to scalar and gaugino masses as indicated above can be thought of as threshold corrections to this RG running.

Messengers with masses far below the GUT scale will affect the running of gauge couplings and might therefore be expected to ruin the apparent unification shown in Figure 1.14. However, if the messengers come in complete multiplets of the $SU(5)$ global symmetry[w] that contains the Standard Model gauge group, and are not very different in mass, then approximate unification of gauge couplings will still occur when they are extrapolated up to the same scale $M_U$ (but with a larger unified value for the gauge couplings at that scale). For this reason, a popular class of models is obtained by taking the messengers to consist of $N_5$ copies of the $\mathbf{5} + \overline{\mathbf{5}}$ of $SU(5)$, resulting in

$$\sum_I n_1(I) = \sum_I n_2(I) = \sum_I n_3(I) = N_5\,. \tag{1.268}$$

[w]This $SU(5)$ may or may not be promoted to a local gauge symmetry at the GUT scale. For our present purposes, it is used only as a classification scheme, since the global $SU(5)$ symmetry is only approximate in the effective theory at the (much lower) messenger mass scale where gauge mediation takes place.

Equations (1.265) and (1.266) then reduce to

$$M_a = \frac{\alpha_a}{4\pi}\Lambda N_5, \tag{1.269}$$

$$m^2_{\phi_i} = 2\Lambda^2 N_5 \sum_{a=1}^{3} C_a(i)\left(\frac{\alpha_a}{4\pi}\right)^2, \tag{1.270}$$

since now there are $N_5$ copies of the minimal messenger sector particles running around the loops. For example, the minimal model in eq. (1.249) corresponds to $N_5 = 1$. A single copy of $\mathbf{10} + \overline{\mathbf{10}}$ of $SU(5)$ has Dynkin indices $\sum_I n_a(I) = 3$, and so can be substituted for 3 copies of $\mathbf{5} + \overline{\mathbf{5}}$. (Other combinations of messenger multiplets can also preserve the apparent unification of gauge couplings.) Note that the gaugino masses scale like $N_5$, while the scalar masses scale like $\sqrt{N_5}$. This means that sleptons and squarks will tend to be lighter relative to the gauginos for larger values of $N_5$ in non-minimal models. However, if $N_5$ is too large, then the running gauge couplings will diverge before they can unify at $M_U$. For messenger masses of order $10^6$ GeV or less, for example, one needs $N_5 \leq 4$.

There are many other possible generalizations of the basic gauge-mediation scenario as described above. An important general expectation in these models is that the strongly interacting sparticles (squarks, gluino) should be heavier than weakly interacting sparticles (sleptons, bino, winos), simply because of the hierarchy of gauge couplings $\alpha_3 > \alpha_2 > \alpha_1$. The common feature that makes all of these models attractive is that the masses of the squarks and sleptons depend only on their gauge quantum numbers, leading automatically to the degeneracy of squark and slepton masses needed for suppression of flavor-changing effects. But the most distinctive phenomenological prediction of gauge-mediated models may be the fact that the gravitino is the LSP. This can have crucial consequences for both cosmology and collider physics, as we will discuss further in sections 1.8.5.

### 1.6.8. *Extra-dimensional and anomaly-mediated supersymmetry breaking*

It is also possible to take the partitioning of the MSSM and supersymmetry breaking sectors shown in Figure 1.15 seriously as geography. This can be accomplished by assuming that there are extra spatial dimensions of the Kaluza-Klein or warped type,[161] so that a physical distance separates the

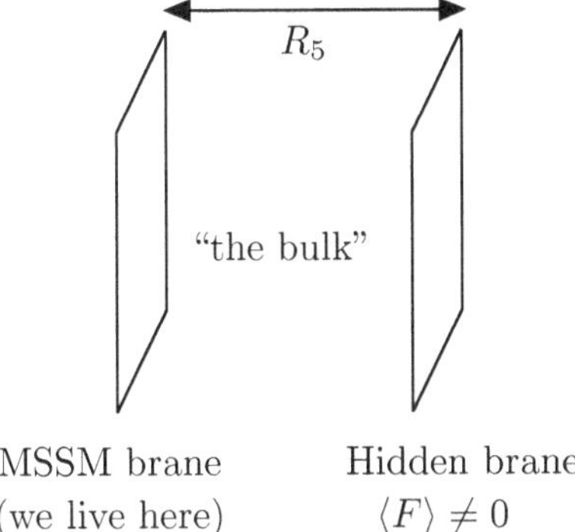


Fig. 1.19. The separation of the susy-breaking sector from the MSSM sector could take place along a hidden spatial dimension, as in the simple example shown here. The branes are 4-dimensional parallel spacetime hypersurfaces in a 5-dimensional spacetime.

visible and hidden[x] sectors. This general idea opens up numerous possibilities, which are hard to classify in a detailed way. For example, string theory suggests six such extra dimensions, with a staggeringly huge number of possible solutions.

Many of the more recently popular models used to explore this extra-dimensional mediated supersymmetry breaking (the acronym XMSB is tempting) use just one single hidden extra dimension with the MSSM chiral supermultiplets confined to one 4-dimensional spacetime brane and the supersymmetry-breaking sector confined to a parallel brane a distance $R_5$ away, separated by a 5-dimensional bulk, as in Figure 1.19. Using this as an illustration, the dangerous flavor-violating terms proportional to $y'^{ijk}$ and $k^i_j$ in eq. (1.242) are suppressed by factors like $e^{-R_5 M_5}$, where $R_5$ is the size of the 5th dimension and $M_5$ is the 5-dimensional fundamental (Planck) scale, and it is assumed that the MSSM chiral supermultiplets are confined to their brane. Therefore, it should be enough to require that $R_5 M_5 \gg 1$, in other words that the size of the 5th dimension (or, more generally, the volume of the compactified space) is relatively large in units of the fundamental length scale. Thus the suppression of flavor-violating effects does not require any fine-tuning or extreme hierarchies, because it is exponential.

One possibility is that the gauge supermultiplets of the MSSM propagate in the bulk, and so mediate supersymmetry breaking.[162–165] This

[x]The name "sequestered" is often used instead of "hidden" in this context.

mediation is direct for gauginos, with

$$M_a \sim \frac{\langle F \rangle}{M_5(R_5 M_5)}, \tag{1.271}$$

but is loop-suppressed for the soft terms involving scalars. This implies that in the simplest version of the idea, often called "gaugino mediation", soft supersymmetry breaking is dominated by the gaugino masses. The phenomenology is therefore quite similar to that of the "no-scale" boundary conditions mentioned in subsection 1.6.6 in the context of PMSB models. Scalar squared masses and the scalar cubic couplings come from renormalization group running down to the electroweak scale. It is useful to keep in mind that gaugino mass dominance is really the essential feature that defeats flavor violation, so it may well turn out to be more robust than any particular model that provides it.

It is also possible that the gauge supermultiplet fields are also confined to the MSSM brane, so that the transmission of supersymmetry breaking is due entirely to supergravity effects. This leads to anomaly-mediated supersymmetry breaking (AMSB),[166] so-named because the resulting MSSM soft terms can be understood in terms of the anomalous violation of a local superconformal invariance, an extension of scale invariance. In one formulation of supergravity,[147] Newton's constant (or equivalently, the Planck mass scale) is set by the VEV of a scalar field $\phi$ that is part of a non-dynamical chiral supermultiplet (called the "conformal compensator"). As a gauge fixing, this field obtains a VEV of $\langle \phi \rangle = 1$, spontaneously breaking the local superconformal invariance. Now, in the presence of spontaneous supersymmetry breaking $\langle F \rangle \neq 0$, for example on the hidden brane, the auxiliary field component also obtains a non-zero VEV, with

$$\langle F_\phi \rangle \sim \frac{\langle F \rangle}{M_{\rm P}} \sim m_{3/2}. \tag{1.272}$$

The non-dynamical conformal compensator field $\phi$ is taken to be dimensionless, so that $F_\phi$ has dimensions of [mass].

In the classical limit, there is still no supersymmetry breaking in the MSSM sector, due to the exponential suppression provided by the extra dimensions.[y] However, there is an anomalous violation of superconformal (scale) invariance manifested in the running of the couplings. This causes supersymmetry breaking to show up in the MSSM by virtue of the non-zero

[y]AMSB can also be realized without invoking extra dimensions. The suppression of flavor-violating MSSM soft terms can instead be achieved using a strongly-coupled conformal field theory near an infrared-stable fixed point.[167]

beta functions and anomalous dimensions of the MSSM brane couplings and fields. The resulting soft terms are[166] (using $F_\phi$ to denote its VEV from now on):

$$M_a = F_\phi \beta_{g_a}/g_a, \tag{1.273}$$

$$(m^2)^i_j = \frac{1}{2}|F_\phi|^2 \frac{d}{dt}\gamma^i_j$$
$$= \frac{1}{2}|F_\phi|^2 \left[\beta_{g_a}\frac{\partial}{\partial g_a} + \beta_{y^{kmn}}\frac{\partial}{\partial y^{kmn}} + \beta_{y^*_{kmn}}\frac{\partial}{\partial y^*_{kmn}}\right]\gamma^i_j, \tag{1.274}$$

$$a^{ijk} = -F_\phi \beta_{y^{ijk}}, \tag{1.275}$$

where the anomalous dimensions $\gamma^i_j$ are normalized as in eqs. (1.166) and (1.177)-(1.183). As in the GMSB scenario of the previous subsection, gaugino masses arise at one-loop order, but scalar squared masses arise at two-loop order. Also, these results are approximately flavor-blind for the first two families, because the non-trivial flavor structure derives only from the MSSM Yukawa couplings.

There are several unique features of the AMSB scenario. First, there is no need to specify at which renormalization scale eqs. (1.273)-(1.275) should be applied as boundary conditions. This is because they hold at every renormalization scale, exactly, to all orders in perturbation theory. In other words, eqs. (1.273)-(1.275) are not just boundary conditions for the renormalization group equations of the soft parameters, but solutions as well. (These AMSB renormalization group trajectories can also be found from this renormalization group invariance property alone,[168] without reference to the supergravity derivation.) In fact, even if there are heavy supermultiplets in the theory that have to be decoupled, the boundary conditions hold both above and below the arbitrary decoupling scale. This remarkable insensitivity to ultraviolet physics in AMSB ensures the absence of flavor violation in the low-energy MSSM soft terms. Another interesting prediction is that the gravitino mass $m_{3/2}$ in these models is actually much larger than the scale $m_{\rm soft}$ of the MSSM soft terms, since the latter are loop-suppressed compared to eq. (1.272).

There is only one unknown parameter, $F_\phi$, among the MSSM soft terms in AMSB. Unfortunately, this exemplary falsifiability is marred by the fact that it is already falsified. The dominant contributions to the first-family squark and slepton squared masses are:

$$m^2_{\tilde{q}} = \frac{|F_\phi|^2}{(16\pi^2)^2}\left(8g_3^4 + \ldots\right), \tag{1.276}$$

$$m^2_{\tilde{e}_L} = -\frac{|F_\phi|^2}{(16\pi^2)^2}\left(\frac{3}{2}g_2^4 + \frac{99}{50}g_1^4\right), \tag{1.277}$$

$$m^2_{\tilde{e}_R} = -\frac{|F_\phi|^2}{(16\pi^2)^2}\frac{198}{25}g_1^4. \tag{1.278}$$

The squarks have large positive squared masses, but the sleptons have negative squared masses, so the AMSB model in its simplest form is not viable. These signs come directly from those of the beta functions of the strong and electroweak gauge interactions, as can be seen from the right side of eq. (1.274).

The characteristic ultraviolet insensitivity to physics at high mass scales also makes it somewhat non-trivial to modify the theory to escape this tachyonic slepton problem by deviating from the AMSB trajectory. There can be large deviations from AMSB provided by supergravity,[169] but then in general the flavor-blindness is also forfeit. One way to modify AMSB is to introduce additional supermultiplets that contain supersymmetry-breaking mass splittings that are large compared to their average mass.[170] Another way is to combine AMSB with gaugino mediation.[171] Some other proposals can be found in.[172] Finally, there is a perhaps less motivated approach in which a common parameter $m_0^2$ is added to all of the scalar squared masses at some scale, and chosen large enough to allow the sleptons to have positive squared masses above LEP bounds. This allows the phenomenology to be studied in a framework conveniently parameterized by just:

$$F_\phi,\ m_0^2,\ \tan\beta,\ \arg(\mu), \tag{1.279}$$

with $|\mu|$ and $b$ determined by requiring correct electroweak symmetry breaking as described in the next section. (Some sources use $m_{3/2}$ or $M_{\rm aux}$ to denote $F_\phi$.) The MSSM gaugino masses at the leading non-trivial order are unaffected by the *ad hoc* addition of $m_0^2$:

$$M_1 = \frac{F_\phi}{16\pi^2}\frac{33}{5}g_1^2, \tag{1.280}$$

$$M_2 = \frac{F_\phi}{16\pi^2}g_2^2, \tag{1.281}$$

$$M_3 = -\frac{F_\phi}{16\pi^2}3g_3^2. \tag{1.282}$$

This implies that $|M_2| \ll |M_1| \ll |M_3|$, so the lightest neutralino is actually mostly wino, with a lightest chargino that is only of order 200 MeV heavier, depending on the values of $\mu$ and $\tan\beta$. The decay $\widetilde{C}_1^\pm \to \widetilde{N}_1\pi^\pm$ produces a very soft pion, implying unique and difficult signatures in colliders.[173–177]

Another large general class of models breaks supersymmetry using the geometric or topological properties of the extra dimensions. In the Scherk-Schwarz mechanism,[178] the symmetry is broken by assuming different boundary conditions for the fermion and boson fields on the compactified space. In supersymmetric models where the size of the extra dimension is parameterized by a modulus (a massless or nearly massless excitation) called a radion, the $F$-term component of the radion chiral supermultiplet can obtain a VEV, which becomes a source for supersymmetry breaking in the MSSM. These two ideas turn out to be often related. Some of the variety of models proposed along these lines can be found in Ref. 179. These mechanisms can also be combined with gaugino-mediation and AMSB. It seems likely that the possibilities are not yet fully explored.

## 1.7. The Mass Spectrum of the MSSM

### 1.7.1. *Electroweak symmetry breaking and the Higgs bosons*

In the MSSM, the description of electroweak symmetry breaking is slightly complicated by the fact that there are two complex Higgs doublets $H_u = (H_u^+,\, H_u^0)$ and $H_d = (H_d^0,\, H_d^-)$ rather than just one in the ordinary Standard Model. The classical scalar potential for the Higgs scalar fields in the MSSM is given by

$$\begin{aligned} V = {} & (|\mu|^2 + m^2_{H_u})(|H_u^0|^2 + |H_u^+|^2) + (|\mu|^2 + m^2_{H_d})(|H_d^0|^2 + |H_d^-|^2) \\ & + [b\,(H_u^+ H_d^- - H_u^0 H_d^0) + \text{c.c.}] \\ & + \frac{1}{8}(g^2 + g'^2)(|H_u^0|^2 + |H_u^+|^2 - |H_d^0|^2 - |H_d^-|^2)^2 \\ & + \frac{1}{2} g^2 |H_u^+ H_d^{0*} + H_u^0 H_d^{-*}|^2. \end{aligned} \tag{7.283}$$

The terms proportional to $|\mu|^2$ come from $F$-terms [see eq. (1.145)]. The terms proportional to $g^2$ and $g'^2$ are the $D$-term contributions, obtained from the general formula eq. (1.135) after some rearranging. Finally, the terms proportional to $m^2_{H_u}$, $m^2_{H_d}$ and $b$ are just a rewriting of the last three terms of eq. (1.152). The full scalar potential of the theory also includes many terms involving the squark and slepton fields that we can ignore here, since they do not get VEVs because they have large positive squared masses.

We now have to demand that the minimum of this potential should break electroweak symmetry down to electromagnetism $SU(2)_L \times U(1)_Y \to U(1)_{\text{EM}}$, in accord with experiment. We can use the freedom to make gauge transformations to simplify this analysis. First, the freedom to make $SU(2)_L$ gauge transformations allows us to rotate away a possible VEV for one of the weak isospin components of one of the scalar fields, so without loss of generality we can take $H_u^+ = 0$ at the minimum of the potential. Then one can check that a minimum of the potential satisfying $\partial V/\partial H_u^+ = 0$ must also have $H_d^- = 0$. This is good, because it means that at the minimum of the potential electromagnetism is necessarily unbroken, since the charged components of the Higgs scalars cannot get VEVs. After setting $H_u^+ = H_d^- = 0$, we are left to consider the scalar potential

$$\begin{aligned} V = &(|\mu|^2 + m_{H_u}^2)|H_u^0|^2 + (|\mu|^2 + m_{H_d}^2)|H_d^0|^2 - (b\, H_u^0 H_d^0 + \text{c.c.}) \\ &+\frac{1}{8}(g^2 + g'^2)(|H_u^0|^2 - |H_d^0|^2)^2. \end{aligned} \tag{7.284}$$

The only term in this potential that depends on the phases of the fields is the $b$-term. Therefore, a redefinition of the phase of $H_u$ or $H_d$ can absorb any phase in $b$, so we can take $b$ to be real and positive. Then it is clear that a minimum of the potential $V$ requires that $H_u^0 H_d^0$ is also real and positive, so $\langle H_u^0 \rangle$ and $\langle H_d^0 \rangle$ must have opposite phases. We can therefore use a $U(1)_Y$ gauge transformation to make them both be real and positive without loss of generality, since $H_u$ and $H_d$ have opposite weak hypercharges ($\pm 1/2$). It follows that CP cannot be spontaneously broken by the Higgs scalar potential, since the VEVs and $b$ can be simultaneously chosen real, as a convention. This means that the Higgs scalar mass eigenstates can be assigned well-defined eigenvalues of CP, at least at tree-level. (CP-violating phases in other couplings can induce loop-suppressed CP violation in the Higgs sector, but do not change the fact that $b$, $\langle H_u^0 \rangle$, and $\langle H_d \rangle$ can always be chosen real and positive.)

In order for the MSSM scalar potential to be viable, we must first make sure that the potential is bounded from below for arbitrarily large values of the scalar fields, so that $V$ will really have a minimum. (Recall from the discussion in sections 1.3.2 and 1.3.4 that scalar potentials in purely supersymmetric theories are automatically non-negative and so clearly bounded from below. But, now that we have introduced supersymmetry breaking, we must be careful.) The scalar quartic interactions in $V$ will stabilize the potential for almost all arbitrarily large values of $H_u^0$ and $H_d^0$. However, for the special directions in field space $|H_u^0| = |H_d^0|$, the quartic contributions to $V$ [the second line in eq. (7.284)] are identically zero. Such directions in field space are called $D$-flat directions, because along them the part of the scalar potential coming from $D$-terms vanishes. In order for the potential to be bounded from below, we need the quadratic part of the scalar potential to be positive along the $D$-flat directions. This requirement amounts to

$$2b < 2|\mu|^2 + m_{H_u}^2 + m_{H_d}^2. \tag{7.285}$$

Note that the $b$-term always favors electroweak symmetry breaking. Requiring that one linear combination of $H_u^0$ and $H_d^0$ has a negative squared mass near $H_u^0 = H_d^0 = 0$ gives

$$b^2 > (|\mu|^2 + m_{H_u}^2)(|\mu|^2 + m_{H_d}^2). \tag{7.286}$$

If this inequality is not satisfied, then $H_u^0 = H_d^0 = 0$ will be a stable minimum of the potential (or there will be no stable minimum at all), and electroweak symmetry breaking will not occur.

Interestingly, if $m_{H_u}^2 = m_{H_d}^2$ then the constraints eqs. (7.285) and (7.286) cannot both be satisfied. In models derived from the minimal supergravity or gauge-mediated boundary conditions, $m_{H_u}^2 = m_{H_d}^2$ is supposed to hold at tree level at the input scale, but the $X_t$ contribution to the RG equation for $m_{H_u}^2$ [eq. (1.201)] naturally pushes it to negative or small values $m_{H_u}^2 < m_{H_d}^2$ at the electroweak scale. Unless this effect is significant, the parameter space in which the electroweak symmetry is broken would be quite small. So in these models electroweak symmetry breaking is actually driven by quantum corrections; this mechanism is therefore known as *radiative electroweak symmetry breaking*. Note that although a negative value for $|\mu|^2 + m_{H_u}^2$ will help eq. (7.286) to be satisfied, it is not strictly necessary. Furthermore, even if $m_{H_u}^2 < 0$, there may be no electroweak symmetry breaking if $|\mu|$ is too large or if $b$ is too small. Still, the large negative contributions to $m_{H_u}^2$ from the RG equation are an important factor in ensuring that electroweak symmetry breaking can occur in models with simple boundary conditions for the soft terms. The realization that this works most naturally with a large top-quark Yukawa coupling provides additional motivation for these models.[151,180]

Having established the conditions necessary for $H_u^0$ and $H_d^0$ to get nonzero VEVs, we can now require that they are compatible with the observed phenomenology of electroweak symmetry breaking, $SU(2)_L \times U(1)_Y \to U(1)_{\text{EM}}$. Let us write

$$v_u = \langle H_u^0 \rangle, \qquad v_d = \langle H_d^0 \rangle. \tag{7.287}$$

These VEVs are related to the known mass of the $Z^0$ boson and the electroweak gauge couplings:

$$v_u^2 + v_d^2 = v^2 = 2m_Z^2/(g^2 + g'^2) \approx (174\,\text{GeV})^2. \tag{7.288}$$

The ratio of the VEVs is traditionally written as

$$\tan\beta \equiv v_u/v_d. \tag{7.289}$$

The value of $\tan\beta$ is not fixed by present experiments, but it depends on the Lagrangian parameters of the MSSM in a calculable way. Since $v_u = v\sin\beta$ and $v_d = v\cos\beta$ were taken to be real and positive by convention, we have $0 < \beta < \pi/2$, a requirement that will be sharpened below. Now one can write down the conditions $\partial V/\partial H_u^0 = \partial V/\partial H_d^0 = 0$ under which the potential eq. (7.284) will have a minimum satisfying eqs. (7.288) and (7.289):

$$m_{H_u}^2 + |\mu|^2 - b\cot\beta - (m_Z^2/2)\cos(2\beta) = 0, \tag{7.290}$$

$$m_{H_d}^2 + |\mu|^2 - b\tan\beta + (m_Z^2/2)\cos(2\beta) = 0. \tag{7.291}$$

It is easy to check that these equations indeed satisfy the necessary conditions eqs. (7.285) and (7.286). They allow us to eliminate two of the Lagrangian parameters $b$ and $|\mu|$ in favor of $\tan\beta$, but do not determine the phase of $\mu$. Taking $|\mu|^2$, $b$, $m_{H_u}^2$ and $m_{H_d}^2$ as input parameters, and $m_Z^2$ and $\tan\beta$ as output parameters obtained by solving these two equations, one obtains:

$$\sin(2\beta) = \frac{2b}{m_{H_u}^2 + m_{H_d}^2 + 2|\mu|^2}, \tag{7.292}$$

$$m_Z^2 = \frac{|m_{H_d}^2 - m_{H_u}^2|}{\sqrt{1-\sin^2(2\beta)}} - m_{H_u}^2 - m_{H_d}^2 - 2|\mu|^2. \tag{7.293}$$

(Note that $\sin(2\beta)$ is always positive. If $m_{H_u}^2 < m_{H_d}^2$, as is usually assumed, then $\cos(2\beta)$ is negative; otherwise it is positive.)

As an aside, eqs. (7.292) and (7.293) highlight the "$\mu$ problem" already mentioned in section 1.5.1. Without miraculous cancellations, all of the input parameters ought to be within an order of magnitude or two of $m_Z^2$. However, in the MSSM, $\mu$ is a supersymmetry-respecting parameter appearing in the superpotential, while $b$, $m_{H_u}^2$, $m_{H_d}^2$ are supersymmetry-breaking parameters. This has lead to a widespread belief that the MSSM must be extended at very high energies to include a mechanism that relates the effective value of $\mu$ to the supersymmetry-breaking mechanism in some way; see Refs. 66–68 for examples.

Even if the value of $\mu$ is set by soft supersymmetry breaking, the cancellation needed by eq. (7.293) is often remarkable when evaluated in specific model frameworks, after constraints from direct searches for the Higgs bosons and superpartners are taken into account. For example, expanding for large $\tan\beta$, eq. (7.293) becomes

$$m_Z^2 = -2(m_{H_u}^2 + |\mu|^2) + \frac{2}{\tan^2\beta}(m_{H_d}^2 - m_{H_u}^2) + \mathcal{O}(1/\tan^4\beta). \quad (7.294)$$

Typical viable solutions for the MSSM have $-m_{H_u}^2$ and $|\mu|^2$ each much larger than $m_Z^2$, so that significant cancellation is needed. In particular, large top squark squared masses, needed to avoid having the Higgs boson mass turn out too small [see eq. (7.307) below] compared to the direct search limits from LEP, will feed into $m_{H_u}^2$. The cancellation needed in the minimal model may therefore be at the several per cent level. It is impossible to objectively characterize whether this should be considered worrisome, but it could be taken as a weak hint in favor of non-minimal models.

The discussion above is based on the tree-level potential, and involves running renormalized Lagrangian parameters, which depend on the choice of renormalization scale. In practice, one must include radiative corrections at one-loop order, at least, in order to get numerically stable results. To do this, one can compute the loop corrections $\Delta V$ to the effective potential $V_{\text{eff}}(v_u, v_d) = V + \Delta V$ as a function of the VEVs. The impact of this is that the equations governing the VEVs of the full effective potential are obtained by simply replacing

$$m_{H_u}^2 \to m_{H_u}^2 + \frac{1}{2v_u}\frac{\partial(\Delta V)}{\partial v_u}, \qquad m_{H_d}^2 \to m_{H_d}^2 + \frac{1}{2v_d}\frac{\partial(\Delta V)}{\partial v_d} \quad (7.295)$$

in eqs. (7.290)-(7.293), treating $v_u$ and $v_d$ as real variables in the differentiation. The result for $\Delta V$ has now been obtained through two-loop order in the MSSM.[181] The most important corrections come from the one-loop diagrams involving the top squarks and top quark, and experience shows that the validity of the tree-level approximation and the convergence of perturbation theory are therefore improved by choosing a renormalization scale roughly of order the average of the top squark masses.

The Higgs scalar fields in the MSSM consist of two complex $SU(2)_L$-doublet, or eight real, scalar degrees of freedom. When the electroweak symmetry is broken, three of them are the would-be Nambu-Goldstone bosons $G^0$, $G^\pm$, which become the longitudinal modes of the $Z^0$ and $W^\pm$ massive vector bosons. The remaining five Higgs scalar mass eigenstates consist of two CP-even neutral scalars $h^0$ and $H^0$, one CP-odd neutral scalar $A^0$, and a charge $+1$ scalar $H^+$ and its conjugate charge $-1$ scalar $H^-$. (Here we define $G^- = G^{+*}$ and $H^- = H^{+*}$. Also, by convention, $h^0$

is lighter than $H^0$.) The gauge-eigenstate fields can be expressed in terms of the mass eigenstate fields as:

$$\begin{pmatrix} H_u^0 \\ H_d^0 \end{pmatrix} = \begin{pmatrix} v_u \\ v_d \end{pmatrix} + \frac{1}{\sqrt{2}} R_\alpha \begin{pmatrix} h^0 \\ H^0 \end{pmatrix} + \frac{i}{\sqrt{2}} R_{\beta_0} \begin{pmatrix} G^0 \\ A^0 \end{pmatrix} \tag{7.296}$$

$$\begin{pmatrix} H_u^+ \\ H_d^{-*} \end{pmatrix} = R_{\beta_\pm} \begin{pmatrix} G^+ \\ H^+ \end{pmatrix} \tag{7.297}$$

where the orthogonal rotation matrices

$$R_\alpha = \begin{pmatrix} \cos\alpha & \sin\alpha \\ -\sin\alpha & \cos\alpha \end{pmatrix}, \tag{7.298}$$

$$R_{\beta_0} = \begin{pmatrix} \sin\beta_0 & \cos\beta_0 \\ -\cos\beta_0 & \sin\beta_0 \end{pmatrix}, \qquad R_{\beta_\pm} = \begin{pmatrix} \sin\beta_\pm & \cos\beta_\pm \\ -\cos\beta_\pm & \sin\beta_\pm \end{pmatrix}, \tag{7.299}$$

are chosen so that the quadratic part of the potential has diagonal squared-masses:

$$\begin{aligned} V = {} & \frac{1}{2} m_{h^0}^2 (h^0)^2 + \frac{1}{2} m_{H^0}^2 (H^0)^2 + \frac{1}{2} m_{G^0}^2 (G^0)^2 + \frac{1}{2} m_{A^0}^2 (A^0)^2 \\ & + m_{G^\pm}^2 |G^+|^2 + m_{H^\pm}^2 |H^+|^2 + \ldots . \end{aligned} \tag{7.300}$$

Then, provided that $v_u, v_d$ minimize the tree-level potential,[z] one finds that $\beta_0 = \beta_\pm = \beta$, and $m_{G^0}^2 = m_{G^\pm}^2 = 0$, and

$$m_{A^0}^2 = 2b/\sin(2\beta) = 2|\mu|^2 + m_{H_u}^2 + m_{H_d}^2, \tag{7.301}$$

$$m_{h^0,H^0}^2 = \frac{1}{2}\Big( m_{A^0}^2 + m_Z^2 \mp \sqrt{(m_{A^0}^2 - m_Z^2)^2 + 4 m_Z^2 m_{A^0}^2 \sin^2(2\beta)} \Big), \tag{7.302}$$

$$m_{H^\pm}^2 = m_{A^0}^2 + m_W^2. \tag{7.303}$$

The mixing angle $\alpha$ is determined, at tree-level, by

$$\frac{\sin 2\alpha}{\sin 2\beta} = -\left( \frac{m_{H^0}^2 + m_{h^0}^2}{m_{H^0}^2 - m_{h^0}^2} \right), \qquad \frac{\tan 2\alpha}{\tan 2\beta} = \left( \frac{m_{A^0}^2 + m_Z^2}{m_{A^0}^2 - m_Z^2} \right), \tag{7.304}$$

and is traditionally chosen to be negative; it follows that $-\pi/2 < \alpha < 0$ (provided $m_{A^0} > m_Z$). The Feynman rules for couplings of the mass eigenstate Higgs scalars to the Standard Model quarks and leptons and the electroweak vector bosons, as well as to the various sparticles, have been worked out in detail in Ref. 182, 183.

[z] It is often more useful to expand around VEVs $v_u, v_d$ that do not minimize the tree-level potential, for example to minimize the loop-corrected effective potential instead. In that case, $\beta$, $\beta_0$, and $\beta_\pm$ are all slightly different.

The masses of $A^0$, $H^0$ and $H^\pm$ can in principle be arbitrarily large since they all grow with $b/\sin(2\beta)$. In contrast, the mass of $h^0$ is bounded above. From eq. (7.302), one finds at tree-level:[184]

$$m_{h^0} < m_Z|\cos(2\beta)|. \tag{7.305}$$

This corresponds to a shallow direction in the scalar potential, along the direction $(H_u^0 - v_u, H_d^0 - v_d) \propto (\cos\alpha, -\sin\alpha)$. The existence of this shallow direction can be traced to the fact that the quartic Higgs couplings are given by the square of the electroweak gauge couplings, via the $D$-term. A contour map of the potential, for a typical case with $\tan\beta \approx -\cot\alpha \approx 10$, is shown in Figure 1.20.If the tree-level inequality (7.305) were robust, the lightest Higgs boson of the MSSM would have been discovered at LEP2. However, the tree-level formula for the squared mass of $h^0$ is subject to quantum corrections that are relatively drastic. The largest such contributions typically come from top and stop loops, as shown[aa] in Figure 1.21. In

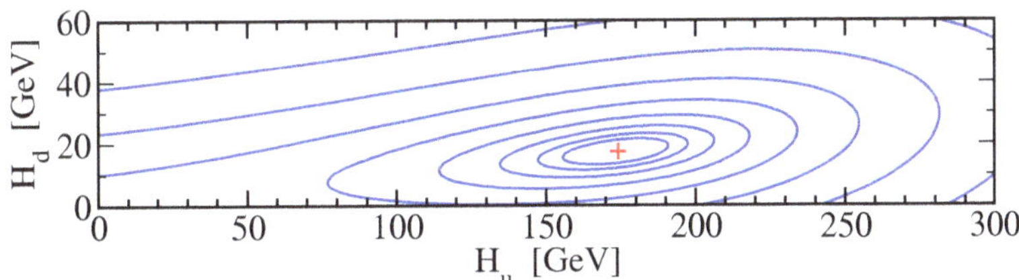


Fig. 1.20. A contour map of the Higgs potential, for a typical case with $\tan\beta \approx -\cot\alpha \approx 10$. The minimum of the potential is marked by +, and the contours are equally spaced equipotentials. Oscillations along the shallow direction, with $H_u^0/H_d^0 \approx 10$, correspond to the mass eigenstate $h^0$, while the orthogonal steeper direction corresponds to the mass eigenstate $H^0$.

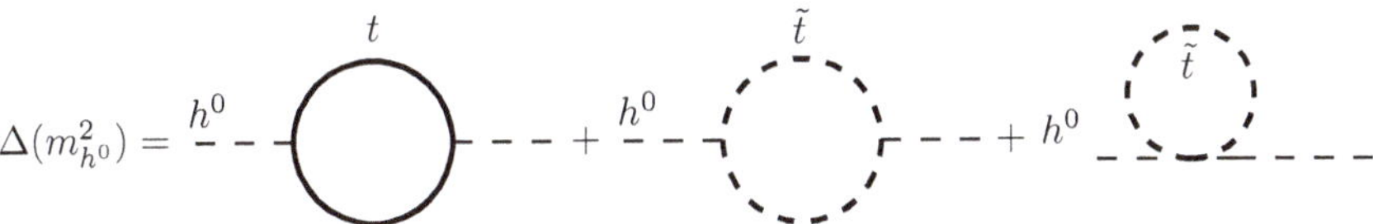


Fig. 1.21. Contributions to the MSSM lightest Higgs mass from top-quark and top-squark one-loop diagrams. Incomplete cancellation, due to soft supersymmetry breaking, leads to a large positive correction to $m^2_{h^0}$ in the limit of heavy top squarks.

---

[aa]In general, one-loop 1-particle-reducible tadpole diagrams should also be included. However, they just cancel against tree-level tadpoles, and so both can be omitted, if the VEVs $v_u$ and $v_d$ are taken at the minimum of the loop-corrected effective potential (see previous footnote).

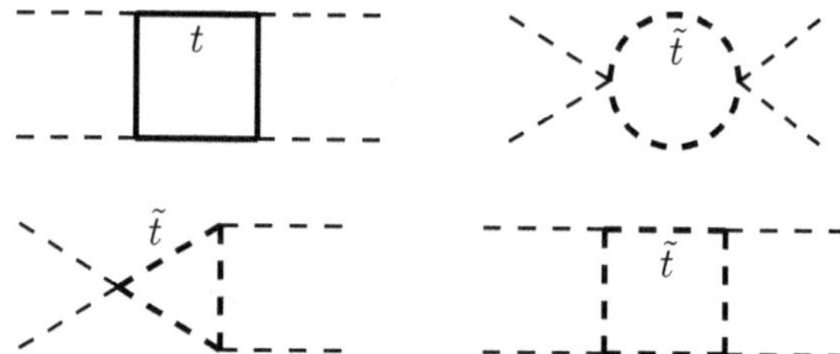


Fig. 1.22. Integrating out the top quark and top squarks yields large positive contributions to the quartic Higgs coupling in the low-energy effective theory, especially from these one-loop diagrams.

the simple limit of top squarks that have a small mixing in the gauge eigenstate basis and with masses $m_{\tilde{t}_1}$, $m_{\tilde{t}_2}$ much greater than the top quark mass $m_t$, one finds a large positive one-loop radiative correction to eq. (7.302):

$$\Delta(m^2_{h^0}) = \frac{3}{4\pi^2}\cos^2\!\alpha\ y^2_t m^2_t \ln\left(m_{\tilde{t}_1} m_{\tilde{t}_2}/m^2_t\right). \tag{7.306}$$

This shows that $m_{h^0}$ can exceed the LEP bounds.

An alternative way to understand the size of the radiative correction to the $h^0$ mass is to consider an effective theory in which the heavy top squarks and top quark have been integrated out. The quartic Higgs couplings in the low-energy effective theory get large positive contributions from the the one-loop diagrams of Figure 1.22. This increases the steepness of the Higgs potential, and can be used to obtain the same result for the enhanced $h^0$ mass.

An interesting case, often referred to as the "decoupling limit", occurs when $m_{A^0} \gg m_Z$. Then $m_{h^0}$ can saturate the upper bounds just mentioned, with $m^2_{h^0} \approx m^2_Z \cos^2(2\beta)+$ loop corrections. The particles $A^0$, $H^0$, and $H^\pm$ will be much heavier and nearly degenerate, forming an isospin doublet that decouples from sufficiently low-energy experiments. The angle $\alpha$ is very nearly $\beta - \pi/2$, and $h^0$ has the same couplings to quarks and leptons and electroweak gauge bosons as would the physical Higgs boson of the ordinary Standard Model without supersymmetry. Indeed, model-building experiences have shown that it is not uncommon for $h^0$ to behave in a way nearly indistinguishable from a Standard Model-like Higgs boson, even if $m_{A^0}$ is not too huge. However, it should be kept in mind that the couplings of $h^0$ might turn out to deviate significantly from those of a Standard Model Higgs boson.

Top-squark mixing (to be discussed in section 1.7.4) can result in a further large positive contribution to $m^2_{h^0}$. At one-loop order, and working in the decoupling limit for simplicity, eq. (7.306) generalizes to:

$$m_{h^0}^2 = m_Z^2 \cos^2(2\beta) + \frac{3}{4\pi^2} \sin^2\beta\, y_t^2 \Big[ m_t^2 \ln\left(m_{\tilde{t}_1} m_{\tilde{t}_2}/m_t^2\right) \\ + c_{\tilde{t}}^2 s_{\tilde{t}}^2 (m_{\tilde{t}_2}^2 - m_{\tilde{t}_1}^2) \ln(m_{\tilde{t}_2}^2/m_{\tilde{t}_1}^2) \\ + c_{\tilde{t}}^4 s_{\tilde{t}}^4 \Big\{ (m_{\tilde{t}_2}^2 - m_{\tilde{t}_1}^2)^2 - \frac{1}{2}(m_{\tilde{t}_2}^4 - m_{\tilde{t}_1}^4) \ln(m_{\tilde{t}_2}^2/m_{\tilde{t}_1}^2) \Big\} / m_t^2 \Big]. \quad (7.307)$$

Here $c_{\tilde{t}}$ and $s_{\tilde{t}}$ are the cosine and sine of a top squark mixing angle $\theta_{\tilde{t}}$, defined more specifically below following eq. (7.353). For fixed top-squark masses, the maximum possible $h^0$ mass occurs for rather large top squark mixing, $c_{\tilde{t}}^2 s_{\tilde{t}}^2 = m_t^2/[m_{\tilde{t}_2}^2 + m_{\tilde{t}_1}^2 - 2(m_{\tilde{t}_2}^2 - m_{\tilde{t}_1}^2)/\ln(m_{\tilde{t}_2}^2/m_{\tilde{t}_1}^2)]$ or $1/4$, whichever is less. It follows that the quantity in square brackets in eq. (7.307) is always less than $m_t^2[\ln(m_{\tilde{t}_2}^2/m_t^2) + 3]$. The LEP constraints on the MSSM Higgs sector make the case of large top-squark mixing noteworthy.

Including these and other important corrections,[185–194] one can obtain only a weaker, but still very interesting, bound

$$m_{h^0} \lesssim 135\,\text{GeV} \quad (7.308)$$

in the MSSM. This assumes that all of the sparticles that can contribute to $m_{h^0}^2$ in loops have masses that do not exceed 1 TeV. By adding extra supermultiplets to the MSSM, this bound can be made even weaker. However, assuming that none of the MSSM sparticles have masses exceeding 1 TeV and that all of the couplings in the theory remain perturbative up to the unification scale, one still has[195]

$$m_{h^0} \lesssim 150\,\text{GeV}. \quad (7.309)$$

This bound is also weakened if, for example, the top squarks are heavier than 1 TeV, but the upper bound rises only logarithmically with the soft masses, as can be seen from eq. (7.306). Thus it is a fairly robust prediction of supersymmetry at the electroweak scale that at least one of the Higgs scalar bosons must be light. (However, if one is willing to extend the MSSM in a completely general way above the electroweak scale, none of these bounds need apply.) For a given set of model parameters, it is always important to take into account the complete set of one-loop corrections and even the dominant two-loop effects in order to get reasonably accurate predictions for the Higgs masses and mixings.[185–194]

In the MSSM, the masses and CKM mixing angles of the quarks and leptons are determined not only by the Yukawa couplings of the superpotential but also the parameter $\tan\beta$. This is because the top, charm and up quark mass matrix is proportional to $v_u = v \sin\beta$ and the bottom, strange, and down quarks and the charge leptons get masses proportional to $v_d = v \cos\beta$. At tree-level,

$$m_t = y_t v \sin\beta, \qquad m_b = y_b v \cos\beta, \qquad m_\tau = y_\tau v \cos\beta. \quad (7.310)$$

These relations hold for the running masses rather than the physical pole masses, which are significantly larger for $t, b$.[196] Including those corrections, one can relate the Yukawa couplings to $\tan\beta$ and the known fermion masses and CKM mixing angles. It is now clear why we have not neglected $y_b$ and $y_\tau$, even though $m_b, m_\tau \ll m_t$. To a first approximation, $y_b/y_t = (m_b/m_t)\tan\beta$ and $y_\tau/y_t = (m_\tau/m_t)\tan\beta$, so that $y_b$ and $y_\tau$ cannot be neglected if $\tan\beta$ is much larger than 1. In fact, there are good theoretical motivations for considering models with large $\tan\beta$. For example, models based on the GUT gauge group $SO(10)$ can unify the running top, bottom and tau Yukawa couplings at the unification scale; this requires $\tan\beta$ to be very roughly of order $m_t/m_b$.[197,198]

Note that if one tries to make $\sin\beta$ too small, $y_t$ will be nonperturbatively large. Requiring that $y_t$ does not blow up above the electroweak scale, one finds that $\tan\beta \gtrsim 1.2$ or so, depending on the mass of the top quark, the QCD coupling, and other details. In principle, there is also a constraint on $\cos\beta$ if one requires that $y_b$ and $y_\tau$ do not become nonperturbatively large. This gives a rough upper bound of $\tan\beta \lesssim 65$. However, this is complicated somewhat by the fact that the bottom quark mass gets significant one-loop non-QCD corrections in the large $\tan\beta$ limit.[198] One can obtain a stronger upper bound on $\tan\beta$ in some models where $m^2_{H_u} = m^2_{H_d}$ at the input scale, by requiring that $y_b$ does not significantly exceed $y_t$. [Otherwise, $X_b$ would be larger than $X_t$ in eqs. (1.201) and (1.202), so one would expect $m^2_{H_d} < m^2_{H_u}$ at the electroweak scale, and the minimum of the potential would have $\langle H^0_d\rangle > \langle H^0_u\rangle$. This would be a contradiction with the supposition that $\tan\beta$ is large.] The parameter $\tan\beta$ also directly impacts the masses and mixings of the MSSM sparticles, as we will see below.

### 1.7.2. *Neutralinos and charginos*

The higgsinos and electroweak gauginos mix with each other because of the effects of electroweak symmetry breaking. The neutral higgsinos ($\widetilde{H}^0_u$ and $\widetilde{H}^0_d$) and the neutral gauginos ($\widetilde{B}$, $\widetilde{W}^0$) combine to form four mass eigenstates called *neutralinos*. The charged higgsinos ($\widetilde{H}^+_u$ and $\widetilde{H}^-_d$) and winos ($\widetilde{W}^+$ and $\widetilde{W}^-$) mix to form two mass eigenstates with charge $\pm 1$ called *charginos*. We will denote[bb] the neutralino and chargino mass eigenstates by $\widetilde{N}_i$ $(i = 1,2,3,4)$ and $\widetilde{C}_i^{\pm}$ $(i = 1,2)$. By convention, these are labeled in ascending order, so that $m_{\widetilde{N}_1} < m_{\widetilde{N}_2} < m_{\widetilde{N}_3} < m_{\widetilde{N}_4}$ and $m_{\widetilde{C}_1} < m_{\widetilde{C}_2}$. The lightest neutralino, $\widetilde{N}_1$, is usually assumed to be the LSP, unless there is a lighter gravitino or unless $R$-parity is not conserved, because it is the only MSSM particle that can make a good dark matter candidate. In this subsection, we will describe the mass spectrum and mixing of the neutralinos and charginos in the MSSM.

In the gauge-eigenstate basis $\psi^0 = (\widetilde{B}, \widetilde{W}^0, \widetilde{H}_d^0, \widetilde{H}_u^0)$, the neutralino mass part of the Lagrangian is

$$\mathcal{L}_{\text{neutralino mass}} = -\frac{1}{2}(\psi^0)^T \mathbf{M}_{\widetilde{N}} \psi^0 + \text{c.c.}, \tag{7.311}$$

where

$$\mathbf{M}_{\widetilde{N}} = \begin{pmatrix} M_1 & 0 & -g'v_d/\sqrt{2} & g'v_u/\sqrt{2} \\ 0 & M_2 & gv_d/\sqrt{2} & -gv_u/\sqrt{2} \\ -g'v_d/\sqrt{2} & gv_d/\sqrt{2} & 0 & -\mu \\ g'v_u/\sqrt{2} & -gv_u/\sqrt{2} & -\mu & 0 \end{pmatrix}. \tag{7.312}$$

The entries $M_1$ and $M_2$ in this matrix come directly from the MSSM soft Lagrangian [see eq. (1.152)], while the entries $-\mu$ are the supersymmetric higgsino mass terms [see eq. (1.144)]. The terms proportional to $g, g'$ are the result of Higgs-higgsino-gaugino couplings [see eq. (1.132) and Figure 1.5g,h], with the Higgs scalars replaced by their VEVs [eqs. (7.288), (7.289)]. This can also be written as

$$\mathbf{M}_{\widetilde{N}} = \begin{pmatrix} M_1 & 0 & -c_\beta\, s_W\, m_Z & s_\beta\, s_W\, m_Z \\ 0 & M_2 & c_\beta\, c_W\, m_Z & -s_\beta\, c_W\, m_Z \\ -c_\beta\, s_W\, m_Z & c_\beta\, c_W\, m_Z & 0 & -\mu \\ s_\beta\, s_W\, m_Z & -s_\beta\, c_W\, m_Z & -\mu & 0 \end{pmatrix}. \tag{7.313}$$

Here we have introduced abbreviations $s_\beta = \sin\beta$, $c_\beta = \cos\beta$, $s_W = \sin\theta_W$, and $c_W = \cos\theta_W$. The mass matrix $\mathbf{M}_{\widetilde{N}}$ can be diagonalized by a unitary matrix $\mathbf{N}$ to obtain mass eigenstates:

$$\widetilde{N}_i = \mathbf{N}_{ij}\psi_j^0, \tag{7.314}$$

so that

$$\mathbf{N}^*\mathbf{M}_{\widetilde{N}}\mathbf{N}^{-1} = \begin{pmatrix} m_{\widetilde{N}_1} & 0 & 0 & 0 \\ 0 & m_{\widetilde{N}_2} & 0 & 0 \\ 0 & 0 & m_{\widetilde{N}_3} & 0 \\ 0 & 0 & 0 & m_{\widetilde{N}_4} \end{pmatrix} \tag{7.315}$$

[bb]Other common notations use $\widetilde{\chi}_i^0$ or $\widetilde{Z}_i$ for neutralinos, and $\widetilde{\chi}_i^{\pm}$ or $\widetilde{W}_i^{\pm}$ for charginos.

has real positive entries on the diagonal. These are the magnitudes of the eigenvalues of $\mathbf{M}_{\widetilde{N}}$, or equivalently the square roots of the eigenvalues of $\mathbf{M}^{\dagger}_{\widetilde{N}}\mathbf{M}_{\widetilde{N}}$. The indices $(i,j)$ on $\mathbf{N}_{ij}$ are (mass, gauge) eigenstate labels. The mass eigenvalues and the mixing matrix $\mathbf{N}_{ij}$ can be given in closed form in terms of the parameters $M_1$, $M_2$, $\mu$ and $\tan\beta$, by solving quartic equations, but the results are very complicated and not illuminating.

In general, the parameters $M_1$, $M_2$, and $\mu$ in the equations above can have arbitrary complex phases. A redefinition of the phases of $\widetilde{B}$ and $\widetilde{W}$ always allows us to choose a convention in which $M_1$ and $M_2$ are both real and positive. The phase of $\mu$ within that convention is then really a physical parameter and cannot be rotated away. [We have already used up the freedom to redefine the phases of the Higgs fields, since we have picked $b$ and $\langle H_u^0\rangle$ and $\langle H_d^0\rangle$ to be real and positive, to guarantee that the off-diagonal entries in eq. (7.313) proportional to $m_Z$ are real.] However, if $\mu$ is not real, then there can be potentially disastrous CP-violating effects in low-energy physics, including electric dipole moments for both the electron and the neutron. Therefore, it is usual [although not strictly mandatory, because of the possibility of nontrivial cancellations involving the phases of the (scalar)$^3$ couplings and the gluino mass] to assume that $\mu$ is real in the same set of phase conventions that make $M_1$, $M_2$, $b$, $\langle H_u^0\rangle$ and $\langle H_d^0\rangle$ real and positive. The sign of $\mu$ is still undetermined by this constraint.

In models that satisfy eq. (1.189), one has the nice prediction

$$M_1 \approx \frac{5}{3}\tan^2\theta_W\, M_2 \approx 0.5 M_2 \tag{7.316}$$

at the electroweak scale. If so, then the neutralino masses and mixing angles depend on only three unknown parameters. This assumption is sufficiently theoretically compelling that it has been made in most phenomenological studies; nevertheless it should be recognized as an assumption, to be tested someday by experiment.

There is a not-unlikely limit in which electroweak symmetry breaking effects can be viewed as a small perturbation on the neutralino mass matrix. If

$$m_Z \ll |\mu \pm M_1|,\ |\mu \pm M_2|, \tag{7.317}$$

then the neutralino mass eigenstates are very nearly a "bino-like" $\widetilde{N}_1 \approx \widetilde{B}$; a "wino-like" $\widetilde{N}_2 \approx \widetilde{W}^0$; and "higgsino-like" $\widetilde{N}_3, \widetilde{N}_4 \approx (\widetilde{H}_u^0 \pm \widetilde{H}_d^0)/\sqrt{2}$, with mass eigenvalues:

$$m_{\widetilde{N}_1} = M_1 - \frac{m_Z^2 s_W^2 (M_1 + \mu\sin 2\beta)}{\mu^2 - M_1^2} + \ldots, \tag{7.318}$$

$$m_{\widetilde{N}_2} = M_2 - \frac{m_W^2(M_2 + \mu \sin 2\beta)}{\mu^2 - M_2^2} + \dots, \tag{7.319}$$

$$m_{\widetilde{N}_3}, m_{\widetilde{N}_4} = |\mu| + \frac{m_Z^2(I - \sin 2\beta)(\mu + M_1 c_W^2 + M_2 s_W^2)}{2(\mu + M_1)(\mu + M_2)} + \dots, \tag{7.320}$$

$$|\mu| + \frac{m_Z^2(I + \sin 2\beta)(\mu - M_1 c_W^2 - M_2 s_W^2)}{2(\mu - M_1)(\mu - M_2)} + \dots, \tag{7.321}$$

where we have taken $M_1$ and $M_2$ real and positive by convention, and assumed $\mu$ is real with sign $I = \pm 1$. The subscript labels of the mass eigenstates may need to be rearranged depending on the numerical values of the parameters; in particular the above labeling of $\widetilde{N}_1$ and $\widetilde{N}_2$ assumes $M_1 < M_2 \ll |\mu|$. This limit, leading to a bino-like neutralino LSP, often emerges from minimal supergravity boundary conditions on the soft parameters, which tend to require it in order to get correct electroweak symmetry breaking.

The chargino spectrum can be analyzed in a similar way. In the gauge-eigenstate basis $\psi^\pm = (\widetilde{W}^+, \widetilde{H}_u^+, \widetilde{W}^-, \widetilde{H}_d^-)$, the chargino mass terms in the Lagrangian are

$$\mathcal{L}_{\text{chargino mass}} = -\frac{1}{2}(\psi^\pm)^T \mathbf{M}_{\widetilde{C}} \psi^\pm + \text{c.c.} \tag{7.322}$$

where, in $2 \times 2$ block form,

$$\mathbf{M}_{\widetilde{C}} = \begin{pmatrix} \mathbf{0} & \mathbf{X}^T \\ \mathbf{X} & \mathbf{0} \end{pmatrix}, \tag{7.323}$$

with

$$\mathbf{X} = \begin{pmatrix} M_2 & g v_u \\ g v_d & \mu \end{pmatrix} = \begin{pmatrix} M_2 & \sqrt{2} s_\beta\, m_W \\ \sqrt{2} c_\beta\, m_W & \mu \end{pmatrix}. \tag{7.324}$$

The mass eigenstates are related to the gauge eigenstates by two unitary $2\times2$ matrices $\mathbf{U}$ and $\mathbf{V}$ according to

$$\begin{pmatrix} \widetilde{C}_1^+ \\ \widetilde{C}_2^+ \end{pmatrix} = \mathbf{V} \begin{pmatrix} \widetilde{W}^+ \\ \widetilde{H}_u^+ \end{pmatrix}, \qquad \begin{pmatrix} \widetilde{C}_1^- \\ \widetilde{C}_2^- \end{pmatrix} = \mathbf{U} \begin{pmatrix} \widetilde{W}^- \\ \widetilde{H}_d^- \end{pmatrix}. \tag{7.325}$$

Note that the mixing matrix for the positively charged left-handed fermions is different from that for the negatively charged left-handed fermions. They are chosen so that

$$\mathbf{U}^* \mathbf{X} \mathbf{V}^{-1} = \begin{pmatrix} m_{\widetilde{C}_1} & 0 \\ 0 & m_{\widetilde{C}_2} \end{pmatrix}, \tag{7.326}$$

with positive real entries $m_{\widetilde{C}_i}$. Because these are only 2×2 matrices, it is not hard to solve for the masses explicitly:

$$m^2_{\widetilde{C}_1}, m^2_{\widetilde{C}_2} = \frac{1}{2}\Big[|M_2|^2 + |\mu|^2 + 2m_W^2 \mp \sqrt{(|M_2|^2 + |\mu|^2 + 2m_W^2)^2 - 4|\mu M_2 - m_W^2 \sin 2\beta|^2}\Big]. \quad (7.327)$$

These are the (doubly degenerate) eigenvalues of the $4 \times 4$ matrix $\mathbf{M}^\dagger_{\widetilde{C}}\mathbf{M}_{\widetilde{C}}$, or equivalently the eigenvalues of $\mathbf{X}^\dagger\mathbf{X}$, since

$$\mathbf{V}\mathbf{X}^\dagger\mathbf{X}\mathbf{V}^{-1} = \mathbf{U}^*\mathbf{X}\mathbf{X}^\dagger\mathbf{U}^T = \begin{pmatrix} m^2_{\widetilde{C}_1} & 0 \\ 0 & m^2_{\widetilde{C}_2} \end{pmatrix}. \quad (7.328)$$

(But, they are *not* the squares of the eigenvalues of $\mathbf{X}$.) In the limit of eq. (7.317) with real $M_2$ and $\mu$, the chargino mass eigenstates consist of a wino-like $\widetilde{C}_1^\pm$ and and a higgsino-like $\widetilde{C}_2^\pm$, with masses

$$m_{\widetilde{C}_1} = M_2 - \frac{m_W^2(M_2 + \mu \sin 2\beta)}{\mu^2 - M_2^2} + \ldots \quad (7.329)$$

$$m_{\widetilde{C}_2} = |\mu| + \frac{m_W^2 I(\mu + M_2 \sin 2\beta)}{\mu^2 - M_2^2} + \ldots. \quad (7.330)$$

Here again the labeling assumes $M_2 < |\mu|$, and $I$ is the sign of $\mu$. Amusingly, $\widetilde{C}_1$ is nearly degenerate with the neutralino $\widetilde{N}_2$ in the approximation shown, but that is not an exact result. Their higgsino-like colleagues $\widetilde{N}_3$, $\widetilde{N}_4$ and $\widetilde{C}_2$ have masses of order $|\mu|$. The case of $M_1 \approx 0.5M_2 \ll |\mu|$ is not uncommonly found in viable models following from the boundary conditions in section 1.6, and it has been elevated to the status of a benchmark framework in many phenomenological studies. However it cannot be overemphasized that such expectations are not mandatory.

The Feynman rules involving neutralinos and charginos may be inferred in terms of $\mathbf{N}$, $\mathbf{U}$ and $\mathbf{V}$ from the MSSM Lagrangian as discussed above; they are collected in Refs. 25, 182. Feynman rules based on two-component spinor notation have also recently been given in Ref. 199. In practice, the masses and mixing angles for the neutralinos and charginos are best computed numerically. Note that the discussion above yields the tree-level masses. Loop corrections to these masses can be significant, and have been found systematically at one-loop order in Ref. 200.

### 1.7.3. *The gluino*

The gluino is a color octet fermion, so it cannot mix with any other particle in the MSSM, even if $R$-parity is violated. In this regard, it is unique among all of the MSSM sparticles. In models with minimal supergravity or gauge-mediated boundary conditions, the gluino mass parameter $M_3$ is related to the bino and wino mass parameters $M_1$ and $M_2$ by eq. (1.189), so

$$M_3 = \frac{\alpha_s}{\alpha} \sin^2\theta_W\, M_2 = \frac{3}{5}\frac{\alpha_s}{\alpha} \cos^2\theta_W\, M_1 \tag{7.331}$$

at any RG scale, up to small two-loop corrections. This implies a rough prediction

$$M_3 : M_2 : M_1 \approx 6 : 2 : 1 \tag{7.332}$$

near the TeV scale. It is therefore reasonable to suspect that the gluino is considerably heavier than the lighter neutralinos and charginos (even in many models where the gaugino mass unification condition is not imposed).

For more precise estimates, one must take into account the fact that $M_3$ is really a running mass parameter with an implicit dependence on the RG scale $Q$. Because the gluino is a strongly interacting particle, $M_3$ runs rather quickly with $Q$ [see eq. (1.188)]. A more useful quantity physically is the RG scale-independent mass $m_{\tilde{g}}$ at which the renormalized gluino propagator has a pole. Including one-loop corrections to the gluino propagator due to gluon exchange and quark-squark loops, one finds that the pole mass is given in terms of the running mass in the $\overline{\rm DR}$ scheme by[118]

$$m_{\tilde{g}} = M_3(Q)\Big(1 + \frac{\alpha_s}{4\pi}\big[15 + 6\ln(Q/M_3) + \sum A_{\tilde{q}}\big]\Big) \tag{7.333}$$

where

$$A_{\tilde{q}} = \int_0^1 dx\, x \ln\big[x m_{\tilde{q}}^2/M_3^2 + (1-x)m_q^2/M_3^2 - x(1-x) - i\epsilon\big]. \tag{7.334}$$

The sum in eq. (7.333) is over all 12 squark-quark supermultiplets, and we have neglected small effects due to squark mixing. [As a check, requiring $m_{\tilde{g}}$ to be independent of $Q$ in eq. (7.333) reproduces the one-loop RG equation for $M_3(Q)$ in eq. (1.188).] The correction terms proportional to $\alpha_s$ in eq. (7.333) can be quite significant, because the gluino is strongly interacting, with a large group theory factor [the 15 in eq. (7.333)] due to its color octet nature, and because it couples to all of the squark-quark pairs. The leading two-loop corrections to the gluino pole mass have also been found,[201] and typically increase the prediction by another 1 or 2%.

### 1.7.4. *The squarks and sleptons*

In principle, any scalars with the same electric charge, $R$-parity, and color quantum numbers can mix with each other. This means that with completely arbitrary soft terms, the mass eigenstates of the squarks and sleptons of the MSSM should be obtained by diagonalizing three $6 \times 6$ squared-mass matrices for up-type squarks ($\widetilde{u}_L$, $\widetilde{c}_L$, $\widetilde{t}_L$, $\widetilde{u}_R$, $\widetilde{c}_R$, $\widetilde{t}_R$), down-type squarks ($\widetilde{d}_L$, $\widetilde{s}_L$, $\widetilde{b}_L$, $\widetilde{d}_R$, $\widetilde{s}_R$, $\widetilde{b}_R$), and charged sleptons ($\widetilde{e}_L$, $\widetilde{\mu}_L$, $\widetilde{\tau}_L$, $\widetilde{e}_R$, $\widetilde{\mu}_R$, $\widetilde{\tau}_R$), and one $3 \times 3$ matrix for sneutrinos ($\widetilde{\nu}_e$, $\widetilde{\nu}_\mu$, $\widetilde{\nu}_\tau$). Fortunately, the general hypothesis of flavor-blind soft parameters eqs. (1.158) and (1.159) predicts that most of these mixing angles are very small. The third-family squarks and sleptons can have very different masses compared to their first- and second-family counterparts, because of the effects of large Yukawa ($y_t$, $y_b$, $y_\tau$) and soft ($a_t$, $a_b$, $a_\tau$) couplings in the RG equations (1.203)-(1.207). Furthermore, they can have substantial mixing in pairs ($\widetilde{t}_L$, $\widetilde{t}_R$), ($\widetilde{b}_L$, $\widetilde{b}_R$) and ($\widetilde{\tau}_L$, $\widetilde{\tau}_R$). In contrast, the first- and second-family squarks and sleptons have negligible Yukawa couplings, so they end up in 7 very nearly degenerate, unmixed pairs $(\widetilde{e}_R, \widetilde{\mu}_R)$, $(\widetilde{\nu}_e, \widetilde{\nu}_\mu)$, $(\widetilde{e}_L, \widetilde{\mu}_L)$, $(\widetilde{u}_R, \widetilde{c}_R)$, $(\widetilde{d}_R, \widetilde{s}_R)$, $(\widetilde{u}_L, \widetilde{c}_L)$, $(\widetilde{d}_L, \widetilde{s}_L)$. As we have already discussed in section 1.5.4, this avoids the problem of disastrously large virtual sparticle contributions to flavor-changing processes.

Let us first consider the spectrum of first- and second-family squarks and sleptons. In many models, including both minimal supergravity [eq. (1.246)] and gauge-mediated [eq. (1.262)] boundary conditions, their running squared masses can be conveniently parameterized, to a good approximation, as:

$$m^2_{Q_1} = m^2_{Q_2} = m_0^2 + K_3 + K_2 + \frac{1}{36}K_1, \tag{7.335}$$

$$m^2_{\overline{u}_1} = m^2_{\overline{u}_2} = m_0^2 + K_3 + \frac{4}{9}K_1, \tag{7.336}$$

$$m^2_{\overline{d}_1} = m^2_{\overline{d}_2} = m_0^2 + K_3 + \frac{1}{9}K_1, \tag{7.337}$$

$$m^2_{L_1} = m^2_{L_2} = m_0^2 + K_2 + \frac{1}{4}K_1, \tag{7.338}$$

$$m^2_{\overline{e}_1} = m^2_{\overline{e}_2} = m_0^2 + K_1. \tag{7.339}$$

A key point is that the same $K_3$, $K_2$ and $K_1$ appear everywhere in eqs. (7.335)-(7.339), since all of the chiral supermultiplets couple to the same gauginos with the same gauge couplings. The different coefficients in front of $K_1$ just correspond to the various values of weak hypercharge squared for each scalar.

In minimal supergravity models, $m_0^2$ is the same common scalar squared mass appearing in eq. (1.246). It can be very small, as in the "no-scale" limit, but it could also be the dominant source of the scalar masses. The contributions $K_3$, $K_2$ and $K_1$ are due to the RG running[cc] proportional to the gaugino masses. Explicitly, they are found at one loop order by solving eq. (1.196):

$$K_a(Q) = \begin{Bmatrix} 3/5 \\ 3/4 \\ 4/3 \end{Bmatrix} \times \frac{1}{2\pi^2} \int_{\ln Q}^{\ln Q_0} dt \; g_a^2(t) \, |M_a(t)|^2 \quad (a = 1, 2, 3). \quad (7.340)$$

Here $Q_0$ is the input RG scale at which the minimal supergravity boundary condition eq. (1.246) is applied, and $Q$ should be taken to be evaluated near the squark and slepton mass under consideration, presumably less than about 1 TeV. The running parameters $g_a(Q)$ and $M_a(Q)$ obey eqs. (1.161) and (1.189). If the input scale is approximated by the apparent scale of gauge coupling unification $Q_0 = M_U \approx 2 \times 10^{16}$ GeV, one finds that numerically

$$K_1 \approx 0.15 m_{1/2}^2, \qquad K_2 \approx 0.5 m_{1/2}^2, \qquad K_3 \approx (4.5 \text{ to } 6.5) m_{1/2}^2. \quad (7.341)$$

for $Q$ near the electroweak scale. Here $m_{1/2}$ is the common gaugino mass parameter at the unification scale. Note that $K_3 \gg K_2 \gg K_1$; this is a direct consequence of the relative sizes of the gauge couplings $g_3$, $g_2$, and $g_1$. The large uncertainty in $K_3$ is due in part to the experimental uncertainty in the QCD coupling constant, and in part to the uncertainty in where to choose $Q$, since $K_3$ runs rather quickly below 1 TeV. If the gauge couplings and gaugino masses are unified between $M_U$ and $M_P$, as would occur in a GUT model, then the effect of RG running for $M_U < Q < M_P$ can be absorbed into a redefinition of $m_0^2$. Otherwise, it adds a further uncertainty roughly proportional to $\ln(M_P/M_U)$, compared to the larger contributions in eq. (7.340), which go roughly like $\ln(M_U/1 \text{ TeV})$.

In gauge-mediated models, the same parameterization eqs. (7.335)-(7.339) holds, but $m_0^2$ is always 0. At the input scale $Q_0$, each MSSM scalar gets contributions to its squared mass that depend only on its gauge interactions, as in eq. (1.262). It is not hard to see that in general these contribute in exactly the same pattern as $K_1$, $K_2$, and $K_3$ in eq. (7.335)-(7.339). The subsequent evolution of the scalar squared masses down to the electroweak scale again just yields more contributions to the $K_1$, $K_2$, and $K_3$

[cc]The quantity $S$ defined in eq. (1.197) vanishes for both minimal supergravity and gauge-mediated boundary conditions, and remains small under RG evolution.

parameters. It is somewhat more difficult to give meaningful numerical estimates for these parameters in gauge-mediated models than in the minimal supergravity models without knowing the messenger mass scale(s) and the multiplicities of the messenger fields. However, in the gauge-mediated case one quite generally expects that the numerical values of the ratios $K_3/K_2$, $K_3/K_1$ and $K_2/K_1$ should be even larger than in eq. (7.341). There are two reasons for this. First, the running squark squared masses start off larger than slepton squared masses already at the input scale in gauge-mediated models, rather than having a common value $m_0^2$. Furthermore, in the gauge-mediated case, the input scale $Q_0$ is typically much lower than $M_P$ or $M_U$, so that the RG evolution gives relatively more weight to RG scales closer to the electroweak scale, where the hierarchies $g_3 > g_2 > g_1$ and $M_3 > M_2 > M_1$ are already in effect.

In general, one therefore expects that the squarks should be considerably heavier than the sleptons, with the effect being more pronounced in gauge-mediated supersymmetry breaking models than in minimal supergravity models. For any specific choice of model, this effect can be easily quantified with a numerical RG computation. The hierarchy $m_{\rm squark} > m_{\rm slepton}$ tends to hold even in models that do not fit neatly into any of the categories outlined in section 1.6, because the RG contributions to squark masses from the gluino are always present and usually quite large, since QCD has a larger gauge coupling than the electroweak interactions.

Regardless of the type of model, there is also a "hyperfine" splitting in the squark and slepton mass spectrum produced by electroweak symmetry breaking. Each squark and slepton $\phi$ will get a contribution $\Delta_\phi$ to its squared mass, coming from the $SU(2)_L$ and $U(1)_Y$ $D$-term quartic interactions [see the last term in eq. (1.135)] of the form (squark)$^2$(Higgs)$^2$ and (slepton)$^2$(Higgs)$^2$, when the neutral Higgs scalars $H_u^0$ and $H_d^0$ get VEVs. They are model-independent for a given value of $\tan\beta$:

$$\begin{aligned}\Delta_\phi &= \frac{1}{2}(T_{3\phi}g^2 - Y_\phi g'^2)(v_d^2 - v_u^2)\\ &= (T_{3\phi} - Q_\phi \sin^2\theta_W)\cos(2\beta)\, m_Z^2,\end{aligned} \tag{7.342}$$

where $T_{3\phi}$, $Y_\phi$, and $Q_\phi$ are the third component of weak isospin, the weak hypercharge, and the electric charge of the left-handed chiral supermultiplet to which $\phi$ belongs. For example, $\Delta_{\tilde{u}_L} = (\frac{1}{2} - \frac{2}{3}\sin^2\theta_W)\cos(2\beta)\, m_Z^2$ and $\Delta_{\tilde{d}_L} = (-\frac{1}{2} + \frac{1}{3}\sin^2\theta_W)\cos(2\beta)\, m_Z^2$ and $\Delta_{\tilde{u}_R} = (\frac{2}{3}\sin^2\theta_W)\cos(2\beta)\, m_Z^2$. These $D$-term contributions are typically smaller than the $m_0^2$ and $K_1$, $K_2$, $K_3$ contributions, but should not be neglected. They split apart the components of the $SU(2)_L$-doublet sleptons and squarks. Including them, the first-family squark and slepton masses are now given by:

$$m^2_{\tilde{d}_L} = m_0^2 + K_3 + K_2 + \frac{1}{36}K_1 + \Delta_{\tilde{d}_L}, \quad (7.343)$$

$$m^2_{\tilde{u}_L} = m_0^2 + K_3 + K_2 + \frac{1}{36}K_1 + \Delta_{\tilde{u}_L}, \quad (7.344)$$

$$m^2_{\tilde{u}_R} = m_0^2 + K_3 + \frac{4}{9}K_1 + \Delta_{\tilde{u}_R}, \quad (7.345)$$

$$m^2_{\tilde{d}_R} = m_0^2 + K_3 + \frac{1}{9}K_1 + \Delta_{\tilde{d}_R}, \quad (7.346)$$

$$m^2_{\tilde{e}_L} = m_0^2 + K_2 + \frac{1}{4}K_1 + \Delta_{\tilde{e}_L}, \quad (7.347)$$

$$m^2_{\tilde{\nu}} = m_0^2 + K_2 + \frac{1}{4}K_1 + \Delta_{\tilde{\nu}}, \quad (7.348)$$

$$m^2_{\tilde{e}_R} = m_0^2 + K_1 + \Delta_{\tilde{e}_R}, \quad (7.349)$$

with identical formulas for the second-family squarks and sleptons. The mass splittings for the left-handed squarks and sleptons are governed by model-independent sum rules

$$m^2_{\tilde{e}_L} - m^2_{\tilde{\nu}_e} = m^2_{\tilde{d}_L} - m^2_{\tilde{u}_L} = g^2(v_u^2 - v_d^2)/2 = -\cos(2\beta)\, m_W^2. \quad (7.350)$$

In the allowed range $\tan\beta > 1$, it follows that $m_{\tilde{e}_L} > m_{\tilde{\nu}_e}$ and $m_{\tilde{d}_L} > m_{\tilde{u}_L}$, with the magnitude of the splittings constrained by electroweak symmetry breaking.

Let us next consider the masses of the top squarks, for which there are several non-negligible contributions. First, there are squared-mass terms for $\widetilde{t}_L^*\widetilde{t}_L$ and $\widetilde{t}_R^*\widetilde{t}_R$ that are just equal to $m^2_{Q_3} + \Delta_{\tilde{u}_L}$ and $m^2_{\overline{u}_3} + \Delta_{\tilde{u}_R}$, respectively, just as for the first- and second-family squarks. Second, there are contributions equal to $m_t^2$ for each of $\widetilde{t}_L^*\widetilde{t}_L$ and $\widetilde{t}_R^*\widetilde{t}_R$. These come from $F$-terms in the scalar potential of the form $y_t^2 H_u^{0*} H_u^0 \widetilde{t}_L^* \widetilde{t}_L$ and $y_t^2 H_u^{0*} H_u^0 \widetilde{t}_R^* \widetilde{t}_R$ (see Figures 1.8b and 1.8c), with the Higgs fields replaced by their VEVs. (Of course, similar contributions are present for all of the squarks and sleptons, but they are too small to worry about except in the case of the top squarks.) Third, there are contributions to the scalar potential from $F$-terms of the form $-\mu^* y_t \widetilde{\overline{t}}\widetilde{t} H_d^{0*}$ + c.c.; see eqs. (1.146) and Figure 1.10a. These become $-\mu^* v y_t \cos\beta\, \widetilde{t}_R^* \widetilde{t}_L$ + c.c. when $H_d^0$ is replaced by its VEV. Finally, there are contributions to the scalar potential from the soft (scalar)$^3$ couplings $a_t \widetilde{\overline{t}} \widetilde{Q}_3 H_u^0$ + c.c. [see the first term of the second line of eq. (1.152), and eq. (1.190)], which become $a_t v \sin\beta\, \widetilde{t}_L \widetilde{t}_R^*$ + c.c. when $H_u^0$ is replaced by its VEV. Putting these all together, we have a squared-mass matrix for the top squarks, which in the gauge-eigenstate basis $(\widetilde{t}_L, \widetilde{t}_R)$ is given by

$$\mathcal{L}_{\text{stop masses}} = -\begin{pmatrix} \widetilde{t}_L^* & \widetilde{t}_R^* \end{pmatrix} \mathbf{m}^2_{\widetilde{\mathbf{t}}} \begin{pmatrix} \widetilde{t}_L \\ \widetilde{t}_R \end{pmatrix} \tag{7.351}$$

where

$$\mathbf{m}^2_{\widetilde{\mathbf{t}}} = \begin{pmatrix} m^2_{Q_3} + m_t^2 + \Delta_{\tilde{u}_L} & v(a_t^* \sin\beta - \mu y_t \cos\beta) \\ v(a_t \sin\beta - \mu^* y_t \cos\beta) & m^2_{\overline{u}_3} + m_t^2 + \Delta_{\tilde{u}_R} \end{pmatrix}. \tag{7.352}$$

This hermitian matrix can be diagonalized by a unitary matrix to give mass eigenstates:

$$\begin{pmatrix} \widetilde{t}_1 \\ \widetilde{t}_2 \end{pmatrix} = \begin{pmatrix} c_{\tilde{t}} & -s_{\tilde{t}}^* \\ s_{\tilde{t}} & c_{\tilde{t}} \end{pmatrix} \begin{pmatrix} \widetilde{t}_L \\ \widetilde{t}_R \end{pmatrix}. \tag{7.353}$$

Here $m^2_{\tilde{t}_1} < m^2_{\tilde{t}_2}$ are the eigenvalues of eq. (7.352), and $|c_{\tilde{t}}|^2 + |s_{\tilde{t}}|^2 = 1$. If the off-diagonal elements of eq. (7.352) are real, then $c_{\tilde{t}}$ and $s_{\tilde{t}}$ are the cosine and sine of a stop mixing angle $\theta_{\tilde{t}}$, which can be chosen in the range $0 \leq \theta_{\tilde{t}} < \pi$. Because of the large RG effects proportional to $X_t$ in eq. (1.203) and eq. (1.204), at the electroweak scale one finds that $m^2_{\overline{u}_3} < m^2_{Q_3}$, and both of these quantities are usually significantly smaller than the squark squared masses for the first two families. The diagonal terms $m_t^2$ in eq. (7.352) tend to mitigate this effect somewhat, but the off-diagonal entries will typically induce a significant mixing, which always reduces the lighter top-squark squared-mass eigenvalue. Therefore, models often predict that $\widetilde{t}_1$ is the lightest squark of all, and that it is predominantly $\tilde{t}_R$.

A very similar analysis can be performed for the bottom squarks and charged tau sleptons, which in their respective gauge-eigenstate bases $(\widetilde{b}_L, \widetilde{b}_R)$ and $(\widetilde{\tau}_L, \widetilde{\tau}_R)$ have squared-mass matrices:

$$\mathbf{m}^2_{\widetilde{\mathbf{b}}} = \begin{pmatrix} m^2_{Q_3} + \Delta_{\tilde{d}_L} & v(a_b^* \cos\beta - \mu y_b \sin\beta) \\ v(a_b \cos\beta - \mu^* y_b \sin\beta) & m^2_{\overline{d}_3} + \Delta_{\tilde{d}_R} \end{pmatrix}, \tag{7.354}$$

$$\mathbf{m}^2_{\widetilde{\boldsymbol{\tau}}} = \begin{pmatrix} m^2_{L_3} + \Delta_{\tilde{e}_L} & v(a_\tau^* \cos\beta - \mu y_\tau \sin\beta) \\ v(a_\tau \cos\beta - \mu^* y_\tau \sin\beta) & m^2_{\overline{e}_3} + \Delta_{\tilde{e}_R} \end{pmatrix}. \tag{7.355}$$

These can be diagonalized to give mass eigenstates $\widetilde{b}_1, \widetilde{b}_2$ and $\widetilde{\tau}_1, \widetilde{\tau}_2$ in exact analogy with eq. (7.353).

The magnitude and importance of mixing in the sbottom and stau sectors depends on how big $\tan\beta$ is. If $\tan\beta$ is not too large (in practice, this usually means less than about 10 or so, depending on the situation under study), the sbottoms and staus do not get a very large effect from the mixing terms and the RG effects due to $X_b$ and $X_\tau$, because $y_b, y_\tau \ll y_t$

from eq. (7.310). In that case the mass eigenstates are very nearly the same as the gauge eigenstates $\widetilde{b}_L$, $\widetilde{b}_R$, $\widetilde{\tau}_L$ and $\widetilde{\tau}_R$. The latter three, and $\widetilde{\nu}_\tau$, will be nearly degenerate with their first- and second-family counterparts with the same $SU(3)_C \times SU(2)_L \times U(1)_Y$ quantum numbers. However, even in the case of small $\tan\beta$, $\widetilde{b}_L$ will feel the effects of the large top Yukawa coupling because it is part of the doublet containing $\widetilde{t}_L$. In particular, from eq. (1.203) we see that $X_t$ acts to decrease $m^2_{Q_3}$ as it is RG-evolved down from the input scale to the electroweak scale. Therefore the mass of $\widetilde{b}_L$ can be significantly less than the masses of $\widetilde{d}_L$ and $\widetilde{s}_L$.

For larger values of $\tan\beta$, the mixing in eqs. (7.354) and (7.355) can be quite significant, because $y_b$, $y_\tau$ and $a_b$, $a_\tau$ are non-negligible. Just as in the case of the top squarks, the lighter sbottom and stau mass eigenstates (denoted $\widetilde{b}_1$ and $\widetilde{\tau}_1$) can be significantly lighter than their first- and second-family counterparts. Furthermore, $\widetilde{\nu}_\tau$ can be significantly lighter than the nearly degenerate $\widetilde{\nu}_e$, $\widetilde{\nu}_\mu$.

The requirement that the third-family squarks and sleptons should all have positive squared masses implies limits on the magnitudes of $a_t^* \sin\beta - \mu y_t \cos\beta$ and $a_b^* \cos\beta - \mu y_b \sin\beta$ and and $a_\tau^* \cos\beta - \mu y_\tau \sin\beta$. If they are too large, then the smaller eigenvalue of eq. (7.352), (7.354) or (7.355) will be driven negative, implying that a squark or charged slepton gets a VEV, breaking $SU(3)_C$ or electromagnetism. Since this is clearly unacceptable, one can put bounds on the $(\text{scalar})^3$ couplings, or equivalently on the parameter $A_0$ in minimal supergravity models. Even if all of the squared-mass eigenvalues are positive, the presence of large $(\text{scalar})^3$ couplings can yield global minima of the scalar potential, with non-zero squark and/or charged slepton VEVs, which are disconnected from the vacuum that conserves $SU(3)_C$ and electromagnetism.[202] However, it is not always immediately clear whether the mere existence of such disconnected global minima should really disqualify a set of model parameters, because the tunneling rate from our "good" vacuum to the "bad" vacua can easily be longer than the age of the universe.[203]

### 1.7.5. *Summary: The MSSM sparticle spectrum*

In the MSSM there are 32 distinct masses corresponding to undiscovered particles, not including the gravitino. In this section we have explained how the masses and mixing angles for these particles can be computed, given an underlying model for the soft terms at some input scale. Assuming only that the mixing of first- and second-family squarks and sleptons is negligible, the

Table 1.5. The undiscovered particles in the Minimal Supersymmetric Standard Model (with sfermion mixing for the first two families assumed to be negligible.

| Names | Spin | $P_R$ | Gauge Eigenstates | Mass Eigenstates |
|---|---|---|---|---|
| Higgs bosons | 0 | +1 | $H_u^0\ H_d^0\ H_u^+\ H_d^-$ | $h^0\ H^0\ A^0\ H^\pm$ |
| squarks | 0 | −1 | $\widetilde{u}_L\ \widetilde{u}_R\ \widetilde{d}_L\ \widetilde{d}_R$ | (same) |
| | | | $\widetilde{s}_L\ \widetilde{s}_R\ \widetilde{c}_L\ \widetilde{c}_R$ | (same) |
| | | | $\widetilde{t}_L\ \widetilde{t}_R\ \widetilde{b}_L\ \widetilde{b}_R$ | $\widetilde{t}_1\ \widetilde{t}_2\ \widetilde{b}_1\ \widetilde{b}_2$ |
| sleptons | 0 | −1 | $\widetilde{e}_L\ \widetilde{e}_R\ \widetilde{\nu}_e$ | (same) |
| | | | $\widetilde{\mu}_L\ \widetilde{\mu}_R\ \widetilde{\nu}_\mu$ | (same) |
| | | | $\widetilde{\tau}_L\ \widetilde{\tau}_R\ \widetilde{\nu}_\tau$ | $\widetilde{\tau}_1\ \widetilde{\tau}_2\ \widetilde{\nu}_\tau$ |
| neutralinos | 1/2 | −1 | $\widetilde{B}^0\ \widetilde{W}^0\ \widetilde{H}_u^0\ \widetilde{H}_d^0$ | $\widetilde{N}_1\ \widetilde{N}_2\ \widetilde{N}_3\ \widetilde{N}_4$ |
| charginos | 1/2 | −1 | $\widetilde{W}^\pm\ \widetilde{H}_u^+\ \widetilde{H}_d^-$ | $\widetilde{C}_1^\pm\ \widetilde{C}_2^\pm$ |
| gluino | 1/2 | −1 | $\widetilde{g}$ | (same) |
| goldstino (gravitino) | 1/2 (3/2) | −1 | $\widetilde{G}$ | (same) |

mass eigenstates of the MSSM are listed in Table 1.5. A complete set of Feynman rules for the interactions of these particles with each other and with the Standard Model quarks, leptons, and gauge bosons can be found in Refs. 25, 182. Feynman rules based on two-component spinor notation have also recently been given in Ref. 199.

Specific models for the soft terms typically predict the masses and the mixing angles angles for the MSSM in terms of far fewer parameters. For example, in the minimal supergravity models, the only free parameters not already measured by experiment are $m_0^2$, $m_{1/2}$, $A_0$, $\mu$, and $b$. In gauge-mediated supersymmetry breaking models, the free parameters include at least the scale $\Lambda$, the typical messenger mass scale $M_{\rm mess}$, the integer number $N_5$ of copies of the minimal messengers, the goldstino decay constant $\langle F\rangle$, and the Higgs mass parameters $\mu$ and $b$. After RG evolving the soft terms down to the electroweak scale, one can demand that the scalar potential gives correct electroweak symmetry breaking. This allows us to trade $|\mu|$ and $b$ (or $B_0$) for one parameter $\tan\beta$, as in eqs. (7.291)-(7.290). So, to a reasonable approximation, the entire mass spectrum in minimal supergravity models is determined by only five unknown parameters: $m_0^2$, $m_{1/2}$, $A_0$, $\tan\beta$, and Arg($\mu$), while in the simplest gauge-mediated supersymmetry breaking models one can pick parameters $\Lambda$, $M_{\rm mess}$, $N_5$, $\langle F\rangle$, $\tan\beta$, and Arg($\mu$). Both frameworks are highly predictive. Of course, it is easy to

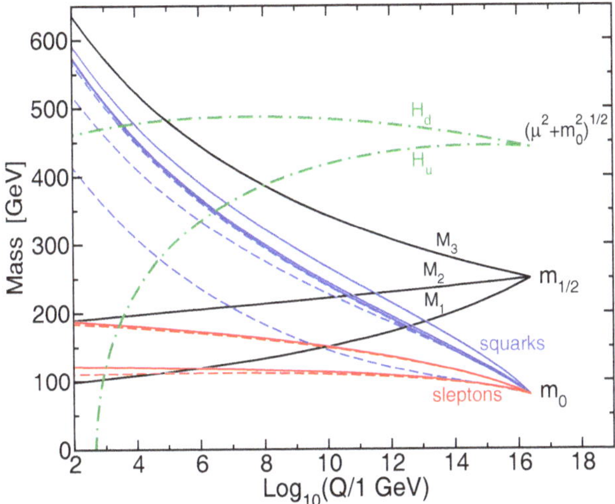


Fig. 1.23. RG evolution of scalar and gaugino mass parameters in the MSSM with typical minimal supergravity-inspired boundary conditions imposed at $Q_0 = 2.5 \times 10^{16}$ GeV. The parameter $\mu^2 + m_{H_u}^2$ runs negative, provoking electroweak symmetry breaking.

imagine that the essential physics of supersymmetry breaking is not captured by either of these two scenarios in their minimal forms. For example, the anomaly mediated contributions could play a role, perhaps in concert with the gauge-mediation or Planck-scale mediation mechanisms.

Figure 1.23 shows the RG running of scalar and gaugino masses in a typical model based on the minimal supergravity boundary conditions imposed at $Q_0 = 2.5 \times 10^{16}$ GeV. [The parameter values used for this illustration were $m_0 = 80$ GeV, $m_{1/2} = 250$ GeV, $A_0 = -500$ GeV, $\tan\beta = 10$, and sign($\mu$)= +.] The running gaugino masses are solid lines labeled by $M_1$, $M_2$, and $M_3$. The dot-dashed lines labeled $H_u$ and $H_d$ are the running values of the quantities $(\mu^2 + m_{H_u}^2)^{1/2}$ and $(\mu^2 + m_{H_d}^2)^{1/2}$, which appear in the Higgs potential. The other lines are the running squark and slepton masses, with dashed lines for the square roots of the third family parameters $m_{\tilde{d}_3}^2$, $m_{Q_3}^2$, $m_{\tilde{u}_3}^2$, $m_{L_3}^2$, and $m_{\tilde{e}_3}^2$ (from top to bottom), and solid lines for the first and second family sfermions. Note that $\mu^2 + m_{H_u}^2$ runs negative because of the effects of the large top Yukawa coupling as discussed above, providing for electroweak symmetry breaking. At the electroweak scale, the values of the Lagrangian soft parameters can be used to extract the physical masses, cross-sections, and decay widths of the particles, and other observables such as dark matter abundances and rare process rates. There are a variety of publicly available programs that do these tasks, including radiative corrections; see for example Refs. 204–213, 194.

Figure 1.24 shows deliberately qualitative sketches of sample MSSM mass spectrum obtained from three different types of models assumptions. The first is the output from a minimal supergravity-inspired model with relatively low $m_0^2$ compared to $m_{1/2}^2$ (in fact the same model parameters as used for Figure 1.23). This model features a near-decoupling limit for the Higgs sector, and a bino-like $\widetilde{N}_1$ LSP, nearly degenerate wino-like $\widetilde{N}_2, \widetilde{C}_1$, and higgsino-like $\widetilde{N}_3, \widetilde{N}_4, \widetilde{C}_2$. The gluino is the heaviest superpartner. The squarks are all much heavier than the sleptons, and the lightest sfermion is a stau. Variations in the model parameters have important and predictable effects. For example, taking larger $m_0^2$ in minimal supergravity models will tend to squeeze together the spectrum of squarks and sleptons and move them all higher compared to the neutralinos, charginos and gluino. Taking larger values of $\tan\beta$ with other model parameters held fixed will usually tend to lower $\widetilde{b}_1$ and $\widetilde{\tau}_1$ masses compared to those of the other sparticles.

The second sample sketch in Figure 1.24 is obtained from a typical minimal GMSB model, with boundary conditions as in eq. (1.269) [with $N_5 = 1$, $\Lambda = 150$ TeV, $\tan\beta = 15$, and sign($\mu$)= + at a scale $Q_0 = M_{\rm mess} = 300$ TeV for the illustration]. Here we see that the hierarchy between strongly interacting sparticles and weakly interacting ones is quite large. Changing the messenger scale or $\Lambda$ does not reduce the relative splitting between squark and slepton masses, because there is no analog of the universal $m_0^2$ contribution here. Increasing the number of messenger fields tends to decrease the squark and slepton masses relative to the gaugino masses, but still keeps the hierarchy between squark and slepton masses intact. In the model shown, the NLSP is a bino-like neutralino, but for larger number of messenger fields it could be either a stau, or else co-NLSPs $\tilde{\tau}_1$, $\tilde{e}_L$, $\tilde{\mu}_L$, depending on the choice of $\tan\beta$.

The third sample sketch in Figure 1.24 is obtained from an AMSB model with an additional universal scalar mass $m_0 = 450$ TeV added at $Q_0 = 2 \times 10^{16}$ GeV to rescue the sleptons, and with $m_{3/2} = 60$ TeV, $\tan\beta = 10$, and sign($\mu$)= + for the illustration. Here the most striking feature is that the LSP is a wino-like neutralino, with $m_{\tilde{C}_1} - m_{\tilde{N}_1}$ only about 160 MeV.

It would be a mistake to rely too heavily on specific scenarios for the MSSM mass and mixing spectrum, and the above illustrations are only a tiny fraction of the available possibilities. However, it is also useful to keep in mind some general lessons that often recur in various different models.

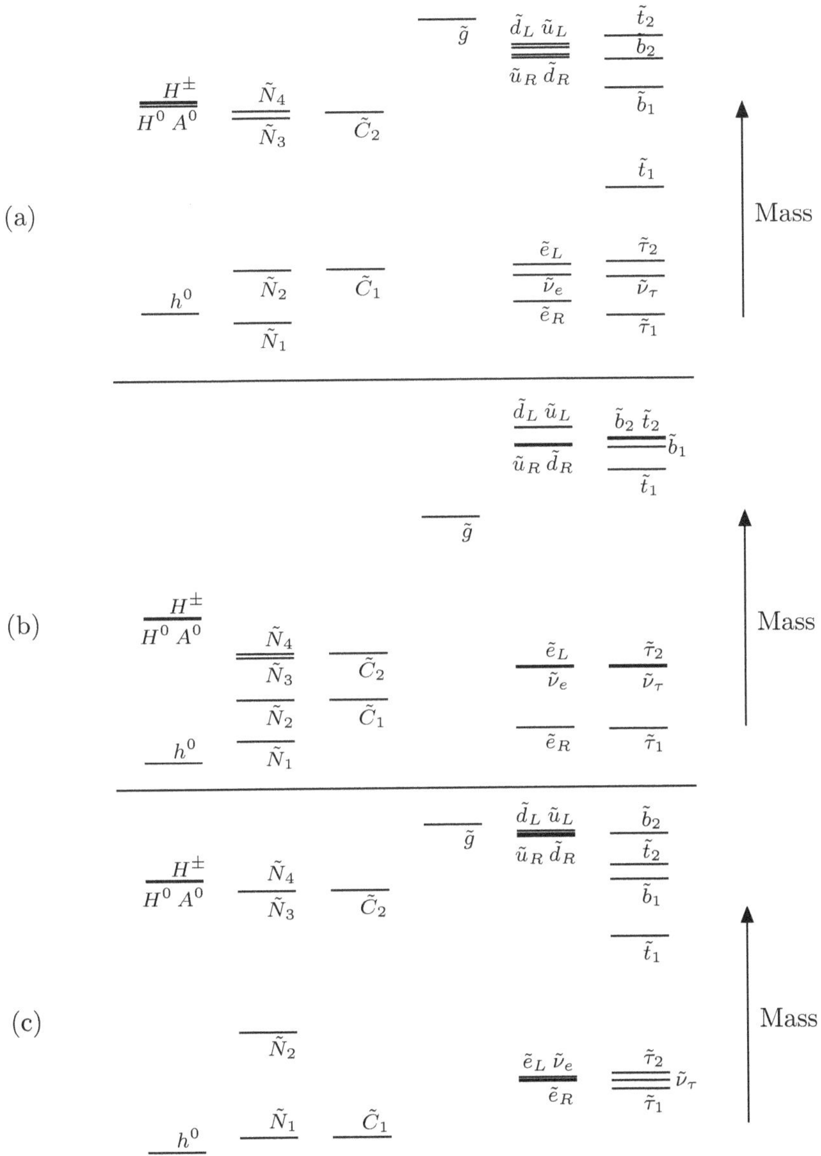


Fig. 1.24. Three sample schematic mass spectra for the undiscovered particles in the MSSM, for (a) minimal supergravity with $m_0^2 \ll m_{1/2}^2$, (b) minimal GMSB with $N_5 = 1$, and (c) AMSB with an extra $m_0^2$ for scalars. These spectra are presented for entertainment purposes only! No warranty, expressed or implied, guarantees that they look anything like the real world.

Indeed, there has emerged a sort of folklore concerning likely features of the MSSM spectrum, partly based on theoretical bias and partly on the constraints inherent in most known viable softly-broken supersymmetric theories. We remark on these features mainly because they represent the prevailing prejudices among supersymmetry theorists, which is certainly a useful thing to know even if one wisely decides to remain skeptical. For example, it is perhaps not unlikely that:

- The LSP is the lightest neutralino $\widetilde{N}_1$, unless the gravitino is lighter or $R$-parity is not conserved. If $M_1 < M_2, |\mu|$, then $\widetilde{N}_1$ is likely to be bino-like, with a mass roughly 0.5 times the masses of $\widetilde{N}_2$ and $\widetilde{C}_1$ in many well-motivated models. If, instead, $|\mu| < M_1, M_2$, then the LSP $\widetilde{N}_1$ has a large higgsino content and $\widetilde{N}_2$ and $\widetilde{C}_1$ are not much heavier. And, if $M_2 \ll M_1, |\mu|$, then the LSP will be a wino-like neutralino, with a chargino only very slightly heavier.
- The gluino will be much heavier than the lighter neutralinos and charginos. This is certainly true in the case of the "standard" gaugino mass relation eq. (1.189); more generally, the running gluino mass parameter grows relatively quickly as it is RG-evolved into the infrared because the QCD coupling is larger than the electroweak gauge couplings. So even if there are big corrections to the gaugino mass boundary conditions eqs. (1.245) or (1.260), the gluino mass parameter $M_3$ is likely to come out larger than $M_1$ and $M_2$.
- The squarks of the first and second families are nearly degenerate and much heavier than the sleptons. This is because each squark mass gets the same large positive-definite radiative corrections from loops involving the gluino. The left-handed squarks $\widetilde{u}_L$, $\widetilde{d}_L$, $\widetilde{s}_L$ and $\widetilde{c}_L$ are likely to be heavier than their right-handed counterparts $\widetilde{u}_R$, $\widetilde{d}_R$, $\widetilde{s}_R$ and $\widetilde{c}_R$, because of the effect parameterized by $K_2$ in eqs. (7.343)-(7.349).
- The squarks of the first two families cannot be lighter than about 0.8 times the mass of the gluino in minimal supergravity models, and about 0.6 times the mass of the gluino in the simplest gauge-mediated models as discussed in section 1.6.7 if the number of messenger squark pairs is $N_5 \leq 4$. In the minimal supergravity case this is because the gluino mass feeds into the squark masses through RG evolution; in the gauge-mediated case it is because the gluino and squark masses are tied together by eqs. (1.265) and (1.266).

- The lighter stop $\widetilde{t}_1$ and the lighter sbottom $\widetilde{b}_1$ are probably the lightest squarks. This is because stop and sbottom mixing effects and the effects of $X_t$ and $X_b$ in eqs. (1.203)-(1.205) both tend to decrease the lighter stop and sbottom masses.
- The lightest charged slepton is probably a stau $\widetilde{\tau}_1$. The mass difference $m_{\widetilde{e}_R} - m_{\widetilde{\tau}_1}$ is likely to be significant if $\tan\beta$ is large, because of the effects of a large tau Yukawa coupling. For smaller $\tan\beta$, $\widetilde{\tau}_1$ is predominantly $\widetilde{\tau}_R$ and it is not so much lighter than $\widetilde{e}_R$, $\widetilde{\mu}_R$.
- The left-handed charged sleptons $\widetilde{e}_L$ and $\widetilde{\mu}_L$ are likely to be heavier than their right-handed counterparts $\widetilde{e}_R$ and $\widetilde{\mu}_R$. This is because of the effect of $K_2$ in eq. (7.347). (Note also that $\Delta_{\widetilde{e}_L} - \Delta_{\widetilde{e}_R}$ is positive but very small because of the numerical accident $\sin^2\theta_W \approx 1/4$.)
- The lightest neutral Higgs boson $h^0$ should be lighter than about 150 GeV, and may be much lighter than the other Higgs scalar mass eigenstates $A^0$, $H^\pm$, $H^0$.

The most important point is that by measuring the masses and mixing angles of the MSSM particles we will be able to gain a great deal of information that can rule out or bolster evidence for competing proposals for the origin and mediation of supersymmetry breaking.

## 1.8. Sparticle Decays

This section contains a brief qualitative overview of the decay patterns of sparticles in the MSSM, assuming that $R$-parity is conserved. We will consider in turn the possible decays of neutralinos, charginos, sleptons, squarks, and the gluino. If, as is most often assumed, the lightest neutralino $\widetilde{N}_1$ is the LSP, then all decay chains will end up with it in the final state. Section 1.8.5 discusses the alternative possibility that the gravitino/goldstino $\widetilde{G}$ is the LSP. For the sake of simplicity of notation, we will often not distinguish between particle and antiparticle names and labels in this section, with context and consistency (dictated by charge and color conservation) resolving any ambiguities.

### 1.8.1. *Decays of neutralinos and charginos*

Let us first consider the possible two-body decays. Each neutralino and chargino contains at least a small admixture of the electroweak gauginos $\widetilde{B}$, $\widetilde{W}^0$ or $\widetilde{W}^\pm$, as we saw in section 1.7.2. So $\widetilde{N}_i$ and $\widetilde{C}_i$ inherit couplings of weak interaction strength to (scalar, fermion) pairs, as shown in

Figure 1.9b,c. If sleptons or squarks are sufficiently light, a neutralino or chargino can therefore decay into lepton+slepton or quark+squark. To the extent that sleptons are probably lighter than squarks, the lepton+slepton final states are favored. A neutralino or chargino may also decay into any lighter neutralino or chargino plus a Higgs scalar or an electroweak gauge boson, because they inherit the gaugino-higgsino-Higgs (see Figure 1.9b,c) and $SU(2)_L$ gaugino-gaugino-vector boson (see Figure 1.5c) couplings of their components. So, the possible two-body decay modes for neutralinos and charginos in the MSSM are:

$$\widetilde{N}_i \to Z\widetilde{N}_j,\ W\widetilde{C}_j,\ h^0\widetilde{N}_j,\ \ell\widetilde{\ell},\ \nu\widetilde{\nu},\ [A^0\widetilde{N}_j,\ H^0\widetilde{N}_j,\ H^\pm\widetilde{C}_j^\mp,\ q\widetilde{q}], \quad (8.356)$$

$$\widetilde{C}_i \to W\widetilde{N}_j,\ Z\widetilde{C}_1,\ h^0\widetilde{C}_1,\ \ell\widetilde{\nu},\ \nu\widetilde{\ell},\ [A^0\widetilde{C}_1,\ H^0\widetilde{C}_1,\ H^\pm\widetilde{N}_j,\ q\widetilde{q}'], \quad (8.357)$$

using a generic notation $\nu$, $\ell$, $q$ for neutrinos, charged leptons, and quarks. The final states in brackets are the more kinematically implausible ones. (Since $h^0$ is required to be light, it is the most likely of the Higgs scalars to appear in these decays. This could even be the best way to discover the Higgs.) For the heavier neutralinos and chargino ($\widetilde{N}_3$, $\widetilde{N}_4$ and $\widetilde{C}_2$), one or more of the two-body decays in eqs. (8.356) and (8.357) is likely to be kinematically allowed. Also, if the decays of neutralinos and charginos with a significant higgsino content into third-family quark-squark pairs are open, they can be greatly enhanced by the top-quark Yukawa coupling, following from the interactions shown in Figure 1.7b,c.

It may be that all of these two-body modes are kinematically forbidden for a given chargino or neutralino, especially for $\widetilde{C}_1$ and $\widetilde{N}_2$ decays. In that case, they have three-body decays

$$\widetilde{N}_i \to ff\widetilde{N}_j, \quad \widetilde{N}_i \to ff'\widetilde{C}_j, \quad \widetilde{C}_i \to ff'\widetilde{N}_j, \quad \text{and} \quad \widetilde{C}_2 \to ff\widetilde{C}_1, \quad (8.358)$$

through the same (but now off-shell) gauge bosons, Higgs scalars, sleptons, and squarks that appeared in the two-body decays eqs. (8.356) and (8.357). Here $f$ is generic notation for a lepton or quark, with $f$ and $f'$ distinct members of the same $SU(2)_L$ multiplet (and of course one of the $f$ or $f'$ in each of these decays must actually be an antifermion). The chargino and neutralino decay widths into the various final states can be found in Refs. 214–216.

The Feynman diagrams for the neutralino and chargino decays with $\widetilde{N}_1$ in the final state that seem most likely to be important are shown in Figure 1.25. In many situations, the decays

$$\widetilde{C}_1^\pm \to \ell^\pm\nu\widetilde{N}_1, \qquad \widetilde{N}_2 \to \ell^+\ell^-\widetilde{N}_1 \quad (8.359)$$

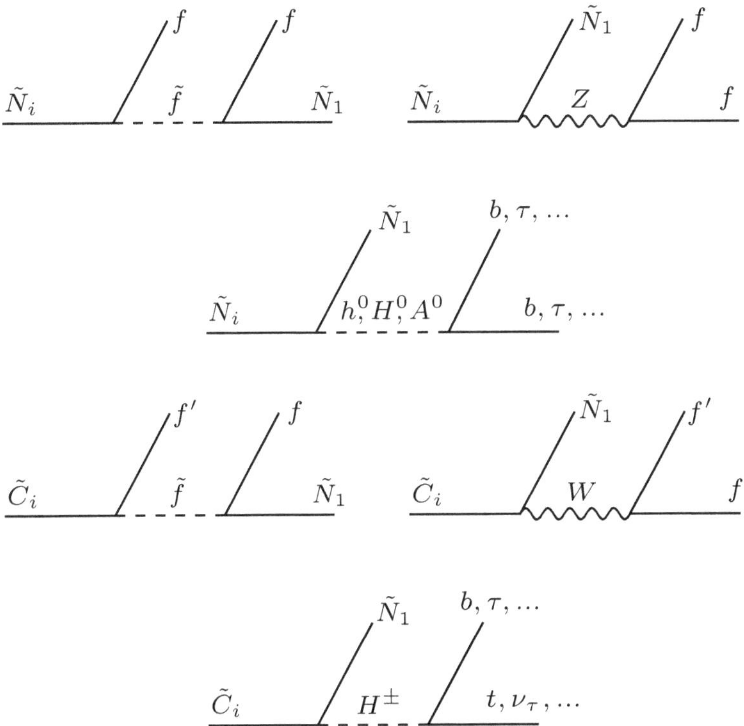


Fig. 1.25. Feynman diagrams for neutralino and chargino decays with $\tilde{N}_1$ in the final state. The intermediate scalar or vector boson in each case can be either on-shell (so that actually there is a sequence of two-body decays) or off-shell, depending on the sparticle mass spectrum.

can be particularly important for phenomenology, because the leptons in the final state often will result in clean signals. These decays are more likely if the intermediate sleptons are relatively light, even if they cannot be on-shell. Unfortunately, the enhanced mixing of staus, common in models, can often result in larger branching fractions for both $\tilde{N}_2$ and $\tilde{C}_1$ into final states with taus, rather than electrons or muons. This is one reason why tau identification may be a crucial limiting factor in attempts to discover and study supersymmetry.

In other situations, decays without isolated leptons in the final state are more useful, so that one will not need to contend with background events with missing energy coming from leptonic $W$ boson decays in Standard Model processes. Then the decays of interest are the ones with quark partons in the final state, leading to

$$\widetilde{C}_1 \to jj\widetilde{N}_1, \qquad \widetilde{N}_2 \to jj\widetilde{N}_1, \tag{8.360}$$

where $j$ means a jet. If the second of these decays goes through an on-shell (or nearly so) $h^0$, then these will usually be $b$-jets.

### 1.8.2. *Slepton decays*

Sleptons can have two-body decays into a lepton and a chargino or neutralino, because of their gaugino admixture, as may be seen directly from the couplings in Figures 1.9b,c. Therefore, the two-body decays

$$\widetilde{\ell} \to \ell\widetilde{N}_i, \quad \widetilde{\ell} \to \nu\widetilde{C}_i, \quad \widetilde{\nu} \to \nu\widetilde{N}_i, \quad \widetilde{\nu} \to \ell\widetilde{C}_i \tag{8.361}$$

can be of weak interaction strength. In particular, the direct decays

$$\widetilde{\ell} \to \ell\widetilde{N}_1 \quad \text{and} \quad \widetilde{\nu} \to \nu\widetilde{N}_1 \tag{8.362}$$

are (almost[dd]) always kinematically allowed if $\widetilde{N}_1$ is the LSP. However, if the sleptons are sufficiently heavy, then the two-body decays

$$\widetilde{\ell} \to \nu\widetilde{C}_1, \quad \widetilde{\ell} \to \ell\widetilde{N}_2, \quad \widetilde{\nu} \to \nu\widetilde{N}_2, \quad \text{and} \quad \widetilde{\nu} \to \ell\widetilde{C}_1 \tag{8.363}$$

can be important. The right-handed sleptons do not have a coupling to the $SU(2)_L$ gauginos, so they typically prefer the direct decay $\widetilde{\ell}_R \to \ell\widetilde{N}_1$, if $\widetilde{N}_1$ is bino-like. In contrast, the left-handed sleptons may prefer to decay as in eq. (8.363) rather than the direct decays to the LSP as in eq. (8.362), if the former is kinematically open and if $\widetilde{C}_1$ and $\widetilde{N}_2$ are mostly wino. This is because the slepton-lepton-wino interactions in Figure 1.9b are proportional to the $SU(2)_L$ gauge coupling $g$, whereas the slepton-lepton-bino interactions in Figure 1.9c are proportional to the much smaller $U(1)_Y$ coupling $g'$. Formulas for these decay widths can be found in Ref. 215.

### 1.8.3. *Squark decays*

If the decay $\widetilde{q} \to q\widetilde{g}$ is kinematically allowed, it will always dominate, because the quark-squark-gluino vertex in Figure 1.9a has QCD strength. Otherwise, the squarks can decay into a quark plus neutralino or chargino: $\widetilde{q} \to q\widetilde{N}_i$ or $q'\widetilde{C}_i$. The direct decay to the LSP $\widetilde{q} \to q\widetilde{N}_1$ is always kinematically favored, and for right-handed squarks it can dominate because $\widetilde{N}_1$ is mostly bino. However, the left-handed squarks may strongly prefer to decay into heavier charginos or neutralinos instead, for example $\widetilde{q} \to q\widetilde{N}_2$

[dd]An exception occurs if the mass difference $m_{\tilde{\tau}_1} - m_{\tilde{N}_1}$ is less than $m_\tau$.

or $q'\widetilde{C}_1$, because the relevant squark-quark-wino couplings are much bigger than the squark-quark-bino couplings. Squark decays to higgsino-like charginos and neutralinos are less important, except in the cases of stops and sbottoms, which have sizable Yukawa couplings. The gluino, chargino or neutralino resulting from the squark decay will in turn decay, and so on, until a final state containing $\widetilde{N}_1$ is reached. This results in numerous and complicated decay chain possibilities called cascade decays.[217]

It is possible that the decays $\widetilde{t}_1 \to t\widetilde{g}$ and $\widetilde{t}_1 \to t\widetilde{N}_1$ are both kinematically forbidden. If so, then the lighter top squark may decay only into charginos, by $\widetilde{t}_1 \to b\widetilde{C}_1$. If even this decay is kinematically closed, then it has only the flavor-suppressed decay to a charm quark, $\widetilde{t}_1 \to c\widetilde{N}_1$, and the four-body decay $\widetilde{t}_1 \to bff'\widetilde{N}_1$. These decays can be very slow,[218] so that the lightest stop can be quasi-stable on the time scale relevant for collider physics, and can hadronize into bound states.

### 1.8.4. *Gluino decays*

The decay of the gluino can only proceed through a squark, either on-shell or virtual. If two-body decays $\widetilde{g} \to q\widetilde{q}$ are open, they will dominate, again because the relevant gluino-quark-squark coupling in Figure 1.9a has QCD strength. Since the top and bottom squarks can easily be much lighter than all of the other squarks, it is quite possible that $\widetilde{g} \to t\widetilde{t}_1$ and/or $\widetilde{g} \to b\widetilde{b}_1$ are the only available two-body decay mode(s) for the gluino, in which case they will dominate over all others. If instead all of the squarks are heavier than the gluino, the gluino will decay only through off-shell squarks, so $\widetilde{g} \to qq\widetilde{N}_i$ and $qq'\widetilde{C}_i$. The squarks, neutralinos and charginos in these final states will then decay as discussed above, so there can be many competing gluino decay chains. Some of the possibilities are shown in Figure 1.26. The cascade decays can have final-state branching fractions that are individually small and quite sensitive to the model parameters.

The simplest gluino decays, including the ones shown in Figure 1.26, can have 0, 1, or 2 charged leptons (in addition to two or more hadronic jets) in the final state. An important feature is that when there is exactly one charged lepton, it can have either charge with exactly equal probability. This follows from the fact that the gluino is a Majorana fermion, and does not "know" about electric charge; for each diagram with a given lepton charge, there is always an equal one with every particle replaced by its antiparticle.

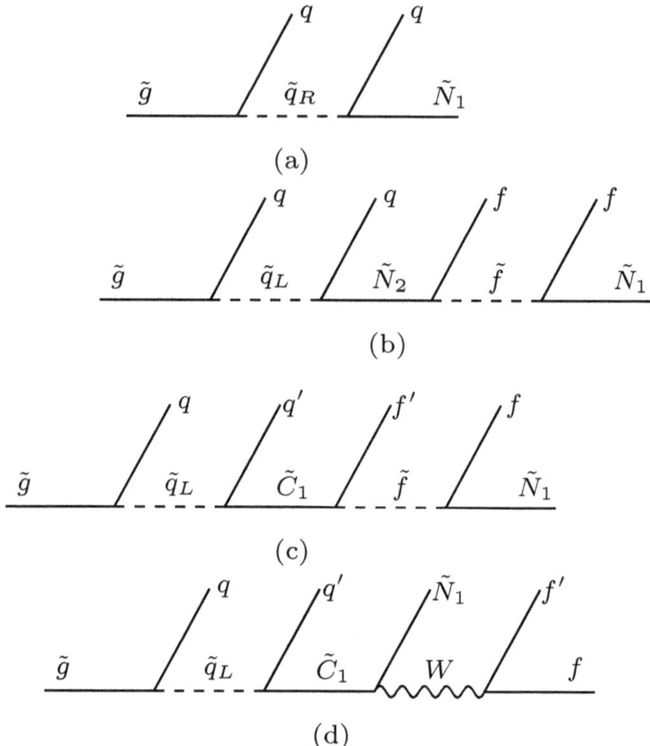


Fig. 1.26. Some of the many possible examples of gluino cascade decays ending with a neutralino LSP in the final state. The squarks appearing in these diagrams may be either on-shell or off-shell, depending on the mass spectrum of the theory.

### 1.8.5. *Decays to the gravitino/goldstino*

Most phenomenological studies of supersymmetry assume explicitly or implicitly that the lightest neutralino is the LSP. This is typically the case in gravity-mediated models for the soft terms. However, in gauge-mediated models (and in "no-scale" models), the LSP is instead the gravitino. As we saw in section 1.6.5, a very light gravitino may be relevant for collider phenomenology, because it contains as its longitudinal component the goldstino, which has a non-gravitational coupling to all sparticle-particle pairs $(\widetilde{X}, X)$. The decay rate found in eq. (1.239) for $\widetilde{X} \rightarrow X\widetilde{G}$ is usually not fast enough to compete with the other decays of sparticles $\widetilde{X}$ as mentioned above, *except* in the case that $\widetilde{X}$ is the next-to-lightest supersymmetric particle (NLSP). Since the NLSP has no competing decays, it should always decay into its superpartner and the LSP gravitino.

In principle, any of the MSSM superpartners could be the NLSP in models with a light goldstino, but most models with gauge mediation of supersymmetry breaking have either a neutralino or a charged lepton playing this role. The argument for this can be seen immediately from eqs. (1.265) and (1.266); since $\alpha_1 < \alpha_2, \alpha_3$, those superpartners with only $U(1)_Y$ interactions will tend to get the smallest masses. The gauge-eigenstate sparticles with this property are the bino and the right-handed sleptons $\widetilde{e}_R$, $\widetilde{\mu}_R$, $\widetilde{\tau}_R$, so the appropriate corresponding mass eigenstates should be plausible candidates for the NLSP.

First suppose that $\widetilde{N}_1$ is the NLSP in light goldstino models. Since $\widetilde{N}_1$ contains an admixture of the photino (the linear combination of bino and neutral wino whose superpartner is the photon), from eq. (1.239) it decays into photon + goldstino/gravitino with a partial width

$$\Gamma(\widetilde{N}_1 \to \gamma\widetilde{G}) = 2 \times 10^{-3}\,\kappa_{1\gamma} \left(\frac{m_{\widetilde{N}_1}}{100\text{ GeV}}\right)^5 \left(\frac{\sqrt{\langle F\rangle}}{100\text{ TeV}}\right)^{-4} \text{eV}. \quad (8.364)$$

Here $\kappa_{1\gamma} \equiv |\mathbf{N}_{11}\cos\theta_W + \mathbf{N}_{12}\sin\theta_W|^2$ is the "photino content" of $\widetilde{N}_1$, in terms of the neutralino mixing matrix $\mathbf{N}_{ij}$ defined by eq. (7.315). We have normalized $m_{\widetilde{N}_1}$ and $\sqrt{\langle F\rangle}$ to (very roughly) minimum expected values in gauge-mediated models. This width is much smaller than for a typical flavor-unsuppressed weak interaction decay, but it is still large enough to allow $\widetilde{N}_1$ to decay before it has left a collider detector, if $\sqrt{\langle F\rangle}$ is less than a few thousand TeV in gauge-mediated models, or equivalently if $m_{3/2}$ is less than a keV or so when eq. (1.238) holds. In fact, from eq. (8.364), the mean decay length of an $\widetilde{N}_1$ with energy $E$ in the lab frame is

$$d = 9.9 \times 10^{-3}\,\frac{1}{\kappa_{1\gamma}}\,(E^2/m^2_{\widetilde{N}_1} - 1)^{1/2} \left(\frac{m_{\widetilde{N}_1}}{100\text{ GeV}}\right)^{-5} \left(\frac{\sqrt{\langle F\rangle}}{100\text{ TeV}}\right)^{4} \text{cm}, \quad (8.365)$$

which could be anything from sub-micron to multi-kilometer, depending on the scale of supersymmetry breaking $\sqrt{\langle F\rangle}$. (In other models that have a gravitino LSP, including certain "no-scale" models,[219] the same formulas apply with $\langle F\rangle \to \sqrt{3}m_{3/2}M_{\text{P}}$.)

Of course, $\widetilde{N}_1$ is not a pure photino, but contains also admixtures of the superpartner of the $Z$ boson and the neutral Higgs scalars. So, one can also have[144] $\widetilde{N}_1 \to Z\widetilde{G}$, $h^0\widetilde{G}$, $A^0\widetilde{G}$, or $H^0\widetilde{G}$, with decay widths given in Ref. 145. Of these decays, the last two are unlikely to be kinematically allowed, and only the $\widetilde{N}_1 \to \gamma\widetilde{G}$ mode is guaranteed to be kinematically allowed for a gravitino LSP. Furthermore, even if they are open, the decays $\widetilde{N}_1 \rightarrow Z\widetilde{G}$ and $\widetilde{N}_1 \rightarrow h^0\widetilde{G}$ are subject to strong kinematic suppressions proportional to $(1 - m_Z^2/m_{\widetilde{N}_1}^2)^4$ and $(1 - m_{h^0}^2/m_{\widetilde{N}_1}^2)^4$, respectively, in view of eq. (1.239). Still, these decays may play an important role in phenomenology if $\langle F \rangle$ is not too large, $\widetilde{N}_1$ has a sizable zino or higgsino content, and $m_{\widetilde{N}_1}$ is significantly greater than $m_Z$ or $m_{h^0}$.

A charged slepton makes another likely candidate for the NLSP. Actually, more than one slepton can act effectively as the NLSP, even though one of them is slightly lighter, if they are sufficiently close in mass so that each has no kinematically allowed decays except to the goldstino. In GMSB models, the squared masses obtained by $\widetilde{e}_R$, $\widetilde{\mu}_R$ and $\widetilde{\tau}_R$ are equal because of the flavor-blindness of the gauge couplings. However, this is not the whole story, because one must take into account mixing with $\widetilde{e}_L$, $\widetilde{\mu}_L$, and $\widetilde{\tau}_L$ and renormalization group running. These effects are very small for $\widetilde{e}_R$ and $\widetilde{\mu}_R$ because of the tiny electron and muon Yukawa couplings, so we can quite generally treat them as degenerate, unmixed mass eigenstates. In contrast, $\widetilde{\tau}_R$ usually has a quite significant mixing with $\widetilde{\tau}_L$, proportional to the tau Yukawa coupling. This means that the lighter stau mass eigenstate $\widetilde{\tau}_1$ is pushed lower in mass than $\widetilde{e}_R$ or $\widetilde{\mu}_R$, by an amount that depends most strongly on $\tan\beta$. If $\tan\beta$ is not too large then the stau mixing effect leaves the slepton mass eigenstates $\widetilde{e}_R$, $\widetilde{\mu}_R$, and $\widetilde{\tau}_1$ degenerate to within less than $m_\tau \approx 1.8$ GeV, so they act effectively as co-NLSPs. In particular, this means that even though the stau is slightly lighter, the three-body slepton decays $\widetilde{e}_R \rightarrow e\tau^\pm\widetilde{\tau}_1^\mp$ and $\widetilde{\mu}_R \rightarrow \mu\tau^\pm\widetilde{\tau}_1^\mp$ are not kinematically allowed; the only allowed decays for the three lightest sleptons are $\widetilde{e}_R \rightarrow e\widetilde{G}$ and $\widetilde{\mu}_R \rightarrow \mu\widetilde{G}$ and $\widetilde{\tau}_1 \rightarrow \tau\widetilde{G}$. This situation is called the "slepton co-NLSP" scenario.

For larger values of $\tan\beta$, the lighter stau eigenstate $\widetilde{\tau}_1$ is more than 1.8 GeV lighter than $\widetilde{e}_R$ and $\widetilde{\mu}_R$ and $\widetilde{N}_1$. This means that the decays $\widetilde{N}_1 \rightarrow \tau\widetilde{\tau}_1$ and $\widetilde{e}_R \rightarrow e\tau\widetilde{\tau}_1$ and $\widetilde{\mu}_R \rightarrow \mu\tau\widetilde{\tau}_1$ are open. Then $\widetilde{\tau}_1$ is the sole NLSP, with all other MSSM supersymmetric particles having kinematically allowed decays into it. This is called the "stau NLSP" scenario.

In any case, a slepton NLSP can decay like $\widetilde{\ell} \rightarrow \ell\widetilde{G}$ according to eq. (1.239), with a width and decay length just given by eqs. (8.364) and (8.365) with the replacements $\kappa_{1\gamma} \rightarrow 1$ and $m_{\widetilde{N}_1} \rightarrow m_{\widetilde{\ell}}$. So, as for the neutralino NLSP case, the decay $\widetilde{\ell} \rightarrow \ell\widetilde{G}$ can be either fast or very slow, depending on the scale of supersymmetry breaking.

If $\sqrt{\langle F \rangle}$ is larger than roughly $10^3$ TeV (or the gravitino is heavier than a keV or so), then the NLSP is so long-lived that it will usually escape a typical collider detector. If $\widetilde{N}_1$ is the NLSP, then, it might as well be the LSP from the point of view of collider physics. However, the decay of $\widetilde{N}_1$ into the gravitino is still important for cosmology, since an unstable $\widetilde{N}_1$ is clearly not a good dark matter candidate while the gravitino LSP conceivably could be. On the other hand, if the NLSP is a long-lived charged slepton, then one can see its tracks (or possibly decay kinks) inside a collider detector.[144] The presence of a massive charged NLSP can be established by measuring an anomalously long time-of-flight or high ionization rate for a track in the detector.

## 1.9. Concluding Remarks

In this primer, I have attempted to convey some of the more essential features of supersymmetry as it is understood so far. One of the most amazing qualities of supersymmetry is that so much is known about it already, despite the present lack of direct experimental data. Even the terms and stakes of many of the important outstanding questions, especially the paramount issue "How is supersymmetry broken?", are already rather clear. That this can be so is a testament to the unreasonably predictive quality of the symmetry itself.

We have seen that sensible and economical models for supersymmetry at the TeV scale can be used as convenient templates for experimental searches. Two of the simplest and most popular possibilities are the "minimal supergravity" scenario with new parameters $m_0^2$, $m_{1/2}$, $A_0$, $\tan\beta$ and $\text{Arg}(\mu)$, and the "gauge-mediated" scenario with new parameters $\Lambda$, $M_{\text{mess}}$, $N_5$, $\langle F \rangle$, $\tan\beta$, and $\text{Arg}(\mu)$. However, one should not lose sight of the fact that the only indispensable idea of supersymmetry is simply that of a symmetry between fermions and bosons. Nature may or may not be kind enough to realize this beautiful idea within one of the specific frameworks that have already been explored well by theorists.

The experimental verification of supersymmetry will not be an end, but rather a revolution in high energy physics. It seems likely to present us with questions and challenges that we can only guess at presently. The measurement of sparticle masses, production cross-sections, and decay modes will rule out some models for supersymmetry breaking and lend credence to others. We will be able to test the principle of $R$-parity conservation, the idea that supersymmetry has something to do with the dark matter, and possibly make connections to other aspects of cosmology including baryogenesis and inflation. Other fundamental questions, like the origin of the

$\mu$ parameter and the rather peculiar hierarchical structure of the Yukawa couplings may be brought into sharper focus with the discovery of the MSSM spectrum. Understanding the precise connection of supersymmetry to the electroweak scale will surely open the window to even deeper levels of fundamental physics.

## Appendix: Non-Renormalizable Supersymmetric Lagrangians

In section 1.3, we discussed only renormalizable supersymmetric Lagrangians. However, like all known theories that include general relativity, supergravity is non-renormalizable as a quantum field theory. It is therefore clear that non-renormalizable interactions must be present in any low-energy effective description of the MSSM. Fortunately, these can be neglected for most phenomenological purposes, because non-renormalizable interactions have couplings of negative mass dimension, proportional to powers of $1/M_{\rm P}$ (or perhaps $1/\Lambda_{\rm UV}$, where $\Lambda_{\rm UV}$ is some other cutoff scale associated with new physics). This means that their effects at energy scales $E$ ordinarily accessible to experiment are typically suppressed by powers of $E/M_{\rm P}$ (or by powers of $E/\Lambda_{\rm UV}$). For energies $E \lesssim 1$ TeV, the consequences of non-renormalizable interactions are therefore usually far too small to be interesting.

Still, there are several reasons why one might be interested in non-renormalizable contributions to supersymmetric Lagrangians. First, some very rare processes (like proton decay) can only be described using an effective MSSM Lagrangian that includes non-renormalizable terms. Second, one may be interested in understanding physics at very high energy scales where the suppression associated with non-renormalizable terms is not enough to stop them from being important. For example, this could be the case in the study of the very early universe, or in understanding how additional gauge symmetries get broken. Third, the non-renormalizable interactions may play a crucial role in understanding how supersymmetry breaking is transmitted to the MSSM. Finally, it is sometimes useful to treat strongly coupled supersymmetric gauge theories using non-renormalizable effective Lagrangians, in the same way that chiral effective Lagrangians are used to study hadron physics in QCD. Unfortunately, we will not be able to treat these subjects in any sort of systematic way. Instead, we will merely sketch a few of the key elements that go into defining a non-renormalizable supersymmetric Lagrangian. More detailed treatments may be found for example in Refs. 12, 16, 18, 20, 26, 28, 29.

Let us consider a supersymmetric theory containing gauge and chiral supermultiplets whose Lagrangian may contain terms that are non-renormalizable. This includes supergravity as a special case, but applies more generally. It turns out that the part of the Lagrangian containing terms with up to two spacetime derivatives is completely determined by specifying three functions of the complex scalar fields (or more formally, of the chiral superfields). They are:

- The superpotential $W(\phi_i)$, which we have already encountered in the case of renormalizable supersymmetric Lagrangians. It must be an analytic function of the superfields treated as complex variables; in other words it depends only on the $\phi_i$ and not on the $\phi^{*i}$. It must be invariant under the gauge symmetries of the theory, and has dimensions of $[\text{mass}]^3$.
- The *Kähler potential* $K(\phi_i, \phi^{*i})$. Unlike the superpotential, the Kähler potential is a function of both $\phi_i$ and $\phi^{*i}$. It is gauge-invariant, real, and has dimensions of $[\text{mass}]^2$. In the special case of renormalizable theories, we did not have to discuss the Kähler potential explicitly, because at tree-level it is always $K = \phi^{i*}\phi_i$ (with $i$ summed over as usual).
- The *gauge kinetic function* $f_{ab}(\phi_i)$. Like the superpotential, it is an analytic function of the $\phi_i$ treated as complex variables. It is dimensionless and symmetric under interchange of its two indices $a, b$, which run over the adjoint representations of the gauge groups of the model. In the special case of renormalizable supersymmetric Lagrangians, it is just a constant (independent of the $\phi_i$), and is equal to the identity matrix divided by the gauge coupling squared: $f_{ab} = \delta_{ab}/g_a^2$. More generally, it also determines the non-renormalizable couplings of the gauge supermultiplets.

The whole Lagrangian with up to two derivatives can now be written down in terms of these. This is a non-trivial consequence of supersymmetry, because many different individual couplings in the Lagrangian are determined by the same three functions.

For example, in supergravity models, the part of the scalar potential that does not depend on the gauge kinetic function can be found as follows. First, one may define the real, dimensionless *Kähler function*:

$$G = K/M_{\text{P}}^2 + \ln(W/M_{\text{P}}^3) + \ln(W^*/M_{\text{P}}^3). \tag{A.1}$$

(Just to maximize the confusion, $G$ is also sometimes referred to as the Kähler potential. Also, many references use units with $M_{\rm P} = 1$, which simplifies the expressions but can slightly obscure the correspondence with the global supersymmetry limit of large $M_{\rm P}$.) From $G$, one can construct its derivatives with respect to the scalar fields and their complex conjugates: $G^i = \delta G/\delta\phi_i$; $G_i = \delta G/\delta\phi^{*i}$; and $G_i^j = \delta^2 G/\delta\phi^{*i}\delta\phi_j$. As in section 1.3.2, raised (lowered) indices $i$ correspond to derivatives with respect to $\phi_i$ ($\phi^{*i}$). Note that $G_i^j = K_i^j/M_{\rm P}^2$, which is sometimes called the Kähler metric, does not depend on the superpotential. The inverse of this matrix is denoted $(G^{-1})_i^j$, or equivalently $M_{\rm P}^2(K^{-1})_i^j$, so that $(G^{-1})_i^k G_k^j = (G^{-1})_k^j G_i^k = \delta_i^j$. In terms of these objects, the generalization of the $F$-term contribution to the scalar potential in ordinary renormalizable global supersymmetry turns out to be, after a complicated derivation:[146,147]

$$V_F = M_{\rm P}^4\, e^G \Big[G^i (G^{-1})_i^j G_j - 3\Big] \tag{A.2}$$

in supergravity. It can be rewritten as

$$V_F = K_i^j F_j F^{*i} - 3e^{K/M_{\rm P}^2} WW^*/M_{\rm P}^2, \tag{A.3}$$

where

$$F_i = -M_{\rm P}^2\, e^{G/2}\, (G^{-1})_i^j G_j = -e^{K/2M_{\rm P}^2}\, (K^{-1})_i^j \Big(W_j^* + W^* K_j/M_{\rm P}^2\Big), \tag{A.4}$$

with $K^i = \delta K/\delta\phi_i$ and $K_j = \delta K/\delta\phi^{*j}$. The $F_i$ are order parameters for supersymmetry breaking in supergravity (generalizing the auxiliary fields in the renormalizable global supersymmetry case). In other words, local supersymmetry will be broken if one or more of the $F_i$ obtain a VEV. The gravitino then absorbs the would-be goldstino and obtains a squared mass

$$m_{3/2}^2 = \langle K_j^i F_i F^{*j}\rangle/3M_{\rm P}^2. \tag{A.5}$$

Now, assuming a minimal Kähler potential $K = \phi^{*i}\phi_i$, then $K_i^j = (K^{-1})_i^j = \delta_i^j$, so that expanding eqs. (A.3) and (A.4) to lowest order in $1/M_{\rm P}$ just reproduces the results $F_i = -W_i^*$ and $V = F_i F^{*i} = W^i W_i^*$, which were found in section 1.3.2 for renormalizable global supersymmetric theories [see eqs. (1.108)-(1.111)]. Equation (A.5) also reproduces the expression for the gravitino mass that was quoted in eq. (1.238).

The scalar potential eq. (A.2) does not include the $D$-term contributions from gauge interactions, which are given by

$$V_D = \frac{1}{2}{\rm Re}[f_{ab}^{-1}\, \widehat{D}_a \widehat{D}_b], \tag{A.6}$$

where

$$\widehat{D}^a \equiv -G^i(T^a)_i{}^j\phi_j = -\phi^{*j}(T^a)_j{}^iG_i = -K^i(T^a)_i{}^j\phi_j = -\phi^{*j}(T^a)_j{}^iK_i, \tag{A.7}$$

are real order parameters of supersymmetry breaking, with the last three equalities following from the gauge invariance of $W$ and $K$. The full scalar potential is

$$V = V_F + V_D, \tag{A.8}$$

and it depends on $W$ and $K$ only through the combination $G$ in eq. (A.1). There are many other contributions to the supergravity Lagrangian, which also turn out to depend only on $G$ and $f_{ab}$, and can be found in Ref. 146, 147. This allows one to consistently redefine $W$ and $K$ so that there are no purely holomorphic or purely anti-holomorphic terms appearing in the latter.

Note that in the tree-level global supersymmetry case $f_{ab} = \delta_{ab}/g_a^2$ and $K^i = \phi^{*i}$, eq. (A.6) reproduces the result of section 1.3.4 for the renormalizable global supersymmetry $D$-term scalar potential, with $\widehat{D}^a = D^a/g^a$ being the $D$-term order parameter for supersymmetry breaking.

Unlike in the case of global supersymmetry, the scalar potential in supergravity is *not* necessarily non-negative, because of the $-3$ term in eq. (A.2). Therefore, in principle, one can have supersymmetry breaking with a positive, negative, or zero vacuum energy. Recent developments in experimental cosmology[220] imply a positive vacuum energy associated with the acceleration of the scale factor of the observable universe,

$$\rho_{\rm vac}^{\rm observed} = \frac{\Lambda}{8\pi G_{\rm Newton}} \approx (2.3 \times 10^{-12}\ {\rm GeV})^4, \tag{A.9}$$

but this is also certainly tiny compared to the scales associated with supersymmetry breaking. Therefore, it is tempting to simply assume that the vacuum energy is 0 within the approximations pertinent for working out the supergravity effects on particle physics at high energies. However, it is notoriously unclear *why* the terms in the scalar potential in a supersymmetry-breaking vacuum should conspire to give $\langle V\rangle \approx 0$ at the minimum. A naive estimate, without miraculous cancellations, would give instead $\langle V\rangle$ of order $|\langle F\rangle|^2$, so at least roughly $(10^{10}\ {\rm GeV})^4$ for Planck-scale mediated supersymmetry breaking, or $(10^4\ {\rm GeV})^4$ for Gauge-mediated supersymmetry breaking. Furthermore, while $\rho_{\rm vac} = \langle V\rangle$ classically, the former is a very large-distance scale measured quantity, while the latter is associated with effective field theories at length scales comparable to and shorter than those familiar to high energy physics. So, in the absence of a compelling explanation for the tiny value of $\rho_{\rm vac}$, it is not at all clear that $\langle V\rangle \approx 0$ is really the right condition to impose.[221] Nevertheless, with $\langle V\rangle = 0$ imposed as a constraint,[ee] eqs. (A.3)-(A.5) tell us that $\langle K^i_j F_i F^{*j}\rangle = 3M_{\rm P}^4 e^{\langle G\rangle} = 3e^{\langle K\rangle/M_{\rm P}^2}|\langle W\rangle|^2/M_{\rm P}^2$, and an equivalent formula for the gravitino mass is therefore $m_{3/2} = e^{\langle G\rangle/2}M_{\rm P}$.

An instructive special case arises if we assume a minimal Kähler potential and divide the fields $\phi_i$ into a visible sector including the MSSM fields $\varphi_i$, and a hidden sector containing a field $X$ that breaks supersymmetry for us (and other fields that we need not treat explicitly). In other words, suppose that the superpotential and the Kähler potential have the form

$$W = W_{\rm vis}(\varphi_i) + W_{\rm hid}(X), \tag{A.10}$$

$$K = \varphi^{*i}\varphi_i + X^*X. \tag{A.11}$$

Now let us further assume that the dynamics of the hidden sector fields provides non-zero VEVs

$$\langle X\rangle = xM_{\rm P}, \qquad \langle W_{\rm hid}\rangle = wM_{\rm P}^2, \qquad \langle \delta W_{\rm hid}/\delta X\rangle = w' M_{\rm P}, \tag{A.12}$$

which define a dimensionless quantity $x$, and $w$, $w'$ with dimensions of [mass]. Requiring $\langle V\rangle = 0$ yields $|w' + x^* w|^2 = 3|w|^2$, and

$$m_{3/2} \;=\; |\langle F_X\rangle|/\sqrt{3}M_{\rm P} \;=\; e^{|x|^2/2}|w|. \tag{A.13}$$

Now we suppose that it is valid to expand the scalar potential in powers of the dimensionless quantities $w/M_{\rm P}$, $w'/M_{\rm P}$, $\varphi_i/M_{\rm P}$, etc., keeping only terms that depend on the visible sector fields $\varphi_i$. It is not a difficult exercise to show that in leading order the result is:

$$\begin{aligned} V = {} & (W^*_{\rm vis})_i (W_{\rm vis})^i + m_{3/2}^2 \varphi^{*i}\varphi_i \\ & + e^{|x|^2/2}\left[w^*\varphi_i (W_{\rm vis})^i + (x^* w'^* + |x|^2 w^* - 3w^*)W_{\rm vis} + {\rm c.c.}\right]. \end{aligned} \tag{A.14}$$

A tricky point here is that we have rescaled the visible sector superpotential $W_{\rm vis} \rightarrow e^{-|x|^2/2}W_{\rm vis}$ everywhere, in order that the first term in eq. (A.14) is the usual, properly normalized, $F$-term contribution in global supersymmetry. The next term is a universal soft scalar squared mass of the form eq. (1.246) with

$$m_0^2 \;=\; |\langle F_X\rangle|^2/3M_{\rm P}^2 \;=\; m_{3/2}^2. \tag{A.15}$$

[ee] We do this only to follow popular example; as just noted we cannot endorse this imposition.

The second line of eq. (A.14) just gives soft (scalar)$^3$ and (scalar)$^2$ analytic couplings of the form eqs. (1.247) and (1.248), with

$$A_0 = -x^* \langle F_X \rangle / M_{\rm P}, \qquad B_0 = \Bigl( \frac{1}{x + w'^*/w^*} - x^* \Bigr) \langle F_X \rangle / M_{\rm P} \quad (A.16)$$

since $\varphi_i (W_{\rm vis})^i$ is equal to $3W_{\rm vis}$ for the cubic part of $W_{\rm vis}$, and to $2W_{\rm vis}$ for the quadratic part. [If the complex phases of $x$, $w$, $w'$ can be rotated away, then eq. (A.16) implies $B_0 = A_0 - m_{3/2}$, but there are many effects that can ruin this prediction.] The Polonyi model mentioned in section 1.6.6 is just the special case of this exercise in which $W_{\rm hid}$ is assumed to be linear in $X$.

However, there is no particular reason why $W$ and $K$ must have the simple form eq. (A.10) and eq. (A.11). In general, the superpotential can be expanded like

$$W = W_{\rm ren} + \frac{1}{M_{\rm P}} w^{ijkn} \phi_i \phi_j \phi_k \phi_n + \frac{1}{M_{\rm P}^2} w^{ijknm} \phi_i \phi_j \phi_k \phi_n \phi_m + \ldots \quad (A.17)$$

where $W_{\rm ren}$ is the renormalizable superpotential with terms up to $\phi^3$. Similarly, the Kähler potential can be expanded like

$$\begin{aligned} K = \phi_i \phi^{*i} &+ \frac{1}{M_{\rm P}} \bigl( k^{ij}_k \phi_i \phi_j \phi^{*k} + {\rm c.c.} \bigr) \\ &+ \frac{1}{M_{\rm P}^2} \bigl( k^{ij}_{kn} \phi_i \phi_j \phi^{*k} \phi^{*n} + k^{ijk}_n \phi_i \phi_j \phi_k \phi^{*n} + {\rm c.c.} \bigr) + \ldots, \end{aligned} \quad (A.18)$$

where terms in $K$ that are analytic in $\phi$ (and $\phi^*$) are assumed to have been absorbed into $W$ (and $W^*$), as explained above. The form of the first term is dictated by the requirement of canonical kinetic terms for the chiral supermultiplet fields. If one now plugs eqs. (A.17) and (A.18) with arbitrary hidden sector fields and VEVs into eq. (A.2), one obtains a general form like eq. (1.242) for the soft terms. It is only when special assumptions are made [like eqs. (A.10), (A.11)] that one gets the phenomenologically desirable results in eqs. (1.244)-(1.248). Thus supergravity by itself does not guarantee universality of the soft terms. Furthermore, there is no guarantee that expansions in $1/M_{\rm P}$ of the form given above are valid or appropriate. In superstring models, the dilaton and moduli fields have Kähler potential terms proportional to $M_{\rm P}^2 \ln[(\phi + \phi^*)/M_{\rm P}]$. (The moduli are massless fields that do not appear in the tree-level perturbative superpotential. The dilaton is a special modulus field whose VEV determines the gauge couplings in the theory.)

Gaugino masses arise from non-renormalizable terms through a non-minimal gauge kinetic function $f_{ab}$. Expanding it in powers of $1/M_{\rm P}$ as

$$f_{ab} = \delta_{ab}\Big[1/g_a^2 + f_a^i \phi_i/M_{\rm P} + \ldots\Big], \tag{A.19}$$

it is possible to show that the gaugino mass induced by supersymmetry breaking is

$$m_{\lambda^a} = {\rm Re}[f_a^i]\langle F_i \rangle / 2M_{\rm P}. \tag{A.20}$$

The assumption of universal gaugino masses therefore follows if the dimensionless quantities $f_a^i$ are the same for each of the three MSSM gauge groups; this is automatic in certain GUT and superstring-inspired models, but not in general.

Finally, let us mention how gaugino condensates can provide supersymmetry breaking in supergravity models. This again requires that the gauge kinetic function has a non-trivial dependence on the scalar fields, as in eq. (A.19). Then eq. (A.4) is modified to

$$F_i = -M_{\rm P}^2\, e^{G/2}\, (G^{-1})_i^j G_j - \frac{1}{4}(K^{-1})_i^j \frac{\partial f_{ab}^*}{\partial \phi^{*j}} \lambda^a \lambda^b + \ldots. \tag{A.21}$$

Now if there is a gaugino condensate $\langle \lambda^a \lambda^b \rangle = \delta^{ab}\Lambda^3$ and $\langle (K^{-1})_j^i \partial f_{ab}/\partial \phi_j \rangle \sim 1/M_{\rm P}$, then $|\langle F_i \rangle| \sim \Lambda^3/M_{\rm P}$. Then as above, the non-vanishing $F$-term gives rise to soft parameters of order $m_{\rm soft} \sim |\langle F_i \rangle|/M_{\rm P} \sim \Lambda^3/M_{\rm P}^2$, as in eq. (1.233).

## Acknowledgments

I am thankful to Gordy Kane and to James Wells for many helpful comments and suggestions on this primer. I am also indebted to my other collaborators on supersymmetry and related matters, Sandro Ambrosanio, Nima Arkani-Hamed, Diego Castaño, Ray Culbertson, Yanou Cui, Michael Dine, Manuel Drees, Herbi Dreiner, Tony Gherghetta, Howie Haber, Ian Jack, Tim Jones, Chris Kolda, Graham Kribs, Stefano Moretti, David Morrissey, Steve Mrenna, Jian-ming Qian, Dave Robertson, Scott Thomas, Kazuhiro Tobe, Mike Vaughn, Graham Wilson, Youichi Yamada, James Younkin, and especially Pierre Ramond, for many illuminating and inspiring conversations. Corrections to previous versions have been provided by Daniel Arnold, Jorge de Blas, Gudrun Hiller, Graham Kribs, Bob McElrath, Verónica Sanz, Shufang Su, John Terning, Keith Thomas,

Scott Thomas, and Sean Tulin. I will be grateful to receive further corrections at spmartin@niu.edu, and a list of them will be maintained at http://zippy.physics.niu.edu/primer.html. I thank the Aspen Center for Physics and the Stanford Linear Accelerator Center for their hospitality, and the students of PHYS 686 at NIU in Spring 2004, the 2005 ICTP Summer School on Particle Physics and the 2008 CERN/Fermilab Hadron Collider Physics Summer School for asking interesting questions. This work was supported in part by the U.S. Department of Energy, and by National Science Foundation grants PHY-9970691, PHY-0140129, PHY-0456635, and PHY-0757325.

## References

1. S. Weinberg, Phys. Rev. D **13**, 974 (1976), Phys. Rev. D **19**, 1277 (1979); E. Gildener, Phys. Rev. D **14**, 1667 (1976); L. Susskind, Phys. Rev. D **20**, 2619 (1979); G. 't Hooft, in *Recent developments in gauge theories*, Proceedings of the NATO Advanced Summer Institute, Cargese 1979, (Plenum, 1980).
2. D.A. Eliezer and R.P. Woodard, Nucl. Phys. B **325**, 389 (1989).
3. S. Dimopoulos and S. Raby, Nucl. Phys. B **192**, 353 (1981); E. Witten, Nucl. Phys. B **188**, 513 (1981); M. Dine, W. Fischler and M. Srednicki, Nucl. Phys. B **189**, 575 (1981); S. Dimopoulos and H. Georgi, Nucl. Phys. B **193**, 150 (1981); N. Sakai, Z. Phys. C **11**, 153 (1981); R.K. Kaul and P. Majumdar, Nucl. Phys. B **199**, 36 (1982).
4. S. Coleman and J. Mandula, Phys. Rev. **159** (1967) 1251; R. Haag, J. Lopuszanski, and M. Sohnius, Nucl. Phys. B **88**, 257 (1975).
5. P. Fayet, Phys. Lett. B **64**, 159 (1976).
6. P. Fayet, Phys. Lett. B **69**, 489 (1977), Phys. Lett. B **84**, 416 (1979).
7. G.R. Farrar and P. Fayet, Phys. Lett. B **76**, 575 (1978).
8. P. Ramond, Phys. Rev. D **3**, 2415 (1971); A. Neveu and J.H. Schwarz, Nucl. Phys. B **31**, 86 (1971); J.L. Gervais and B. Sakita, Nucl. Phys. B **34**, 632 (1971).
9. Yu. A. Gol'fand and E. P. Likhtman, JETP Lett. **13**, 323 (1971).
10. J. Wess and B. Zumino, Nucl. Phys. B **70** (1974) 39.
11. D.V. Volkov and V.P. Akulov, Phys. Lett. B **46**, 109 (1973).
12. J. Wess and J. Bagger, *Supersymmetry and Supergravity*, (Princeton Univ. Press, 1992).
13. G.G. Ross, *Grand Unified Theories*, (Addison-Wesley, 1985).
14. P.P. Srivastava, *Supersymmetry and Superfields and Supergravity; an Introduction*, (Adam-Hilger, 1986).
15. P.G.O. Freund, *Introduction to Supersymmetry*, (Cambridge University Press, 1986).
16. P.C. West, *Introduction to Supersymmetry and Supergravity*, (World Scientific, 1990).

17. R.N. Mohapatra, *Unification and Supersymmetry: The Frontiers of Quark-Lepton Physics*, Springer-Verlag, New York 1992.
18. D. Bailin and A. Love, *Supersymmetric Gauge Field Theory and String Theory*, (Institute of Physics Publishing, 1994).
19. P. Ramond, *Journeys Beyond the Standard Model*, (Frontiers in Physics, Perseus Books 1999).
20. S. Weinberg, *The Quantum Theory of Fields, Vol. 3: Supersymmetry*, (Cambridge University Press, 2000).
21. M. Drees, R. Godbole and P. Roy, *Theory and Phenomenology of Sparticles*, (World Scientific, 2004).
22. H. Baer and X. Tata, *Weak Scale Supersymmetry*, (Cambridge University Press, 2006).
23. P. Binetruy, *Supersymmetry*, (Oxford University Press, 2006).
24. J. Terning, *Modern Supersymmetry: Dynamics and Duality* (Oxford University Press, 2006).
25. H.E. Haber and G.L. Kane, Phys. Rept. **117**, 75 (1985).
26. H.P. Nilles, Phys. Rept. **110**, 1 (1984).
27. M.F. Sohnius, Phys. Rept. **128**, 39 (1985).
28. S.J. Gates, M.T. Grisaru, M. Rocek and W. Siegel, *Superspace or One Thousand and One Lessons in Supersymmetry*, [hep-th/0108200].
29. P. Van Nieuwenhuizen, Phys. Rept. **68** (1981) 189.
30. R. Arnowitt, A. Chamseddine and P. Nath, *N=1 Supergravity*, (World Scientific, 1984).
31. D.R.T. Jones, "Supersymmetric gauge theories", in *TASI Lectures in Elementary Particle Physics 1984*, ed. D.N. Williams, TASI publications, Ann Arbor 1984.
32. H.E. Haber, "Introductory low-energy supersymmetry", TASI-92 lectures, [hep-ph/9306207].
33. P. Ramond, "Introductory lectures on low-energy supersymmetry", TASI-94 lectures, [hep-th/9412234].
34. J.A. Bagger, "Weak-scale supersymmetry: theory and practice", TASI-95 lectures, [hep-ph/9604232].
35. M.E. Peskin, "Duality in supersymmetric Yang-Mills theory," TASI-96 lectures, [hep-th/9702094].
36. H. Baer et al., "Low energy supersymmetry phenomenology", in *Report of the Working Group on Electroweak Symmetry Breaking and New Physics* of the 1995 study of the future of particle physics in the USA, (World Scientific, 1996), [hep-ph/9503479],
37. M. Drees and S.P. Martin, "Implications of SUSY model building", as in,[36] [hep-ph/9504324].
38. J.D. Lykken, "Introduction to Supersymmetry", TASI-96 lectures, [hep-th/9612114].
39. S. Dawson, "SUSY and such", Lectures given at NATO Advanced Study Institute on Techniques and Concepts of High-energy Physics, [hep-ph/9612229].

40. M. Dine, "Supersymmetry Phenomenology (With a Broad Brush)", [hep-ph/9612389].
41. J.F. Gunion, "A simplified summary of supersymmetry", [hep-ph/9704349].
42. M.A. Shifman, "Nonperturbative dynamics in supersymmetric gauge theories", Prog. Part. Nucl. Phys. **39**, 1 (1997) [hep-th/9704114].
43. X. Tata, "What is supersymmetry and how do we find it?", [hep-ph/9706307].
44. M.A. Luty, "2004 TASI lectures on supersymmetry breaking," [hep-th/0509029].
45. I. Aitchison,"Supersymmetry and the MSSM: an elementary introduction," [hep-ph/0505105].
46. D.J.H. Chung et al., Phys. Rept. **407**, 1 (2005) [hep-ph/0312378].
47. J. Terning, "Non-perturbative supersymmetry," TASI 2002 lectures [hep-th/0306119].
48. H. Murayama, "Supersymmetry phenomenology," ICTP Summer School Lectures 1999, [hep-ph/0002232].
49. S. Ferrara, editor, *Supersymmetry*, (World Scientific, 1987).
50. A. Salam and J.A. Strathdee, Nucl. Phys. B **76**, 477 (1974); S. Ferrara, J. Wess and B. Zumino, Phys. Lett. B **51**, 239 (1974). See[12] for a pedagogical introduction to superfields and superspace.
51. J. Wess and B. Zumino, Phys. Lett. B **49**, 52 (1974); J. Iliopoulos and B. Zumino, Nucl. Phys. B **76**, 310 (1974).
52. J. Wess and B. Zumino, Nucl. Phys. B **78**, 1 (1974).
53. L. Girardello and M.T. Grisaru Nucl. Phys. B **194**, 65 (1982).
54. J. Polchinski and L. Susskind, Phys. Rev. D **26**, 3661 (1982).
55. D.R.T. Jones, L. Mezincescu and Y.P. Yao, Phys. Lett. B **148**, 317 (1984);
56. L.J. Hall and L. Randall, Phys. Rev. Lett. **65**, 2939 (1990).
57. M. Dine and D. MacIntire, Phys. Rev. D **46**, 2594 (1992) [hep-ph/9205227].
58. I. Jack and D.R.T. Jones, Phys. Lett. B **457**, 101 (1999) [hep-ph/9903365].
59. P.J. Fox, A.E. Nelson and N. Weiner, JHEP **0208**, 035 (2002) [hep-ph/0206096].
60. Z. Chacko, P.J. Fox and H. Murayama, Nucl. Phys. B **706**, 53 (2005) [hep-ph/0406142].
61. I. Antoniadis, K. Benakli, A. Delgado and M. Quiros, Adv. Stud. Theor. Phys. **2**, 645 (2008) [hep-ph/0610265].
62. G.D. Kribs, E. Poppitz and N. Weiner, "Flavor in Supersymmetry with an Extended R-symmetry," [hep-ph/0712.2039].
63. A.E. Blechman and S.P. Ng, JHEP **0806**, 043 (2008) [hep-ph/0803.3811].
64. S.D.L. Amigo, A.E. Blechman, P.J. Fox and E. Poppitz, "R-symmetric gauge mediation," [hep-ph/0809.1112].
65. T. Plehn and T.M.P. Tait, "Seeking Sgluons," [hep-ph/0810.3919].
66. J.E. Kim and H. P. Nilles, Phys. Lett. B **138**, 150 (1984); J.E. Kim and H. P. Nilles, Phys. Lett. B **263**, 79 (1991); E.J. Chun, J.E. Kim and H.P. Nilles, Nucl. Phys. B **370**, 105 (1992).
67. G.F. Giudice and A. Masiero, Phys. Lett. B **206**, 480 (1988); J.A. Casas and C. Muñoz, Phys. Lett. B **306**, 288 (1993) [hep-ph/9302227].

68. G. Dvali, G.F. Giudice and A. Pomarol, Nucl. Phys. B **478**, 31 (1996) [hep-ph/9603238].
69. F. Zwirner, Phys. Lett. B **132**, 103 (1983); S. Dawson, Nucl. Phys. B **261**, 297 (1985); R. Barbieri and A. Masiero, Nucl. Phys. B **267**, 679 (1986); S. Dimopoulos and L. Hall, Phys. Lett. B **207**, 210 (1988); V. Barger, G. Giudice, and T. Han, Phys. Rev. D **40**, 2987 (1989); R. Godbole, P. Roy and X. Tata, Nucl. Phys. B **401**, 67 (1993) [hep-ph/9209251]; G. Bhattacharyya and D. Choudhury, Mod. Phys. Lett. A **10**, 1699 (1995) [hep-ph/9503263];
70. For reviews, see G. Bhattacharyya, Nucl. Phys. Proc. Suppl. **52A**, 83 (1997) [hep-ph/9608415]; H.K. Dreiner, "An introduction to explicit R-parity violation," in *Perspectives on Supersymmetry*, ed. G.L. Kane (World Scientific, 1998), [hep-ph/9707435]; R. Barbier *et al.*, "Report of the GDR working group on the R-parity violation," [hep-ph/9810232]; B. Allanach *et al.*, "Searching for R-parity violation at Run-II of the Tevatron," [hep-ph/9906224]; B. Allanach, A. Dedes and H.K. Dreiner, Phys. Rev. D **69**, 115002 (2004) [hep-ph/0309196]; M. Chemtob, Prog. Part. Nucl. Phys. **54**, 71 (2005) [hep-ph/0406029].
71. G. 't Hooft, Phys. Rev. Lett. **37**, 8 (1976).
72. S. Dimopoulos and H. Georgi, Nucl. Phys. B **193**, 150 (1981); S. Weinberg, Phys. Rev. D **26**, 287 (1982); N. Sakai and T. Yanagida, Nucl. Phys. B **197**, 533 (1982); S. Dimopoulos, S. Raby and F. Wilczek, Phys. Lett. B **112**, 133 (1982).
73. H. Goldberg, Phys. Rev. Lett. **50**, 1419 (1983); J. Ellis, J. Hagelin, D.V. Nanopoulos, K. Olive, and M. Srednicki, Nucl. Phys. B **238**, 453 (1984).
74. L. Krauss and F. Wilczek, Phys. Rev. Lett. **62**, 1221 (1989).
75. L.E. Ibáñez and G. Ross, Phys. Lett. B **260**, 291 (1991); T. Banks and M. Dine, Phys. Rev. D **45**, 1424 (1992) [hep-th/9109045]; L.E. Ibáñez, Nucl. Phys. B **398**, 301 (1993) [hep-ph/9210211].
76. R.N. Mohapatra, Phys. Rev. D **34**, 3457 (1986); A. Font, L.E. Ibáñez and F. Quevedo, Phys. Lett. B **228**, 79 (1989);
77. S.P. Martin Phys. Rev. D **46**, 2769 (1992) [hep-ph/9207218], Phys. Rev. D **54**, 2340 (1996) [hep-ph/9602349].
78. R. Kuchimanchi and R.N. Mohapatra, Phys. Rev. D **48**, 4352 (1993) [hep-ph/9306290], Phys. Rev. Lett. **75**, 3989 (1995) [hep-ph/9509256]; C.S. Aulakh, K. Benakli and G. Senjanović, Phys. Rev. Lett. **79**, 2188 (1997) [hep-ph/9703434]; C.S. Aulakh, A. Melfo and G. Senjanović, Phys. Rev. D **57**, 4174 (1998) [hep-ph/9707256].
79. S. Dimopoulos and D. Sutter, Nucl. Phys. B **452**, 496 (1995) [hep-ph/9504415].
80. J. Ellis and D.V. Nanopoulos, Phys. Lett. B **110**, 44 (1982); R. Barbieri and R. Gatto, Phys. Lett. B **110**, 211 (1982); B.A. Campbell, Phys. Rev. D **28**, 209 (1983).
81. M.J. Duncan, Nucl. Phys. B **221**, 285 (1983); J.F. Donahue, H.P. Nilles and D. Wyler, Phys. Lett. B **128**, 55 (1983); A. Bouquet, J. Kaplan and

C.A. Savoy, Phys. Lett. B **148**, 69 (1984); M. Dugan, B. Grinstein and L.J. Hall, Nucl. Phys. B **255**, 413 (1985); F. Gabbiani and A. Masiero, Nucl. Phys. B **322**, 235 (1989); J. Hagelin, S. Kelley and T. Tanaka, Nucl. Phys. B **415**, 293 (1994).
82. L.J. Hall, V.A. Kostalecky and S. Raby, Nucl. Phys. B **267**, 415 (1986); F. Gabbiani and A. Masiero, Phys. Lett. B **209**, 289 (1988); R. Barbieri and L.J. Hall, Phys. Lett. B **338**, 212 (1994) [hep-ph/9408406]; R. Barbieri, L.J. Hall and A. Strumia, Nucl. Phys. B **445**, 219 (1995) [hep-ph/9501334].
83. J. Ellis, S. Ferrara and D.V. Nanopoulos, Phys. Lett. B **114**, 231 (1982); W. Buchmüller and D. Wyler, Phys. Lett. B **121**, 321 (1983); J. Polchinski and M.B. Wise, Phys. Lett. B **125**, 393 (1983); F. del Aguila, M.B. Gavela, J.A. Grifols and A. Méndez, Phys. Lett. B **126**, 71 (1983) [Erratum-ibid. B **129**, 473 (1983)]; D.V. Nanopoulos and M. Srednicki, Phys. Lett. B **128**, 61 (1983).
84. S. Bertolini, F. Borzumati, A. Masiero and G. Ridolfi, Nucl. Phys. B **353**, 591 (1991); R. Barbieri and G.F. Giudice, Phys. Lett. B **309**, 86 (1993) [hep-ph/9303270].
85. J. Hisano et al., Phys. Lett. B **357**, 579 (1995) [hep-ph/9501407].
86. F. Gabbiani, E. Gabrielli, A. Masiero and L. Silvestrini, Nucl. Phys. B **477**, 321 (1996) [hep-ph/9604387].
87. J.L. Hewett and J.D. Wells, Phys. Rev. D **55**, 5549 (1997) [hep-ph/9610323].
88. Y. Grossman, Y. Nir and R. Rattazzi, [hep-ph/9701231].
89. S. Pokorski, J. Rosiek and C. A. Savoy, Nucl. Phys. B **570**, 81 (2000) [hep-ph/9906206];
90. S. Abel, S. Khalil and O. Lebedev, Nucl. Phys. B **606**, 151 (2001) [hep-ph/0103320].
91. M. Misiak, S. Pokorski and J. Rosiek, "Supersymmetry and FCNC effects," [hep-ph/9703442].
92. J.A. Bagger, K.T. Matchev and R.J. Zhang, Phys. Lett. B **412**, 77 (1997) [hep-ph/9707225].
93. Y. Nir and M.P. Worah, Phys. Lett. B **423**, 319 (1998) [hep-ph/9711215].
94. M. Ciuchini et al., Nucl. Phys. B **523**, 501 (1998) [hep-ph/9711402].
95. M. Ciuchini et al., JHEP **9810**, 008 (1998) [hep-ph/9808328].
96. A. Masiero and H. Murayama, Phys. Rev. Lett. **83**, 907 (1999) [hep-ph/9903363].
97. S. Khalil, T. Kobayashi and A. Masiero, Phys. Rev. D **60**, 075003 (1999) [hep-ph/9903544].
98. A.J. Buras et al., Nucl. Phys. B **566**, 3 (2000) [hep-ph/9908371].
99. F. Borzumati, C. Greub, T. Hurth and D. Wyler, Phys. Rev. D **62**, 075005 (2000) [hep-ph/9911245].
100. A.J. Buras, P. Gambino, M. Gorbahn, S. Jager and L. Silvestrini, Nucl. Phys. B **592**, 55 (2001) [hep-ph/0007313].
101. P.H. Chankowski and L. Slawianowska, Phys. Rev. D **63**, 054012 (2001) [hep-ph/0008046].
102. T. Besmer, C. Greub and T. Hurth, Nucl. Phys. B **609**, 359 (2001) [hep-ph/0105292].

103. G. Burdman, E. Golowich, J. Hewett and S. Pakvasa, Phys. Rev. D **66**, 014009 (2002) [hep-ph/0112235].
104. A.J. Buras, P.H. Chankowski, J. Rosiek and L. Slawianowska, Nucl. Phys. B **659**, 3 (2003) [hep-ph/0210145].
105. M. Ciuchini, E. Franco, A. Masiero and L. Silvestrini, Phys. Rev. D **67**, 075016 (2003) [Erratum-ibid. D **68**, 079901 (2003)] [hep-ph/0212397].
106. M.L. Brooks *et al.* [MEGA Collaboration], Phys. Rev. Lett. **83**, 1521 (1999) [hep-ex/9905013], Phys. Rev. D **65**, 112002 (2002) [hep-ex/0111030].
107. A.G. Cohen, D.B. Kaplan and A.E. Nelson, Phys. Lett. B **388**, 588 (1996) [hep-ph/9607394].
108. J.D. Wells, "Implications of supersymmetry breaking with a little hierarchy between gauginos and scalars," [hep-ph/0306127], Phys. Rev. D **71**, 015013 (2005) [hep-ph/0411041]. "Split supersymmetry", as in N. Arkani-Hamed and S. Dimopoulos, JHEP **0506**, 073 (2005) [hep-th/0405159]; G. F. Giudice and A. Romanino, Nucl. Phys. B **699**, 65 (2004) [Erratum-ibid. B **706**, 65 (2005)] [hep-ph/0406088], abandons the motivation of supersymmetry as a solution to the hierarchy problem. The most logical and sublime refinement of this line of thinking is found in P.J. Fox *et al.*, "Supersplit supersymmetry," [hep-th/0503249].
109. Y. Nir and N. Seiberg, Phys. Lett. B **309**, 337 (1993) [hep-ph/9304307].
110. P. Langacker, "Precision Tests Of The Standard Model," in Proceedings of the PASCOS90 Symposium, (World Scientific, 1990); J. Ellis, S. Kelley, and D. Nanopoulos, Phys. Lett. B **260**, 131 (1991); U. Amaldi, W. de Boer, and H. Furstenau, Phys. Lett. B **260**, 447 (1991); P. Langacker and M. Luo, Phys. Rev. D **44**, 817 (1991); C. Giunti, C.W. Kim and U.W. Lee, Mod. Phys. Lett. A **6**, 1745 (1991);
111. W. Siegel, Phys. Lett. B **84**, 193 (1979); D.M. Capper, D.R.T. Jones and P. van Nieuwenhuizen, Nucl. Phys. B **167**, 479 (1980).
112. W. Siegel, Phys. Lett. B **94**, 37 (1980); L. V. Avdeev, G. A. Chochia and A. A. Vladimirov, Phys. Lett. B **105**, 272 (1981); L.V. Avdeev and A.A. Vladimirov, Nucl. Phys. B **219**, 262 (1983).
113. I. Jack and D.R.T. Jones, "Regularisation of supersymmetric theories", in *Perspectives on Supersymmetry*, ed. G.L. Kane (World Scientific, 1998), [hep-ph/9707278]
114. D. Stöckinger, JHEP **0503**, 076 (2005) [hep-ph/0503129]; A. Signer and D. Stöckinger, Phys. Lett. B **626**, 127 (2005) [hep-ph/0508203].
115. D. Evans, J.W. Moffat, G. Kleppe, and R.P. Woodard, Phys. Rev. D **43**, 499 (1991); G. Kleppe and R.P. Woodard, Phys. Lett. B **253**, 331 (1991), Nucl. Phys. B **388**, 81 (1992) [hep-th/9203016]; G. Kleppe, Phys. Lett. B **256**, 431 (1991).
116. V. Novikov, M. Shifman, A. Vainshtein and V. Zakharov, Nucl. Phys. B **229**, 381 (1983), Phys. Lett. B **166**, 329 (1986); J. Hisano and M. Shifman, Phys. Rev. D **56**, 5475 (1997) [hep-ph/9705417].
117. I. Antoniadis, C. Kounnas and K. Tamvakis, Phys. Lett. B **119**, 377 (1982); G.A. Schuler, S. Sakakibara and J.G. Korner, Phys. Lett. B **194**, 125 (1987); Y. Yamada, Phys. Lett. B **316**, 109 (1993) [hep-ph/9307217].

118. S.P. Martin and M.T. Vaughn, Phys. Lett. B **318**, 331 (1993) [hep-ph/9308222].
119. I. Jack, D.R.T. Jones and K.L. Roberts, Z. Phys. C **62**, 161 (1994) [hep-ph/9310301], Z. Phys. C **63**, 151 (1994) [hep-ph/9401349].
120. A. Salam and J. Strathdee, Phys. Rev. D **11**, 1521 (1975); M.T. Grisaru, W. Siegel and M. Rocek, Nucl. Phys. B **159**, 429 (1979).
121. D.R.T. Jones, Nucl. Phys. B **87**, 127 (1975); D.R.T. Jones and L. Mezincescu, Phys. Lett. B **136**, 242 (1984).
122. P.C. West, Phys. Lett. B **137**, 371 (1984); D.R.T. Jones and L. Mezincescu, Phys. Lett. B **138**, 293 (1984); J.E. Bjorkman and D.R.T. Jones, Nucl. Phys. B **259**, 533 (1985).
123. K. Inoue, A. Kakuto, H. Komatsu and H. Takeshita, Prog. Theor. Phys. **68**, 927 (1982) [Erratum-ibid. **70**, 330 (1983)], Prog. Theor. Phys. **71**, 413 (1984); N.K. Falck Z. Phys. C **30**, 247 (1986).
124. S.P. Martin and M.T. Vaughn, Phys. Rev. D **50**, 2282 (1994) [hep-ph/9311340]; Y. Yamada, Phys. Rev. D **50**, 3537 (1994) [hep-ph/9401241]; I. Jack and D.R.T. Jones, Phys. Lett. B **333**, 372 (1994) [hep-ph/9405233]; I. Jack, D.R.T. Jones, S.P. Martin, M.T. Vaughn and Y. Yamada, Phys. Rev. D **50**, 5481 (1994) [hep-ph/9407291].
125. I. Jack, D.R.T. Jones and C.G. North, Nucl. Phys. B **473**, 308 (1996) [hep-ph/9603386], Phys. Lett. B **386**, 138 (1996) [hep-ph/9606323]; P.M. Ferreira, I. Jack, D.R.T. Jones, Phys. Lett. B **387**, 80 (1996) [hep-ph/9605440].
126. P.M. Ferreira, I. Jack, D.R.T. Jones and C.G. North, Nucl. Phys. B **504**, 108 (1997) [hep-ph/9705328]; I. Jack, D.R.T. Jones and A. Pickering, Phys. Lett. B **435**, 61 (1998) [hep-ph/9805482].
127. I. Jack and D.R.T. Jones, Phys. Lett. B **415**, 383 (1997) [hep-ph/9709364]; L.V. Avdeev, D.I. Kazakov and I.N. Kondrashuk, Nucl. Phys. B **510**, 289 (1998) [hep-ph/9709397]; N. Arkani-Hamed, G. F. Giudice, M. A. Luty and R. Rattazzi, Phys. Rev. D **58**, 115005 (1998) [hep-ph/9803290]; I. Jack, D.R.T. Jones and A. Pickering, Phys. Lett. B **432**, 114 (1998) [hep-ph/9803405].
128. J.R. Ellis, C.H. Llewellyn Smith and G.G. Ross, Phys. Lett. B **114**, 227 (1982).
129. S. Dimopoulos, G.R. Dvali, R. Rattazzi and G.F. Giudice, Nucl. Phys. B **510**, 12 (1998) [hep-ph/9705307]. M.A. Luty and J. Terning, Phys. Rev. D **62**, 075006 (2000) [hep-ph/9812290]. T. Banks, [hep-ph/0510159].
130. K.A. Intriligator, N. Seiberg and D. Shih, JHEP **0604**, 021 (2006) [hep-th/0602239].
131. S.A. Abel, C.S. Chu, J. Jaeckel and V.V. Khoze, JHEP **0701**, 089 (2007) [hep-th/0610334]. S.A. Abel, J. Jaeckel and V. V. Khoze, JHEP **0701**, 015 (2007) [hep-th/0611130]. N.J. Craig, P.J. Fox and J.G. Wacker, Phys. Rev. D **75**, 085006 (2007) [hep-th/0611006]. W. Fischler, V. Kaplunovsky, C. Krishnan, L. Mannelli and M.A.C. Torres, JHEP **0703**, 107 (2007) [hep-th/0611018].
132. P. Fayet and J. Iliopoulos, Phys. Lett. B **51**, 461 (1974); P. Fayet, Nucl. Phys. B **90**, 104 (1975).

133. However, see for example P. Binétruy and E. Dudas, Phys. Lett. B **389**, 503 (1996) [hep-th/9607172]; G. Dvali and A. Pomarol, Phys. Rev. Lett. **77**, 3728 (1996) [hep-ph/9607383]; R.N. Mohapatra and A. Riotto, Phys. Rev. D **55**, 4262 (1997) [hep-ph/9611273]; N. Arkani-Hamed, M. Dine and S.P. Martin, Phys. Lett. B **431**, 329 (1998) [hep-ph/9803432]. A non-zero Fayet-Iliopoulos term for an anomalous $U(1)$ symmetry is commonly found in superstring models: M. Green and J. Schwarz, Phys. Lett. B **149**, 117 (1984); M. Dine, N. Seiberg and E. Witten, Nucl. Phys. B **289**, 589 (1987); J. Atick, L. Dixon and A. Sen, Nucl. Phys. B **292**, 109 (1987). This may even help to explain the observed structure of the Yukawa couplings: L.E. Ibáñez, Phys. Lett. B **303**, 55 (1993) [hep-ph/9205234]; L.E. Ibáñez and G.G. Ross, Phys. Lett. B **332**, 100 (1994) [hep-ph/9403338]; P. Binétruy, S. Lavignac, P. Ramond, Nucl. Phys. B **477**, 353 (1996) [hep-ph/9601243]. P. Binétruy, N. Irges, S. Lavignac and P. Ramond, Phys. Lett. B **403**, 38 (1997) [hep-ph/9612442]; N. Irges, S. Lavignac, P. Ramond, Phys. Rev. D **58**, 035003 (1998) [hep-ph/9802334].
134. L. O'Raifeartaigh, Nucl. Phys. B **96**, 331 (1975).
135. M.A. Luty and W.I. Taylor, Phys. Rev. D **53**, 3399 (1996) [hep-th/9506098]. M. Dine, L. Randall and S.D. Thomas, Nucl. Phys. B **458**, 291 (1996) [hep-ph/9507453]. T. Gherghetta, C.F. Kolda and S.P. Martin, Nucl. Phys. B **468**, 37 (1996) [hep-ph/9510370].
136. S. Coleman and E. Weinberg, Phys. Rev. D **7**, 1888 (1973).
137. E. Witten, Nucl. Phys. B **202**, 253 (1982);
138. I. Affleck, M. Dine and N. Seiberg, Nucl. Phys. B **241**, 493 (1984), Nucl. Phys. B **256**, 557 (1985);
139. A.E. Nelson and N. Seiberg, Nucl. Phys. B **416**, 46 (1994) [hep-ph/9309299].
140. L. Randall, "Models of dynamical supersymmetry breaking," [hep-ph/9706474]; A. Nelson, Nucl. Phys. Proc. Suppl. **62**, 261 (1998) [hep-ph/9707442]; G.F. Giudice and R. Rattazzi, Phys. Rept. **322**, 419 (1999) [hep-ph/9801271]; E. Poppitz and S.P. Trivedi, Ann. Rev. Nucl. Part. Sci. **48**, 307 (1998) [hep-th/9803107]; Y. Shadmi and Y. Shirman, Rev. Mod. Phys. **72**, 25 (2000) [hep-th/9907225]. K.A. Intriligator and N. Seiberg, Class. Quant. Grav. **24**, S741 (2007) [hep-ph/0702069].
141. For example, see: S. Franco and A.M. Uranga, JHEP **0606**, 031 (2006) [hep-th/0604136]. H. Ooguri and Y. Ookouchi, Nucl. Phys. B **755**, 239 (2006) [hep-th/0606061]. T. Banks, [hep-ph/0606313]. S. Franco, I. Garcia-Etxebarria and A.M. Uranga, JHEP **0701**, 085 (2007) [hep-th/0607218]. R. Kitano, Phys. Lett. B **641**, 203 (2006) [hep-ph/0607090]. A. Amariti, L. Girardello and A. Mariotti, JHEP **0612**, 058 (2006) [hep-th/0608063]. M. Dine, J.L. Feng and E. Silverstein, Phys. Rev. D **74**, 095012 (2006) [hep-th/0608159]. E. Dudas, C. Papineau and S. Pokorski, JHEP **0702**, 028 (2007) [hep-th/0610297]. H. Abe, T. Higaki, T. Kobayashi and Y. Omura, Phys. Rev. D **75**, 025019 (2007) [hep-th/0611024]. M. Dine and J. Mason, Phys. Rev. D **77**, 016005 (2008) [hep-ph/0611312]. R. Kitano, H. Ooguri and Y. Ookouchi, Phys. Rev. D **75**, 045022 (2007) [hep-ph/0612139]. H. Murayama and Y. Nomura, Phys. Rev. Lett. **98**, 151803 (2007) [hep-

ph/0612186]. C. Csaki, Y. Shirman and J. Terning, JHEP **0705**, 099 (2007) [hep-ph/0612241]. O. Aharony and N. Seiberg, JHEP **0702**, 054 (2007) [hep-ph/0612308]. D. Shih, JHEP **0802**, 091 (2008) [hep-th/0703196]. K.A. Intriligator, N. Seiberg and D. Shih, JHEP **0707**, 017 (2007) [hep-th/0703281]. A. Giveon, A. Katz, Z. Komargodski and D. Shih, JHEP **0810**, 092 (2008) [hep-th/0808.2901].

142. P. Fayet, Phys. Lett. B **70**, 461 (1977), Phys. Lett. B **86**, 272 (1979), and in *Unification of the fundamental particle interactions* (Plenum, 1980).
143. N. Cabibbo, G.R. Farrar and L. Maiani, Phys. Lett. B **105**, 155 (1981); M.K. Gaillard, L. Hall and I. Hinchliffe, Phys. Lett. B **116**, 279 (1982) [Erratum-ibid. B **119**, 471 (1982)]; J. Ellis and J.S. Hagelin, Phys. Lett. B **122**, 303 (1983); D.A. Dicus, S. Nandi and J. Woodside, Phys. Lett. B **258**, 231 (1991); D.R. Stump, M. Wiest, and C.P. Yuan, Phys. Rev. D **54**, 1936 (1996) [hep-ph/9601362]. S. Ambrosanio, G. Kribs, and S.P. Martin, Phys. Rev. D **56**, 1761 (1997) [hep-ph/9703211].
144. S. Dimopoulos, M. Dine, S. Raby and S.Thomas, Phys. Rev. Lett. **76**, 3494 (1996) [hep-ph/9601367]; S. Dimopoulos, S. Thomas and J. D. Wells, Phys. Rev. D **54**, 3283 (1996) [hep-ph/9604452].
145. S. Ambrosanio et al., Phys. Rev. Lett. **76**, 3498 (1996) [hep-ph/9602239], Phys. Rev. D **54**, 5395 (1996) [hep-ph/9605398].
146. P. Nath and R. Arnowitt, Phys. Lett. B **56**, 177 (1975); R. Arnowitt, P. Nath and B. Zumino, Phys. Lett. B **56**, 81 (1975); S. Ferrara, D.Z. Freedman and P. van Nieuwenhuizen, Phys. Rev. D **13**, 3214 (1976); S. Deser and B. Zumino, Phys. Lett. B **62**, 335 (1976); D.Z. Freedman and P. van Nieuwenhuizen, Phys. Rev. D **14**, 912 (1976); E. Cremmer et al., Nucl. Phys. B **147**, 105 (1979); J. Bagger, Nucl. Phys. B **211**, 302 (1983).
147. E. Cremmer, S. Ferrara, L. Girardello, and A. van Proeyen, Nucl. Phys. B **212**, 413 (1983).
148. S. Deser and B. Zumino, Phys. Rev. Lett. **38**, 1433 (1977); E. Cremmer et al., Phys. Lett. B **79**, 231 (1978).
149. H. Pagels, J.R. Primack, Phys. Rev. Lett. **48**, 223 (1982); T. Moroi, H. Murayama, M. Yamaguchi, Phys. Lett. B **303**, 289 (1993).
150. A.H. Chamseddine, R. Arnowitt and P. Nath, Phys. Rev. Lett. **49**, 970 (1982); R. Barbieri, S. Ferrara and C. A. Savoy, Phys. Lett. B **119**, 343 (1982); L.E. Ibáñez, Phys. Lett. B **118**, 73 (1982); L.J. Hall, J.D. Lykken and S. Weinberg, Phys. Rev. D **27**, 2359 (1983); N. Ohta, Prog. Theor. Phys. **70**, 542 (1983).
151. J. Ellis, D.V. Nanopoulos and K. Tamvakis, Phys. Lett. B **121**, 123 (1983); L. Alvarez-Gaumé, J. Polchinski, and M. Wise, Nucl. Phys. B **221**, 495 (1983).
152. P. Moxhay and K. Yamamoto, Nucl. Phys. B **256**, 130 (1985); K. Grassie, Phys. Lett. B **159**, 32 (1985); B. Gato, Nucl. Phys. B **278**, 189 (1986); N. Polonsky and A. Pomarol, N. Polonsky and A. Pomarol, Phys. Rev. Lett. **73**, 2292 (1994) [hep-ph/9406224].
153. G.G. Ross and R.G. Roberts, G. G. Ross and R. G. Roberts, Nucl. Phys. B **377**, 571 (1992); H. Arason et al., Phys. Rev. Lett. **67**, 2933 (1991);

R. Arnowitt and P. Nath, Phys. Rev. Lett. **69**, 725 (1992); Phys. Rev. D **46**, 3981 (1992); D.J. Castaño, E.J. Piard, P. Ramond, Phys. Rev. D **49**, 4882 (1994) [hep-ph/9308335]; V. Barger, M. Berger and P. Ohmann, Phys. Rev. D **49**, 4908 (1994) [hep-ph/9311269]; G.L. Kane, C. Kolda, L. Roszkowski, J.D. Wells, Phys. Rev. D **49**, 6173 (1994) [hep-ph/9312272]; B. Ananthanarayan, K.S. Babu and Q. Shafi, Nucl. Phys. B **428**, 19 (1994) [hep-ph/9402284]; M. Carena, M. Olechowski, S. Pokorski, C.E.M. Wagner, Nucl. Phys. B **419**, 213 (1994) [hep-ph/9311222]; W. de Boer, R. Ehret and D. Kazakov, Z. Phys. C **67**, 647 (1995) [hep-ph/9405342]; M. Carena et al., Nucl. Phys. B **491**, 103 (1997) [hep-ph/9612261].

154. V. Kaplunovsky and J. Louis, Phys. Lett. B **306**, 269 (1993) [hep-th/9303040]; R. Barbieri, J. Louis and M. Moretti, Phys. Lett. B **312**, 451 (1993) [Erratum-ibid. B **316**, 632 (1993)] [hep-ph/9305262]; A. Brignole, L.E. Ibáñez and C. Muñoz, Nucl. Phys. B **422**, 125 (1994) [Erratum-ibid. B **436**, 747 (1995)] [hep-ph/9308271].

155. J. Polonyi, Hungary Central Research Institute report KFKI-77-93 (1977) (unpublished). See[18] for a pedagogical account.

156. For a review, see A.B. Lahanas and D.V. Nanopoulos, Phys. Rept. **145**, 1 (1987).

157. For a review, see A. Brignole, L.E. Ibáñez and C. Muñoz, "Soft supersymmetry-breaking terms from supergravity and superstring models", in *Perspectives on Supersymmetry*, ed. G.L. Kane (World Scientific, 1998), [hep-ph/9707209].

158. M. Dine and W. Fischler, Phys. Lett. B **110**, 227 (1982); C.R. Nappi and B.A. Ovrut, Phys. Lett. B **113**, 175 (1982); L. Alvarez-Gaumé, M. Claudson and M. B. Wise, Nucl. Phys. B **207**, 96 (1982).

159. M. Dine, A. E. Nelson, Phys. Rev. D **48**, 1277 (1993) [hep-ph/9303230]; M. Dine, A.E. Nelson, Y. Shirman, Phys. Rev. D **51**, 1362 (1995) [hep-ph/9408384]; M. Dine, A.E. Nelson, Y. Nir, Y. Shirman, Phys. Rev. D **53**, 2658 (1996) [hep-ph/9507378].

160. S. Dimopoulos, G.F. Giudice and A. Pomarol, Phys. Lett. B **389**, 37 (1996) [hep-ph/9607225]; S.P. Martin Phys. Rev. D **55**, 3177 (1997) [hep-ph/9608224]; E. Poppitz and S.P. Trivedi, Phys. Lett. B **401**, 38 (1997) [hep-ph/9703246].

161. V.A. Rubakov and M.E. Shaposhnikov, Phys. Lett. B **125**, 136 (1983), Phys. Lett. B **125**, 139 (1983); L. Randall and R. Sundrum, Phys. Rev. Lett. **83**, 3370 (1999) [hep-ph/9905221]; Phys. Rev. Lett. **83**, 4690 (1999) [hep-th/9906064], For accessible reviews of recent work on extra dimensions in general, see C. Csaki, "TASI lectures on extra dimensions and branes," [hep-ph/0404096]; A. Pérez-Lorenzana, J. Phys. Conf. Ser. **18**, 224 (2005) [hep-ph/0503177]; R. Sundrum, "To the fifth dimension and back. (TASI 2004)," [hep-th/0508134]. G.D. Kribs, "TASI 2004 Lectures on the Phenomenology of Extra Dimensions," [hep-ph/0605325].

162. E.A. Mirabelli and M.E. Peskin, Phys. Rev. D **58**, 065002 (1998) [hep-th/9712214].

163. D.E. Kaplan, G.D. Kribs and M. Schmaltz, Phys. Rev. D **62**, 035010 (2000) [hep-ph/9911293]; Z. Chacko, M.A. Luty, A.E. Nelson and E. Ponton, JHEP **0001**, 003 (2000) [hep-ph/9911323].
164. M. Schmaltz and W. Skiba, Phys. Rev. D **62**, 095005 (2000) [hep-ph/0001172]; Phys. Rev. D **62**, 095004 (2000) [hep-ph/0004210].
165. The gaugino mediation mechanism can be "deconstructed", replacing the extra dimension by a lattice of gauge groups, as in C. Csaki, J. Erlich, C. Grojean and G.D. Kribs, Phys. Rev. D **65**, 015003 (2002) [hep-ph/0106044]; H.C. Cheng, D.E. Kaplan, M. Schmaltz and W. Skiba, Phys. Lett. B **515**, 395 (2001) [hep-ph/0106098].
166. L. Randall and R. Sundrum, Nucl. Phys. B **557**, 79 (1999) [hep-th/9810155]; G.F. Giudice, M.A. Luty, H. Murayama and R. Rattazzi, JHEP **9812**, 027 (1998) [hep-ph/9810442].
167. M.A. Luty and R. Sundrum, Phys. Rev. D **65**, 066004 (2002) [hep-th/0105137], Phys. Rev. D **67**, 045007 (2003) [hep-th/0111231].
168. I. Jack, D.R.T. Jones and A. Pickering, Phys. Lett. B **426**, 73 (1998) [hep-ph/9712542]. I. Jack and D.R.T. Jones, Phys. Lett. B **465**, 148 (1999) [hep-ph/9907255],
169. J.A. Bagger, T. Moroi and E. Poppitz, JHEP **0004**, 009 (2000) [hep-th/9911029]; M.K. Gaillard and B.D. Nelson, Nucl. Phys. B **588**, 197 (2000) [hep-th/0004170]; A. Anisimov, M. Dine, M. Graesser and S. D. Thomas, Phys. Rev. D **65**, 105011 (2002) [hep-th/0111235], JHEP **0203**, 036 (2002) [hep-th/0201256]; M. Dine et al., Phys. Rev. D **70**, 045023 (2004) [hep-ph/0405159].
170. A. Pomarol and R. Rattazzi, JHEP **9905**, 013 (1999) [hep-ph/9903448].
171. D.E. Kaplan and G.D. Kribs, JHEP **0009**, 048 (2000) [hep-ph/0009195]; Z. Chacko and M.A. Luty, JHEP **0205**, 047 (2002) [hep-ph/0112172]; R. Sundrum, Phys. Rev. D **71**, 085003 (2005) [hep-th/0406012].
172. Z. Chacko, M.A. Luty, I. Maksymyk and E. Ponton, JHEP **0004**, 001 (2000) [hep-ph/9905390]; E. Katz, Y. Shadmi and Y. Shirman, JHEP **9908**, 015 (1999) [hep-ph/9906296]; I. Jack and D.R.T. Jones, Phys. Lett. B **482**, 167 (2000) [hep-ph/0003081]; M. Carena, K. Huitu and T. Kobayashi, Nucl. Phys. B **592**, 164 (2001) [hep-ph/0003187]; B.C. Allanach and A. Dedes, JHEP **0006**, 017 (2000) [hep-ph/0003222]; N. Arkani-Hamed, D.E. Kaplan, H. Murayama and Y. Nomura, JHEP **0102**, 041 (2001) [hep-ph/0012103]; N. Okada, Phys. Rev. D **65**, 115009 (2002) [hep-ph/0202219]; R. Harnik, H. Murayama and A. Pierce, JHEP **0208**, 034 (2002) [hep-ph/0204122]; A.E. Nelson and N.T. Weiner, [hep-ph/0210288]; O.C. Anoka, K.S. Babu and I. Gogoladze, Nucl. Phys. B **686**, 135 (2004) [hep-ph/0312176]; R. Kitano, G.D. Kribs and H. Murayama, Phys. Rev. D **70**, 035001 (2004) [hep-ph/0402215].
173. C.H. Chen, M. Drees and J.F. Gunion, Phys. Rev. D **55**, 330 (1997) [Erratum-ibid. D **60**, 039901 (1999)] [hep-ph/9607421].
174. S.D. Thomas and J.D. Wells, Phys. Rev. Lett. **81**, 34 (1998) [hep-ph/9804359].

175. J.L. Feng et al., Phys. Rev. Lett. **83**, 1731 (1999) [hep-ph/9904250]; J.L. Feng and T. Moroi, Phys. Rev. D **61**, 095004 (2000) [hep-ph/9907319].
176. T. Gherghetta, G.F. Giudice and J.D. Wells, Nucl. Phys. B **559**, 27 (1999) [hep-ph/9904378].
177. J.F. Gunion and S. Mrenna, Phys. Rev. D **62**, 015002 (2000) [hep-ph/9906270]. Phys. Rev. D **64**, 075002 (2001) [hep-ph/0103167], R. Rattazzi, A. Strumia and J. D. Wells, Nucl. Phys. B **576**, 3 (2000) [hep-ph/9912390]. F.E. Paige and J.D. Wells, "Anomaly mediated SUSY breaking at the LHC," [hep-ph/0001249]; H. Baer, J.K. Mizukoshi and X. Tata, Phys. Lett. B **488**, 367 (2000) [hep-ph/0007073].
178. J. Scherk and J.H. Schwarz, Phys. Lett. B **82**, 60 (1979); Nucl. Phys. B **153**, 61 (1979).
179. I. Antoniadis, Phys. Lett. B **246**, 377 (1990); I. Antoniadis, C. Muñoz and M. Quirós, Nucl. Phys. B **397**, 515 (1993) [hep-ph/9211309]; G.R. Dvali and M. A. Shifman, Nucl. Phys. B **504**, 127 (1997) [hep-th/9611213]; A. Pomarol and M. Quirós, Phys. Lett. B **438**, 255 (1998) [hep-ph/9806263]; I. Antoniadis, S. Dimopoulos, A. Pomarol and M. Quirós, Nucl. Phys. B **544**, 503 (1999) [hep-ph/9810410]; A. Delgado, A. Pomarol and M. Quirós, Phys. Rev. D **60**, 095008 (1999) [hep-ph/9812489]; Z. Chacko and M. A. Luty, JHEP **0105**, 067 (2001) [hep-ph/0008103]; T. Gherghetta and A. Pomarol, Nucl. Phys. B **602**, 3 (2001) [hep-ph/0012378]; D. Marti and A. Pomarol, Phys. Rev. D **64**, 105025 (2001) [hep-th/0106256]; R. Barbieri, L.J. Hall and Y. Nomura, Nucl. Phys. B **624**, 63 (2002) [hep-th/0107004]; D.E. Kaplan and N. Weiner, [hep-ph/0108001]; J. Bagger, F. Feruglio and F. Zwirner, JHEP **0202**, 010 (2002) [hep-th/0108010]; T. Gherghetta and A. Riotto, Nucl. Phys. B **623**, 97 (2002) [hep-th/0110022].
180. L.E. Ibáñez and G.G. Ross, Phys. Lett. B **110**, 215 (1982).
181. S.P. Martin, Phys. Rev. D **65**, 116003 (2002) [hep-ph/0111209], Phys. Rev. D **66**, 096001 (2002) [hep-ph/0206136].
182. J.F. Gunion and H.E. Haber, Nucl. Phys. B **272**, 1 (1986), Nucl. Phys. B **278**, 449 (1986), Nucl. Phys. B **307**, 445 (1988). (Errata in [hep-ph/9301205].)
183. J.F. Gunion, H.E. Haber, G.L. Kane and S. Dawson, *The Higgs Hunter's Guide*, (Addison-Wesley 1991), (Errata in [hep-ph/9302272].)
184. K. Inoue, A. Kakuto, H. Komatsu and S. Takeshita, Prog. Theor. Phys. **67**, 1889 (1982); R.A. Flores and M. Sher, Annals Phys. **148**, 95 (1983).
185. H.E. Haber and R. Hempfling, Phys. Rev. Lett. **66**, 1815 (1991); Y. Okada, M. Yamaguchi and T. Yanagida, Prog. Theor. Phys. **85**, 1 (1991), Phys. Lett. B **262**, 54 (1991); J. Ellis, G. Ridolfi and F. Zwirner, Phys. Lett. B **257**, 83 (1991), Phys. Lett. B **262**, 477 (1991).
186. G. Gamberini, G. Ridolfi and F. Zwirner, Nucl. Phys. B **331**, 331 (1990); R. Barbieri, M. Frigeni and F. Caravaglio, Phys. Lett. B **258**, 167 (1991); A. Yamada, Phys. Lett. B **263**, 233 (1991), Z. Phys. C **61**, 247 (1994); J.R. Espinosa and M. Quirós, Phys. Lett. B **266**, 389 (1991); A. Brignole, Phys. Lett. B **281**, 284 (1992); M. Drees and M.M. Nojiri, Nucl. Phys. B **369**, 54 (1992), Phys. Rev. D **45**, 2482 (1992).

H.E. Haber and R. Hempfling, Phys. Rev. D **48**, 4280 (1993) [hep-ph/9307201]; P.H. Chankowski, S. Pokorski and J. Rosiek, Phys. Lett. B **274**, 191 (1992); Phys. Lett. B **281** (1992) 100; R. Hempfling and A.H. Hoang, Phys. Lett. B **331**, 99 (1994) [hep-ph/9401219]; J. Kodaira, Y. Yasui and K. Sasaki, Phys. Rev. D **50**, 7035 (1994) [hep-ph/9311366]; J.A. Casas, J.R. Espinosa, M. Quirós and A. Riotto, Nucl. Phys. B **436**, 3 (1995) [Erratum-ibid. B **439**, 466 (1995)] [hep-ph/9407389]; M. Carena, M. Quirós and C. Wagner, Nucl. Phys. B **461**, 407 (1996) [hep-ph/9508343]. H.E. Haber, R. Hempfling and A.H. Hoang, Z. Phys. C **75**, 539 (1997) [hep-ph/9609331].

187. S. Heinemeyer, W. Hollik and G. Weiglein, Phys. Rev. D **58**, 091701 (1998) [hep-ph/9803277]; Phys. Lett. B **440**, 296 (1998) [hep-ph/9807423]; Eur. Phys. J. C **9**, 343 (1999) [hep-ph/9812472].
188. R.J. Zhang, Phys. Lett. B **447**, 89 (1999) [hep-ph/9808299]. J.R. Espinosa and R.J. Zhang, JHEP **0003**, 026 (2000) [hep-ph/9912236]; Nucl. Phys. B **586**, 3 (2000) [hep-ph/0003246].
189. A. Pilaftsis and C.E.M. Wagner, Nucl. Phys. B **553**, 3 (1999) [hep-ph/9902371]. M. Carena, J.R. Ellis, A. Pilaftsis and C.E.M. Wagner, Nucl. Phys. B **586**, 92 (2000) [hep-ph/0003180]; Nucl. Phys. B **625**, 345 (2002) [hep-ph/0111245].
190. M. Carena et al., Nucl. Phys. B **580**, 29 (2000) [hep-ph/0001002].
191. J.R. Espinosa and I. Navarro, Nucl. Phys. B **615**, 82 (2001) [hep-ph/0104047].
192. G. Degrassi, P. Slavich and F. Zwirner, Nucl. Phys. B **611**, 403 (2001) [hep-ph/0105096]; A. Brignole, G. Degrassi, P. Slavich and F. Zwirner, Nucl. Phys. B **631**, 195 (2002) [hep-ph/0112177]; Nucl. Phys. B **643**, 79 (2002) [hep-ph/0206101].
193. S.P. Martin, Phys. Rev. D **67**, 095012 (2003) [hep-ph/0211366]; Phys. Rev. D **70**, 016005 (2004) [hep-ph/0312092]; Phys. Rev. D **71**, 016012 (2005) [hep-ph/0405022].
194. S. Heinemeyer, W. Hollik and G. Weiglein, "FeynHiggs: A program for the calculation of the masses of the neutral CP-even Higgs bosons in the MSSM," Comput. Phys. Commun. **124**, 76 (2000) [hep-ph/9812320]; T. Hahn, W. Hollik, S. Heinemeyer and G. Weiglein, "Precision Higgs masses with FeynHiggs 2.2," [hep-ph/0507009].
195. G.L. Kane, C. Kolda and J.D. Wells, Phys. Rev. Lett. **70**, 2686 (1993) [hep-ph/9210242]; J.R. Espinosa and M. Quirós, Phys. Lett. B **302**, 51 (1993) [hep-ph/9212305].
196. R. Tarrach, Nucl. Phys. B **183**, 384 (1981); H. Gray, D.J. Broadhurst, W. Grafe and K. Schilcher, Z. Phys. C **48**, 673 (1990); H. Arason et al., Phys. Rev. D **46**, 3945 (1992).
197. L.E. Ibáñez and C. Lopez, Phys. Lett. B **126**, 54 (1983); V. Barger, M.S. Berger and P. Ohmann, Phys. Rev. D **47**, 1093 (1993) [hep-ph/9209232]; P. Langacker and N. Polonsky, Phys. Rev. D **49**, 1454 (1994) [hep-ph/9306205]; P. Ramond, R.G. Roberts, G.G. Ross, Nucl. Phys. B **406**, 19 (1993) [hep-ph/9303320]; M. Carena, S. Pokorski and C. Wagner,

Nucl. Phys. B **406**, 59 (1993) [hep-ph/9303202]; G. Anderson et al., Phys. Rev. D **49**, 3660 (1994) [hep-ph/9308333].

198. L.J. Hall, R. Rattazzi and U. Sarid, Phys. Rev. D **50**, 7048 (1994) [hep-ph/9306309]; M. Carena, M. Olechowski, S. Pokorski and C.E.M. Wagner, Nucl. Phys. B **426**, 269 (1994) [hep-ph/9402253]. R. Hempfling, Phys. Rev. D **49**, 6168 (1994); R. Rattazzi and U. Sarid, Phys. Rev. D **53**, 1553 (1996) [hep-ph/9505428].
199. H.K. Dreiner, H.E. Haber, and S.P. Martin, "Two-component spinor techniques and Feynman rules for quantum field theory and supersymmetry", arXiv:0812.1594 [hep-ph], submitted to Physics Reports.
200. D. Pierce and A. Papadopoulos, Phys. Rev. D **50**, 565 (1994) [hep-ph/9312248], Nucl. Phys. B **430**, 278 (1994) [hep-ph/9403240]; D. Pierce, J.A. Bagger, K. Matchev, and R.-J. Zhang, Nucl. Phys. B **491**, 3 (1997) [hep-ph/9606211].
201. Y. Yamada, Phys. Lett. B **623**, 104 (2005) [hep-ph/0506262], S.P. Martin, Phys. Rev. D **72**, 096008 (2005) [hep-ph/0509115].
202. M. Claudson, L.J. Hall and I. Hinchliffe, Nucl. Phys. B **228**, 501 (1983); J.A. Casas, A. Lleyda, C. Muñoz, Nucl. Phys. B **471**, 3 (1996) [hep-ph/9507294]; T. Falk, K.A. Olive, L. Roszkowski, and M. Srednicki Phys. Lett. B **367**, 183 (1996) [hep-ph/9510308]; H. Baer, M. Brhlik and D.J. Castaño, Phys. Rev. D **54**, 6944 (1996) [hep-ph/9607465]. For a review, see J.A. Casas, in *Perspectives on Supersymmetry*, ed. G.L. Kane (World Scientific, 1998). [hep-ph/9707475].
203. A. Kusenko, P. Langacker and G. Segre, Phys. Rev. D **54**, 5824 (1996) [hep-ph/9602414].
204. F.E. Paige, S.D. Protopescu, H. Baer and X. Tata, "ISAJET 7.69: A Monte Carlo event generator for p p, anti-p p, and e+ e- reactions," [hep-ph/0312045].
205. B.C. Allanach, "SOFTSUSY: A C++ program for calculating supersymmetric spectra," Comput. Phys. Commun. **143**, 305 (2002) [hep-ph/0104145].
206. A. Djouadi, J.L. Kneur and G. Moultaka, "SuSpect: A Fortran code for the supersymmetric and Higgs particle spectrum in the MSSM," [hep-ph/0211331].
207. W. Porod, "SPheno, a program for calculating supersymmetric spectra, SUSY particle decays and SUSY particle production at e+ e- colliders," Comput. Phys. Commun. **153**, 275 (2003) [hep-ph/0301101].
208. M. Muhlleitner, A. Djouadi and Y. Mambrini, "SDECAY: A Fortran code for the decays of the supersymmetric particles in the MSSM," Comput. Phys. Commun. **168**, 46 (2005) [hep-ph/0311167].
209. A. Djouadi, J. Kalinowski and M. Spira, "HDECAY: A program for Higgs boson decays in the standard model and its supersymmetric extension," Comput. Phys. Commun. **108**, 56 (1998) [hep-ph/9704448].
210. U. Ellwanger, J.F. Gunion and C. Hugonie, JHEP **0502**, 066 (2005) [hep-ph/0406215], U. Ellwanger and C. Hugonie, "NMHDECAY 2.0: An updated

program for sparticle masses, Higgs masses, couplings and decay widths in the NMSSM," [hep-ph/0508022].
211. J.S. Lee et al., "CPsuperH: A computational tool for Higgs phenomenology in the minimal supersymmetric standard model with explicit CP violation," Comput. Phys. Commun. **156**, 283 (2004) [hep-ph/0307377].
212. P. Gondolo et al., "DarkSUSY: A numerical package for dark matter calculations in the MSSM," [astro-ph/0012234]. JCAP **0407**, 008 (2004) [astro-ph/0406204].
213. G. Belanger, F. Boudjema, A. Pukhov and A. Semenov, Comput. Phys. Commun. **149**, 103 (2002) [hep-ph/0112278], "MicrOMEGAs: Version 1.3," [hep-ph/0405253].
214. A. Bartl, H. Fraas and W. Majerotto, Nucl. Phys. B **278**, 1 (1986); Z. Phys. C **30**, 441 (1986); Z. Phys. C **41**, 475 (1988); A. Bartl, H. Fraas, W. Majerotto and B. Mösslacher, Z. Phys. C **55**, 257 (1992). For large $\tan\beta$ results, see H. Baer, C.-h. Chen, M. Drees, F. Paige and X. Tata, Phys. Rev. Lett. **79**, 986 (1997) [hep-ph/9704457];
215. H. Baer et al., Int. J. Mod. Phys. A **4**, 4111 (1989).
216. H.E. Haber and D. Wyler, Nucl. Phys. B **323**, 267 (1989); S. Ambrosanio and B. Mele, Phys. Rev. D **55**, 1399 (1997) [Erratum-ibid. D **56**, 3157 (1997)] [hep-ph/9609212]; S. Ambrosanio et al., Phys. Rev. D **55**, 1372 (1997) [hep-ph/9607414].
217. H. Baer et al., Phys. Lett. B **161**, 175 (1985); G. Gamberini, Z. Phys. C **30**, 605 (1986); H.A. Baer, V. Barger, D. Karatas and X. Tata, Phys. Rev. D **36**, 96 (1987); R.M. Barnett, J.F. Gunion and H.A. Haber, Phys. Rev. D **37**, 1892 (1988).
218. K. Hikasa and M. Kobayashi, Phys. Rev. D **36**, 724 (1987).
219. J. Ellis, J.L. Lopez, D.V. Nanopoulos Phys. Lett. B **394**, 354 (1997) [hep-ph/9610470]; J.L. Lopez, D.V. Nanopoulos Phys. Rev. D **55**, 4450 (1997) [hep-ph/9608275].
220. D.N. Spergel *et al.* [WMAP Collaboration], Astrophys. J. Suppl. **148**, 175 (2003) [astro-ph/0302209], and [astro-ph/0603449]. M. Tegmark *et al.* [SDSS Collaboration], Phys. Rev. D **69**, 103501 (2004) [astro-ph/0310723];
221. K. Choi, J.E. Kim and H.P. Nilles, Phys. Rev. Lett. **73**, 1758 (1994) [hep-ph/9404311]; K. Choi, J.E. Kim and G.T. Park, Nucl. Phys. B **442**, 3 (1995) [hep-ph/9412397]. See also N.C. Tsamis and R.P. Woodard, Phys. Lett. B **301**, 351 (1993); Nucl. Phys. B **474**, 235 (1996) [hep-ph/9602315]; Annals Phys. **253**, 1 (1997) [hep-ph/9602316]; Annals Phys. **267**, 145 (1998) [hep-ph/9712331]. and references therein, for discussion of nonperturbative quantum gravitational effects on the effective cosmological constant. This work implies that requiring the tree-level vacuum energy to vanish may not be correct or meaningful. Moreover, perturbative supergravity or superstring predictions for the vacuum energy may not be relevant to the question of whether the observed cosmological constant is sufficiently small.

# Twenty Open Questions and a Postscript: Supersymmetry Enters the Era of the LHC

Keith R. Dienes[*,†,‡] and Christopher Kolda[§]

[*]*Physics Division, National Science Foundation, Arlington, VA 22230 USA*

[†]*Department of Physics, University of Maryland, College Park, MD 20742 USA*

[‡]*Department of Physics, University of Arizona, Tucson, AZ 85721 USA*

[§]*Department of Physics, University of Notre Dame, South Bend, IN 46556 USA*

[‡]*dienes@physics.arizona.edu*

[§]*ckolda@nd.edu*

We give a brief overview of twenty open theoretical questions in supersymmetric particle physics. The twenty questions we have chosen range from the GeV scale to the Planck scale, and include issues pertaining to the Minimal Supersymmetric Standard Model and its extensions, SUSY-breaking, cosmology, grand unified theories, and string theory. Throughout, our goal is to address those topics in which supersymmetry plays a fundamental role, and which are areas of active research in the field. This survey is written at an introductory level and is aimed at people who are not necessarily experts in the field.

At first glance, supersymmetry appears to be a theoretical success story writ large. With one simple idea, such diverse issues as extending the Lorentz group, solving the gauge hierarchy problem, coupling gauge theories to gravity, and generating gauge coupling unification all seem to fall into place. The lack of direct experimental evidence for supersymmetry (SUSY) dampens our enthusiasm somewhat, but only now (and over the next few years) are experiments really beginning to probe the domain where SUSY should be expected to be manifest.

[‡]Permanent address.

Nonetheless, there is a price to be paid for the successes of SUSY, and we mean more than simply the doubling of the Standard Model (SM) particle spectrum. SUSY introduces into physics a host of new questions which must be addressed, and hopefully, answered. Most of these questions were once real problems — problems which seemed to detract from, or perhaps even invalidate, SUSY as a viable fundamental symmetry of nature. Although some of the questions presented here do not have attractive solutions, none of them, nor any others that we are aware of, threaten to rule out SUSY. Instead, as competing answers to these questions have been found, these questions become opportunities not only to simply discover SUSY, but also to probe physics well beyond the scale of SUSY.

The questions that we will consider in this work have been chosen because they satisfy a number of important criteria. First and foremost, each defines an area of active research in the field; in many ways, the list of questions that follows forms a summary of current topics of interest in applying SUSY to the physics of the SM.

Second, these are questions which are intrinsically supersymmetric and may not even arise in the SM alone. Thus, generic questions in the SM (such as the cosmological constant, inflation, baryogenesis, and fermion mass hierarchies, to name a few) are not included here. This is not to say that SUSY does not have implications for these subjects, for it usually does, and when it does it typically recasts the terms of the debate completely by redefining the spectrum of possible solutions. However, in this article we will concentrate on those issues which are intrinsically supersymmetric and whose solutions one would hope to find in a complete description of a supersymmetric SM. After all, it would be wonderful if SUSY could explain why the electron is so much lighter than the top quark, but there is no obvious reason to suppose that it will since the question itself is intrinsically non-supersymmetric.

Third, these are questions whose answers may tell us a great deal about physics beyond the MSSM, yet which might be probed using only a combination of experimental measurements of the MSSM and theoretical constraints. In this sense, these questions are windows through which insight far beyond the weak scale may be sought.

Because we focus primarily on questions that have direct applicability to the physics of the SM or MSSM, a large number of interesting topics will receive only abbreviated attention. For example, recent results on the dynamics of SUSY gauge theories may have profound effects on how we approach the MSSM in coming years (as they already have had on the question of SUSY-breaking), but their current applications are limited. Thus we will only touch upon this and related topics of a more formal nature.

The open questions we have chosen to address are as follows. First, at the lowest energy scales, we have selected a number of open questions that pertain to the MSSM itself:

- *Question #1:* Why doesn't the proton decay in $10^{-17}$ years?
- *Question #2:* How is flavor-changing suppressed?
- *Question #3:* Why isn't CP violation ubiquitous?
- *Question #4:* Where does the $\mu$-term come from?
- *Question #5:* Why does the MSSM conserve color and charge?

Next, we consider open questions pertaining to SUSY-breaking:

- *Question #6:* How is SUSY broken?
- *Question #7:* Once SUSY is broken, how do we find out?

Then, we consider two open questions pertaining to natural extensions of the MSSM:

- *Question #8:* Can singlets and SUSY coexist?
- *Question #9:* How do extra $U(1)$'s fit into SUSY?

Our next set of open questions addresses the interplay between supersymmetry and cosmological issues:

- *Question #10:* How does SUSY shed light on dark matter?
- *Question #11:* Are gravitinos dangerous to cosmology?
- *Question #12:* Are moduli cosmologically dangerous?

We then turn our attention to supersymmetric GUT's:

- *Question #13:* Does the MSSM unify into a SUSY GUT?
- *Question #14:* Proton decay again: Why doesn't the proton decay in $10^{32}$ years?
- *Question #15:* Can SUSY GUT's explain the masses of fermions?

Next, we discuss some recent formal developments concerning SUSY and gauge theory:

- *Question #16:* N=1 SUSY duality: How has SUSY changed our view of gauge theory?

Finally, our last set of questions addresses supersymmetry at the very highest scales, in the context of string theory:

- *Question #17:* Why strings?
- *Question #18:* What roles does SUSY play in string theory?
- *Question #19:* How is SUSY broken in string theory?
- *Question #20:* Making ends meet: How can we understand gauge coupling unification from string theory?

Finally, many of the topics that we shall discuss here will be covered in much greater detail in the topical chapters of this book, and we will try to indicate the relevant chapters as we go along.

## Section I: Open Questions in the MSSM

The Standard Model forms the bedrock of modern high-energy physics, and accurately describes all physical phenomena down to scales of $10^{-16}$ cm. However, there are many possible ways of extending the SM down to smaller length scales. These include extra gauge interactions, new matter, new levels of compositeness (technicolor), and supersymmetry. While supersymmetry does not succeed as an *explanation* of the features of the SM, it provides a remarkably robust extension to the SM which is in agreement with all experimental data. This cannot be said for many other possible extensions (such as, *e.g.*, the simplest versions of technicolor). Moreover, it is quite possible and perhaps even likely that other forms of potential new physics might appear at the same energy scale as supersymmetry. Thus, as a first step, it is important to investigate how the structure of supersymmetry might be joined with that of the SM in a cohesive framework.

Although there are various ways in which SUSY might be joined with the SM, for simplicity one can pursue a *minimal* construction, and attempt to write down a Lagrangian which is the most *general* effective Lagrangian for the *minimal* extension of the SM which is invariant under SUSY transformations up to *soft-breaking* terms. This then results in the Lagrangian of the Minimal Supersymmetric Standard Model (MSSM). We will not attempt to fully define the MSSM, its field content, or its Lagrangian, but instead we refer the reader to any of several standard references,[1] or to the article of S. Martin in this volume. We will therefore say only a few words of a general nature.

The minimal extension of the SM with unbroken SUSY is a simple model with fewer free parameters than the SM itself, despite the large number of new fields. It is only in the breaking of SUSY that the number of free parameters becomes large — but of course, it is only in the breaking of SUSY that the model can even attempt to describe nature as we observe it. In the SM, the field content along with the gauge symmetries serve to provide a number of accidental symmetries at the renormalizable level. These include, for example, baryon number $B$, and lepton number $L$. These symmetries also serve to forbid flavor-changing neutral currents (FCNC's) up to small corrections arising from Yukawa couplings in loops. The MSSM shares neither of these properties. As we will discuss, the most general MSSM would have the proton decay with a weak-interaction lifetime and large FCNC's. The reason is this same proliferation of fields and free parameters. Even after imposing symmetries to forbid the fast proton decay (see Question #1 below), one finds[2] that the MSSM contains 106 new, independent (real) parameters above and beyond those of the SM. These consist of 26 masses (resulting from 12 squark, 9 slepton, and 3 gaugino masses, plus $\mu$- and $B_\mu$-terms), 37 mixing angles, and 43 CP-violating phases. Understanding, constraining, and ultimately measuring these parameters is one of the primary goals of the SUSY program in particle physics.

Of course, the MSSM is unlikely to be the end of the story. Therefore, although we will begin this article by considering open questions within the MSSM, we will later allow this structure to expand by considering new singlets, extra gauge symmetries, grand unification, and ultimately embeddings within string theory.

One recurring feature in many of the open questions in this article is the question of "naturalness": why is some coupling (or mass) very small or even zero when it need not have been zero *a priori*? More precisely, this is a question of "Dirac naturalness." The idea of Dirac naturalness is built on the supposition that in any physical system, all couplings and interactions which are not otherwise forbidden should be allowed, and that all ratios of couplings, as well as all ratios of masses, should be $\mathcal{O}(1)$. We generally find that a theory is Dirac natural if some exact symmetry exists which forbids the undesired couplings. Note that Dirac naturalness is not to be confused with "'t Hooft naturalness" or "technical naturalness," which is the problem besetting the Higgs sector of the Standard Model. In the latter case one seeks to understand small numbers or ratios, such as the ratio of the weak to Planck scale, in terms of approximate symmetries. The role of radiative corrections is very important for 't Hooft naturalness, because even if one could choose the ratio of two couplings or two masses to be far from unity at tree level, without some approximate symmetry at play there would be no reason for this ratio to persist beyond tree level. The chiral symmetry of the SM fermions is a classic example of this phenomenon: an approximate symmetry protects the fermion masses from receiving corrections proportional to heavy mass scales.

Without SUSY, there is nothing like a chiral symmetry to protect scalar masses from heavy mass scales. But with SUSY, the chiral symmetry in the fermionic sector protects the scalars too. This is a general feature of SUSY — couplings and masses which are SUSY-preserving are automatically natural *à la* 't Hooft even if they are unnatural *à la* Dirac. Thus when we speak of naturalness in the context of SUSY, we will generally be referring to Dirac naturalness, for which SUSY provides no automatic solutions.

## Question #1 Why doesn't the proton decay in $10^{-17}$ years?

As discussed in the article of S. Martin in this volume, the most general, renormalizable superpotential for the MSSM can be organized into three pieces, $W = W_0 + W_{\not B} + W_{\not L}$, where

$$W_0 = y^U_{ij} Q^i u^j H_U + y^D_{ij} Q^i d^j H_D + y^E_{ij} L^i e^j H_D + \mu H_U H_D \tag{2.1}$$

$$W_{\not L} = \lambda_{ijk} Q^i d^j L^k + \lambda'_{ijk} e^i L^j L^k + \mu'_i H_U L^i \tag{2.2}$$

$$W_{\not B} = \lambda''_{ijk} u^i d^j d^k \ . \tag{2.3}$$

The first piece preserves the global $B$ and $L$ quantum numbers of the SM, while the other two each violate one of either $B$ or $L$. This is to be contrasted with the case of the SM in which one can obtain $B$-violating operators only by going to dimension six, or $L$-violating operators only by going to dimension five. Allowing simultaneous $B$- and $L$-violation would be a disaster. For example, given non-zero $\lambda$ and $\lambda''$, one can form a four-fermion operator $QudL$ which can mediate proton decay and is only suppressed by squark masses,[3,4] not $M_{\rm GUT}$ or $M_{\rm Pl}$.

There is a simple remedy for this: one can set $W_{\not B} = W_{\not L} = 0$ by hand. In a non-supersymmetric theory, this would be 't Hooft unnatural unless there existed some exact global or gauge symmetry to ensure that these couplings remain zero. In a supersymmetric theory, however, couplings in the superpotential are always 't Hooft natural due to the "non-renormalization" theorem, which says that any couplings in $W$ that are set to zero will remain zero to orders in perturbation theory. Nonetheless, setting otherwise-allowed couplings to zero is always Dirac unnatural and so we might look for a symmetry that arises when these couplings vanish. The obvious candidates, global $B$ and $L$ symmetries, are probably not suitable to play this role since we expect from a number of arguments (*e.g.*, involving GUT's, $\nu$ masses, cosmological baryon asymmetry) that they will be broken by non-renormalizable or non-perturbative terms.

Fortunately it is easy to invent a new discrete symmetry which simultaneously forbids all the unwanted terms and which allows those that are phenomenologically necessary. This is "matter parity," a $\mathbb{Z}_2$ symmetry under which all the matter fields $(L, e, Q, u, d)$ are odd and the Higgs fields are even.[5] An added benefit of such a symmetry is the exact stability of the lightest supersymmetric particle (LSP), even against higher-order interactions, thereby providing a candidate for the (cold) dark matter in the universe.

Though matter parity may provide a "Dirac natural" solution to the proton decay problem, there is a more recent definition of naturalness which matter parity does not necessarily satisfy. We shall refer to this new definition, which comes out of ideas in string theory and quantum gravity, as "local naturalness." Whereas Dirac naturalness would allow any exact global symmetry to forbid the unwanted couplings, local naturalness requires that these symmetries be *gauge* symmetries, or at least the indirect consequence of an underlying gauge symmetry.[6] This requirement follows from the result/belief that in a theory of quantum gravity, there may be effects (potentially arising from Planck-scale physics) which violate any and all symmetries of the theory which are not gauged or somehow protected by a gauge symmetry. These protected symmetries include global and discrete subgroups of gauge symmetries. And though these Planck-scale effects might be small, suppressed by powers of $M_{\rm Pl}$, in some cases this may be enough to violate known constraints. In the case of proton decay, such a violation occurs. It would be desirable, then, to embed matter parity into a gauge symmetry.

In order to see how this might be done, note that the matter parity $P_M$ of any MSSM field with baryon and lepton numbers $B$ and $L$ respectively can be written as

$$P_M = (-1)^{3(B-L)}. \tag{2.4}$$

Matter parity can therefore easily be accommodated as a discrete subgroup of a gauged $U(1)_{B-L}$ symmetry, something which appears in many extensions of the SM. In fact, the action of the full $U(1)_{B-L}$ symmetry would be to forbid exactly the unwanted terms in $W$. If this selection rule is to survive the breakdown of the $U(1)_{B-L}$ symmetry, then one need only require that all order parameters (*e.g.*, Higgs VEV's) carry *even* integer values of $3(B-L)$. This restriction can also be generalized to groups containing $U(1)_{B-L}$; for example, in $SO(10)$ one finds[7] that matter parity occurs as a discrete gauge symmetry after GUT-breaking if that breaking is done without giving VEV's to spinor representations (*e.g.*, $\mathbf{16}, \mathbf{144}, \mathbf{560}\ldots$).

The question of how one can embed such a discrete symmetry into a gauge symmetry can be further generalized by asking whether or not the set of discrete charges in the MSSM is "anomaly-free," *i.e.*, whether or not these charges obey certain constraints which allow them to be interpreted as arising from a non-anomalous gauge symmetry (as in the example above). One can in fact show that matter parity is the only (generation-independent) $\mathbb{Z}_2$ symmetry which is anomaly-free and prevents dimension-four proton decay.[8]

Of course, the original argument against allowing non-zero $W_{\not B}$ and $W_{\not L}$ was that both could not be allowed without leading to rapid nucleon decay. This argument suffices to forbid one or the other, but not both. In fact, much work has been done considering extensions of the MSSM with so-called "$R$-parity violation" in which either $W_{\not B}$ *or* $W_{\not L}$ is non-zero, but not both. The price one pays is that the LSP is no longer stable and so there is no easy candidate for the dark matter. However, the phenomenology and motivations of $R$-parity violation are too complicated to discuss here; for details, see the contribution of H. Dreiner to this volume.

Thus far, we have said nothing about proton decay from higher-dimension operators. In particular there are many terms of dimension five *which are invariant under matter parity* which can mediate proton decay. For example, let us consider the superpotential term $W = (\frac{\eta}{M})QQQL$. Even if one identifies $M$ with the Planck mass, current bounds on the proton lifetime require that $\eta$ be unnaturally small: $\eta \lesssim 10^{-7}$. One could consider alternative $\mathbb{Z}_N$ symmetries for the superpotential, and in fact one particular alternative[8] stands out: a $\mathbb{Z}_3$ symmetry called "baryon parity." Under baryon parity the MSSM fields $(\widetilde{Q}, \widetilde{u}, \widetilde{d}, \widetilde{L}, \widetilde{e}, H_U, H_D)$ have $\mathbb{Z}_3$ charges (0,2,1,2,2,1,2) respectively. Baryon parity has many interesting properties: it is the only (generation-independent) $\mathbb{Z}_3$ symmetry of the MSSM without discrete gauge anomalies; it prevents dimension-four proton decay by forbidding $W_{\not B}$; it allows the $\mu$-term; it allows $W_{\not L}$ and therefore neutrino masses; it forbids dimension-five proton decay; but it also allows the decay of the LSP.

Is matter parity, or any of its competitors, a real symmetry of nature? This is essentially an experimental question, but its implications go far beyond questions of detection signals, for these symmetries teach us about dark matter and the ultimate fate of the universe, and also provide insights into the symmetry structure of the MSSM at very high energies.

## Question #2 How is flavor-changing suppressed?

There is no reason to expect that the mass and interaction eigenstates of the SM fermions coincide; that is, the quark and lepton mass matrices need not be diagonal in the interaction eigenbasis. In fact we know that they are not diagonal (since $\theta_c \neq 0$). One obviously expects the same to be true of the scalars of the MSSM. But before SUSY-breaking, one can expect that at least the mass matrices of fermions and their scalar superpartners will be diagonal in the same basis. This need not be true after SUSY-breaking.

In the interaction basis, non-diagonal mass matrices would seem to lead to large flavor-changing neutral currents (FCNC's). Within the SM, it is the GIM mechanism which prevents this from occurring. Flavor-independent neutral current (NC) gauge interactions couple gauge bosons to propagating fermions and their conjugates, each rotated from their interaction basis by matrices $U$ and $U^\dagger$ respectively. As long as $U$ is unitary, then $U^\dagger U = \mathbf{1}$ and the NC gauge interaction conserves flavor. The same holds for the scalar partners which are rotated by unitary $\widetilde{U}$ and $\widetilde{U}^\dagger$. Thus gauge bosons do not induce FCNC's for particles or their superpartners. The Problem arises with gauginos, which couple to both a particle *and* its superpartner simultaneously. In this case the coupling has the form $\widetilde{U}^\dagger U$, which need not be diagonal. Thus gauginos can generate FCNC's.[3,9] (This is not a complete disaster because for flavor-changing processes involving only fermions on external lines, gaugino contribution can arise only beyond the tree-level.)

Because FCNC's are suppressed by GIM in the SM, the dominant source for FCNC's in low-energy processes may come from SUSY. The tightest bounds on FCNC's presently come from $K^0 - \overline{K^0}$ mixing. Requiring that the MSSM prediction for the $K^0 - \overline{K^0}$ mass difference not exceed the measured value yields limits of the form:[11]

$$\frac{\mathcal{A}^2}{m^2_{\widetilde{Q}}}\left(\frac{\delta m^2_{\widetilde{Q}}}{m^2_{\widetilde{Q}}}\right)^2 \leq 5\times 10^{-9}\ \mathrm{GeV}^{-2} \tag{2.5}$$

where $\mathcal{A}^2$ is a product of angles which rotate from the quark mass basis to the squark mass basis, and $\delta m^2_{\widetilde{Q}}$ is the mass difference between the $\widetilde{d_L}$ and $\widetilde{s_L}$ squarks. There are corresponding limits on $\widetilde{d_R} - \widetilde{s_R}$ splittings as well as mixed left-right limits.

Given the above constraint, it is clear that if the mass splittings are $\mathcal{O}(1)$ and the angles take average values ($\mathcal{A}^2 \sim 1/20$), the squark mass scale must be $\gtrsim 3\,\mathrm{TeV}$. Similar bounds exist from $D^0 - \overline{D^0}$ and $B^0 - \overline{B^0}$ mixing, and in the slepton sector from processes such as $\mu \to e\gamma$. However it is worth noting that bounds on FCNC's tend to constrain only the first two generations of squarks and sleptons; the third generation is rather weakly constrained at present.

There are three primary proposals for solving the SUSY flavor problem: degeneracy, alignment, and decoupling.

Degeneracy attempts to solve the flavor problem by positing that squarks and sleptons of a given flavor are mass-degenerate,[10,11] *i.e.*, $\delta m^2_{\widetilde{Q}} = 0$. (One way to see that this would forbid FCNC's is to note that if, for example, the $\widetilde{d}$ and $\widetilde{s}$ were to have equal mass, they would exactly cancel each other's contributions to any flavor-changing process.) This extension of the GIM mechanism to the SUSY sector is often referred to as "super-GIM": $m^2_{\widetilde{u}} = m^2_{\widetilde{c}} \simeq m^2_{\widetilde{t}}$ for L,R squarks separately, likewise $m^2_{\widetilde{e}} = m^2_{\widetilde{\mu}} \simeq m^2_{\widetilde{\tau}}$. Given the weaker bounds on FCNC's involving the third generation, the final equalities need only be approximate.

Degeneracy models come in a variety of flavors themselves (for more details see Question #7). For many years, models based on supergravity (SUGRA) as a mediator of SUSY-breaking were considered members of this family. Now it is understood that there exist a number of rather generic phenomena which break degeneracy in SUGRA models, including non-minimal Kähler potentials, GUT effects, and superstring thresholds. Even if all of these could be ruled out, it is important to realize that *sources of flavor physics between the Planck and weak scales tend to violate the degeneracy.* This is a generic feature: if flavor physics occurs below the scale at which SUSY-breaking is communicated to the SM, scalar mass degeneracies will tend to be spoiled. An alternative to SUGRA mediation is gauge-mediated SUSY-breaking (GMSB). In GMSB two important things happen: the scalars of the MSSM receive their soft masses from SM gauge interactions, which are by definition flavor-universal, and these masses are communicated at scales often very close to the weak scale so that there is little room for new physics to spoil the degeneracy. Finally it has also been suggested that the flavor physics itself arises from non-abelian global symmetries ("horizontal" symmetries) under which the families of the SM/MSSM form non-trivial representations. These models usually predict some combination of degeneracy among the flavors and alignment, the next mechanism. One perhaps noteworthy aspect of non-abelian horizontal symmetries is that they can be gauged only in special cases,[13] because the broken Cartan generators of a local symmetry typically generate $D$-terms which destroy the mass degeneracy that one worked so hard to obtain in the first place.

Alignment solves the flavor problem not by setting $\delta\widetilde{m}^2 = 0$ as with degeneracy, but by enforcing $U = \widetilde{U}$ (or equivalently $\mathcal{A} = 0$) to a very high accuracy.[12] Here the flavor physics is typically generated by one or more $U(1)$ gauge interactions. Models of abelian horizontal symmetries seek to tie the generation of the scalar soft masses to that of the fermion mass/Yukawa matrices. All of these models generate the hierarchies in the fermion masses as powers of a small expansion parameter, usually the ratio between some flavor-violating VEV's and the UV scale of the theory. Particularly interesting among these models are those in which the flavor $U(1)$ is pseudo-anomalous[14] — the anomalies in its fermionic sector are cancelled by non-linear transformations of the dilaton/axion superfield as in the Green-Schwarz mechanism of string theory. Such a $U(1)$ must be broken just below the string scale via a one-loop induced Fayet-Iliopoulos term, and it is the ratio of this breaking scale to the string scale that provides the expansion parameter for building the mass hierarchies.

Finally, decoupling has been proposed as a solution to the flavor problem.[15] Here one simply makes the first two generations of sparticles heavy enough (typically $10 - 100\,\text{TeV}$) so that their contributions to FCNC processes vanish. But what about the 't Hooft naturalness problem in the SM Higgs sector that SUSY can solve only if the superpartners are light ($\lesssim 1\,\text{TeV}$)? Recall that the troublesome diagrams involving scalars were all suppressed by Yukawa couplings. If we make the reasonable assumption that naturalness requires the gauginos, higgsinos, *and the third generation of squarks and sleptons* to lie below 1 TeV, then all the other squarks and sleptons can have masses as large as $(m_t/m_{q,\ell}) \times (1\,\text{TeV})$ without violating naturalness constraints. (Here $m_{q,\ell}$ is the appropriate partner quark/lepton mass.) Once again one takes advantage of the fact that the bounds on third generation FCNC's are weak in order to allow large mass splittings between the third generation and the first two. One should note, however, that there exist problems with the behavior of the decoupling scenario at two-loop order;[16] it is not clear at present that such a scenario can be made to work without inadvertently breaking QCD and/or QED.

Once SUSY is discovered, it may not take long to discern which of these paths nature has chosen to follow. Observation of scalar partners should quickly tell us whether they are degenerate or not (*i.e.*, super-GIM or aligned); likewise, if only the third generation is found, we learn that the other spartners must be very heavy (*i.e.*, decoupled). However, it will take much more experimental and theoretical work to determine just how each of these choices is concretely realized.

## Question #3 Why isn't CP violation ubiquitous?

In the MSSM there are 43 physical CP-violating phases above and beyond that of the SM. (In this discussion, we will ignore $\theta_{\rm QCD}$.) Unlike the single SM phase, which does not typically engender large CP-violating effects[a] because in physical processes it always comes in proportional to the small Jarlskog parameter $J$, the phases of the MSSM can show up in large, easily observed, and easily constrained experimental processes. The tightest constraints come from CP-violating in the kaon system and the electric dipole moment of the neutron. The former receives SUSY contributions only if there are also flavor-changing SUSY contributions such as those discussed above; the latter exists even if SUSY is flavor-preserving. (For a review, see Ref. 17 and the contribution of A. Masiero and L. Silvestrini to this volume.)

Let us consider the second case first, with all SUSY flavor-changing effects put to zero through universal soft masses. For simplicity let us make the usual assumption that the trilinear couplings are all proportional to the Yukawas, and that the gaugino masses are universal at some scale. Then there remain only two physical phases beyond the SM associated with some combination of the $\mu$-term, the $B_\mu$-term, the $A$-terms, and the gaugino masses:

$$\phi_A = \arg(A^* M_3) \qquad\qquad \phi_B = \arg(B_\mu^* \mu M_3). \tag{2.6}$$

At one-loop order, gluinos and squarks can contribute to the electric dipole moment (EDM) of quarks, and then in turn nuclei. The EDM of the

[a] Even in the SM it is not always true that large CP-violating observables are lacking; for example, in $B-\overline{B}$ mixing, CP-violating effects can be $\mathcal{O}(1)$ since $J$ appears divided by small quark mixing angles.

neutron can be calculated to be[18] (for $M_3 \simeq m_{\tilde{q}} \equiv \tilde{m}$):

$$d_N \simeq 2 \left( \frac{100\,\mathrm{GeV}}{\tilde{m}} \right)^2 \sin(\phi_A - \phi_B) \times 10^{-23}\, e\,\mathrm{cm}, \tag{2.7}$$

where experimentally $d_N < 1.1 \times 10^{-25}\, e\,\mathrm{cm}$. Clearly, if the phases $\phi_{A,B}$ take values of $\mathcal{O}(1)$, then the sparticles must be heavier than 1 TeV; or for sparticles around 100 GeV, the phases must be $\lesssim 10^{-2}$. In either case (heavy sparticles or small phases) some degree of unnaturalness is introduced.

In the most general case where flavor violations are also allowed, not only does the number of physical CP-violating phases increase, but the bounds from observables become much stronger. For example, gluinos and squarks can appear in the internal lines of box diagrams contributing to $\epsilon_K$ (*i.e.*, the imaginary part of the $K^0 - \overline{K^0}$ mixing amplitude). One finds for $M_3 \simeq m_{\tilde{Q}} \simeq m_{\tilde{d}} \equiv \tilde{m}$ that[11]

$$\epsilon_K \simeq \left( \frac{10^{10}\,\mathrm{GeV}^2}{\tilde{m}^2} \right) \mathrm{Im} \left( \frac{\delta m^2_{\tilde{Q}}}{m^2_{\tilde{Q}}} \frac{\delta m^2_{\tilde{d}}}{m^2_{\tilde{d}}} \right) . \tag{2.8}$$

Experimentally, $\epsilon_K$ is known to be about $2.3 \times 10^{-3}$. Thus, in order to allow $\mathcal{O}(1)$ mass splittings and phases, the squarks and sleptons must have masses $\gtrsim 1000$ TeV! Conversely, in order to allow masses below 1 TeV, either the mass splittings or the CP-violating angles must be made unnaturally small.

The CP problem is solved in ways that are similar to those that solve the flavor problem. For the $\epsilon_K$ problem, degeneracy appears to be an attractive solution (decoupling from $\epsilon_K$ would seem to put too heavy a burden on any theory, pushing squarks to 1000 TeV). The EDM problem is not so simply solved, but then it is not quite as serious. Decoupling *can* eliminate this problem; it is also solved in some classes of very minimal gauge-mediated supersymmetry-breaking (GMSB) models.[20] One particularly interesting line of inquiry has involved attempts to use this CP-violation as the source needed during baryogenesis;[21] in this case one obtains strong constraints on the parameters on the MSSM and thus a highly predictive model that will soon be tested.

## Question #4 Where does the $\mu$-term come from?

The $\mu$-term of the MSSM prompts another question of Dirac naturalness: how do we induce a parameter to take a value far below its "natural" scale?

For the $\mu$-parameter of the MSSM,

$$W = \cdots + \mu H_U H_D, \tag{2.9}$$

the natural value is the UV cutoff of the MSSM. This is not a statement about radiative corrections *à la* 't Hooft, for in SUSY models the non-renormalization theorem will protect a weak-scale $\mu$-parameter from any large corrections. Rather, it is a question of why a mass parameter which is SUSY-invariant and $SU(3) \times SU(2) \times U(1)$-invariant would have a value typical of SUSY-breaking, or SM-breaking, masses. *A priori*, we would expect it to have a value of order the scale at which $H_U H_D$ no longer forms a gauge singlet, or the Planck scale, whichever is smaller. Within the context of a GUT model, the problem is exacerbated. In $SU(5)$ parlance, the $\mu$-term provides the mass for the complete **5** and $\overline{\mathbf{5}}$ of Higgs, thereby becoming entangled in the famous doublet-triplet splitting problem.

Phenomenologically, we know that $\mu \sim m_Z$ because minimization of the MSSM Higgs potential yields the result

$$\mu^2 = \frac{m_{H_D}^2 - m_{H_U}^2 \tan^2\beta}{\tan^2\beta - 1} - \tfrac{1}{2} m_Z^2 \tag{2.10}$$

where all the masses on the right side are $\sim m_Z$. Only a gross fine-tuning would allow values of $\mu$ very different than $m_Z$. We could in principle set $\mu \equiv 0$ by invoking a Peccei-Quinn symmetry, but the by-product of this would be a standard axion, already ruled out by direct searches.

There is an almost-default solution to the $\mu$-problem which goes by the acronym NMSSM (Next-to-Minimal...). In this model, a gauge singlet $N$ is introduced whose role is to produce a $\mu$-term through its VEV. The superpotential would have the form:

$$W = \lambda N H_U H_D + \lambda' N^3 \tag{2.11}$$

while $N$ itself is presumed to have a soft $(\text{mass})^2$ term which is negative. (Such a negative $(\text{mass})^2$ can actually arise naturally if $\lambda$ or $\lambda'$ is large, since it will drive the soft mass term negative in the infrared, just as occurs for the $H_U$ mass term in radiative electroweak symmetry breaking.) Obviously a $\mu$-term arises: $\mu = \lambda \langle N \rangle$. Several terms contribute to a $B_\mu$-term, including the trilinear soft term $\lambda A_N \langle N \rangle H_U H_D$, and $F_N$ via $B_\mu \sim F_N H_U H_D \sim \lambda' \langle N^2 \rangle H_U H_D$. However, the singlet solution to the $\mu$-problem is not without problems, as we will discuss further after Question #8.

Within SUGRA there is actually a more attractive solution, known as the Giudice-Masiero mechanism.[22] If the $\mu$-term is forbidden by some symmetry (say a discrete symmetry) which is violated in the hidden sector, then a $\mu$-term can arise through a non-minimal Kähler potential: $K = \cdots + (S/M_{\rm Pl})H_U H_D$, where the ellipsis represents the canonical terms and $S$ is a hidden sector field with $F$-term $\langle F_S \rangle = m_Z M_{\rm Pl}$. Then

$$\int d^4\theta \frac{S}{M_{\rm Pl}} H_U H_D = \int d^2\theta \frac{F_S}{M_{\rm Pl}} H_U H_D \equiv \int d^2\theta \, \mu H_U H_D. \qquad (2.12)$$

In this way the $\mu$-term, which is SUSY-preserving, is actually tied to the breaking of SUSY and thus naturally $\sim m_Z$. The corresponding value of $B_\mu$ will then also be $\sim m_Z^2$.

Within non-SUGRA models, there are no such simple solutions. In particular, GMSB models struggle with a severe $\mu$-problem. In generic models one typically finds $B_\mu/\mu^2 \gg 1$ where one needs $B_\mu \sim \mu^2$ phenomenologically. In some special cases one finds $B_\mu/\mu^2 \ll 1$. Here one would expect an axion, but one-loop corrections to $B_\mu$ pull the axion mass above $m_Z$. The hallmark of such a scenario[20] is a very large value for $\tan\beta$ ($\sim 50$).

## Question #5 Why does the MSSM conserve color and charge?

In the SM, the only field which can receive a VEV is the Higgs field, and because of its quantum numbers, a Higgs VEV uniquely breaks $SU(3) \times SU(2) \times U(1) \to SU(3) \times U(1)$, thereby preserving QCD and QED. In the MSSM there are a large number of charged and colored scalars in addition to the Higgs, any of which could receive a VEV and break the gauge group even further. Whether or not this occurs is a function of the potential felt by these scalars. Unfortunately, minimizing this potential can be an arduous task, complicated by the large numbers of scalar fields. (For a full discussion of efforts on this front, see Ref. 23 and the contribution of A. Casas to this volume.)

Scalar potentials in SUSY receive contributions from three sources: $D$-terms, $F$-terms, and soft-breaking terms. The first of these provides the quartic terms $V \sim \lambda\varphi^4$ with $\lambda \geq 0$. For fields which carry some gauge charge, $\lambda = 0$ can occur along only special directions in field space known as "$D$-flat directions" or "$D$-moduli." Along a flat direction the quartic potential takes the form $V \sim (\varphi_1^2 - \varphi_2^2)^2 \to 0$ as $|\varphi_1| \to |\varphi_2|$, and so the potential far from the origin may not be well-behaved (*i.e.*, $\varphi$ may run off to infinity). Whether or not this occurs will depend on the $F$-terms and soft terms.

The $F$-terms in turn contribute quadratic, cubic, and quartic terms to the potential. Because these terms are supersymmetric, the portion of the potential due to $F$-terms is positive semi-definite. This does not mean, however, that the minimum of the potential lies at the origin. Rather, it means that directions in field space with non-zero quartic $F$ contributions will be well-behaved far away from the origin. Still, there will be a subset of the $D$-flat directions which are also $F$-flat and whose behavior will be completely controlled by soft-breaking terms.

But the soft mass contributions are problematic, as they can have either sign. Because they contribute to only the quadratic and cubic pieces of $V$, one can analyze their structure most readily along flat directions in which the quartic pieces all vanish. Then one finds two distinct types of problems which may arise: potentials which are charge- and color-breaking (CCB) at their minima, or potentials which are unbounded from below (UFB).

CCB most readily occurs along directions which are $D$-flat, though not necessarily $F$-flat; then the $\varphi^4$ contributions to the potential are suppressed by Yukawas. The canonical example[24] of CCB involves only the fields $H_U$, $\widetilde{Q}$ and $\widetilde{u}$ which have a $D$-flat direction in which $|H_U| = |\widetilde{Q}_u| = |\widetilde{u}|$. The potential along this direction receives dangerous cubic contributions from the soft trilinear terms $\lambda_u A_u \widetilde{Q} H_U \widetilde{u}$ which can dominate over the small residual quartic terms (proportional to Yukawa couplings since this is not an $F$-flat direction) out to large field values. The condition that a secondary (and deeper) minimum not be generated away from the origin then results in the famous bound:

$$|A_u|^2 \leq 3(m^2_{\widetilde{Q}} + m^2_{\widetilde{u}} + m^2_{H_U} + \mu^2). \tag{2.13}$$

In principle, bounds such as this can be derived along every $D$-flat direction of the MSSM. It is clear that rigorously analyzing such a possibility is hopeless when one considers that the space of all $D$-flat directions is itself 37-complex dimensional!

The appearance of UFB directions is also common, occurring along directions which are both $D$- and $F$-flat. The usual example is found right in the Higgs potential along the direction $|H_U| = |H_D|$. As with all UFB potentials, there is no quartic contribution to the potential along this direction (nor in this example is there a cubic piece). Stability of the potential then requires that the quadratic pieces be positive semi-definite, *i.e.*,

$$m^2_{H_U} + m^2_{H_D} + 2\mu^2 \geq 2|B_\mu|. \tag{2.14}$$

Once again, bounds such as these are fairly generic — the space of directions which are both $D$- and $F$-flat in the MSSM is 29-complex dimensional.[25]

Can these considerations prove useful beyond providing bounds on soft parameters? The answer to this question depends somewhat on the values of the soft masses which are measured experimentally. One can turn around the above analysis and ask: What would it mean if the measured masses violated a CCB/UFB constraint? In the case of UFB directions, the stability of the potential must be rescued, either by non-renormalizable operators coming from new physics (*e.g.*, $V \sim \varphi^6/M^2$) or by one-loop contributions to the effective potential (*e.g.*, $V \sim \mathrm{STr}\, m^4(\varphi) \log[m^2(\varphi)/Q^2]/64\pi^2$). In either case, the result is generally a bounded potential with CCB VEV's well above the weak scale.

However, the existence of a global CCB minimum below that of the SM does not necessarily imply that we should be in it. It is entirely possible that, at the end of inflation and reheating, the universe found itself in the current vacuum even though there is a deeper vacuum elsewhere. If the barrier between the two vacua is high, the time scale for our universe to tunnel to the new vacuum may be much larger than its current age. But why would the universe end up in the "wrong" vacuum to begin with? Presumably the answer lies in how the CCB minima are lifted by finite-temperature effects relative to the SM-like minimum.[26] It may also depend on whether or not some late (weak-scale) inflation occurred and what its reheating temperature was. Once the universe ended up in the SM-like vacuum, transitions to the CCB vacuum would be exponentially suppressed by the height of the intervening barrier, easily leading to a metastable (but very long-lived) universe. Thus the appropriate question raised by discovering violation of the CCB bounds may well be cosmological rather than directly experimental.

## Section II: Open Questions on SUSY-Breaking

All of the questions we have discussed thus far have begun with the structure of the MSSM. Implicit in that construction is the fact that as a symmetry of nature, SUSY makes some profoundly (and obviously) wrong predictions. In supersymmetric theories, there is an absolute correspondence between fermions and bosons that is not manifest experimentally. Specifically, SUSY requires a spectrum in which every fermionic degree of freedom has a bosonic counterpart with identical mass and quantum numbers. Therefore, in order for SUSY to play any role in low-energy physics, it must clearly be broken (or hidden), just as is the $SU(2)$ gauge symmetry of weak interactions.

## Question #6 How is SUSY broken?

Once it is clear that SUSY must be broken, the next logical step would seem to be to find some way in which to break SUSY spontaneously. Why spontaneously? Our experience with gauge interactions teaches us that Ward identities, and with them all of the desirable properties of symmetries, are preserved after the symmetry is broken *only if* the symmetry-breaking is done spontaneously. For gauge symmetries, this means that renormalizability and unitarity are lost for explicit breaking; for SUSY, it is the cancellation of the quadratic divergences that would be lost. There is also another, more philosophical, reason for demanding spontaneous SUSY-breaking: symmetries which are explicitly broken are not symmetries at all (even if they may be useful for classification purposes), while symmetries which are broken spontaneously are still symmetries of the theory — they are just not symmetries of the vacuum state of the theory.

It is clear that breaking SUSY spontaneously entails adding to the SM some fields and their interactions to act as a SUSY-breaking sector, just as the Higgs and its potential are added to the SM. Can the fields of the MSSM play this role themselves? For a number of reasons, it turns out that they cannot. First, of the MSSM fields that could receive a SUSY-breaking VEV, it is only the Higgs and sneutrinos which can do so without breaking too many gauge symmetries. But after explicit calculation, one finds that the Higgs/sneutrino potential is minimized at the origin with $V = 0$, so no SUSY-breaking occurs. It thus becomes quickly apparent that some new fields must be added to do the job of SUSY-breaking.

By putting in a set of new fields, it is easy enough to break SUSY (for example, through an O'Raifeartaigh-type superpotential). But how, if at all, can these fields couple to the usual MSSM fields? One of the properties of SUSY which is preserved after spontaneous breaking[3] is the famous supertrace formula, applied separately to each individual supermultiplet:

$$\mathrm{STr}\, M^2 = 0. \tag{2.15}$$

This formula is phenomenologically untenable, for it predicts that the masses of scalars are distributed evenly above and below the fermion masses. For example, one of the up-type squarks must be no heavier than an up quark!

Eq. (2.15) is true only at tree-level. Even so, this constraint requires that a SUSY-breaking sector must not have renormalizable tree-level couplings to ordinary matter. Because the SUSY-breaking must be kept at some distance from the SM sector, it has been called variously a "hidden" or "secluded" sector, while the SM is said to live in the "visible" sector.

Because the actual breaking of SUSY is far removed from the SM, the question of how SUSY is broken is also far removed from experimental probes. Thus, this question becomes subsidiary to the next issue we will consider, namely that of communicating SUSY-breaking to the SM. Indeed, of the multitude of models which we now know to break SUSY and which could live in the hidden sector, it is often the case that the resulting visible-sector phenomenology is much less sensitive to model-specific details than to the method itself by which the SUSY-breaking is communicated. We therefore leave further discussion on the topic of SUSY-breaking to the contribution of M. Peskin to this volume.

Note that there is one aspect of SUSY-breaking that may nevertheless have universal applicability, even if we do not know the details of the SUSY-breaking sector itself. Within a field theory, SUSY-breaking generically occurs because $F$-terms receive VEV's (the case of $D$-term VEV's rarely drives SUSY-breaking in realistic models). The $F$-VEV's are controlled by the superpotential as in the original O'Raifeartaigh model, which always requires that a mass scale be present to set the scale of the VEV's. In order to obtain weak-scale SUSY, we expect that this mass scale, regardless of the method of SUSY-breaking, will be far below the Planck scale. Is this natural *à la* Dirac?

Witten realized that in fact this *is* natural if the scale in the superpotential comes from strong-coupling dynamics in some asymptotically free gauge theory.[27] Indeed, in strongly coupled gauge theories, potentials of the form

$$V = \frac{\Lambda^n}{\varphi^{n-4}} + \lambda\varphi^4 \tag{2.16}$$

are typical: the first term might arise from instantons, gaugino condensation, or other strong dynamics, while the second term occurs at tree-level. Such a potential breaks SUSY, with $F_\varphi \sim \Lambda^2$. Furthermore, since $\Lambda$ is the strong-coupling scale of the gauge group, it can be expressed via dimensional transmutation as

$$\Lambda = M_{\rm Pl}\, e^{8\pi^2/bg^2(M_{\rm Pl})} \tag{2.17}$$

where the one-loop $\beta$-function coefficient $b$ is negative. Thus $\Lambda$ is a new scale in the theory, exponentially far from the Planck scale. In this way, SUSY-breaking may hold the key to understanding the fundamental hierarchy problem of particle physics, explaining why the Planck scale is so far from the weak scale.

**Question #7** **Once SUSY is broken, how do we find out?**

We have said that the dynamics of SUSY-breaking must occur far from the sector of the SM. But this begs the question: how does the SM "learn" that SUSY has been broken? Remember that at tree-level, $\mathrm{STr}\, M^2 = 0$, a constraint that must be violated in the visible sector. We have already hinted at how this can be done: *Eq. (2.15) is true only at tree-level and for renormalizable interactions.* Two routes therefore seem open to us for communicating SUSY-breaking to the visible sector: loops and non-renormalizable interactions. Both have found application in realistic models.

### 2.7.1. *Supergravity mediation*

The "default" mechanism for communicating SUSY-breaking is through the non-renormalizable interactions found in supergravity theories. (A full introduction to supergravity as a mediator of SUSY-breaking can be found in the contribution of R. Arnowitt and P. Nath to this volume, or in the standard references.[1]) Local SUSY, *i.e.*, supergravity, is automatically a non-renormalizable field theory for gravity, containing the spin-two graviton and its spin-3/2 partner, the gravitino. The Lagrangian for a supergravity model is determined in terms of three arbitrary functions of the superfields: the superpotential $W(\varphi)$, the Kähler potential $K(\varphi, \varphi^\dagger)$, and the gauge kinetic function $f(\varphi)$. Note that $W$ and $f$ are holomorphic in $\varphi$, while $K$ is real.

The minimal supergravity-mediation model relies on the following assumptions:

- The superpotential can be written in the form $W = W_H(X) + W_V(\varphi)$ where $W_H(X)$ is the superpotential for the hidden sector fields $X$, and $W_V(\varphi)$ is the superpotential for the visible sector fields $\varphi$.
- The Kähler potential is the minimal one: $K = \sum_i X_i^\dagger X_i + \sum_i \varphi_i^\dagger \varphi_i$.
- The gauge kinetic function is given as $f = cX/M_{\mathrm{Pl}}$ for some constant $c \sim \mathcal{O}(1)$.

- In the hidden sector, SUSY breaks such that $\langle F_X \rangle \neq 0$, $\langle W_H \rangle \sim \langle F_X \rangle M_{\rm Pl}$, and $\langle V \rangle = 0$.

After SUSY-breaking, the scalar potential of supergravity

$$V(\chi) = e^{K/M_{\rm Pl}^2} \left( \left| \frac{\partial W}{\partial \chi_i} + \frac{\partial K}{\partial \chi_i} \frac{W}{M_{\rm Pl}^2} \right|^2 - 3\frac{|W|^2}{M_{\rm Pl}^2} \right) \quad \text{for } \chi = X, \varphi \tag{2.18}$$

reduces in the visible sector to

$$V(\varphi) = e^{\langle K \rangle / M_{\rm Pl}^2} \frac{|\langle W_H \rangle|^2}{M_{\rm Pl}^4} \sum_i |\varphi_i|^2 , \tag{2.19}$$

giving masses to all visible sector fields $\sim F_X/M_{\rm Pl}$. One of the more well-known features of this mechanism is that all the visible-sector scalars receive exactly the same mass, leading to the well-advertised mass universalities of supergravity. Similar analyses give universal trilinear and bilinear terms, also $\sim F_X/M_{\rm Pl}$. The gaugino masses arise from

$$\mathcal{L} = \frac{1}{4} e^{K/2M_{\rm Pl}^2} \sum_i \frac{\partial f_a}{\partial \chi_i} M_{\rm Pl} \left( \chi_i + \frac{M_{\rm Pl}^2}{W} \frac{\partial W}{\partial \chi_i} \right) \lambda^a \lambda^a , \tag{2.20}$$

again yielding (universal) masses $\sim F_X/M_{\rm Pl}$.

If supergravity is the *dominant* mediator of SUSY-breaking, then the weak scale can be *defined* to be $F_X/M_{\rm Pl}$, *i.e.*, $F_X \sim m_Z M_{\rm Pl}$. In fact, even if supergravity is *not* the dominant mediator, it will still communicate SUSY-breaking to the visible sector and the masses induced will always be $\sim F_X/M_{\rm Pl}$. Many of these results can be summarized neatly by writing the effective operators for the soft masses in superspace:

$$\begin{aligned} m^2 \varphi^* \varphi &\sim \int d^4\theta \, \frac{X^\dagger X}{M^2} \varphi^\dagger \varphi = \frac{|F_X|^2}{M^2} \varphi^* \varphi \\ M_a \lambda^a \lambda^a &\sim \int d^2\theta \, \frac{X}{M} W^{\alpha a} W^a_\alpha = \frac{F_X}{M} \lambda^a \lambda^a \\ A_{ijk} \varphi_i \varphi_j \varphi_k &\sim \int d^2\theta \, \frac{X}{M} \varphi_i \varphi_j \varphi_k = \frac{F_X}{M} \varphi_i \varphi_j \varphi_k \end{aligned} \tag{2.21}$$

where $M$ is the scale of the messenger interactions/particles. In supergravity, $M = M_{\rm Pl}$. Note that all masses are of the same order, $m \sim F_X/M$, as we found in supergravity.

The final outputs of minimal supergravity are four mass parameters which describe all the soft masses of the MSSM: a universal scalar mass $m_0$, a universal gaugino mass $M_{1/2}$, a universal $A$-term $A_0$, and a universal $B$-term $B_0$. (In addition, one must also specify the SUSY-preserving $\mu$-term in order to fully describe the MSSM Lagrangian.) These four parameters can then be evolved from the GUT or Planck scale down to the weak scale in order to form the basis for realistic SUSY phenomenology studies.[28]

It is obvious why such a scenario has been the favored mechanism for communicating SUSY-breaking since its inception: it simplifies the spectrum of the MSSM considerably; it is automatic in the sense that any theory that connects gravity to a supersymmetric field theory would seemingly have to include supergravity; and, at lowest order, it produces the kind of universal masses necessary to solve the FCNC problem described previously. Why should we even consider anything else?

It is by now well-known that, beyond the lowest order, there are many effects in supergravity models that can significantly disrupt the mass universality at the weak scale. For example, there is no way to forbid all terms of the form $y_{ij}(X^\dagger, X)\varphi_i^\dagger\varphi_j$ from appearing in the Kähler potential $K$. Though such terms are suppressed by powers of $1/M_{\rm Pl}$, the hidden sector fields receive VEV's $\sim m_Z M_{\rm Pl}$ so that terms of this type contribute to the scalar masses with size $\sim m_Z$. Furthermore, because gravity does not "know about" the mass basis choice imposed by the Higgs Yukawa interactions, there is no reason for $y_{ij}$ to be diagonal in the same basis as the fermions. There are also other effects, including RGE running in the third generation, which can spoil universality and lead to observable FCNC effects. More generally, as we have already emphasized, any source of flavor physics between the Planck and weak scales will tend to violate the degeneracy.

### 2.7.2. *Gauge mediation*

One might hope for some way of communicating SUSY-breaking that yields mass universality more robustly. In that vein, there has been much recent interest in so-called "gauge-mediated" models.[29] The basic principle for these models is rather simple: If the scalar soft masses are functions only of the gauge charges of the individual sparticles, universality is automatic. (Remember that universality in this context only refers to sparticles with identical quantum numbers, such as $\widetilde{d}_L, \widetilde{s}_L, \widetilde{b}_L$-squarks.) Furthermore, if the scale at which the communication of SUSY-breaking takes place is well below the Planck scale, then the Planckian "corrections" discussed above cannot disrupt the universality ($F_X/M_{\rm Pl} \ll m_Z$).

We will not say much here about the details of gauge-mediation; interested readers should see the contributions of M. Peskin and S. Dimopoulos to this volume. The effective mass operators are changed from those in Eq. (2.21) in two ways: first, the messenger scale is now $M \ll M_{\rm Pl}$, and second, the soft masses arise through loops, so each operator experiences an additional $n$-loop suppression $\sim (\alpha/\pi)^n$. Specifically, we obtain

$$\begin{aligned} m^2\varphi^*\varphi &\sim \left(\frac{\alpha}{\pi}\right)^2 \frac{|F_X|^2}{M^2}\varphi^*\varphi \\ M_a\lambda^a\lambda^a &\sim \left(\frac{\alpha}{\pi}\right) \frac{F_X}{M}\lambda^a\lambda^a \\ A_{ijk}\varphi_i\varphi_j\varphi_k &\sim \left(\frac{\alpha}{\pi}\right)^2 \frac{F_X}{M}\varphi_i\varphi_j\varphi_k. \end{aligned} \tag{2.22}$$

If the scalar mass is identified to be $\sim m_Z$, then the gaugino mass will also be $\sim m_Z$, but the $A$-term will be $\sim (\alpha/\pi)m_Z \ll m_Z$. So non-zero $A$-terms essentially do not arise in gauge-mediated models, though they may reappear through renormalization group flow.

Can experiments differentiate this type of mediation from supergravity mediation? Perhaps most significantly for phenomenology, these models predict that the lightest SUSY particle will be the gravitino and that other SUSY particles can decay into it with observable lifetimes. For an interesting portion of the parameter space of these models, the missing-energy signal typical of SUSY models is augmented by two hard photons. However, the rest of the phenomenology of these models is very similar to that of the supergravity models.

### 2.7.3. *Mediation via pseudo-anomalous $U(1)$*

Finally, there also exists one additional method for communicating SUSY-breaking that we shall mention. In string theories, there is often one $U(1)$ gauge group factor (typically denoted $U(1)_X$) whose fermionic matter content appears to be anomalous but under which the string axion field transforms non-linearly, cancelling the anomaly. (This will be discussed in more detail after Question #18.) The $U(1)_X$ gauge fields by necessity have interactions in both the visible *and* hidden sectors, interactions which can communicate SUSY-breaking.[19] Because of the anomalous matter content, the $U(1)_X$ gauge superfield acquires a Fayet-Iliopoulos term at one-loop order which breaks the gauge symmetry at a scale one to two orders of magnitude below the Planck scale: $\epsilon \equiv M/M_{\rm Pl} \simeq 10^{-(1-2)}$. The effective visible sector mass operators are analogous to those in Eq. (2.21), except that the $X$ fields are not singlets but are instead charged under the anomalous $U(1)_X$. Thus the scalar masses can still arise as they do in Eq. (2.21), but the gaugino masses and $A$-terms cannot. For these latter cases, we must make the replacement

$$\int d^2\theta \, \frac{X}{M} \longrightarrow \int d^2\theta \, \frac{X^+X^-}{M^2} = \frac{F_{X^+}X^-}{M^2} \sim \epsilon M. \tag{2.23}$$

Thus the gauginos are generically much lighter than the scalars. In order to satisfy experimental bounds on the gauginos, the scalars must then be very heavy ($> 1\,\mathrm{TeV}$). This would reintroduce the naturalness problem of the SM. One solution is that the third generation scalars have no $U(1)_X$ charge and so they and the gauginos receive masses $F_X/M_{\mathrm{Pl}} \equiv m_Z$, while the first and second generation scalars are charged under $U(1)_X$, giving them masses $\sim m_Z/\epsilon$ where they would be somewhat immune to the supergravity corrections which could lead to FCNCs. (There is also the possibility that for $\epsilon$ small enough, such states could decouple from flavor-changing processes; the problems here would be the same ones that have been noticed for the decoupling scenarios discussed after Question #2.)

The phenomenology of these models has not been explored in any great detail. Since these models offer the chance to combine the best parts of the universality and decoupling solutions to the SUSY flavor problem, they may deserve more attention.

## Section III: Open Questions in Simple Extensions of the MSSM

Having considered the various issues that can arise in the MSSM, we now turn our attention to two of its simplest and best-motivated extensions. Probably the simplest extension of any gauge theory is to add to the spectrum one or more states which are complete gauge singlets. In specific SUSY models, gauge singlets are often introduced whose VEV's can provide mass scales without breaking gauge symmetries; the best example is the case of a singlet solution to the $\mu$-problem already discussed after Question #4. The simplest possible extension of the SM gauge group is the addition of extra $U(1)$ factors. Such models have been considered in the past for many reasons: extra $U(1)$'s arise naturally in higher-rank GUT groups, they are useful for communicating SUSY-breaking as in gauge-mediated models, *etc.* Depending on the model, these $U(1)$'s can arise in either the hidden or the visible sector, and as we shall see, each case has its own set of open questions.

## Question #8 Can gauge singlets and SUSY coexist?

We have already argued that, at least for the $\mu$-problem, it might be useful to add a gauge singlet to the spectrum of the MSSM. But there are two primary barriers to doing so, one cosmological and the other fundamental. If a singlet $S$ appears in the MSSM coupled to $H_U H_D$ in place of an explicit $\mu$-term, the action of the MSSM possesses a $\mathbb{Z}_3$ discrete symmetry under which all superfields are singly charged. When $S$ receives a VEV at the weak scale, it breaks the $\mathbb{Z}_3$ symmetry and could, in principle, precipitate the formation of domain walls in the universe. Such walls would dominate the energy density of the universe, yielding $\Omega \gg 1$. Solutions to this problem usually involve either a period of late inflation or breaking the $\mathbb{Z}_3$ symmetry explicitly[30] through non-renormalizable terms in the superpotential or Kähler potential.

However, the more fundamental problem arising for fields which are gauge and global singlets is that the tadpoles associated with them can reintroduce quadratic divergences which destabilize the gauge hierarchy.[31] These tadpoles can arise at one-loop order for non-minimal Kähler potential $K$, or at two-loop order even if $K$ is minimal. Since tadpole diagrams arise only for gauge singlets, the loop will be cut off by the scale of some new physics, usually new physics under which the singlet accrues gauge charges. Similar arguments can also be made for global symmetries; in this case, however, we should demand "local naturalness" since gravitational corrections to the Kähler potential may violate the global symmetry and reintroduce the tadpoles with $\mathcal{O}(1)$ coefficients.

As a simple example, let us consider the Kähler potential

$$K = \cdots + (N + N^\dagger)\Phi^\dagger\Phi/M_{\rm Pl} \tag{2.24}$$

where $N$ is the singlet and $\Phi$ is any light chiral superfield in the theory. At one-loop order, the resulting contribution to the Lagrangian in a theory with supergravity is given by

$$\begin{aligned} \delta\mathcal{L} &\sim \frac{1}{16\pi^2}\frac{\Lambda^2}{M_{\rm Pl}}\int d^4\theta\, e^K (N + N^\dagger) \\ &\sim m_{3/2}^2 M_{\rm Pl} N + m_{3/2} M_{\rm Pl} F_N \ . \end{aligned} \tag{2.25}$$

Here $\Lambda$ is the cutoff for the loop integration and can be taken to be the scale at which the singlet picks up some charge; for true singlets we take $\Lambda = M_{\rm Pl}$. The final equality in Eq. (2.25) follows from the fact that the superspace density $e^K$ receives a VEV of the form $\langle e^K \rangle \sim 1 + m_{3/2}\theta^2 +$

$m_{3/2}^2\theta^2\overline{\theta}^2$ in the process of SUSY-breaking. Unless the gravitino mass is exceedingly small (as can happen in some models of low-energy SUSY-breaking[32]), this contribution destabilizes the singlet VEV, pulling it — and whatever it couples to — up to large values. However, it may be possible to build realistic, and very generic, models of GUT- and intermediate-scale symmetry breaking which are actually driven by the tadpole contributions rather than upset by them.[33] Thus, what may have seemed a problem may indeed become a virtue.

## Question #9 **How do extra $U(1)$'s fit into SUSY?**

There are two primary ways of extending the gauge structure of the MSSM: we can embed the MSSM gauge groups into a large simple group as with GUT models (see Question #13 below), or into a larger direct-product group structure. In the second case, which we shall discuss here, it is difficult to build realistic models in which this additional gauge-group structure is non-abelian, for such extensions typically require extending the multiplet structure of the MSSM $SU(3) \times SU(2)$ gauge groups. On the other hand, additional *abelian* gauge groups are relatively simple to introduce, and thus they find their way into many possible extensions of the MSSM. (For a full discussion, see the contribution of M. Cvetič and P. Langacker to this volume.)

In non-supersymmetric models, the scale at which the additional gauge group $U(1)'$ breaks is arbitrary. This is partially due to the fact that it is difficult (and perhaps impossible) to stabilize the gauge hierarchy in such theories. Within the context of supersymmetric theories, however, the scales of extra gauge interactions are tightly constrained by the form of the SUSY scalar potential. There are two primary cases that one can consider. The first possibility is that the $U(1)'$ breaks along a direction in the potential which is $D$- and $F$-flat, so that the scale of symmetry breaking is set by non-renormalizable operators and/or radiative corrections to the potential.[34] The second possibility is that the breaking of the additional $U(1)'$ gauge symmetry is not along a flat direction. In this case the symmetry-breaking scale is constrained by SUSY to be very close to the weak scale. The first possibility is difficult to rule out, and is in fact highly model-dependent. The second possibility, by contrast, is in many ways more natural, but begs the question: if the new interactions should lie near the weak scale, where are they?

If the extra gauge interactions live in the hidden sector, the argument against non-abelian groups disappears. However, the only interesting scale for symmetry-breaking from the point of view of the visible sector is the scale at which SUSY is also broken. For non-abelian groups, any of the previously discussed methods for communicating SUSY-breaking to the visible sector would now play their role, and the physics of the hidden sector itself would become difficult to probe. But in the case of an extra abelian symmetry, something else can occur.

If SUSY is broken in the hidden sector at a scale $\Lambda \gg m_Z$, then it is expected that a VEV of size $\sim \Lambda^2$ for the $U(1)$ $D$-term will be generated. This in itself is not undesirable. However there is another generic feature of models with multiple $U(1)$'s which, when combined with such large $D$-terms, can become dangerous.[35] Since for a $U(1)$ interaction the gauge field strength tensor $F_{\mu\nu}$ is gauge-invariant, the Lagrangian can contain terms which mix the field strengths of two different $U(1)$'s. Specifically, we can have

$$\mathcal{L} = -\frac{1}{4}F^{(a)}_{\mu\nu}F^{(a)\mu\nu} - \frac{1}{4}F^{(b)}_{\mu\nu}F^{(b)\mu\nu} - \chi\, F^{(a)}_{\mu\nu}F^{(b)\mu\nu} + \cdots \tag{2.26}$$

for a $U(1)_a \times U(1)_b$ theory. Even if $\chi = 0$ at tree-level, it can be generated by loops. And because the mixing operator is dimension-four, contributions from massive (*e.g.*, stringy) states do not decouple since they are not suppressed by $M_{\rm Pl}^{-1}$.

When this "gauge kinetic mixing" is generalized to the SUSY case, mixing of the field strengths $F^{(a)}_{\mu\nu}$ implies mixing of the field strength spinors $W_\alpha$ which in turn implies mixing of their $D$-components. On integrating out the auxiliary $D$-fields, the scalar potential of each sector is sensitive to the SUSY-breaking $D$-VEV's that are present in the other sector. Thus the squarks, sleptons, and Higgs bosons of the MSSM, all of which are charged under $U(1)_Y$, learn about the SUSY-breaking scale in the hidden sector. Such contributions, if present, destabilize the gauge hierarchy in the MSSM.

Are there ways out of this disaster? Several options exist.[35] First, this result is special for extra $U(1)$'s; such gauge kinetic mixing cannot occur for non-abelian gauge symmetries. Second, there are discrete symmetries which can forbid such mixing; these are essentially charge-conjugation symmetries which act on one $U(1)$ but not the other: $C(A^{(a)}_\mu) = -A^{(a)}_\mu$ but $C(A^{(b)}_\mu) = +A^{(b)}_\mu$. Such symmetries can arise naturally if, for example, the two groups are unified into some non-abelian group $G_N$ whose central $\mathbb{Z}_N$ is left unbroken after $G_N \to [U(1)]^2$.

## Section IV: Open Questions on SUSY Cosmology

The interplay of particle physics and cosmology has never been stronger. It has always been clear that particle physics provides important inputs into models of cosmology, but as the field of cosmology has matured, the opposite has become just as true. SUSY opens exciting new avenues for cosmology: it can provide the needed dark matter of the universe, it may provide a natural mechanism for inflation, it provides several new possibilities for baryogenesis, and so forth. But the cosmological sword is two-edged, and we find that cosmology can also serve to constrain SUSY — for example, a given particle physics model might overclose the universe, or dissociate light nuclei after nucleosynthesis, or worse. The next few questions address some of these important questions concerning the interplay between SUSY and cosmology.

### Question #10 How does SUSY shed light on dark matter?

In supersymmetric models with $R$-parity (or matter-parity) conservation, sparticles can only interact in pairs, thereby guaranteeing that the lightest SUSY particle (LSP) is absolutely stable. This provides both an important constraint and the exciting possibility that SUSY may produce stable, cosmological relics. It is known that non-luminous matter is needed to explain the rotation curves of galaxies (and galactic clusters) at large radii where luminous matter densities have fallen near zero. And even larger densities of "dark matter" (*i.e.*, $\rho = \rho_c$ where $\rho_c$ is the closure density) are needed in order to place the universe in a stable evolutionary trajectory such that its current age and density are not fine-tuned. Such larger densities are also needed in order to reproduce the otherwise successful predictions of inflation.

However a number of constraints place strong limitations on the form of the dark matter: nucleosynthesis does not allow very much of the dark matter to be baryons; heavy isotope searches constrain the ability of strongly or electromagnetically interacting matter from acting as the dark matter; and structure formation simulations generally rule out neutrinos and other particle species which are relativistic when they fall out of thermal equilibrium. Of all the classes of particles, those which remain as good candidates for the dark matter are the so-called WIMP's — Weakly Interacting Massive Particles.

The MSSM provides two ideal candidates for the dark matter, both of them WIMP's: the sneutrino and the neutralino. A detailed discussion of how well each of these particles serves as a dark matter candidate, as well as a complete list of references, can be found in the contribution of J. Wells to this volume; for now let us simply summarize the results.

After detailed calculations,[36] one finds that the sneutrino, though a WIMP, does not provide a good source of dark matter. First, in most models it is not the LSP. Second, even in those models where it is the LSP, its relic densities tend to be far from those needed for a dark matter candidate. Current experimental bounds from direct detection also serve to limit the densities of sneutrinos allowed in the solar neighborhood too severely.

The neutralino can be either a good candidate or a bad one, because the neutralino is itself an admixture of the bino, wino and the higgsinos, each with very different properties. Bino neutralinos (*i.e.*, neutralinos which are mostly composed of binos) are the most common in realistic models and also provide the best source of dark matter.[38] They can provide reasonable relic densities throughout a broad mass range from tens to hundreds of GeV (and even thousands of GeV in some parts of parameter space). Winos, because they interact more strongly, usually provide much smaller densities for the same range of masses. Higgsinos are in general poor dark matter candidates. They tend to interact too strongly and therefore stay in thermal equilibrium until their densities are depleted. Even if they could somehow provide the galactic dark matter, they are very easily detected by a variety of searches. Only if $\tan\beta$ is very close to one and the Higgsino neutralino decouples from the $Z$ would the Higgsino neutralino be a good dark matter candidate.[37]

What is most inspiring about the possibility of SUSY dark matter is that models of particles physics devised solely to satisfy particle physics constraints and prejudices nevertheless simultaneously provide a candidate for the long-sought-after dark matter. (In fact, SUSY had predicted stable relics even before it was understood that *non-baryonic* dark matter was cosmologically useful.) Furthermore, it may be possible to study the dark matter candidate both at accelerators and in dark matter detectors, hopefully verifying that the properties observed in one match those seen in the other. This would close the dark matter question once and for all.

### Question #11 Are gravitinos dangerous to cosmology?

When global SUSY is broken (at a scale $\sqrt{F}$), there is always a spin-1/2 goldstino state $\widetilde{G}_\alpha$ in the massless spectrum. When SUSY is promoted to a local symmetry (supergravity), the goldstino is eaten by the massless spin-3/2 gravitino. The resulting fermion has mass $m_{3/2} \sim F/M_{\rm Pl}$, "transverse" components which interact with matter gravitationally, and "longitudinal" components which couple derivatively to the SUSY current. As is typical for a Goldstone field,[39] this coupling is suppressed by $1/F$:

$$\mathcal{L} = \frac{1}{F}\widetilde{G}_\alpha \partial_\mu J_{\alpha\mu} \sim \bar{\lambda}^A \gamma^\rho \sigma^{\mu\nu} \partial_\rho \widetilde{G} F^A_{\mu\nu} + \bar{\psi}_L \gamma^\mu \gamma^\nu \partial_\mu \widetilde{G} D_\nu \phi. \tag{2.27}$$

A lower bound on the gravitino mass is provided by the requirement that $F \gtrsim m_Z^2$ so that $m_{3/2} \gtrsim 10^{-5}\,\text{eV}$; similarly, an upper bound comes from demanding that $F \lesssim m_Z M_{\rm Pl}$ so that $m_{3/2} \lesssim 1\,\text{TeV}$.

In the early universe, gravitinos are believed to have existed in thermal equilibrium with a plasma of hidden- and visible-sector fields. As the universe cooled, the annihilation rate for gravitinos eventually fell below the expansion rate, and they decoupled, effectively locking in their relic density.

Calculating the relevant cross-sections and solving the Boltzmann equation allows one to put an upper bound on the mass of a stable gravitino in order to avoid overclosing the universe. If there exists no mechanism for diluting the gravitino densities, then one finds[40] that $m_{3/2} \lesssim 2h^2\,\text{keV}$ where $h$ is the Hubble constant in units of 100 km/sec/Mpc. On the other hand, if the gravitino is very, very light, then its interactions are quite strong and it can stay in equilibrium below the QCD phase transition. From standard Big Bang nucleosynthesis (BBN) results, we know that the number of neutrino species allowed is $\lesssim 3$, while a coupled gravitino would behave as an additional species, violating the bounds. Thus, we find that $M_{3/2} \gtrsim 10^{-6}\,\text{eV}$ so that $\widetilde{G}$ decouples before nucleosynthesis.[41] Slightly stronger bounds (such as $m_{3/2} \gtrsim 10^{-5}\,\text{eV}$) can be derived from limits on the cooling rate for supernova SN1987A via gravitino production and emission from the core of the star.[42]

As with any unwanted relic, the gravitino excess can be diluted away by a period of inflation. However it is important that the reheating after inflation not produce a new population of gravitinos. This places upper bounds on the reheating temperature $T_R$.

Let us consider the case where the gravitino is the LSP (such as in gauge-mediated models or in no-scale supergravity).[43] In the mass range $1\,\text{keV} \lesssim m_{3/2} \lesssim 100\,\text{keV}$, large densities of gravitinos will be produced if *any* of the MSSM superpartners are produced, since such superpartners will in time decay to gravitinos. Thus, assuming that the superpartners are at the weak scale, we have $T_R \lesssim m_Z$. However, if $100\,\text{keV} \lesssim m_{3/2} \lesssim (3 -$

300) GeV (where the upper bound depends on the SUSY masses), then the principal source of gravitinos is not through SUSY decays, but rather through scattering processes $A+B \to C+\widetilde{G}$; here the reheating temperature must not exceed $T_R \lesssim 10^8\, m_{3/2}$. Finally, for $m_{3/2} \gtrsim (3-300)$ GeV, the decay rate of MSSM superpartners into gravitinos is so small that the decays take place after nucleosynthesis, disastrously photo-dissociating light nuclei.

For the first two cases, a period of late inflation (perhaps thermal inflation — see Question #12) can dilute away the gravitino problem. However, any baryon densities present before this late inflation would also be diluted away, requiring mechanisms for baryogenesis which operate at temperatures below $T_R$. This is particularly difficult if $T_R \lesssim m_Z$, perhaps requiring electroweak baryogenesis or use of the Affleck-Dine mechanism. Note that a period of inflation cannot help for the last case of $m_{3/2} \gtrsim (3-300)$ GeV.

## Question #12 Are moduli cosmologically dangerous?

The "moduli problem", as we have chosen to call it, is actually a large class of problems corresponding to the physics of the different moduli which occur in the MSSM and its extensions. The essence of the moduli problem is that fields with extremely flat potentials and weak couplings to other fields tend to be cosmologically dangerous.

The original example (the "Polonyi problem") is provided by that hidden-sector field (the "Polonyi field") which is a gauge singlet, has a nearly flat potential and no renormalizable couplings to other matter, and whose $F$-component is responsible for SUSY-breaking in supergravity-mediated models. After receiving a scalar VEV $\sim M_{\rm Pl}$ and $F$-VEV $\sim m_Z M_{\rm Pl}$, the physical Polonyi (scalar) field $\Phi$ emerges as the scalar partner of the goldstino/gravitino and thus has a mass $\sim m_Z$. The Polonyi VEV, which sets the natural scale of its oscillations, is much larger than its mass, and so at temperatures far above the weak scale, $\Phi$ feels only the potential induced by the (SUSY-breaking) temperature and vacuum energy. In particular, the minimum in which $\Phi$ finds itself at finite temperature $T$ and finite Hubble constant $H$ need not correspond to the minimum of the $T = H = 0$ potential.

Somewhat more precisely, the problem can be stated like this: Once $T$ and $H$ fall below $\langle\Phi\rangle$, $\Phi$ finds itself far from its true minimum and begins to oscillate with amplitude $\sim \sqrt{m_Z M_{\rm Pl}}$. However, since it is only weakly coupled to matter, there is little friction to damp the oscillations (*i.e.*, $\Gamma_\Phi$ is small), which allows the oscillations to continue for times approaching minutes. During this time, $\Phi$ will dominate the energy density of the universe; if $\Phi$ is too long-lived, it will in fact overclose the universe. But even if $\Phi$ is not so long-lived, it is the decays of $\Phi$, occurring long after baryogenesis and perhaps even nucleosynthesis, that are the principal concern. After several minutes of slow $\Phi$ oscillations, the temperature of the universe is far below the weak scale, but $\Phi$ decays are still occurring, dumping entropy into the universe in quantities which are more than sufficient to overdilute the baryon density (and/or dissociate light nuclei) without increasing the temperature enough to restart the original $B$-violating processes that could replenish it.

The above argument can be generalized and refined by identifying $\Phi$ with other moduli of the theory. These include (but are not limited to): string moduli whose VEV's $\sim M_{\rm Pl}$ parametrize the size of compact dimensions; the string dilaton whose VEV $\sim M_{\rm Pl}$ parametrizes the string coupling constant; $D$- and $F$-flat directions of the MSSM which have no potential before SUSY-breaking; and Higgs fields responsible for breaking GUT's to the MSSM.

A *partial* resolution of the moduli problem may rest in the fact that for many types of moduli, there is no reason for the couplings of the moduli to other matter to be particularly small. Thus, once $T, H \lesssim m_\Phi$ and the oscillations begin, $\Gamma_\Phi$ can be sizable, leading to fast decays which do not appreciably dilute the baryon density. This is the generalization of one of the early suggestions for solving the Polonyi problem — namely, introducing extra fields in the hidden sector to mediate decays of the Polonyi field into gravitinos.[44] Denoting $\langle\Phi\rangle$ at $T = H = 0$ as $\langle\Phi\rangle_0$ and at $T, H \neq 0$ as $\langle\Phi\rangle_{T,H}$, we find that there are then four classes of cosmological histories:[45]

- $\langle\Phi\rangle_0 = \langle\Phi\rangle_{T,H} = 0$: In this case there is no moduli problem and no interesting cosmology in the moduli sector.
- $\langle\Phi\rangle_0 = 0$ but $\langle\Phi\rangle_{T,H} \neq 0$: In this case oscillations begin after $H \lesssim m_\Phi$ but are quickly damped by sizable $\Gamma_\Phi$ so there is no moduli problem. Note that if $\Phi$ carries non-zero $B$ or $L$, this case can lead to Affleck-Dine baryogenesis.[46]
- $\langle\Phi\rangle_{T,H} = 0$ but $\langle\Phi\rangle_0 \neq 0$: In this case oscillations again begin once $H, T \lesssim m_\Phi$. However, the moduli cannot have large couplings to light particles (otherwise they would not be light), and thus $\Gamma_\Phi$ is small and the oscillations last a long time. During this time, the universe can "thermally" inflate[45] due to the energy stored in $\Phi$. In

general such a period of inflation does not solve the moduli problem; however, for $\langle\Phi\rangle_0 \sim (m_Z M_{\rm Pl})^{1/2}$, the problems associated with the inflating moduli do not occur *and* there is sufficient inflation to dilute any other moduli fields.

- $\langle\Phi\rangle_{T,H} \neq 0$ and $\langle\Phi\rangle_0 \neq 0$: In this case the moduli problem arises again since the $\Phi$ decays must be suppressed, but no period of thermal inflation occurs.

For stringy moduli (usually denoted $T$), the conditions necessary to avoid washing out the baryon density are harder to fulfill. For one thing, when string moduli begin oscillating, their amplitudes are generally $\sim M_{\rm Pl}$. Secondly, their couplings to matter are always suppressed by powers of $m_{3/2}/M_{\rm Pl}$, so that $\Gamma_T$ is very small and the oscillations last a long time. Thus, more generally than the case discussed above, one can expect a moduli problem to arise if $\langle T\rangle_{T,H} \neq \langle T\rangle_0$.

If we identify the modulus in question to be the dilaton $S$, many of the same problems arise, along with a new one.[48] Because the dilaton couples to the vacuum energy density, the non-zero vacuum energies which are supposed to drive inflation lose much of their energy into driving oscillations of $S$, thereby slowing the expansion rate. Thus, inflation must wait until $S$ settles into the minimum of its potential. This itself is difficult, since the potential for $S$, which arises non-perturbatively, goes to zero as $S \to \infty$. Thus, a successful inflationary model must force $S$ into some local minimum of the full potential without pushing it over the barrier which separates finite, time-independent dilaton VEV's from infinite, time-dependent ones. This appears to be a difficult problem, with no simple solutions currently available.

There is one other independent mechanism which may help lessen the moduli problem. If $\Phi$ couples to some other field $\psi$, the equation of motion for $\psi$ is that of a harmonic oscillator with periodic driving force, the period being that of the oscillating moduli. Such an equation, known as Mathieu's equation, is known to have regions of instability in which the solutions are exponentially growing. Physically, these solutions correspond to coherent decays of the moduli at rates far exceeding those for single particle decays. This phenomenon is known as parametric resonance.[47] If parametric resonance occurs, the moduli will quickly dump most of their energy into particles, instead of slowing over several minutes. There are many questions still to be answered about parametric resonance, the conditions under which it will occur, and the means by which the decay products thermalize, but this idea appears to be an exciting advance in our understanding of cosmological moduli physics.

Finally, it would seem natural to use the moduli themselves as the inflatons. We shall not comment on the successes or difficulties in using any of the moduli, stringy or not, for inflation, but leave that discussion for the contribution of L. Randall to this volume.

## Section V: Open Questions on SUSY Grand Unification

One particularly attractive idea for physics at high energy scales concerns the possible appearance of a grand unified theory, or GUT, with a single symmetry group that is large enough to incorporate the $SU(3) \times SU(2) \times U(1)$ gauge group of the Standard Model as a subgroup.[49] The idea of grand unification has a long history independent of SUSY, but once SUSY is included in the picture, a number of new results and predictions arise. We will consider some of these issues in this section.

### Question #13 Does the MSSM unify into a supersymmetric GUT?

There are several profound attractions to the idea of grand unification. Perhaps the most obvious is that GUT's have the potential to unify the diverse set of particle representations and parameters found in the MSSM into a single, comprehensive, and hopefully predictive framework. For example, through the GUT symmetry one might hope to explain the quantum numbers of the fermion spectrum, or even the origins of fermion mass. Moreover, by unifying all $U(1)$ generators within a non-abelian theory, GUT's would also provide an explanation for the quantization of electric charge; note that this is a puzzle in the Standard Model due to the abelian $U(1)_Y$ hypercharge group factor whose allowed eigenvalues are arbitrary. Furthermore, because they generally lead to baryon-number violation, GUT's have the potential to explain the cosmological baryon/anti-baryon asymmetry. By combining GUT's with supersymmetry in the context of SUSY GUT's,[50] we then hope to realize the attractive features of GUT's simultaneously with those of supersymmetry in a single theory.

There is also one compelling piece of *experimental* evidence for the existence of *supersymmetric* GUT's. It is a straightforward matter to extrapolate the strong, electroweak, and hypercharge gauge couplings of either the Standard Model or the MSSM to higher energies by using their one-loop renormalization group equations.[51] The results are shown in Fig. 2.1. One sees that if this extrapolation is performed within the non-supersymmetric Standard Model, these couplings fail to unify at any scale. However, performing this extrapolation within the context of the supersymmetric MSSM — *i.e.*, assuming only the minimal MSSM particle content with superpartners near the $Z$ scale — one obtains an apparent gauge coupling unification[52,53] of the form

$$\frac{5}{3}\,\alpha_Y(M_{\rm GUT}) \;=\; \alpha_2(M_{\rm GUT}) \;=\; \alpha_3(M_{\rm GUT}) \;\approx\; \frac{1}{25} \qquad (2.28)$$

at the scale $M_{\rm GUT} \approx 2\times 10^{16}$ GeV. This single unified gauge coupling is then easy to interpret as that of a single GUT group $G_{\rm GUT}$ which breaks at the scale $M_{\rm GUT}$ down to $SU(3)\times SU(2)\times U(1)$. Note that it is the introduction of *supersymmetry* which enables this gauge coupling unification to take place without any further intermediate-scale structure.

While there are *a priori* many choices for such possible groups $G_{\rm GUT}$, the list can be narrowed down by requiring groups of rank $\geq 4$ that have complex representations. The smallest possibilities are then $SU(5)$, $SU(6)$, $SO(10)$, and $E_6$. Amongst these choices, $SO(10)$ is particularly attractive because $SO(10)$ is the smallest simple Lie group for which a single anomaly-free irreducible representation (namely the spinor **16** representation) can accommodate the entire MSSM fermion content of each generation. Specifically, under the decomposition $SO(10) \supset SU(5)\times U(1)' \supset SU(3)\times SU(2)\times U(1)_Y \times U(1)'$, the **16** representation decomposes as

$$\begin{aligned} \mathbf{16} &\to \mathbf{10}_{-1} \oplus \overline{\mathbf{5}}_3 \oplus \mathbf{1}_{-5} \\ &\to \{(\mathbf{3},\mathbf{2})_{1/6} \oplus (\overline{\mathbf{3}},1)_{-2/3} \oplus (\mathbf{1},\mathbf{1})_1\}_{-1} \\ &\qquad \oplus \{(\overline{\mathbf{3}},\mathbf{1})_{1/3} \oplus (\mathbf{1},\mathbf{2})_{-1/2}\}_3 \oplus \{(\mathbf{1},\mathbf{1})_0\}_{-5}\,. \end{aligned} \qquad (2.29)$$

These representations are respectively identified as the left-handed quark $Q$, the right-handed up quark $u^c_R$, the right-handed electron $e^c_R$, the right-handed down quark $d^c_R$, the left-handed lepton $L$, and the right-handed neutrino $\nu^c_R$. Note that all Standard Model particles are incorporated, with all of their correct quantum numbers, and no extraneous particles are introduced. Furthermore, such $SU(5)$-based unification scenarios provide a natural explanation for the normalization factor 5/3 which appears in Eq. (2.28): this is simply the group-theoretic factor by which the Standard Model hypercharge generator must be rescaled in order to join with the non-abelian generators into a single $SU(5)$ non-abelian multiplet.

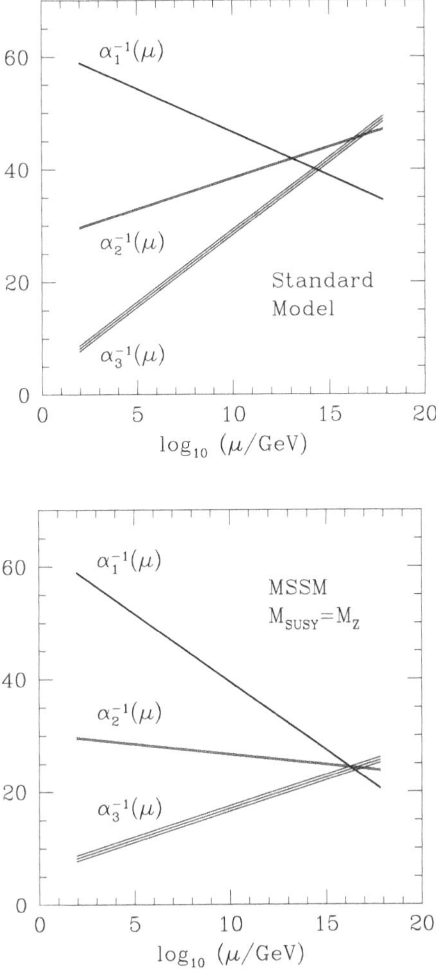


Fig. 2.1. One-loop evolution of the gauge couplings within the non-supersymmetric Standard Model and within the Minimal Supersymmetric Standard Model (MSSM). In both cases $\alpha_1 \equiv (5/3)\alpha_Y$ where $\alpha_Y$ is the hypercharge coupling in the conventional normalization. The relative width of each line reflects current experimental uncertainties.

The apparent gauge coupling unification of the MSSM is strong circumstantial evidence in favor of the emergence of a SUSY GUT near $10^{16}$ GeV. However, GUT theories naturally lead to a variety of outstanding questions. Understanding the answers to these questions therefore provides a window into high-scale physics.

**Question #14** **Proton decay again: Why doesn't the proton decay in $10^{32}$ years?**

Perhaps the most important problem that SUSY GUT's must address is the proton-lifetime problem. In general, GUT's lead to a number of processes that can mediate proton decay. For example, proton decay can be mediated via the off-diagonal $SU(5)$ $X$ gauge bosons that connect quarks to quarks and quarks to leptons. Such gauge bosons arise, along with the Standard Model gauge bosons, in the decomposition of the $SU(5)$ adjoint representation; they transform in the $(\mathbf{3},\mathbf{2})_{-5/6}$ and $(\overline{\mathbf{3}},\mathbf{2})_{5/6}$ representations of $SU(3)\times SU(2)\times U(1)_Y$, and thus have fractional electric charges $-4/3$ and $-1/3$ respectively. Because interactions via these $X$ gauge bosons violate baryon-number ($B$) and lepton-number ($L$) symmetries, such gauge bosons can mediate proton decay via processes such as $p\to\pi^0 e^+$. However, in supersymmetric GUT's this is not the dominant source of proton decay because the $X$ gauge boson must have a mass $M_X\approx M_{\rm GUT}\approx 2\times 10^{16}$ GeV. This is two orders of magnitude higher than the expected "unification" scale of non-supersymmetric GUT's. Furthermore, since this decay is gauge-mediated, the contribution to the branching ratio for proton decay via this process goes as $\Gamma\sim g^4 m_p^5/M_X^4$ where $g\approx 0.7$ is the unified gauge coupling and $m_p$ is the proton mass. The factor $M_X^{-4}$ is a typical suppression for a dimension-six operator, and results in an expected lifetime $\tau(p\to\pi^0 e^+)\approx 10^{36}$ years.

A much more problematic dimension-five contribution arises in supersymmetric GUT's via mediation by colored Higgsino triplets.[54] Because the MSSM requires two electroweak Higgs doublets, and because a minimal $SU(5)$ GUT gauge structure forces these doublets to be part of larger (*e.g.*, five-dimensional) Higgs representation, the electroweak doublet Higgs will necessarily have a colored triplet Higgs counterpart which contains a fermionic colored Higgsino component. *A priori*, a given $SU(5)$-invariant mass term for this Higgs multiplet will tend to give the same mass to the doublet Higgs as to the triplet Higgs(ino). Therefore, since the electroweak doublet Higgs is expected to have a mass $\sim 100$ GeV, it is generally quite difficult to give the color triplet Higgs(ino) a large mass. However, a large mass is precisely what we need if we are to avoid rapid proton decay, for this fermionic Higgsino component of the color Higgs triplet can mediate decay processes such as $p\to K^+\overline{\nu}$. In this case, the branching ratios go as $\Gamma\sim h^4 m_p^5/M_{\tilde H}^2 M_{\rm SUSY}^2$ where $h\approx 10^{-5}$ is the Higgs(ino) Yukawa coupling to the light generation and $M_{\rm SUSY}$ is the scale of SUSY-breaking. Despite the fact that this branching ratio is Yukawa-suppressed (by the factor of $h^4$) relative to the dimension-six case, we have only a factor of $M_{\tilde{H}}^{-2}$ mass suppression because the Higgsino mediator is a fermion. Thus, in order to protect against proton decay (and also to preserve gauge coupling unification), the color triplet Higgs must be substantially heavier than the electroweak doublet Higgs. Indeed, in order not to violate current experimental bounds, we must ensure that $\tau(p \to K^+\overline{\nu}) \gtrsim 10^{32}$ years. This is the problem of "doublet-triplet splitting". Once the doublets and triplets are somehow split, supersymmetric non-renormalization theorems should protect this splitting against radiative corrections.

It is striking that the dominant proton decay mode depends so crucially on whether or not supersymmetry is present. Discovery of the $p \to K^+\overline{\nu}$ decay mode can thus serve as a clear signal for supersymmetry.

There are a number of potential solutions to the doublet-triplet splitting problem. Proposals include the so-called "sliding singlet"[55] and "missing partner"[56] mechanisms which apply in the case of $SU(5)$, and also a "Higgs-as-pseudo-Goldstone" mechanism[57] which applies in the case of $SU(6)$. Perhaps the most attractive proposal, however, is the "missing VEV" solution for $SO(10)$, originally proposed by Dimopoulos and Wilczek.[58]

The basic idea behind this mechanism is as follows. One way to break the $SO(10)$ GUT gauge symmetry down to that of the MSSM is to give a vacuum expectation value (VEV) to the adjoint **45** representation. However, because the **45** representation contains two Standard-Model singlets, there are *a priori* many ways in which this can be done without breaking the Standard-Model gauge group. The Dimopoulos-Wilczek mechanism entails giving a VEV to only one of these singlets, and keeping the other VEV fixed at zero. In order to see this explicitly, it is most useful to consider the Pati-Salam decomposition under which $SO(10)$ breaks to the Standard Model gauge group via the pattern

$$SO(10) \supset SU(4)\times SU(2)_L\times SU(2)_R \supset \{SU(3)\times U(1)_C\}\times SU(2)_L\times U(1)_R\,. \tag{2.30}$$

The hypercharge $U(1)$ is then identified as a linear combination of $U(1)_C$ and $U(1)_R$; note that $U(1)_C = U(1)_{B-L}$. Under the first decomposition $SO(10) \supset SU(4) \times SU(2)_L \times SU(2)_R$, the **45** representation decomposes as

$$\mathbf{45} \;\to\; (\mathbf{15},\mathbf{1},\mathbf{1}) \oplus (\mathbf{6},\mathbf{2},\mathbf{2}) \oplus (\mathbf{1},\mathbf{3},\mathbf{1}) \oplus (\mathbf{1},\mathbf{1},\mathbf{3})\ . \tag{2.31}$$

However, under $SU(4) \supset SU(3) \times U(1)_C$, we have $\mathbf{15} \to \mathbf{8}_0 \oplus \mathbf{3}_1 \oplus \overline{\mathbf{3}}_{-1} \oplus \mathbf{1}_0$, while under $SU(2)_R \supset U(1)_R$ we have $\mathbf{3} \to \{q_C = \pm 1, 0\}$. Thus, only the first and fourth terms in Eq. (2.31) contain singlets. The Dimopoulos-Wilczek mechanism consists of giving a VEV to the first singlet *but not the second.* This works because the $SO(10)$ Higgs decomposes as $\mathbf{10} \to (\mathbf{6}, \mathbf{1}, \mathbf{1}) \oplus (\mathbf{1}, \mathbf{2}, \mathbf{2})$, where the first representation contains the triplet Higgs and the second contains the doublet Higgs. By giving a VEV to the $(\mathbf{15}, \mathbf{1}, \mathbf{1})$ representation within the $SO(10)$ adjoint but withholding it from the $(\mathbf{1}, \mathbf{1}, \mathbf{3})$ representation, we see that the effective superpotential term $\Phi_{\mathbf{10}}\Phi_{\mathbf{45}}\Phi'_{\mathbf{10}}$ gives a mass to the triplet Higgs but not the doublet Higgs. Constructing a fully consistent $SO(10)$ model in which this mechanism is implemented in a natural way remains an active area of research, and many proposals exist.[58–60]

### Question #15 Can SUSY GUT's explain the masses of fermions?

In general, the GUT structure imposes not only a unification of gauge couplings, but also a unification of Yukawa couplings. Thus, fermion masses are another generic issue that SUSY GUT's must address. How, through a SUSY GUT, can we explain in a simple way the many free MSSM parameters that describe the fermion masses?

Just as we did for the gauge couplings, it is straightforward to use one-loop renormalization group equations (RGE's) along with the Yukawa couplings in order to extrapolate the observed fermion masses up to the GUT scale. In terms of the generic fermion Yukawa couplings $\lambda_i$, we then find the approximate relations at the GUT scale

$$\begin{aligned} &\lambda_d(M_{\rm GUT}) \approx 3\lambda_e(M_{\rm GUT}) \ , \quad \lambda_s(M_{\rm GUT}) \approx \frac{1}{3}\lambda_\mu(M_{\rm GUT}) \ , \\ &\lambda_b(M_{\rm GUT}) \approx \lambda_\tau(M_{\rm GUT}) \ . \end{aligned} \tag{2.32}$$

Note that because the fundamental GUT idea relates quarks and leptons within a single multiplet, we are particularly interested in such mass relations between quarks and leptons.[61] The issue, then, is to "explain" these relations within the context of a consistent GUT model. Ideally, we would also like to explain additional features of the fermion mass spectrum, such as the inter-generation mass hierarchy, the masses of the *up*-type quarks, and the (near?)-masslessness of the neutrinos. Reviews of the fermion mass problem within GUT scenarios can be found in Ref. 62.

Certain features are easy to explain. For example, the factors of three that appear in Eq. (2.32) can be understood, as first suggested by Georgi and Jarlskog,[63] as Clebsch-Gordon coefficients of the GUT gauge group (which in turn ultimately stem from the fact that there are three quarks for every lepton). This requires the appearance of certain *textures* (*i.e.*, patterns of zero and non-zero entries) in the fermion mass matrices. Likewise, the inter-generation mass hierarchy might be explained if the first-generation mass terms are of higher dimension in the effective superpotential than those of the second and third generations. Indeed, within $SO(10)$, even small neutrino masses can be accommodated via the see-saw mechanism.[64]

The goal, however, is to realize all of these mechanisms simultaneously within the context of a self-consistent supersymmetric GUT model. There are many ways in which such mechanisms can be implemented. For example, judicious use of a **126** representation of $SO(10)$ can give rise to a heavy Majorana right-handed neutrino mass, the proper Georgi-Jarlskog factors of three in the light quark/lepton mass ratios, and GUT symmetry-breaking with automatic $R$-parity conservation. Use of the **120** and **144** representations can also accomplish some (but not all) of the same goals. General studies of the classes of operators that can explain the fermion masses can be found in Ref. 65, and recent GUT models in which such mechanisms are employed can be found in Refs. 59, 60, 66, 67.

Recently, much attention has focused on deriving GUT models that are consistent with the additional constraints that come from string theory. String theory, in particular, tends to severely restrict not only the GUT representations that might be available for model-building, but also their couplings (see, *e.g.*, Refs. 68, 101). It turns out that large representations are often entirely excluded, and only very minimal sets of representations and couplings are allowed. Recent attempts to build field-theoretic models that are consistent with these sorts of constraints can be found in Ref. 60. We shall discuss the recent progress in string GUT model-building in Question #20.

## Section VI: SUSY Duality

Another set of exciting recent developments, made possible in large part due to supersymmetry, concerns the notion of SUSY gauge theory *duality*. Duality in this context refers to the fact that two seemingly dissimilar theories can actually describe the same physics. A wide variety of exact dualities are known to occur for $N \geq 2$ extended supersymmetries, but these are somewhat removed from our immediate world which seems to have at most $N = 1$ SUSY. Therefore we will confine ourselves in this section to a short discussion of $N = 1$ duality.

## Question #16 $N=1$ SUSY duality: How has SUSY changed our view of gauge theory?

$N = 1$ dualities relate two seemingly different theories in the sense that both flow to the same fixed point in the infrared. Such theories may be either both asymptotically free, or one asymptotically free and the other infrared-free. Supersymmetry plays a fundamental role in uncovering these duality relations. By gathering all possible interaction terms into a superpotential that must be holomorphic in the chiral superfields as well as in their couplings, supersymmetry imposes extraordinarily tight constraints on the possible forms of the effective superpotentials that are generated both perturbatively and *non-perturbatively* as one flows from higher to lower energy scales. Indeed, in many cases one is able to determine the effective superpotential exactly. These exact expressions for the effective superpotentials have been used for many purposes: to give new, simpler proofs of some standard supersymmetric non-renormalization theorems that hold beyond perturbation theory; to determine the situations under which strong-coupling dynamics can break supersymmetry; and also to uncover the phase structure of supersymmetric gauge theories.

$N = 1$ dualities come into play when describing the results of this phase structure analysis. As a simple example, let us consider $N = 1$ supersymmetric $SU(N_c)$ gauge theory with $N_f$ flavors transforming in the fundamental representation. Such a theory is asymptotically free if $N_f < 3N_c$; note that this is the supersymmetric generalization of the famous constraint $N_f < (11/2)N_c$ which holds for non-supersymmetric $SU(N_c)$ theories. Using the powerful constraints imposed by the $N = 1$ supersymmetry, the infrared limit of this theory has been determined as a function of the parameters $N_c$ and $N_f$. One finds that if $N_f \geq 3N_c$, the theory flows in the infrared to a (free) theory of non-interacting quarks and gluons, while if $N_f$ is in the range $(3/2)N_c \leq N_f \leq 3N_c$, then the theory flows to a nontrivial interacting fixed point. But what is the infrared limit of the theory if $N_f < (3/2)N_c$? Evidence suggests that if $N_f > N_c+2$, the infrared limit in this case is the *same* as that for the $N_f \geq (3/2)N_c$ case: one again has a free theory of non-interacting elementary constituents. Indeed, the entire phase diagram seems to have a symmetry under $N_c \to N'_c \equiv N_f - N_c$, so that we may identify

$$SU(N_c) \text{ with } N_f \text{ flavors} \iff SU(N_f - N_c) \text{ with } N_f \text{ flavors} \quad (2.33)$$

as "dual" theories. The elementary constituents of the infrared limit of one theory are then identified as the "dual quarks" of its dual theory, and so forth. It is remarkable that two very different theories can be related in this way. In fact, this is only the first in a long list of such duality relations, and examples exist for many other gauge groups and matter representations (including, most interestingly, duals between chiral and non-chiral theories). Anomaly matching conditions provide highly non-trivial checks of these duality conjectures. Recent reviews of this subject can be found in Ref. 69.

The existence of such duality conjectures immediately prompts a number of outstanding questions. First, can one *prove* these conjectures? Such a proof would seem to require the construction of a procedure for passing from a given gauge theory to its dual, and perhaps also include an explicit mapping from the degrees of freedom of one theory to the degrees of freedom of the other. Second, what does the existence of such dualities tell us about the fundamental nature of supersymmetric gauge theories? These dualities suggest, for example, that the particular gauge symmetry itself may not be the crucial defining characteristic of such theories. Finally, of more practical relevance, however, is a third question: to what extent do these duality relations survive supersymmetry breaking? This will ultimately determine the extent to which such duality conjectures may be useful for low-energy phenomenology.

Finally, we mention that there exist other sorts of dualities which also rely heavily on the presence of supersymmetry. Perhaps the best-known among these is Montonen-Olive duality,[70] which is an exact strong/weak coupling duality of finite, interacting, non-abelian gauge theories. At the present time, this proposed duality can be understood only in the context of $N = 4$ supersymmetric field theories and finite $N = 2$ supersymmetric theories, although there do exist extensions to asymptotically free $N = 1$ theories. Once again, a crucial question is whether this duality has an analogue in (or implication for) non-supersymmetric theories. It is fair to say that we are only at the beginning stages of understanding for all of these dualities, and their precise connections and interpretations await further developments.

## Section VII: Open Questions on SUSY and String Theory

In this final section, we will discuss some of the phenomenological connections between supersymmetry and string theory. We will only focus on general themes and basic introductory ideas, since many details will be provided in the subsequent chapters. For an overall introduction to string theory, we recommend Ref. 71. There are also a number of review articles that deal with more specific aspects of string theory. For example, recent discussions concerning string phenomenology and string model-building can be found in Ref. 72. Likewise, reviews of methods of supersymmetry-breaking in string theory can be found in Ref. 73, and a review of gauge coupling unification in string theory can be found in Ref. 74. Note that in these sections we will *not* be discussing some of the more formal aspects of string theory such as string duality; reviews of this topic can be found in Ref. 75.

### Question #17: Why strings?

One of the primary goals of high-energy physics in this century has been to unify the different observed forces and particles within the framework of a single, comprehensive theory. In recent years, this goal has given rise to tremendous interest in *string theory*. The fundamental tenet underlying string theory is that all of the known elementary particles and gauge bosons can be realized as the different excitation modes of a single fundamental closed string of size $\sim 10^{-33}$ cm. Thus, within string theory, the physics of zero-dimensional points is replaced by the physics of one-dimensional strings, and likewise the spacetime physics of one-dimensional worldlines is replaced by the physics of two-dimensional worldsheets.

There are several profound attractions to this idea. First, one finds that among the excitations of such a string there exists a spin-two massless excitation that is naturally identified as the graviton. Thus string theory is a theory of quantized gravity. Indeed, it is this identification which sets the fundamental scale for string theory to be the Planck scale. Second, it turns out that string theory enjoys a measure of finiteness that is not found in ordinary point-particle field theories, and therefore many of the divergences associated with field theory are absent in string theory. Third, it is found that string theory in some sense *requires* gauge symmetry for its internal consistency, and moreover predicts gauge coupling unification. But for the purposes of this review, it turns out that the most intriguing aspect of string theory may be that it seems to predict supersymmetry.

## Question #18: What roles does SUSY play in string theory?

The histories of string theory and supersymmetry are closely intertwined. Indeed, a form of supersymmetry itself was originally discovered[76] in the context of explaining how strings can have fermionic excitations. We shall now briefly sketch several remarkable inter-relations between supersymmetry and string theory, focusing on those special roles that supersymmetry plays in string theory. We shall mostly restrict our attention to *perturbative* string theory, as this is far better understood than recent developments in possible non-perturbative formulations of string theory.

### 2.18.1. *Worldsheet SUSY, spacetime SUSY, and the dimension of spacetime*

Perhaps the most intriguing aspect of supersymmetry in string theory concerns the connections between *worldsheet* supersymmetry, *spacetime* supersymmetry, and the dimension of spacetime. In general, a closed one-dimensional string sweeps out a two-dimensional worldsheet with coordinates $(\sigma_1, \sigma_2)$, and the simplest action for such a string is given by

$$S = \int d^2\sigma \, g_{\mu\nu} \, \partial_\alpha X^\mu(\sigma) \, \partial^\alpha X^\nu(\sigma) \, . \tag{2.34}$$

Here $X^\mu(\sigma)$ indicate the spacetime coordinates of the string as a function of its worldsheet coordinates, the derivatives are with respect to the worldsheet coordinates, and $g_{\mu\nu}$ is the spacetime metric. The spacetime indices run over the range $\mu, \nu = 1, ..., D_c$. From a spacetime perspective, this action is equivalent to the area of the worldsheet embedded in a $D_c$-dimensional spacetime. From the worldsheet perspective, by contrast, this is the action of a two-dimensional field theory in which the coordinates $X^\mu$ appear as a collection of $D$ bosonic worldsheet fields with couplings $g_{\mu\nu}$. Each different excitation state of the string is then interpreted as a different particle in spacetime. Since the fundamental string energy scale is the Planck scale, only the lowest-lying (massless) excitations are observable, and the remaining states are all at the Planck scale.

Eq. (2.34) is the action of the bosonic string, and it turns out that the quantum consistency of this two-dimensional action requires that $D_c = 26$. All of the states of this string have integer spin in spacetime, and are therefore bosons. However, in order to introduce spacetime fermions, a natural idea is to supersymmetrize this worldsheet action, introducing superpartner fermionic fields $\psi^\mu(\sigma)$ on the worldsheet,

$$S = \int d^2\sigma\, g_{\mu\nu} \left[ \partial_\alpha X^\mu(\sigma)\, \partial^\alpha X^\nu(\sigma) - i\overline{\psi}^\mu \rho_\alpha \partial^\alpha \psi^\nu \right] , \qquad (2.35)$$

where $\rho_\alpha$ are the corresponding *two-dimensional* Dirac matrices. We then gauge this worldsheet supersymmetry. This procedure yields the action of the superstring, and a slight variation on this idea (one involving a mixture of both supersymmetrized and non-supersymmetrized actions) yields the action of the heterotic string. However, in either case, one finds that the spectra of these theories contain spacetime fermions as well as bosons. Moreover, the spacetime dimension required for the quantum consistency of this theory falls to $D_c = 10$.

In fact, it turns out that there is an additional remarkable result. If, in addition to the above gauged worldsheet supersymmetry, we introduce an additional *global* worldsheet supersymmetry subject to certain constraints,[77] then the spacetime spectrum of the string not only consists of bosons and fermions, but actually is itself $N = 1$ supersymmetric! Thus in string theory, $N = 1$ supersymmetry in spacetime is realized as the *consequence of two supersymmetries on the worldsheet, one local and one global*! This is a profound observation, implying that $N = 1$ supersymmetry in spacetime can emerge as (and thereby be explained as) the result of a more fundamental *worldsheet* symmetry (in this case, $N = 2$ worldsheet supersymmetry). This is only the first of a number of such profound connections between worldsheet and spacetime supersymmetries, as indicated in Table 2.1.

Table 2.1. Relations between the total number of worldsheet supersymmetries ($N_t$), the number of gauged worldsheet supersymmetries ($N_g$), the resulting critical spacetime dimension $D_c$ before compactification, and the properties of the resulting spacetime spectrum. The asterisk indicates *complex* dimensions.

| $N_g$ | $N_t$ | $D_c$ | spectrum |
|---|---|---|---|
| 0 | 0 | 26 | bosons only |
| 1 | 1 | 10 | bosons and fermions |
| 1 | 2 | 10 | N=1 SUSY |
| 1 | 4 | 10 | N=2 SUSY |
| 2 | 2 | 2* | |
| 4 | 4 | −2 | |

### 2.18.2. *Supersymmetry, strings, and vacuum stability*

Another intriguing connection between supersymmetry and string theory concerns vacuum stability. In field theory, supersymmetry is a very attractive feature, but it is certainly not *required* for consistency. For example, while the non-supersymmetric Standard Model may suffer from a variety of unappealing technical problems (foremost among them the gauge hierarchy problem), it suffers from no fundamental inconsistency. In string theory, however, the situation appears to be entirely different. In general, string theories with non-supersymmetric spacetime spectra (henceforth to be referred to as non-supersymmetric strings) have a non-vanishing one-loop tadpole amplitude for a certain light scalar state called the *dilaton*. As we discussed in Question #12, such a light dilaton causes a variety of phenomenological and cosmological problems. However, the existence of such a dilaton tadpole implies that the dilaton experiences a linear potential — *i.e.*, that the ground state of the string is unstable. Such a non-supersymmetric string model is then presumed to flow (in the space of all possible string models) to another point at which stability is restored and the one-loop dilaton tadpole is cancelled. A recent study of this question can be found in Ref. 78.

Spacetime supersymmetry is an elegant way of cancelling this dilaton tadpole. Although it is not known whether all stable string models must be supersymmetric, this fact is commonly assumed. If this assumption is true, then supersymmetry plays a more profound role in string theory than it does even in field theory, for the fundamental consistency of the string theory would seem to require it. This might then be the best explanation for "why" the world should be supersymmetric, at least at sufficiently high energies. However, as we stated, it is not known whether this assumption is true, and we shall see below that certain non-supersymmetric string models also manage to have a remarkable degree of finiteness and stability.

### 2.18.3. *SUSY and pseudo-anomalous $U(1)$'s*

A closely related issue, one with deep ramifications for string phenomenology, concerns the connection between spacetime supersymmetry and the extra "pseudo-anomalous" gauge symmetries that often appear in realistic string models.

In field theory, consistency requires that there be no anomalies, and indeed all triangle anomalies are cancelled in the Standard Model and its supersymmetric extensions. In string theory, by contrast, there can be $U(1)$ gauge symmetries (typically denoted $U(1)_X$) which are "pseudo-anomalous". This means that $\text{Tr}\, Q_X \neq 0$ where the trace is evaluated over the massless (observable) string states. The reason this is allowed to occur in string theory is that string theory provides a different mechanism, the Green-Schwarz mechanism,[79] which cancels such triangle anomalies even if this trace is non-zero. The Green-Schwarz mechanism works by ensuring that any anomalous variation of the *field-theoretic* $U(1)_X$ triangle diagram is always cancelled by a corresponding non-trivial $U(1)_X$ transformation of the string axion field. This axion field arises generically in string theory as the pseudo-scalar partner of the dilaton, and couples universally to all gauge groups. Thus, the existence of such a mechanism in string theory implies that anomaly cancellation in string theory does not require cancellation of $\text{Tr}\, Q_X$ by itself, and consequently a given string model can remain non-anomalous even while having $\text{Tr}\, Q_X \neq 0$. Indeed, this is the generic case for most realistic string models.

What does this have to do with supersymmetry? It turns out that *even though* the anomalies caused by having $\text{Tr}\, Q_X \neq 0$ are cancelled by the Green-Schwarz mechanism, there is still another danger: such a non-vanishing trace leads to the breaking of spacetime supersymmetry at one-loop order through the appearance of a one-loop Fayet-Iliopoulos $D$-term of the form[80]

$$\frac{g^2_{\text{string}}\, \text{Tr}\, Q_X}{192\, \pi^2}\, M^2_{\text{Pl}} \tag{2.36}$$

in the low-energy superpotential. This in turn destabilizes the string ground state by generating a dilaton tadpole at the two-loop level, and signals that our original string theory (or string model) in which $\text{Tr}\, Q_X \neq 0$ cannot be consistent.

The standard solution to this problem is to give non-vanishing vacuum expectation values (VEV's) to certain scalar fields $\phi$ in the string model in such a way that the offending $D$-term in Eq. (2.36) is cancelled and spacetime supersymmetry is restored. In string moduli space, this procedure is equivalent to moving to a nearby point at which the string ground state is stable, and consequently this procedure is referred to as *vacuum shifting*. The specific VEV's that parametrize this vacuum shift are determined by solving the various $F$- and $D$-term flatness constraints, and one finds that they are typically quite small, of the order $\langle\phi\rangle/M_{\text{string}} \sim \mathcal{O}(1/10)$.

Such vacuum shifting has important consequences for the phenomenology of the string theory. For example, vacuum shifting clearly requires that those scalar fields receiving VEV's be charged under $U(1)_X$. Thus, the act of vacuum shifting breaks $U(1)_X$, with the $U(1)_X$ gauge boson "eating" the axion to become massive. In fact, since the scalars $\phi$ which are charged under $U(1)_X$ are also often charged under other gauge symmetries, giving VEV's to these scalars typically causes further gauge symmetry breaking. Perhaps most importantly, however, vacuum shifting can generate effective superpotential mass terms for vector-like states $\Psi$ that would otherwise be massless. Indeed, upon replacing the scalar fields $\phi$ by their VEV's in the low-energy superpotential, one finds that higher-order non-renormalizable couplings can become lower-order effective mass terms:

$$\frac{1}{M_{\text{string}}^{n-1}}\,\phi^n\,\overline{\Psi}\Psi \;\to\; \frac{1}{M_{\text{string}}^{n-1}}\,\langle\phi\rangle^n\,\overline{\Psi}\Psi\ . \tag{2.37}$$

Moreover, it often turns out that various string selection rules prohibit these types of effective mass terms from appearing in the tree-level superpotential until rather high order. For example, we often have $n \gtrsim 5$ in Eq. (2.37). Since one typically has $\langle\phi\rangle/M_{\text{string}} \sim \mathcal{O}(1/10)$, the effective mass terms that are generated after the vacuum shift are schematically of the order $\langle\phi\rangle^n/M_{\text{string}}^{n-1} \sim (1/10)^n M_{\text{string}}$. Thus, we see that vacuum shifting in string theory provides an economical mechanism for generating intermediate mass scales.

It is remarkable that in string theory, the need to protect supersymmetry against the effects of pseudo-anomalous $U(1)$'s can have all of these important effects. This once again underlines the key idea that supersymmetry plays a profound role in string theory — in some ways, even more profound than the role it plays in field theory.

## Question #19 How is SUSY broken in string theory?

Given the unique role of supersymmetry in string theory, and given that our low-energy world is non-supersymmetric, the next issue that arises is the means by which supersymmetry can be *broken* in string theory. Although there are many different proposals, these can be grouped into essentially three methods: one can break SUSY within perturbative string theory itself (so that one obtains a non-supersymmetric string); one can break SUSY within the low-energy effective field theory derived from a supersymmetric string; and one can break SUSY via a new scenario (the Hořava-Witten scenario) that makes use of certain features of non-perturbative string theory.

### 2.19.1. *Within string theory itself*

Perhaps the most direct way of breaking supersymmetry in string theory is within the full string theory itself. Thus, one would obtain a string that has no spacetime supersymmetry at *any* scale, not even the Planck scale. As we stated above, such strings are generally not stable (due to their non-vanishing dilaton tadpoles), but it is not known whether there might exist a special subset of non-supersymmetric strings which *are* stable. In many ways, this question is the stringy analogue of the cosmological constant problem: how can one find a non-supersymmetric ground state which preserves a near-exact (if not absolutely exact) cancellation of the cosmological constant? Indeed, in string theory these two questions are actually related in a deep way, and various proposals exist for solving this problem.[81]

Breaking supersymmetry within the string theory itself can be done in a variety of different ways. In all cases, however, the basic idea is to implement a carefully chosen "twist" when compactifying the string so that all superpartner states (including the gravitinos themselves) suffer so-called "GSO projections" and are removed from the string spectrum. In many (but not all) cases, this method is equivalent to the well-known Scherk-Schwarz mechanism[82] in which supersymmetry is broken through the special dependence that compactified fields have on the coordinates of compactified dimensions. This procedure was introduced into string theory in Ref. 83, and has since been pursued in a number of contexts.[84–88]

Breaking supersymmetry this way offers a number of distinct advantages. The most important may be that it *preserves the string itself*. Specifically, because this method results in another *string* theory, it preserves the string symmetries (such as modular invariance) that underlie many of the properties of string theory (such as finiteness) that we would like to preserve even after SUSY-breaking. For example, it has been shown that even though spacetime supersymmetry is broken in such scenarios, there is always a hidden "misaligned supersymmetry"[84] that remains in the string spectrum. This misaligned supersymmetry tightly constrains the distribution of bosonic and fermionic states throughout the string spectrum in such a way that even though SUSY is broken, bosons and fermions nevertheless provide cancelling contributions to string amplitudes, and certain mass supertraces continue to vanish.[84,85] Indeed, the phenomenology of misaligned supersymmetry ensures that these supertraces cancel not in the usual scale-by-scale manner, multiplet-by-multiplet, but rather through subtle simultaneous conspiracies between physics at different energy scales. This may have important phenomenological applications.

There is also another important phenomenological aspect of such theories. In some sense, since SUSY is being directly broken at the string scale, one might suspect that all gravitinos must have Planck-scale masses. However, this is not the case: it turns out that one can often "dial" the gravitino mass $m_{3/2}$ in such scenarios. But various string consistency constraints then imply[89] that such theories will essentially have an extra dimension whose radius is $R \sim m_{3/2}^{-1}$. Thus, in such string models, the existence of a TeV-scale gravitino implies the existence a TeV-scale extra dimension, which in turn implies the existence of infinite towers of TeV-scale string states with TeV-scale mass separations. The phenomenology of such scenarios is discussed in Ref. 86.

Finally, we remark that even though breaking SUSY through the string itself does not provide supersymmetry at any scale below the Planck scale, this need not be in conflict with gauge coupling unification. A discussion of this point can be found in Ref. 74.

### 2.19.2. *Within the low-energy effective theory*

The second way of breaking SUSY in string theory is to start with a supersymmetric string at the Planck scale, and then break SUSY within the low-energy effective field theory that is derived from the massless (observable) modes of the string. Since this method is essentially field-theoretic in nature, occurring purely within the language of the effective field theory, it does not necessarily result in a particle spectrum that can be interpreted as the low-energy limit of a non-supersymmetric string. This method therefore presumably breaks some or all of the consistency constraints that underlie string theory, and destroys the fundamental finiteness properties of the string. However, it offers the advantage that a purely field-theoretic treatment of SUSY-breaking will suffice.

Because this method of SUSY-breaking is field-theoretic, all of the SUSY-breaking mechanisms we have outlined in Question #6 apply to this case as well. The most commonly assumed scenario is that the dynamics of extra "hidden" string sectors will break supersymmetry through some mechanism (*e.g.*, gaugino condensation[90]) which is then communicated to the observable sector through either gravitational or gauge interactions. The ensuing phenomenologies are then analyzed in purely field-theoretic terms, and will be discussed in upcoming chapters.

### 2.19.3. *SUSY-breaking in strongly coupled strings*

Finally, there also exists a third scenario for SUSY-breaking within string theory. At strong coupling, it has been proposed[91] that the ten-dimensional $E_8 \times E_8$ heterotic string can be described as the compactification of an *eleven*-dimensional theory known as 'M-theory' on a line segment of finite length $\rho$. The two $E_8$ gauge factors are presumed to exist at opposite endpoints of this line segment. In order to incorporate GUT-scale gauge coupling unification within this scenario, it turns out[92] that the length $\rho$ of the eleventh dimension must be substantially larger than the eleven-dimensional Planck length. Thus, one has a situation in which the two $E_8$ gauge factors communicate primarily with their own ten-dimensional worlds located at opposite ends of an eleven-dimensional bulk, and only gravitational interactions connect these two "ends of the world" with each other.

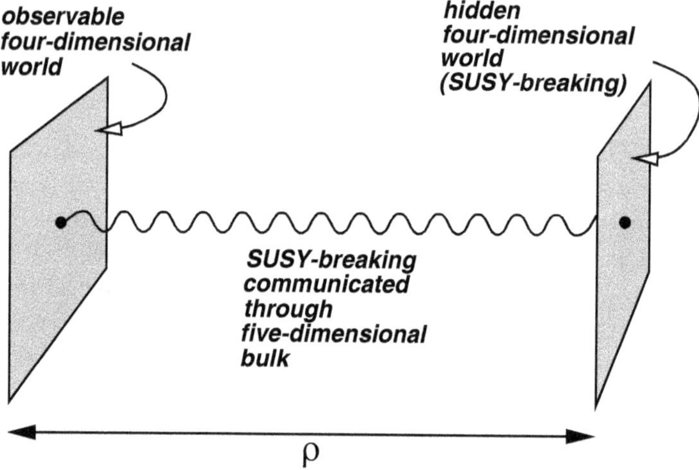


Fig. 2.2. The Hořava-Witten scenario for communicating SUSY-breaking from a hidden world across a five-dimensional bulk of length $\rho$ to the observable world.

If we now imagine further compactifying this picture to four dimensions, we obtain a scenario in which one four-dimensional world is the "observable" world that descends from one of the $E_8$ gauge factors, while the other four-dimensional world represents the "hidden" sector that descends from the other $E_8$ gauge factor. Since the radii of the additional six-dimensional compactification must be considerably smaller than the length of the eleventh dimension, one obtains an effective situation in which two four-dimensional worlds are connected through a five-dimensional bulk. This situation is illustrated in Fig. 2.2. Most interestingly, if strong-coupling dynamics in the hidden $E_8$ gauge factor causes SUSY-breaking to occur in that sector (such as via gaugino condensation), the effects of such SUSY-breaking will be communicated gravitationally to the observable world through the five-dimensional interior bulk. This scenario thereby places the question of SUSY-breaking in an entirely new geometric context. For example, in some circumstances the resulting gravitino mass can be identified with the radius $\rho$ of the fifth dimension. Various phenomenological consequences of this picture of SUSY-breaking are currently being explored,[93] in the context of both gaugino condensation and Scherk-Schwarz compactification.

## Question #20 Making ends meet: How can we understand gauge coupling unification from string theory?

In this section we will discuss another important issue connected with strings and supersymmetry, namely gauge coupling unification. As we have seen, the strong, electroweak, and hypercharge gauge couplings appear to unify at approximately $M_{\text{MSSM}} \approx 2 \times 10^{16}$ GeV when extrapolated within the framework of the MSSM. Indeed, this observation is often taken as evidence for supersymmetry, which also provides elegant solutions for the finiteness and gauge hierarchy problems. Thus, the currently accepted field-theoretic scenario calls for some sort of grand unified group (GUT) above $M_{\text{MSSM}}$; the MSSM gauge group and spectrum between $M_{\text{MSSM}}$ and the scale of SUSY-breaking $M_{\text{SUSY}}$ at which the superpartners decouple; and simply the Standard Model gauge group and spectrum below $M_{\text{SUSY}}$.

This is a compelling picture, except for various problems. First, $M_{\text{MSSM}}$ is close to the Planck scale, but gravity is not incorporated. Second, one would in principle like to explain the spectrum of the MSSM — *e.g.*, to explain why there are three generations, or to derive the fermion mass matrices. Third, if there is a GUT theory above $M_{\text{MSSM}}$, what about proton-lifetime problems? One requires some sort of doublet-triplet splitting mechanism, as was discussed in Question #14. Finally, why should we require gauge coupling unification at all? This is, after all, only a theoretical prejudice, and is not required for the consistency of the model.

### 2.20.1. *The predictions from string theory*

Of course, string theory can solve these problems. First, as we have seen, it naturally incorporates quantized gravity, in the sense that a spin-two massless particle (the graviton) always appears in the string spectrum. Second, $N = 1$ supersymmetric field theories with non-abelian gauge groups naturally appear as the limits of a certain class of string models (the heterotic strings). Third, string theory can provide, in principle, a uniform framework for understanding three generations, fermion matrices, doublet-triplet splitting mechanism, *etc.* — in principle, there are no free parameters! Finally, it also turns out that *independently of the existence of a unified gauge symmetry*, heterotic string theories always give rise to a natural unification of gauge couplings. Indeed, in heterotic string theory, the gauge and gravitational couplings automatically unify[94] to form a single coupling constant $g_{\text{string}}$:

$$8\pi \frac{G_N}{\alpha'} = g_i^2\, k_i = g_{\text{string}}^2 \ . \tag{2.38}$$

Here $G_N$ is the gravitational (Newton) coupling; $\alpha'$ is the Regge slope (which sets the mass scale for string theory); $g_i$ are the gauge couplings; and the normalization constants $k_i$ are the *affine levels* (also sometimes called *Kač-Moody* levels) at which the different group factors are realized. For non-abelian group factors we have $k_i \in \mathbb{Z}^+$, while for $U(1)$ gauge factors the $k_i$ are arbitrary. Thus, string theory appears to give us precisely the features we want.

There are, however, some crucial differences between string theory and field theory. First, string theory is a *finite* theory: the gauge couplings run only within the framework of the string low-energy *effective* theory. Second, in string theory all couplings are ultimately dynamical variables, related to the expectation values of scalar moduli fields. The third difference is the dependence on the affine levels $k_i$. These levels are essentially normalizations, and are therefore analogous to the hypercharge normalization $k_Y = 5/3$ which appears in $SU(5)$ or $SO(10)$ embeddings, but in string theory such normalizations also appear for the *non*-abelian gauge couplings as well. It turns out that the most easily constructed string models have $k_i = 1$ for non-abelian factors.

The most important difference concerns the scale of the unification. The string unification scale is set by $\alpha'$ (which in turn is set by Planck scale), and at the one-loop level one finds:[95]

$$M_{\text{string}} \approx g_{\text{string}} \times 5 \times 10^{17} \text{ GeV} \ . \tag{2.39}$$

Since extrapolation of low-energy data suggests that $g_{\text{string}} \approx \mathcal{O}(1)$, we thus find that $M_{\text{string}} \approx 5 \times 10^{17}$ GeV — a factor of 20 discrepancy relative to the MSSM prediction!

Is this is a major problem? A factor of 20 sounds large, but this is only a 10% effect in the *logarithms* of the mass scales. Unfortunately, however, this discrepancy leads to wildly incorrect values for the low-energy observables $\sin^2\theta_W$ and $\alpha_{\text{strong}}$ at the weak scale. In other words, if we start our MSSM running of gauge couplings down from $M_{\text{string}}$ rather than from $M_{\text{MSSM}}$, we find that string theory predicts values for these quantities which differ from their experimentally observed values by many standard deviations. This is the problem of gauge coupling unification in string theory. Essentially, given the high-energy predictions of string theory and our low-energy experimental couplings, we face the classic question: how can we make the two ends meet?

### 2.20.2. *Overview of possible solutions*

Over the past decade, a number of solutions to this question have been proposed. We shall here outline only six possible classes of solutions. The reader should consult Ref. 74 for a more complete discussion of these and other solutions.

The first solution reconciles $M_{\text{MSSM}}$ and $M_{\text{string}}$ by assuming that the three low-energy gauge couplings indeed unify at $M_{\text{MSSM}}$ because of the presence of a unifying gauge symmetry group $G$ at that scale, whereupon the new unified gauge coupling $g_G$ runs upwards to $M_{\text{string}}$ where it unifies with the gravitational coupling. Thus, at the string scale, we are essentially realizing the GUT group $G$ as our gauge symmetry: these are "string GUT models". Note that in this context we therefore consider only those unification groups $G$ such as $SU(5)$ or $SO(10)$ which are *simple.* An essential property of such groups is that they require a Higgs scalar representation in the adjoint of $G$ in order to break $G$ down to the MSSM gauge group.

The second possible solution makes use of the affine levels $k_i$ that appear in the string unification relation in Eq. (2.38). Indeed, in string theory, these levels $k_i$ need not take the values $(k_Y, k_2, k_3) = (5/3, 1, 1)$ that we naïvely expect them to have in the MSSM. It is then possible that non-standard values for these levels could alter the runnings in such a way as to reconcile the string unification scale with the MSSM unification scale. This would clearly be a stringy effect.

The third solution supposes that there can be large "heavy string threshold corrections" at the string scale. These corrections represent the contributions from the infinite towers of massive (Planck-scale) string states that are otherwise neglected in an analysis of the purely massless string spectrum. This would also be an intrinsically stringy effect.

A fourth solution involves "light SUSY thresholds" — the corrections that arise due to the breaking of supersymmetry — and are typically analyzed in field theory.

A fifth solution involves extra matter beyond the MSSM at intermediate mass scales. While introducing such matter may seem *ad hoc* from the field-theory perspective, it turns out that certain exotic non-MSSM states appear in, and are actually *required for the self-consistency* of, many realistic string models.

Finally, a sixth solution[92] involves possible effects due to non-perturbative string physics. For example, as we have seen, recent developments in string duality suggest that at strong coupling, the behavior of heterotic strings can be modelled by other theories for which the heterotic string prediction in Eq. (2.38) is no longer valid. This then effectively loosens the tight constraints between the gauge couplings and the gravitational coupling, which in turn enables one to separate the gauge-coupling unification scale from the gravitationally-determined string scale.

Thus, we are faced with one over-riding question: Which solution(s) to the problem — *i.e.*, which "path to unification" — does string theory actually take?

It is perhaps worth emphasizing that this a much more difficult question in string theory than it would be in field theory. In field theory, one can imagine rather easily building a model that realizes any one of the above proposals. In string theory, however, there are deeper string consistency constraints which arise due to the fact that four-dimensional (spacetime) physics is ultimately derived from two-dimensional (worldsheet) physics. Thus four-dimensional spectra, gauge symmetries, couplings, *etc.*, are all ultimately determined or constrained by worldsheet symmetries. This tends to make it difficult, when string model-building, to realize one given desirable phenomenological feature in one sector of a string model without upsetting some other desired feature in a different sector of the model.

The question, then, is to determine which of the above potential solutions to the string unification problem are self-consistent in string theory, and can be realized in *actual realistic string models.*

### 2.20.3. *Current status*

We shall now give a quick summary of the current status of some of these proposed solutions. A more detailed review (along with appropriate references) can be found in Ref. 74.

*String GUT models*: As mentioned above, the goal in this approach is to construct realistic string GUT models — *i.e.*, string models whose low-energy limits reproduce standard $SU(5)$ or $SO(10)$ unification scenarios. The major problem that one faces, however, is that while it is generally easy to obtain the required gauge group, obtaining the required *matter representations* has proven to be very difficult. The fundamental reason for this difficulty is that: (i) the string requirement that the worldsheet conformal field theory be unitary ends up restricting the allowed massless matter representations that the string model can produce; and (ii) for GUT symmetry breaking, one requires a Higgs scalar transforming in the adjoint of the GUT gauge group. Together, these two requirements imply that one must realize the GUT gauge symmetry at an affine level $k_{\rm GUT} \geq 2$, and historically it has proven to be a highly non-trivial task to construct such a higher-level string GUT model with three generations.[96–98]

At present, the three-generation problem has been solved at level two only in the case of $SU(5)$.[97–99] At level three, however, there currently exist three-generation models for $SU(5)$, $SU(6)$, $SO(10)$, and $E_6$.[100] However, much phenomenological analysis of these models still remains to be done. In some cases, these models tend to have extra chiral matter, or unsuitable couplings. Doublet-triplet splitting also remains a problem, and appears to require fine-tuning. There are also rather tight constraints[101] concerning the allowed representations and couplings for these models which restrict their phenomenologies significantly. However, the important point is that the issues concerning string GUT model-building now seem to be more of a technical rather than fundamental nature, and further progress can be expected.

*Non-Standard Levels and Hypercharge Normalizations:* In this solution to the unification problem, one attempts to realize the MSSM gauge group and particle content in a given model, but to reconcile the discrepancy between $M_{\rm MSSM}$ and $M_{\rm string}$ by having non-standard values for the levels $(k_Y, k_2, k_3)$. A straightforward analysis[102,103] shows that in order to do the job, the required levels would be:

$$k_2 = k_3 = 1, 2 \; ; \qquad k_Y/k_2 \approx 1.45 - 1.5 \; . \tag{2.40}$$

Thus, restricting our attention the simpler level-one models, the question arises: can one even realize realistic string models with $k_Y$ in this range?

This question is motivated by the observation that the standard $SO(10)$ hypercharge embedding naturally leads to the MSSM value $k_Y = 5/3$, and most trivial modifications or extensions to this embedding tend to increase $k_Y$. Thus, more generally, we ask whether it is even possible to realize hypercharge embeddings with $k_Y < 5/3$, and whether this would cause undesirable effects on the rest of such a string model. Note that one always must have $k_Y \geq 1$ in any string model containing at least the MSSM spectrum.[104]

The current status of this approach is as follows. In general, it is very difficult to arrange to have $k_Y < 5/3$ in string theory.[98,103] However, some self-consistent string models with $k_Y < 5/3$ have been constructed.[98] Unfortunately, all of these models have unwanted fractionally charged states that could survive in their light spectra. This is to be expected, since there is a general result[104] that if a string model is to completely avoid fractionally-charged color-neutral string states, then its affine levels must obey the relation

$$3\,k_Y + 3\,k_2 + 4\,k_3 \;=\; 0 \pmod{12}\ . \tag{2.41}$$

For $k_2 = k_3 = 1, 2$, this implies $k_Y/k_2 \geq 5/3$. Of course, it is possible that fractionally charged states appear but are extremely massive, or that they might bind together into color-neutral objects under the influence of extra hidden-sector interactions. A general classification of the binding scenarios that can eliminate such fractionally charged states has been performed,[103] but no string model has yet been constructed which realizes these scenarios.

*Heavy string threshold corrections:* Heavy string threshold corrections are the contributions due to the infinite towers of massive Planck-scale string states that are otherwise neglected when deriving a low-energy effective action from the string. In order to reconcile the values of the three low-energy gauge couplings $g_i$ with string-scale unification, it turns out that such corrections $\Delta_i$ must have the relative sizes

$$\Delta_{\hat{Y}} - \Delta_2 \approx -28\ , \quad \Delta_{\hat{Y}} - \Delta_3 \approx -58\ , \quad \Delta_2 - \Delta_3 \approx -30\ , \tag{2.42}$$

where $\hat{Y} \equiv Y/\sqrt{5/3}$ is the renormalized hypercharge. These corrections are quite sizable, and the fundamental question is then how to obtain corrections of this size.

The formalism for calculating these corrections was first derived in Ref. 95 and more recently refined in Ref. 105. From these results, a number of theoretical mechanisms were identified for making these corrections sufficiently large. Perhaps the most obvious mechanism[106] is to construct a string model with a large modulus (such as a large compactified dimension), for as the size of such a radius is increased, various momentum states become lighter and lighter. The contributions of such states to the threshold corrections therefore become more substantial, ultimately leading to a decompactification of the theory. Unfortunately, it is not known why a given string model should be expected to have such a large modulus. Indeed, the general expectation is that in realistic string models, moduli should settle at or near the self-dual point for which moduli are of order one.[107]

Explicit calculations of these threshold corrections have been carried out within several realistic string models. Here the term "realistic" denotes string models with the following properties: $N = 1$ spacetime SUSY; appropriate gauge groups [such as $SU(3) \times SU(2) \times U(1)$, Pati-Salam $SO(6) \times SO(4)$, or flipped $SU(5) \times U(1)$]; the proper massless observable spectrum (including three complete chiral MSSM generations with correct quantum numbers, hypercharges, and Higgs scalar representations); and anomaly cancellation. Many realistic models also exhibit additional attractive features, such as a semi-stable proton, proper fermion mass hierarchy, and a heavy top quark. A collection of such models, all of which are realized in the free-fermionic construction with an underlying $\mathbb{Z}_2 \times \mathbb{Z}_2$ orbifold structure, can be found in Ref. 108.

Unfortunately, the results found within these models are not encouraging: in each of the realistic string models of Ref. 108, it is found[109] that threshold corrections are unexpectedly small, and moreover they have the wrong sign. For example, in one such string model it was found that

$$\Delta_{\hat{Y}} - \Delta_2 \approx 1.6\,, \quad \Delta_{\hat{Y}} - \Delta_3 \approx 5 \tag{2.43}$$

which does not fare well against Eq. (2.42). This behavior seems to be generic to the entire class of realistic string models in Ref. 108. Thus it seems that threshold corrections by themselves are not able to resolve the discrepancy with the low-energy couplings in these realistic string models. Indeed, despite some interesting proposals,[110] there do not presently exist any realistic string models with small moduli for which the threshold corrections are sufficiently large.

*Light SUSY thresholds and intermediate-scale gauge structure:* Light SUSY thresholds are the effects that arise from SUSY-breaking at some intermediate scale: they can be parametrized in terms of the usual soft SUSY-breaking parameters $\{m_0, m_{1/2}, m_h, m_{\tilde{h}}\}$, or one can take non-universal boundary terms for the sparticle masses. Similarly, the effects from intermediate-scale gauge structure arise whenever there is a gauge symmetry, such as $SO(6) \times SO(4)$ or flipped $SU(5) \times U(1)$, which is broken at some intermediate scale $M_I$. Such effects are then parametrized in terms of $M_I$. Both of these effects are analyzed purely in terms of the low-energy field theory derived from the string, and consequently their evaluation proceeds exactly as in field theory. A detailed calculation of these effects must also include two-loop corrections, the effects of Yukawa couplings, and even the effects of scheme conversion (from the $\overline{\text{DR}}$ scheme in which the string scale is evaluated to the $\overline{\text{MS}}$ scheme through which the low-energy couplings are extracted from experiment). Within the context of the low-energy effective theories derived from the realistic string models in Ref. 108, such a calculation has been performed.[109] The results indicate that the light SUSY thresholds are generally insufficient to resolve the discrepancies, and that the effects of intermediate gauge structure in the realistic string models only *enlarge* the disagreement with experiment! This latter result is surprising, given that $M_I$ can be tuned in principle to *any* value below $M_{\text{string}}$, and serves to illustrate the rather tight (and predictive) constraints that a given string model provides.

*Extra Matter Beyond the MSSM:* Finally, there is the possibility of extra matter beyond the MSSM. While all of the above results assumed only the MSSM spectrum, string theory often *requires* that additional exotic states appear in the massless spectrum. Their effects must therefore be included. Such states appear in a majority of the realistic string models, usually appear in vector-like representations, and ultimately have masses determined by cubic and higher-order terms in the superpotential (which are determined in turn by the specific SUSY-breaking mechanism employed, as well as by a host of additional factors). In one string model, for example, it has been estimated[111] that such extra states will naturally sit at an intermediate scale $\approx 10^{11}$ GeV. In the realistic string models[108] with $SU(3) \times SU(2) \times U(1)$ gauge groups, such matter typically arises in rather specific $SU(3) \times SU(2) \times U(1)_Y$ representations such as $(\mathbf{3},\mathbf{2})_{1/6}$, $(\overline{\mathbf{3}},\mathbf{1})_{1/3}$, $(\overline{\mathbf{3}},\mathbf{1})_{1/6}$, and $(\mathbf{1},\mathbf{2})_0$. While the first two representations can be fit into standard $SO(10)$ mutliplets, the remaining two cannot, and are truly exotic.

What is remarkable, however, is that this extra matter is just what is needed: because of their unusual hypercharge assignments, these representations have one-loop beta-function coefficients $b_i$ where $b_1$ turns out to be much smaller than $b_2$ or $b_3$. These representations therefore have the potential to modify the running of the $SU(2)$ and $SU(3)$ couplings without seriously affecting the $U(1)$ coupling. Moreover, in some string models, these extra non-MSSM matter representations also appear in precisely the right *combinations* to do the job. Details can be found in Ref. 109. Similar scenarios using such extra non-MSSM matter can also be found, *e.g.*, in Ref. 112. Thus, on the basis of this evidence, it appears that extra intermediate-scale matter beyond the MSSM may turn out to be the string-preferred route to string-scale unification. It is remarkable that string theory, which predicts an unexpectedly high unification scale, often also simultaneously predicts precisely the extra exotic matter necessary to reconcile this higher scale with the observed low-energy couplings.

## Conclusions

In this article, we have surveyed a number of questions and issues that arise in supersymmetric particle physics, ranging from the MSSM at the lowest scales to string theory at the highest scales. It is remarkable that supersymmetry not only provides a window into physics at so many different energy regimes, but also has such a profound impact in all of these areas. Indeed, at the very least it either refines old questions or proposes new ones, and in most cases it actually changes the language of the debate. Supersymmetry is perhaps the only extension of the Standard Model which has such a direct impact on so many types of new phenomena, including gravity. Moreover, as we have seen, supersymmetry has global applications, ranging from high-energy accelerator experiments to astrophysics and cosmology. The questions that supersymmetry prompts therefore provide unique opportunities for studying all sorts of new physics, and finding the answers to any of these questions — from the most phenomenological to the most theoretical — will undoubtedly teach us much about the physics that we expect to be exploring in the twenty-first century.

## Postscript

This essay is adapted from a review article that was originally written more than a decade ago.[113] However, as this book goes to publication, the world's physics community has just learned of the first collisions at the long-awaited Large Hadron Collider (LHC) at CERN. Though the data and analyses required to discover new physics, and supersymmetry in particular, are still at least a year or two in the future, there is a palpable excitement among particle physicists that many of the questions addressed in this essay will soon be examined and our ideas tested.

In fact, the last decade has already seen a number of theoretical and experimental developments that constrain SUSY and sharpen our thinking about how it might manifest itself in the real world. For brevity's sake, we will mention only three broad areas here.

First, the experimental particle physics program over the past decade has seen a number of important experiments brought to successful completion. In Japan and the US, the two B-factories, with their dedicated experiments, BaBar and Belle, probed CP violation and flavor-changing in the heavy quark sector, while simultaneously studying rare decays of $\tau$-leptons. Their most famous result was that CP violation in $b \to d$ transitions appears to be consistent with the CKM paradigm,[114] meaning that another opportunity for non-canonical SUSY flavor physics to show itself has passed us by. In a sense, the SUSY flavor problem is now even more difficult than it was a decade ago. Similarly, neither the B-factories nor Fermilab found any evidence for rare flavor-changing neutral currents mediated by SUSY Higgs bosons,[115] a signature that was suggested less than ten years ago.[116] On the other hand, the Brookhaven E821 experiment found a discrepancy in the anomalous magnetic moment of the muon that stands now slightly more than $3\sigma$ from the Standard Model.[117] Could this be the first real evidence for weak-scale SUSY?

A second area in which a great deal has happened to impact SUSY in the last decade is astrophysics. Studies of type-Ia supernovae by two groups found that the universe's expansion has sped up in the recent past,[118] contrary to previous belief. This result, coupled with precision measurements of the cosmic microwave background radiation by the WMAP experiment,[119] has resolved a long-standing feud between theoretical and observational astrophysicists about the abundance of dark matter in the universe. Current best fits indicate that dark matter has a density of about a quarter of critical density ($\Omega_{CDM} \simeq 0.23$), and that the total energy of the universe is roughly equal to critical density ($\Omega \simeq 1$), the remainder mostly composed of "dark energy" or a cosmological constant. This dark matter density is much smaller than previously supposed by most particle theorists, and this result, coupled with better codes for calculating neutralino relic abundances, has severely constrained the parameter space of SUSY models, forcing more and more fine-tuning on the theory.[120]

Thirdly, the last decade saw the end of the LEP physics program, and at the end of that program, a bound on the Higgs boson mass of 114 GeV.[121] In time theorists began to understand that Higgs masses above 114 GeV created their own "little hierarchy" problem.[122] The problem can be understood simply: In order to push Higgs masses above 114 GeV, we must invoke top squarks that are nearly, or above, the TeV mass range (since the effect of the heavy stops is only logarithmic). But top squarks with TeV masses pull up the soft Higgs masses quadratically, which drives up the electroweak scale in turn. While a number of partial solutions have been proposed in the literature, this remains very much an open issue.

With the start of the LHC, we may finally see many of these, and other, questions addressed. Will the LHC see long-lived neutralinos? Will there be other sparticles in the theory to help push up the Higgs mass? Are the squarks all degenerate, or are other flavor-preserving mechanisms at work? In terms of the original set of open questions, as well as those newer questions mentioned just above, the final word on SUSY belongs to experiment, most particularly the imminent LHC.

## Acknowledgments

We would like to thank J. Bagger, A. Kusenko, P. Langacker, J. March-Russell, and N. Polonsky for many helpful discussions, and especially K.S. Babu, T. Gherghetta, G. Kane, and M. Peskin for their comments on this article. We would also like to acknowledge the hospitality of the Aspen Center for Physics, where parts of this work were completed. KRD was supported in part by the U.S. Department of Energy under Grant DE-FG02-04ER-41298.

## References

1. J. Wess and J. Bagger, *Introduction to Supersymmetry, 2nd Ed.* (Princeton University Press, 1992); H.P. Nilles, *Phys. Rep.* **110** (1984) 1; H. Haber and G. Kane, *Phys. Rep.* **117** (1985) 75.
2. S. Dimopoulos and D. Sutter, *Nucl. Phys.* **B452** (1995) 496.
3. S. Ferrara, L. Girardello, and F. Palumbo, *Phys. Rev.* **D20** (1979) 403; S. Dimopoulos and H. Georgi, *Nucl. Phys.* **B193** (1981) 150.
4. S. Weinberg, *Phys. Rev.* **D26** (1982) 287.
5. G. Farrar and P. Fayet, *Phys. Lett.* **B76** (1978) 575; S. Dimopoulos, S. Raby, and F. Wilczek, *Phys. Lett.* **B112** (1982) 133.
6. T. Banks and L. Dixon, *Nucl. Phys.* **B307** (1988) 93; M. Kamionkowski and J. March-Russell, *Phys. Rev. Lett.* **69** (1992) 1485; R. Holman *et al.*, *Phys. Rev. Lett.* **69** (1992) 1489.
7. S. Martin, *Phys. Rev.* **D46** (1992) 2769.
8. L. Ibáñez and G. Ross, *Nucl. Phys.* **B368** (1991) 3.
9. L. Hall, V. Kostelecky, and S. Raby, *Nucl. Phys.* **B267** (1986) 415; H. Georgi, *Phys. Lett.* **B169** (1986) 231.

10. J. Donoghue, H.P. Nilles, and D. Wyler, *Phys. Lett.* **B128** (1983) 55.
11. F. Gabbiani and A. Masiero, *Nucl. Phys.* **B322** (1989) 235; M. Dine, A. Kagan, and S. Samuel, *Phys. Lett.* **B243** (1990) 250; F. Gabbiani, E. Gabrielli, A. Masiero, and L. Silvestrini, *Nucl. Phys.* **B477** (1996) 321.
12. Y. Nir and N. Seiberg, *Phys. Lett.* **B309** (1993) 337.
13. K.S Babu and S. Barr, *Phys. Lett.* **B387** (1996) 87.
14. L. Ibáñez and G. Ross, *Phys. Lett.* **B332** (1994) 100; P. Binétruy and P. Ramond, *Phys. Lett.* **B350** (1995) 49; P. Binétruy, S. Lavignac, and P. Ramond, *Nucl. Phys.* **B477** (1996) 353; R. Zhang, *Phys. Lett.* **B402** (1997) 101; A. Nelson and D. Wright, *Phys. Rev.* **D56** (1997) 1598.
15. A. Cohen, D. Kaplan, and A. Nelson, *Phys. Lett.* **B388** (1996) 588.
16. N. Arkani-Hamed and H. Murayama, hep-ph/9703259.
17. For a nice review, see: Y. Grossman, Y. Nir, and R. Rattazzi, hep-ph/9701231.
18. W. Fischler, S. Paban, and S. Thomas, *Phys. Lett.* **B289** (1992) 373.
19. P. Binétruy and E. Dudas, *Phys. Lett.* **B389** (1996) 503; G. Dvali and A. Pomarol, *Phys. Rev. Lett.* **77** (1996) 3728.
20. K.S. Babu, C. Kolda, and F. Wilczek, *Phys. Rev. Lett.* **77** (1996) 3070; M. Dine, Y. Nir, and Y. Shirman, *Phys. Rev.* **D55** (1997) 1501; J. Bagger *et al.*, *Phys. Rev.* **D55** (1997) 3188; R. Rattazzi and U. Sarid, hep-ph/9612464; F. Borzumati, hep-ph/9702307.
21. M. Carena *et al.*, hep-ph/9702409.
22. G. Guidice and A. Masiero, *Phys. Lett.* **B206** (1988) 480.
23. J. Casas, A. Lleyda and C. Muñoz, *Nucl. Phys.* **B471** (1996) 3.
24. J. Gunion, H. Haber, and M. Sher, *Nucl. Phys.* **B306** (1988) 1.
25. T. Gherghetta, C. Kolda, and S. Martin, *Nucl. Phys.* **B468** (1996) 37.
26. A. Kusenko, P. Langacker, and G. Segrè, *Phys. Rev.* **D54** (1996) 5824.
27. E. Witten, *Nucl. Phys.* **B188** (1981) 513.
28. R. Roberts and G. Ross, *Nucl. Phys.* **B377** (1992) 571; S. Kelley *et al.*, *Nucl. Phys.* **B398** (1993) 3; M. Drees and M. Nojiri, *Phys. Rev.* **D47** (1993) 376; G. Kane, L. Roszkowski, C. Kolda, and J. Wells, *Phys. Rev.* **D49** (1994) 6173; M. Carena, M. Olechowski, S. Pokorski, and C. Wagner, *Nucl. Phys.* **B419** (1994) 213.
29. Recent work was inspired by the papers of M. Dine and A. Nelson, *Phys. Rev.* **D48** (1993) 1277; M. Dine, A. Nelson, and Y. Shirman, *Phys. Rev.* **D51** (1995) 1362; M. Dine, A. Nelson, Y. Nir, and Y. Shirman, *Phys. Rev.* **D53** (1996) 2658.
30. For recent work, see: S. Abel, S. Sarkar, and P. White, *Nucl. Phys.* **B454** (1995) 663.
31. U. Ellwanger, *Phys. Lett.* **B133** (1983) 187; J. Bagger and E. Poppitz, *Phys. Rev. Lett.* **71** (1993) 2380; J. Bagger, E. Poppitz, and L. Randall, *Nucl. Phys.* **B455** (1995) 59.
32. H.P. Nilles and N. Polonsky, hep-ph/9707249.
33. C. Kolda, S. Pokorski, and N. Polonsky, *in preparation.*
34. M. Cvetič and P. Langacker, *Phys. Rev.* **D54** (1996) 3570; M. Cvetič *et al.*, hep-ph/9703317; hep-ph/9705391.

35. K.R. Dienes, C. Kolda, and J. March-Russell, *Nucl. Phys.* **B492** (1997) 104 [hep-ph/9610479].
36. T. Falk, K. Olive, and M. Srednicki, *Phys. Lett.* **B339** (1994) 248.
37. G. Kane and J. Wells, *Phys. Rev. Lett.* **76** (1996) 4458.
38. For a review, see: G. Jungman, M. Kamionkowski, and K. Griest, *Phys. Rep.* **267** (1996) 195.
39. P. Fayet, *Phys. Rep.* **105** (1984) 21.
40. H. Pagels and J. Primack, *Phys. Rev. Lett.* **48** (1982) 223.
41. T. Gherghetta, *Nucl. Phys.* **B485** (1997) 25.
42. M. Luty and E. Ponton, hep-ph/9706268.
43. T. Moroi, H. Murayama, and M. Yamaguchi, *Phys. Lett.* **B303** (1993) 289.
44. M. Dine, W. Fischler, and D. Nemeschansky, *Phys. Lett.* **B136** (1984) 169.
45. D. Lyth and E. Stewart, *Phys. Rev.* **D53** (1996) 1784.
46. I. Affleck and M. Dine, *Nucl. Phys.* **B249** (1985) 361. M. Dine, L. Randall, and S. Thomas, *Nucl. Phys.* **B458** (1996) 291.
47. L. Kofman, A. Linde, and A. Starobinskii, *Phys. Rev. Lett.* **73** (1994) 3195; Y. Shtanov, J. Traschen, and R. Brandenberger, *Phys. Rev.* **D51** (1995) 5438.
48. R. Brustein and P. Steinhardt, *Phys. Lett.* **B302** (1993) 196.
49. J.C. Pati and A. Salam, *Phys. Rev.* **D8** (1973) 1240; H. Georgi and S.L. Glashow, *Phys. Rev. Lett.* **32** (1974) 438.
50. S. Dimopoulos, S. Raby, and F. Wilczek, *Phys. Rev.* **D24** (1981) 1681; S. Dimopoulos and H. Georgi, *Nucl. Phys.* **B193** (1981) 150; N. Sakai, *Zeit. Phys.* **C11** (1981) 153; L.E. Ibáñez and G.G. Ross, *Phys. Lett.* **B105** (1981) 439.
51. H. Georgi, H.R. Quinn, and S. Weinberg, *Phys. Rev. Lett.* **33** (1974) 451.
52. M.B. Einhorn and D.R.T. Jones, *Nucl. Phys.* **B196** (1982) 475; W.J. Marciano and G. Senjanovic, *Phys. Rev.* **D25** (1982) 3092.
53. For recent precision calculations, see, *e.g.*: U. Amaldi, W. de Boer, and H. Fürstenau, *Phys. Lett.* **B260** (1991) 447; J. Ellis, S. Kelley, and D.V. Nanopoulos, *Phys. Lett.* **B260** (1991) 131; P. Langacker and M. Luo, *Phys. Rev.* **D44** (1991) 817; P. Langacker and N. Polonsky, *Phys. Rev.* **D47** (1993) 4028; M. Carena, S. Pokorski, and C.E.M. Wagner, *Nucl. Phys.* **B406** (1993) 59; J. Bagger, K. Matchev, and D. Pierce, *Phys. Lett.* **B348** (1995) 443.
54. N. Sakai and T. Yanagida, *Nucl. Phys.* **B197** (1982) 533; S. Weinberg, *Phys. Rev.* **D26** (1982) 287.
55. E. Witten, *Phys. Lett.* **B105** (1981) 267; D.V. Nanopoulos and K. Tamvakis, *Phys. Lett.* **B113** (1982) 151; S. Dimopoulos and H. Georgi, *Phys. Lett.* **B117** (1982) 287; L. Ibáñez and G. Ross, *Phys. Lett.* **B110** (1982) 215.
56. A. Buras *et al.*, *Nucl. Phys.* **B135** (1985) 66; H. Georgi, *Phys. Lett.* **B108** (1982) 283; A. Masiero *et al.*, *Phys. Lett.* **B115** (1982) 380; B. Grinstein, *Nucl. Phys.* **B206** (1982) 387.
57. K. Inoue, A. Kakuto, and T. Takano, *Prog. Theor. Phys.* **75** (1984) 664; A. Anselm and A. Johansen, *Phys. Lett.* **B200** (1988) 331; Z. Berezhiani and G. Dvali, *Sov. Phys. Lebedeev Inst. Reps.* **5** (1989) 55; R. Barbieri, G. Dvali, and M. Moretti, *Phys. Lett.* **B312** (1993) 137.

58. S. Dimopoulos and F. Wilczek, ITP preprint NSF-ITP-82-07 (1982), unpublished. For a recent discussion, see: K.S. Babu and S.M. Barr, *Phys. Rev.* **D48** (1993) 5354; *Phys. Rev.* **D50** (1994) 3529.
59. D. Lee and R.N. Mohapatra, *Phys. Rev.* **D51** (1995) 1353; L.J. Hall and S. Raby, *Phys. Rev.* **D51** (1995) 6524.
60. J. Hisano, H. Murayama, and T. Yanagida, *Phys. Rev.* **D49** (1994) 4966; K.S. Babu and S.M. Barr, *Phys. Rev.* **D51** (1995) 2463; G. Dvali and S. Pokorski, *Phys. Lett.* **B379** (1996) 126; S.M. Barr and S. Raby, hep-ph/9705366.
61. A.J. Buras, J. Ellis, M.K. Gaillard, and D.V. Nanopoulos, *Nucl. Phys.* **B135** (1978) 66.
62. S. Raby, hep-ph/9501349; Z. Berezhiani, hep-ph/9602325.
63. H. Georgi and C. Jarlskog, *Phys. Lett.* **B86** (1979) 297.
64. M. Gell-Mann, P. Ramond, and R. Slansky, in *Supergravity* (edited by P. van Nieuwenhuizen and D.Z. Freedman), North-Holland, Amsterdam (1979); T. Yanagida, in *Proceedings of the Workshop on Unified Theories and Baryon Number in the Universe* (edited by A. Sawada and A. Sugamoto), K.E.K. preprint 79-18 (1979); R.N. Mohapatra and G. Senjanovic, *Phys. Rev. Lett.* **44** (1980) 912.
65. H. Arason *et al.*, *Phys. Rev.* **D47** (1993) 232; G. Anderson *et al.*, *Phys. Rev.* **D49** (1994) 3660.
66. X.G. He and S. Meljanac, *Phys. Rev.* **D41** (1990) 1620; S. Dimopoulos, L.J. Hall, and S. Raby, *Phys. Rev. Lett.* **68** (1992) 1984; *Phys. Rev.* **D45** (1992) 4192; G. Anderson *et al.*, *Phys. Rev.* **D47** (1993) 3702.
67. K.S. Babu and R.N. Mohapatra, *Phys. Rev. Lett.* **70** (1993) 2845; *Phys. Rev. Lett.* **74** (1995) 2418; K.S. Babu and S.M. Barr, hep-ph/9512389; *Phys. Rev. Lett.* **75** (1995) 2088.
68. J. Ellis, J. Lopez, and D.V. Nanopoulos, *Phys. Lett.* **B245** (1990) 375.
69. N. Seiberg, hep-th/9506077; K. Intriligator and N. Seiberg, *Nucl. Phys. Proc. Suppl.* **55B** (1997) 157.
70. C. Montonen and D. Olive, *Phys. Lett.* **B72** (1977) 117.
71. For introductions and reviews, see: M.B. Green, J.H. Schwarz, and E. Witten, *Superstring Theory, Vols. 1 & 2* (Cambridge University Press, Cambridge, 1987); M. Dine, ed., *String Theory in Four Dimensions* (North-Holland, Amsterdam, 1988); B. Schellekens, ed., *Superstring Construction* (North-Holland, Amsterdam, 1989); J. Polchinski, *What is String Theory?, 1994 Les Houches Summer School Lectures*, hep-th/9411028.
72. For recent general discussions on various aspects of string phenomenology and string model-building, see, *e.g.*: L.E. Ibáñez, hep-th/9112050; hep-th/9505098; I. Antoniadis, hep-th/9307002; M. Dine, hep-ph/9309319; J.L. Lopez, *Surveys H.E. Phys.* **8** (1995) 135; J.L. Lopez and D.V. Nanopoulos, hep-ph/9511266; J.D. Lykken, hep-ph/9511456; hep-th/9607144; F. Quevedo, hep-th/9603074; hep-ph/9707434; A.E. Faraggi, hep-ph/9707311; G. Cleaver, hep-th/9708023; Z. Kakushadze, G. Shiu, S.-H.H. Tye, and Y. Vtorov-Karevsky, hep-th/9710149; and references therein.

73. T.R. Taylor, `hep-ph/9510281`; F. Quevedo, `hep-th/9511131`.
74. K.R. Dienes, *Phys. Reports* **287** (1997) 447 [`hep-th/9602045`].
75. J. Polchinski, *Prog. Theor. Phys. Suppl.* **123** (1996) 9; *Rev. Mod. Phys.* **68** (1996) 1245; `hep-th/9611050`; J. Schwarz, `hep-th/9607201`; M. Duff, *Int. J. Mod. Phys.* **A11** (1996) 5623; M. Dine, `hep-th/9609051`; A. Sen, `hep-th/9609176`; M.R. Douglas, `hep-th/9610041`; P.K. Townsend, `hep-th/9612121`; S. Förste and J. Louis, `hep-th/9612192`; S. Kachru, `hep-th/9705173`.
76. P. Ramond, *Phys. Rev.* **D3** (1971) 2415.
77. L.J. Dixon and J.A. Harvey, *Nucl. Phys.* **B274** (1986) 93; A Sen, *Nucl. Phys.* **B278** (1986) 289; T. Banks, L. Dixon, D. Friedan, and E. Martinec, *Nucl. Phys.* **B299** (1988) 613.
78. J.D. Blum and K.R. Dienes, Phys. Lett. B **414**, 260 (1997) [arXiv:hep-th/9707148]; Nucl. Phys. B **516**, 83 (1998) [arXiv:hep-th/9707160].
79. M.B. Green and J. Schwarz, *Phys. Lett.* **B149** (1984) 117.
80. M. Dine, N. Seiberg, and E. Witten, *Nucl. Phys.* **B289** (1987) 589.
81. See, *e.g.*: G. Moore, *Nucl. Phys.* **B293** (1987) 139; Erratum: *ibid.* **B299** (1988) 847; T.R. Taylor, *Nucl. Phys.* **B303** (1988) 543; J. Balog and M.P. Tuite, *Nucl. Phys.* **B319** (1989) 387; K.R. Dienes, *Phys. Rev.* **D42** (1990) 2004, *Phys. Rev. Lett.* **65** (1990) 1979; T. Gannon and C.S. Lam, *Phys. Rev.* **D46** (1992) 1710.
82. J. Scherk and J.H. Schwarz, *Phys. Lett.* **B82** (1979) 60.
83. R. Rohm, *Nucl. Phys.* **B237** (1984) 553.
84. K.R. Dienes, *Nucl. Phys.* **B429** (1994) 533; `hep-th/9409114`; `hep-th/9505194`.
85. K.R. Dienes, M. Moshe, and R.C. Myers, *Phys. Rev. Lett.* **74** (1995) 4767; `hep-th/9506001`.
86. I. Antoniadis, *Phys. Lett.* **B246** (1990) 377; I. Antoniadis, C. Muñoz, and M. Quirós, *Nucl. Phys.* **B397** (1993) 515; I. Antoniadis and K. Benakli, *Phys. Lett.* **B326** (1994) 69.
87. S. Ferrara, C. Kounnas, and M. Porrati, *Phys. Lett.* **B197** (1987) 135; *Phys. Lett.* **B206** (1988) 25; *Nucl. Phys.* **B304** (1988) 500; S. Ferrara, C. Kounnas, M. Porrati, and F. Zwirner, *Nucl. Phys.* **B318** (1989) 75.
88. C. Bachas, `hep-th/9503030`; J.G. Russo and A.A. Tseytlin, *Nucl. Phys.* **B461** (1996) 131; A.A. Tseytlin, `hep-th/9510041`; M. Spalinski and H.P. Nilles, *Phys. Lett.* **B392** (1997) 67; I. Shah and S. Thomas, `hep-th/9705182`.
89. I. Antoniadis, C. Bachas, D.C. Lewellen, and T. Tomaras, *Phys. Lett.* **B207** (1988) 441.
90. See, *e.g.*, H.P. Nilles, *Phys. Lett.* **B115** (1982) 193; *Nucl. Phys.* **B217** (1983) 366; *Int. J. Mod. Phys.* **AA5** (1990) 4199; J.P. Derendinger, L.E. Ibáñez, and H.P. Nilles, *Phys. Lett.* **B155** (1985) 65; M. Dine, R. Rohm, N. Seiberg, and E. Witten, *Phys. Lett.* **B156** (1985) 55; T.R. Taylor, *Phys. Lett.* **B164** (1985) 43; *Phys. Lett.* **B252** (1990) 59; C. Kounnas and M. Porrati, *Phys. Lett.* **B191** (1987) 91; P. Binétruy and M.K. Gaillard, *Phys. Lett.* **B232** (1989) 83; *Phys. Lett.* **B253** (1991) 119; *Nucl. Phys.* **B358** (1991) 121; A. Font, L.E. Ibáñez, D. Lüst, and F. Quevedo, *Phys. Lett.* **B249** (1990) 35;

J.A. Casas, Z. Lalak, C. Muñoz, and G.G. Ross, *Nucl. Phys.* **B347** (1990) 243; V.S. Kaplunovsky and J. Louis, *Phys. Lett.* **B306** (1993) 269; *Nucl. Phys.* **B422** (1994) 57; P. Binétruy, M.K. Gaillard, and Y.Y. Wu, *Nucl. Phys.* **B493** (1997) 27; and references therein.

91. P. Hořava and E. Witten, *Nucl. Phys.* **B460** (1996) 506.
92. E. Witten, *Nucl. Phys.* **B471** (1996) 135; P. Hořava and E. Witten, *Nucl. Phys.* **B475** (1996) 94.
93. P. Hořava, *Phys. Rev.* **D54** (1996) 7561; T. Banks and M. Dine, *Nucl. Phys.* **B479** (1996) 173; hep-th/9609046; E. Caceres, V.S. Kaplunovsky, and I.M. Mandelberg, hep-th/9606036; T.J. Li, J.L. Lopez, and D.V. Nanopoulos, hep-ph/9702237; hep-ph/9704247; E. Dudas and C. Grojean, hep-th/9704177; I. Antoniadis and M. Quirós, hep-th/9705037; hep-th/9707208; H.P. Nilles, M. Olechowski, and M. Yamaguchi, hep-th/9707143; Z. Lalak and S. Thomas, hep-th/9707223. E. Dudas, hep-th/9709043; K. Choi, H.B. Kim, and C. Muñoz, hep-th/9711158; A. Lukas, B.A. Ovrut, and D. Waldram, hep-th/9711197.
94. P. Ginsparg, *Phys. Lett.* **B197** (1987) 139.
95. V.S. Kaplunovsky, *Nucl. Phys.* **B307** (1988) 145; Erratum: *ibid.* **B382** (1992) 436.
96. D.C. Lewellen, *Nucl. Phys.* **B337** (1990) 61; A. Font, L.E. Ibáñez, and F. Quevedo, *Nucl. Phys.* **B345** (1990) 389; G. Aldazabal, A. Font, L.E. Ibáñez, and A.M. Uranga, *Nucl. Phys.* **B452** (1995) 3; S. Chaudhuri, S.-W. Chung, G. Hockney, and J.D. Lykken, *Nucl. Phys.* **B456** (1995) 89; A.A. Maslikov, I. Naumov, and G.G. Volkov, *Int. J. Mod. Phys.* **A11** (1996) 1117; J. Erler, hep-th/9602032; G. Cleaver, hep-th/9604183.
97. G. Aldazabal, A. Font, L.E. Ibáñez, and A.M. Uranga, *Nucl. Phys.* **B465** (1996) 34.
98. S. Chaudhuri, G. Hockney, and J.D. Lykken, *Nucl. Phys.* **B469** (1996) 357.
99. D. Finnell, *Phys. Rev.* **D53** (1996) 5781.
100. Z. Kakushadze and S.-H.H. Tye, *Phys. Rev. Lett.* **77** (1996) 2612; *Phys. Lett.* **B392** (1997) 335; *Phys. Rev.* **D55** (1997) 7878; *Phys. Rev.* **D55** (1997) 7896; hep-ph/9705202.
101. K.R. Dienes and J. March-Russell, *Nucl. Phys.* **B479** (1996) 113; K.R. Dienes, *Nucl. Phys.* **B488** (1997) 141.
102. J.A. Casas and C. Muñoz, *Phys. Lett.* **B214** (1988) 543; L. Ibáñez, *Phys. Lett.* **B318** (1993) 73; K. Benakli and G. Senjanovic, *Phys. Rev.* **D54** (1996) 5734.
103. K.R. Dienes, A.E. Faraggi, and J. March-Russell, *Nucl. Phys.* **B467** (1996) 44.
104. A.N. Schellekens, *Phys. Lett.* **B237** (1990) 363.
105. E. Kiritsis and C. Kounnas, *Nucl. Phys. Proc. Suppl.* **41** (1995) 331; *Nucl. Phys.* **B442** (1995) 472; hep-th/9507051; *Nucl. Phys. Proc. Suppl.* **45** (1996) 207; E. Kiritsis, C. Kounnas, P.M. Petropoulos, and J. Rizos, *Phys. Lett.* **B385** (1996) 87.
106. L.J. Dixon, V.S. Kaplunovsky, and J. Louis, *Nucl. Phys.* **B355** (1991) 649.
107. A. Font, L.E. Ibáñez, D. Lüst, and F. Quevedo, *Phys. Lett.* **B245** (1990)

401; M. Cvetič, A. Font, L.E. Ibáñez, D. Lüst, and F. Quevedo, *Nucl. Phys.* **B361** (1991) 194; B. de Carlos, J.A. Casas, and C. Muñoz, *Nucl. Phys.* **B399** (1993) 623.

108. I. Antoniadis, J. Ellis, J. Hagelin, and D.V. Nanopoulos, *Phys. Lett.* **B231** (1989) 65; A.E. Faraggi, D.V. Nanopoulos, and K. Yuan, *Nucl. Phys.* **B335** (1990) 347; J.L. Lopez, D.V. Nanopoulos, and K. Yuan, *Nucl. Phys.* **B399** (1993) 654; I. Antoniadis, G.K. Leontaris, and J. Rizos, *Phys. Lett.* **B245** (1990) 161; A.E. Faraggi, *Phys. Lett.* **B278** (1992) 131; *Nucl. Phys.* **B387** (1992) 239; *Phys. Lett.* **B302** (1993) 202.
109. K.R. Dienes and A.E. Faraggi, *Phys. Rev. Lett.* **75** (1995) 2646; *Nucl. Phys.* **B457** (1995) 409.
110. P. Mayr, H.P. Nilles, and S. Stieberger, *Phys. Lett.* **B317** (1993) 53; H.P. Nilles and S. Stieberger, *Phys. Lett.* **B367** (1996) 126; *Nucl. Phys.* **B499** (1997) 3.
111. A.E. Faraggi, *Phys. Rev.* **D46** (1992) 3204.
112. I. Antoniadis, J. Ellis, S. Kelley, and D.V. Nanopoulos, *Phys. Lett.* **B272** (1991) 31; S. Kelley, J.L. Lopez, and D.V. Nanopoulos, *Phys. Lett.* **B278** (1992) 140; D. Bailin and A. Love, *Phys. Lett.* **B280** (1992) 26; M.K. Gaillard and R. Xiu, *Phys. Lett.* **B296** (1992) 71; S.P. Martin and P. Ramond, *Phys. Rev.* **D51** (1995) 6515; B.C. Allanach and S.F. King, `hep-ph/9601391`; C. Bachas, C. Fabre, and T. Yanagida, *Phys. Lett.* **B370** (1996) 49; J.L. Lopez and D.V. Nanopoulos, *Phys. Rev. Lett.* **76** (1996) 1566.
113. K. R. Dienes and C. F. Kolda, arXiv:hep-ph/9712322.
114. B. Aubert *et al.* [BABAR Collaboration], Phys. Rev. Lett. **89**, 201802 (2002); K. Abe *et al.* [Belle Collaboration], Phys. Rev. D **66**, 071102 (2002).
115. V. M. Abazov *et al.* [D0 Collaboration], Phys. Rev. D **76**, 092001 (2007); T. Aaltonen *et al.* [CDF Collaboration], Phys. Rev. Lett. **100**, 101802 (2008); B. Aubert *et al.* [BABAR Collaboration], Phys. Rev. D **73**, 092001 (2006); I. Adachi *et al.* [Belle Collaboration], arXiv:0810.0335 [hep-ex].
116. S. R. Choudhury and N. Gaur, Phys. Lett. B **451**, 86 (1999); K. S. Babu and C. Kolda, Phys. Rev. Lett. **84**, 228 (2000).
117. G. W. Bennett *et al.* [Muon g-2 Collaboration], Phys. Rev. Lett. **92**, 161802 (2004).
118. A. G. Riess *et al.* [Supernova Search Team Collaboration], Astron. J. **116**, 1009 (1998); S. Perlmutter *et al.* [Supernova Cosmology Project Collaboration], Astrophys. J. **517**, 565 (1999).
119. E. Komatsu *et al.* [WMAP Collaboration], Astrophys. J. Suppl. **180**, 330 (2009).
120. See, for example, N. Arkani-Hamed, A. Delgado and G. F. Giudice, Nucl. Phys. B **741**, 108 (2006).
121. R. Barate *et al.* [LEP Working Group for Higgs Boson Searches], Phys. Lett. B **565**, 61 (2003).
122. A. Birkedal, Z. Chacko and M. K. Gaillard, JHEP **0410**, 036 (2004).

# Developments in Supergravity Unified Models

Richard Arnowitt* and Pran Nath†

*Center for Theoretical Physics, Department of Physics, Texas A&M University, College Station, TX 77843-4242

†Department of Physics, Northeastern University, Boston, MA 02115, USA

*arnowitt@physics.tamu.edu

†nath@neu.edu

A review is given of developments in supergravity unified models proposed in 1982 and their implications for current and future experiment are discussed.

## 3.1. Introduction

Supersymmetry (SUSY) was initially introduced as a global symmetry[1,2] on purely theoretical grounds that nature should be symmetric between bosons and fermions. It was soon discovered, however, that models of this type had a number of remarkable properties.[3] Thus the bose-fermi symmetry led to the cancellation of a number of the infinities of conventional field theories, in particular the quadratic divergences in the scalar Higgs sector of the Standard Model (SM). Thus SUSY could resolve the gauge hierarchy problem that plagued the SM. Further, the hierarchy problem associated with grand unified models[4] (GUT), where without SUSY, loop corrections gave all particles GUT size masses[5,6] was also resolved. In addition, SUSY GUT models with minimal particle spectrum raised the value for the scale of grand unification, $M_G$, to $M_G \cong 2 \times 10^{16}$ GeV, so that the predicted proton decay rate[6,7] was consistent with existing experimental bounds. Thus in spite of the lack of any direct experimental evidence for the existence of SUSY particles, supersymmetry became a highly active field among particle theorists.

However, by about 1980, it became apparent that global supersymmetry was unsatisfactory in that a phenomenologically acceptable picture of spontaneous breaking of supersymmetry did not exist. Thus the success of the SUSY grand unification program discussed above was in a sense spurious in that the needed SUSY threshold $M_S$ (below which the SM held) could not be theoretically constructed. In order to get a phenomenologically viable model, one needed "soft breaking" masses[8] (i.e. supersymmetry breaking terms of dimenison $\leq 3$ which maintain the gauge hierarchy) and these had to be introduced in an ad hoc fashion by hand. In the minimal SUSY model, the MSSM,[9] where the particle spectrum is just that of the supersymmeterized SM, one could introduce as many as 105 additional parameters (62 new masses and mixing angles and 43 new phases) leaving one with a theory with little predictive power.

A resolution of the problem of how to break supersymmetry spontaneously was achieved by promoting supersymmetry to a local symmetry,[10] and specifically supergravity.[11] Here gravity is included into the dynamics. One can then construct supergravity [SUGRA] grand unified models[12,13] where the spontaneous breaking of supersymmetry occurs in a "hidden" sector via supergravity interactions in a fashion that maintains the gauge hierarchy. In such theories there remains, however, the question of at what scale does supersymmetry break, and what is the "messenger" that communicates this breaking from the hidden to the physical sector. In this chapter we consider models where supersymmetry breaks at a scale $Q > M_G$ with gravity being the messenger.[12,14–17] Such models are economical in that both the messenger field and the agency of supersymmetry breaking are supersymmetrized versions of fields and interactions that already exist in nature (i.e. gravity). Alternately within gravity mediation, supersymmetry could be broken by gaugino condensation.[18] This mechanism is a likely possibility within string theory.

The strongest direct evidence supporting supergravity GUT models is the apparent experimental grand unification of the three gauge coupling constants.[19] This result is non-trivial not only because three lines do not ordinarily intersect at one point, but also because there is only a narrow acceptable window for $M_G$. Thus one requires $M_G \gtrsim 5 \times 10^{15}\ GeV$ so as not to violate current experimental bounds on proton decay for the $p \to e^+ + \pi^0$ channel (which occurs in almost all GUT models) and one requires $M_G \lesssim 5 \times 10^{17}$ GeV $\cong M_{string}$ (the string scale) so that gravitational effects do not become large invalidating the analysis. Further, assuming an MSSM type of particle spectrum between the electroweak scale $M_Z$ and $M_G$, acceptable grand unification occurs only with one pair of Higgs doublets and at most four generations. Finally, naturalness requires that SUSY thresholds be at $M_S \lesssim 1$ TeV which turns out to be the case. Thus the possibility of grand unification is tightly constrained.

As discussed in Sec. 3.5 below, the grand unified models with R parity invariance produce a natural candidate for the dark matter observed astronomically. Further, the amount of dark matter produced in the early universe can be calculated, and remarkably the theory naturally predicts a relic density of dark matter today of size seen by WMAP and other observations. Thus SUGRA GUTS allows the construction of models valid from mass $M_G$ down to the electroweak scale, and backwards in time to $\sim 10^{-8}$ sec after the Big Bang (when the dark matter was created), a unification of particle physics and early universe cosmology. At present, grand unification in SUGRA GUTs can be obtained to within about 2-3 std.[20,21] However, the closeness of $M_G$ to the Planck scale, $M_{P\ell} = (\hbar c/8\pi G_N)^{1/2} \cong 2.4 \times 10^{18}$ GeV, suggests the possibility that there are $O(M_G/M_{P\ell})$ corrections to these models. One might, in fact, expect such structures to arise in string theory as nonrenormalizable operators (NROs) obtained upon integrating out the tower of Planck mass states. Such terms would produce $\approx 1\%$ corrections at $M_G$ which might grow to $\approx 5\%$ corrections at $M_Z$. Indeed, as will be seen in Sec. 3.2, it is just such NRO terms involving the hidden sector fields that give rise to the soft breaking masses, and so it would not be surprising to find such structures in the physical sector as well. Thus SUGRA GUTs should be viewed as an effective theory and, as will be discussed in Sec. 3.8, with small deviations between theory and experiment perhaps opening a window to Planck scale physics.

One of the fundamental aspects of the SM, not explained by that theory, is the origin of the spontaneous breaking of SU(2) x U(1). SUGRA GUTS offers an explanation of this due to the existence of soft SUSY breaking masses at $M_G$. Thus as long as at least one of the soft breaking terms are present at $M_G$, breaking of SU(2) x U(1) can occur at a lower energy,[12,22] providing a natural Higgs mechanism. Further, radiative breaking occurs at the electroweak scale provided the top quark is heavy ie. 100 GeV $\lesssim m_t \lesssim$ 200 GeV. The minimal SUGRA model[12,15,16] (mSUGRA), which assumes universal soft breaking terms, requires only four additional parameters and one sign to describe all the interactions and masses of the 32 SUSY particles. Thus the mSUGRA is predictive model producing many sum rules among the sparticle masses,[23] and for that reason the model is used in much of the phenomenological analysis of the past decades. However, we will see in

Sec. 3.2 that there are reasons to consider non-universal extensions of the mSUGRA, and inclusion of the nonuniversalities can produce significant modifications of the sparticle masses and their signatures.

### 3.2. Soft Breaking Masses

Supergravity interactions with chiral matter fields, $\{\chi_i(x), \phi_i(x)\}$ (where $\chi_i(x)$ are left (L) Weyl spinors and $\phi_i(x)$ are complex scalar fields) depend upon three functions of the scalar fields: the superpotential $W(\phi_i)$, the gauge kinetic function $f_{\alpha\beta}(\phi_i, \phi_i^\dagger)$ (which enters in the Lagrangian as $f_{\alpha\beta}F^\alpha_{\mu\nu}F^{\mu\nu\beta}$ with $\alpha, \beta$ = gauge indices) and the Kahler potential $K(\phi_i, \phi_i^\dagger)$ (which appears in the scalar kinetic energy as $K^i_j\partial_\mu\phi_i\partial^\mu\phi_j^\dagger$, $K^i_j \equiv \partial^2K^2/\partial\phi_i\partial\phi_j^\dagger$ and elsewhere). W and K enter only in the combination

$$G(\phi_i, \phi_i^\dagger) = \kappa^2 K(\phi_i, \phi_i^+) + \ell n[\kappa^6 \mid W(\phi_i) \mid^2] \tag{3.1}$$

where $\kappa = 1/M_{P\ell}$. Writing $\{\phi_i\} = \{\phi_a, z\}$ where $\phi_a$ are physical sector fields (squarks, sleptons, higgs) and z are the hidden sector fields whose VEVs $\langle z \rangle = \mathcal{O}(M_{P\ell})$ break supersymmetry, one assumes that the superpotential decomposes into a physical and a hidden part,

$$W(\phi_i) = W_{phy}(\phi_a, \kappa z) + W_{hid}(z) \tag{3.2}$$

Supersymmetry breaking is scaled by requiring $\kappa^2 W_{hid} = \mathcal{O}(M_S)\tilde{W}_{hid}(\kappa z)$ and the gauge hierarchy is then guaranteed by the additive nature of the terms in Eq.(3.2). Thus only gravitational interactions remain to transmit SUSY breaking from the hidden sector to the physical sector.

A priori, the functions W, K and $f_{\alpha\beta}$ are arbitrary. However, they are greatly constrained by the conditions that the model correctly reduce to the SM at low energies, and that non-renormalizable corrections be scaled by $\kappa$ (as would be expected if they were the low energy residue of string physics of the Planck scale). Thus one can expand these functions in polynomials of the physical fields $\phi_a$

$$\begin{aligned} f_{\alpha\beta}(\phi_i) &= c_{\alpha\beta} + \kappa d^a_{\alpha\beta}(x,y)\phi_a + \cdots, \\ W_{phys}(\phi_i) &= \frac{1}{6}\lambda^{abc}(x)\phi_a\phi_b\phi_c + \frac{1}{24}\kappa\lambda^{abcd}(x)\phi_a\phi_b\phi_c\phi_d + \cdots, \\ K(\phi_i, \phi_i^\dagger) &= \kappa^{-2}c(x,y) + c^a_b(x,y)\phi_a\phi_b^\dagger \\ &\quad + (c^{ab}(x,y)\phi_a\phi_b + h.c.) + \kappa(c^a_{bc}\phi_a\phi_b^\dagger\phi_c^\dagger + h.c.) + \cdots. \end{aligned} \tag{3.3}$$

Here x=$\kappa z$ and y =$\kappa z^{\dagger}$, so that $\langle x\rangle$, $\langle y\rangle = \mathcal{O}$ (1). The scaling hypothesis for the NRO's imply then that the VEVs of the coefficients $c_{\alpha\beta}$, $c^a_{\alpha\beta}$, $\lambda^{abc}$, $c$, $c^a_b$, $c^{ab}$ etc. are all $\mathcal{O}(1)$. The holomorphic terms in K labeled by $c^{ab}$ can be transformed from K to W by a Kahler transformation, $K \to K - (c^{ab}\phi_a\phi_b + h.c.)$ and

$$W \to W exp[\kappa^2 c^{ab}\ \phi_a\phi_b] = W + \tilde{\mu}^{ab}\phi_a\phi_b + \cdots \tag{3.4}$$

where $\tilde{\mu}^{ab}(x,y) = \kappa^2 W c^{ab}$. Hence $\langle\tilde{\mu}^{ab}\rangle = \mathcal{O}(\mathcal{M}_S)$, and one obtains a $\mu$-term with the right order of magnitude after SUSY breaking provided only that $c^{ab}$ is not zero.[24] The cubic terms in W are just the Yukawa couplings with $\langle\lambda^{abc}(x)\rangle$ being the Yukawa coupling constants. Also $\langle c_{\alpha\beta}\rangle = \delta_{\alpha\beta}$, $\langle c^a_b\rangle = \delta^a_b$ and $\langle c_{xy}\rangle = 1$ ($c_x \equiv \partial c/\partial x$ etc.) so that the field kinetic energies have canonical normalization.

The breaking of SUSY in the hiddden sector leads to the generation of a series of soft breaking terms.[12,14–17] We consider here the case where $\langle x\rangle = \langle y\rangle$ (i.e. the hidden sector SUSY breaking does not generate any CP violation) and state the leading terms. Gauginos gain a soft breaking mass term at $M_G$ of $\tilde{m}_{\alpha\beta}\lambda^\alpha\gamma^0\lambda^\beta$ ($\lambda^\alpha$ = gaugino Majorana field) where

$$\tilde{m}_{\alpha\beta} = \kappa^{-2}\langle G^i(K^{-1})^i_j Re f^{\dagger}_{\alpha\beta j}\rangle m_{3/2} \tag{3.5}$$

Here $G^i \equiv \partial G/\partial\phi_i$, $(K^{-1})^i_j$ is the matrix universe of $K^i_j$, $f_{\alpha\beta j} = \partial f_{\alpha\beta}/\partial\phi^{\dagger}_j$ and $m_{3/2}$ is the gravitino mass: $m_{3/2} = \kappa^{-1}\langle exp[G/2]\rangle$ In terms of the expansion of Eq.(3.3) one finds

$$\tilde{m}_{\alpha\beta} = [c + \ell n(W_{hid})]_x Re\ c^*_{\alpha\beta y} m_{3/2}, \tag{3.6}$$

and $m_{3/2} = (\exp\frac{1}{2}c)\kappa^2 W_{hid}$ where it is understood from now on that x is to be replaced by its VEV in all functions (e.g. c(x)$\to c(\langle x\rangle) = \mathcal{O}(1)$) so that $m_{3/2} = \mathcal{O}(M_S)$. One notes the following about Eq.(3.6): (i) For a simple GUT group, gauge invariance implies that $c_{\alpha\beta} \sim \delta_{\alpha\beta}$ and so gaugino masses are universal (labeled by $m_{1/2}$) at mass scales above $M_G$. (ii) While $m_{1/2}$ is scaled by $m_{3/2} = \mathcal{O}(M_S)$, it can differ from it by a significant amount. (iii) From Eq.(3.6) one sees that it is the NRO such as $\kappa z m_{3/2}\lambda^\alpha\gamma^0\lambda^\alpha$ that gives rise to $m_{1/2}$. Below $M_G$, where the GUT group is broken, the second term in the $f_{\alpha\beta}$ of Eq.(3.3) would also contribute yielding a NRO of size $\kappa d^a_{\alpha\beta}\phi_a m_{3/2}\ \lambda^\alpha\gamma^0\lambda^\beta \sim (M_G/M_{P\ell})\ m_{3/2}\lambda\gamma^0\lambda$[25] for fields with VEV $\langle\phi_a\rangle = \mathcal{O}(M_G)$ which break the GUT group. Such terms give small corrections to the universality of the gaugino masses and affect grand unification. They are discussed in Sec. 3.8.

The effective potential for the scalar components of chiral multiplets is given by[12,26]

$$V = e^{\kappa^2 K}\ [(K^{-1})^j_i (W^i + \kappa^2 K^i W)\ (W^j + \kappa^2 K^j W)^\dagger - 3\kappa^2 \mid W \mid^2] + V_D,$$
$$V_D = \frac{1}{2} g_\alpha g_\beta\ (Ref^{-1})_{\alpha\beta}\ (K^i (T^\alpha)_{ij} \phi_j)\ (K^k (T^\beta)_{k\ell} \phi_\ell), \tag{3.7}$$

where $W^i = \partial W / \partial \phi_i$ etc., and where $g_\alpha$ are the gauge coupling constants. Eqs.(3.2-3.4) then lead to the following soft breaking terms at $M_G$:

$$V_{soft} = (m_0^2)^a_b\ \phi_a \phi^\dagger_b + \left[\frac{1}{3}\tilde{A}^{abc} \phi_a \phi_b \phi_c + \frac{1}{2}\tilde{B}^{ab} \phi_a \phi_b + h.c.\right] \tag{3.8}$$

In the following, we impose for simplicity the condition that the cosmological constant vanish after SUSY breaking, i.e. $\langle V \rangle = 0$. [One of course could accommodate the tiny cosmological constant suggested by the supernova observation.] This is a fine tuning of $\mathcal{O}(M_S^2 M_{P\ell}^2)$. From Eq.(3.7) one notes that the soft breaking terms are in general not universal unless one assumes that the fields z couple universally to the physical sector.

### 3.3. Radiative Breaking and the Low Energy Theory

In Sec. 3.2, the SUGRA GUT model above the GUT scale i.e. at $Q > M_G$ was discussed. Below $M_G$ the GUT group is spontaneously broken, and we will assume here that the SM group, SU(3) x SU(2) x U(1), holds for $Q < M_G$. Contact with accelerator physics at low energy can then be achieved using the renormalization group equations (RGE)[27] running from $M_G$ to the electroweak scale $M_Z$. As one proceeds downward from $M_G$, the coupling constants and masses evolve, and provided at least one soft breaking parameter and also the $\mu$ parameter at $M_G$ is not zero, the large top quark Yukawa can turn the $H_2$ running $(\text{mass})^2$, $m^2_{H_2}(Q)$, negative at the electroweak scale.[22] Thus the spontaneous breaking of supersymmetry at $M_G$ triggers the spontaneous breaking of SU(2) x U(1) at the electroweak scale. In this fashion all the masses and coupling constants at the electroweak scale can be determined in terms of the fundamental parameters (Yukawa coupling constants and soft breaking parameters) at the GUT scale, and the theory can be subjected to experimental tests.

The conditions for electroweak symmetry breaking arise from minimizing the effective potential V at the electroweak scale with respect to the Higgs VEVs $v_{1,2} = \langle H_{1,2} \rangle$. This leads to the equations[22]

$$\mu^2 = \frac{\mu_1^2 - \mu_2^2 tan^2\beta}{tan^2\beta - 1} - \frac{1}{2} M_Z^2; \quad sin^2\beta = \frac{-2B\mu}{2\mu^2 + \mu_1^2 + \mu_2^2} \tag{3.9}$$

where $tan\beta = v_2/v_1$, B is the quadratic soft breaking parameter ($V^B_{soft} = B\mu H_1 H_2$), $\mu_i = m^2_{H_i} + \Sigma_i$, and $\Sigma_i$ are loop corrections.[28] All parameters are running parameters at the electroweak scale which one takes for convenience to be $Q \simeq \sqrt{m_{\tilde{t}_1} m_{\tilde{t}_2}}$ to minimize loop corrections. Eq.(3.9) then determines the $\mu$ parameter and allows the elimination of B in terms of $tan\beta$. This determination of $\mu$ greatly enhances the predictive power of the model.

In general there are two broad regions of electroweak symmetry breaking implied by the soft parameters appearing in Eq.(3.9). One region is where the soft parameters can be arranged to lie on the surface of an ellipsoid with their radii fixed by the value of $\mu$. In this case for fixed $\mu$, $m_0$ and $m_{\frac{1}{2}}$ cannot get too large since the surface of an ellipsoid is a closed surface. However, it turns out that when loop corrections[28] to the effective potential are included the nature of electroweak symmetry breaking can change rather drastically. Then the soft parameters instead of lying on the surface of an ellipsoid, lie on the surface of a hyperboloid and this branch may appropriately be called the hyperbolic branch (HB)(see the first paper of[29]). Since the surface of a hyperboloid in open, the soft parameters can get large with $\mu$ fixed. Specifically, the HB allows TeV size scalars with small values of $\mu$ and thus small fine tunings. The region of TeV size scalars is also known as the Focus Point (FP) region (see the second paper of[29]).

The renormalization group equations evolve the universal gaugino mass $m_{1/2}$ at $M_G$ to separate masses for SU(3), SU(2) and U(1) at $M_Z$

$$\tilde{m}_i = (\alpha_i(M_Z)/\alpha_G)m_{1/2}; \quad i = 1, 2, 3 \tag{3.10}$$

where at 1-loop, the gluino mass $m_{\tilde{g}} = \tilde{m}_3$.[30] The simplest model is the one with universal soft breaking masses. This model depends on only the four SUSY parameters and one sign at the GUT scale[12,15,16]

$$m_0, \quad m_{1/2}, \quad A_0, \quad B_0, \quad sign(\mu_0) \tag{3.11}$$

Alternately, at the electroweak scale one may choose $m_0$ , $m_{\frac{1}{2}}$, $A_t$ , $tan\beta$ , and sign($\mu$) as the independent parameters. Universality can be derived in a variety of ways. From a string view point it could arise, for example, from dilaton dominance, or when modular weights are all equal, and in both GUTS and in strings it could arise from a family symmetry at the GUT/string scale. This case has been extensively discussed in the literature.[31–35] The deviations from universality can be significant, however, and affect $\mu^2$ and sparticle masses, and these parameters play a crucial role in predictions of the theory. Several analyses exist where the SUGRA models have been extended to include non-universalities.[36–38] These extensions include non-universalities in the gaugino sector, in the Higgs sector and in the third generation sector consistent with flavor changing neutral current constraints.

### 3.4. Supersymmetric Corrections to Electroweak Phenomena

SUGRA models make contributions to all electroweak processes at the loop level through the exchange of sparticles. We discuss here some of the more prominent ones which include the muon anomalous magnetic moment $g_\mu - 2$, and the the flavor changing neutral current processes $b \to s\gamma$ and $B_s^0 \to \mu^+\mu^-$. These processes are all probes of new physics. Thus in $g_\mu - 2$ the sparticle loops at one loop make contributions (see 3.1) which are comparable to the electroweak contributions from the Standard Model.[39–41] The most recent evaluations of the difference between experiment and theory give for $\delta a_\mu = (g_\mu^{exp} - g_\mu^{SM})/2$, the result[42]

$$\delta a_\mu = (24.6 \pm 8.0) \times 10^{-10} \tag{3.12}$$

If the above result holds up it would imply upper limits on sparticle masses within the range of the LHC energies [a]. These conclusions were already drawn from the result of the Brookhaven experiment E821 in 2001.[44]

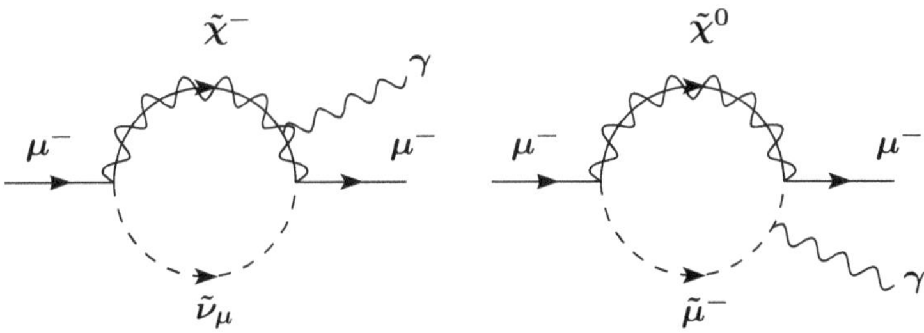


Fig. 3.1. Supersymmetric electroweak contributions to $g_\mu - 2$.

Flavor changing neutral current processes also provide an important constraint on supergravity unified models. A process of great interest here is the decay $b \to s+\gamma$. Further, it is well known from the early days that $b \to s+\gamma$ experiment imposes an important constraint on the parameter space of supergravity models[45] and specifically on the analysis of dark matter. The

[a]In most extra dimension models the corrections to $g_\mu - 2$ are rather small[43] and it is difficult to accommodate a deviation of size Eq.(3.12) .

current experimental value for this branching ratio from the the Heavy Flavor Averaging Group (HFAG)[46] along with the BABAR, Belle and CLEO experimental results gives $\mathcal{B}r(B \to X_s\gamma) = (352 \pm 23 \pm 9) \times 10^{-6}$. In the SM this decay proceeds at the loop level with the exchange of W and Z bosons and the most recent evaluation including the next to next leading order (NNLO) QCD corrections is given by[47] $\mathcal{B}r(b \to s\gamma) = (3.15 \pm 0.23) \times 10^{-4}$. In supersymmetry there are additional diagrams which contribute to this process.[48] Thus in SUGRA unification one has contributions from the exchange of the charged Higgs, the charginos, the neutralinos and from the gluino. It is well known that the contribution from the charged Higgs exchange is always positive[49] while the contribution from the exchange of the other SUSY particles can be either positive or negative with the contribution of the charginos being usually the dominant one.[50]

A comparison of the experimental and theoretical evaluations in the SM point to the possibility that a positive correction to the SM value is needed. As noted above such a positive correction can arise from supersymmetry specifically from the exchange of the charged higgs which implies the possibility of a relatively light charged Higgs. Further, over most of the parameter space the chargino exchange contributions are often negative pointing to a cancellation between the charged Higgs and the chargino exchange contributions and also hint at the possibility of a relatively light chargino and possibly of a relatively light stop.

The rare process $B_s \to \mu^+\mu^-$ is of interest as it is a probe of phyiscs beyond the standard model.[51,52] The branching ratio for this process in the SM is $\mathcal{B}r(B_s \to \mu^+\mu^-) = (3.1 \pm 1.4) \times 10^{-9}$ (for $V_{ts} = 0.04 \pm 0.002$)). In supersymmetric models it can get large for large $\tan\beta$ since decay branching ratio increases as $\tan^6\beta$. The current experimental limit at 95% (90%) C.L. reported by CDF is $\mathcal{B}r(B_s \to \mu^+\mu^-) = 5.8 \times 10^{-8}$ $(4.7 \times 10^{-8})$.[53] Since in supersymmetric theories this branching ratio can increase as $\tan^6\beta$ the experimental data does constrain the analysis at least for large $\tan\beta$ and the implications of this constraint have been analyzed in several works.[54,55] Additionally, this specific decay is very sensitive to CP phases and thus the experiment also constrains the CP phases in SUGRA models in certain regions of the parameter space.[56]

### 3.5. Dark Matter in SUGRA Unification

As mentioned earlier one of the remarkable results of supergravity grand unification with R parity invariance is the prediction that the lightest neutralino $\chi_1^0$ is the LSP over most of the parameter space.[57] In this part of the parameter space the $\chi_1^0$ is a candidate for cold dark matter (CDM). We discuss now the relic density of $\chi_1$ within the framework of the Big Bang Cosmology. The quantity that is computed theoretically is $\Omega_{\chi_1}h^2$ where $\Omega_{\chi_1} = \rho_{\chi_1}/\rho_c$, $\rho_{\chi_1}$ is the neutralino relic density and $\rho_c$ is the critical relic density needed to close the universe, $\rho_c = 3H^2/8\pi G_N$, and $H = h100km/sMpc$ is the Hubble constant. One of the important elements in the computation of the relic density concerns the correct thermal averaging of the quantity $(\sigma v)$ where $\sigma$ is the neutralino annihilation cross section in the early universe and $v$ is the relative neutralino velocity. Normally the thermal average is calculated by first making the approximation $\sigma v = a + bv^2$ and then evaluating its thermal average.[58] However, it is known that such an approximation breaks down in the vicinity of thresholds and poles.[59] Precisely such a situation exists for the case of the annihilation of the neutralino through the Z and Higgs poles. An accurate analysis of the neutralino relic density in the presence of Z and Higgs poles was given in ref.[60] [b] and similar analyses have also been carried out since by other authors.[62] There are a number of possibilites for the detection of dark matter both direct and indirect.[63] We discuss first the direct method which involves the scattering of incident neutralino dark matter in the Milky Way from nuclei in terrestial targets. The event rates consist of two parts:[64] one involves an axial interaction and the other a scalar interaction. The axial (spin dependent) part $R_{SD}$ falls off as $R_{SD} \sim 1/M_N$ for large $M_N$ where $M_N$ is the mass of the target nucleus, while the scalar (spin independent) part behaves as $R_{SI} \sim M_N$ and increases with $M_N$. Thus for heavy target nuclei the spin independent part $R_{SI}$ dominates over most of the parameter space of the model.

In recent years the direct detection dark matter experiments have begun to provide significant bounds on the spin independent neutralino -proton cross section $\sigma_{\tilde{\chi}^0 p}$ and thus theoretical computations of this quantity can be directly compared with the data. The predictions of the SUGRA models lie over a wide range. With typical asssumptions of naturalness on sparticle masses below 1 TeV, $\sigma_{\tilde{\chi}^0 p}$ can lie in the range $10^{-43}$ cm$^2$ to $10^{-48}$ cm$^2$. The current experiments such as CDMS[65] and XENON[66] have already begun to constrain a part of the SUGRA parameter space and improved experiments[67] are expected to probe the parameter space further. Inclusion of

[b]The analysis of Ref. 60 has been used to show that annihilation near a Breit-Wigner pole generates a significant enhancement of $< \sigma v >_H$ in the halo of the galaxy relative to $< \sigma v >_{X_f}$ at the freezeout.[61]

non-universalities is seen to produce definite signatures in the event rate analysis.[37,68] The satisfaction of the relic density in SUGRA models can occur via coannihilations. One of the most studied coannihilation is with the coannihilation of the neutralino with the stau. However, with the inclusion of non-universalities of soft parameters many other coannihilaitons become possible such as coannihilations of the LSP with charginos, stops, gluinos, and heavier neutralinos etc. The coannihilation with the gluino which occurs most dominantly when the gluino is the NLSP and exhibits the interesting phenomenon that the sparticle production cross sections are dominated by the gluino production making the observation of other sparticles challenging.[69]

In addition to the direct detection dark matter experiments, there are also indirect signatures for dark matter. Thus, e.g., neutrinos arising from the annihilation of neutralinos in the center of the Earth and Sun can produce detectable signals. Further, it was suggested quite sometime ago[70] that the annihilation process $\tilde{\chi}^0\tilde{\chi}^0 \to W^+W^-$ with the subsequent decays of the W's, e.g., $W^+ \to e^+\nu$ could generate a detectable positron excess in anti-matter probes. One of the typical problems encountered in most theoretical analyses of the positron excess is the following: in order to have the appropriate positron signal in PAMELA[71] one needs to have a $\chi^0\chi^0$ annihilation cross section to $WW$ with $< \sigma v >_{WW} \simeq 10^{-24} cm^3/s$. However, the relic density has an inverse proportionality to the annihilation cross section at the freeze out, i.e., $\Omega_{\tilde{\chi}^0} h^2 \propto [\int_{x_f}^{\infty} < \sigma_{eff} v > \frac{dx}{x^2}]^{-1}$ which leads to too low a relic density. To overcome this problem most works typically resort to large so called boost factors. Effectively, what this implies is that the annihilation cross section of dark matter is taken to satisfy the relic density and then to get the right strength positron signal a boost factor is assumed. It is argued that such boost factors can arise from clumping of dark matter in the galaxy. However, while a boost factor of O(2-10) could arise arise from clumping of dark matter in the galaxy, it appears unreasonable to assume large clumping factors (sometimes as large as $10^3$ or even larger) to fit the data. In the context of the minimal supergravity model, one simple solution arises due to the automatic suppression of the relic density from coannihilation effects with hidden sector matter in SUGRA models with an extended $U(1)^n$ sector.[72] In this case with n=3, one finds good fits to the positron excess from PAMELA while maintaining the neutralino relic density in the WMAP[73] error corridor. At the same time one can maintain compatibility with the anti-proton flux[71] and the photon flux from the FERMI-LAT experiment.[74]

### 3.6. Signatures at Colliders

Sparticle decays produce missing energy signals since at least one of the carriers of missing energy will be the neutralino. Signals of this type were studied early on after the advent of supergravity models in the supersymmetric decays of the W and Z bosons[75] and such analyses have since been extended to the decays of all of the supersymmetric particles (For a review of sparticle decays see Ref. 13). Using these decay patterns one finds a variety of supersymmetric signals for SUSY particles at colliders where SUSY particles are expected to be pair produced when sufficient energies in the center of mass system are achieved. One signal of special interest in the search for supersymmetry is the trileptonic signal through off shell $W^*$ production as well as via other production and decay chains.[76] For example in $p\bar{p}$ collisions one can have $p\bar{p} \to \tilde{\chi}_1^{\pm} + \tilde{\chi}_2^0 + X \to (l_1\bar{\nu}_1\tilde{\chi}_1^0) + (l_2\bar{l}_2\tilde{\chi}_1^0) + X$ which gives a signal of three leptons and missing energy. In addition to the trileptonic signal there is a long list of possible signatures for the discovery of supersymmetry and test of SUGRA models. These include multileptons and multijet and missing energy. Thus one can devise a variety of combinations with n number of leptons ($e$ or $\mu$), m number of $\tau$'s, k number of jets ($m, n = 0, 1, 2, 3, ..; k \geq 2$) leading to a large number of possibilities. Further, one can add to this list tagged b jet signals and kinematical signatures such as missing transverse momentum $P_T^{miss}$, effective mass $P_T^{miss} + \sum_j P_T^j$, invariant mass of $e^+e^-$, $\mu^+\mu^-$, $\tau^+\tau^-$, and invariant mass of all jets which provide important signatures.

An important result concerns the fact that one can utilize measurements at the LHC to predict phenomena related to dark matter, showing the unification of particle physics and cosmology. Within the mSUGRA framework, existing constraints on the parameter space combined with the cold dark matter constraints pick out three regions: (i) the $\tilde{\chi}_1 - \tilde{\tau}_1$ coannihilation (CA) region where $m_0$ is small but $m_{1/2}$ can rise to 1 TeV, (ii) the hyperbolic (HB)/focus (FP) region, where $m_{1/2}$ is relatively small but $m_0$ is large, and (iii) the pole region[57] (alternately called the funnel region) where annihilation in the early universe goes via heavy Higgs poles. For the CA region it can be shown[77] that purely from the measurements at the LHC one can predict the dark matter relic density with an uncertainty of 6% with $30\text{fb}^{-1}$ of data which is comparable to the uncertainty in the determination of the relic density by WMAP. The relevant signal here consists of low energy $\tau$ leptons from $\tilde{\chi}_2^0 \to \tau\tilde{\tau}_1 \to \tau\tau\tilde{\chi}_1^0$ where the mass difference of the $\tilde{\tau}_1$ and $\tilde{\chi}_1^0$ is constrained to lie within in 5-15 GeV by the current experimental bounds.[80] In addition it is possible to test experimentally the universality of the gaugino masses and if not, measure the amount of non-universality as well as obtain precision measurements of the gaugino mass, squark and lighter stau masses. A similar analysis may be carried out in the HB/FP region[78] and very likely can be done for the ILC.[79]

We discuss now an approach by which the LHC data can be used to discriminate among a variety of models. This approach utilizes the idea of sparicle mass hierarchies which we now describe. Thus as mentioned already in MSSM there are 32 supersymmetric particles including the Higgs fields. In general they can generate a large number of mass hierarchies. Assuming the lightest sparticle is the lightest neutralino, there are still in excess of $10^{25}$ possible mass hierarchies in which the sparticles can arrange themselves. Of these only one will eventually be realized if all the sparticle masses are finally measured at the LHC or in other future collider experiments. The question then is how predictive are SUGRA models in pinning down the mass hierarchical patterns. The above question can be answered within the SUGRA framework including the REWSB constraints, the WMAP and other relevant experimental constraints.[81] An analysis along these lines but limited to four particle mass hierarchies aside from the lightest Higgs boson mass would in general lead to roughly $O(10^4)$ such mass hierarchical patterns. However, within the mSUGRA framework with the constraints mentioned above one finds that the number of possibilities reduces to just 16 for $\mu$ positive and 6 more for $\mu$ negative. These possibilities are labeled as minimal supergravity patterns mSP1-mSP16 for $\mu > 0$ and mSP17-mSP22 for $\mu < 0$. These allowed set of models can be further subdivided into classes according to their next to the lowest mass particle (NLSP). Thus one finds the dominant sub classes of patterns among mSP1-mSP16 to be the Chargino Pattern, Stau Pattern, Stop Pattern and Higgs Pattern. In addition for $\mu < 0$ one finds additional Stau and Stop Patterns and also a Neutralino Pattern where the second neutralino is the NLSP.[81]

A similar analysis can also be carried out for the non-universal SUGRA case. Here allowing for non-universalities in the Higgs sector, gaugino sector and in the third generation sector one finds 22 new sparticle patterns for the first four sparticles (excluding the lightest Higgs boson). These are labeled NUSP1-NUSP22. It is found that no new patterns arise from non-universalities in the Higgs sector and all the new patterns are from non-universalties in the gaugino sector and in the third generation sectors. It is shown in[81] that these new patterns have distinctive signatures and can be discriminated by appropriate choices of events with leptons, jets and missing energy. Specifically using leptons, jets and missing energy events one can discriminate between the stau coannihilation branch and the hyperbolic branch/focus point region. Thus it is also found that one can identify the origin of dark matter using LHC data.

### 3.7. CP Violation

The minimal supergravity model with universal soft breaking has two independent CP violating phases which can be chosen to be the phase of the $\mu$ parameter ($\theta_\mu$) and the phase of the trilinear coupling $A_0$ ($\alpha_{A_0}$). In the more general soft breaking of the non-universal supergravity model, one may have many more phases. For instance, with non-universal gaugino masses each of the gaugino masses in the $U(1)_Y$, $SU(2)_L$ and $SU(3)_C$ may have a phase, i.e., $\tilde{m}_i = |\tilde{m}_i|e^{i\xi_i}$ (i=1,2, 3) of which two are independent. Similarly, the trilinear couplings may be complex and flavor dependent, so that $A_a = |A_a|e^{i\alpha_{A_i}}$. For the most general allowed soft breaking in MSSM, the list of allowed phases is much larger, and even after field redefinitions many CP phases remain. In general the CP phases lead to large supersymmetric contributions to the electric dipole moments (EDMs) of the neutron and of the leptons leading to their EDMs far in excess of experiment. These EDMs can be made compatible by a variety of means, such as through choice of small CP phases,[82,83] large sparticle masses,[84] via the cancellation mechanism,[85,86] or via the CP phases arising only in the third generation.[87] If the CP phases are large they will lead to mixings [88] of CP even Higgs and CP odd Higgs states and there would be many phenomenological implications at colliders and elsewhere.[89,90]

### 3.8. Planck Scale Corrections and Further Tests of SUGRA GUT and Post GUT Physics

Because of the proximity of the GUT scale to the Planck scale one can expect corrections of size O($M_G/M_{Pl}$) to grand unification where $M_{Pl}$ is the Planck mass. For example, Planck scale corrections can modify the gauge kinetic energy function so that one has for the gauge kinetic energy term -$(1/4)f_{\alpha\beta}F_\alpha^{\mu\nu}F_{\beta\mu\nu}$. For the minimal SU(5) theory, $f_{\alpha\beta}$ in SUGRA models can assume the form $f_{\alpha\beta} = \delta_{\alpha\beta} + (c/2M_{Pl})d_{\alpha\beta\gamma}\Sigma^\gamma$ where $\Sigma$ is the scalar field in the 24 plet of SU(5). After the spontaneous breaking of SU(5) and a re-diagonalization of the gauge kinetic energy function, one finds a splitting of the $SU(3) \times SU(2) \times U(1)$ gauge coupling constants at the

GUT scale. These splittings generate a corrections to $\alpha_i(M_Z)$, and using the LEP data one can put constraints on c. One finds that[21] $-1 \leq c \leq 3$. The Planck scale correction also helps relax the stringent constraint on $tan\beta$ imposed by $b-\tau$ unification. Thus in the absence of Planck scale correction one has that $b-\tau$ unification requires $tan\beta$ to lie in two rather sharply defined corridors.[91] One of these corresponds to a small value of $tan\beta$, i.e., $tan\beta \sim 2$ and the second a large value $tan\beta \sim 50$. This stringent constraint is somewhat relaxed by the inclusion of Planck scale corrections.[21]

SUSY grand unified models contain many sources of proton instability. Thus in addition to the p decay occuring via the exchange of superheavy vector lepto-quarks, one has the possibility of p dacay from dimension (dim) 4 (dim 3 in the superpotential) and dim 5 (dim 4 in the superpotential) operators.[92] The lepto-quarks exchange would produce $p \to e^+\pi^0$ as its dominant mode with an expected lifetime[93] of $\sim 1 \times 10^{35\pm1}[M_X/10^{16}]^4 y$ where $M_X \cong 1.1 \times 10^{16}$ while the current lower limit on this decay mode from Super Kamiokande is $\sim 2 \times 10^{33}$ yr. Thus the $e^+\pi^0$ mode may be at the edge of being accessible in proposed experiments[94] such as at DUSEL which will have improved sensitivities for this decay mode. Proton decay from dim 4 operators is much too rapid but is easily forbidden by the imposition of R parity invariance. The p decay from dim 5 operators is more involved. It depends on both the GUT physics as well as on the low energy physics such as the masses of the squarks and of the gauginos. Analysis in supergravity unified models[95] shows that one can make concrete predictions of the p decay modes within these models once the sparticle spectrum is determined.

Precision determination of soft SUSY breaking parameters can be utilized as a vehicle for the test of the predictions of supergravity grand unification. Specifically it has been proposed that precision measurement of the soft breaking parameters can also act as a test of physics at the post GUT and string scales.[96] Thus, for example, if one has a concrete model of the soft breaking parameters at the string scale then these parameters can be evolved down to the grand unification scale leading to a predicted set of non-universalities there. If the SUSY particle spectra and their interactions are known with precision at the electro-weak scale, then this data can be utilized to test a specific model at the post GUT or string scales. Future colliders such as the LHC[97] and the NLC[98] will allow one to make mass measurements with significant accuracy. Thus accuracies of up to a few percent in the mass measurements will be possible at these colliders allowing a test of post GUT and string physics up to an accuracy of $\sim 10\%$.[96]

### 3.9. Conclusion

Supergravity grand unification provides a framework for the supersymmetric unification of the electro-weak and the strong interactions where supersymmetry is broken spontaneously by a super Higgs effect in the hidden sector and the breaking communicated to the visible sector via gravitational interactions. The minimal version of the model based on a generation independent Kahler potential contains only four additional arbitrary parameters and one sign in terms of which all the SUSY mass spectrum and all the SUSY interaction structure is determined. This model is thus very predictive. A brief summary of the predictions and the phenomenological implications of the model were given. Many of the predictions of the model can be tested at current collider energies and at energies that would be achievable at the LHC. We also discussed here extensions of the minimal supergravity model to include non-unversalities in the soft SUSY breaking parameters. Some of the implications of these non-universalities on predictions of the model were discussed. Future experiments should be able to see if the predictions of supergravity unification are indeed verified in nature.

## References


1. P. Ramond, Phys. Rev. D **3**, 2415 (1971).
2. Yu A. Golfand and E.P. Likhtman, JETP Lett. **13**, 452 (1971); D. Volkov and V.P. Akulov, JETP Lett. **16**, 438 (1972).
3. J. Wess and B. Zumino, Nucl. Phys. **B78**, 1 (1974).
4. J. C. Pati and A. Salam, Phys. Rev. D **10**, 275 (1974) [Erratum-ibid. D **11**, 703 (1975)]; H. Georgi and S.L. Glashow, Phys. Rev. Lett. **32**, 438 (1974).
5. E. Gildener, Phys. Rev. **D14**, 1667 (1976).
6. S. Dimopoulos and H. Georgi, Nucl. Phys. **B193**, 150 (1981).
7. S. Dimopoulos, S. Raby and F. Wilczek, Phys. Rev. **D24**, 1681 (1981); N. Sakai, Z. Phys. **C11**, 153 (1981).
8. L. Giradello and M.T. Grisaru, Nucl. Phys. **B194**, 65 (1982).
9. For a review of the properties of the MSSM see H. Haber and G. Kane, Phys. Rep. **117**, 75 (1985).
10. P. Nath and R. L. Arnowitt, Phys. Lett. B **56**, 177 (1975); R. L. Arnowitt, P. Nath and B. Zumino, Phys. Lett. B **56**, 81 (1975).
11. D.Z. Freedman, P. van Nieuwenhuizen and S. Ferrara, Phys. Rev. **D14**, 912 (1976); S. Deser and B. Zumino, Phys. Lett. **B62**, 335 (1976).
12. A.H. Chamseddine, R. Arnowitt and P. Nath, Phys. Rev. Lett. **49**, 970 (1982).
13. For reviews see P. Nath, R. Arnowitt and A.H. Chamseddine, *Applied N*

=*1 Supergravity* (World Scientific, Singapore, 1984); H.P. Nilles, Phys. Rep. **110**, 1 (1984); R. Arnowitt and P. Nath, Proc. of VII J.A. Swieca Summer School ed. E. Eboli (World Scientific, Singapore, 1994); S. P. Martin, "A Supersymmetry Primer," arXiv:hep-ph/9709356; H. Baer and X. Tata, "Weak scale supersymmetry: From superfields to scattering events," *Cambridge, UK: Univ. Pr. (2006) 537 p*; M. Drees, R. Godbole and P. Roy, "Theory and phenomenology of sparticles: An account of four-dimensional N=1 supersymmetry in high energy physics," *Hackensack, USA: World Scientific (2004) 555 p.*

14. R. Barbieri, S. Ferrara and C. A. Savoy, Phys. Lett. B **119**, 343 (1982).
15. L. J. Hall, J. D. Lykken and S. Weinberg, Phys. Rev. D **27** (1983) 2359.
16. P. Nath, R. L. Arnowitt and A. H. Chamseddine, Nucl. Phys. B **227**, 121 (1983).
17. Alternate possibilities are considered elsewhere in this book.
18. H. P. Nilles, Phys. Lett. B **115**, 193 (1982); H. P. Nilles, Nucl. Phys. B **217**, 366 (1983).
19. J. Ellis, S. Kelley and D.V. Nanopoulos, Phys. Lett. **B249**, 441 (1990); **B260**, 131 (1991); U. Amaldi, W. de Boer and H. Furstenau, Phys. Lett. **B260**, 447 (1991); P. Langacker and M. X. Luo, Phys. Rev. D **44**, 817 (1991); F. Anselmo, L. Cifarelli, A. Peterman and A. Zichichi, Nuovo Cim. A **104** (1991) 1817.
20. P.H. Chankowski, Z. Pluciennik, and S. Pokorski, Nucl. Phys. **B439**, 23 (1995); J. Bagger, K. Matchev and D. Pierce, Phys. Lett **B348**, 443 (1995); R.Barbieri and L.J. Hall, Phys. Rev. Lett. **68**, 752 (1992); L.J. Hall and U. Sarid, Phys. Rev. Lett. **70**, 2673 (1993); P. Langacker and N. Polonsky, Phys. Rev. **D47**, 4028 (1993).
21. T. Dasgupta, P. Mamales and P. Nath, Phys. Rev. **D52**, 5366 (1995); D. Ring, S. Urano and R. Arnowitt, Phys. Rev. **D52**, 6623 (1995); S. Urano, D. Ring and R. Arnowitt, Phys. Rev. Lett. **76**, 3663 (1996); P. Nath, Phys. Rev. Lett. **76**, 2218 (1996).
22. K. Inoue et al., Prog. Theor. Phys. **68**, 927 (1982); L. Ibañez and G.G. Ross, Phys. Lett. **B110**, 227 (1982); L. Alvarez-Gaumé, J. Polchinski and M.B. Wise, Nucl. Phys. **B221**, 495 (1983); J. Ellis, J. Hagelin, D.V. Nanopoulos and K. Tamvakis, Phys. Lett. **B125**, 2275 (1983); L. E. Ibañez and C. Lopez, Phys. Lett. **B128**, 54 (1983); Nucl. Phys. **B233**, 545 (1984); L.E. Ibañez, C. Lopez and C. Muñoz, Nucl. Phys. **B256**, 218 (1985).
23. S. P. Martin and P. Ramond, Phys. Rev. D **48**, 5365 (1993) [arXiv:hep-ph/9306314].
24. S. Soni and A. Weldon, Phys. Lett. **B126**, 215 (1983).
25. C.T. Hill, Phys. Lett. **B135**, 47 (1984); Q. Shafi and C. Wetterich, Phys. Rev. Lett. **52**, 875 (1984).
26. E. Cremmer, S. Ferrara, L.Girardello and A. van Proeyen, Phys. Lett. **116B**, 231(1982).
27. S. P. Martin and M. T. Vaughn, Phys. Rev. D **50**, 2282 (1994) [Erratum-ibid. D **78**, 039903 (2008)] [arXiv:hep-ph/9311340].
28. R. L. Arnowitt and P. Nath, Phys. Rev. D **46**, 3981 (1992).

29. K. L. Chan, U. Chattopadhyay and P. Nath, Phys. Rev. D **58** (1998) 096004; J. L. Feng, K. T. Matchev and T. Moroi, Phys. Rev. Lett. **84**, 2322 (2000); H. Baer, C. Balazs, A. Belyaev, T. Krupovnickas and X. Tata, JHEP **0306**, 054 (2003); For a review see, A. B. Lahanas, N. E. Mavromatos and D. V. Nanopoulos, Int. J. Mod. Phys. D **12**, 1529 (2003).
30. QCD corrections to this result are discussed in S.P. Martin and M.T. Vaughn, Phys. Lett. **B318**, 331 (1993); D. Pierce and A. Papadopoulos, Nucl. Phys. **B430**, 278 (1994).
31. G. Ross and R.G. Roberts, Nucl. Phys. **B377**, 571 (1992); R. L. Arnowitt and P. Nath, Phys. Rev. Lett. **69**, 725 (1992); M. Drees and M.M. Nojiri, Nucl. Phys. **B369**, 54 (1993); S. Kelley *et. al.*, Nucl. Phys. **B398**, 3 (1993); M. Olechowski and S. Pokorski, Nucl. Phys. **B404**, 590 (1993); G. Kane, C. Kolda, L. Roszkowski and J. Wells, Phys. Rev. **D49**, 6173 (1994); D.J. Castaño, E. Piard and P. Ramond, Phys. Rev. **D49**, 4882 (1994); W. de Boer, R. Ehret and D. Kazakov, Z.Phys. **C67**, 647 (1995); V. Barger, M.S. Berger, and P. Ohmann, Phys. Rev. **D49**, 4908 (1994); H. Baer, M. Drees, C. Kao, M. Nojiri and X. Tata, Phys. Rev. **D50**, 2148 (1994); H. Baer, C.-H. Chen, R. Munroe, F. Paige and X. Tata, Phys. Rev. **D51**, 1046 (1995).
32. B. C. Allanach, J. P. J. Hetherington, M. A. Parker and B. R. Webber, JHEP **0008**, 017 (2000); B. C. Allanach, G. Belanger, F. Boudjema, A. Pukhov and W. Porod, arXiv:hep-ph/0402161; B. C. Allanach, C. G. Lester and A. M. Weber, JHEP **0612**, 065 (2006).
33. A. Djouadi, M. Drees and J. L. Kneur, JHEP **0108**, 055 (2001); JHEP **0603**, 033 (2006); arXiv:hep-ph/0112026.
34. R. R. de Austri, R. Trotta and L. Roszkowski, JHEP **0605**, 002 (2006); R. R. de Austri, R. Trotta and L. Roszkowski, JHEP **0605**, 002 (2006).
35. D. Zerwas, AIP Conf. Proc. **1078**, 90 (2009) [arXiv:0808.3506 [hep-ph]]; R. Lafaye, T. Plehn, M. Rauch and D. Zerwas, Eur. Phys. J. C **54**, 617 (2008) [arXiv:0709.3985 [hep-ph]].
36. J.R. Ellis, K. Enqvist, D.V. Nanopoulos and K. Tamvakis, Phys. Lett. B **155**, 381 (1985); M. Drees, Phys. Lett. B **158**, 409 (1985); G. Anderson, C.H. Chen, J.F. Gunion, J.D. Lykken, T. Moroi and Y. Yamada, [hep-ph/9609457].
37. P. Nath and R. L. Arnowitt, Phys. Rev. D **56**, 2820 (1997); R. L. Arnowitt, B. Dutta and Y. Santoso, Nucl. Phys. B **606**, 59 (2001); J. R. Ellis, T. Falk, K. A. Olive and Y. Santoso, Nucl. Phys. B **652**, 259 (2003); D. G. Cerdeno and C. Munoz, JHEP **0410**, 015 (2004).
38. A. Corsetti and P. Nath, Phys. Rev. D **64**, 125010 (2001); U. Chattopadhyay and P. Nath, Phys. Rev. D **65**, 075009 (2002); A. Birkedal-Hansen and B. D. Nelson, Phys. Rev. D **67**, 095006 (2003); U. Chattopadhyay and D. P. Roy, Phys. Rev. D **68**, 033010 (2003); G. Belanger, F. Boudjema, A. Cottrant, A. Pukhov and A. Semenov, Nucl. Phys. B **706**, 411 (2005); H. Baer, A. Mustafayev, E. K. Park, S. Profumo and X. Tata, JHEP **0604**, 041 (2006); K. Choi and H. P. Nilles JHEP **0704** (2007) 006; I. Gogoladze, R. Khalid, N. Okada and Q. Shafi, arXiv:0811.1187 [hep-ph]; S. Bhattacharya, A. Datta and B. Mukhopadhyaya, Phys. Rev. D **78**,

115018 (2008); M. E. Gomez, S. Lola, P. Naranjo and J. Rodriguez-Quintero, arXiv:0901.4013 [hep-ph]; B. Altunkaynak, P. Grajek, M. Holmes, G. Kane and B. D. Nelson, arXiv:0901.1145 [hep-ph]; U. Chattopadhyay, D. Das and D. P. Roy, arXiv:0902.4568 [hep-ph]; S. Bhattacharya and J. Chakrabortty, arXiv:0903.4196 [hep-ph].

39. T. C. Yuan, R. Arnowitt, A.H. Chamseddine and P. Nath, Z. Phys. **C26**, 407(1984); D. A. Kosower, L. M. Krauss, N. Sakai, Phys. Lett. **133B**, 305(1983).
40. J. Lopez, D.V. Nanopoulos, and X. Wang, Phys. Rev. **D49**, 366(1994).
41. U. Chattopadhyay and P. Nath, Phys. Rev. **D53**, 1648(1996); T. Moroi, Phys. Rev. D **53**, 6565 (1996).
42. M. Davier, A. Hoecker, B. Malaescu, C. Z. Yuan and Z. Zhang, arXiv:0908.4300 [hep-ph].
43. P. Nath and M. Yamaguchi, Phys. Rev. D **60**, 116006 (1999) [arXiv:hep-ph/9903298].
44. J. L. Feng and K. T. Matchev, Phys. Rev. Lett. **86**, 3480 (2001); E. A. Baltz and P. Gondolo, Phys. Rev. Lett. **86**, 5004 (2001); L. L. Everett, G. L. Kane, S. Rigolin and L. T. Wang, Phys. Rev. Lett. **86**, 3484 (2001); U. Chattopadhyay and P. Nath, Phys. Rev. Lett. **86**, 5854 (2001); T. Ibrahim, U. Chattopadhyay and P. Nath, Phys. Rev. **D64**, 016010(2001); J. Ellis, D.V. Nanopoulos, K. A. Olive, Phys. Lett. B **508**, 65 (2001); R. Arnowitt, B. Dutta, B. Hu, Y. Santoso, Phys. Lett. B **505**, 177 (2001); S. P. Martin, J. D. Wells, Phys. Rev. D **64**, 035003 (2001); H. Baer, C. Balazs, J. Ferrandis, X. Tata, Phys.Rev.**D64**: 035004, (2001).
45. P. Nath and R. L. Arnowitt, Phys. Lett. B **336**, 395 (1994); F. Borzumati, M. Drees and M. M. Nojiri, Phys. Rev. D **51**, 341 (1995); L. Bergstrom and P. Gondolo, Astropart. Phys. **5**, 263 (1996)
46. E. Barberio *et al.* [Heavy Flavor Averaging Group], arXiv:0808.1297 [hep-ex].
47. M. Misiak *et al.*, Phys. Rev. Lett. **98** (2007) 022002.
48. S. Bertolini, F. Borzumati and A. Masiero, Phys. Rev. Lett. **59**, 180(1987); G. Degrassi, P. Gambino and G. F. Giudice, JHEP **0012** (2000) 009; F. Borzumati, C. Greub, T. Hurth and D. Wyler, Phys. Rev. D **62**, 075005 (2000); M. E. Gomez, T. Ibrahim, P. Nath and S. Skadhauge, Phys. Rev. D **74** (2006) 015015; G. Degrassi, P. Gambino and P. Slavich, Phys. Lett. B **635** (2006) 335.
49. J.L. Hewett, Phys. Rev. Lett. **70**, 1045(1993); V. Barger, M. Berger, P. Ohmann. and R.J.N. Phillips, Phys. Rev. Lett. **70**, 1368(1993).
50. M. Diaz, Phys. Lett. **B304**, 278(1993); J. Lopez, D.V. Nanopoulos, and G. Park, Phys. Rev. **D48**, 974(1993); R. Garisto and J.N. Ng, Phys. Lett. **B315**, 372(1993); J. Wu, R. Arnowitt and P. Nath, Phys. Rev. **D51**, 1371(1995); V. Barger, M. Berger, P. Ohman and R.J.N. Phillips, Phys. Rev. **D51**, 2438(1995); H. Baer and M. Brhlik, Phys. Rev. D **55**, 3201 (1997) [arXiv:hep-ph/9610224].
51. S. R. Choudhury and N. Gaur, Phys. Lett. B **451**, 86 (1999); K. S. Babu and C. Kolda, Phys. Rev. Lett. **84**, 228 (2000).

52. C. Bobeth, T. Ewerth, F. Kruger and J. Urban, Phys. Rev. D **64**, 074014 (2001).
53. T. Aaltonen *et al.* [CDF Collaboration], Phys. Rev. Lett. **100**, 101802 (2008).
54. A. Dedes, H. K. Dreiner, and U. Nierste, Phys. Rev. Lett. **87**, 251804 (2001); A. Dedes, H. K. Dreiner, U. Nierste and P. Richardson, arXiv:hep-ph/0207026.
55. R. L. Arnowitt, B. Dutta, T. Kamon and M. Tanaka, Phys. Lett. B **538**, 121 (2002) [arXiv:hep-ph/0203069].
56. T. Ibrahim and P. Nath, Phys. Rev. D **67**, 016005 (2003) [arXiv:hep-ph/0208142].
57. R. Arnowitt and P. Nath, Phys. Rev. Lett. **69**, 725(1992); P. Nath and R. Arnowitt, Phys. Lett. **B289**, 368(1992).
58. For a review see, E. W. Kolb and M. S. Turner, *The Early Universe* (Addison-Wesley, Redwood City, CA, 1989).
59. K. Greist and D. Seckel, Phys. Rev. **D43**, 3191 (1991); P. Gondolo and G. Gelmini, Nucl. Phys. **B360**, 145(1991).
60. R. Arnowitt and P. Nath, Phys. Lett. **B299**, 58(1993); **303**, 403(E)(1993); P. Nath and R. Arnowitt, Phys. Rev. **70**, 3696(1993).
61. D. Feldman, Z. Liu and P. Nath, Phys. Rev. D **79**, 063509 (2009); See also, M. Ibe, H. Murayama and T. T. Yanagida, Phys. Rev. D **79**, 095009 (2009).
62. H. Baer and M. Brhlik, Phys. Rev. D **53**, 597 (1996); V. D. Barger and C. Kao, Phys. Rev. D **57**, 3131 (1998).
63. For a review see, G. Jungman, M. Kamionkowski and K. Greist, Phys. Rep. **267**,195(1995); C. Munoz, Int. J. Mod. Phys. A **19**, 3093 (2004).
64. M.W. Goodman and E. Witten, Phys. Rev. **D31**, 3059(1983); K.Greist, Phys. Rev. **D38**, (1988)2357; **D39**,3802(1989)(E); J.Ellis and R.Flores, Phys. Lett. **B300**,175(1993); R. Barbieri, M. Frigeni and G.F. Giudice, Nucl. Phys. **B313**,725(1989); M.Srednicki and R.Watkins, Phys. Lett. **B225**,140(1989); R.Flores, K.Olive and M.Srednicki, Phys. Lett. **B237**,72(1990); A. Bottino etal, Astro. Part. Phys. **1**, 61 (1992); **2**, 77 (1994).
65. Z. Ahmed *et al.* [CDMS Collaboration], Phys. Rev. Lett. **102**, 011301 (2009).
66. J. Angle *et al.* [XENON Collaboration], Phys. Rev. Lett. **100**, 021303 (2008).
67. SuperCDMS (Projected) 2-ST@Soudan,SuperCDMS Proposal, SuperCDMS (Projected) 25kg (7-ST@Snolab),SuperCDMS Proposal, LUX Proposal, 300 kg LXe Projection, R. Gaitskell, T. Shut, et al.
68. V. Berezinsky, A. Bottino, J. Ellis, N. Forrengo, G. Mignola, and S. Scopel, Astropart.Phys.**5**, 1(1996).
69. D. Feldman, Z. Liu and P. Nath, Phys. Rev. D **80**, 015007 (2009); I. Gogoladze, R. Khalid and Q. Shafi, Phys. Rev. D **80**, 095016 (2009) [arXiv:0908.0731 [hep-ph]].
70. M. S. Turner and F. Wilczek, Phys. Rev. D **42**, 1001 (1990).
71. O. Adriani *et al.* [PAMELA Collaboration], Nature **458**, 607 (2009); Phys. Rev. Lett. **102**, 051101 (2009).
72. D. Feldman, Z. Liu, P. Nath and B. D. Nelson, Phys. Rev. D **80**, 075001 (2009); D. Feldman, arXiv:0908.3727 [hep-ph].

73. D. N. Spergel *et al.* [WMAP Collaboration], Astrophys. J. Suppl. **170**, 377 (2007); E. Komatsu *et al.* [WMAP Collaboration], Astrophys. J. Suppl. **180**, 330 (2009).
74. A. A. Abdo *et al.* [The Fermi LAT Collaboration], Phys. Rev. Lett. **102**, 181101 (2009); See also, J. Chang *et al.*, Nature **456**, 362 (2008).
75. S. Weinberg, Phys. Rev. Lett. **50**, 387 (1983); R. Arnowitt, A. H. Chamseddine and P. Nath, Phys. Rev. Lett. **50** (1983) 232; A. H. Chamseddine, P. Nath and R. Arnowitt, Phys. Lett. B **129**, 445 (1983); P. Nath, R. Arnowitt and A. H. Chamseddine, HUTP-83/A077; D. A. Dicus, S. Nandi, W. W. Repko and X. Tata, Phys. Rev. Lett. **51**, 1030 (1983); Phys. Rev. D **29**, 67 (1984); Phys. Rev. D **29**, 1317 (1984); D. A. Dicus, S. Nandi and X. Tata, Phys. Lett. B **129**, 451 (1983); J. M. Frere and G. L. Kane, Nucl. Phys. B **223**, 331 (1983); J. R. Ellis, J. M. Frere, J. S. Hagelin, G. L. Kane and S. T. Petcov, Phys. Lett. B **132**, 436 (1983); H. Goldberg, Phys. Rev. Lett. **50**, 1419 (1983); J. Ellis, J.S. Hagelin, D.V. Nanopoulos, and M. Srednicki, Phys. Rev. Lett. **127B**, 233(1983).
76. P. Nath and R. Arnowitt, Mod. Phys. Lett. **A2**, 331(1987); R. Barbieri, F. Caravaglio, M. Frigeni, and M. Mangano, Nucl. Phys. **B367**, 28(1991); H. Baer and X. Tata, Phys. Rev. **D47**, 2739(1992); J.L. Lopez, D.V. Nanopoulos, X. Wang and A. Zichichi, Phys. Rev. **D48**, 2062(1993); H. Baer, C. Kao and X. Tata, Phys. Rev. **D48**, 5175(1993).
77. R. L. Arnowitt, B. Dutta, A. Gurrola, T. Kamon, A. Krislock and D. Toback, Phys. Rev. Lett. **100**, 231802 (2008) [arXiv:0802.2968 [hep-ph]].
78. B. Dutta and T. Kamon, talk at PASCOS2009, DESY Hamburg, July 9, 2009.
79. V. Khotilovich, R. L. Arnowitt, B. Dutta and T. Kamon, Phys. Lett. B **618** (2005) 182 [arXiv:hep-ph/0503165].
80. R. L. Arnowitt *et al.*, Phys. Lett. B **649**, 73 (2007); R. L. Arnowitt, B. Dutta, T. Kamon, N. Kolev and D. A. Toback, Phys. Lett. B **639**, 46 (2006) [arXiv:hep-ph/0603128].
81. D. Feldman, Z. Liu and P. Nath, Phys. Rev. Lett. **99**, 251802 (2007); Phys. Lett. B **662**, 190 (2008); JHEP **0804**, 054 (2008); Phys. Rev. D **78**, 083523 (2008); N. Chen, D. Feldman, Z. Liu and P. Nath, arXiv:0911.0217 [hep-ph].
82. S. Weinberg, Phys. Rev. Lett. **63** (1989) 2333.
83. R. L. Arnowitt, J. L. Lopez and D. V. Nanopoulos, Phys. Rev. D **42**, 2423 (1990); R. L. Arnowitt, M. J. Duff and K. S. Stelle, Phys. Rev. D **43**, 3085 (1991).
84. P. Nath, Phys. Rev. Lett. **66**, 2565 (1991); Y. Kizukuri and N. Oshimo, Phys. Rev. **D46**, 3025 (1992).
85. T. Ibrahim and P. Nath, Phys. Lett. B **418**, 98 (1998); Phys. Rev. **D57**, 478(1998); T. Falk and K Olive, Phys. Lett. **B 439**, 71(1998); M. Brhlik, G.J. Good, and G.L. Kane, Phys. Rev. **D59**, 115004 (1999); M. Brhlik, L. Everett, G.L. Kane and J. Lykken, Phys. Rev. Lett. **83** (1999) 2124; A. Bartl, T. Gajdosik, W. Porod, P. Stockinger, and H. Stremnitzer, Phys. Rev. **60**, 073003(1999); S. Pokorski, J. Rosiek and C.A. Savoy, Nucl.Phys. **B570**, 81(2000). V. D. Barger, T. Falk, T. Han, J. Jiang, T. Li and T. Plehn, Phys.

Rev. D **64**, 056007 (2001); S.Abel, S. Khalil, O.Lebedev, Phys. Rev. Lett. **86**, 5850(2001).
86. E. Accomando, R. Arnowitt and B. Dutta, Phys. Rev. D **61**, 115003 (2000); R. L. Arnowitt, B. Dutta and Y. Santoso, Phys. Rev. D **64**, 113010 (2001).
87. D. Chang, W. Y. Keung and A. Pilaftsis, Phys. Rev. Lett. **82** (1999) 900.
88. A. Pilaftsis, Phys. Lett. B **435** (1998) 88.
89. J. R. Ellis, J. S. Lee and A. Pilaftsis, Phys. Rev. D **70** (2004) 075010.
90. T. Ibrahim and P. Nath, Rev. Mod. Phys. **80** (2008) 577.
91. V. Barger, M.S. Berger, and P. Ohman, Phys. Lett. **B314**, 351(1993); W. Bardeen, M. Carena, S. Pokorski, and C.E.M. Wagner, Phys. Lett. **B320**, 110(1994).
92. S.Weinberg, Phys.Rev.**D26**,287(1982); N.Sakai and T.Yanagida, Nucl.Ph ys.**B197**, 533(1982); S.Dimopoulos, S.Raby and F.Wilcek, Phys.Lett. **112B**, 133(1982); J.Ellis, D.V.Nanopoulos and S.Rudaz, Nucl.Phys. **B202**,43(1982).
93. W.J. Marciano, talk at SUSY-97, Philadelphia, May 1997.
94. A. Rubbia, arXiv:0908.1286 [hep-ph]; S. Raby *et al.*, arXiv:0810.4551 [hep-ph]; P. Nath and P. Fileviez Perez, Phys. Rept. **441**, 191 (2007).
95. R. L. Arnowitt, A. H. Chamseddine and P. Nath, Phys. Lett. B **156**, 215 (1985); P. Nath, A. H. Chamseddine and R. L. Arnowitt, Phys. Rev. D **32**, 2348 (1985); J. Hisano, H. Murayama and T. Yanagida, Nucl. Phys. B **402**, 46 (1993); R. L. Arnowitt and P. Nath, Phys. Rev. D **49**, 1479 (1994).
96. R. L. Arnowitt and P. Nath, Phys. Rev. D **56**, 2833 (1997).
97. H. Baer, C. Chen, F. Paige, and X. Tata, Phys. Rev. **DD52**, 2746(1995); I. Hinchliffe, F.E. Paige, M.D. Shapiro, J. Soderqvist and W. Yao, Phys. Rev. **D55**, 5520(1997).
98. T. Tsukamoto, K. Fujii, H. Murayama, M. Yamaguchi, and Y. Okada, Phys. Rev. **D. 51**, 3153(1995); J.L. Feng, M.E. Peskin, H. Murayama, and X. Tata, Phys. Rev. **D52**, 1418(1992); S. Kuhlman et. al., "Physics and Technology of the NLC: Snowmass 96", hep-ex/9605011.

## Soft Supersymmetry-Breaking Terms from Supergravity and Superstring Models

A. Brignole*, L. E. Ibáñez† and C. Muñoz†

*Theory Division, CERN, CH-1211 Geneva 23, Switzerland

†Departamento de Física Teórica, Universidad Autónoma de Madrid, Cantoblanco, 28049 Madrid, Spain

We review the origin of soft supersymmetry-breaking terms in $N = 1$ supergravity models of particle physics. We first consider general formulae for those terms in general models with a hidden sector breaking supersymmetry at an intermediate energy scale. The results for some simple models are given. We then consider the results obtained in some simple superstring models in which particular assumptions about the origin of supersymmetry breaking are made. These are models in which the seed of supersymmetry breaking is assumed to be originated in the dilaton/moduli sector of the theory.

### 4.1. Introduction

A phenomenological implementation of the idea of supersymmetry (SUSY) in the standard model requires the presence of SUSY breaking. There are essentially two large families of models in this context, depending on whether the scale of spontaneous SUSY breaking is high (of order $10^{10}$–$10^{13}$ GeV) or low (of order 1–$10^2$ TeV). We will focus on the former possibility. The latter possibility, considered in other chapters of this book, has only recently received sufficient attention, since it was realized from the very first days of SUSY phenomenology that the existence of certain supertrace constraints in spontaneously broken SUSY theories made the building of realistic models quite complicated. Possible solutions to these early difficulties are discussed elsewhere and we are not going to discuss them further here.

A more pragmatic attitude to the issue of SUSY breaking is the addition of explicit soft SUSY-breaking terms of the appropriate size (of order

$10^2$–$10^3$ GeV) in the Lagrangian and with appropriate flavour symmetries to avoid dangerous flavour-changing neutral currents (FCNC) transitions. The problem with this pragmatic attitude is that, taken by itself, lacks any theoretical explanation. Supergravity theories provide an attractive context that can justify such a procedure. Indeed, if one considers the SUSY standard model and couples it to $N = 1$ supergravity, the spontaneous breaking of local SUSY in a hidden sector generates explicit soft SUSY-breaking terms of the required form in the effective low-energy Lagrangian.[1,2] If SUSY is broken at a scale $\Lambda_S$, the soft terms have a scale of order $\Lambda_S^2/M_{Planck}$. Thus one obtains the required size if SUSY is broken at an intermediate scale $\Lambda_S \sim 10^{10}$ GeV, as mentioned above. Large classes of supergravity models, as we discuss in section 2, give rise to universal soft SUSY-breaking terms, providing for an understanding of FCNC supression. In the last few years it has often been stated in the literature that this class of supergravity models have a flavour-changing problem. We think more appropriate to say that some particular models get interesting *constraints* from FCNC bounds. A generic statement like that seems unjustified, since it is usually based on a strong assumption, i.e. the existence of a region in between the grand unified theory (GUT) scale and the Planck (or superstring) scale in which important flavour non-diagonal renormalization effects take place.

Recently there have been studies of supergravity models obtained in particularly simple classes of superstring compactifications.[3] Such heterotic models have a natural hidden sector built-in: the complex dilaton field $S$ and the complex moduli fields $T_i$. These gauge singlet fields are generically present in four-dimensional models: the dilaton arises from the gravitational sector of the theory and the moduli parametrize the size and shape of the compactified variety. Assuming that the auxiliary fields of those multiplets are the seed of SUSY breaking, interesting predictions for this simple class of models are obtained. These are reviewed in section 3. The analysis does not assume any specific SUSY-breaking mechanism. We leave section 4 for some final comments and additional references to recent work.

## 4.2. Soft Terms from Supergravity

### 4.2.1. *General computation of soft terms*

The full N=1 supergravity Lagrangian[1] (up to two derivatives) is specified in terms of two functions which depend on the chiral superfields $\phi_M$ of the theory (denoted by the same symbol as their scalar components): the analytic gauge kinetic function $f_a(\phi_M)$ and the real gauge-invariant Kähler function $G(\phi_M, \phi_M^*)$. $f_a$ determines the kinetic terms for the fields in the vector multiplets and in particular the gauge coupling constant, $Ref_a = 1/g_a^2$. The subindex $a$ is associated with the different gauge groups of the theory since in general $\mathcal{G} = \prod_a \mathcal{G}_a$. For example, in the case of the pure SUSY standard model coupled to supergravity, $a$ would correspond to $SU(3)_c$, $SU(2)_L$, $U(1)_Y$. $G$ is a combination of two functions

$$G(\phi_M, \phi_M^*) = K(\phi_M, \phi_M^*) + \log |W(\phi_M)|^2 \; , \tag{4.1}$$

where $K$ is the Kähler potential, $W$ is the complete analytic superpotential, and we use from now on the standard supergravity mass units where $M_P \equiv M_{Planck}/\sqrt{8\pi} = 1$. $W$ is related with the Yukawa couplings (which eventually determine the fermion masses) and also includes possibly nonperturbative effects

$$W = \hat{W}(h_m) + \frac{1}{2}\mu_{\alpha\beta}(h_m)C^\alpha C^\beta + \frac{1}{6}Y_{\alpha\beta\gamma}(h_m)C^\alpha C^\beta C^\gamma + \dots \; , \tag{4.2}$$

where we assume two different types of scalar fields $\phi_M = h_m, C^\alpha$: $C^\alpha$ correspond to the observable sector and in particular include the SUSY standard model fields, while $h_m$ correspond to a hidden sector. The latter fields may develop large ($\gg M_W$) vacuum expectation values (VEVs) and are responsible for SUSY breaking if some auxiliary components $F^m$ (see below) develop nonvanishing VEVs. The ellipsis indicates terms of higher order in $C^\alpha$ whose coefficients are suppressed by negative powers of $M_P$. The second derivative of $K$ determines the kinetic terms for the fields in the chiral supermultiplets and is thus important for obtaining the proper normalization of the fields. Expanding in powers of $C^\alpha$ and $C^{*\overline{\alpha}}$ we have

$$\begin{aligned} K = \hat{K}(h_m, h_m^*) + \; & \tilde{K}_{\overline{\alpha}\beta}(h_m, h_m^*)C^{*\overline{\alpha}}C^\beta \\ + & \left[\frac{1}{2}Z_{\alpha\beta}(h_m, h_m^*)C^\alpha C^\beta \; + \; h.c.\right] + \dots \; , \end{aligned} \tag{4.3}$$

where the ellipsis indicates terms of higher order in $C^\alpha$ and $C^{*\overline{\alpha}}$. Notice that the coefficients $\tilde{K}_{\overline{\alpha}\beta}$, $Y_{\alpha\beta\gamma}$, $\mu_{\alpha\beta}$, and $Z_{\alpha\beta}$ which appear in (4.2) and (4.3) may depend on the hidden sector fields in general. The bilinear terms associated with $\mu_{\alpha\beta}$ and $Z_{\alpha\beta}$ are often forbidden by gauge invariance in specific models, but they may be *relevant* in order to solve the so-called $\mu$ problem in the context of the minimal supersymmetric standard model (MSSM), as we will discuss below. In this case the two Higgs doublets, which are necessary to break the electroweak symmetry, have opposite hypercharges. Therefore those terms are allowed and may generate both the $\mu$ parameter and the corresponding soft bilinear term.

The (F part of the) tree-level supergravity scalar potential, which is crucial to analyze the breaking of SUSY, is given by

$$V(\phi_M, \phi_M^*) = e^G \left(G_M K^{M\bar{N}} G_{\bar{N}} \ - \ 3\right) = \left(\bar{F}^{\bar{N}} K_{\bar{N}M} F^M \ - \ 3e^G\right) \ , \quad (4.4)$$

where $G_M \equiv \partial_M G \equiv \partial G/\partial\phi_M$ and the matrix $K^{M\bar{N}}$ is the inverse of the Kähler metric $K_{\bar{N}M} \equiv \partial_{\bar{N}}\partial_M K$. We have also written $V$ as a function of the $\phi_M$ auxiliary fields, $F^M = e^{G/2}K^{M\bar{P}}G_{\bar{P}}$. When, at the minimum of the scalar potential, some of the hidden sector fields $h_m$ acquire VEVs in such a way that at least one of their auxiliary fields ($\hat{K}^{m\overline{n}}$ is the inverse of the hidden field metric $\hat{K}_{\overline{n}m}$)

$$F^m = e^{G/2}\hat{K}^{m\overline{n}}G_{\overline{n}} \quad (4.5)$$

is non-vanishing, then SUSY is spontaneously broken and soft SUSY-breaking terms are generated in the observable sector. Let us remark that, for simplicity, we are assuming vanishing D-term contributions to SUSY breaking. When this is not the case, their effects on soft terms can be found e.g. in Ref. 4. The goldstino, which is a combination of the fermionic partners of the above fields, is swallowed by the gravitino via the superHiggs effect. The gravitino becomes massive and its mass

$$m_{3/2} = e^{G/2} \quad (4.6)$$

sets the overall scale of the soft parameters.

*General results*

Using the above information, the soft SUSY-breaking terms in the observable sector can be computed. They are obtained by replacing $h_m$ and their auxiliary fields $F^m$ by their VEVs in the supergravity Lagrangian and taking the so-called flat limit where $M_P \to \infty$ but $m_{3/2}$ is kept fixed. Then the non-renormalizable gravity corrections are formally eliminated and one is left with a global SUSY Lagrangian plus a set of soft SUSY-breaking terms. On the one hand, from the fermionic part of the supergravity Lagrangian, soft gaugino masses for the canonically *normalized* gaugino fields can be obtained

$$M_a = \frac{1}{2}\left(Ref_a\right)^{-1} F^m \partial_m f_a \ , \quad (4.7)$$

as well as the *un-normalized* Yukawa couplings of the observable sector fermions and the SUSY *un-normalized* masses of some of them (those with bilinear terms either in the superpotential or in the Kähler potential, e.g. the Higgsinos in the case of the MSSM)

$$Y'_{\alpha\beta\gamma} = \frac{\hat{W}^*}{|\hat{W}|} e^{\hat{K}/2} Y_{\alpha\beta\gamma} \,, \tag{4.8}$$

$$\mu'_{\alpha\beta} = \frac{\hat{W}^*}{|\hat{W}|} e^{\hat{K}/2} \mu_{\alpha\beta} + m_{3/2} Z_{\alpha\beta} - \overline{F}^{\overline{m}} \partial_{\overline{m}} Z_{\alpha\beta} \,. \tag{4.9}$$

On the other hand, scalar soft terms arise from the expansion of the supergravity scalar potential (4.4)

$$V_{soft} = m'^2_{\overline{\alpha}\beta} C^{*\overline{\alpha}} C^\beta + \left( \frac{1}{6} A'_{\alpha\beta\gamma} C^\alpha C^\beta C^\gamma + \frac{1}{2} B'_{\alpha\beta} C^\alpha C^\beta + h.c. \right) . \tag{4.10}$$

In the most general case, when hidden and observable sector matter metrics are not diagonal, the *un-normalized* soft scalar masses, trilinear and bilinear parameters are given respectively by[5]

$$\begin{aligned} m'^2_{\overline{\alpha}\beta} &= \left( m^2_{3/2} + V_0 \right) \tilde{K}_{\overline{\alpha}\beta} \\ &\quad - \overline{F}^{\overline{m}} \left( \partial_{\overline{m}} \partial_n \tilde{K}_{\overline{\alpha}\beta} - \partial_{\overline{m}} \tilde{K}_{\overline{\alpha}\gamma} \tilde{K}^{\gamma\overline{\delta}} \partial_n \tilde{K}_{\overline{\delta}\beta} \right) F^n \,, \end{aligned} \tag{4.11}$$

$$\begin{aligned} A'_{\alpha\beta\gamma} &= \frac{\hat{W}^*}{|\hat{W}|} e^{\hat{K}/2} F^m \Big[ \hat{K}_m Y_{\alpha\beta\gamma} + \partial_m Y_{\alpha\beta\gamma} \\ &\quad - \left( \tilde{K}^{\delta\overline{\rho}} \partial_m \tilde{K}_{\overline{\rho}\alpha} Y_{\delta\beta\gamma} + (\alpha \leftrightarrow \beta) + (\alpha \leftrightarrow \gamma) \right) \Big] \,, \end{aligned} \tag{4.12}$$

$$\begin{aligned} B'_{\alpha\beta} &= \frac{\hat{W}^*}{|\hat{W}|} e^{\hat{K}/2} \Big\{ F^m \Big[ \hat{K}_m \mu_{\alpha\beta} + \partial_m \mu_{\alpha\beta} \\ &\quad - \left( \tilde{K}^{\delta\overline{\rho}} \partial_m \tilde{K}_{\overline{\rho}\alpha} \mu_{\delta\beta} + (\alpha \leftrightarrow \beta) \right) \Big] - m_{3/2} \mu_{\alpha\beta} \Big\} \\ &\quad + \left( 2m^2_{3/2} + V_0 \right) Z_{\alpha\beta} - m_{3/2} \overline{F}^{\overline{m}} \partial_{\overline{m}} Z_{\alpha\beta} \\ &\quad + m_{3/2} F^m \Big[ \partial_m Z_{\alpha\beta} - \left( \tilde{K}^{\delta\overline{\rho}} \partial_m \tilde{K}_{\overline{\rho}\alpha} Z_{\delta\beta} + (\alpha \leftrightarrow \beta) \right) \Big] \\ &\quad - \overline{F}^{\overline{m}} F^n \Big[ \partial_{\overline{m}} \partial_n Z_{\alpha\beta} - \left( \tilde{K}^{\delta\overline{\rho}} \partial_n \tilde{K}_{\overline{\rho}\alpha} \partial_{\overline{m}} Z_{\delta\beta} + (\alpha \leftrightarrow \beta) \right) \Big] \end{aligned} \tag{4.13}$$

where $\tilde{K}^{\alpha\overline{\beta}}$ is the inverse of the observable matter metric $\tilde{K}_{\overline{\beta}\gamma}$. $V_0$ is the VEV of the scalar potential (4.4), i.e. the tree-level cosmological constant

$$V_0 = \overline{F}^{\overline{m}} \hat{K}_{\overline{m}n} F^n - 3m^2_{3/2} \,. \tag{4.14}$$

It has a bearing on measurable quantities like scalar masses and therefore it is preferable to leave it undetermined (see Refs. 6–8 for a discussion on this point). Notice that, after normalizing the fields to get canonical kinetic terms, the first piece in (4.11) will lead to universal diagonal soft masses but the second piece will generically induce *off-diagonal* contributions. Actually, universality is a desirable property not only to reduce the number of independent parameters in SUSY models, but also for phenomenological reasons, particularly to avoid FCNC. The latter is a *low-energy* phenomenological constraint that must be satisfied by any supergravity model. It is worth mentioning in this context that one–loop corrections to the soft parameters (4.7), (4.11), (4.12) and (4.13) have recently been computed in Ref. 9. They may induce FCNC phenomena even when the tree-level computation gives a universal soft mass. Also a discussion about the loop effects on the contribution of the cosmological constant to the soft terms can be found there. Concerning the $A$ and $B$ parameters, notice that we have not factored out the Yukawa couplings and mass terms respectively as usual, since proportionality is not guaranteed in (4.12) and (4.13). Finally, from (4.5) and (4.6), one can easily see that the (tree-level) soft parameters in (4.7), (4.11) and (4.12) are generically $\mathcal{O}(m_{3/2})$, as mentioned above. A departure from this result is only possible when some of them vanish. We will mention below some examples of this type, i.e. the case of no-scale supergravity models and special limits of superstring models.

*The $\mu$ problem*

The set of mass parameters in the MSSM Higgs potential includes, besides $\mathcal{O}(m_{3/2}^2)$ soft masses, the square of $\mu'_{\alpha\beta}$ (4.9) and the $B$ parameter (4.13), which is $\mathcal{O}(m_{3/2}\mu'_{\alpha\beta})$. In order to have correct electroweak symmetry breaking, the SUSY mass term $\mu'_{\alpha\beta}$ should also be $\mathcal{O}(m_{3/2})$. This is the so-called $\mu$ problem.[2] In this respect, notice that $\mu'_{\alpha\beta} = \mathcal{O}(m_{3/2})$ is *naturally* achieved in the presence of a non-vanishing $Z_{\alpha\beta}$ in the Kähler potential.[10,11] The other possible source of the mass $\mu'_{\alpha\beta}$ is the SUSY mass $\mu_{\alpha\beta}$ in the superpotential.[12] This case is more involved since in principle the natural scale of $\mu_{\alpha\beta}$ would be $M_P$. However, a possible solution can be obtained if the superpotential contains e.g. a non-renormalizable term[11,12]

$$\lambda(h_m)\hat{W}(h_m)H_1H_2 \ , \tag{4.15}$$

characterized by the coupling $\lambda$, which mixes the observable sector with the hidden sector. Since $m_{3/2} = e^{G/2} = e^{\hat{K}/2}|\hat{W}|$, if that term exists then an effective $\mu$ parameter $\mathcal{O}(m_{3/2})$ is generated *dynamically* when $h_m$ acquire VEVs:

$$\mu = \lambda(h_m)\hat{W}(h_m) \ . \tag{4.16}$$

We should add that both mechanisms to generate $\mu'_{\alpha\beta}$, a bilinear term in the Kähler potential or in the superpotential, could be present simultaneously. Notice also that the two mechanisms are equivalent if $Z$ depends only on $h_m$ (not on $h_m^*$). Indeed, in that case the supergravity theory is equivalent to the one with a Kähler potential $K$ without the terms $ZH_1H_2 + h.c.$ and a superpotential $We^{ZH_1H_2}$, since the $G$ function (4.1) is the same for both. After expanding the exponential, the superpotential will have a contribution $Z\hat{W}H_1H_2$, i.e. a term of the type (4.15). Finally, let us mention that several new sources of the $\mu$ term due to loop effects on (4.9), which are naturally of order the weak scale, have recently been computed in Ref. 9.

We recall that the solutions mentioned here in order to solve the $\mu$ problem are *naturally* present in superstring models. For instance, in large classes of superstring models the Kähler potential does contain bilinear terms analytic in the observable fields as in (4.3), with specific coefficients $Z_{\alpha\beta}$,[13–15] so that a $\mu$ parameter may be naturally generated. Concerning superpotential contributions, we recall that a 'direct' $\mu H_1H_2$ term in $W$ (4.2) is naturally absent (otherwise the natural scale for $\mu$ would be $M_P$), since in supergravity models deriving from superstring theory mass terms for light fields are forbidden in the superpotential by scale invariance of the theory. However, the superpotential (4.2) may well contain an 'effective' $\mu H_1H_2$ term, e.g. a term of the type (4.15)[11,15] induced by non-perturbative SUSY-breaking mechanisms like gaugino-squark condensation in the hidden sector.

### *The low-energy spectrum*

The results (4.7), (4.8), (4.9), (4.11), (4.12) and (4.13) should be understood as being valid at some high scale $\mathcal{O}(M_P)$ and the standard RGEs must be used to obtain the low-energy values. Although the SUSY spectrum will depend in general on the details of $SU(2)_L \times U(1)_Y$ breaking, there are several particles whose mass is rather independent of those details and is mostly given by the boundary conditions and the renormalization group running. In particular, neglecting all Yukawa couplings except the one of the top, that is the case of the gluino $g$, all the squarks (except stops and left sbottom) $Q_L = (u_L, d_L), u_L^c, d_L^c$ and all the sleptons $L_L = (v_L, e_L), e_L^c$. For all these particles one can write explicit expressions for the masses in terms of the soft parameters (after normalizing the fields to get canonical kinetic terms). For instance, assuming that gauginos have a common initial mass (e.g. due to a universal $f$ function) and that there is nothing but the MSSM in between the weak scale and the Planck scale, one obtains the approximate numerical expressions:

$$\begin{aligned}
M_g^2(M_Z) &\simeq 9.8\ M^2\ , \\
m_{Q_L}^2(M_Z) &\simeq m_{Q_L}^2 + 8.3\ M^2\ , \\
m_{u_L^c,d_L^c}^2(M_Z) &\simeq m_{u_L^c,d_L^c}^2 + 8\ M^2\ , \\
m_{L_L}^2(M_Z) &\simeq m_{L_L}^2 + 0.7\ M^2\ , \\
m_{e_L^c}^2(M_Z) &\simeq m_{e_L^c}^2 + 0.23\ M^2\ ,
\end{aligned} \tag{4.17}$$

where the second term in the expression of the scalar masses is the effect of gaugino loop contributions. In the above formulae we have neglected the scalar potential D-term contributions, which are normally small compared to the terms above, and the contribution of the $U(1)_Y$ D-term in the RGEs of scalar masses. These may be found e.g. in Ref. 16.

### 4.2.2. *Supergravity models*

We now specialize the above general discussion to the case of supergravity models where the observable (here MSSM) matter fields have diagonal metric:

$$\tilde{K}_{\overline{\alpha}\beta}(h_m, h_m^*) = \delta_{\overline{\alpha}\beta}\tilde{K}_\alpha(h_m, h_m^*)\ . \tag{4.18}$$

This possibility is particularly interesting due to its simplicity and also for phenomenological reasons related to the absence of FCNC in the effective low-energy theory (see Refs. 6, 9, 17–20 for a discussion on this point). Besides, the supergravity models that will be studied below correspond to this situation. Then the Kähler potential (4.3), to lowest order in the observable fields $C^\alpha$, and the superpotential (4.2) have the form

$$K = \hat{K}(h_m, h_m^*) + \tilde{K}_\alpha(h_m, h_m^*)C^{*\overline{\alpha}}C^\alpha + [Z(h_m, h_m^*)H_1H_2 + h.c.] \tag{4.19}$$

$$\begin{aligned}
W = \hat{W}(h_m) + \mu(h_m)H_1H_2\ + &\sum_{generations} [Y_u(h_m)Q_LH_2u_L^c \\
&+ Y_d(h_m)Q_LH_1d_L^c + Y_e(h_m)L_LH_1e_L^c]\ ,
\end{aligned} \tag{4.20}$$

where $C^\alpha = Q_L, u^c_L, d^c_L, L_L, e^c_L, H_1, H_2$, and we have taken for simplicity diagonal Yukawa couplings ($Y_{\alpha\beta\gamma} = Y_u, Y_d, Y_e$, in a self-explanatory notation). Now the form of the effective soft Lagrangian obtained from (4.7) and (4.10) is given by

$$\mathcal{L}_{soft} = \frac{1}{2}(M_a \widehat{\lambda}^a \widehat{\lambda}^a + h.c.) - m^2_\alpha \widehat{C}^{*\overline{\alpha}} \widehat{C}^\alpha - \left(\frac{1}{6} A_{\alpha\beta\gamma} \widehat{Y}_{\alpha\beta\gamma} \widehat{C}^\alpha \widehat{C}^\beta \widehat{C}^\gamma + B\widehat{\mu}\widehat{H}_1\widehat{H}_2 + h.c.\right) , \quad (4.21)$$

with

$$m^2_\alpha = \left(m^2_{3/2} + V_0\right) - \overline{F}^{\overline{m}} F^n \partial_{\overline{m}} \partial_n \log \tilde{K}_\alpha , \quad (4.22)$$

$$A_{\alpha\beta\gamma} = F^m \left[\hat{K}_m + \partial_m \log Y_{\alpha\beta\gamma} - \partial_m \log(\tilde{K}_\alpha \tilde{K}_\beta \tilde{K}_\gamma)\right] , \quad (4.23)$$

$$\begin{aligned} B = \widehat{\mu}^{-1} (\tilde{K}_{H_1} \tilde{K}_{H_2})^{-1/2} \Bigg\{ & \frac{\hat{W}^*}{|\hat{W}|} e^{\hat{K}/2} \mu \left(F^m \left[\hat{K}_m + \partial_m \log \mu \right.\right. \\ & \left.\left. - \partial_m \log(\tilde{K}_{H_1} \tilde{K}_{H_2})\right] - m_{3/2}\right) \\ & + \left(2m^2_{3/2} + V_0\right) Z - m_{3/2} \overline{F}^{\overline{m}} \partial_{\overline{m}} Z \\ & + m_{3/2} F^m \left[\partial_m Z - Z \partial_m \log(\tilde{K}_{H_1} \tilde{K}_{H_2})\right] \\ & - \overline{F}^{\overline{m}} F^n \left[\partial_{\overline{m}} \partial_n Z - \partial_{\overline{m}} Z \partial_n \log(\tilde{K}_{H_1} \tilde{K}_{H_2})\right] \Bigg\} , \end{aligned} \quad (4.24)$$

where $\widehat{C}^\alpha$ and $\widehat{\lambda}^a$ are the scalar and gaugino canonically *normalized* fields respectively

$$\widehat{C}^\alpha = \tilde{K}^{1/2}_\alpha C^\alpha , \quad (4.25)$$

$$\widehat{\lambda}^a = (Ref_a)^{1/2} \lambda^a , \quad (4.26)$$

and the rescaled Yukawa couplings and $\mu$ parameter

$$\widehat{Y}_{\alpha\beta\gamma} = Y_{\alpha\beta\gamma} \frac{\hat{W}^*}{|\hat{W}|} e^{\hat{K}/2} (\tilde{K}_\alpha \tilde{K}_\beta \tilde{K}_\gamma)^{-1/2} , \quad (4.27)$$

$$\widehat{\mu} = \left(\frac{\hat{W}^*}{|\hat{W}|} e^{\hat{K}/2} \mu + m_{3/2} Z - \overline{F}^{\overline{m}} \partial_{\overline{m}} Z\right) (\tilde{K}_{H_1} \tilde{K}_{H_2})^{-1/2} , \quad (4.28)$$

have been factored out in the $A$ and $B$ terms as usual.

Now we are ready to study specific supergravity models. As follows from the above discussion, the particular values of the soft parameters depend on the type of supergravity theory from which the MSSM derives and, in general, on the mechanism of SUSY breaking (through the presence of $\hat{W}(h_m)$ in $m_{3/2}$ and F terms). However, it is still possible to learn things about soft parameters without knowing all the details of SUSY breaking. In order to show this, let us consider two simple and interesting supergravity models studied extensively in the literature: minimal supergravity and no-scale supergravity.

(i) *Minimal supergravity*

This model corresponds to use the form of $K$ that leads to minimal (canonical) kinetic terms in the supergravity Lagrangian, namely

$$\tilde{K}_\alpha(h_m, h_m^*) = 1 \tag{4.29}$$

in (4.19). Then, irrespective of the SUSY-breaking mechanism, the scalar masses and the $A$, $B$ parameters can be straightforwardly computed using (4.22), (4.23) and (4.24)

$$m_\alpha^2 = m_{3/2}^2 + V_0 \ , \tag{4.30}$$

$$A_{\alpha\beta\gamma} = F^m \left(\hat{K}_m + \partial_m \log Y_{\alpha\beta\gamma}\right) \ , \tag{4.31}$$

$$\begin{aligned} B = \widehat{\mu}^{-1} \Bigg\{ & \frac{\hat{W}^*}{|\hat{W}|} e^{\hat{K}/2} \mu \left[F^m \left(\hat{K}_m + \partial_m \log \mu\right) - m_{3/2}\right] \\ & + \left(2m_{3/2}^2 + V_0\right) Z + \ m_{3/2} \left(F^m \partial_m Z - \overline{F}^{\overline{m}} \partial_{\overline{m}} Z\right) \\ & - \overline{F}^{\overline{m}} F^n \partial_{\overline{m}} \partial_n Z \Bigg\} \ , \end{aligned} \tag{4.32}$$

where

$$\widehat{\mu} = \frac{\hat{W}^*}{|\hat{W}|} e^{\hat{K}/2} \mu + m_{3/2} Z - \overline{F}^{\overline{m}} \partial_{\overline{m}} Z \ . \tag{4.33}$$

Notice that the scalar masses are automatically *universal* in this model. Further simplifications can be obtained if $Z = 0$ and if the superpotential parameters $Y_{\alpha\beta\gamma}$ and $\mu$ do not depend on the hidden sector fields. Under such assumptions, which are common in the literature, (4.31) and (4.32) generate universal $A$ parameters, as well as the relation

$$B = A - m_{3/2} \ . \tag{4.34}$$

Furthermore, if we assume $V_0 = 0$, then $m \equiv m_\alpha = m_{3/2}$ and the well known result for the $B$ parameter, $B = A - m$, is recovered. This supergravity model is attractive for its simplicity and for the natural explanation that it offers to the universality of the soft scalar masses.

We remark that although minimal (canonical) kinetic terms for hidden matter, $\hat{K}(h_m, h_m^*) = \sum_m h_m h_m^*$, are also usually assumed, we have seen that it is not a necessary condition in order to obtain the above results. Concerning the kinetic terms for vector multiplets, it can be seen from (4.7) that the minimal (canonical) choice $f_a = const.$ is not phenomenologically interesting, since it implies $M_a = 0$. Nonvanishing and universal gaugino masses can be obtained if all the $f_a$ have the same dependence on the hidden sector fields, i.e. $f_a(h_m) = c_a f(h_m)$ for the different gauge group factors of the theory. This is in fact what happens, at tree level, in supergravity models deriving from superstring theory, as we will see in the next section. As an additional comment, we stress that relation (4.34) depends on the particular mechanism that is used to generate the $\mu$ parameter. As a counter-example, notice that if one takes e.g. an $h_m$–dependent $\mu$ as in (4.16) with $\lambda = const.$, instead of taking $\mu = const.$, then (4.32) gives

$$B = 2m_{3/2} + \frac{V_0}{m_{3/2}} \ , \tag{4.35}$$

with $\widehat{\mu} = m_{3/2}\lambda$ from (4.33). Thus the relation (4.34) does not hold. The above result (4.35) can be obtained also if one takes $\mu = 0$ in the superpotential (4.20) and $Z = const.$ in the Kähler potential (4.19). This also follows from our discussion above about the equivalence between the two mechanisms when $Z$ is an analytic function.

(ii) *No-scale supergravity*

In no-scale supergravity models,[21] after the spontaneous breaking of SUSY, the tree-level potential vanishes identically along some directions. A simple example of this type of models has just one hidden field $h$, a Kähler potential (4.19) with

$$\hat{K} = -3 \ \log(h + h^*) \ , \ \tilde{K}_\alpha = (h + h^*)^{-1} \ , \tag{4.36}$$

and a superpotential (4.20) with a hidden field independent $\hat{W}$, i.e. $\hat{W} = const.$ Then, the attractive result of a *vanishing* (flat) tree-level effective potential for the hidden sector (4.14) is obtained

$$V_0 = 0 \ , \tag{4.37}$$

for all VEVs of $h$. On the other hand, the soft parameters, using (4.22), (4.23) and (4.24), are given by

$$m_\alpha^2 = 0 \,, \tag{4.38}$$

$$A_{\alpha\beta\gamma} = -m_{3/2}(h+h^*)\partial_h \log Y_{\alpha\beta\gamma} \,, \tag{4.39}$$

$$\begin{aligned} B = -\hat{\mu}^{-1} m_{3/2}(h+h^*)^2 \Bigg\{ & \frac{\hat{W}^*}{|\hat{W}|}(h+h^*)^{-3/2}\partial_h \mu \\ & + m_{3/2}\left[\partial_{h^*} Z + \partial_h Z + (h+h^*)\partial_h \partial_{h^*} Z\right]\Bigg\} \,, \end{aligned} \tag{4.40}$$

where

$$\widehat{\mu} = (h+h^*)\left[\frac{\hat{W}^*}{|\hat{W}|}(h+h^*)^{-3/2}\mu + m_{3/2}Z + m_{3/2}(h+h^*)\partial_{h^*} Z\right] . \tag{4.41}$$

Assuming now that the $\mu$ and $Z$ coefficients and the Yukawa couplings are hidden field independent, the well known result for the soft parameters is recovered:

$$m_\alpha = A_{\alpha\beta\gamma} = B = 0 \,. \tag{4.42}$$

Although the above parameters are vanishing at the high scale, gaugino masses (4.7) can induce non-vanishing values at the electroweak scale due to radiative corrections.

In conclusion, both supergravity models considered in this section are interesting and give rise to concrete predictions for the soft parameters. However, one can think of many possible supergravity models (with different $K$, $W$ and $f$) leading to *different* results for the soft terms. This arbitrariness, as we will see in the next section, can be ameliorated in supergravity models deriving from superstring theory, where $K$, $f$, and the hidden sector are more constrained. We can already anticipate, however, that in such a context the kinetic terms are generically *not* canonical. Besides, although Kähler potentials of the no-scale type may appear at tree-level, the superpotentials are in general hidden field *dependent*. Moreover, the Yukawa couplings $Y_{\alpha\beta\gamma}$ and the bilinear coefficients $\mu$ and $Z$ are also generically hidden field *dependent*.

Finally, we remark that further constraints on the soft parameter space of the MSSM can be obtained if one wishes to avoid low-energy charge and color breaking minima deeper than the standard vacuum.[22] On these grounds, and assuming also radiative symmetry breaking with nothing but the MSSM in between the weak scale and the Planck scale, e.g. large regions in the parameter space ($m$, $M$, $A$, $B$) of the minimal supergravity model i) are forbidden. In the limiting case $m = 0$ the whole parameter space turns out to be excluded. This has obvious implications, e.g. for the no-scale supergravity model ii). If the same kind of analysis is applied to the soft parameters of superstring models, again strong constraints can be obtained, as we will comment below.

### 4.3. Soft Terms from Superstring Theory

#### 4.3.1. *General parametrization of SUSY breaking*

We are going to consider N=1 four-dimensional superstrings where the rôle of hidden sector fields is effectively played by $r$ moduli fields $T_i$, $i = 1, ..., r$ and the dilaton field $S$, i.e. $h_m = S, T_i$ following the notation of the previous section. We recall that we are denoting the $T$- and $U$-type (Kähler class and complex structure in the Calabi-Yau language) moduli collectively by $T_i$. The associated effective N=1 supergravity Kähler potentials (4.3), to lowest order in the matter fields, are of the type:

$$K = \hat{K}(S,S^*,T_i,T_i^*) \;+\; \tilde{K}_{\overline{\alpha}\beta}(T_i,T_i^*)C^{*\overline{\alpha}}C^{\beta} \\ + \;\left[\frac{1}{2}Z_{\alpha\beta}(T_i,T_i^*)C^{\alpha}C^{\beta} + h.c.\right] \;, \tag{4.43}$$

where at the superstring tree level

$$\hat{K}(S,S^*,T_i,T_i^*) = -\log(S+S^*) \;+\; \hat{K}(T_i,T_i^*) \;. \tag{4.44}$$

The first piece in (4.44) is the usual term corresponding to the complex dilaton $S$ that is present for any compactification. The second piece is the Kähler potential of the moduli fields, which in general depends on the compactification scheme and can be a complicated function. For the moment we leave it generic, but in the next subsection we will analyze some specific classes of superstring models where it has been computed. The same comment applies to $\tilde{K}_{\overline{\alpha}\beta}(T_i,T_i^*)$ and $Z_{\alpha\beta}(T_i,T_i^*)$. In the case of the superpotential (4.2), $Y_{\alpha\beta\gamma}(T_i)$ is also independent of $S$, but the non-perturbative contributions $\hat{W}(S,T_i)$ and $\mu_{\alpha\beta}(S,T_i)$ may depend in general on both $S$ and $T_i$. Finally, for any four-dimensional superstring the tree-level gauge kinetic function is independent of the moduli sector and is simply given by

$$f_a = k_a S \;, \tag{4.45}$$

where $k_a$ is the Kac–Moody level of the gauge factor. Usually (level one case) one takes $k_3 = k_2 = \frac{3}{5}k_1 = 1$, but this is irrelevant for our tree-level computation since $k_a$ will not contribute to the soft parameters.

As we will show below, it is important to know what fields, either $S$ or $T_i$, play the predominant role in the process of SUSY breaking. This will have relevant consequences in determining the pattern of soft parameters, and therefore the spectrum of physical particles.[6] That is why it is very useful to introduce the following parametrization, consistent with (4.14), for the VEVs of dilaton and moduli auxiliary fields

$$\begin{aligned} F^S &= \sqrt{3} C m_{3/2} K_{\overline{S}S}^{-1/2} \sin\theta e^{-i\gamma_S} \quad , \\ F^i &= \sqrt{3} C m_{3/2} \cos\theta P^{i\overline{j}} \Theta_{\overline{j}} \ , \end{aligned} \tag{4.46}$$

where the constant $C$ is defined as follows

$$C^2 = 1 + \frac{V_0}{3m_{3/2}^2} \ . \tag{4.47}$$

This parametrization is valid for the general case of off-diagonal moduli metric, since $P$ is a matrix canonically normalizing the moduli fields, i.e. $P^\dagger \hat{K} P = 1$ where $\hat{K} \equiv \hat{K}_{\overline{i}j}$ and 1 stands for the unit matrix. The angle $\theta$ and the complex parameters $\Theta_{\overline{j}}$ just parametrize the direction of the goldstino in the $S$, $T_i$ field space (see below (4.5)) and $\sum_j \Theta_j^* \Theta_{\overline{j}} = 1$. We have also allowed for the possibility of some complex phases which could be relevant for the CP structure of the theory (see Refs. 6, 17, 18, 20, 23, 24 for a discussion on this point). Notice that if the tree-level cosmological constant $V_0$ is assumed to vanish, one has $C = 1$, but we prefer for the moment to leave it undetermined as we did in the previous section (see below (4.14)).

Notice that such a phenomenological approach allows us to 'reabsorb' (or circumvent) our ignorance about the (nonperturbative) $S$- and $T_i$- dependent part of the superpotential (4.2), $\hat{W}(S, T_i)$, which is responsible for SUSY breaking.

It is now a straightforward exercise, plugging (4.43), (4.44), (4.45) and (4.46) into (4.7), (4.11), (4.12) and (4.13), to compute the soft SUSY-breaking parameters as functions of $\theta$ and $\Theta_{\overline{j}}$. On the one hand, since the tree-level gauge kinetic function is given for any four-dimensional superstring by (4.45), the tree-level gaugino masses are universal, independent of the moduli sector, and simply given by:

$$M_a = \sqrt{3} C m_{3/2} \sin\theta e^{-i\gamma_S} \ . \tag{4.48}$$

On the other hand, the bosonic soft parameters depend in general on the moduli sector (i.e. on the functions $\tilde{K}_{\overline{\alpha}\beta}, Z_{\alpha\beta}(T_i, T_i^*), \ldots$ and on the parameters $\cos\theta$ and $\Theta_{\overline{j}}$) and therefore they should be studied in the context of specific classes of superstring models. However, we will first focus on the very interesting limit $\cos\theta = 0$, which corresponds to the case where the dilaton sector is the source of all the SUSY breaking (see (4.46)) and the results are compactification independent.

*Dilaton SUSY breaking*

Since the dilaton couples in a universal manner to all particles, this limit is quite model *independent.*[6,13] Indeed, the expressions for all the soft parameters (except $B$) are quite simple and independent of the four-dimensional superstring considered. After canonically normalizing the fields, one obtains:

$$m_\alpha^2 = m_{3/2}^2 + V_0 , \tag{4.49}$$

$$M_a = \sqrt{3} C m_{3/2} e^{-i\gamma_S} , \tag{4.50}$$

$$A_{\alpha\beta\gamma} = -M_a , \tag{4.51}$$

$$\begin{aligned} B = \widehat{\mu}^{-1}(\tilde{K}_{H_1}\tilde{K}_{H_2})^{-1/2} \left\{ \frac{\hat{W}^*}{|\hat{W}|} e^{\hat{K}/2} \mu m_{3/2} \left(-1 \right.\right. \\ \left.\left. - \sqrt{3} C e^{-i\gamma_S} \left[1 - (S+S^*)\partial_S \log\mu\right]\right) + Z\left(2m_{3/2}^2 + V_0\right) \right\} , \end{aligned} \tag{4.52}$$

where

$$\widehat{\mu} = \left( \frac{\hat{W}^*}{|\hat{W}|} e^{\hat{K}/2} \mu + m_{3/2} Z \right) (\tilde{K}_{H_1}\tilde{K}_{H_2})^{-1/2} . \tag{4.53}$$

Although the general expression for $B$ is more involved than the ones of the other soft parameters, a considerable simplification occurs if $Z$ is the only source of the $\mu$ term. In this case $B$ reduces to

$$B = 2m_{3/2} + \frac{V_0}{m_{3/2}} . \tag{4.54}$$

and thus becomes independent of the four-dimensional superstring considered, as the other parameters. It is easy to check that the same result (4.54) is also obtained if $Z = 0$ and the superpotential contains a $\mu$ coefficient of the form (4.16), where now $\mu = \lambda(T_i)\hat{W}(S, T_i)$. Notice that the expressions for $m_\alpha$ (4.49) and $B$ (4.54) coincide with the corresponding ones obtained in the minimal supergravity model i), (4.30) and (4.35) respectively. Furthermore, if $Z = 0$ and $\partial_S \mu = 0$, the expression for $B$ obtained from (4.52) coincides with the corresponding one (4.34) of the minimal supergravity model.

This dilaton-dominated scenario is attractive for its simplicity and for the natural explanation that it offers to the *universality* of the soft terms. For possible explicit SUSY–breaking mechanisms where this limit might be obtained see Ref. 25. Because of the simplicity of this scenario, the low-energy predictions are quite precise.[6,26–28] Assuming a vanishing cosmological constant and imposing, e.g. from the limits on the electric dipole moment of the neutron, $\gamma_S = 0 \bmod \pi$ (4.49), (4.50) and (4.51) give [a]

$$m_\alpha = m_{3/2} \ , \ M_a = \pm\sqrt{3} m_{3/2} \ , \ A_{\alpha\beta\gamma} = -M_a \ . \qquad (4.55)$$

Since scalars are lighter than gauginos at the high scale and all mass ratios are fixed, at low-energy ($\sim M_Z$) one finds the following mass ratios for the gluino, slepton and squark (except stops and left sbottom) masses

$$M_g : m_{Q_L} : m_{u^c_L} : m_{d^c_L} : m_{L_L} : m_{e^c_L} \simeq 1 : 0.94 : 0.92 : 0.92 : 0.32 : 0.24 \ , \qquad (4.56)$$

as can be computed e.g. from (4.17). Although squarks and sleptons have the same soft mass at the high scale, at low-energy the former are much heavier than the latter because of the gluino contribution to the renormalization of their masses. The rest of the spectrum is very dependent on the details of $SU(2)_L \times U(1)_Y$ breaking and therefore on the values of $B$ and $\widehat{\mu}$. For $B = 2m_{3/2}$ (see(4.54)) and $\mu$ treated as a free parameter this analysis can be found in Ref. 26. Modifications to this scenario due to the effect of possible superstring non–perturbative corrections to the Kähler potential can be found in Ref. 30.

It is worth noticing here that, although the value of $\widehat{\mu}$ (4.53) is compactification dependent even in this dilaton-dominated scenario, the simple result $\widehat{\mu} = m_{3/2}$ can be obtained in any compactification scheme where the source of $\widehat{\mu}$ is a $Z$ term in the Kähler potential fulfilling the property $Z = (\tilde{K}_{H_1} \tilde{K}_{H_2})^{1/2}$. In fact, we will see in the next subsection that this is the case of some classes of orbifold models. Notice that, when such a property holds, the whole SUSY spectrum depends only on one parameter,

[a] It is worth remarking that these particular boundary conditions have also interesting finiteness properties. In particular, they preserve one-loop finiteness of $N = 1$ finite theories.[29]

$m_{3/2}$, since

$$m_\alpha = m_{3/2} \ , \ M_a = \pm\sqrt{3}m_{3/2} \ , \ A_{\alpha\beta\gamma} = -M_a \ , \ B = 2m_{3/2} \ , \ \widehat{\mu} = m_{3/2} \ . \tag{4.57}$$

Besides, this parameter can be fixed from the phenomenological requirement of correct electroweak breaking $2M_W^2/g_2^2 = \langle |H_1|^2\rangle + \langle |H_2|^2\rangle$. Thus at the end of the day we are left essentially with no free parameters. In Ref. 31 the consistency of the above boundary conditions with the appropriate radiative electroweak symmetry breaking was explored. Unfortunately, it was found that they are not consistent with the measured value of the top-quark mass, namely the mass obtained in this scheme turns out to be too small. A possible way-out to this situation is to assume that also the moduli fields contribute to SUSY breaking, since the soft terms are then modified. Of course, this amounts to a departure of the pure dilaton-dominated scenario. This possibility will be discussed in the context of orbifold models in the next subsection.

Finally, we recall that the phenomenological problem of the pure dilaton-dominated limit mentioned above is also obtained in a different context, namely from requiring the absence of low-energy charge and color breaking minima deeper than the standard vacuum.[32] In fact, on these grounds, the dilaton-dominated limit is excluded not only for a $\mu$ term generated through the Kähler potential but for any possible mechanism solving the $\mu$ problem. The results indicate that the whole free parameter space ($m_{3/2}$, $B$, $\mu$) is excluded after imposing the present experimental data on the top mass. Again this rests on the assumption of radiative symmetry breaking with nothing but the MSSM in between the weak scale and the superstring scale.

*Dilaton/Moduli SUSY breaking*

In general the moduli fields $T_i$ may also contribute to SUSY breaking, i.e. $F^i \neq 0$ in (4.46), and therefore their effects on soft parameters must also be included.[6,8,19,33,34] In this sense it is interesting to note that explicit possible scenarios of SUSY breaking by gaugino condensation in superstrings, when analyzed at the one–loop level, lead to the mandatory inclusion of the moduli in the game (in fact the moduli are the main source of SUSY breaking in these cases).[35] Since different compactification schemes give rise to different expressions for the moduli-dependent part of the Kähler potential (4.43), the computation of the bosonic soft parameters will be model *dependent*. The results are discussed below in the context of some specific superstring models.

### 4.3.2. *Superstring models*

To illustrate the main features of mixed dilaton/moduli SUSY breaking, we will concentrate mainly on the case of diagonal moduli and matter metrics. For instance, under this assumption the parametrization (4.46) is simplified to

$$\begin{aligned} F^S &= \sqrt{3} C m_{3/2} \hat{K}_{\overline{S}S}^{-1/2} \sin\theta e^{-i\gamma_S} \quad , \\ F^i &= \sqrt{3} C m_{3/2} \hat{K}_{\overline{i}i}^{-1/2} \cos\theta \Theta_i e^{-i\gamma_i} \ , \end{aligned} \tag{4.58}$$

where $\sum_i \Theta_i^2 = 1$. Although this is the generic case e.g. in most orbifolds, off–diagonal metrics are present in general in Calabi–Yau compactifications. This may lead to FCNC effects in the low–energy effective N=1 softly broken Lagrangian. The analysis of soft SUSY-breaking parameters in Calabi–Yau compactifications is therefore more involved and can be found in Ref. 36 using parametrization (4.46). A similar analysis for the few orbifolds with off–diagonal metrics was carried out in Ref. 34. Some comments about the "off-diagonal" results will be made below. Also in the case of orbifold compactifications with continuous Wilson lines off-diagonal moduli metrics arise, due to the moduli–Wilson line mixing. However, this analysis turns out to be simple[37] and the results are similar to the ones studied below in the diagonal case.

Since the moduli part of the Kähler potential (4.43) has been computed for $(0, 2)$ symmetric Abelian orbifolds, we will concentrate here on these models. They contain generically three $T$-type moduli (the exceptions are the orbifolds $Z_3$, $Z_4$ and $Z_6'$, which have 9, 5 and 5 respectively, and are precisely the ones with off-diagonal metrics) and, at most, three $U$-type moduli. We will denote them collectively by $T_i$, where e.g. $T_i = U_{i-3}$; $i = 4, 5, 6$. For this class of models the Kähler potential has the form

$$K = -\log(S + S^*) - \sum_i \log(T_i + T_i^*) + \sum_\alpha |C^\alpha|^2 \Pi_i (T_i + T_i^*)^{n_\alpha^i} \ . \tag{4.59}$$

Here $n_\alpha^i$ are (zero or negative) fractional numbers usually called "modular weights" of the matter fields $C^\alpha$. For each given Abelian orbifold, independently of the gauge group or particle content, the possible values of the modular weights are very restricted. For a classification of modular weights for all Abelian orbifolds see Ref. 38. The piece proportional to $Z_{\alpha\beta}$

in (4.43) has been shown to be present in Calabi–Yau compactifications and orbifolds. In particular, in the case of orbifolds, such a term arises when the untwisted sector has at least one complex–structure field $U$ and has been explicitly computed. We will analyze separately this case below, as well as the associated $\mu$ and $B$ parameters, whereas we will concentrate here on the other bosonic soft parameters. Plugging the particular form (4.59) of the Kähler potential and the parametrization (4.58) in (4.22) and (4.23) we obtain the following results for the scalar masses and trilinear parameters:[19,33,34]

$$m_\alpha^2 = m_{3/2}^2 \left(1 + 3C^2 \cos^2\theta \; \vec{n_\alpha}.\vec{\Theta^2}\right) \; + \; V_0 \; , \tag{4.60}$$

$$A_{\alpha\beta\gamma} = -\sqrt{3} C m_{3/2} \Bigg( \sin\theta e^{-i\gamma_S} + \cos\theta \sum_{i=1}^{6} e^{-i\gamma_i} \Theta_i \big[ 1 + n_\alpha^i + n_\beta^i + n_\gamma^i - (T_i + T_i^*) \partial_i \log Y_{\alpha\beta\gamma} \big] \Bigg) \, . \tag{4.61}$$

It is easy to check that the results (4.49) and (4.51) are recovered in the limit where $\cos\theta \to 0$. Notice that neither the scalar (4.60) nor the gaugino masses (4.48) have any explicit dependence on $S$ or $T_i$: they only depend on the gravitino mass and the goldstino angles. This is one of the advantages of a parametrization in terms of such angles. Although in the case of the $A$-parameter an explicit $T_i$-dependence may appear in the term proportional to $\partial_i \log Y_{\alpha\beta\gamma}$, it disappears in several interesting cases.[34] Using the above information, we can now analyze the structure of soft parameters available in Abelian orbifolds.

In the dilaton-dominated case ($\cos\theta = 0$) the soft parameters are universal, as already studied in the previous section. However, in general, they show a lack of universality due to the modular weight dependence (see (4.60) and (4.61)). So, even with diagonal matter metrics, FCNC effects may appear. However, we recall that the low-energy running of the scalar masses has to be taken into account. In particular, in the squark case, for gluino masses heavier than (or of the same order as) the scalar masses at the boundary scale, there are large flavour-independent gluino loop contributions which are the dominant source of scalar masses (see (4.17)). We will show below that this situation is very common in orbifold models. The above effect can therefore help in fulfilling the FCNC constraints.

Another feature of the case under study is that, depending on the goldstino direction, tachyons may appear. For $\cos^2\theta \geq 1/3$, the goldstino direction cannot be chosen arbitrarily if one is interested in avoiding tachyons (see (4.60)). Nevertheless, having a tachyonic sector is not necessarily a problem, it may even be an advantage.[34] In the case of superstring GUTs (or the standard model with extra U(1) interactions), the negative squared mass may just induce gauge symmetry breaking by forcing a VEV for a particular scalar, GUT-Higgs field, in the model. The latter possibility provides us with interesting phenomenological consequences: the breaking of SUSY could directly induce further gauge symmetry breaking.

Finally, let us consider three particles $C^\alpha$, $C^\beta$ and $C^\gamma$, coupled through a Yukawa $Y_{\alpha\beta\gamma}$. They may belong both to the untwisted (**U**) sector or to a twisted (**T**) sector, i.e. we consider couplings of the type **UUU**, **UTT**, **TTT**. Then, using the above formulae (4.60) and (4.48), with negligible $V_0$, one finds[34] that in general for *any* choice of goldstino direction

$$m_\alpha^2 + m_\beta^2 + m_\gamma^2 \leq |M_a|^2 = 3m_{3/2}^2 \sin^2\theta \ . \tag{4.62}$$

Remarkably, the same sum rule is fulfilled even in the presence of off-diagonal metrics, as it is the case of the orbifolds $Z_3$, $Z_4$ and $Z_6'$. The three scalar mass eigenvalues will be in general non-degenerate, which in turn may induce FCNC. This can be automatically avoided in the dilaton dominated limit or under special conditions (for instance, when $\hat{W}$ does not depend on the moduli, a no-scale scenario arises and the mass eigenvalues vanish). The same problem is present in Calabi-Yau compactifications, where again the mass eigenvalues are typically non-degenerate. Besides, the sum rule (4.62) is violated in general.[36] Coming back to the orbifold case, notice that the above sum rule implies that on average scalars are lighter than gauginos. For small $\sin\theta$, some particular scalar mass may become bigger than the gaugino mass, but in that case at least one of the other scalars involved in the sum rule would be forced to have a *negative* squared mass. This situation is quite dangerous in the context of standard model four-dimensional superstrings, since some observable particles, like Higgses, squarks or sleptons, could be forced to acquire large VEVs (of order the superstring scale). If the above sum rule is applied and squared soft masses are (conservatively) required to be non-negative in order to avoid instabilities of the scalar potential, then the tree level soft masses of observable scalars are constrained to be always smaller than gaugino masses at the boundary scale:

$$m_\alpha < M_a \ . \tag{4.63}$$

In turn, this implies that, at low-energy ($\sim M_Z$), the masses of gluinos, sleptons and squarks (except stops and left sbottom) are ordered as

$$m_l \ < \ m_q \ \simeq M_g \ , \qquad (4.64)$$

where gluinos are slightly heavier than scalars. Therefore, in spite of the different set of (non-universal) soft scalar masses, the low-energy phenomenological predictions of the mixed dilaton/moduli SUSY breaking become qualitatively similar to those of the pure dilaton-dominated SUSY breaking. This holds especially for the squark masses, as follows e.g. from (4.63) and (4.17). In the case of sleptons, which do not feel gluino loop effects, the boundary values of the soft masses (4.63) can be relatively more important at low-energy, and larger deviations from the numerical results of (4.56) can be obtained. Analyses of the low-energy predictions of the dilaton/moduli scenario taking account of the radiative symmetry breaking can be found in Refs. 6, 28, 39.

Before concluding, we recall that exceptions to the above pattern (4.63), (4.64) can arise in several situations.[34] For instance, since the total squared Higgs masses receive a positive contribution $\mu^2$, the corresponding soft masses may be allowed to be negative: in this case the restrictions from the sum rule would be relaxed. Another example concerns MSSM Yukawa couplings that arise effectively from higher dimension operators: in this case the three-particle sum rule itself may not hold. Finally, a departure from relations (4.63) and (4.64) can also arise when both scalar and gaugino masses vanish at tree level. Such a vanishing can happen in the fully moduli-dominated SUSY breaking, e.g. if SUSY breaking is equally shared among $T_1, T_2, T_3$ and one consider untwisted particles: then superstring loop effects become important and tend to make scalars heavier than gauginos.[6] In any event, we stress again that potential violations of the pattern in (4.63) and (4.64) can occur only when SUSY breaking is mainly moduli dominated (specifically, $\cos^2\theta \geq 2/3$), since only in this case the gaugino masses can decrease below $m_{3/2}$ and possibly become lighter than some scalar mass.

*The B parameter and the $\mu$ problem*

As already discussed in section 2.1, the two mechanisms to solve the $\mu$ problem in the context of supergravity are *naturally* present in superstring models. We will concentrate here on the case in which $\mu$ arises from a bilinear term in the Kähler potential (4.3). The alternative mechanism which generates $\mu$ from the superpotential, as in (4.16), may also be present in orbifolds, but the results are more model dependent. They can be found in Ref. 34. We recall that, in any orbifold with at least one complex-structure field $U$, the Kähler potential of the untwisted sector possesses the structure $Z(T_i, T_i^*)C_1C_2 + h.c.$[14,15] and can therefore generate a $\mu$ term. Specifically, the $Z_N$ orbifolds based on $Z_4$, $Z_6$, $Z_8$, $Z_{12}'$ and the $Z_N \times Z_M$ orbifolds based on $Z_2 \times Z_4$ and $Z_2 \times Z_6$ do all have a $U$-type field in (say) the third complex plane. In addition, the $Z_2 \times Z_2$ orbifold has $U$ fields in the three complex planes. In all these models the piece of the Kähler potential involving the moduli and the untwisted matter fields $C_{1,2}$ in the third complex plane has the form

$$K_3 = -\log\left[(T_3 + T_3^*)(U_3 + U_3^*) - (C_1 + C_2^*)(C_1^* + C_2)\right] \tag{4.65}$$

$$\simeq -\log(T_3 + T_3^*) - \log(U_3 + U_3^*) + \frac{(C_1 + C_2^*)(C_1^* + C_2)}{(T_3 + T_3^*)(U_3 + U_3^*)} . \tag{4.66}$$

Now, from the expansion shown in (4.66), one can easily read off the functions $Z$, $\tilde{K}_1$, $\tilde{K}_2$ associated to $C_1$ and $C_2$:

$$Z = \tilde{K}_1 = \tilde{K}_2 = \frac{1}{(T_3 + T_3^*)(U_3 + U_3^*)} . \tag{4.67}$$

Let us assume that the MSSM can be obtained from a superstring model of the kind mentioned above and let us identify the fields $C_1$ and $C_2$ with the electroweak Higgs fields $H_1$ and $H_2$. Plugging back the expressions (4.67) in (4.24) and (4.28) with $\mu = 0$, and using the parametrization (4.58), one can compute $\mu$ and $B$ for this interesting class of models:[34]

$$\widehat{\mu} = m_{3/2}\left[1 + \sqrt{3}C\cos\theta(e^{i\gamma_3}\Theta_3 + e^{i\gamma_6}\Theta_6)\right] , \tag{4.68}$$

$$\begin{aligned} B\widehat{\mu} = 2m_{3/2}^2 &\Big[1 + \sqrt{3}C\cos\theta(\cos\gamma_3\Theta_3 + \cos\gamma_6\Theta_6) \\ &+ 3C^2\cos^2\theta\cos(\gamma_3 - \gamma_6)\Theta_3\Theta_6\Big] + V_0 . \end{aligned} \tag{4.69}$$

Notice that, in the limit where $\cos\theta \to 0$, the results in (4.57) are recovered. In addition, we recall from (4.60) that the soft masses are

$$m_{H_1}^2 = m_{H_2}^2 = m_{3/2}^2\left[1 - 3C^2\cos^2\theta(\Theta_3^2 + \Theta_6^2)\right] + V_0 . \tag{4.70}$$

In general, the quadratic part of the Higgs potential after SUSY breaking has the form (see (4.21))

$$V_2 = (m_{H_1}^2 + |\widehat{\mu}|^2)|\widehat{H}_1|^2 + (m_{H_2}^2 + |\widehat{\mu}|^2)|\widehat{H}_2|^2 + (B\widehat{\mu}\widehat{H}_1\widehat{H}_2 + h.c.), \tag{4.71}$$

where we recall that $\hat{H}_1$ and $\hat{H}_2$ are the canonically normalized Higgs fields. In the specific case under consideration, from (4.68), (4.69) and (4.70) we find the remarkable result that the three coefficients in $V_2$ are equal, i.e.

$$m_{H_1}^2 + |\widehat{\mu}|^2 = m_{H_2}^2 + |\widehat{\mu}|^2 = B\widehat{\mu} . \tag{4.72}$$

so that $V_2$ has the simple form

$$V_2 = B\widehat{\mu}\,(\widehat{H}_1 + \widehat{H}_2^*)(\widehat{H}_1^* + \widehat{H}_2)\ , \tag{4.73}$$

and therefore $\tan\beta = \frac{<\hat{H}_2>}{<\hat{H}_1>} = -1$. Of course, this corresponds to the boundary condition on the scalar potential at the superstring scale: at lower energies the renormalization group equations should be used. Although the common value of the three coefficients in (4.72) depends on the Goldstino direction via the parameters $\cos\theta$, $\Theta_3$, $\Theta_6, \ldots$ (see e.g. the expression of $B\widehat{\mu}$ in (4.69)), we stress that the equality itself and the form of $V_2$ hold *independently* of the Goldstino direction.

Starting from such 'superstringy' boundary conditions for the MSSM parameters, one can explore their consistency with radiative electroweak-symmetry breaking[31] (see also Ref. 40). One finds that consistency with the measured value of the top-quark mass cannot be achieved in the dilaton-dominated scenario (as already mentioned in section 3.1). The only SUSY-breaking scenario compatible with such constraints requires a suppressed dilaton contribution and important (often dominant) contributions from the $T_3$, $U_3$ moduli.

### 4.4. Final Comments and Outlook

It is worth remarking that most of the above results for soft terms in superstring models refer to certain simple *perturbative* heterotic compactifications. In addition, it is assumed that the goldstino is a fermionic partner of some combination of the dilaton and/or the moduli fields. Recently some information about the non-perturbative regime in superstring theory has been obtained in terms of the S-dualities[41] of the theory. All superstring theories seem to correspond to some points in the parameter space of a unique eleven-dimensional underlying theory, M-theory.[42] Although the structure of this theory is largely unknown, some preliminary attempts have been made to extract some information of phenomenological interest. A scenario to understand the difference between the GUT scale and the superstring scale has been put forward.[43] Supersymmetry breaking and other phenomenological issues have also been explored within this context in Ref. 44. Studies in progress concerning *non-perturbative* superstring vacua with $N = 1$ SUSY will certainly bring us new surprises.

## References


1. For a review, see: H.P. Nilles, *Phys. Rep.* **110** (1984) 1, and references therein.
2. For a recent review, see: C. Muñoz, *hep-th/9507108*, and references therein.
3. For a review, see: C. Muñoz, *hep-ph/9601325*, and references therein.
4. Y. Kawamura, T. Kobayashi and T. Komatsu, *hep-ph/9609462*, and references therein.
5. S.K. Soni and H.A. Weldon, *Phys. Lett.* **B126** (1983) 215.
6. A. Brignole, L.E. Ibáñez and C. Muñoz, *Nucl. Phys.* **B422** (1994) 125 [Erratum: **B436** (1995) 747].
7. K. Choi, J.E. Kim and H.P. Nilles, *Phys. Rev. Lett.* **73** (1994) 1758; K. Choi, J.E. Kim and G.T. Park, *Nucl. Phys.* **B442** (1995) 3.
8. S. Ferrara, C. Kounnas and F. Zwirner, *Nucl. Phys.* **B429** (1994) 589 [Erratum: **B433** (1995) 255].
9. K. Choi, J.S. Lee and C. Muñoz, *hep-ph/9709250.*
10. G.F. Giudice and A. Masiero, *Phys. Lett.* **B206** (1988) 480.
11. J.A. Casas and C. Muñoz, *Phys. Lett.* **B306** (1993) 288.
12. J.E. Kim and H.P. Nilles, *Phys. Lett.* **B138** (1984) 150, *Phys. Lett.* **B263** (1991) 79; E.J. Chun, J.E. Kim and H.P. Nilles, *Nucl. Phys.* **B370** (1992) 105.
13. V.S. Kaplunovsky and J. Louis *Phys. Lett.* **B306** (1993) 269.
14. G. Lopes-Cardoso, D. Lüst and T. Mohaupt, *Nucl. Phys.* **B432** (1994) 68.
15. I. Antoniadis, E. Gava, K.S. Narain and T.R. Taylor, *Nucl. Phys.* **B432** (1994) 187.
16. A. Lleyda and C. Muñoz, *Phys. Lett.* **B317** (1993) 82.
17. D. Choudhury, F. Eberlein, A. Köning, J. Louis and S. Pokorski, *Phys. Lett.* **B342** (1995) 180.
18. J. Louis and Y. Nir, *Nucl. Phys.* **B447** (1995) 18.
19. P. Brax and M. Chemtob, *Phys.Rev.* **D51** (1995) 6550.
20. P. Brax and C.A. Savoy, *Nucl. Phys.* **B447** (1995) 227.
21. For a review, see: A.B. Lahanas and D.V. Nanopoulos, *Phys. Rep.* **145** (1987) 1, and references therein.
22. J.A. Casas, A. Lleyda and C. Muñoz, *Nucl. Phys.* **B471** (1996) 3, *Phys. Lett.* **B389** (1996) 305.
23. K. Choi, *Phys. Rev. Lett.* **72** (1994) 1592.
24. B. Acharya, D. Bailin, A. Love, W.A. Sabra and S. Thomas, *Phys. Lett.* **B357** (1995) 387; D. Bailin, G.V. Kraniotis and A. Love, *hep-th/9705244*, *hep-th/9707105.*
25. A. de la Macorra and G.G. Ross, *Nucl. Phys.* **B404** (1993) 321; V. Halyo and E. Halyo, *Phys. Lett.* **B382** (1996) 89.
26. R. Barbieri, J. Louis and M. Moretti, *Phys. Lett.* **B312** (1993) 451 [Erratum: **B316** (1993) 632].

27. J.L. Lopez, D.V. Nanopoulos and A. Zichichi, *Phys. Lett.* **B319** (1993) 451.
28. S. Khalil, A. Masiero and F. Vissani *Phys. Lett.* **B375** (1996) 154.
29. See e.g. L.E. Ibáñez, *hep-th/9505098*, Proc. of Strings 95, World Scientific (1995), and references therein.
30. J.A. Casas, *Phys. Lett.* **B384** (1996) 103.
31. A. Brignole, L.E. Ibáñez and C. Muñoz, *Phys. Lett.* **B387** (1996) 305.
32. J.A. Casas, A. Lleyda and C. Muñoz, *Phys. Lett.* **B380** (1996) 59.
33. T. Kobayashi, D. Suematsu, K. Yamada and Y. Yamagishi, *Phys. Lett.* **B348** (1995) 402.
34. A. Brignole, L.E. Ibáñez, C. Muñoz and C. Scheich, *Z. Phys.* **C74** (1997) 157.
35. A. Font, L.E. Ibañez, D. Lüst and F. Quevedo, *Phys. Lett.* **B245** (1990) 401; S. Ferrara, N. Magnoli, T.R. Taylor and G. Veneziano, *Phys. Lett.* **B245** (1990) 409; M. Cvetic, A. Font, L.E. Ibañez, D. Lüst and F. Quevedo, *Nucl. Phys.* **B361** (1991) 194; B. de Carlos, J.A. Casas and C. Muñoz, *Phys. Lett.* **B299** (1993) 234, *Nucl. Phys.* **B399** (1993) 623; A. de la Macorra and G.G. Ross, *Phys. Lett.* **B325** (1994) 85.
36. H.B. Kim and C. Muñoz, *Z. Phys.* **C75** (1997) 367.
37. H.B. Kim and C. Muñoz, *Mod. Phys. Lett.* **A12** (1997) 315.
38. L.E. Ibáñez and D. Lüst, *Nucl. Phys.* **B382** (1992) 305.
39. C.-H. Chen, M. Drees and J.F. Gunion, *Phys. Rev.* **D55** (1997) 330; Y. Kawamura, S. Khalil and T. Kobayashi, *hep-ph/9703239*; A. Love and P. Stadler, *hep-ph/9709234*.
40. Y. Kawamura, T. Kobayashi and M. Watanabe, *DPSU-97-5, hep-ph/9609462*.
41. A. Font, L.E. Ibáñez, D. Lüst and F. Quevedo, *Phys. Lett.* **B249** (1990) 35; A. Sen, *Int.J. Mod.Phys.* **A9** (1994) 3707.
42. See e.g. J.H. Schwarz, *hep-th/9607201*, P.K. Townsend, *hep-th/9612121*, and references therein.
43. E. Witten, *Nucl. Phys.* **B471** (1996) 135.
44. T. Banks and M. Dine, *hep-th/9605136*, *hep-th/9608197*, *hep-th/9609046*; E. Caceres, V.S. Kaplunovsky and I.M. Mandelberg, *hep-th/9606036*; P. Horava, *hep-th/9608019*; T. Li, J. Lopez and D.V. Nanopoulos, *hep-ph/9702237*, *hep-ph/9704247*; E. Dudas and C. Grojean, *hep-th/9704177*; I. Antoniadis and M. Quirós, *hep-th/9705037*, *hep-th/9707208*; K. Choi, *hep-th/9706171*; H.P. Nilles, M. Olechowski and M. Yamaguchi, *hep-th/9707143*; Z. Lalak and S. Thomas, *hep-th/9707223*; V. Kaplunovsky and J. Louis, *hep-th/9708049*; E. Dudas, *hep-th/9709043*.

# Mass Density of Neutralino Dark Matter

James D. Wells

*CERN, Theory Group (PH-TH),*
*CH-1211 Geneva 23, Switzerland*

*Michigan Center for Theoretical Physics,*
*University of Michigan, Ann Arbor, USA*

The lightest supersymmetric particle (LSP) is stable in an $R$-parity conserving theory. In this article the steps needed to calculate the present day mass density of such a particle are detailed. It is shown that there can be a significant amount of LSP dark matter in the universe. Furthermore, relic abundance considerations put an upper bound on how large supersymmetry breaking masses can be without resorting to finetuning arguments.

## 5.1. Introduction

The most general gauge invariant superpotential will generally lead to unacceptable fast proton decay. For this reason, a discrete symmetry must be posited that banishes baryon and/or lepton violating operators which contribute to this decay. The simplest such discrete symmetry is $R$-parity.[1] Exact $R$-parity conservation makes the lightest supersymmetric partner (LSP) stable, thereby introducing many interesting phenomena. The most studied phenomenon is the large missing energy signature associated with production of superpartners and subsequent decay into the LSP plus jets, photons, leptons, etc. For experimental reasons[2] and theoretical reasons[3] the LSP is now expected to be the lightest neutralino.

A stable LSP has more than collider physics consequences. If they are created in the hot and violent early days of the universe, then there should be some left over today, and perhaps they could have significant cosmological and astrophysical consequences. If it turns out that the LSP constitutes a significant mass fraction of the universe then it should be possible to witness large-scale gravitational effects of these particles on galaxies and clusters. Experiments have been testing large-scale gravity for decades now, and the current consensus states that there are non-luminous sources of gravitational import beyond the ordinary baryonic matter that makes up planets and stars.[4] Part of the evidence of additional mass-energy beyond our luminous matter includes rotation curves of galaxies and infall of clusters. The evidence for non-baryonic dark matter can be attributed to the successful agreement between measurement and theory in big bang nucleosynthesis (BBN). BBN tells us that baryonic mass fraction of the universe is probably less than 10%.[5] (I am making the usual assumption that the total energy density of the universe is the critical energy density in accord with a general inflation scenario.)

One is left wondering what the rest of the universe is made of. One suggestion is a weakly interacting massive object, or WIMP. The acronym WIMP is a good general label, but it is a bit misleading. Upon closer inspection a particle of $\mathcal{O}(m_W)$ mass which has full-strength $SU(2)$ (weak) interactions is generally not a good dark matter candidate, and yields a relic abundance less than is required to have astrophysical significance. Some additional suppressions are generally needed in the annihilation cross-section. As we will see in subsequent sections, the LSP generally has the right suppressions to make it a good dark matter candidate while at the same time having mass near the weak-scale. For this reason, I will not use the generic word WIMP, and instead refer to the dark matter candidate as the LSP. LSPs are excellent candidates for the dark matter because *if enough can be around* they allow conformance to experimentally determined properties of galactic rotation curves, structure formation and big bang nucleosynthesis. Furthermore, calculation of their relic abundance indicates that LSPs could contribute most of the energy density of the universe. This is a non-trivial separate test that confirms that *enough can be around* to solve the observational problems.

In the lightest neutralino, supersymmetry provides a natural dark matter candidate. In fits of optimism one could even declare the galactic rotation curve anomalies, etc. as positive experimental evidences for supersymmetry. However, in this chapter the focus will be on two main topics. First, and foremost, I will provide the details on how to calculate the relic abundance of a weakly interacting particle. Since obtaining the correct relic abundance is a somewhat involved calculation, but also an important one, it is useful to have detailed discussion. Second, I will tailor some additional remarks about the calculation to the lightest neutralino, and show how the results constrain *other* supersymmetric particle masses. In general it can be shown that there is an upper limit to superpartner masses due to relic abundance considerations alone. The limit is based on physical necessity and not on arbitrary fine-tuning considerations. This insight is as important as the realization that supersymmetry can cure the "dark matter problem."

The subsequent sections reflect the goals presented in the previous paragraph. Much of the emphasis will be placed in the techniques of calculating the relic abundance. I have attempted to include in this one source all the necessary general relativity, statistical mechanics, and particle physics knowledge needed to follow a precise calculation. I have also included a section which derives an accurate and approximate solution to the Boltzmann equation. These results will then be used to analyze quantitatively how the supersymmetric spectrum is affected by the relic abundance constraint.

## 5.2. Solving the Boltzmann Equation

The starting point is the Boltzmann equation,

$$\frac{dn}{dt} = -3Hn - \langle\sigma v\rangle(n^2 - n_{eq}^2) \tag{5.1}$$

where $H$ is the Hubble constant, $n$ is the particle number density in question, $n_{eq}$ is the particle number equilibrium density, and $\langle\sigma v\rangle$ is the thermal averaged cross-section. The Boltzmann equation is simple in form but somewhat subtle to solve. The differentiating parameter is time $t$, however $n$ and $n_{eq}$ are most easily characterized by temperature. Furthermore, the Hubble constant evolution is best traced by the relative time change in the scale parameter $H = \dot{a}/a$. Only one parameter is independent, and so the first step will be to cast the Boltzmann equation into a purely temperature dependent relation. The two equations that will allow us to do that are the "Friedmann equation" and the "conservation of entropy equation."

The use of Einstein's equation of general relativity is necessary to reveal the explicit time dependence of $\dot{a}/a$. This familiar equation states that

$$R_{\mu\nu} - \frac{1}{2}Rg_{\mu\nu} = 8\pi G_N T_{\mu\nu}. \tag{5.2}$$

To do explicit calculations the metric and stress-energy must be defined. We use the flat Robertson-Walker metric

$$ds^2 = dt^2 - a^2(t)\left[dx^2 + dy^2 + dz^2\right] \tag{5.3}$$

which assumes the universe to be flat, homogeneous, and isotropic. This is equivalent to the metric tensor

$$g_{\mu\nu} = \mathrm{diag}(1, -a^2(t), -a^2(t), -a^2(t)). \quad (5.4)$$

The general stress-energy consistent with an homogeneous and isotropic universe is

$$T_{\mu\nu} = \mathrm{diag}(\rho, -p, -p, -p). \quad (5.5)$$

Einstein's equation is of course valid for each $\{\mu\nu\}$, however we only need the 00 component. From the metric tensor it is straightforward[6] to compute the Ricci tensor component $R_{00}$ and the Ricci scalar $R$:

$$R_{00} = -3\frac{\ddot{a}}{a} \quad (5.6)$$

$$R = -6\left(\frac{\ddot{a}}{a} + \frac{\dot{a}^2}{a^2}\right). \quad (5.7)$$

(Dots indicate time derivative.) The 00 component of the Einstein equation under the flat Robertson-Walker metric is then simply,

$$\frac{\dot{a}}{a} = \frac{\sqrt{\rho}}{\kappa} \quad \text{where} \quad \kappa = \sqrt{\frac{3}{8\pi G_N}}. \quad (5.8)$$

This equation is often called the Friedmann equation. This will be quite useful to get rid of the scale factor in the Boltzmann equation. The value of $\rho$ on the right hand side of the equation will be calculated later and is conveniently parameterized by its temperature dependence. Thus, the Friedmann equation provides a nice connection between the time dependent relative scale factor (the Hubble constant) and the temperature.

In thermal equilibrium entropy is conserved, and using the first law of thermodynamics we can identify the entropy as $S(T) = (\rho + p)V/T$ up to an irrelevant constant. With $V = a^3$ we define the entropy density $s(T)$ as

$$s(T) = \frac{S(T)}{a^3} = \frac{\rho + p}{T}. \quad (5.9)$$

The conservation of entropy means that $S(T) = s(T)a^3$ is time independent:

$$\frac{d}{dt}(s(T)a^3) = 0 \implies \dot{T} = -3\frac{\dot{a}}{a}\frac{s(T)}{s'(T)}. \quad (5.10)$$

(The prime on $s'(T)$ indicates a temperature derivative.) This equation is a direct result of the conservation of entropy in thermal equilibrium and so is called the "conservation of entropy equation". Its utility is relating the time derivative of the temperature ($\dot{T}$) to the scale factor time derivative ($\dot{a}/a$).

We actually have enough information from the above paragraphs to construct the temperature dependent Boltzmann equation. First, we rewrite $dn/dt = (dn/dT)\dot{T}$ and use the "conservation of entropy" equation to replace $\dot{T}$ in favor of $\dot{a}/a$. Then we use the "Friedmann equation" to replace $\dot{a}/a$ in favor of the energy density $\rho(T)$. The result is

$$\frac{dn}{dT} = \frac{s'(T)}{s(T)} \left\{ n + \frac{\kappa J(t)}{3\sqrt{\rho(T)}} [n^2 - n_{eq}^2(T)] \right\} \tag{5.11}$$

where $J(T) = \langle \sigma v \rangle (T)$ is the thermal averaged cross-section.

This equation might not look like much progress; however, all nontrivial dependences from $\rho(T)$ and $s(T)$ are easily calculated as functions of temperature. This will be demonstrated below. At sufficiently high temperature ($T_H$), where all relevant particles are in thermal equilibrium, then we know as a boundary condition that $n(T_H) = n_{eq}(T_H)$. One then need only integrate down to today's temperature ($T \simeq 0$) to obtain the current number density in the universe $n(0)$. The mass density is then just $\rho_\chi = m_\chi n(0)$ where $m_\chi$ is the mass of the relic particle of interest.

Often it is of interest to compare the mass density of our relic particle to the critical density $\rho_c = \kappa^2 H^2$ needed for a flat universe. The relevant formula is

$$\Omega_\chi = \frac{m_\chi n(0)}{\kappa^2 H^2}. \tag{5.12}$$

For a flat universe the sum of all contributing $\Omega$'s (baryons, neutrinos, cosmological constant, neutralinos, etc.) must be equal to 1. If a massive stable particle makes up a significant fraction of the total critical density then it is an interesting cold dark matter candidate. If the calculated critical density is too high, then it is said to "overclose the universe", meaning that the universe became matter dominated too early and it is impossible to reconcile the current mass density and the Hubble constant with the age of the universe.

In order to effectively solve the Boltzmann equation we must have an understanding of the thermodynamic quantities which enter the equation. These are the equilibrium number density $n_{eq}(T)$, the energy density $\rho(T)$, and the entropy density $s(T)$. Since $s(T) = (\rho + p)/T$ we can focus on calculating the pressure $p(T)$ rather than computing $s(T)$ directly. Calculating thermodynamic quantities is standard statistical mechanics and can be found in numerous sources. Here I will merely argue the most salient points that will lead to workable equations.

The density of states in a phase space volume $d^3\vec{x}d^3\vec{k}$ is $1/(2\pi)^3$. Integrating over the volume, the phase space volume density of states is

$$d\xi = \frac{V}{(2\pi)^3}d^3\vec{k} = \frac{V}{(2\pi)^3}(4\pi)k^2dk, \tag{5.13}$$

where $k = |\vec{k}|$ here. We are interested to begin with in the number of particles per unit volume (number density). It will be necessary to multiply the density of states times the mean occupation number for a given momentum state $|\vec{k}\rangle$. The mean occupation number is expressed by the Fermi and Bose distribution functions,

$$f_\eta(k,T) = \frac{1}{e^{\sqrt{k^2+m^2}/T}+\eta} \tag{5.14}$$

where $\eta = -1, 1$ for bosons and fermions respectively. Thus, the total number density integrated over all momentum modes is

$$n_{eq}(T) = \frac{g}{V}\int d\xi f_\eta(k,T) = \frac{g}{2\pi^2}\int_0^\infty dk k^2 f_\eta(k,T) \tag{5.15}$$

where $g$ is the number of internal spin degrees of freedom.

Similarly, we can calculate the energy density and pressure as

$$\rho(T) = \frac{g}{2\pi^2}\int_0^\infty dk k^2\sqrt{k^2+m^2}f_\eta(k,T) \tag{5.16}$$

$$p(T) = \frac{g}{2\pi^2}\int_0^\infty dk\frac{k^4}{\sqrt{k^2+m^2}}f_\eta(k,T). \tag{5.17}$$

Since $s(T) = (\rho(T)+p(T))/T$ we are done. We should keep in mind that $n_{eq}(T)$ in the Boltzmann equation only applies for the one relic particle. However, the $\rho(T)$ and $s(T)$ in the equation are the total energy density and entropy density summed over all particles.

We always want $T$ to be identified with the photon temperature. The thermodynamic quantities technically should be calculated at the temperature of each particle $T_i$ and then the contributions of each particle to the density and pressure should be summed. This creates a subtlety when a stable particle decouples from the photon thermal bath yet still contributes to the mass density, pressure, and entropy of the universe. When a particle decouples its entropy density is separately conserved from that of the photon bath. However, at the decoupling temperature $T_d$ the sum of the two entropies for $T < T_d$ must be equal to the total entropy for

$T > T_d$. This is just entropy conservation. Mathematically this is expressed as $s(T_A)T_A^3 = s(T_B)T_B^3$ where $T_B = T_d + \delta$ before decoupling of particle $\chi$ and $T = T_\gamma = T_\chi$, and $T_A = T_d - \delta$ is the temperature after decoupling and $T = T_\gamma \neq T_\chi$. However when $\delta \to 0^+$ we can identify $T_A = T$ and $T_B = T_\chi$ to obtain

$$T_i = T\left[\frac{s(T_d - \delta)}{s(T_d + \delta)}\right]^{1/3}. \tag{5.18}$$

For stable particles which have decoupled then $T_i$ should be substituted into the formulas for $\rho(T)$ and $s(T)$.

When additional particles "distribution decouple" from the thermal bath ($W$, $h$, etc. whose density goes to zero as $T \ll m$) the $T$ in Eq. (5.18) is no longer the current photon temperature but rather the photon temperature before the "distribution decoupling". Using conservation of entropy again at the decoupling boundary, we can find the relation between $T_\chi$ and the current photon temperature ($T = T_\gamma$):

$$T_\chi = \lim_{\delta \to 0^+} T\left[\frac{s(T_d - \delta)}{s(T_d + \delta)}\right]^{1/3} \prod_{T_\gamma < m_i < T_d} \left[\frac{s(T_d^i - \delta)}{s(T_d^i + \delta)}\right]^{1/3} \tag{5.19}$$

$$= T\left[\frac{s(T_\gamma + \delta)}{s(T_d + \delta)}\right]^{1/3}. \tag{5.20}$$

The thermodynamic quantity integrals are straightforward to calculate numerically for any mass and any temperature. They also can be solved analytically in the limits that $T \gg m$ and $T \ll m$. For example, if $T \gg m$ then the boson energy density is $gT^4\pi^2/30$, and if $T \ll m$ then energy density decreases rapidly as $\rho \sim T^{3/2}e^{-m/T}$. Doing the same for fermions we can construct the following approximation for $\rho(T)$ and $s(T)$:

$$\rho(T) = \frac{\pi^2}{30}\bar{\rho}(T)T^4 \tag{5.21}$$

$$s(T) = \frac{2\pi^2}{45}\bar{s}(T)T^3 \tag{5.22}$$

where

$$\bar{\rho}(T) \simeq \sum_{i=bosons} g_i\theta(T_i - m_i)\left(\frac{T_i}{T}\right)^4 + \frac{7}{8}\sum_{i=fermions} g_i\theta(T_i - m_i)\left(\frac{T_i}{T}\right)^4 \tag{5.23}$$

$$\bar{s}(T) \simeq \sum_{i=bosons} g_i\theta(T_i - m_i)\left(\frac{T_i}{T}\right)^3 + \frac{7}{8}\sum_{i=fermions} g_i\theta(T_i - m_i)\left(\frac{T_i}{T}\right)^3. \tag{5.24}$$

Again, for most particles $T_i = T$; stable particles decoupled from the photon will have $T_i \neq T$. The equilibrium number density is $n_{eq}(T) \simeq 1.2gT^3/\pi^2$, $0.9gT^3/\pi^2$ (bosons, fermions) for $T \gg m$ and decreases rapidly for $T \ll m$.

### 5.3. Approximating the Relic Abundance

Often it is useful to employ approximation techniques to calculate efficiently and reliably the relic abundance.[6–9] It is based primarily on dividing thermal history into two distinct eras: before freeze-out, and after freeze-out. Freeze-out means that the annihilation rate of the particle, $\Gamma = n_{eq}\langle\sigma v\rangle$, is less than the Hubble expansion rate. Once this occurs the particle can no longer remain in equilibrium.

From previous sections we have the tools to solve for this freeze-out temperature given the condition that $\Gamma = H$. Massive weakly interacting particles are non-relativistic when they freeze-out, and so the equilibrium number density can be well approximated as

$$n_{eq} = g\left(\frac{mT}{2\pi}\right)^{3/2} e^{-m/T}, \tag{5.25}$$

where $g$ is the number of degrees of freedom of the relic particle (2 for a neutralino). It is convenient to reëxpress all temperatures as the dimensionless variable $x \equiv T/m_\chi$ such that

$$n_{eq} = \frac{gm^3}{(2\pi)^{3/2}} x^{3/2} e^{-1/x}. \tag{5.26}$$

Furthermore, from our discussion earlier we know that the Hubble constant is

$$H = \frac{\dot{a}}{a} = \sqrt{\frac{8\pi\rho_\chi G}{3}} = \sqrt{N_F}T^2 A \tag{5.27}$$

$$= \sqrt{N_F}m^2x^2 A \tag{5.28}$$

where $A = \sqrt{8\pi^3 G/45}$. The freeze-out condition that $n_{eq}\langle\sigma v\rangle = H$ at $x = x_F$ yields a transcendental equation for $x_F$,

$$x_F^{-1} = \ln\left[\frac{gm\langle\sigma v\rangle(x_F)}{\sqrt{N_F}\sqrt{x_F}A(2\pi)^{3/2}}\right]. \tag{5.29}$$

Eq. (5.29) is only logarithmically dependent on $\langle\sigma v\rangle$ and yields a value of $x_F \simeq 1/20$ quite consistently for massive weakly interacting particles.

After freeze-out the actual number density remains high above the subsequent would-be equilibrium number density. It is therefore appropriate the approximation $n^2 - n_{eq}^2 \simeq n^2$. The Boltzmann equation can then be conveniently rewritten as

$$\frac{df}{dx} = \frac{m_\chi}{A\sqrt{N_F}}\langle\sigma v\rangle f^2 \tag{5.30}$$

where $f = n/T^3$ subject to the boundary condition $f(x_F) = n_{eq}/T_F^3$. To find the current number density we must integrate this equation from $x_F$ (freeze-out) down to $x \simeq 0$ (today):

$$f(0) = \frac{A\sqrt{N_F}}{m_\chi J(x_F) + A\sqrt{N_F} f^{-1}(x_F)} \simeq \frac{A\sqrt{N_F}}{m_\chi J(x_F)} \tag{5.31}$$

where

$$J(x_F) = \int_0^{x_F} dx \langle\sigma v\rangle(x). \tag{5.32}$$

The calculated relic abundance is then simply

$$\rho_\chi = m_\chi T_\chi^3 f(0) = \frac{A\sqrt{N_F}}{J(x_F)}\left(\frac{T_\chi}{T_\gamma}\right)^3 T_\gamma^3. \tag{5.33}$$

The ratio $T_\chi^3/T_\gamma^3 < 1$ is the photon reheating effect from entropy conservation, which was calculated earlier. Scaling this to the critical density $\rho_c = 3H_0^2/8\pi G$ yields the relic's fraction of critical density. It is customary to define the measured value of the Hubble constant as $H_0 = 100\,h\,\mathrm{km\ s^{-1}\,Mpc^{-1}}$. Then

$$\Omega_\chi h^2 = \frac{A\sqrt{N_F}}{(\rho_c/h^2)J(x_F)}\left(\frac{T_\chi}{T_\gamma}\right)^3 T_\gamma^3. \tag{5.34}$$

We can further reduce the expression for $\Omega_\chi h^2$ by noting that

$$\left(\frac{T_\chi}{T_\gamma}\right)^3 = \frac{s(T_\gamma)}{s(T_d)} \simeq \frac{N_F(T_\gamma)}{N_F(T_d)} = \frac{2}{N_F}. \tag{5.35}$$

Using this relation and evaluating all numerical constants we get the convenient form

$$\Omega_\chi h^2 = \frac{1}{\mu^2\sqrt{N_F}J(x_F)} \tag{5.36}$$

where $\mu = 1.2 \times 10^5\,\mathrm{GeV}$. The above formula is an accurate approximation to the Boltzmann equation solution to within about 15%. Table 1 lists the value of $N_F$ for various ranges of the freeze-out temperature $T_F = x_F m_\chi$.

Table 5.1. Degrees of freedom corresponding to freeze-out temperature $T_F = x_F m_\chi \simeq m_\chi/20$.

| $T_F = m_\chi x_F$ | $N_F$ |
|---|---|
| $m_s - m_c$ | 494/8 |
| $m_c - m_\tau$ | 578/8 |
| $m_\tau - m_b$ | 606/8 |
| $m_b - m_W$ | 690/8 |
| $m_W - m_Z$ | 738/8 |
| $m_Z - m_t$ | 762/8 |
| $> m_t$ | 846/8 |

Since a massive weakly interacting particle decouples in the non-relativistic regime, one is able to expand the annihilation cross section in powers of the relative velocity

$$\sigma v = a + \frac{b}{6}v^2 + \cdots . \tag{5.37}$$

The thermal average[10] of this expansion yields

$$\langle \sigma v \rangle = a + \left( b - \frac{3}{2}a \right) x + \cdots . \tag{5.38}$$

With $x_F \simeq 1/20$ this is a quickly converging expansion. The constants $a$ and $b$ can be evaluated straight-forwardly with knowledge of the squared-matrix element of the annihilation process.[11] Helicity amplitudes also exist for all MSSM processes.[12]

The power series expansion of the thermal averaged cross-section breaks down, however, when the relic particle annihilates into a pole (e.g., a $\chi\chi \to Z$ resonance).[13,14] One then needs to use more careful techniques. Also, the Boltzmann equation becomes more complicated if there are other particles around with mass slightly higher than the stable relic (within a "thermal distance" $|m_\chi - m_i| \lesssim T_F$).[14] In this case, the close-by particles can help restore thermal equilibrium through coannihilation channels, effectively lowering the relic abundance. These potentially important cases will not be considered further here.

## 5.4. Neutralino Dark Matter

From the previous sections we have found that $\Omega_\chi h^2$ depends on $1/\langle \sigma v \rangle$. By dimensional analysis we can identify $\langle \sigma v \rangle \sim 1/m^2_{susy}$, which implies that $\Omega_\chi h^2 \propto m^2_{susy}$. Since the age of the universe constraints are compatible only with $\Omega_\chi h^2 < 1$ (very conservative requirement) then it should not surprise us that relic abundance considerations put a upper limit to how large $m_{susy}$ can be. We shall see quantitatively the results of these constraints in the following paragraphs. Before going straight to that, a few general comments about the lightest neutralino are useful to review.

The neutralino is a majorana particle, meaning that the particle and conjugate are the same. Therefore when two of them come together to annihilate, Fermi statistics requires each to be in a different helicity state.[15] Thus $\chi\chi \to f\bar{f}$ requires a final state helicity flip for the external fermions, and the total annihilation rate in the $s$-wave is suppressed by $m_f^2/m_W^2$ compared to processes that do not require helicity flips. This is often referred to as "p-wave suppression": the p-wave configuration doesn't have this Fermi statistics requirement, although it is suppressed by powers of $v^2$. However, at higher values of $m_\chi$ other channels start opening up, such as $W^+W^-$ and $t\bar{t}$ which do not have this suppression (although they may still have other coupling constant suppressions).

The majorana nature of the neutralino is one important property of supersymmetric dark matter. The other important realization is that numerous other supersymmetric particles play a role in the neutralino annihilation channels. The lightest neutralino, being the LSP, only annihilates into standard model particles when the temperature is near freeze-out. However, many important diagrams of these annihilation processes depend on intermediate supersymmetric states. For example, $\chi\chi \to e^+e^-$ depends not only on an intermediate $s$-channel $Z$ boson, but also on $t$-channel slepton exchange. Similarly, $\chi\chi \to q\bar{q}$ depends on $t$-channel squark exchange, and $\chi\chi \to W^-W^+$ can depend crucially on the chargino spectrum. In general, the neutralino annihilation rate and therefore the LSP relic abundance depends on the entire supersymmetric low energy spectrum – this includes particle content, masses, and mixing angles.

Upon closer inspection most choices of the supersymmetric spectrum ultimately give rise to an LSP relic abundance dependent on only a few parameters in the theory. Relic abundance in the minimal model depends most sensitively on only two parameters, the bino mass and the right-handed slepton mass. As discussed in Martin's introductory article in this volume, this minimal model is described partly by a common scalar mass $m_0$ at the high scale and a common gaugino mass $m_{1/2}$. As long as $m_0$ is not much greater than $m_{1/2}$ one typically finds in the low energy spectrum a bino (superpartner of the hypercharge gauge boson) as the lightest supersymmetric particle. One also typically finds the squarks significantly heavier than the sleptons, with the right-handed sleptons being lighter than the left-handed sleptons.

Taking our hints from the minimal model, a useful initial exercise is to declare the LSP as a *pure* bino. Then, the largest contribution to the neutralino annihilation rate will be from $t$-channel $\tilde{l}_R$ exchange in $\chi\chi \to l^- l^+$. This case has been studied in detail,[12] where it was shown that

$$\Omega_\chi h^2 = \frac{\Sigma^2}{M^2 m_\chi^2}\left[\left(1 - \frac{m_\chi^2}{\Sigma}\right)^2 + \frac{m_\chi^4}{\Sigma^2}\right]^{-1} \tag{5.39}$$

where $M \simeq 1\,\text{TeV}$ and $\Sigma = m_\chi^2 + m_{\tilde{l}_R}^2$. A good approximation to the requirement that $\Omega_\chi h^2 < 1$ yields

$$\frac{m_{\tilde{l}_R}^2}{m_\chi} \lesssim 200\,\text{GeV}. \tag{5.40}$$

This result has two important consequences. One, it satisfies the general observation that the relic abundance constraint puts an upper limit on at least one superpartner. And second, in this important example of pure bino, both the bino and the right-handed lepton superpartner must have masses below about 200 GeV. *A priori* the upper bound could have been any number. The result is compatible with weak scale supersymmetry ($m_{susy} \sim m_W$), long thought to be important in electroweak symmetry breaking.

It is not expected that the lightest supersymmetric particle is a pure bino, but rather mostly bino. The coupling of the neutralino to the $Z$ boson depends on the higgsino components. If the LSP is partly Higgsino then efficient annihilations through the $Z$ boson could be possible, thereby reducing the relic abundance for a particular $m_\chi$ value. Limits on the superpartner spectrum still exist; they just happen to be at different scales than the 200 GeV we found for the pure bino case.

Our example realistic particle spectrum is a precise formulation of the minimal model with electroweak symmetry breaking enforced. The full $4\times4$ neutralino mass matrix is diagonalized numerically, and the relic abundance is calculated. The four contour plots[16] of Fig. 5.1 demonstrate the effects of the relic abundance on the parameter space in the $m_0$ vs. $m_{1/2}$ plane. The value of the top quark mass for this contour plot is 170 GeV and the sign of the supersymmetric $\mu$ parameter is chosen to be negative. In (a) $\tan\beta = 5$, and $A_0/m_0 = 0$, (b) $\tan\beta = 5$ and $A_0/m_0 = -2$, in (c) $\tan\beta = 20$ and $A_0/m_0 = 3$, and in (d) $\tan\beta = 10$, $A_0/m_0 = -2$, and the sign of $\mu$ has been changed to positive. The region inside the solid curve is allowed by all constraints. Letter labels are placed outside the solid curve to demonstrate which constraint makes the region phenomenologically unacceptable: `A` indicates $\Omega_\chi h^2 > 1$; `B` $B(B \to s\gamma) > 5.4 \times 10^{-4}$; `C` $m_{\chi_1^+} > 47\,\text{GeV}$; `E` electroweak symmetry breaking problems (full one loop effective potential becomes unbounded from below); and, `L` the LSP becomes charged ($m_{\tilde{l}_R} < m_\chi$). Current limits on $m_{\chi_1^+}$ and $B(b \to s\gamma)$ are somewhat more restrictive than the limits quoted above, however they will only shrink the allowed parameter space a little bit near `B` and `C`. Inside the dotted line corresponds to one particular definition of acceptable finetuning allowed in the electroweak symmetry breaking solutions.[16] It is not important for our further discussion.

The peak or spike in the allowed region for $m_{1/2} \simeq 130\,\text{GeV}$ corresponds to $\chi\chi$ annihilations through a $Z$ pole. Since the LSP is not just a pure bino, and has some higgsino component to it, the $Z$ pole annihilation rate becomes extremely important in this region. Also, it appears obvious from the figures that the relic abundance constraint has the most effect in the increasing $m_0$ direction. This is understandable for several reasons. Away from the $Z$ and Higgs poles, the LSP annihilations occur most efficiently through the $t$-channel scalar diagrams. For a fixed $m_\chi$ value, which is roughly equivalent to fixed $m_{1/2}$, the annihilation cross-section decreases rapidly as the scalar mass $m_0$ increases. Hence, a cutoff in the allowed region must occur as $m_0$ increases.

Forays into the large $m_{1/2}$ region with $m_0$ fixed usually end up with other problems, the most common of which is $\tilde{l}_R$ becoming the LSP. This is simply a result of the renormalization group equation dependences on the $U(1)_Y$ couplings. A quick inspection of the renormalization group equations for $m_{\tilde{B}}$ and for $m_{\tilde{l}_R}$ proves that if $m_{1/2} \gg m_0, m_W$, then the right-handed slepton must be lighter than the bino. Ruling out this region of parameter space can be considered a relic abundance constraint, since the dark matter probably cannot charged.[17]

Realization that a weak-scale higgsino could be a legitimate dark matter particle is a rather recent development. One way to obtain an higgsino as the lightest neutralino is to make $|\mu|$ much less than the gaugino parameters in the neutralino mass matrix. A very low value of $\mu$ will create a roughly degenerate triplet of higgsinos. The charged higgsino and the neutral higgsinos can all coannihilate together with full $SU(2)$ strength, allowing the LSP to stay in thermal contact with the photons more effectively, thereby lowering the relic abundance of the higgsino LSP to an insignificant level.

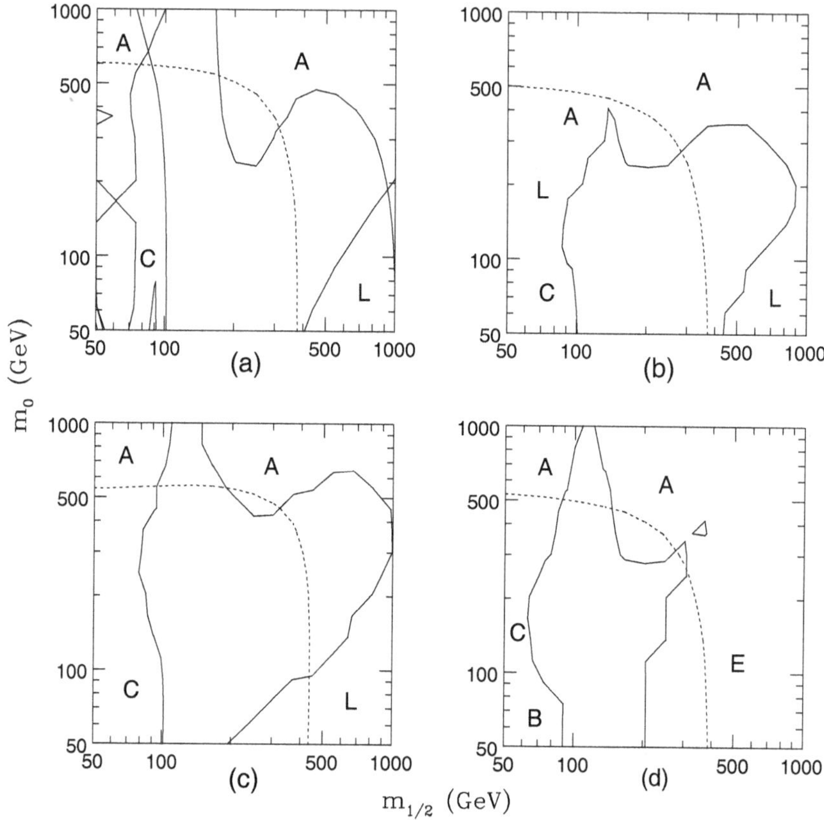


Fig. 5.1. Contours of allowed parameter space. $m_t = 170\,\text{GeV}$ for each frame. In frame (a) $\tan\beta = 5$, $\text{sgn}(\mu) = -$, and $A_0/m_0 = 0$, (b) $\tan\beta = 5$, $\text{sgn}(\mu) = -$ and $A_0/m_0 = -2$, in (c) $\tan\beta = 20$, $\text{sgn}(\mu) = -$ and $A_0/m_0 = 3$, and in (d) $\tan\beta = 10$, $\text{sgn}(\mu) = +$, and $A_0/m_0 = -2$. See the text for explanation of letter labels.

These coannihilation channels are often cited as the reason why higgsinos are not viable dark matter candidates. This claim is true in general, but there are two specific cases that I would like to summarize below that allow the higgsino to be a good dark matter candidate.

Drees *et al.* have pointed out that potentially large one-loop splittings among the higgsinos can render the coannihilations less relevant.[18] Under some conditions with light top squark masses, one-loop corrections to the neutralino mass matrix will split the otherwise degenerate higgsinos. If the mass difference can be more than about 5% of the LSP mass, then the

LSP will decouple from the photons alone and not with its other higgsino partners, thereby increasing its relic abundance.

Another possibility[19] relating to a higgsino LSP is to set equal the bino and wino mass to approximately $m_Z$. Then set the $-\mu$ term to less than $m_W$. This non-universality among the gauginos and particular choice for the higgsino mass parameter, produces a light higgsino with mass approximately equal to $\mu$, a photino with mass at about $m_Z$, and the rest of the neutralinos and both charginos with mass above $m_W$. There are no coannihilation channels to worry about with this higgsino dark matter candidate since no other chargino or neutralino mass is near it. The value of $\tan\beta$ is also required to be near one so that the lightest neutralino is an almost pure symmetric combination of $\tilde{H}_u$ and $\tilde{H}_d$ higgsino states. The exactly symmetric combination does not couple to $Z$ boson (at tree level). The annihilation cross section near $\tan\beta \sim 1$ is proportional to $\cos^2 2\beta$. The relic abundance scales inversely proportional to this, and so the nearly symmetric higgsino in this case is a very good dark matter candidate. Note that there are no $t$-channel slepton or squark diagrams since higgsinos couple to sfermions proportional to the fermion mass. Because the higgsino mass is below $m_W$, the top quark final state is kinematically inaccessible, and so the large top Yukawa cannot play a direct role in the higgsino annihilations.

This non-minimal higgsino dark matter candidate described in the previous paragraph was motivated by the $e^+e^-\gamma\gamma$ event reported by the CDF collaboration at Fermilab.[20] The non-minimal parameters[21] which leads to a radiative decay of the second lightest neutralino (photino) into the lightest neutralino (symmetric higgsino) and photon also miraculously yield a model with a good higgsino dark matter candidate.

### 5.5. Nonthermal Sources

Over the last decade there have been two issues that have gained prominence in the area of neutralino relic abundance: coannihilations and nonthermal sources. Coannihilation is discussed above, but its importance has taken on increased urgency given that the lower bounds on superpartner masses have increased due to LEP2 and Tevatron not finding superpartners. Although the direct bounds on slepton masses are only about 100 GeV, indirectly we expect them to be higher due to theory correlations with other parameters that are constrained more tightly by the data. The most important such constraint comes from the Higgs boson mass. The lightest Higgs boson must be greater than 114 GeV if it acts like a SM Higgs, which is generically expected. The one-loop corrections to the Higgs mass must be quite substantial in order to reach that limit, which correlates with an overall heavy superpartner mass spectrum (see, e.g., Ref. 29).

It is by no means guaranteed that a heavy squark superpartner spectrum needed for heavier Higgs boson is directly requiring the sleptons to be similarly heavy. However, if the heavy superpartner spectrum correlation is strong among all sleptons and squarks, this would mean that the 'bulk region' of thermal relic abundance, which is the region of primary consideration in the discussion above, is not accessible. Instead, additional suppressions of the relic abundance needs to be found. One additional suppression factor is annihilating through a heavy Higgs boson pole, and another suppression factor is coannihilation channels. These were discussed above briefly, but the current situation suggests these ideas may be required in nature. We do not discuss it further here, but refer to the reader to one of the first places that especially emphasizes this issue in so-called focus point supersymmetry,[30] and the references to that work.

There are other candidates for dark matter that have thermal relic abundance that is well below what is needed to be the cold dark matter of the universe. This is especially true for sub-TeV Higgsino or Wino dark matter. The annihilations and coannihilations of these states are so efficient that a cosmologically uninteresting remnant of them results if their masses are less than a TeV (Higgsino) or two (Wino). Nevertheless they are stable neutralino superpartners and it behooves us to ask if there is a way that they could be promoted to dark matter candidates. The answer is yes, and the mechanism is non-thermal production.

By non-thermal production we simply mean that the universe cools and there are remnants of other objects in addition to the neutralinos, and the late decay of those other objects can be into neutralinos which ultimately can have a large relic abundance. There are several examples of this mechanism in the literature. Heavy gravitinos and moduli in anomaly mediated supersymmetry breaking, or any supersymmetry transmission mechanism that gives a similar hierarchy between gauginos and gravitinos, are excellent illustrations. For example, normal thermal processes can produce heavy gravitinos in the early universe which decay after the neutralino freezeout but before big bang nucleosynthesis begins. This can simultaneously solve the gravitino problem and the dark matter problem of higgsino or wino relics.[31] One can also get dark matter from decaying cosmic strings, or other topological defects in the early universe.[32] A review of non-thermal dark matter can be found in Ref. 33 and by S. Watson in this volume.

## 5.6. Conclusion

The minimal model bino and the higgsino described above work as dark matter candidates both qualitatively and quantitatively. Nature, of course, might not conform to either of these specific possibilities, but it is straightforward to catalog the non-minimal possibilities. There are, of course, numerous ways to go beyond what is presented here.[22] Scalar mass universality could be relaxed.[23] One could in fact suppose that $R$-parity is not exactly conserved and the LSP is not a stable particle. Perhaps one could allow small $R$-parity violations which create meta-stable LSPs whose lifetimes are greater than the age of the universe to solve the dark matter dilemma. However, it is not enough to just make the lifetime greater than the age of the universe. Remarkably, the measurements of positrons and photons in cosmic rays require that the LSP lifetimes into these particles be many orders of magnitude beyond the age of the universe.[24] In other words, it is not easy to make the LSP a meta-stable dark matter candidate. If $R$-parity were broken in nature, it appears necessary to look elsewhere for the dark matter candidate.

One common theme exists in non-minimal models which accommodate a supersymmetric solution to the dark matter problem. It is the requirement that $\chi\chi$ not couple to the $Z$ boson. A full strength $SU(2)$ coupling to the $Z$ boson generally allows too efficient LSP annihilation, and therefore too small relic abundance to be relevant to the dark matter problem. This theme can be found in several non-minimal examples such as the symmetric Higgsino[18,19] described above, the $\tilde{Z}$,[25] and the sterile neutralino[26] dark matter candidates. Of course, the minimal model also provides a low-strength coupling to the $Z$ naturally with the mostly bino dark matter candidate.[27] The older supersymmetry dark matter literature considered the photino, which also does not couple to the $Z$, as the primary dark matter candidate.[15]

Theoretical ideas about the low-energy superpartner spectrum *always* have dark matter implications. The supersymmetric solution to the dark matter problem is perhaps only superseded by gauge coupling unification in experimentally based indications that nature might be described by softly broken supersymmetry. It is mainly for this reason that much experimental effort is expended on the search for supersymmetric relics.[28] Likewise, theoretical efforts on understanding the origin of LSP stability and the prediction of dark matter properties should continue to enlighten us.

## References


1. G. Farrar, P. Fayet, Phys. Lett. B **76**, 575 (1978); S. Dimopoulos, H. Georgi, Nucl. Phys. B **193**, 150 (1981); N. Sakai, T. Yanagida, Nucl. Phys. B **197**, 83 (1982); S. Weinberg, Phys. Rev. D **26**, 287 (1982).
2. T. Falk, K. Olive, M. Srednicki, Phys. Lett. B **339**, 248 (1994).
3. E. Diehl, G.L. Kane, C. Kolda, J.D. Wells, Phys. Rev. D. **52**, 4223 (1995).
4. V. Trimble, Ann. Rev. Astron. Astrophys. **25**, 425 (1987); P. Sikivie, Nucl. Phys. Proc. Suppl. **43**, 90 (1995).
5. D. Schramm, M. Turner, astro-ph/9706069.
6. R. Wald, *General Relativity*, Chicago: University of Chicago Press (1984); E. Kolb, M. Turner, *The Early Universe*, Redwood City, USA: Addison-Wesley (1990).
7. B. Lee, S. Weinberg, Phys. Rev. Lett. **39**, 165 (1977); M. Vysotskii, A. Dolgov, Ya. Zeldovich, Pisma Zh.Eksp.Teor.Fiz. **26**, 200 (1977); P. Hut, Phys. Lett. B **69**, 85 (1977).
8. K. Olive, D. Schramm, G. Steigman, Nucl. Phys. B **180**, 497 (1981).
9. J. Ellis, J. Hagelin, D. Nanopoulos, M. Srednicki, Nucl. Phys. B **238**, 453 (1984).
10. M. Srednicki, R. Watkins, K. Olive, Nucl. Phys. B **310**, 693 (1988); P. Gondolo, G. Gelmini, Nucl. Phys. B **360**, 145 (1991).
11. J.D. Wells, hep-ph/9404219; L. Roszkowski, Phys. Rev. D **50**, 4842 (1994).
12. M. Drees, M. Nojiri, Phys. Rev. D **47**, 376 (1993).
13. G. Kane, I. Kani, Nucl. Phys. B **277**, 525 (1986).
14. K. Griest, D. Seckel, Phys. Rev. D **43**, 3191 (1991).
15. H. Goldberg, Phys. Rev. Lett. **50**, 1419 (1983).
16. G.L. Kane, C. Kolda, L. Roszkowski, J.D. Wells, Phys. Rev. D **49**, 6173 (1994).
17. P. Smith, in *Proceedings of the First International Symposium on Sources of Dark Matter in the Universe*, edited by D. Cline, World Scientific, 1995. J. Basdevant, R. Mochkovitch, J. Rich, M. Spiro, A. Vidal-Madjar, Phys. Lett. B **234**, 395 (1990).
18. M. Drees, M. Nojiri, D. Roy, Y. Yamada, Phys. Rev. D **56**, 276 (1997).
19. G.L. Kane, J.D. Wells, Phys. Rev. Lett. **76**, 4458 (1996); K. Freese, M. Kamionkowski, Phys. Rev. D **55**, 1771 (1997).
20. S. Park, "Search for new phenomena at CDF," 10th Topical Workshop on Proton-Antiproton Collider Physics, edited by Rajendran Raja and John Yoh, AIP Press, 1995.
21. S. Ambrosanio, G. Kane, G. Kribs, S. Martin, S. Mrenna, Phys. Rev. Lett. **76**, 3498 (1996); Phys. Rev. D **55**, 1372 (1997).
22. I do not discuss models of low-energy supersymmetry breaking which could have the gravitino as the LSP, or even exotic messenger particles as the cold dark matter. See for example, S. Dimopoulos, G. Giudice, A. Pomarol, Phys. Lett. B **389**, 37 (1996).
23. P. Nath, R. Arnowitt, hep-ph/9701301.

24. J. Ellis, G. Gelmini, J. Lopez, D. Nanopoulos, S. Sakar, Nucl. Phys. B **373**, 399 (1992); G. Kribs, I. Rothstein, Phys. Rev. D **55**, 4435 (1997).
25. A. Gabutti, M. Olechowski, S. Cooper, S. Pokorski, L. Stodolsky, Astropart. Phys. **6**, 1 (1996).
26. B. de Carlos, J.R. Espinosa, hep-ph/9705315.
27. L. Roszkowski, Phys. Lett. B **262**, 59 (1991).
28. G. Jungman, M. Kamionkowski, K. Griest, Phys. Rept. **267**, 195 (1996).
29. K. Tobe and J. D. Wells, Phys. Rev. D **66**, 013010 (2002) [arXiv:hep-ph/0204196].
30. J. L. Feng, K. T. Matchev and F. Wilczek, Phys. Lett. B **482**, 388 (2000) [arXiv:hep-ph/0004043].
31. T. Gherghetta, G. F. Giudice and J. D. Wells, Nucl. Phys. B **559**, 27 (1999) [arXiv:hep-ph/9904378].
32. R. Jeannerot, X. Zhang and R. H. Brandenberger, JHEP **9912**, 003 (1999) [arXiv:hep-ph/9901357]. Y. Cui and D. E. Morrissey, Phys. Rev. D **79**, 083532 (2009) [arXiv:0805.1060 [hep-ph]].
33. B. S. Acharya, G. Kane, S. Watson and P. Kumar, Phys. Rev. D **80**, 083529 (2009) [arXiv:0908.2430 [astro-ph.CO]].

# A Wino-Like LSP World: Theoretical and Phenomenological Motivations

Daniel Feldman and Gordon Kane

*Michigan Center for Theoretical Physics, University of Michigan, Ann Arbor, MI 48109, USA*

Central to the identification of the lightest superpartner (LSP), and the accompanying spectra of heavier supersymmetric particles, is to understand the particle content of the LSP mass eigenstate. There are strong arguments, both theoretical and phenomenological, that point to a LSP with a significant wino component that can contribute substantially to the relic density of dark matter and even accommodate the entire relic abundance. Data from the Large Hadron Collider (LHC) is expected to test signals of such models. Early signatures from the cascades of colored superpartners may point to the presence and form of softly broken supersymmetry, and of electroweak symmetry breaking.

## 6.1. Annihilating Dark Matter in the Halo

From the standpoint of annihilating dark matter, among the mechanisms that can simultaneously explain the PAMELA[1] and WMAP[2] relic density data, some are based on a wino-like LSP. These include a non-thermally produced wino LSP[3] with the relic abundance explained via moduli decay[4] which arises in a concrete framework of fluxless string compactifications,[6,7] or a thermal wino-like LSP with a weakly interacting co-annihilating hidden sector with mass terms for the Majoranas arising from Stueckelberg fields and/or Kinetic Mixing.[8] Here we will focus on the collider and cosmological implications of the wino dominated dark matter. Other dark matter annihilation mechanisms capable of explaining both data sets are a Sommerfeld enhancement in the halo,[9] a Breit-Wigner enhancement of dark matter annihilations in the halo,[10] and Leptonic Asymmetric Dark Matter,[11] and we refer the reader to the original papers for details on these proposals.

An early approach that gave a motivation for a wino LSP was the proposal that that the mechanism of anomaly mediation might be the dominant mechanism of supersymmetry breaking[14,15] (for early foundational work see Ref. 16). If a nearly pure wino LSP is to account for the entire relic density it has become more clear that the universe likely has a non-thermal cosmological history. This has received increasing attention as a generic feature of a comprehensive underlying theory; we comment briefly on this and a more detailed discussion is given in Ref. 12. If however the LSP has a relatively large wino component but has a suitable admixture of other eigen-components, a thermal mechanism for explaining its relic abundance is quite possible. We review this idea as well. Our goal in this chapter is simply to emphasize that in the past decade a wino-like LSP has been recognized to be both theoretically and phenomenologically well motivated. We proceed by describing several examples for readers who wish to see more details.

## 6.2. Non-Thermal Winos from Moduli Stabilized on a $G_2$ Manifold

### 6.2.1. *Soft breaking from the $G_2$*

A recent string model of interest that *predicts* a LSP that is dominantly wino is that of the $G_2$-MSSM. The $G_2$-MSSM was formally introduced in Ref. 6 and was motivated by the work of Acharya and Witten,[5] where it was demonstrated that $M$-theory compactifications on a manifold $X$ of $G_2$ holonomy can give rise to chiral fermions in four dimensions only if $X$ is not smooth; specifically the fermions need to be localized at conical singularities. Thus in Ref. 6 it was shown that moduli stabilization is manifest in a large class of $M$-theory compactifications in the zero-flux sector, with minimally two non-abelian asymptotically free gauge groups which gives rise to a moduli potential. The simultaneous breaking of supersymmetry becomes possible when at least one of the hidden sectors contains charged matter, and the cosmological constant can be tuned tuned towards zero.

The underlying framework is described by $\mathcal{N} = 1$ supergravity where a generalized sector of soft susy breaking is derived from both a tree level supergravity contribution[17] and an anomalous supergravity contribution,[14,16] as in Ref. 15. The soft parameters can be parametrized at the unification scale as $m_0 = s \cdot m_{3/2}$, $m_a = f_a \cdot m_{3/2}$, $A_3 = a_3 \cdot m_{3/2}$ where $m_{3/2} \sim \mathcal{O}(10 - 100)$ TeV is the gravitino mass, $m_0$ is a universal scalar mass, $m_a$ are the gaugino masses, $A_3$ are the tri-linear couplings of the third generation, and $\tan\beta$ is found to lie in the range $\tan\beta \sim 1.5 - 2.0$. Here the parameters $(s, f_a, a_3)$ are functions of the microscopic theory which are determined by specification of the Kähler potential $K$, superpotential $W$ and gauge kinetic function $f$ (see below). The soft parameters are well approximated by (for the complete analytical expressions see Ref. 6) $s \sim 1$, $f_a = f'_a \alpha_G - \epsilon\, \eta$, where $f'_{1,2,3} = (0.35, 0.58, 0.64)$, and $\eta = 1 - \alpha_G \delta$ parametrizes gauge coupling corrections in the tree level sector of the gaugino masses, and $\epsilon \sim 0.024$ depends on the parameters of the hidden sector potential which are responsible for tuning the cosmological constant to zero. The terms entering for the tri-linears of the third generation are well approximated by $a_3 = 1.67 - 3\epsilon\, ln(4\pi^{1/3} V_7) - 2\epsilon\, ln(|Y_{t,b,\tau}|)$, where $Y$ are normalized Yukawas and where $V_7$ is the normalized volume of the $G_2$ manifold which also enters in the determination of the gravitino mass. Here the largeness of the gravitino mass decouples the scalars while the gaugino masses are suppressed relative to gravitino mass, where the suppression enters via the volume of hidden sector three cycles. The physical values of the soft parameters are sensitive to the precise value of the unified gauge coupling and threshold corrections. The largeness of the gravitino mass drives the $\mu$ term to be order $m_{3/2}$ for electroweak symmetry breaking, which in turn induces a relatively large self energy correction[18] to the electroweak gaugino masses.[7]

The above arises from coupling the moduli and hidden sector matter to the visible SUGRA sector with the following Kähler potential $K$, superpotential $W$ and gauge kinetic function $f$ at the compactification scale $(\sim M_{\text{unif}})$:[6,7]

$$
\begin{aligned}
K/\overline{M}_{\text{pl}}^2 &= -3\ln(4\pi^{1/3} V_7) + \bar{\phi}\phi, \quad V_7 = \prod_{i=1}^{N} s_i^{a_i}\,, \; a_i \in Q^+ \\
W &= \overline{M}_{\text{pl}}^3 \left( C_1\, P\, \phi^{-(2/P)}\, e^{ib_1 f_1} + C_2\, Q\, e^{ib_2 f_2} \right); \; b_1 = \frac{2\pi}{P},\, b_2 = \frac{2\pi}{Q} \\
f_1 &= f_2 \equiv f_{\text{hid}} = \sum_{i=1}^{N} N_i\, z_i;\; z_i = t_i + i s_i. \qquad (6.1)
\end{aligned}
$$

Here, $V_7$ is the volume of the $G_2$ manifold $X$ (in units of the eleven-dimensional Planck length $l_{11}$), the composite scalar $\phi$ is the effective meson field; there is one for each pair of massless quarks which forms the composite. $f_{1,2}$ are the gauge kinetic functions of two hidden sectors at tree-level, $s_i$ are the $N$ geometric moduli of the $G_2$ manifold and $t_i$ are axionic scalars. The parameters $P$ and $Q$ are proportional to the beta function coefficients of the gauge groups. The difference $Q - P$ is fixed by the solution corresponding to a metastable minimum with spontaneously broken supersymmetry and one finds $Q - P \sim 3$.[6] The effective parameter $P_{\rm eff} \equiv Pln(C_1/C_2)$ takes a value of 84 for $Q - P = 3$ as governed by the tuning of the cosmological to zero. It is this relative smallness of the factor $1/P_{\rm eff}$, which enters for example in the tree-level gaugino sector, that allows the anomaly terms and tree terms to contribute to the soft masses on the same footing.

### 6.2.2. *Moduli masses*

The system of moduli fields which participate in reheating the universe at various epochs are the $N$ geometric moduli $s_i$, and the hidden sector meson field $\phi$. Since these moduli and meson will mix in general, the physical moduli correspond to mass eigenstates. There is one heavy eigenstate with mass[6] $m_{X_N} = (7K_1/3 + K_2)^{1/2} m_{3/2}$, along with $(N-1)$ degenerate light eigenstates with mass $m_{X_j} = (K_2)^{1/2} m_{3/2}$, and an eigenstate with mass $m_\phi = (K_4 - \frac{K_3^2}{K_1})^{1/2} m_{3/2}$. The $K$ factors are uniquely determined in terms of the VEV of the meson field and the parameters discussed above. A remarkable result is that[6,7]

$$m_{X_j} \approx 1.96\, m_{3/2},\ j = 1, \cdots, N-1\ . \tag{6.2}$$

That is, the light moduli are phase space restricted and the decay into gravitinos are suppressed which essentially eliminates the moduli induced gravitino problem.[6,7]

### 6.2.3. *The right halo cross section and just about the right abundance from non-thermal winos*

For a pure wino in the large $\mu$ limit such as that which arises in the $G_2$ models, one has $\tilde{N}_1^0 = \widetilde{W}$, $\tilde{N}_2^0 = \tilde{B}$, $\tilde{N}_{3,4}^0 = \frac{1}{\sqrt{2}}(\tilde{H}_1 \mp \tilde{H}_2)$, where $\tilde{N}_n^0 = \mathcal{N}_{nm}\widetilde{Z}_m$, $\widetilde{Z}^T = (\widetilde{W}, \widetilde{B}, \tilde{H}_1, \tilde{H}_2)$ and the annihilation cross section of the winos into $W$ Bosons proceeds through an s-wave via chargino exchange

$$\langle \sigma v \rangle = \frac{g_2^4}{2\pi m_{\widetilde{W}}^2} \frac{(1 - m_W^2/m_{\widetilde{W}}^2)^{3/2}}{(2 - m_W^2/m_{\widetilde{W}}^2)^2} \tag{6.3}$$

with $\langle \sigma v \rangle \sim (1.6 - 2.2) \times 10^{-7} \mathrm{GeV}^{-2}$ for $m_{\widetilde{W}} \in (170, 200)$ GeV. The lack of velocity dependence and any non-pertubative effects in this limit fixes the halo cross section with the wino mass determined. As far as the relic abundance, quite generally, in the absence of co-annihilations for s-wave annihilations the relic abundance follows from solving the Boltzmann equations. This is equivalent to

$$\Omega h^2 = h^2 \frac{m_{\widetilde{W}} n_{\widetilde{W}}/s}{\rho_c/s_0} = \frac{h^2}{4(\rho_c/s_0)} \left(\frac{90}{\pi^2 g_*}\right)^{1/2} \frac{1}{\overline{M}_{\rm pl}} \frac{1}{\langle\sigma v\rangle} \frac{m_{\widetilde{W}}}{T} \quad (6.4)$$

where $n_{\widetilde{W}} = H/\langle\sigma v\rangle$, $H^2 = T^4 g_* \pi^2/(90\overline{M}_{\rm pl}^2)$, $s = (2\pi^2/45)g_* T^3$ and $\rho_c/s_0 = 3.6 \times 10^{-9}$GeV $h^2$, with $g_* = g_*(T)$. Note that the above holds for both the thermal case[20] and the non-thermal case; the difference for the non-thermal case arises in that the neutralino density can become re-populated well after freeze out (where $m/T \sim 20$ at $T = T_{\rm freeze}$). Instead, for the non-thermal case, the reheat temperature $T_R$ occurs at temperatures on the scale (5-100) MeV, depending on the particular model.

Thus the presence of moduli (denoted here now more generally by $M$) are ubiquitous in string theories whose low energy limit are supergravity models (as we have just discussed in the previous section). If such a scalar decays after freeze out it can decay into susy particles repopulating the neutralino abundance. Here we discuss the phenomena in some generality.

Assuming the energy density during modulus decay is transferred completely to radiation one has $H = \Gamma_M$, and thus $T_R = (90\pi^{-2}/g_*)^{1/4}\sqrt{\Gamma_M \overline{M}_{\rm pl}}$, where the modulus decay is parametrized as $\Gamma_M = c_M M^3/\Lambda^2$. The scale $\Lambda$, the modulus mass $M$ and its coupling are determined by the underlying model. In order to avoid the gravitino problem (for recent work see Ref. 19), the modulus is taken to satisfy $M < 2m_{3/2}$ (as mentioned previously this is remarkably manifest in the $G_2$ models[6]), where $m_{3/2}$ is the gravitino mass. The relic density calculation of the abundance of winos is then entirely straight forward with the temperature $T \to T_R$ and one obtains

$$\Omega_{\tilde{W}} h^2 \simeq 0.32 \frac{1}{\alpha\sqrt{c_M}} \left(\frac{3 \times 10^{-7}\text{GeV}^{-2}}{\langle\sigma v\rangle}\right) \left(\frac{m_{\tilde{W}}}{200\text{GeV}}\right) \left(\frac{m_{3/2}}{100\text{TeV}}\right)^{-3/2} \quad (6.5)$$

were $\langle\sigma v\rangle$ is to be velocity averaged with a Boltzmann distribution and we have used $g_* = 10.75$ (the appropriate degrees of freedom for MeV scale temperatures), $M \sim 2m_{3/2}$ and $\Lambda = \overline{M}_{\rm pl}/\alpha$,where $\alpha$ parametrizes deviations from the reduced Planck scale. For example, moduli couplings at a string scale would correspond to $\alpha \sim \sqrt{V}$, where $V$ is the (dimensionless) volume of compactification and thus for a pure wino with mass of $\sim 170$ GeV, the WMAP relic density ($\Omega_{\tilde{W}} h^2 \sim 0.1$ ) is then achieved for

$c_M \sim 1$, $\Lambda = M_{\text{String}} \sim 2 \times 10^{17}$GeV, (which is similar to the heterotic string scale) for gravitino mass of $\sim 40$ TeV.

For the case of the $G_2$-MSSM $\alpha = 1$ and $c_M \in (0.5, 4)$ over the part of the model parameter space investigated. Thus a gravitino mass on order of 100 TeV is needed for a pure wino with mass of 170 GeV to give the observed relic density within the WMAP error band. A more detailed understanding of the underlying string theory may allow for a lighter modulus mass.

### 6.3. Wino-Like Dark Matter in the Stueckelberg Extensions and with Kinetic Mixings

#### 6.3.1. *The right relic abundance and just about the right halo cross section from extra $U(1)_X$ factors*

The presence of matter in the hidden sector can have observable effects on the visible (MSSM) sector. One such effect is that extra Majorana matter very weakly coupled to the MSSM neutralinos can coannihilate with the LSP which has the effect of enhancing the relic density for the LSP by as much as an order of magnitude or more.[8,40] This enhancement of the relic abundance can occur through the presence of $n$ $U(1)_X$ gauge symmetries in the hidden sector and correspondingly $n$ new scalars (axions) each being absorbed through a Stueckelberg mechanism[40,43] generating masses for the $n$ $U(1)_X$ gauge Bosons.[8]

Thus, in top-down approaches to building realistic models based on D-Branes the SM gauge group can be produced but one also encounters residual Abelian group factors where the extra $U(1)$s usually correspond to massive vector fields. In particular frameworks they lead to terms in the action of the form $B \wedge F$, i.e. terms of the form $\frac{1}{2}\epsilon^{\mu\nu\alpha\beta}B_{\mu\nu}F_{\alpha\beta}$ which are needed for anomaly cancellation via a [four dimensional] Green-Schwarz (GS) mechanism.[35] In other frameworks the $B \wedge F$ couplings can arise for the non-anomalous cases as well;[34,38–40] for the anomalous case see i.e. Refs. 41, 42. These types of couplings can give rise to Stueckelberg mass terms. For example, under the duality transformation $\partial_\mu \sigma \sim \epsilon_{\mu\nu\rho\sigma}\partial^\nu B^{\rho\sigma}$, as illustrated in Ref. 34, vector fields gain mass via the GS mechanism leading to Stueckelberg mass terms in the Lagrangian of the form

$$\mathcal{L}_{\text{St}} = -\frac{1}{2}\sum_I (\partial^\mu \sigma^I - k^{Ia} C_a^\mu)^2, \tag{6.6}$$

with the index $I$ running over all the 4D fields and thus runs over each of psuedoscalars $\sigma^I$ (Ramond axions) and over the killing vector coefficients

$k^{Ia}$. Here $a$ indexes a Brane stack with a supporting gauge group (for example: $U(N_a) = SU(N_a) \times U(1)_a$) with each Abelian vector field denoted as $C_a^\mu$. It is then clear that the quadratic term for $C_a^\mu$ in Eq.(6.6) gives rise to mass terms for the Abelian vector fields. The orthogonality of the killing vectors determines which Bosonic states will be rendered massive or if they remain massless. Upon addition of gauge fixing terms, the cross terms in Eq.(6.6) cancel and thus the psuedoscalars and vector fields decouple. The mechanism outlined here is distinct from a Higgs mechanism for mass generation as there is no extra dynamical scalar field.

In the supersymmetric case, a Stueckelberg extension of the MSSM[40] arises from an extended Lagrangian with $N_V$ Abelian gauge fields and $N_S$ axions

$$\mathcal{L}_{\rm St} = \int d^2\theta d^2\bar{\theta} \sum_{j=1}^{N_S} \left( S_j + \bar{S}_j + \sum_{i=1}^{N_V} M_{ij} C_i \right)^2 , \tag{6.7}$$

where $(S_j,\, C_i)$ are (chiral,vector) supermultiplets.

The case of interest to us here is when $N_S = N_V - 1$ so generically all $N_V - 1$ number of axions are absorbed by $N_V - 1$ number of vector bosons making them massive and leaving one vector boson massless. To generate communication between the hidden and the visible sectors we thus identify one of the fields in the $N_V$ set to be the hypercharge, e.g., $B_Y = C_1$. When one includes the electroweak sector of MSSM, there would then be an automatic coupling between the hidden and the visible sectors through $B_Y$ which enters both in the MSSM sector and in the Stueckelberg sector. Consequently a spontaneous breaking in the MSSM sector with inclusion of the above interactions will lead to a massless photon, a weakly coupled $Z$ boson and a system of $N_V - 1$ extra $Z'$ bosons. Additionally, beyond the MSSM, there will be $N_V - 1$ number of CP even Higgs fields.

In the fermionic sector we gain a Stueckelberg chiral fermion for each $S_j$ and a Stueckelberg gaugino for each $C_i$. These mix with the neutral fermions of the MSSM sector producing a neutralino mass matrix of extended dimension. Thus for $n$ number of $U(1)_X$, there are $2n+4$ Majorana states:

$$(N_1^0, (n_1^0, n_2^0 \ldots n_{2n}^0), N_2^0, N_3^0, N_4^0) \tag{6.8}$$

where $N_i^0$ $(i = 1, 2, 3, 4)$ are essentially the four neutralino states of the MSSM and $n_\alpha^0$, $(\alpha = 1, ..., 2n)$ are the additional states.[8] With the Majorana fields of the hidden sector interacting extra weakly, their interaction cross sections are suppresed, and one finds that there is an enhancement of the relic density by a factor $B_{Co}$ through coannihilation effects entering through the number of degrees of freedom supplied by the hidden sector Majoranas. *This enhancement to the relic density* (to be distinguished physically from a boost in the halo!) is given by[8]

$$B_{\rm Co} \simeq \frac{\sum_{a,b}\int_{x_f}^{\infty}\langle\sigma_{ab}v\rangle\gamma_a\gamma_b\frac{dx}{x^2}}{\sum_{A,B}\int_{x_f}^{\infty}\langle\sigma_{AB}v\rangle\Gamma_A\Gamma_B\frac{dx}{x^2}},$$

$$\gamma_a = \frac{g_a(1+\Delta_a)^{3/2}e^{-\Delta_a x}}{\sum_b g_b(1+\Delta_b)^{3/2}e^{-\Delta_b x}}, \text{ MSSM}$$

$$\Gamma_A = \frac{g_A(1+\Delta_A)^{3/2}e^{-\Delta_A x}}{\sum_A g_A(1+\Delta_A)^{3/2}e^{-\Delta_A x}}, \text{ MSSM} \otimes \text{Hid.}$$

Here $a$ runs over the channels which coannihilate in the MSSM sector, while $A$ runs over channels both in the MSSM sector and in the hidden sector (*i.e.*, $A=1,\ldots,n_v+n_h$). For the case when these fields that coannihilate are completely degenerate with the LSP, and the hidden sector interactions are suppressed, one has $B_{\rm Co}=(1+d_h/d_v)^2$, where $d_s=\sum_s g_s$, for $s=(v,h)$, i.e.

$$(\Omega h^2)_{N^0} \simeq (1+\frac{d_h}{d_v})^2(\Omega h^2)_{\rm MSSM}, \tag{6.9}$$

and one has for the case of $n$ hidden sector $U(1)$s the result

$$B_{\rm Co} = (1+2n)^2 \qquad (\text{no MSSM particle co}-\text{annihilations}). \tag{6.10}$$

As studied in Refs. 8, 43, such an extra weakly interacting set of hidden sector neutralinos can be sourced not only by Stueckelberg mass mixings but also through kinetic mixings. This has recently been observed in other stringy environments where the extra weak Majoranas, such as Stinos[43] have been dubbed string photini.[44] In fact such a situation can arise from both Stueckelberg mass and kinetic mixings operating at comparable scales.[43,45]

Quite generally, in order for the above models to produce a significant flux in the halo they must lead to the LSP having a significant wino component. Such eigen content arises in non-universal SUGRA models and string models with non-universalites in the gaugino sector (for early work see Ref. 30). A pure wino can be obtained by driving down $M_2(GUT)$ and halo cross sections of the size $2\times10^{-7}\text{GeV}^{-2}$ can manifest.[8] However in this case, the degenerate set of $U(1)$ factors must be large, since MSSM co-annihilations reduce the maximal $B_{\rm Co}$ to $B_{\rm Co}=(1+2n/3)^2$. Yet when MSSM co-annihilations are reduced, the wino component can also remain large $\mathcal{N}_{1,2} \sim 0.70$, and there can be non-negligible mixtures of bino and Higgsino components. In this wino mixed case, the annihilation cross section in the halo can be about a factor of 5 lower than the pure wino case, while the $B_{\mathrm{Co}}$ enhancement of the relic density is large enough (maximally about 45 in practice with a $U(1)^3_X$ gauge symmetry) to bring the mixed wino neutralino abundance up to the lower limit on the WMAP constraint. Fits to the PAMELA data require a small boost in the halo on the order of (3-5). Thus such a mixed wino model is less susceptible to potenial overproduction of protons and photons but can still produce a good fit to the PAMELA data and the WMAP data. Nevertheless, a large wino component is needed.

### 6.4. Directly Detecting Wino-Like Dark Matter

Experiments are actively attempting to detect dark matter via their spin dependant and spin independent scattering with nuclei. The LSPs have a velocity distribution near the earth and in the local galactic halo, and they are travelling with non relativistic speed order $0.001c$. As such their momentum transfer is rather small (order 100 MeV for LSP masses of order 100 GeV). Therefore, the relevant interactions for the direct detection of dark matter (LSP collisions with nuclei) may be calculated in the limit of zero momentum transfer. For the case of the MSSM, over most of the viable parameter space, the relevant piece of the interaction Lagrangian is[25,26]

$$\begin{aligned}\mathcal{L} = &\bar{\chi}\gamma^\mu\gamma^5\chi\bar{q}_i\gamma_\mu(\alpha_{1i}+\alpha_{2i}\gamma^5)q_i + \alpha_{3i}\bar{\chi}\chi\bar{q}_i q_i \\ &+ \alpha_{4i}\bar{\chi}\gamma^5\chi\bar{q}_i\gamma^5 q_i + \alpha_{5i}\bar{\chi}\chi\bar{q}_i\gamma^5 q_i + \alpha_{6i}\bar{\chi}\gamma^5\chi\bar{q}_i q_i\,.\end{aligned} \tag{6.11}$$

Indeed, the spin independent (SI) cross section is currently being probed in experimental searches.[28] For the cross section of neutralinos scattering elastically off target nuclei, in terms of the reduced mass of the neutralino and the target system ($\mu_{\chi T}$), one has

$$\sigma_{\chi(T)} = \frac{4\mu^2_{\chi T}}{\pi}(Zf_p + (A-Z)f_n)^2\,, \tag{6.12}$$

where $(Z, A)$ are the atomic (number,mass) of the nucleus, and the interactions with the quarks in the target nuclei are dominated by $t$-channel CP-even Higgs exchange, and $s$-channel squark exchange and are housed in

$$f_{p/n} = \sum_{q=u,d,s} f^{(p/n)}_{T_q} a_q \frac{m_{p/n}}{m_q} + \frac{2}{27} f^{(p/n)}_{TG} \sum_{q=c,b,t} a_q \frac{m_{p/n}}{m_q}\,. \tag{6.13}$$

Here $f_{TG}^{(p/n)}$ is given by $1 - f_{T_u}^{(p/n)} - f_{T_d}^{(p/n)} - f_{T_s}^{(p/n)}$ and arises via gluon exchange with the nucleon and the $f_{T_q}^{(p/n)}$ are determined from light quark masses obtained from baryon masses via matrix elements and from the value of the pion-nucleon sigma-term. Numerical values and further details are given in, for example, in Ref. 27. Under constraints from collider data on the sparticle masses and mixings, the dominant couplings that enter in the spin independent cross section are[25,26]

$$\begin{aligned} a_q \equiv a_{3i} = & -\frac{1}{2(m_{1i}^2 - m_\chi^2)} \Re\left[(X_i)(Y_i)^*\right] - \frac{1}{2(m_{2i}^2 - m_\chi^2)} \Re\left[(W_i)(V_i)^*\right] \\ & -\frac{g_2 m_q}{4 m_W B}\left[\Re\left(\delta_1[g_2 n_{12} - g_Y n_{11}]\right) DC\left(-\frac{1}{m_H^2} + \frac{1}{m_h^2}\right)\right. \\ & \left. +\Re\left(\delta_2[g_2 n_{12} - g_Y n_{11}]\right)\left(\frac{D^2}{m_h^2} + \frac{C^2}{m_H^2}\right)\right] . \end{aligned} \tag{6.14}$$

The first term arises from squark exchange and is typically much suppressed, while the dominant effects enter through the exchange of Higgs Bosons. The parameters $\delta_{1,2}$ depend on eigen components of the LSP wave function and $B, C, D$ depend on VEVs of the Higgs fields and the Higgs mixing parameter $\alpha$ and are given by

$$\begin{aligned} &\text{for u quarks:} \quad \delta_1 = n_{13} \quad \delta_2 = n_{14} \quad B = \sin\beta \quad C = \sin\alpha \quad D = \cos\alpha \\ &\text{for d quarks:} \quad \delta_1 = n_{14} \quad \delta_2 = -n_{13} \quad B = \cos\beta \quad C = \cos\alpha \quad D = -\sin\alpha\ . \end{aligned} \tag{6.15}$$

In order for wino-like dark matter to give rise to a detectable spin independent cross section, at current sensitivities, a mixture of Higgsino content is needed, where the spin independent nucleon cross section of $\sim 5 \times 10^{-44}$ cm$^2$ at dark matter mass of $\sim 60$ GeV is currently the minimum of the confidence limits.[28] The LSP can have a large wino component and produce a large spin independent cross section as well as a large flux in the halo. Such can be achieved, for example in a model with split gaugino masses at the GUT scale [a] such as $M_a = m_{1/2}(1 + \Delta_a)$, $a = 1, 2, 3$ with $(m_0, m_{1/2}, A_0, \tan\beta, (\Delta_1, \Delta_2, \Delta_3)) = ((1000, 800, 0)\text{GeV}, 10, (-.4 -.7 -.4))$, with sign$(\mu) > 0$, where $(m_0, A_0)$ are the universal scalar mass and trilinear coupling, respectively, and $\tan\beta$ is the rato of the Higgs VEVS and $\mu$ enters

[a] a natural class of models where this happens are grand unified models such as $SU(5)$, $SO(10)$, and $E_6$ where the GUT symmetry is broken by a non-singlet $F$ term (for recent work in this direction see Refs. 31–33).

as the bilinear term in the superpotential (see i.e. Ref. 29). After radiative electroweak symmetry breaking the produced spectrum and mixing leads to a wino-like eigenstate with both a strong halo cross section and scattering cross sections:

$$(N_{\tilde{B}}, N_{\widetilde{W}}, N_{\tilde{H}_1}, N_{\tilde{H}_2}) = (0.234, -0.957, 0.161, -0.064), \qquad (6.16)$$

$$\sigma_{\rm SI} = 1 \times 10^{-8} \ \text{pb}, \qquad \sigma_{\rm SD} = 2 \times 10^{-5} \ \text{pb}, \qquad (6.17)$$

$$\langle \sigma v \rangle_{\widetilde{W}\widetilde{W} \to W^+W^-} = 2 \times 10^{-24} \ \text{cm}^3/\text{s}, \quad m_{\widetilde{W}} = 185 \ \text{GeV} \ . \qquad (6.18)$$

Thus, this class of model produces positrons in the halo which describe the PAMELA data (see next section), and produces a spin independent scattering cross section within reach of the CDMS and Xenon experiments.

## 6.5. Positron Flux from Wino-Like Dark Matter

A wino-like LSP can annihilate in the galaxy producing $W^{\pm}$ bosons. The resulting positrons diffuse through the galaxy to satellite detectors. Several analyses have shown that the positron flux ratio from the annihilations of wino-like dark matter provides a relatively good description of the PAMELA data.[3,8,22]

The positron flux can be described semi-analytically. The flux enters as a solution to the diffusion loss equation, which is solved in a region with a cylindrical boundary, and is well approximated under steady state conditions by

$$\Phi_{\bar{e}}(E) = \frac{B_{\bar{e}} v_{\bar{e}}}{8\pi b(E)} \frac{\rho_{\odot}^2}{m_{\tilde{\chi}^0}^2} F(E) \ , \quad [\text{GeV} \cdot \text{cm}^2 \cdot \text{s} \cdot \text{sr}]^{-1} \qquad (6.19)$$

$$F(E) = \int_E^{M_{\tilde{\chi}^0}} dE' \sum_k \langle \sigma v \rangle_{\rm halo}^k \frac{dN_{\bar{e}}^k}{dE'} \cdot \mathcal{I}(E, E'). \qquad (6.20)$$

The particle physics depends on $\langle \sigma v \rangle_{\rm halo}$, the velocity averaged cross section in the halo of the galaxy, and $dN/dE$, the positron fragmentation functions. The astrophysics depends on $b(E) = E^2/(\text{GeV}\cdot\tau_E)$ with $\tau_E = (1\text{–}3)\times 10^{16}\,\text{s}$ which parametrizes the energy loss in the flux from the presence of magnetic fields and from scattering off galactic photons. $\mathcal{I}(E, E')$ is the a-dimensional halo function and the minimal parameters needed to describe it are the diffusion/propagation parameters $\delta$, $K_0$ and $L$, with diffusion coefficient $K(E) = K_0(E/\text{GeV})^{\delta}$ and $L$ being the half height of the cylinder, (for some fits see with various halo profiles see i.e. Ref. 21). The local halo density, $\rho_{\odot}$, lies in the range $\sim(0.3-0.6)\mathrm{GeV/cm^3}$ (with recent result pointing towards a value closer to $\sim 0.4\ \mathrm{GeV/cm^3}$) in the vicinity of galactic plane and at a distance of $r_\odot = 8.5\,\mathrm{kpc}$. $B_{\bar{e}}$ is a so-called boost factor which parametrizes the possible local inhomogeneities of the dark matter distribution. Large boost factors have been used in the literature, even as large as 10,000, to explain the PAMELA data, however analyses of simulations increasingly suggest boost factors near unity. A pure wino needs no multiplicative boost factor,[3] while a wino with some mixture requires small halo boosts.[8]

Thus, for wino dominated dark matter, $\frac{dN_{\bar{e}}^k}{dE'} \to \frac{dN_{\bar{e}}^{WW}}{dE'}$ with $\langle\sigma v\rangle_{\widetilde{W}\widetilde{W}\to W^+W^-}$ the overwhelmingly dominant source of positrons. The relative strength of the halo cross section is governed then by the dark matter mass, and the strength wino component (see Eq.(6.3) for the dominant contribution of a pure wino) however the shape of the flux distribution is controlled by the fragmentation function and the halo/profile model. The fragmentation function for the $W$ boson falls off at the mass of the dark matter, and therefore one expects to see a significant dip in the positron fraction at threshold. The fall off may however be compensated by an astrophysical flux from shock waves/pulsars or other astrophysical remnants.[3]

Of course, not only are positrons produced from the decays of the $W$ bosons, but hadrons and photons as well. Separate analyses[3,8] have shown, with both semi-analytical fits and fits using Galprop,[23] that the hadronic fluxes are in close accord with the PAMELA data. The precise nature of the fits depend sensitively on the halo/diffusion models and the proper handling of cosmic backgrounds. Recent preliminary analyses on the photon spectrum indicates that a pure wino may be constrained by the production of photons.[24] If this is the case, this may represent a test of pure wino dark matter, however, as discussed in the previous sections non-neglible non-wino components can play an important role in the size of the $\langle\sigma v\rangle$ which is a sensitive factor in the flux predictions.

## 6.6. Dark Matter and the LHC

Our understanding of what we expect to observe at the LHC is tied closely to the relations between the LHC space of signatures and their possible degeneracies.[47] In addition, there has been recent effort to connect the LHC space of signals with dark matter signatures in broad classes of models.[48] The above connection, dark matter and the LHC, becomes rather relevant in the context of models which predict light gauginos and in particular light gluinos (see i.e. Refs. 33, 50). In such cases, the production of colored sparticles via gluino production[49] becomes the central source of high PT jets and missing energy and can lead to early discovery prospects at low luminosity.

Specifically in the context of a wino LSP, collider implications have been discussed in some depth in the literature (see i.e. Refs. 8, 46, 52–59). The near degeneracy of a pure wino with the lightest chargino yields a mass spliting of order 160 MeV making its discovery prospects at hadron colliders challenging. The chargino decay can lead a wino and pion giving rise to a displaced vertex of a track length of a few centimetres. In association, dilepton signals with displaced vertices have been emphasized in Ref. 51. Typically however the role of colored sparticle production in models with a wino-like LSP has been put on a back burner. It has recently been observed however, that light gluinos can arise when the LSP has a significant wino component and when the scalars of the theory decouple.[6] Here one finds dominant LHC production modes are

$$pp \to [(\tilde{g}\tilde{g}), (\widetilde{W}\widetilde{C}_1), (\widetilde{C}_1^{\pm}, \widetilde{C}_1^{\mp})] \tag{6.21}$$

and the 3 body decay modes lead to rich jet and missing $E_T$ signatures translating into early discovery reach of the gluino as low as 500 GeV with $\sqrt{s} = 7$ TeV for integrated luminosity in the neighbourhood of 800 pb$^{-1}$ with LSP winos in the $m_{\widetilde{W}} \in [170, 200]$ GeV (as motivated by the reported PAMELA data).[46] In recent works, the role of strongly produced superpartners at the LHC in models with a wino-like LSP has been emphasized[8] inspired by these recent observations that a wino can fit the PAMELA data[3,8] and it has been found that a wino-like LSP can occur with a compressed spectrum, in particular for both the squarks and gluino.[8] The lighted colored superpartners can produce stunning monojet, multijet and lepton signals with less than fb$^{-1}$ of data when the mass splitting between the LSP and chargino opens up due in part to a non-negligible Higgsino component. The tagging of bjets in these models becomes very relevant as the three body decay of the $\tilde{g} \to \widetilde{W}b\bar{b}$ can dominate and the reconstruction of the invariant mass of the 2 bjet system has been shown to reveal a potential clue to the size of the gluino mass from its inferred kink in the spectrum. Indeed in greater generality, the gluino three body decay modes

$$\tilde{g} \to b\bar{b}\widetilde{W}, \quad \tilde{g} \to q\bar{q}\widetilde{W} \quad \tilde{g} \to t\bar{t}\widetilde{N}_2, \tag{6.22}$$

$$\tilde{g} \to t\bar{b}\widetilde{C}^- + h.c., \quad \tilde{g} \to q_u\bar{q}_d\widetilde{C}^- + h.c, \tag{6.23}$$

can all become important depending on the part of the parameter space and loop decays of gauginos can also play a role.[60]

## 6.7. Concluding Remarks

The implications of a wino-like LSP in connection with PAMELA[3,8] and in connection with the LHC have recently been studied in some detail.[8,46] In particular, the rather exciting possibility of the observable effects from the production of light gluinos at the LHC, and a possible interpretation of dark matter in the PAMELA Satellite data, represent two prominent discovery channels for the identification of the existence of superpartners. A third related indication of wino-like dark matter could come from an enhanced spin independent cross section when the wino content is supplemented by non-neglible sources of Higgsino and bino content[8] producing cross sections in the physically interesting region of $\sim \mathcal{O}(10^{-44})\text{cm}^2$. If the relic density is indeed composed of a wino-like LSP, the two possibilities discussed here, namely an extended neutralino sector, or a non-thermal cosmological history, must be taken very seriously.

The LHC could open a paradigm shift in our understanding of the nature of physics beyond the Standard Model of particle physics. Within the plethora of models that have been proposed, the clearest and most well motived generalization of the Standard Model is clearly softly broken supersymmetry. The eigen content of the LSP plays a central role in determining observable manifestations of SUSY in forthcoming experiments. The LSP wavefunction indeed plays a central role in governing the spin independent cross section of LSPs scattering off of target nuclei, as well as the size of the annihilation cross sections of the LSPs in the early universe and in the galactic halo. The eigen content also implies certain relations on the amount of LSP missing energy produced in hadron collisions, as well as relations amongst the spectrum of the heavier unstable supersymmetric states under the constraints of radiative electroweak symmetry breaking and therefore the possible decays and productions modes of superpartners.

We have entered a data-rich period and some answers to the nature of the LSP are being directly tested. Shortly, it may indeed be possible to answer the seemingly simple question: what is the nature of the lightest supersymmetric particle? Here we have emphasized several theoretical and phenomenological motivations that indicate the distinct possibility that the LSP has a sizeable wino component; and should this be the case, we can expect to constrain or discover models which do produce a wino-like LSP in the very near future.

## Acknowledgments

We collectively thank collaborators who have worked on many of the topics discussed here including: Bobby Acharya, Konstantin Bobkov, Piyush Kumar, Boris Kors, Zuowei Liu, Ran Lu, Pran Nath, Brent Nelson, Aaron Pierce, Jing Shao, Scott Watson, and Liantao Wang.

## References

1. [PAMELA Collaboration], Nature **458**, 607 (2009).
2. [WMAP Collaboration], Astrophys. J. Suppl. **180**, 330 (2009).
3. G. Kane, R. Lu and S. Watson, Phys. Lett. B **681**, 151 (2009); P. Grajek, G. Kane, D. Phalen, A. Pierce, S. Watson, Phys. Rev. D **79**, 043506 (2009).
4. T. Moroi and L. Randall, Nucl. Phys. B **570**, 455 (2000).
5. B. S. Acharya and E. Witten, arXiv:hep-th/0109152.
6. B. S. Acharya, K. Bobkov, G. L. Kane, P. Kumar and J. Shao, Phys. Rev. D **76**, 126010 (2007); Phys. Rev. D **78**, 065038 (2008).
7. B. S. Acharya, P. Kumar, K. Bobkov, G. Kane, J. Shao and S. Watson, JHEP **0806**, 064 (2008).
8. D. Feldman, Z. Liu, P. Nath and B. D. Nelson, Phys. Rev. D **80**, 075001 (2009); D. Feldman, arXiv:0908.3727 [hep-ph].
9. J. Hisano, S. Matsumoto, M. M. Nojiri and O. Saito, Phys. Rev. D **71**, 063528 (2005); N. Arkani-Hamed, D. P. Finkbeiner, T. R. Slatyer and N. Weiner, Phys. Rev. D **79**, 015014 (2009); S. Cassel, D. M. Ghilencea and G. G. Ross, Nucl. Phys. B **827**, 256 (2010).
10. D. Feldman, Z. Liu and P. Nath, Phys. Rev. D **79**, 063509 (2009); M. Ibe, H. Murayama and T. T. Yanagida, Phys. Rev. D **79**, 095009 (2009); Y. Bai, M. Carena and J. Lykken, Phys. Rev. D **80**, 055004 (2009).
11. D. E. Kaplan, M. A. Luty and K. M. Zurek, Phys. Rev. D **79**, 115016 (2009); T. Cohen and K. M. Zurek, arXiv:0909.2035 [hep-ph].
12. For a detailed discussion see the Chapter by S. Watson.
13. H. E. Haber and G. L. Kane, Phys. Rept. **117**, 75 (1985); D. J. H. Chung, L. L. Everett, G. L. Kane, S. F. King, J. D. Lykken and L. T. Wang, Phys. Rept. **407**, 1 (2005).
14. L. Randall and R. Sundrum, Nucl. Phys. B **557**, 79 (1999); G. F. Giudice, M. A. Luty, H. Murayama and R. Rattazzi, JHEP **9812**, 027 (1998).
15. M. K. Gaillard, B. D. Nelson and Y. Y. Wu, Phys. Lett. B **459**, 549 (1999); J. A. Bagger, T. Moroi and E. Poppitz, JHEP **0004**, 009 (2000).
16. L. E. Ibanez and D. Lust, Nucl. Phys. B **382**, 305 (1992); V. Kaplunovsky and J. Louis, Nucl. Phys. B **422**, 57 (1994).
17. A. H. Chamseddine, R. Arnowitt and P. Nath, Phys. Rev. Lett. **49** (1982) 970; L. Hall, J. Lykken and S. Weinberg, Phys. Rev. **D27**, 2359 (1983); H. P. Nilles, Phys. Rept. **110**, 1 (1984); V. S. Kaplunovsky and J. Louis, Phys. Lett. B **306**, 269 (1993); A. Brignole, L. E. Ibanez and C. Munoz,

Nucl. Phys. B **422**, 125 (1994); A. Brignole, L. E. Ibanez and C. Munoz, arXiv:hep-ph/9707209.
18. D. M. Pierce, J. A. Bagger, K. T. Matchev and R. j. Zhang, Nucl. Phys. B **491**, 3 (1997).
19. S. Nakamura and M. Yamaguchi, Phys. Lett. B **638**, 389 (2006).
20. For a detailed discussion see the Chapter by J. Wells.
21. M. Cirelli, R. Franceschini and A. Strumia, Nucl. Phys. B **800**, 204 (2008).
22. J. Hisano, M. Kawasaki, K. Kohri and K. Nakayama, Phys. Rev. D **79**, 063514 (2009).
23. A. W. Strong and I. V. Moskalenko, Astrophys. J. **509**, 212 (1998).
24. S. Murgia, "2009 Fermi Symposium", Washington DC, on behalf of the FERMI-LAT Collaboration; and for related analysis see also, Winer, Plenary Talk at SUSY 09, Northeastern University, Boston; T. A. Porter and f. t. F. Collaboration, arXiv:0907.0294 [astro-ph.HE]; T. L. Collaboration, arXiv:0912.0973 [astro-ph.HE], to appear in Phys. Rev. Letters.
25. U. Chattopadhyay, T. Ibrahim and P. Nath, Phys. Rev. D **60**, 063505 (1999).
26. J. R. Ellis, A. Ferstl and K. A. Olive, Phys. Lett. B **481**, 304 (2000); J. R. Ellis, K. A. Olive and C. Savage, Phys. Rev. D **77**, 065026 (2008).
27. G. Belanger, F. Boudjema, A. Pukhov and A. Semenov, Comput. Phys. Commun. **180**, 747 (2009).
28. Z. Ahmed *et al.* [CDMS Collaboration], Phys. Rev. Lett. **102**, 011301 (2009) [arXiv:0802.3530 [astro-ph]]; J. Angle *et al.* [XENON Collaboration], Phys. Rev. Lett. **100**, 021303 (2008) [arXiv:0706.0039 [astro-ph]]. E. Armengaud *et al.*, arXiv:0912.0805 [astro-ph.CO].
29. See S.P. Martin's Chapter for a review
30. A. Corsetti and P. Nath, Phys. Rev. D **64**, 125010 (2001).
31. S. P. Martin, Phys. Rev. D **79**, 095019 (2009),
32. P. Athron, S. F. King, D. J. Miller, S. Moretti and R. Nevzorov, Phys. Rev. D **80**, 035009 (2009).
33. D. Feldman, Z. Liu and P. Nath, Phys. Rev. D **80**, 015007 (2009).
34. R. Blumenhagen, B. Kors, D. Lust, S. Stieberger, Phys. Rept. **445**, 1 (2007).
35. M. B. Green and J. H. Schwarz, Phys. Lett. B **149**, 117 (1984).
36. B. Kors and P. Nath, JHEP **0412**, 005 (2004); JHEP **0507**, 069 (2005).
37. D. Feldman, B. Kors and P. Nath, Phys. Rev. D **75**, 023503 (2007).
38. D. Ghilencea, L. Ibanez, N. Irges, F. Quevedo, JHEP **0208** (2002) 016.
39. B. Körs and P. Nath, Phys. Lett. B **586**, 366 (2004).
40. D. Feldman, Z. Liu and P. Nath, Phys. Rev. Lett. **97**, 021801 (2006).
41. C. Coriano', N. Irges and E. Kiritsis, Nucl. Phys. B **746**, 77 (2006).
42. J. Kumar, A. Rajaraman and J. D. Wells, Phys. Rev. D **77**, 066011 (2008); Y. Mambrini, arXiv:0907.2918 [hep-ph].
43. D. Feldman, B. Kors and P. Nath, Phys. Rev. D **75**, 023503 (2007).
44. A. Arvanitaki, N. Craig, S. Dimopoulos, S. Dubovsky and J. March-Russell, arXiv:0909.5440 [hep-ph].
45. D. Feldman, Z. Liu and P. Nath, Phys. Rev. D **75**, 115001 (2007); S. A. Abel, M. D. Goodsell, J. Jaeckel, V. V. Khoze and A. Ringwald, JHEP **0807**, 124

(2008); C. P. Burgess, J. P. Conlon, L. Y. Hung, C. H. Kom, A. Maharana and F. Quevedo, JHEP **0807**, 073 (2008).
46. G. Kane, D. Feldman, R. Lu and B.D. Nelson, to appear.
47. N. Arkani-Hamed, G. L. Kane, J. Thaler and L. T. Wang, JHEP **0608**, 070 (2006); G. L. Kane, P. Kumar and J. Shao, J. Phys. G **34**, 1993 (2007); J. P. Conlon, C. H. Kom, K. Suruliz, B. C. Allanach and F. Quevedo, JHEP **0708**, 061 (2007); D. Feldman, Z. Liu and P. Nath, Phys. Rev. Lett. **99**, 251802 (2007); JHEP **0804**, 054 (2008).
48. D. Feldman, Z. Liu and P. Nath, Phys. Lett. B **662**, 190 (2008); B. Altunkaynak, M. Holmes and B. D. Nelson, JHEP **0810**, 013 (2008); D. Feldman, Z. Liu and P. Nath, Phys. Rev. D **78**, 083523 (2008); N. Bhattacharyya, A. Datta and S. Poddar, Phys. Rev. D **78**, 075030 (2008); G. Kane and S. Watson, Mod. Phys. Lett. A **23**, 2103 (2008).
49. G. L. Kane and J. P. Leveille, Phys. Lett. B **112**, 227 (1982); P. R. Harrison and C. H. Llewellyn Smith, Nucl. Phys. B **213**, 223 (1983); E. Reya and D. P. Roy, Phys. Lett. B **141**, 442 (1984); S. Dawson, E. Eichten and C. Quigg, Phys. Rev. D **31**, 1581 (1985); W. Beenakker, R. Hopker, M. Spira and P. M. Zerwas, Nucl. Phys. B **492**, 51 (1997).
50. B. S. Acharya, P. Grajek, G. L. Kane, E. Kuflik, K. Suruliz and L. T. Wang, arXiv:0901.3367 [hep-ph].
51. T. Gherghetta, G. F. Giudice and J. D. Wells, Nucl. Phys. B **559**, 27 (1999).
52. C. H. Chen, M. Drees and J. F. Gunion, Phys. Rev. Lett. **76**, 2002 (1996); Phys. Rev. D **55**, 330 (1997).
53. J. L. Feng, T. Moroi, L. Randall, M. Strassler and S. f. Su, Phys. Rev. Lett. **83**, 1731 (1999) [arXiv:hep-ph/9904250].
54. S. Mrenna and J. F. Gunion, Int. J. Mod. Phys. A **16S1B**, 822 (2001).
55. G. L. Kane, J. D. Lykken, S. Mrenna, B. D. Nelson, L. T. Wang and T. T. Wang, Phys. Rev. D **67**, 045008 (2003); Phys. Lett. B **551**, 146 (2003). P. Binetruy, A. Birkedal-Hansen, Y. Mambrini and B. D. Nelson, Eur. Phys. J. C **47**, 481 (2006).
56. M. Ibe, T. Moroi and T. T. Yanagida, Phys. Lett. B **644**, 355 (2007).
57. U. Chattopadhyay, D. Das, P. Konar and D. P. Roy, Phys. Rev. D **75**, 073014 (2007).
58. P. Langacker, G. Paz, L. T. Wang and I. Yavin, Phys. Rev. Lett. **100**, 041802 (2008); P. Langacker, G. Paz, L. T. Wang and I. Yavin, Phys. Rev. D **77**, 085033 (2008).
59. M. R. Buckley, L. Randall and B. Shuve, arXiv:0909.4549 [hep-ph].
60. H. E. Haber and G. L. Kane, Nucl. Phys. B **232**, 333 (1984); E. Ma and G. G. Wong, Mod. Phys. Lett. A **3**, 1561 (1988); H. Baer, R. M. Barnett, M. Drees, J. F. Gunion, H. E. Haber, D. L. Karatas and X. R. Tata, Int. J. Mod. Phys. A **2**, 1131 (1987); H. Baer, X. Tata and J. Woodside, Phys. Rev. D **42**, 1568 (1990); M. Toharia and J. D. Wells, JHEP **0602**, 015 (2006). H. Baer, A. Mustafayev, E. K. Park, S. Profumo and X. Tata, JHEP **0604**, 041 (2006); D. Feldman, Z. Liu and P. Nath, Phys. Rev. D **80**, 015007 (2009); M. A. Diaz, B. Panes and P. Urrejola, arXiv:0910.1554 [hep-ph].

# Reevaluating the Cosmological Origin of Dark Matter

Scott Watson*

*Department of Physics, University of Michigan,*
*450 Church Street, Ann Arbor, MI 48109*
*gswatson@syr.edu*

The origin of dark matter as a thermal relic offers a compelling way in which the early universe was initially populated by dark matter. Alternative explanations typically appear exotic compared to the simplicity of thermal production. However, recent observations and progress from theory suggest that it may be necessary to be more critical. This is important because ongoing searches probing the microscopic properties of dark matter typically rely on the assumption of dark matter as a single, unique, thermal relic. On general grounds I will argue that non-thermal production of dark matter seems to be a robust prediction of physics beyond the standard model. However, if such models are to lead to realistic phenomenology, the theoretical framework is highly restrictive, and we find that viable models would result in concrete and testable predictions. Although many challenges remain, the non-thermal component of such models may offer a new way to test string theories that are formulated to provide realistic particle physics near the electroweak scale.

## 7.1. Cosmological Evidence for Dark Matter

The first hint for the existence of dark matter came from observations of the nearby Coma cluster of galaxies by Fritz Zwicky in 1933.[1] Zwicky found that by assuming the galaxies comprising the cluster were in equilibrium, their velocity distribution implied a cluster mass far exceeding that inferred from the luminous matter contained within the cluster. Today, through a number of complementary and more sophisticated techniques, cluster studies suggest a relative abundance of dark matter $\Omega_{cdm} = 0.2$ to 0.3, where $\Omega_{cdm} = \rho_{cdm}/\rho_c$ is the fractional amount of dark matter

*On leave from the Department of Physics, Syracuse University, Syracuse, NY 13244.

as compared to the critical density for collapse which today is given by $\rho_c = 3H_0^2/(8\pi G) \approx 10^{-29}\text{g}\cdot\text{cm}^{-3}$ with a Hubble parameter of around 70 $\text{km}\cdot\text{s}^{-1}\cdot\text{Mpc}^{-1}$ and $G = 6.67 \times 10^{-11}$ is Newton's gravitational constant.

A more precise measure of dark matter can be obtained from less direct observations, such as the temperature anisotropies of the cosmic microwave background (CMB), and the evolution and formation of the large scale structure (LSS) of the universe. This is because the evolution of density inhomogeneities that eventually grow to form LSS is quite sensitive to the properties of the primordial bath of particles from which they evolve. At the time the CMB photons last scattered, by mass the particles were primarily composed of dark matter. Combining probes of the CMB, structure formation, and distance probes such as supernovae the amount of dark matter is found to be[2]

$$\Omega_{cdm} = 0.233 \pm 0.013, \tag{7.1}$$

implying that the total energy budget of the universe is comprised of a little less than a quarter dark matter.

In addition to determining the dark matter abundance, these observations, along with the above mentioned galaxy and cluster observations, also tell us that the dark matter must be 'cold' – meaning non-relativistic at the time of structure formation, stable (at least until recently), and 'dark' meaning without significant electromagnetic interactions. The latter, when combined with constraints from Big Bang Nucleosynthesis (BBN), suggests the particles are at most weakly interacting with themselves and with other particles. Combining all of these cosmological observations, we find that what is expected is a WIMP, that is a Weakly Interacting Massive Particle.

## 7.2. Reevaluating the WIMP Miracle

### 7.2.1. *WIMPs as thermal relics*

Big Bang cosmology predicts that as the universe expands it cools[a]. Thus, if we consider the expansion in reverse, we expect at some point in the early universe that the cosmic temperature would have exceeded the mass of the dark matter particles rendering them relativistic. At this temperature the particles are relatively light and easy to produce from the primordial plasma so that their creation and annihilation would be near thermal equilibrium.

[a]In this subsection we briefly review the scenario of thermal production of WIMPs. For a more detailed treatment we refer the reader to Ref. 3.

In equilibrium, the rate at which particles annihilate in a fixed comoving volume $a^3$ is $n_x a^3 \times n_x \langle \sigma_x v \rangle$, where $n_x$ is the number density of dark matter particles of mass $m_x$, $\sigma_x$ is their annihilation cross section, and $\langle \sigma_x v \rangle$ is the thermally averaged cross-section and relative velocity of the particles. In equilibrium, particle annihilations should be balanced by particle pair-creation and the rate is given by $(n_x^{eq})^2 a^3 \langle \sigma_x v \rangle$, so that the number in a comoving volume is constant. This is expressed by the Boltzmann equation

$$\frac{d(n_x a^3)}{dt} = -a^3 \langle \sigma_x v \rangle \left[ n_x^2 - (n_x^{eq})^2 \right], \tag{7.2}$$

where the first term on the right is dilution due to particle annihilations ($XX \to \gamma\gamma$), and the second term is the reverse process of particle creation from the thermal bath ($\gamma\gamma \to XX$). At high temperatures when $T \gg m_x$, we have $n_x^{eq} \sim T^3$ and since $T \sim 1/a$ the last two terms cancel and the particle density simply scales with the expansion. Once the particles become non-relativistic ($m_x \ll T$) then $n_x^{eq} \sim e^{-m_x/T}$ becomes Boltzmann suppressed and particle production becomes negligible, so that the density of particles rapidly drops due to both the expansion and annihilations. Finally, once the number density drops to the point where the cosmic expansion exceeds the annihilation rate per particle $H \gtrsim n_x \langle \sigma_x v \rangle$, the particles 'freezeout' and their number per comoving volume is

$$\frac{n_x}{s} = \left. \frac{H}{s \langle \sigma_x v \rangle} \right|_{T=T_f}, \tag{7.3}$$

where all parameters appearing in this expression are to be evaluated at the freeze-out temperature $T_f$ and we have introduced the entropy density $s = (2\pi^2/45) g_* T^3 \sim 1/a^3$, which gives a more convenient way to define the comoving frame and $g_*$ is the number of relativistic degrees of freedom at the time of freeze-out. The freeze-out temperature can be found from the number density, since one finds that it closely tracks the equilibrium density near freeze-out. Thus, at freeze-out $n_x \sim n_x^{eq} \sim e^{-m_x/T_f}$, and the mass to temperature ratio at this time is only logarithmically sensitive to changes in the parameters appearing in (7.3). In fact, for thermally produced dark matter associated with weak-scale physics this ratio is typically $m_x/T = 25$, with corrections up to at most a factor of two[b].

[b] Of course, the Boltmann equation can always be solved numerically and one finds good agreement with the analytic argument given above. See Ref. 5 for a more thorough discussion.

Assuming no significant entropy production following freeze-out, the number of dark matter particles per comoving volume (7.3) will be preserved until today resulting in a density of dark matter

$$\Omega_{cdm}(T) \equiv \frac{\rho_{cdm}(T)}{\rho_c} = \frac{m_x n_x(T)}{\rho_c} = \frac{m_x}{\rho_c}\left(\frac{n_x(T_f)}{s(T_f)}\right) s(T)$$
$$= \frac{m_x}{\rho_c}\left(\frac{H}{s\langle\sigma_x v\rangle}\right)_{T=T_f} s(T). \qquad (7.4)$$

Making the additional assumption that the universe is entirely radiation dominated at freeze-out so that $H \sim T^2$ and using $s \sim T^3$ we find that the critical density in dark matter evaluated today is

$$\Omega_{cdm}(T_0) = \frac{45}{2\pi\sqrt{10}}\left(\frac{s_0}{\rho_c m_p}\right)\left(\frac{m_x}{g_*^{1/2}\langle\sigma_x v\rangle T_f}\right),$$
$$\approx 0.23 \times \left(\frac{10^{-26}\,\mathrm{cm}^3 \cdot \mathrm{s}^{-1}}{\langle\sigma_x v\rangle}\right), \qquad (7.5)$$

where $g_* = 106.75$ is the number of relativistic degrees of freedom around the typical temperature of dark matter freezeout (see Ref. 3 for a more detailed discussion), and the entropy density today is $s_0 = 2970\ \mathrm{cm}^{-3}$,

This result is interesting for several reasons. First, we note that the abundance only depends on the self annihilation cross section of the dark matter particles, and we saw that any changes in the theory enter as logarithmic corrections – i.e. this scenario is robust. Thus, measurements of the thermal relic density won't lead to any deeper understanding of physics beyond the standard model or the evolution of the universe prior to freeze-out. This will be an important difference from the non-thermal case that we will discuss below. Another interesting fact about the result above is that if we compare this result for the abundance of dark matter produced thermally with the precision cosmological measurement given in (7.1), we find that $\langle\sigma_x v\rangle \approx 10^{-26}\mathrm{cm}^3{\cdot}\mathrm{s}^{-1}$ (or $\sigma_x \approx 1$ picobarn) and thus we are lead to expect a new particle with weak scale interactions. Of course, we already expect new physics to appear near the electroweak scale to properly account for a light Higgs. Such theories for an extension of the standard model postulate new symmetries above the electroweak scale, and at low energy their breaking results in a lightest stable particle associated with the new physics. One example is provided by the supersymmetric (SUSY) neutralino, which after the spontaneous breaking of SUSY remains stable under a residual discrete symmetry, i.e. R-parity. That the weak scale cross section naturally emerges when comparing the cosmological observations with the thermal prediction (7.5), and the fact that this was independently expected from theoretical considerations related to the Higgs has lead some to refer to this coincidence as the 'WIMP Miracle'. To summarize,

Assuming:

- The WIMPs were at some point relativistic and reached chemical equilibrium.
- At the time of freeze-out, the universe was radiation dominated (all other contributions to the energy density were negligible).
- Following freeze-out there was no significant entropy production.
- There were no other late-time sources of dark matter particles (e.g. decays from other particles).
- There is only one species of dark matter particle and any other new particles are unstable or have significantly larger mass.

we find that

- The relic density does not depend on the expansion history, only on the temperature at freeze-out.
- The relic density does not depend on any high scale physics, only on the low-energy cross-section.
- The answer is very robust to changes in the cross-section and mass of the particles.
- When combined with cosmological observations – we expect new physics at the electroweak scale.

Although all the assumptions listed above are well motivated – and the resulting model is quite simple and compelling – it is important to proceed with caution when attempting to promote any candidate signature coming from particle experiments to a claim that one has gained a complete understanding of cosmological dark matter. In addition to the challenge of reconstructing the properties of dark matter from signatures at colliders, direct, and indirect detection, there are also a number of challenges associated with the reconstruction of the relic density of dark matter itself. These include that the relic density could be comprised of more than one kind of particle, that the expansion history prior to BBN could be more complicated than expected, or that the late decay of particles could alter the abundance of dark matter particles. These are just a couple possibilities that could stymy the extrapolation of a confirmed particle detection to an accurate picture of cosmological dark matter.

### 7.2.2. *Other dark matter*

One key assumption underlying the connection between the thermal relic abundance (7.5) and LHC, is that the WIMP is a unique dark matter candidate and that its mass is far below the next to lightest particle associated with new physics. As an example of the latter, in supersymmetric theories it is common that the next to lightest SUSY particle (NLSP) can be nearly degenerate in mass with the LSP. If this is the case, not only could the NLSP be mistaken as a stable WIMP (LSP) in the LHC detector – as the lifetime of a particle in the detector is only $10^{-8}$ s, or the NLSP and its decay products might both be neutral – but cosmologically, coannihilations[13] between the NLSP and LSP will significantly reduce the thermal relic density estimated in (7.5).

Another important possibility is that there is more than one type of dark matter. Thus, the total dark matter abundance should always be thought of as

$$\Omega_{cdm}^{total} = \sum_i \Omega_{cdm}^{(i)}, \tag{7.6}$$

where the sum is over all contributions to the dark matter energy budget. In fact, because we now know that neutrinos have mass, we also know that they must make up some part of the dark matter. However, we also know that neutrinos are relativistic at the onset of structure formation, i.e. they are 'warm' dark matter, requiring that they must represent a small fraction of the total dark matter. In fact, combining the recent WMAP5 data with other cosmological observations a bound of $\Omega_\nu h^2 \lesssim 0.006$ was obtained in Ref. 2. Of course, in addition to neutrinos there are a number of other possible contributions to the cosmological dark matter, including axions. The QCD axion provides an elegant solution to the strong CP problem, and although tightly constrained, still remains a viable dark matter candidate for some regions of its parameter space (see e.g. Ref. 12). It is also expected that additional axions will generically arise at low energies from effective theories with ultraviolet completions in string theory (see Ref. 14 and references within).

### 7.2.3. Modified expansion history at freeze-out

For the calculation of the thermal relic density one assumption was that the universe was radiation dominated at the time of freeze-out, so that $H \sim T^2$ allowing for the simplification in going from (7.4) to (7.5). This assumption agrees with the observational predictions of BBN occurring a few minutes after the Big Bang. However, there is no cosmological evidence for this assumption prior to the time of BBN.

There are both theoretical and observational indications that this assumption may be too naive. Indeed, given the rich particle phenomenology that occurs at energies above the scale of BBN (energies around a MeV), we might expect this to complicate the simple picture of a purely radiation dominated universe. Moreover, relics from early universe phase transitions, such as scalar condensates or rolling inflatons that didn't completely decay, would also be expected to alter the expansion history.

In fact, theories beyond the standard model generically predict the existence of scalar fields. Many of these fields have little or no potential – so called moduli, so they are often light. Examples include the sizes and shapes of extra dimensions, or flat directions in the complicated SUSY field space of the scalar partners to standard model fermions. In the early universe these moduli will generically be displaced from their low energy minima during phase transitions, such as inflation.[15] Energy can then become stored in the form of coherent oscillations forming a scalar condensate. The cosmological scaling of the condensate depends on which term in the potential is dominant. For a potential with a dominant term $V \sim \phi^\gamma$ one finds that the pressure depends on the energy density as

$$p = \left(\frac{2\gamma}{2+\gamma} - 1\right)\rho, \tag{7.7}$$

where $\rho$ scales as

$$\rho = \rho_0 a^{-6\gamma/(2+\gamma)}. \tag{7.8}$$

Two examples are a massive scalar with negligible interactions for which $\gamma = 2$ and the condensate scales as pressure-less matter $p = 0$, whereas if physics at the high scale is dominant – in the form of non-renormalizable operators – then $\gamma > 4$ and the condensate evolves as a stiff fluid $p \approx \rho$ for large $\gamma$. Whatever the behavior of the condensate, if it contributes appreciably to the total energy density prior to freeze-out the abundance (7.5) will be altered. This is because the presence of addition matter will increase the cosmic expansion rate allowing less time for particle annihilations prior to freeze-out[c]. The expansion rate at the time of freeze-out is then given by

$$H_f = H_{rdu}\left(1+\frac{\rho_\phi}{\rho_r}\right)^{1/2}, \tag{7.9}$$

where $H_{rdu}$ is the expansion rate in a radiation dominated universe and $\rho_\phi$ and $\rho_r$ are the energy density of the scalar condensate and radiation, respectively. Using that at freeze-out $\rho_r = (\pi^2/30)g(T_f)T_f^4$, $\rho_\phi = \rho_{osc}\,(T_f/T_{osc})^p$ where $p \equiv 6\gamma/(2+\gamma)$, and $\rho_{osc}$ is the energy initially in the condensate which began coherent oscillations at temperature $T_{osc}$ we find that the new dark matter abundance is

$$\Omega_{cdm} \to \Omega_{cdm}\sqrt{1+r_0 T_f^{2(\gamma-4)/(2+\gamma)}}, \tag{7.10}$$

where we have used $a \sim 1/T$ for an adiabatic expansion, and the constant

$$r_0 \equiv \frac{30}{\pi^2}\left(\frac{\rho_\varphi(T_{osc})}{g(T_f)T_{osc}^{6\gamma/(2+\gamma)}}\right),$$

where $g(T_f)$ is the number of relativistic degrees of freedom at freeze-out. In practice, typically one finds that for moduli in the early universe $r_0 \gg 1$.[15,32] We see that especially for high energy effects in the potential this can have a significant effect on the resulting relic density. As a simple example, if we consider a massive scalar with negligible interactions ($\gamma = 2$) displaced after a period of inflation we expect $\rho_\varphi(T_{osc}) \simeq m_\varphi^2 m_p^2$ so that $r_0 \simeq (m_\varphi m_p)^2/(g(T_f)T_{osc}^3) \gg 1$ leading to a large enhancement of the relic density. One can also show that there is a significant effect for scalars which are dominated by their kinetic terms (e.g. kination models[16–19] ), which behave like the stiff fluid models discussed above (i.e. $p = \rho$). In fact, this modification to the expansion history was considered in Ref. 20, where it was shown that this would loosen constraints on axionic dark matter. In these examples, the relic density is found to be enhanced compared to that of a purely radiation dominated universe. Of course scalar condensates are not the only additional sources of energy one might expect in the early universe and it is important to note that any additional, significant component

[c]Here we have assumed that radiation contributes substantially to the total energy density or that whatever the primary source of energy density it scales at least as fast as radiation. However, if instead the universe were completely dominated by a massive, non-interacting scalar condensate then this would actually decrease the amount of dark matter. In either situation, the point is that the standard thermal relic density (7.5) will not give the correct result.

will alter the standard thermal abundance of the cosmological dark matter in a way similar to that discussed for scalars above.

### 7.2.4. *Late production of dark matter and entropy*

Two more crucial assumptions that went into the dark matter abundance (7.5) were that there were no other sources of dark matter and/or entropy production following freeze-out. An example of how this can fail is if there is a late period of thermal inflation,[21] which has been argued to be quite natural and necessary for resolving issues with some models coming from string compactifications (see e.g. Ref. 22). Another example is provided by the condensate formation we discussed above. That is, because the moduli have very weak couplings – typically of gravitational strength – the condensate will decay late producing additional particles and entropy. This decay must occur before BBN, which requires the modulus to have a mass larger than around 10 TeV in order to avoid the so-called cosmological moduli problem.[23–26] If the condensate contributes appreciably to the total energy density at the time it decays it will not only produce relativistic particles – and significant entropy – but could also give rise to additional dark matter particles. The former will act to reduce the thermal relic density of dark matter particles $\Omega_{cdm} \rightarrow \Omega_{cdm} \left(T_r/T_f\right)^3$, where $T_r$ is the temperature after the decay and $T_f$ is the freeze-out temperature of the dark matter particles. The factor by which the abundance is diluted can be understood from the scaling of the volume $a^3$ and we have $T \sim 1/a$. As an example, for a 10 TeV scalar the decay to relativistic particles will 'reheat' the universe to a temperature of around an MeV, whereas a 100 GeV WIMP freezes out at a temperature near a GeV. Thus, the scalar decay will dilute the preexisting relic density in dark matter by a factor of about $(T_r/T_f)^3 \simeq 10^9$.

As we have mentioned, in addition to the scalar decaying to relativistic particles it could also decay to WIMPs below their freeze-out temperature. In this case there are two possible results for the relic density, depending on the resulting density of the WIMPs that are produced.[27] If the number density of WIMPs exceeds the critical value

$$n_x^c = \left.\frac{H}{\langle\sigma_x v\rangle}\right|_{T=T_r}, \tag{7.11}$$

then the WIMPs will quickly annihilate down to this value, which acts as an attractor. It is important to note that the fixed point value is evaluated at the time of reheating, in contrast to the freeze-out result (7.3). The other possibility is that the WIMPs produced in the decay do not exceed the fixed point value. In this case their density is just given by $n_x \sim B_x n_\varphi$, where $B_x$ is the branching ratio for scalar decay to WIMPs and $n_\varphi$ is the number density of the scalar condensate. We see that in both these cases the thermal relic density (7.5) would give the wrong answer for the true abundance of dark matter, unless the entropy diluted thermal density of dark matter still manages to exceed the amount coming from the scalar decay. We see that in the case of fixed point production, comparing (7.11) with the calculation for thermal production (7.5) results in a parametric enhancement proportional to the ratio of the freezeout to the reheat temperature, i.e.

$$\Omega_{cdm} \to \Omega_{cdm} \left( \frac{T_f}{T_r} \right), \tag{7.12}$$

which for the example of a 10 TeV scalar results in an overall enhancement of about three orders of magnitude.

Fixing the relic density by the cosmological data (7.1) implies that particles need a larger cross section in order to get the right amount of dark matter. For example, in the case of neutralino dark matter, Winos and Higgsinos annihilate well and have been seen as giving too little dark matter given a thermal history. However, in the theoretically constructed models of Refs. 27, 32 it is found that Winos, Higgsinos, or some mixture can yield the right amount of dark matter due to non-thermal production, which results naturally by requiring consistency of the theory.

It is important to note that even though the naive thermal freezeout calculation no longer determines the relic density, in the fixed point case the answer is still given in terms of the weak scale cross section, and gives a result of the correct order of magnitude for WIMPs with masses of order 100 GeV. The dark matter scale and the electroweak symmetry breaking scale still remain related, and the "WIMP Miracle" survives.

---

In this section, we have seen three possible ways in which the prediction for the amount of cosmological dark matter – and the constraints on microphysics that would result – can be altered. Although the case for more than one type of dark matter, or a more complicated expansion history prior to BBN might seem plausible, the case for a scalar with a mass light enough to decay after dark matter freezeout, but heavy enough to avoid BBN constraints naively would seem quite contrived. In the next section we will argue that this is not the case, and that hints from model building in a way that is consistent with UV physics might predict that non-thermal production of dark matter is the rule rather than an exotic exception.

## 7.3. Non-thermal Production of WIMPs

Non-thermal production of dark matter is not a new idea.[27,32–35,41] However, recent results and future expectations from both theory and experiment suggest that such an origin for dark matter might need to be seriously reconsidered. On the observational side, cosmic ray experiments such as PAMELA and FERMI have reported an excess in both cosmic ray positrons and gamma rays above anticipated astrophysical backgrounds. Although a dark matter explanation seems somewhat unlikely, if the dark matter had a larger cross-section – as made possible by non-thermal production – then candidates like the Wino neutralino may be capable of addressing the excesses through the self annihilations of dark matter.[36–39] Conversely, current and future data from experiments like PAMELA and FERMI can be used to put important constraints on the dark matter cross section and therefore the non-thermal production process.[36–39] By itself these results are certainly not a compelling argument for non-thermal production, but another motivation could be provided in the very near future by the Large Hadron Collider (LHC) or other future colliders. That is, if dark matter was non-thermally produced resulting in a larger self annihilation cross-section, then cross sections for dark matter particles deduced from LHC – when used to calculate the thermal relic density – would result in an unacceptably low cosmological abundance and would be in surprising disagreement with e.g. the WMAP data.[40] Of course, the explanation could also lie elsewhere, e.g. as a consequence of more than one dark matter particle. Thus, we are lead to the possibility of a 'dark matter inverse problem'[40] – stressing the importance of combining collider, astrophysical (direct/indirect detection), and cosmological probes in order to obtain a complete understanding of both the microscopic and cosmological nature of dark matter.

### 7.3.1. *Considerations from fundamental theory*

Given the possible observational consequences of non-thermal production, it is important to ask if such a scenario makes sense from a fundamental viewpoint, or whether such models represent exotic physics. We saw in the last section that interesting (meaning leading to a situation different from thermal production) and viable cases of non-thermal production rely on

**three crucial assumptions** in order for the WIMP Miracle to survive:

- A scalar condensate composed of particles with masses of about 10 – 100 TeV
- Gravitational coupling to all matter
- A new symmetry that when broken leads to a stable dark matter candidate

All of these requirements are a natural consequence of physics beyond the standard model. However, the very particular choice of an approximately 10 TeV scale mass for the decaying scalar – though mandatory – seems quite artificial. That is, if the scalar is lighter than about 10 TeV then it threatens the successes of BBN, whereas if it is much heavier it would decay before dark matter freezeout and we would have the usual thermal dark matter scenario. It is this apparent tuning of the scalar mass that makes the scenario of non-thermal production much less aesthetically appealing than the thermal case which appears quite robust. Indeed, from a phenomenological point of view it is hard to motivate such a scalar mass except in special cases (see e.g. Ref. 27), however the scenario does have the advantage of being testable in current and near term experiments as discussed above.

This picture drastically changes if one considers constructing phenomenological models which are theoretically consistent in the presence of gravity and at high energies, i.e. for models which have a UV completion in quantum gravity. At first, decoupling of scales would seem to suggest that high energy physics – far beyond the scale of electroweak symmetry breaking – should be irrelevant for the low energy physics of dark matter and the standard model. However, string theories, while providing a consistent UV completion, also provide a very rigid set of constraints that must be applied to low energy effective field theories (EFTs) that would otherwise seem perfectly consistent at low energies and in the absence of gravity.[47] In this way one can hope to highly constrain the number of possible phenomenological models, using added constraints resulting from demanding consistency conditions, such as the absence of anomalies in the presence of gravity.[47] String theory provides a framework to build such models, however, whether one uses string theory or some other consistent UV completion a successful top-down approach must **at least** provide:

- A four dimensional effective theory containing a perturbative limit in which we recover the standard model and Einstein gravity.

- An explanation for the hierarchy between the Planck scale and the scale of electroweak symmetry breaking.
- Additional symmetries must be spontaneously broken – as to not reintroduce the hierarchy problem.
- The vacuum should contain a small and positive cosmological constant (or equivalent) today.

Although at this time no single theory has been shown to accomplish all of these goals in a convincing and natural way, it is interesting that in string theories all these problems can be related to the problem of stabilizing light scalars – moduli. These moduli parameterize the structure of the vacuum of the theory. They describe the size and shape of extra dimensions, as well as the location and orientation of any strings and/or branes that are present. In addition, at the phenomenological level, scalars will also appear as the superpartners to the standard model fermions and many of these scalars lead to flat directions in the potential, i.e. directions in field space where no forces act.

Given the expectation of a large number of scalars with little or no potential, it has been an important program in string model building to find ways in which these scalars may have been stabilized, or at least ways in which the formation of scalar condensates might have been prevented[d]. This is crucial to avoid the cosmological moduli problem discussed above.

An essential step in the program to stabilize the vacuum was the inclusion of additional string theoretic ingredients, which were naturally expected to appear in the theory, but had been neglected initially for computational simplicity. It was later found that the inclusion of branes, strings, and generalizations of Maxwell fields (fluxes) lead to stabilizing effects that ultimately lead to string scale masses for many of the scalars. It then follows that these extremely heavy particles would quickly decay in the early universe to lighter particles, and we have an effective decoupling of string scale physics as one would naively expect.

The low energy, four dimensional scalar potential is then given by

$$V = e^{K/m_p^2}\left(\sum_{\alpha}|D_\alpha W|^2 - 3\frac{|W|^2}{m_p^2}\right) \tag{7.13}$$

where the sum runs over all fields present in the low energy theory, $W$ is the superpotential, $K$ is the Kahler potential, and the condition for SUSY is that $D_\alpha W \equiv \partial_\alpha W + W\partial_\alpha K{=}0$. The stabilization of the moduli at high

[d]See e.g. Ref. 48 for a guide to the literature.

energy leads to a constant term in the low energy superpotential $W = W_0$. For a generic choice of flux, SUSY will be broken explicitly and at the string scale. However, if we choose flux that preserves SUSY, then (7.13) with $D_\alpha W = 0$ implies a deep, negative potential leading to an AdS or negative cosmological constant vacuum. In order to break SUSY and lift the potential we must add an energy contribution to the potential that is parametrically of the form[25]

$$\Delta V(\Phi) = m_{3/2}^2 m_p^2 f\left(\frac{\Phi}{m_p}\right), \tag{7.14}$$

where $m_{3/2}$ is the gravitino mass, related to the scale of SUSY breaking by $\Lambda_{SUSY}^2 = m_{3/2} m_p$, and $\Phi$ is the field leading to the symmetry breaking. The inclusion of physics that would lead to a term like that above is restricted if we hope to achieve a realistic and successful theory. It must lead to *spontaneous* SUSY breaking and a gravitino mass of $m_{3/2} \approx$ TeV, if it is to preserve the success of SUSY in explaining the scale of electroweak symmetry breaking. This is important since $m_{3/2}$ sets the mass of the superpartners and these can not be far above the electroweak scale. It must also cancel the contribution on the right side of (7.13) arranging for a small positive cosmological constant.

It might be difficult to understand how a string based model could ever accomplish this given the discrepancy of scales. However, in addition to the stabilized scalars, the presence of additional symmetries in the theory generically leads to the situation that at least one (if not many) of the scalars are not stabilized at the perturbative level[e]. For these scalars it was shown that non-perturbative effects, such as the condensation of fermions (gauginos in a strongly coupled hidden sector),[29] or the presence of additional branes[30] or additional hidden sector matter fields[31] can be used to stabilize the remaining scalars, providing them with a mass. This leads to an additional contribution in the superpotential and we have

$$W = W_0 + m_p^3 e^{-X}, \tag{7.15}$$

where for simplicity we consider the case of a single scalar $X$ and we take the string scale to lie near the Planck scale – these assumptions however are not crucial to the arguments to follow. The Kahler potential is then of

[e]In some cases this is tied to the requirement that the resulting low energy theory must be perturbative (small coupling), and since the expectation values of many of these scalars determine the low energy couplings, this forces their stabilization away from the string scale. (see e.g. Ref. 50).

the form

$$K = -nm_p^2 \log\left(X + \bar{X}\right). \tag{7.16}$$

The SUSY minimum corresponds to

$$D_X W = 0 \rightarrow \langle X \rangle = \log\left(\frac{m_p}{nm_{3/2}}\right) \tag{7.17}$$

and using this in (7.13) we again find the AdS minimum

$$V_{\mathrm{AdS}} = -3m_{3/2}^2 m_p^2, \tag{7.18}$$

which although SUSY preserving, we choose to write in terms of the gravitino mass in anticipation of SUSY breaking. The authors of Ref. 30 then argued that one could break SUSY and lift the vacuum to contain a small cosmological constant by the addition of another brane leading to a contribution to the potential

$$\Delta V \sim m_{3/2}^2 m_p^2. \tag{7.19}$$

It is important to mention that such an addition must meet rigid constraints coming from the high energy theory that are required for the consistency of the theory (tadpole/anomaly cancelation). Given the full potential we can canonically normalize the scalar field

$$\delta X \rightarrow \delta X_c = \frac{\sqrt{n}}{\langle ReX \rangle}\delta X \tag{7.20}$$

and we find that its mass is then given by

$$m_X = \frac{1}{\sqrt{n}} \log\left(\frac{m_p}{nm_{3/2}}\right) m_{3/2}. \tag{7.21}$$

This scaling and its relation to phenomenology was first stressed in Ref. 49. We see that in order to preserve the hierarchy one would need $m_{3/2} \approx TeV$ and so the scalar mass would naturally lie near the TeV scale. Of course this result just demonstrates that if a scalar of string origin is protected under a symmetry until SUSY breaking occurs its mass should be on the order of the gravitino mass, which must be near a TeV for naturalness. This is precisely the result needed for the non-thermal production of dark matter to be natural, suggesting a new 'non-thermal' WIMP miracle.[41]

Of course the scenario mentioned above is very far from realistic. First, the model of Ref. 30 would seem to explicitly break SUSY by the addition of the brane, where a realistic model should spontaneously break the symmetry. However, this point is moot, because the model contains two tunings – one for the cosmological constant and one for the gravitino mass. The latter implies that the phenomenological successes of SUSY are lost. To see this, consider the gravitino mass in the theory which is given parametrically by

$$m_{3/2} = \frac{|W_0|}{m_p^2 V_6}, \tag{7.22}$$

where $V_6$ is the overall volume of the extra dimensions. In the models of Ref. 30, one then tunes the values of the flux to yield a small value for the superpotential ($W_0 \ll 1$) and thus the scale of SUSY breaking. Another class of models, so-called Large Volume models,[51] take the natural value $W_0 \approx 1$, but then tune the volume[f] $V_6 \approx 10^{14} \gg 1$ so as to obtain the correct scale of SUSY breaking. Another possibility arises from considering M-theory compactifications[52] where it is argued that all moduli are stabilized by non-perturbative physics, so that there is no constant contribution to the superpotential ($W_0 = 0$). The geometry of these compactifications is quite complicated and offers a substantial challenge, however if the expectation holds this would realize SUSY breaking dynamically[53] and preserve the hierarchy. These models also predict the existence of a TeV scale scalar mass, which has been shown to give rise to a non-thermal scenario.[32]

It must be stressed that all of these models contain shortcomings and substantial challenges to address, but with our current understanding of moduli stabilization and SUSY breaking, it would seem that a scalar with TeV mass is an inevitable prediction of the theory. Of course, this was also the original motivation for the cosmological moduli problem, which was argued to be very robust given the arguments presented above.

## 7.4. Conclusions

In this review we have seen that dark matter as a thermal relic remains a simplistic and convincing explanation for the cosmological origin of dark matter. We have also seen that there are a number of possible ways in which this paradigm could turn out to be too naive. Recent observations from dark matter experiment suggest that this might be the case, but taken alone are not especially compelling. However, when combined with theoretical expectations, the possibility of non-thermal dark matter seems worthy of serious consideration. This is especially true since it would make concrete predictions for LHC – if we calculate the thermal relic density from the self annihilation cross section of dark matter deduced from LHC alone we

[f]The authors of Ref. 51 argue that this is not a tuning but the natural location when considering higher order corrections to the theory. This remains to be seen however, since it is difficult to systematically calculate all corrections to the theory.

would get disagreement with cosmological observations. We also saw that the existence of light scalars associated with physics beyond the standard model naturally predicts the existence of a scalar with TeV scale mass – the essential ingredient for non-thermal production. This is intimately tied to the cosmological moduli problem, and progress in string theories in addressing this problem suggests that a non-thermal origin of dark matter may be inevitable. However, model building is in an early stage and there are many challenges that remain in building more realistic models that are compatible with both the standard model and at higher energy with quantum gravity. Regardless of the outcome of the theoretical effort, if we are to achieve a complete understanding of dark matter (both microscopic and macroscopic) this will require combining collider, astrophysics (direct and indirect), and cosmological observations with theoretical approaches.

## Acknowledgments

I would like to thank Gordy Kane for discussions, collaboration, and initially suggesting to me to explore many of the ideas presented in this review. I would also like to thank Bobby Acharya, Konstantin Bobkov, Sera Cremonini, Dan Feldman, Phill Grajek, Piyush Kumar, Ran Lu, Dan Phalen, Aaron Pierce, and Jing Shao for discussions and collaboration. The research of S.W. is supported in part by the Michigan Society of Fellows. S.W. would also like to thank Cambridge University – DAMTP and the Mitchell Institute at Texas A&M for hospitality and financial assistance.

## References

1. F. Zwicky, *Helv. Phys. Acta* **6**, 110 (1933).
2. E. Komatsu *et al.* [WMAP Collaboration], "Five-Year Wilkinson Microwave Anisotropy Probe (WMAP) Observations:Cosmological Interpretation," arXiv:0803.0547 [astro-ph].
3. See the Chapter by J. Wells
4. D. Clowe, M. Bradac, A. H. Gonzalez, M. Markevitch, S. W. Randall, C. Jones and D. Zaritsky, "A direct empirical proof of the existence of dark matter," Astrophys. J. **648**, L109 (2006) [arXiv:astro-ph/0608407].
5. E. W. . Kolb and M. S. . Turner, "THE EARLY UNIVERSE." *REDWOOD CITY, USA: ADDISON-WESLEY (1988) 719 P. (FRONTIERS IN PHYSICS, 70)*
6. S. Weinberg, "Cosmology." *NEW YORK, USA: OXFORD UNIVERSITY PRESS (2008) 593 P.*
7. A. M. Green, "Determining the WIMP mass from a single direct detection ex-

periment, a more detailed study," JCAP **0807**, 005 (2008) [arXiv:0805.1704 [hep-ph]].
8. E. A. Baltz, M. Battaglia, M. E. Peskin and T. Wizansky, "Determination of dark matter properties at high-energy colliders," Phys. Rev. D **74**, 103521 (2006) [arXiv:hep-ph/0602187].
9. R. Arnowitt, B. Dutta, A. Gurrola, T. Kamon, A. Krislock and D. Toback, "Determining the Dark Matter Relic Density in the mSUGRA Stau-Neutralino Co-Annhiliation Region at the LHC," arXiv:0802.2968 [hep-ph].
10. M. M. Nojiri, G. Polesello and D. R. Tovey, "Constraining dark matter in the MSSM at the LHC," JHEP **0603**, 063 (2006) [arXiv:hep-ph/0512204].
11. B. Altunkaynak, M. Holmes and B. D. Nelson, "Solving the LHC Inverse Problem with Dark Matter Observations," arXiv:0804.2899 [hep-ph].
12. M. Hertzberg, M. Tegmark, and F. Wilczek, "Axion Cosmology and the Energy Scale of Inflation," [arXiv:0807.1726 (astro-ph)]
13. P. Binetruy, G. Girardi, and P. Salati, Nucl. Phys. **B237** 285 (1984).
14. P. Svrcek and E. Witten, "Axions in string theory," JHEP **0606**, 051 (2006) [arXiv:hep-th/0605206].
15. M. Dine, L. Randall and S. D. Thomas, "Baryogenesis From Flat Directions Of The Supersymmetric Standard Model," Nucl. Phys. B **458**, 291 (1996) [arXiv:hep-ph/9507453].
16. P. Salati, "Quintessence and the relic density of neutralinos," Phys. Lett. B **571**, 121 (2003) [arXiv:astro-ph/0207396].
17. R. Catena, N. Fornengo, A. Masiero, M. Pietronoi, and F. Rosati, "Dark matter relic abundance and scalar-tensory dark energy," Phys. Rev. D **70**, 063519 (2004) [arXiv:astro-ph/0403614].
18. D. J. H. Chung, L. L. Everett and K. T. Matchev, "Inflationary Cosmology Connecting Dark Energy and Dark Matter," Phys. Rev. D **76**, 103530 (2007) [arXiv:0704.3285 [hep-ph]].
19. D. J. H. Chung, L. L. Everett, K. Kong and K. T. Matchev, "Connecting LHC, ILC, and Quintessence," JHEP **0710**, 016 (2007) [arXiv:0706.2375 [hep-ph]].
20. D. Grin, T. L. Smith and M. Kamionkowski, "Axion constraints in non-standard thermal histories," Phys. Rev. D **77**, 085020 (2008) [arXiv:0711.1352 [astro-ph]].
21. D. H. Lyth and E. D. Stewart, "Thermal Inflation And The Moduli Problem," Phys. Rev. D **53**, 1784 (1996) [arXiv:hep-ph/9510204].
22. J. P. Conlon and F. Quevedo, "Astrophysical and Cosmological Implications of Large Volume String Compactifications," JCAP **0708**, 019 (2007) [arXiv:0705.3460 [hep-ph]].
23. G. D. Coughlan, W. Fischler, E. W. Kolb, S. Raby and G. G. Ross, "Cosmological Problems For The Polonyi Potential," Phys. Lett. B **131**, 59 (1983).
24. J. R. Ellis, D. V. Nanopoulos and M. Quiros, "On the Axion, Dilaton, Polonyi, Gravitino and Shadow Matter Problems in Supergravity and Superstring Models," Phys. Lett. B **174**, 176 (1986).
25. B. de Carlos, J. A. Casas, F. Quevedo and E. Roulet, "Model independent properties and cosmological implications of the dilaton and moduli sectors of

4-d strings," Phys. Lett. B **318**, 447 (1993) [arXiv:hep-ph/9308325].
26. T. Banks, D. B. Kaplan and A. E. Nelson, "Cosmological implications of dynamical supersymmetry breaking," Phys. Rev. D **49**, 779 (1994) [arXiv:hep-ph/9308292].
27. T. Moroi and L. Randall, "Wino cold dark matter from anomaly-mediated SUSY breaking," Nucl. Phys. B **570**, 455 (2000) [arXiv:hep-ph/9906527].
28. S. B. Giddings, S. Kachru and J. Polchinski, "Hierarchies from fluxes in string compactifications," Phys. Rev. D **66**, 106006 (2002) [arXiv:hep-th/0105097].
29. H. P. Nilles, "Dynamically Broken Supergravity And The Hierarchy Problem," Phys. Lett. B **115**, 193 (1982).
30. S. Kachru, R. Kallosh, A. Linde and S. P. Trivedi, "De Sitter vacua in string theory," Phys. Rev. D **68**, 046005 (2003) [arXiv:hep-th/0301240].
31. O. Lebedev, H. P. Nilles and M. Ratz, "de Sitter vacua from matter superpotentials," Phys. Lett. B **636**, 126 (2006) [arXiv:hep-th/0603047].
32. B. S. Acharya, P. Kumar, K. Bobkov, G. Kane, J. Shao and S. Watson, "Non-thermal Dark Matter and the Moduli Problem in String Frameworks," JHEP **06**, 064 (2008) arXiv:0804.0863 [hep-ph].
33. G. F. Giudice, E. W. Kolb and A. Riotto, "Largest temperature of the radiation era and its cosmological implications," Phys. Rev. D **64**, 023508 (2001) [arXiv:hep-ph/0005123].
34. M. Kamionkowski and M. S. Turner, Phys. Rev. D **42**, 3310 (1990); R. Jeannerot, X. Zhang and R. H. Brandenberger, JHEP **9912**, 003 (1999) [arXiv:hep-ph/9901357]; M. Fujii and K. Hamaguchi, Phys. Rev. D **66**, 083501 (2002) [arXiv:hep-ph/0205044]; M. Fujii and K. Hamaguchi, Phys. Lett. B **525**, 143 (2002) [arXiv:hep-ph/0110072] G. B. Gelmini and P. Gondolo, Phys. Rev. D **74**, 023510 (2006) [arXiv:hep-ph/0602230]; G. Gelmini, P. Gondolo, A. Soldatenko and C. E. Yaguna, Phys. Rev. D **74**, 083514 (2006) [arXiv:hep-ph/0605016]; K. Olive and J. Silk, Phys. Rev. Lett. **55** (1985) 2362; J. Ellis, D.V. Nanopoulos and S. Sarkar, Nucl. Phys. **B259** (1985); J. Ellis, J.E. Kim and D.V. Nanopoulos, Phys. Lett. **145B** (1984) 181; K. Rajagopal, M. Turner and F. Wilczek, Nucl. Phys. **B358** (1991) 447; D. J. H. Chung, E. W. Kolb and A. Riotto, Phys. Rev. Lett. **81**, 4048 (1998) [arXiv:hep-ph/9805473]; W. B. Lin, D. H. Huang, X. Zhang and R. H. Brandenberger, Phys. Rev. Lett. **86**, 954 (2001) [arXiv:astro-ph/0009003]; X. J. Bi, R. Brandenberger, P. Gondolo, T. j. Li, Q. Yuan and X. m. Zhang, arXiv:0905.1253 [hep-ph]; D. Grin, T. Smith and M. Kamionkowski, arXiv:0812.4721 [astro-ph]; D. Grin, T. L. Smith and M. Kamionkowski, Phys. Rev. D **77**, 085020 (2008) [arXiv:0711.1352 [astro-ph]]; B. Dutta, L. Leblond and K. Sinha, arXiv:0904.3773 [hep-ph]; J. J. Heckman, A. Tavanfar and C. Vafa, arXiv:0812.3155 [hep-th].
35. R. Jeannerot, X. Zhang and R. H. Brandenberger, "Non-thermal production of neutralino cold dark matter from cosmic string decays," JHEP **9912**, 003 (1999) [arXiv:hep-ph/9901357].
36. G. Kane, R. Lu and S. Watson, "PAMELA Satellite Data as a Signal of Non-Thermal Wino LSP Dark Matter," Phys. Lett. B **681**, 151 (2009) [arXiv:0906.4765 [astro-ph.HE]].

37. P. Grajek, G. Kane, D. Phalen, A. Pierce and S. Watson, "Is the PAMELA Positron Excess Winos?," Phys. Rev. D **79**, 043506 (2009) [arXiv:0812.4555 [hep-ph]].
38. P. Grajek, G. Kane, D. J. Phalen, A. Pierce and S. Watson, "Neutralino Dark Matter from Indirect Detection Revisited," arXiv:0807.1508 [hep-ph].
39. M. Nagai and K. Nakayama, "Direct/indirect detection signatures of non-thermally produced dark matter." [arxiv:0807.1634 [hep-ph]]
40. G. Kane and S. Watson, "Dark Matter and LHC: What is the Connection?," Mod. Phys. Lett. A **23**, 2103 (2008) [arXiv:0807.2244 [hep-ph]].
41. B. S. Acharya, G. Kane, S. Watson and P. Kumar, "A Non-thermal WIMP Miracle," Phys. Rev. D **80**, 083529 (2009) [arXiv:0908.2430 [astro-ph.CO]].
42. Y. Cui, S. P. Martin, D. E. Morrissey and J. D. Wells, "Cosmic Strings from Supersymmetric Flat Directions," Phys. Rev. D **77**, 043528 (2008) [arXiv:0709.0950 [hep-ph]].
43. M. Dine and A. Kusenko, "The origin of the matter-antimatter asymmetry," Rev. Mod. Phys. **76**, 1 (2004) [arXiv:hep-ph/0303065].
44. D. Hooper and E. A. Baltz, "Strategies for Determining the Nature of Dark Matter," arXiv:0802.0702 [hep-ph].
45. A. Helmi, S.D.M. White, and V. Springel, Phys. Rev. D **66**, 063502 (2002) [arXiv:astro-ph/0201289].
46. D. Maurin, F. Donato, R. Taillet, and P. Salati, Astrophys. J. **55**, 585 (2001) [arXiv:astro-ph/0101231].
47. C. Vafa, "The string landscape and the swampland," arXiv:hep-th/0509212; N. Arkani-Hamed, L. Motl, A. Nicolis and C. Vafa, JHEP **0706**, 060 (2007) [arXiv:hep-th/0601001]; A. Adams, N. Arkani-Hamed, S. Dubovsky, A. Nicolis and R. Rattazzi, "Causality, analyticity and an IR obstruction to UV completion," JHEP **0610**, 014 (2006) [arXiv:hep-th/0602178].
48. L. McAllister and E. Silverstein, "String Cosmology: A Review," Gen. Rel. Grav. **40**, 565 (2008) [arXiv:0710.2951 [hep-th]].
49. O. Loaiza-Brito, J. Martin, H. P. Nilles and M. Ratz, "log(M(Pl/m(3/2)))," AIP Conf. Proc. **805**, 198 (2006) [arXiv:hep-th/0509158].
50. M. Dine, Y. Nir and Y. Shadmi, Enhanced symmetries and the ground state of string theory, Phys.Lett. B 438, 61 (1998) [arXiv:hep-th/9806124]; S. Cremonini and S. Watson, Dilaton dynamics from production of tensionless membranes, Phys.Rev. D 73, 086007 (2006) [arXiv:hep-th/0601082].
51. J. P. Conlon, F. Quevedo and K. Suruliz, "Large-volume flux compactifications: Moduli spectrum and D3/D7 soft supersymmetry breaking," JHEP **0508**, 007 (2005) [arXiv:hep-th/0505076].
52. B. S. Acharya, K. Bobkov, G. Kane, P. Kumar and D. Vaman, "An M theory solution to the hierarchy problem," Phys. Rev. Lett. **97**, 191601 (2006) [arXiv:hep-th/0606262].
53. E. Witten, "Dynamical Breaking Of Supersymmetry," Nucl. Phys. B **188**, 513 (1981).

# $Z'$ Physics and Supersymmetry

M. Cvetič

*Department of Physics and Astronomy, University of Pennsylvania, Philadelphia, PA 19104-6396*

P. Langacker

*School of Natural Sciences, Institute for Advanced Study, Princeton, NJ 08540*

We review the status of heavy neutral gauge bosons, $Z'$, with emphasis on constraints that arise in supersymmetric models, especially those motivated from superstring compactifications. After elaborating on the status and (lack of) predictive power for general models with an additional $Z'$, we concentrate on motivations and successes for $Z'$ physics in supersymmetric theories in general and in a class of superstring models in particular. We review phenomenologically viable scenarios and their implications with the $Z'$ mass in the electroweak or in the intermediate scale region.

## 8.1. Introduction

The existence of heavy neutral ($Z'$) vector gauge bosons are a feature of many extensions of the standard model (SM). In particular, one (or more) additional $U(1)'$ gauge factors provide one of the simplest extensions of the SM. Additional $Z'$ gauge bosons appear in grand unified theories (GUT's), superstring compactifications, alternative models of electroweak symmetry breaking, in models involving an (almost) hidden dark sector, and other classes of models.[1]

However, for those models which do not incorporate constraints due to supersymmetry, supergravity, or superstring theory, the masses of additional gauge bosons are usually free parameters which can range from the electroweak scale $\mathcal{O}(1\ \text{TeV})$ to the Planck scale $M_{Pl}$ [a]. In addition, models with extended gauge symmetry generically contain exotic particles, e.g., new heavy quarks or leptons which are non-chiral under $SU(2)_L$, or new SM singlets, with masses that are tied to those of the new $Z'$s, but are otherwise unconstrained. Thus, such models lack predictive power for $Z'$ physics, and much of the phenomenological work in this context is of the "searching under the lamp-post" variety. In particular, there is no strong motivation to think that an extra $Z'$ would actually be light enough to be observed at future colliders. Even within ordinary GUT's, there is no robust prediction for the mass of a $Z'$. (There *are*, however, concrete predictions for the relative couplings of the ordinary and exotic particles to the $Z'$ in each ordinary or supersymmetric GUT.)

On the other hand $N = 1$ supersymmetric models typically provide more constraints. First, the scalar potential is determined by the superpotential, Kähler potential, $D$-terms, and soft supersymmetry breaking terms, and is generally more restrictive. In particular, the $U(1)'$ $D$-term, which is typically of the order of $M^2_{Z'}$, breaks supersymmetry, so the $U(1)'$ breaking scale should usually not be much larger than the electroweak scale, i.e., no more than a few TeV.[4,5] (Exceptions involving flat or almost flat directions are mentioned below.) Secondly, superstring models often imply constraints on the superpotential, such as the absence of mass terms and the existence of large (order one) Yukawa couplings, which can determine the mechanism and scale of $U(1)'$ breaking.

There are two promising classes of theoretical structures in which the minimal supersymmetric standard model (MSSM) and its extensions are likely to be embedded. One are superstring models which compactify directly to a gauge group consisting of the SM and possibly additional $U(1)'$ factors.[6–10] Some of the superstring models are based on $E_6 \times E_8$ Calabi-Yau compactifications of the heterotic string, in which $E_6$ is already broken at the string scale, e.g., to the SM gauge group and an additional $U(1)'$ factor, via the Wilson line (Hosotani) mechanism. Such $Z'$ models, primarily employing the quantum numbers associated with a particular $E_6$ breaking pattern, were addressed in Refs. 11–13. The $Z'$ phenomenology of superstring models with a true GUT symmetry at the string scale has not been explored. Most of the recent work on supersymmetric $Z'$s has been

[a]Major exceptions are alternative models of electroweak symmetry breaking,[2] which often involve new gauge symmetries broken at the TeV scale, and models connecting to a hidden dark matter sector by a light (MeV-GeV) weakly-coupled $Z'$.[3] Neither class will be discussed in detail here.

from the point of view of the first type, non-GUT models, and it will be emphasized in this contribution.

In the models studied in Refs. 6–10, 12 the $U(1)'$ breaking is radiative. This is analogous to radiative electroweak breaking, in which the (running) Higgs mass-squared terms are positive at the Planck scale, but one of them is driven negative at a low or intermediate scale by the large Yukawa coupling associated with the top quark. Similarly, in radiative $U(1)'$ breaking one or more SM scalars that carry $U(1)'$ charges have positive mass-squared terms at the Planck scale. However, if these scalars have large Yukawa couplings to exotic multiplets or to Higgs doublets their mass-squares can be driven negative at lower scales so that the scalar develops a vacuum expectation value and breaks the $U(1)'$ symmetry. Typically, the initial (Planck scale) values of the Higgs and SM scalar mass-squares are comparable and given by the scale of soft supersymmetry breaking. In a class of models in which the magnitudes of the Yukawa couplings in the superpotential are motivated to be of $\mathcal{O}(1)$, the radiative breaking can occur, and it depends on the exotic particle content and on the boundary conditions for the soft supersymmetry breaking terms at the large scale. The symmetry breaking usually takes place at the electroweak scale,[6–8,10,12] so that the $Z'$ mass is comparable to the ordinary $Z$ and to the scale of supersymmetry breaking, and is typically less than a few TeV.[7,8,10] However, the breaking can instead occur at an intermediate scale[6,7,9,14] if the minimum occurs along an $F$- and $D$-flat direction [b]. Secluded sector models,[15] in which the flatness of the symmetry breaking direction is lifted by a small $F$ term, interpolate between these cases, and allow $M_{Z'}$ in the multi-TeV range.

A class of superstring compactifications, based on free fermionic constructions[16–20] [c], contains all of these ingredients, including the general particle content and gauge group of the MSSM, and in general additional non-anomalous $U(1)'$ gauge symmetry factors and vector pairs of exotic chiral supermultiplets [d]. In this class of models there are no bilinear (mass) terms in the superpotential, and the non-vanishing trilinear (Yukawa) terms are of order one. These conditions usually suffice to require and allow ra-

[b]The actual minimum is typically shifted slightly away from the $D$-flat direction by soft supersymmetry breaking terms, leading to finite shifts in the sparticle and Higgs masses of order of the soft breaking, even for a large $U(1)'$ breaking scale.

[c]Certain models based on orbifold constructions with Wilson lines[21] also possess the gauge structure and the particle content of the MSSM.

[d]Related classes of models based on higher level Kač-Moody algebra constructions yield[19,22,23] GUT gauge symmetry with adjoint representations. These models have not yet been explored for $Z'$ physics.

diative $U(1)'$ breaking, respectively.[7] Thus, supersymmetric models (with constraints on couplings from a class of superstring models) *have predictive power* for the masses of $Z'$ and the accompanying exotic particles. In that sense they are superior to general models with extended gauge structures. For these reasons, we would like to advocate that within supersymmetry (and superstring theory constraints), perhaps the best motivated physics for future experiments, next to the Higgs and sparticle searches, are searches for $Z'$ and the accompanying exotic particles. On the other hand, at present we have little theoretical control over the type of dynamically preferred superstring compactification, or of the soft supersymmetry breaking parameters, i.e., the origin of supersymmetry breaking in superstring models; it is therefore hard to make general predictions for the specific $U(1)'$ charges and other details of the models.

In supersymmetric models additional $U(1)'$s would have important theoretical implications. For example, an extra $U(1)'$ breaking at the electroweak scale in a supersymmetric extension of the SM could solve the $\mu$ problem,[24,25] by forbidding an elementary superpotential $\mu$-term $W = \mu\hat{H}_1 \cdot \hat{H}_2$, but allowing a trilinear coupling $W = h_s\hat{S}\hat{H}_1 \cdot \hat{H}_2$, where $\hat{S}$ is a SM singlet. The VEV $\langle S\rangle$ not only breaks the $U(1)'$ gauge symmetry but induces an effective $\mu_{eff} = h_s\langle S\rangle$ which is typically at the electroweak or soft supersymmetry breaking scale.[6–8] This mechanism is similar to the NMSSM (or can be considered as one of a larger class of NMSSM-like models),[26] but because of the continuous $U(1)'$ symmetry it is automatically free of domain wall and induced tadpole effects. Other possible implications, to be briefly discussed in Sections 8.4 and 8.5, include the Higgs, scalar partner, chargino, and neutralino masses and couplings, and thus the properties of the lightest supersymmetric particle (LSP); mechanisms for the generation of small Dirac or Majorana neutrino masses (or both); possible flavor changing neutral currents (FCNC); the hierarchies of small masses for quarks and charged leptons; a possible role in the mediation of supersymmetry breaking; the production of superpartners at colliders; and the possibility of electroweak baryogenesis (EWBG).

This review is organized as follows. In Section 8.2 we describe representative $Z'$ models, introduce notation, and summarize the experimental situation. In Section 8.3, we review the features of $Z'$ models with or without supersymmetry and with or without GUT embedding. In Section 8.4 we discuss both the electroweak and the intermediate scale scenarios for $U(1)'$ symmetry breaking in detail. In Section 8.5 we briefly discuss other implications of a $Z'$, especially at the electroweak-TeV scale.

## 8.2. $Z'$ Physics

### 8.2.1. *Overview of $Z'$ models*

The most commonly studied $Z'$ couplings are sequential, GUT, $T_{3R}$ and $B - L$, string-motivated, and phenomenological.[1] In each case, the $Z'$ couples to $g'Q$, with $g'$ the $U(1)'$ gauge coupling and $Q$ the charge. The sequential models have the same couplings to fermions as the SM $Z$, and are mainly introduced as a convenient reference model for describing experimental constraints.

In the GUT-motivated cases[e] $g' = \sqrt{5/3}\sin\theta_W G\lambda_g^{1/2}$, with $G \equiv \sqrt{g_L^2 + g_Y^2} = g_L/\cos\theta_W$, where $g_L$, $g_Y$ are the gauge couplings of $SU(2)_L$ and $U(1)_Y$, and $\lambda_g$ depends on the symmetry breaking pattern.[27] If the GUT group breaks directly to $SU(3)_C \times SU(2)_L \times U(1)_Y \times U(1)'$, then $\lambda_g = 1$. Standard GUT-type examples include: (i) $Z_\chi$, which occurs in $SO_{10} \to SU(5) \times U(1)_\chi$; (ii) $Z_\psi$, from $E_6 \to SO_{10} \times U(1)_\psi$; (iii) $Z_\eta = \sqrt{3/8}Z_\chi - \sqrt{5/8}Z_\psi$, which occurs in Calabi-Yau compactifications of the heterotic string if $E_6$ breaks directly to a rank 5 group[28] via the Wilson line (Hosotani) mechanism; (iv) the general $E_6$ boson $Z(\theta_{E_6}) = \cos\theta_{E_6} Z_\chi + \sin\theta_{E_6} Z_\psi$, where $0 \leq \theta_{E_6} < \pi$ is a mixing angle. The $U(1)'$ charges for other specific examples are given in Refs. 1, 29.

Another class of models is based on linear combinations of $T_{3R}$ and $B - L$, where $T_{3R} = \frac{1}{2}$ $(u_R, \nu_R)$, $-\frac{1}{2}$ $(d_R, e_R)$, and 0 $(\psi_L)$. They are economical in that they are the unique family-universal $U(1)'$ extensions of the SM with nontrivial couplings that do not require any new fermions other than right-handed neutrinos to cancel anomalies.[30] An important subclass, $Z_{LR}$, occurs in left-right symmetric (LR) models,[31] which contain a right-handed charged boson as well as an additional neutral boson. The ratio $\kappa = g_R/g_L$ of the gauge couplings $g_{L,R}$ for $SU(2)_{L,R}$, respectively, parametrizes the whole class of models. $\kappa = 1$ corresponds to manifest or pseudo-manifest left-right symmetry. In this case $\lambda_g = 1$ by construction and $\kappa > 0.55$ for consistency.[32,33]

Superstring constructions that compactify directly to the SM often contain additional non-anomalous $U(1)'$ factors, some of which may survive to the TeV scale.[34,35] Heterotic constructions often descend through an underlying $SO(10)$ or $E_6$ in the higher-dimensional space, and may therefore

[e]The models in this class are often viewed as plausible examples of anomaly-free charge assignments for the ordinary and exotic particles, rather than as full grand unified theories, to avoid some of the difficulties mentioned in Section 8.3.2.

lead to the $T_{3R}$ and $B-L$ or the $E_6$-type charges. Additional or alternative $U(1)'$ structures may emerge that do not have any GUT-type interpretation and therefore have very model dependent charges, as often occurs in free fermionic constructions, for example. Intersecting brane models often involve $Q_{LR}$ because of an underlying Pati-Salam structure.[36] Other branes can lead to other types of $U(1)'$ charges. For example, the construction in Ref. 37 involves two extra $Z'$ s, one coupling to $Q_{LR}$ and the other only to the Higgs and the right-handed fermions.

String constructions usually also involve anomalous $U(1)'$ symmetries,[38] which must be cancelled by a generalized Green-Schwarz mechanism. Heterotic constructions typically have one anomalous combination of $U(1)'$s. In the Type II string constructions with D-branes, the intersecting D-brane models[35] are particularly suited for the study of particle physics, since they provide a geometric origin of the gauge symmetry and the chiral spectrum. Note, however, that T-duality maps to Type I models with magnetized branes, and Type IIB constructions with D-branes at singularities yield analogous results. Specifically, a stack of $N$ D-branes typically yields a gauge symmetry $U(N) \sim SU(N) \times U(1)$ where $U(1)$ is usually anomalous, with the anomaly cancelled via a generalized Green-Schwarz mechanism. The associated $Z'$ typically acquires a string-scale mass due to Chern-Simons terms, a string realization of the Stückelberg mechanism.[39] However, a particular linear combination of multiple $U(1)'$s, associated with multiple stacks of intersecting D-branes, can in principle remain massless, provided specific conditions on the cycles wrapped by the D-branes are satisfied.[40] (For recent implementations of these constraints within bottom-up MSSM constructions with intersecting D-branes, see Ref. 41.) Such massless $Z'$s have to acquire masses at scales below the string scale, as studied in Section 8.4.

Another way to obtain massless $Z'$s in compactifications with D-branes is by the brane-splitting mechanism, where a non-abelian gauge symmetry of a stack of D-branes is broken down to non-anomalous abelian factors after the splitting of a stack of D-branes to wrap different cycles in the same homology. For a detailed analysis of this symmetry breaking mechanism both from the string theory and the field theory perspective, see Ref. 42.

As for anomalous $Z'$s, effective trilinear vertices may be generated between the $Z'$ and the SM gauge bosons.[43] If there are large extra dimensions the string scale and therefore the $Z'$ mass may be very low, e.g., at the

TeV scale, with anomalous decays into $ZZ$, $WW$, and $Z\gamma$.[44] An anomalous $U(1)'$ survives as a global symmetry on the perturbative low energy theory, restricting the possible couplings and having possible implications, e.g., for baryon or lepton number. There has been considerable recent work involving the nonperturbative generation of such otherwise forbidden couplings by string instanton effects in intersecting brane constructions,[45] with implications for Majorana or Dirac neutrino masses, top-Yukawa couplings, and the $\mu$ parameter.

There are many other types of $Z'$ models, often motivated on phenomenological grounds.[1] For example, many authors have considered couplings based on the cancellation of anomalies, with various assumptions concerning the types of exotic particles that are allowed,[1,30,46] whether the minimal MSSM-type gauge coupling unification is maintained,[47] and other conditions. Other classes of models include:

- Those emerging from Little Higgs models; un-unified models; dynamical symmetry breaking models, e.g., with strong $t\bar{t}$ coupling; or other types of new TeV scale dynamics or symmetry breaking mechanisms.[2]
- Kaluza-Klein excitations of the SM gauge bosons in models with large and/or warped extra dimensions.
- Models in which the $Z'$ is decoupled from some or all of the SM particles, such as leptophobic, fermiophobic or weak coupling models.[5] These may have a low scale or even massless $Z'$.
- Models in which the $Z'$ couples to a hidden sector, e.g., associated with dark matter or supersymmetry breaking.[3] Such a $Z'$ may (almost) decouple from the SM particles and serve as a weakly coupled "portal" to the hidden sector (with small mixings due to kinetic mixing, higher-dimensional operators, etc), or may couple to both sectors, e.g., to mediate supersymmetry breaking.
- Supersymmetric models with a secluded or intermediate scale $Z'$, e.g., associated with (approximately) flat directions, small Dirac $m_\nu$ from higher-dimensional operators, etc.
- Models with family-nonuniversal couplings, leading to $Z'$-mediated FCNC.
- Stückelberg models,[39,48] which allow a $Z'$ mass without spontaneous symmetry breaking.

### 8.2.2. *Mass and kinetic mixing*

The $Z - Z'$ mass matrix takes the form

$$M^2 = \begin{pmatrix} M_Z^2 & \gamma M_Z^2 \\ \gamma M_Z^2 & M_{Z'}^2 \end{pmatrix}, \quad (8.1)$$

where $\gamma$ is determined within each model once the Higgs sector is specified. The physical (mass) eigenstates of mass $M_{1,2}$ are then

$$Z_1 = +Z\cos\phi + Z'\sin\phi, \qquad Z_2 = -Z\sin\phi + Z'\cos\phi, \quad (8.2)$$

where $Z_1$ is the known boson and $\tan 2\phi = 2\gamma M_Z^2/(M_{Z'}^2 - M_Z^2)$. The mass $M_1$ is shifted from the SM value $M_Z$ by the mixing, so that $M_Z^2 - M_1^2 = \tan^2\phi(M_2^2 - M_Z^2)$. For $M_Z \ll M_{Z'}$ one has $M_2 \sim M_{Z'}$ and $\phi \sim \gamma M_1^2/M_2^2$.

There may also be a gauge invariant mixing of the $U(1)$ gauge boson kinetic energy terms.[49] Such terms are usually absent initially, but a (usually small) effect may be induced by loops,[49–52] e.g., from nondegenerate heavy particles, or in running couplings when $Tr(Q_Y Q) \neq 0$, with the trace restricted to the light degrees of freedom. This can occur in GUTs, for example, when multiplets are split into light and heavy sectors. Kinetic mixing can also be due to higher genus effects in superstring theory.[53] However, such effects are small for superstring vacua based on the free fermionic construction [f], on the order of at most a %. An important implication[g] of kinetic mixing is that the effective charge of a TeV-scale $Z'$ at low energies may contain a component proportional to the ordinary weak hypercharge [h].

### 8.2.3. *Precision electroweak and collider limits and prospects*

Constraints can be placed on the existence of $Z'$'s either indirectly from fits to high precision electroweak data in weak neutral current experiments, at the $Z$-pole (LEP and SLC), at LEP 2, and at a future International Linear Collider (ILC), or from direct searches at hadron colliders, i.e., the Tevatron and LHC. The results depend on the chiral couplings of the $Z'$,

---

[f] If one of the $U(1)$'s is associated with a large supersymmetry-breaking $D$-term in a "hidden" sector, the kinetic mixing could propagate this large scale to the observable sector.[53]

[g] The effects of kinetic mixing for a light or massless $Z'$ are quite different. For example, kinetic mixing of a massless $Z'$ with the photon could induce tiny effective electric charges for hidden sector particles coupled to the $Z'$.[49]

[h] The $Z'$ may also contain a component of hypercharge in models which are not based on $U(1)_Y \times U(1)'$, such as the LR.

on the relation between the $Z'$ mass and the $Z - Z'$ mixing angle, and on whether there are open decay channels to superpartners and exotics.[54] The GUT based $Z'$ models, described at the beginning of Section 8.2.1, are fairly representative, with lower limits on the $Z'$ mass in the range 800-1200 GeV,[1,29] and the mixing restricted to $|\phi| <$ few $\times 10^{-3}$. The limits may be considerably weaker in nonstandard models, e.g., with reduced couplings to ordinary fermions.[3]

The LHC has a discovery potential to $\sim 4 - 5$ TeV through $pp \to Z' \to e^+e^-, \mu^+\mu^-, jj, \bar{b}b, \bar{t}t, e\mu, \tau^+\tau^-$, and should be able to make diagnostic studies of the $Z'$ couplings up to $M_{Z'} \sim 2 - 2.5$ TeV. Possible probes include relative branching ratios; forward-backward asymmetries; rapidity distributions; lineshape variables; associated production of $Z'Z, Z'W, Z'\gamma$; rare (but enhanced) decays such as $Z' \to W\bar{f}_1 f_2$ involving a radiated $W$; and decays such as $Z' \to W^+W^-, Zh, 3Z$, or $W^{\pm}H^{\mp}$, in which the small mixing is compensated by the longitudinal $W, Z$ enhancement. The ILC would perhaps have an even larger reach for $Z'$ discovery or diagnostics associated with interference effects in cross sections and asymmetries. Reviews and recent references include.[1,55–62]

## 8.3. $Z'$s — Theoretical Considerations

$Z'$ models fall into different classes, depending on whether they are embedded into a GUT and whether supersymmetry is included.

### 8.3.1. *$Z'$ models in GUT's without supersymmetry*

In a general class of models with extended gauge structure which do not incorporate constraints of supersymmetry, supergravity or superstring theory, the mass and couplings of additional gauge bosons are free parameters in general. However, one class of models of special interest is based on the GUT gauge structure.[63] At $M_U$ the GUT gauge group is broken to a smaller one which includes the SM gauge group and may also include additional $U(1)'$ factors. As opposed to general models the gauge couplings of the additional $Z'$ are now determined within each GUT model, and the quantum numbers of additional exotic particles are also fixed.[5] However, even within GUT models, there is *no robust prediction for the $Z'$ mass and the mass of the accompanying exotic particles*; without fine-tuning of parameters their masses are likely to be at $M_U$, while with fine-tuning their masses can be anywhere between $M_U$ and $M_Z$.

### 8.3.2. $Z'$ models in supersymmetric GUT's

Supersymmetric GUT models possess the advantages of the ordinary GUT models, e.g., gauge coupling unification, and they may provide more constraints on the parameters of the theory. The GUT models with the MSSM below $M_U$ have a gauge coupling unification[64] that is consistent with current experiments. Taking the observed $\alpha$ and weak angle $\sin^2\theta_W$ as inputs and extrapolating assuming the particle content of the MSSM, one finds[65] that the running $SU(2)_L$ and $U(1)_Y$ couplings meet at a scale $M_U \sim 3 \times 10^{16}$ GeV. One can then predict $\alpha_s(M_Z) \sim 0.130 \pm 0.010$ for the strong coupling, which is roughly consistent with, but slightly above, most experimental determinations, $\sim 0.12$. To ensure correct electroweak symmetry breaking, fine-tuning of the superpotential parameters or a specific (higher-dimensional) Higgs representation, e.g., the "missing partner" mechanism,[66,67] is needed.

Within a symmetry breaking scheme in which the GUT group is broken down to the SM and additional $U(1)'$ factors, the success of gauge coupling unification can be ensured only for certain exotic particle spectra. This in general involves fine-tuning of parameters in the superpotential. For example, in the $E_6$ models each SM family can be embedded in a 27-plet of $E_6$, along with a right-handed neutrino, Higgs pair $H_{1,2}$, a SM singlet $S$ (which can break the $U(1)'$), and an exotic pair of ($SU(2)$-singlet) $D$-type quarks, along with their superpartners.[13,63] However, maintaining the MSSM-type unification for the SM gauge couplings requires the addition of an extra $H_1 + H_1^*$ or $H_2 + H_2^*$ from an incomplete (and therefore fine-tuned) $27 + 27^*$.[10] The parameters should be further constrained to ensure additional symmetry breaking. In particular, the breaking of $U(1)'$ within the full GUT context may involve large Higgs representations and/or fine-tuning to ensure $M_{Z'} \ll M_U$ and $D$- and $F$-flatness up to $\mathcal{O}(M_{Z'})$. Further fine-tuning of the superpotential parameters would be needed to allow $M_{Z'} \gg M_Z$ [i]. A final and major difficulty is that, in many cases, if one does achieve a low $U(1)'$ breaking scale there will be exotic particles of comparable mass that can mediate unacceptably rapid proton decay unless the GUT relations relating their Yukawa couplings to those of the Higgs are broken.

Thus, in spite of the constraining structure of supersymmetric GUT models, the mass of $Z'$ and the accompanying exotic particles again gener-

[i] A somewhat analogous analysis of the symmetry breaking pattern of supersymmetric $SO(10)$ to the left-right and then SM gauge group was addressed in Ref. 68.

ically tends to be of order $M_U$. Other mass scales could be achieved by choosing specific Higgs chiral supermultiplets and adjusting the amount of fine-tuning for the superpotential parameters.

### 8.3.3. *Supersymmetric $Z'$ models without GUT embedding*

Supersymmetric models with additional $U(1)'$ factors, but without GUT embedding, provide a promising class of models with more predictive power, and are a possible *minimal extension of the MSSM.* If mass parameters are absent in the superpotential, the soft supersymmetry breaking masses and the type of the SM singlets responsible for the $U(1)'$ breaking determine the mass scale of the $Z'$ without additional (excessive) fine-tuning. Furthermore, the dangerous Yukawa couplings that can lead to rapid proton decay may be absent.

Such a class of models can be derived within certain classes of superstring compactifications. In these models the massless particle spectrum and couplings in the superpotential are calculable, and there are *no mass parameters in the superpotential.* Thus, supersymmetric $Z'$ models with built in constraints on couplings from superstring models *provide a class of models with a predictive power* for $Z'$ physics and that of the accompanying exotic particles.

In the following Section we review the properties of $Z'$s based on that type of model. It is primarily based on Refs. 7–9. An important related analysis was given in Ref. 6[j]. Earlier work, which addresses $Z'$s in models with softly broken $N = 1$ supergravity with no direct connection to superstring models, but with $Z'$ quantum numbers obtained from $E_6$ embeddings (i.e., with $E_6$ broken already at the string scale by the Wilson line (Hosotani) mechanism), was given in Ref. 11. More recent analyses in this context appeared in Refs. 12, 70. $Z'$s which arise from the (intersecting) D-brane models of Type II string theory and remain massless at the string scale would also have to obtain mass at the level of the effective theory, along the lines discussed in the next Section.

## 8.4. $U(1)'$ Symmetry Breaking Scenarios

In this section we describe two important $U(1)'$-breaking scenarios in some detail. These were originally discussed in the context of a class of

[j]Some aspects of $Z'$s in superstring models were also addressed in Ref. 69.

superstring compactifications that are based on free fermionic constructions[16–20,22,23] (see also the extended discussion in Ref. 5). This class provides the necessary ingredients for radiative $U(1)'$ breaking, either at the electroweak or at an intermediate scale. The particle content and the gauge group structure contain, along with the MSSM, additional $U(1)'$ gauge symmetry factors. The massless particle spectrum and the superpotential couplings (which do not have mass terms) are calculable. Importantly, in this class of superstring models the trilinear (Yukawa) couplings in the superpotential are either absent or of order one. However, most of the considerations are valid for other classes of string models as well.

In the following we shall concentrate on phenomenological consequences of an additional non-anomalous $U(1)'$ symmetry. Potential additional phenomenological problems [k] will not be addressed.

For the sake of simplicity we consider only one additional $U(1)'$. The symmetry breaking of the additional $U(1)'$ must take place via the Higgs mechanism, in which the scalar components of chiral supermultiplets $S_i$ which carry non-zero charges under the $U(1)'$ acquire non-zero VEV's and spontaneously break the symmetry. The low energy effective action, responsible for symmetry breaking, is specified by the superpotential, Kähler potential, $D$-terms, and soft supersymmetry breaking terms.

We focus on two symmetry breaking scenarios:

*(i) electroweak scale breaking.*[8] When an additional $U(1)'$ is broken by a *single* SM singlet $S$, the mass scale of the $U(1)'$ breaking should be[7] in the electroweak range (and not larger than a few TeV). The $U(1)'$ breaking may be radiative due to the large (of order one) Yukawa couplings of $S$ to exotic particles. The scale of the symmetry breaking is set by the soft supersymmetry breaking scale, in analogy to the radiative breaking of the electroweak symmetry due to the large top Yukawa coupling. The analysis was generalized[6,8] by including the coupling of the two SM Higgs doublets to the SM singlet $S$.

*(ii) Intermediate scale breaking.*[9] This scenario takes place[7] if there are couplings in the renormalizable superpotential of exotic particles to two or more mirror-like singlets $S_i$ with opposite signs of their $U(1)'$ charges. In this case, the potential may have $D$- and $F$-flat directions, along which it consists only of the quadratic soft supersymmetry breaking mass terms. If

[k] For example: (i) there may be additional color triplets which could mediate a too fast proton decay; (ii) the mass spectrum of the ordinary fermions may not be realistic; and (iii) the light exotic particle spectrum may not be consistent with gauge coupling unification.

there is a mechanism to drive a particular linear combination negative at $\mu_{RAD} \gg M_Z$, the $U(1)'$ breaking is at *an intermediate scale* of order $\mu_{RAD}$, or in the case of dominant non-renormalizable terms in the superpotential the $U(1)'$ takes place at an intermediate scale governed by these terms.

We will also comment briefly on the secluded models,[15] which are somewhat intermediate between cases (i) and (ii).

### 8.4.1. *Electroweak scale breaking*

The superpotential in this model is of the form:

$$W = h_s \hat{S} \hat{H}_1 \cdot \hat{H}_2 + h_Q \hat{U}_3^c \hat{Q}_3 \cdot \hat{H}_2, \tag{8.3}$$

where the Higgs doublet superfield $\hat{H}_2$ has only a Yukawa coupling to a single (third) quark family [1]. For simplicity, the Kähler potential is written in a canonical form, providing canonical kinetic energy terms for the matter superfields. Supersymmetry breaking is parametrized with the most general soft mass parameters, with $m^2_{1,2,S,Q,U}$ corresponding to the soft mass-squared terms of the scalar fields $H_1$, $H_2$, $S$, $Q_3$, $U_3^c$, respectively, and $A$ and $A_Q$ are the soft mass parameters associated with the first and the second trilinear terms in the superpotential (8.3).

Gauge symmetry breaking is now driven by the VEV's of the doublets $H_1$, $H_2$ and the singlet $S$, obtained by minimizing the Higgs potential. By an appropriate choice of the global phases of the fields, one can choose the VEV's such that $\langle H^0_{1,2}\rangle = v_{1,2}/\sqrt{2}$ and $\langle S \rangle = s/\sqrt{2}$ are positive in the true minimum. Whether the obtained local minimum of the potential is acceptable will depend on the location and depth of the other possible minima and of the barrier height and width between the minima.[72]

A nonzero value (in the electroweak range) for $s$ renders the first term of the superpotential (8.3) into an effective $\mu$-parameter, i.e., $\mu \hat{H}_1 \hat{H}_2$, with $\mu \equiv h_s s/\sqrt{2}$ in the electroweak range, thus providing an elegant solution to the $\mu$ problem.

The $Z - Z'$ mass-squared matrix is given by (8.1), where

$$M_Z^2 = \frac{1}{4} G^2 (v_1^2 + v_2^2), \qquad M_{Z'}^2 = g'^2 (v_1^2 Q_1^2 + v_2^2 Q_2^2 + s^2 Q_S^2), \tag{8.4}$$

$$\gamma = \frac{2g'}{G} \frac{(v_1^2 Q_1 - v_2^2 Q_2)}{v_1^2 + v_2^2}. \tag{8.5}$$

[1] The masses of other quarks and leptons are obtained in this class of models through non-renormalizable terms.[71] The fermion masses obtained in this way may not be realistic.

Here $g'$ is the gauge coupling for $U(1)'$, $G = \sqrt{g_L^2 + g_Y^2}$, and $Q_{1,2}$ and $Q_S$ are the $U(1)'$ charges for the $\hat{H}_{1,2}$ and $\hat{S}$ superfields.

*Electroweak Scale Conditions*

This symmetry breaking scenario can be classified in three different categories according to the value of the singlet $s$ VEV:

$s = 0$. In this case the breaking is driven only by the two Higgs doublets (this would be the typical case if the soft mass of the singlet remains positive). The $Z'$ boson would generically acquire mass of the same order as the $Z$, and some particles (Higgses, charginos and neutralinos) would tend to be dangerously light [m]. There is a possibility of a small $Z - Z'$ mixing due to cancellations, and by considerable fine-tuning one may be able to arrange the parameters to barely satisfy experimental constraints.

$s \sim v_{1,2}$. This case would naturally give $M_{Z'} \gtrsim M_Z$ (if $g'Q_{1,2}$ is not too small) and the effective $\mu$-parameter may be small. Thus, some sparticles may be expected to be light. One requires $Q_1 = Q_2$ to have negligible $Z - Z'$ mixing [n]. A particularly interesting example is the "Large Trilinear Coupling Scenario", in which a large trilinear soft supersymmetry breaking term dominates the symmetry breaking pattern, and relative signs and the magnitudes of the soft mass-squared terms are not important since they contribute negligibly to the location of the minimum. The three Higgs fields assume approximately equal VEV's: $v_1 \sim v_2 \sim s \sim 174$ GeV. In this scenario, the electroweak phase transition may be first order, with potentially interesting cosmological implications.

$s \gg v_{1,2}$. Unless $g'Q_S$ is large, $M_{Z'} \gg M_Z$ requires $s \gg v_{1,2}$ and the effective $\mu$-parameter is naturally large. In this case the breaking of the $U(1)'$ is triggered effectively by the running of the soft mass-squared term $m_S^2$ towards negative values in the infrared, yielding $s^2 \simeq -(2m_S^2)/(g'^2 Q_S^2)$ and $M_{Z'}^2 \sim -2m_S^2$, with $m_S^2$ evaluated at $s$. The presence of this large singlet VEV influences $SU(2)_L \times U(1)_Y$ breaking already at tree level. The hierarchy $M_{Z'} \gg M_Z$ results from a cancellation of different mass terms of order $M_{Z'}$; the fine-tuning involved is roughly given by $M_{Z'}/M_Z$. The

[m] In the $Q_1 + Q_2 = 0$ (which allows an elementary $\mu$-parameter) or large $\tan\beta = v_2/v_1 \gg 1$ cases one of the neutral gauge bosons becomes massless.[73] This does not provide a viable hierarchy since the $W^\pm$ mass is non-zero and related to $v_{1,2}$ and $G$ in the usual way.

[n] Such models are allowed for, e.g., leptophobic couplings.

$Z - Z'$ mixing is suppressed by the large $Z'$ mass (in addition to any accidental cancellation for particular choices of charges). Excessive fine-tuning would be needed for $M_{Z'} \gg 1$ TeV. More details are given in Ref. 8 (see also Refs. 6, 70).

*String Scale Conditions*

A detailed discussion of the relation of the electroweak scale parameters to the boundary conditions at $M_{string}$ is given in Refs. 5, 8, 10 (see also Refs. 6, 7, 12), assuming that the non-zero Yukawa couplings are given by $h_s = h_Q = \sqrt{2}g_U$ at $M_{string}$, as determined in a class of superstring models. It was shown that the symmetry breaking scenarios described above can be obtained, but that one requires either nonuniversal soft breaking terms at $M_{string}$, or additional exotic particles, e.g., additional $SU(3)_C$ triplets $\hat{D}_{1,2}$, which couple to $\hat{S}$ in the superpotential.

*The Spectra of Other Particles*

The spectrum of physical Higgses after symmetry breaking consists of three neutral CP even scalars ($h_i^0$, $i = 1, 2, 3$), one CP odd pseudoscalar ($A^0$) and a pair of charged Higgses ($H^\pm$), i.e., it has one scalar more than in the MSSM (for a detailed analysis see Refs. 8, 74, 75, for an earlier discussion see Refs. 4, 11). In the neutralino sector, there is an extra $U(1)'$ zino and the higgsino $\tilde{S}$ as well as the four MSSM neutralinos. In these models the LSP is usually mostly $\tilde{B}$. For large gaugino masses, however, the lightest neutralino is the singlino $\tilde{S}$, whose mass is of the order of $M_Z$. It provides a viable dark matter candidate.[76,77]

Masses for the squarks and sleptons can be obtained directly from the MSSM formulae by setting $\mu \equiv h_s s/\sqrt{2}$ and adding the pertinent $D$-term diagonal contributions from the $U(1)'$.[4,78] This extra term can produce significant mass deviations with respect to the minimal model and plays an important role in the connection between parameters at the electroweak and string scales.

### 8.4.2. *Intermediate scale breaking*

A mechanism to generate intermediate scale $U(1)'$ breaking in supersymmetric theories utilizes the $D$-flat directions,[14] which are in general present in the case of two or more SM singlets with opposite signs of the $U(1)'$ charges. For simplicity, consider *two* chiral multiplets $\hat{S}_{1,2}$ that are singlets under the SM gauge group, with the $U(1)'$ charges $Q_{1S}$ and $Q_{2S}$, respectively. If these charges have opposite signs ($Q_{1S}Q_{2S} < 0$), there is a $D$-flat direction:

$$Q_{1S}\langle S_1\rangle^2 + Q_{2S}\langle S_2\rangle^2 = 0. \tag{8.6}$$

In the absence of the self-coupling of $\hat{S}_1$ and $\hat{S}_2$ in the superpotential there is an $F$-flat direction in the $S_{1,2}$ scalar field space as well. Then the only contribution to the scalar potential along this $D$- and $F$-flat direction is due to the soft mass-squared terms $m_{1S}^2|S_1|^2 + m_{2S}^2|S_2|^2$. For the (real) component along the flat direction $s \equiv (\sqrt{2|Q_{2S}|}\text{Re}S_1 + \sqrt{2|Q_{1S}|}\text{Re}S_2)/\sqrt{|Q_{1S}| + |Q_{2S}|}$ the potential is simply

$$V(s) = \frac{1}{2}m^2 s^2, \qquad m^2 \equiv \frac{|Q_{2S}|m_{1S}^2 + |Q_{1S}|m_{2S}^2}{|Q_{1S}| + |Q_{2S}|}, \tag{8.7}$$

where $m_{1S,2S}^2$ are respectively the $S_{1,2}$ soft mass-squared terms. $m^2$ is evaluated at the scale $s$. For $m^2$ positive at the string scale it can be driven to negative values at the electroweak scale if $\hat{S}_1$ and/or $\hat{S}_2$ have a large Yukawa coupling to other fields, which is in general the case for this class of superstring models. In this case, $V(s)$ develops a minimum along the flat direction and $s$ acquires a VEV[o].

From the minimization condition for (8.7) the VEV $\langle s\rangle$ occurs very close to the scale $\mu_{RAD}$ at which $m^2$ crosses zero. This scale is fixed by the RGE evolution of parameters from $M_{string}$ down to the electroweak scale and lies in general at an intermediate scale. It can be achieved with universal boundary conditions at $M_{string}$ as long as there is a large Yukawa coupling of SM singlets to other matter. The precise value depends on the type of couplings of $\hat{S}_{1,2}$ and the particle content of the model. Examples with $\mu_{RAD}$ in the range $10^{15}$ GeV – $10^4$ GeV are discussed in detail in Ref. 9 (see also Refs. 6, 7).

### *Competition with Non-Renormalizable Operators*

The stabilization of the minimum along the $D$-flat direction can also be due to non-renormalizable terms in the superpotential, which lift the $F$-flat direction for sufficiently large $s$. If these terms are important below $\mu_{RAD}$, they will determine $\langle s\rangle$. The relevant non-renormalizable terms are of the

[o]If $m^2$ remains positive at the electroweak scale, but, e.g., $m_{1S}^2 < 0$, then one recovers the electroweak scale case with $\langle S_1\rangle \neq 0$ and $\langle S_2\rangle = 0$.

form [p]

$$W_{\rm NR} = \left(\frac{\alpha_K}{M_{Pl}}\right)^K \hat{S}^{3+K}, \tag{8.8}$$

where $\hat{S}$ is the effective chiral superfield (i.e., $W_{\rm NR}$ may actually involve the product of distinct superfields $\hat{S}_i$) with the (real) scalar component along the $D$-flat direction, $K = 1, 2, \cdots$, and $M_{Pl}$ is the Planck scale.

Including the $F$-term from (8.8), the potential along $s$ is

$$V(s) = \frac{1}{2}m^2 s^2 + \frac{1}{2(K+2)}\left(\frac{s^{2+K}}{M^K}\right)^2, \tag{8.9}$$

where $M \equiv C_K M_{Pl}/\alpha_K$, and $C_K$ is a coefficient of order unity. The VEV of $s$ is then [q]

$$\langle s \rangle = \left(|m| M^K\right)^{\frac{1}{K+1}} \sim (m_{soft} M^K)^{\frac{1}{K+1}}, \tag{8.10}$$

where $m_{soft} = \mathcal{O}(|m|) = \mathcal{O}(M_Z)$ is a typical soft supersymmetry breaking scale. $m^2$ is evaluated at the scale $\langle s \rangle$ and has to satisfy the necessary condition $m^2(\langle s \rangle) < 0$. If non-renormalizable terms are negligible below $\mu_{RAD}$, no solution to (8.10) exists and $\langle s \rangle$ is fixed solely by the running $m^2$.

The coefficients $\alpha_K$ in (8.8) and thus $M$ in (8.9) are in principle calculable, at least in the free fermionic models discussed above. Depending on the $U(1)'$ charges and world-sheet symmetries of the superstring models, not all values of $K$ are allowed. It is expected that the world-sheet integrals that determine the couplings of non-renormalizable terms are such that $M$ increases as $K$ increases. For $K = 1$ one obtains (in a class of models): $M \sim 3 \times 10^{17}$ GeV, and for $K = 2$: $M \sim 7 \times 10^{17}$ GeV.

*Higgs and Higgsino Mass Spectrum*

The mass of the physical field $s$ in the vacuum $\langle s \rangle$ is either $\sim m_{soft}/4\pi$ for pure radiative breaking, or $\sim m_{soft}$ in the case of stabilization by non-renormalizable terms. Thus, the potential is very flat, with possible important cosmological consequences. The physical excitations along the transverse direction have an intermediate mass scale. There remains one

[p] One can also have terms of the form $\alpha_K^K \hat{S}^{2+K}\hat{\Phi}/M_{Pl}^K$, where $\Phi$ is a SM singlet that does not acquire a VEV. These have similar implications as the terms in (8.8).

[q] For simplicity, soft-terms of the type $(AW_{\rm NR} + \text{H.c.})$ with $A \sim m_{soft}$ are not included in (8.9). Such terms do not affect the order of magnitude estimates.

massless pseudoscalar, which can acquire a mass from a soft supersymmetry breaking term of the type $AW_{\rm NR}$ or from loop corrections.

The fermionic part of the $Z' - S_1 - S_2$ sector consists of three neutralinos $(\tilde{B}', \tilde{S}_1, \tilde{S}_2)$. One combination is light, with mass of order $m_{soft}$ (if the minimum is fixed by non-renormalizable terms) or of order $m_{soft}/4\pi$ (obtained at one-loop order when the minimum is instead determined by the running of $m^2$). The two other neutralinos have intermediate scale masses.

Other fields that couple to $\hat{S}_{1,2}$ in the renormalizable superpotential acquire intermediate scale masses, and those that do not remain light.

*μ-Parameter*

The flat direction $S$ can have a set of non-renormalizable couplings to MSSM states that offer a solution to the $\mu$ problem. The non-renormalizable $\mu$-generating terms are of the form,

$$W_\mu \sim \hat{H}_1 \hat{H}_2 \hat{S} \left( \frac{\hat{S}}{M} \right)^P . \tag{8.11}$$

For breaking due to non-renormalizable terms, $K = P$ yields an effective $\mu$-parameter $\mu \sim m_{soft}$, while for pure radiative breaking $\mu \sim \mu_{RAD}^{P+1}/M^P$, which depends crucially on the value of $\mu_{RAD}$.

*Fermion Masses*

Non-renormalizable couplings can also yield mass hierarchies between family generations. Generational up, down, and electron mass terms appear, via

$$W_{u_i} \sim \hat{H}_2 \hat{Q}_i \hat{U}_i^c \left( \frac{\hat{S}}{M} \right)^{P'_{u_i}} ; \quad W_{d_i} \sim \hat{H}_1 \hat{Q}_i \hat{D}_i^c \left( \frac{\hat{S}}{M} \right)^{P'_{d_i}} ; \quad W_{e_i} \sim \hat{H}_1 \hat{L}_i \hat{E}_i^c \left( \frac{\hat{S}}{M} \right)^{P'_{e_i}} , \tag{8.12}$$

with $i$ the family number. $K$ and $P'_i$ can be chosen to yield a realistic hierarchy for the first two generations.[9] Presumably the top mass is associated with a renormalizable coupling ($P'_{u_3} = 0$). The other third family masses do not fit as well; it is possible that $m_b$ and $m_\tau$ are associated with some other mechanism, such as non-renormalizable operators involving the VEV of a different singlet.

There may also be non-renormalizable Majorana and Dirac neutrino terms. They can produce small (non-seesaw) Dirac masses. It is also possible to have small Majorana masses, providing an interesting possibility for oscillations of the ordinary into sterile neutrinos.[79,80] A traditional seesaw can also be obtained, depending on the nature of the non-renormalizable operators.[9]

### 8.4.3. *Secluded models*

In the weak scale model in (8.3) there is some tension between the electroweak scale and developing a large enough $M_{Z'}$. These can be decoupled without tuning when there are several $S$ fields. For example, in the *secluded sector* model[15] there are four standard model singlets $S, S_{1,2,3}$ that are charged under a $U(1)'$, with

$$W = h_s \hat{S} \hat{H}_1 \cdot \hat{H}_2 + \lambda \hat{S}_1 \hat{S}_2 \hat{S}_3. \tag{8.13}$$

(Structures similar to this are often encountered in heterotic string constructions.) $\mu_{eff}$ is given by $h_s \langle S \rangle$, but all four VEVs contribute to $M_{Z'}$. The only couplings between the ordinary $(S, H_{1,2})$ and secluded $(S_{1,2,3})$ sectors are from the $U(1)'$ $D$ term and the soft masses (special values of the $U(1)'$ charges, which allow soft mixing terms, are required to avoid unwanted additional global symmetries[81]). It is straightforward to choose the soft parameters so that there is a runaway (i.e., $F$- and $D$-flat) direction in the limit $\lambda \to 0$, for which the ordinary sector VEVs remain finite while the $S_i$ VEVs become large. For $\lambda$ finite but small, the flatness is lifted and the $\langle S_i \rangle$ and $M_{Z'}$ scale as $1/\lambda$. For example, one can find $M_{Z'}$ in the TeV range for $\lambda \sim 0.05 - 0.1$. Other implications of the secluded models are described in Ref. 82.

## 8.5. Other Implications

The discovery of a TeV-scale $Z'$ with electroweak-strength couplings would have implications far beyond the existence of a new vector particle. The role of a new $U(1)'$ gauge symmetry in solving the $\mu$ problem was already discussed in the Introduction, and a number of other implications in Section 8.4. Here, we list some of the other possibilities (see Ref. 1 for more complete references).

- TeV scale $U(1)'$ models generally involve an extended Higgs sector, requiring at least a SM singlet $S$ to break the $U(1)'$ symmetry. New $F$ and $D$-term contributions can relax the theoretical upper limit of $\sim 130$ GeV on the lightest Higgs scalar in the MSSM up to $\sim 150$

GeV, and smaller values of $\tan\beta$, e.g. $\sim 1$, become possible. Conversely, doublet-singlet mixing can allow a Higgs lighter than the direct SM and MSSM limits. Such mixing as well as the extended neutralino sector can lead to non-standard collider signatures.[75]

- $U(1)'$ models also have extended neutralino sectors,[77] involving at least the $\tilde{Z}'$ gaugino and the $\tilde{S}$ singlino, allowing non-standard couplings (e.g., light singlino-dominated), extended cascades, and modified possibilities for cold dark matter, $g_\mu - 2$, etc.
- Most family-universal $U(1)'$ models (with the exception of those involving $T_{3R}$ and$B - L$) require new exotic fermions to cancel anomalies. These are usually non-chiral with respect to the SM (to avoid precision electroweak constraints) but chiral under the $U(1)'$. A typical example is a pair of $SU(2)$-singlet colored quarks $D_{L,R}$ with charge $-1/3$. Such exotics may decay by mixing, although that is often forbidden by $R$-parity. They may also decay by diquark or leptoquark couplings, or they be quasi-stable, decaying by higher-dimensional operators.[83,84]
- A heavy $Z'$ may decay efficiently into sparticles, exotics, etc., constituting a "SUSY factory".[54,85]
- The $U(1)'$ charges may be family non-universal[r], leading to FCNC due to fermion mixings. The limits from $K$ and $\mu$ decays and interactions are sufficiently strong that only the third family is likely to be non-universal.[87] Third family non-universality may lead to significant tree-level effects,[88] e.g., in $B_s - \bar{B}_s$ mixing or in charmless $B_d$ decays, competing with SM loops, or with enhanced loops in the MSSM with large $\tan\beta$.
- A TeV-scale $U(1)'$ symmetry places new constraints on neutrino mass generation. Various versions allow or exclude Type I or II seesaws, extended seesaws, or small Dirac masses by higher-dimensional operators;[9,79,84,89] small Dirac masses by non-holomorphic soft terms;[90] and either Majorana (seesaw)[45] or small Dirac masses by string instantons.[91]
- Electroweak baryogenesis[92] is relatively easy to implement in $U(1)'$ (and other NMSSM-type) models because the cubic $A$ term associated with the effective $\mu$ parameter can lead to a strongly first order electroweak phase transition,[93,94] and because of possible tree-level $CP$ violation in the Higgs sector (which is not significantly con-

[r]This frequently occurs in string constructions.[34,35] For an $E_6$ GUT example, see Ref. 86.

strained by EDMs).[94] However, the alternative leptogenesis model of baryogenesis, in which a lepton asymmetry is first created by the out of equilibrium decay of a heavy Majorana neutrino, and then converted to a baryon asymmetry by electroweak effects,[95] is forbidden by a light $Z'$, at least in its simplest form, unless the heavy neutrino carries no $U(1)'$ charge.[96]

- $Z'$ gauge bosons may couple only to an otherwise hidden sector associated with dark matter, etc., except for small effects associated, e.g., with kinetic mixing or higher-dimensional operators, thereby serving as a "portal" to that sector.[3] Examples include a massless $Z'$;[49,97] a MeV-GeV "$U$-boson",[98] which is suggested by some dark matter models;[99] hidden valley models, which connect to a strongly coupled hidden sector;[100] and Stückelberg models.[48] Such models are not necessarily supersymmetric.
- Similarly, it often occurs in string constructions that a $Z'$ couples to both the ordinary and "hidden" sectors. This allows the possibility of $Z'$-mediation of supersymmetry breaking,[101] in which a mass difference between the $Z'$ and the associated gaugino is generated in the hidden sector and is communicated to the ordinary sector by loop effects. (See Ref. 102 for specific string realizations.) Scalar masses are generated at one loop and SM gaugino masses at two loops. If the latter are to be $\gtrsim$ 100 GeV, then the scalar masses should be in the 100 TeV range for electroweak-size couplings, requiring a fine-tuning to obtain the electroweak scale (a mini form of split supersymmetry[103]). Alternatively, $Z'$-mediation can be combined with some other mediation mechanism for the SM gauginos to allow lower scalar masses. For example, the combination of $Z'$ and anomaly mediation cures the problems of each and allows a realistic spectrum.[104] Other implications of a $U(1)'$ for supersymmetry breaking and mediation are reviewed in Ref. 101.

## 8.6. Conclusions

Supersymmetric $Z'$ models motivated from a class of superstring models provide a "minimal" extension of the MSSM. The electroweak breaking scenario yields phenomenologically acceptable symmetry breaking patterns which involve a certain but not excessive amount of fine-tuning. It predicts interesting new phenomena testable at future colliders. The intermediate scale scenario provides a framework in which intermediate scales can naturally occur, with implications for the $\mu$-parameter and the fermion masses. There are also possibilities for other mass ranges, reduced couplings, etc. There are important possible implications for the Higgs, sparticle, exotic, and neutrino sectors; the $\mu$ problem; dark matter; baryogenesis; FCNC; and the mediation of supersymmetry breaking.

## Acknowledgments

This work was supported by Department of Energy Grant DOE-EY-76-02-3071 (MC) and by NSF PHY-0503584 and the IBM Einstein Fellowship (PL). We would like to thank F. del Aguila, D. Demir, G. Cleaver, J. R. Espinosa and L. Everett and our many other collaborators for fruitful collaboration.

## References

1. For general reviews, see A. Leike, *Phys. Rept.* **317**, 143 (1999); T. G. Rizzo, arXiv:hep-ph/0610104; P. Langacker, *Rev. Mod. Phys.* **81**, 1199-1228 (2009).
2. *E.g.*, C. T. Hill and E. H. Simmons, *Phys. Rept.* **381**, 235 (2003) [E-**390**, 553 (2004)].
3. For reviews, see P. Langacker, arXiv:0909.3260 [hep-ph]; M. Goodsell, J. Jaeckel, J. Redondo and A. Ringwald, *JHEP* **0911**, 027 (2009).
4. See C. Kolda and S. Martin, *Phys. Rev.* **D53**, 3871 (1996), and references theirin.
5. For a more detailed earlier version of this review, see M. Cvetič and P. Langacker, hep-ph/9707451.
6. D. Suematsu and Y. Yamagishi, *Int. J. Mod. Phys.* **A10**, 4521 (1995).
7. M. Cvetič and P. Langacker, *Phys. Rev.* **D54**, 3570 (1996) and *Mod. Phys. Lett.* **11A**, 1247 (1996).
8. M. Cvetič, D. A. Demir, J. R. Espinosa, L. Everett and P. Langacker, *Phys. Rev.* **D56**, 2861 (1997).
9. G. Cleaver, M. Cvetič, J. R. Espinosa, L. Everett and P. Langacker, *Phys. Rev.* **D57**, 2701 (1998).
10. P. Langacker and J. Wang, *Phys. Rev.* **D58**, 115010 (1998).
11. J. F. Gunion, L. Roszkowski, and H. E. Haber, *Phys. Lett.* **B189**, 409 (1987); *Phys. Rev.* **D38**, 105 (1988).
12. E. Keith, E. Ma, and B. Mukhopadhyaya, *Phys. Rev.* **D55**, 3111 (1997); E. Keith and E. Ma, *Phys. Rev.* **D56**, 7155 (1997).
13. For a review, see J. Hewett and T. Rizzo, *Phys. Rept.* **183**, 193 (1989).
14. M. Dine, V. Kaplunovsky, M. Mangano, C. Nappi and N. Seiberg, *Nucl. Phys.* **B259**, 549 (1985); M. Mangano, *Zeit. Phys.* **C28**, 613 (1985); G. Costa, F. Feruglio, F. Gabbiani and F. Zwirner, *Nucl. Phys.* **B286**,

325 (1987); J. Ellis, K. Enqvist, D. V. Nanopoulos and K. Olive, *Phys. Lett.* **B188**, 415 (1987); R. Arnowitt and P. Nath, *Phys. Rev. Lett.* **60**, 1817 (1988).
15. J. Erler, P. Langacker and T. j. Li, *Phys. Rev.* **D66**, 015002 (2002).
16. I. Antoniadis, C. Bachas and C. Kounnas, *Nucl. Phys.* **B289**, 87 (1987); H. Kawai, D. Lewellen and S. H. H. Tye, *Phys. Rev. Lett.* **57**, 1832 (1986) and *Phys. Rev.* **D34**, 3794 (1986).
17. I. Antoniadis, J. Ellis, J. Hagelin and D. V. Nanopoulos, *Phys. Lett.* **B231**, 65 (1989).
18. A. Faraggi, D. V. Nanopoulos and K. Yuan, *Nucl. Phys.* **B335**, 347 (1990); A. Faraggi, *Phys. Lett.* **B278**, 131 (1992).
19. S. Chaudhuri, S.-W. Chung, G. Hockney and J. Lykken, *Nucl. Phys.* **B456**, 89 (1995); S. Chaudhuri, G. Hockney and J. Lykken, *Nucl. Phys.* **B469**, 357 (1996).
20. For a review, see K. R. Dienes, *Phys. Rept.* **287**, 447 (1997).
21. L. E. Ibáñez, J. E. Kim, H. P. Nilles, and F. Quevedo, *Phys. Lett.* **B191**, 282 (1987); A. Font, L. E. Ibáñez, H. P. Nilles and F. Quevedo, *Phys. Lett.* **B210**, 101 (1988), [E-**B213**, 564 (1988)].
22. G. Aldazabal, A. Font, L. E. Ibáñez, and A.M. Uranga, *Nucl. Phys.* **B452**, 3 (1995) and *Nucl. Phys.* **B465**, 34 (1996).
23. Z. Kakushadze and S. H. H. Tye, *Phys. Lett.* **B392**, 335 (1997), *Phys. Rev.* **D55**, 7878 (1997) and *Phys. Rev.* **D55**, 7896 (1997).
24. J. E. Kim and H. P. Nilles, *Phys. Lett.* **B138**, 150 (1984).
25. Alternative solutions include the string instanton and higher-dimensional operator mechanisms mentioned in Sections 8.2.1 and 8.4.2, respectively, and the Giudice-Masiero mechanism, G. F. Giudice and A. Masiero, *Phys. Lett.* **B206**, 480 (1988),
26. For a review, see M. Maniatis, arXiv:0906.0777 [hep-ph].
27. R. Robinett, *Phys. Rev.* **D26**, 2388 (1982); R. Robinett and J. Rosner, *Phys. Rev.* **D25**, 3036 (1982) and **D26**, 2396 (1982).
28. E. Witten, *Nucl. Phys.* **B258**, 75 (1985).
29. J. Erler, P. Langacker, S. Munir and E. R. Pena, *JHEP* **0908**, 017 (2009).
30. See, e.g., T. Appelquist, B. A. Dobrescu and A. R. Hopper, *Phys. Rev.* **D68**, 035012 (2003).
31. For a review see R.N. Mohapatra, *Unification and Supersymmetry* (Springer, Berlin, 2003); for constraints on general LR models, see P. Langacker and S. Uma Sankar, *Phys. Rev.* **D40**, 1569 (1989).
32. U. Amaldi *et al.*, *Phys. Rev.* **D36**, 1385 (1987).
33. M. Cvetič, B. Kayser, and P. Langacker, *Phys. Rev. Lett.* **68**, 2871 (1992).
34. See, for example, A. E. Faraggi and D. V. Nanopoulos, *Mod. Phys. Lett.* **A6**, 61 (1991); G. Cleaver, M. Cvetič, J. R. Espinosa, L. L. Everett, P. Langacker and J. Wang, *Phys. Rev.* **D59**, 055005 (1999) and *Phys. Rev.* **D59**, 115003 (1999); J. Giedt, *Annals Phys.* **289**, 251 (2001); V. Braun, Y. H. He, B. A. Ovrut and T. Pantev, *JHEP* **0506**, 039 (2005); P. Anastasopoulos, T. P. T. Dijkstra, E. Kiritsis and A. N. Schellekens, *Nucl. Phys.* **B759**, 83 (2006); C. Coriano, A. E. Faraggi and M. Guzzi, *Eur. Phys. J.* **C53**,

421 (2008); O. Lebedev, H. P. Nilles, S. Raby, S. Ramos-Sanchez, M. Ratz, P. K. S. Vaudrevange and A. Wingerter, *Phys. Rev.* **D77**, 046013 (2008).

35. For reviews, see R. Blumenhagen, M. Cvetič, P. Langacker and G. Shiu, *Ann. Rev. Nucl. Part. Sci.* **55**, 71 (2005); R. Blumenhagen, B. Kors, D. Lust and S. Stieberger, *Phys. Rept.* **445**, 1 (2007); F. Marchesano, *Fortsch. Phys.* **55**, 491 (2007).
36. J. C. Pati and A. Salam, *Phys. Rev.* **D10**, 275 (1974) [E-**D11**, 703 (1975)].
37. M. Cvetič, G. Shiu and A. M. Uranga, *Nucl. Phys.* **B615**, 3 (2001).
38. For a review, see E. Kiritsis, *Phys. Rept.* **421**, 105 (2005).
39. E. C. G. Stueckelberg, *Helv. Phys. Acta* **11**, 225 (1938).
40. G. Aldazabal, S. Franco, L. E. Ibanez, R. Rabadan and A. M. Uranga, *J. Math. Phys.* **42**, 3103 (2001).
41. M. Cvetič, J. Halverson and R. 2. Richter, arXiv:0905.3379 [hep-th].
42. M. Cvetic, P. Langacker, T. j. Li and T. Liu, *Nucl. Phys.* **B709**, 241 (2005).
43. C. Coriano', N. Irges and E. Kiritsis, *Nucl. Phys.* **B746**, 77 (2006); P. Anastasopoulos, M. Bianchi, E. Dudas and E. Kiritsis, *JHEP* **0611**, 057 (2006).
44. See, for example, R. Armillis, C. Coriano', M. Guzzi and S. Morelli, *Nucl. Phys.* **B814**, 15679 (2009); J. Kumar, A. Rajaraman and J. D. Wells, *Phys. Rev.* **D77**, 066011 (2008); P. Anastasopoulos, F. Fucito, A. Lionetto, G. Pradisi, A. Racioppi and Y. S. Stanev, *Phys. Rev.* **D78**, 085014 (2008)
45. R. Blumenhagen, M. Cvetič and T. Weigand, *Nucl. Phys.* **B771**, 113 (2007); L. E. Ibanez and A. M. Uranga, *JHEP* **0703**, 052 (2007). For a review, see R. Blumenhagen, M. Cvetič, S. Kachru and T. Weigand, arXiv:0902.3251 [hep-th].
46. See, for example, M. S. Carena, A. Daleo, B. A. Dobrescu and T. M. P. Tait, *Phys. Rev.* **D70**, 093009 (2004); D. A. Demir, G. L. Kane and T. T. Wang, *Phys. Rev.* **D72**, 015012 (2005); P. Batra, B. A. Dobrescu and D. Spivak, *J. Math. Phys.* **47**, 082301 (2006); H. S. Lee, K. T. Matchev and T. T. Wang, *Phys. Rev.* **D77**, 015016 (2008); P. Langacker, G. Paz, L. T. Wang and I. Yavin, *Phys. Rev. Lett.* **100**, 041802 (2008).
47. J. Erler, *Nucl. Phys.* **B586**, 73 (2000); D. E. Morrissey and J. D. Wells, *Phys. Rev.* **D74**, 015008 (2006).
48. P. Nath, arXiv:0812.0958 [hep-ph]; D. Feldman, Z. Liu and P. Nath, *Phys. Rev. Lett.* **97**, 021801 (2006).
49. B. Holdom, *Phys. Lett.* **B166**, 196 (1986).
50. F. del Aguila, G. A. Blair, M. Daniel and G. G. Ross, *Nucl. Phys.* **B283**, 50 (1987); F. del Aguila, G. D. Coughlan and M. Quirós; *Nucl. Phys.* **B307**, 633 (1988), [E-**B312**, 751 (1989)]; F. del Aguila, M. Masip and M. Pérez-Victoria, *Nucl. Phys.* **B456**, 531 (1995).
51. F. del Aguila, M. Cvetič and P. Langacker, *Phys. Rev.* **D52**, 37 (1995).
52. K. S. Babu, C. Kolda, and J. March-Russell, *Phys. Rev.* **D54**, 4635 (1996); *Phys. Rev.* **D57**, 6788 (1998).
53. K. R. Dienes, C. Kolda and J. March-Russell, *Nucl. Phys.* **B492**, 104 (1997).
54. J. Kang and P. Langacker, *Phys. Rev.* **D71**, 035014 (2005).
55. F. del Aguila, M. Cvetič and P. Langacker, *Phys. Rev.* **D48**, 969 (1993); *Phys. Rev.* **D52**, 37 (1995).

56. M. Cvetič and S. Godfrey, hep-ph/9504216.
57. M. Dittmar, A. S. Nicollerat and A. Djouadi, *Phys. Lett.* **B583**, 111 (2004).
58. G. Weiglein *et al.* [LHC/LC Study Group], *Phys. Rept.* **426**, 47 (2006).
59. F. Petriello and S. Quackenbush, *Phys. Rev.* **D77**, 115004 (2008); Y. Li, F. Petriello and S. Quackenbush, *Phys. Rev.* **D80**, 055018 (2009).
60. R. Diener, S. Godfrey and T. A. W. Martin, arXiv:0910.1334 [hep-ph].
61. E. Salvioni, G. Villadoro and F. Zwirner, *JHEP* **0911**, 068 (2009).
62. P. Langacker, arXiv:0911.4294 [hep-ph].
63. For a review, see P. Langacker, *Phys. Rept.* **72**, 185 (1981).
64. S. Dimopoulos, S. Raby and F. Wilczek, *Phys. Rev.* **D24**, 1681 (1981); L. E. Ibáñez and G. G. Ross, *Phys. Lett.* **105B**, 439 (1981).
65. P. Langacker and N. Polonsky, *Phys. Rev.* **D52**, 3081 (1995); J. Bagger, K. T. Matchev and D. Pierce, *Phys. Lett.* **B348**, 443 (1995); and references therein.
66. S. Dimopoulos and F. Wilczek, Santa Barbara preprint 81-0600 (1981).
67. For a review, see, e.g., H. P. Nilles, *Phys. Rept.* **110**, 1 (1984).
68. C. S. Aulakh and R. N. Mohapatra, *Phys. Rev.* **D28**, 217 (1983).
69. A. E. Faraggi and M. Masip, *Phys. Lett.* **B388**, 524 (1996); J. L. Lopez and D. V. Nanopoulos, *Phys. Rev.* **D55**, 397 (1997).
70. T. Gherghetta, T. A. Kaeding, and G. L. Kane, *Phys. Rev.* **D57**, 3178 (1998).
71. A. E. Faraggi, *Phys. Lett.* **B274**, 47 (1992); *Nucl. Phys.* **B403**, 102 (1993).
72. J. A. Casas, A. Lleyda and C. Munoz, *Nucl. Phys.* **B471**, 3 (1996); A. Kusenko, P. Langacker and G. Segrè, *Phys. Rev.* **D54**, 5824 (1996); A. Kusenko and P. Langacker, *Phys. Lett.* **B391**, 29 (1997).
73. We thank M.-X. Luo for emphasizing that to us.
74. V. Barger, P. Langacker, H. S. Lee and G. Shaughnessy, *Phys. Rev.* **D73**, 115010 (2006); S. W. Ham and S. K. OH, arXiv:0906.5526 [hep-ph].
75. E. Accomando *et al.*, arXiv:hep-ph/0608079.
76. B. de Carlos and J. R. Espinosa, *Phys. Lett.* **B407**, 12 (1997).
77. V. Barger, P. Langacker, I. Lewis, M. McCaskey, G. Shaughnessy and B. Yencho, *Phys. Rev.* **D75**, 115002 (2007); S. Y. Choi, H. E. Haber, J. Kalinowski and P. M. Zerwas, *Nucl. Phys.* **B778**, 85 (2007).
78. J. D. Lykken, hep-ph/9610218.
79. P. Langacker, *Phys. Rev.* **D58**, 093017 (1998).
80. For other mechanisms, see E. Ma, *Mod. Phys. Lett.* **A11**, 1893 (1996); E. Keith and E. Ma, *Phys. Rev.* **D54**, 3587 (1996).
81. P. Langacker, G. Paz and I. Yavin, *Phys. Lett.* **B671**, 245 (2009).
82. T. Han, P. Langacker and B. McElrath, *Phys. Rev.* **D70**, 115006 (2004); C. W. Chiang and E. Senaha, *JHEP* **0806**, 019 (2008).
83. J. Kang, P. Langacker and B. D. Nelson, *Phys. Rev.* **D77**, 035003 (2008).
84. S. F. King, S. Moretti and R. Nevzorov, *Phys. Rev.* **D73**, 035009 (2006); P. Athron, S. F. King, D. J. Miller, S. Moretti and R. Nevzorov, *Phys. Rev.* **D80**, 035009 (2009).
85. M. Baumgart, T. Hartman, C. Kilic and L. T. Wang, *JHEP* **0711**, 084 (2007); T. Cohen and A. Pierce, *Phys. Rev.* **D78**, 055012 (2008); H. S. Lee,

*Phys. Lett.* **B674**, 87 (2009); A. Ali, D. A. Demir, M. Frank and I. Turan, *Phys. Rev.* **D79**, 095001 (2009).
86. E. Nardi, *Phys. Rev.* **D48**, 3277 (1993); L. L. Everett, J. Jiang, P. G. Langacker and T. Liu, arXiv:0911.5349 [hep-ph].
87. P. Langacker and M. Plumacher, *Phys. Rev.* **D62**, 013006 (2000).
88. V. Barger, L. Everett, J. Jiang, P. Langacker, T. Liu and C. Wagner, *Phys. Rev.* **D80**, 055008 (2009) and arXiv:0906.3745 [hep-ph].
89. J. h. Kang, P. Langacker and T. j. Li, *Phys. Rev.* **D71**, 015012 (2005).
90. D. A. Demir, L. L. Everett and P. Langacker, *Phys. Rev. Lett.* **100**, 091804 (2008).
91. M. Cvetič and P. Langacker, *Phys. Rev.* **D78**, 066012 (2008).
92. M. Trodden, *Rev. Mod. Phys.* **71**, 1463 (1999).
93. M. Pietroni, *Nucl. Phys.* **B402**, 27 (1993); A. T. Davies, C. D. Froggatt and R. G. Moorhouse, *Phys. Lett.* **B372**, 88 (1996); S. J. Huber and M. G. Schmidt, *Eur. Phys. J.* **C10**, 473 (1999) and *Nucl. Phys.* **B606**, 183 (2001); J. M. Cline, S. Kraml, G. Laporte and H. Yamashita, *JHEP* **0907**, 040 (2009).
94. J. Kang, P. Langacker, T. j. Li and T. Liu, *Phys. Rev. Lett.* **94**, 061801 (2005) and arXiv:0911.2939 [hep-ph].
95. M. Fukugita and T. Yanagida, *Phys. Lett.* **B174**, 45 (1986); *Phys. Rev.* **D42**, 1285 (1990); W. Buchmüller and M. Plümacher, *Phys. Lett.* **B389**, 73 (1996). S. Davidson, E. Nardi and Y. Nir, *Phys. Rept.* **466**, 105 (2008).
96. W. Buchmüller and T. Yanagida, *Phys. Lett.* **B302**, 240 (1993).
97. B. A. Dobrescu, *Phys. Rev. Lett.* **94**, 151802 (2005).
98. P. Fayet, *Phys. Rev.* **D74**, 054034 (2006); M. Pospelov, *Phys. Rev.* **D80**, 095002 (2009).
99. N. Arkani-Hamed, D. P. Finkbeiner, T. R. Slatyer and N. Weiner, *Phys. Rev.* **D79**, 015014 (2009).
100. M. J. Strassler and K. M. Zurek, *Phys. Lett.* **B651**, 374 (2007).
101. P. Langacker, G. Paz, L. T. Wang and I. Yavin, *Phys. Rev.* **D77**, 085033 (2008) and in Ref. 46.
102. H. Verlinde, L. T. Wang, M. Wijnholt and I. Yavin, *JHEP* **0802**, 082 (2008); T. W. Grimm and A. Klemm, *JHEP* **0810**, 077 (2008).
103. N. Arkani-Hamed and S. Dimopoulos, *JHEP* **0506**, 073 (2005).
104. T. Kikuchi and T. Kubo, *Phys. Lett.* **B669**, 81 (2008); J. de Blas, P. Langacker, G. Paz and L. T. Wang, arXiv:0911.1996 [hep-ph].

# Searches for Supersymmetry at High-Energy Colliders*

Jonathan L. Feng, Jean-François Grivaz and Jane Nachtman

*Department of Physics and Astronomy,*
*University of California, Irvine, CA 92697, USA*
*Laboratoire de l'Accélérateur Linéaire, Orsay, France*
*University of Iowa, Iowa City, Iowa 52242, USA*

This review summarizes the state of the art in searches for supersymmetry at colliders on the eve of the LHC era. Supersymmetry is unique among extensions of the standard model in being motivated by naturalness, dark matter, and force unification, both with and without gravity. At the same time, weak-scale supersymmetry encompasses a wide range of experimental signals that are also found in many other frameworks. We recall the motivations for supersymmetry and review the various models and their distinctive features. We then comprehensively summarize searches for neutral and charged Higgs bosons and standard model superpartners at the high energy frontier, considering both canonical and non-canonical supersymmetric models, and including results from LEP, HERA, and the Tevatron.

## 9.1. Introduction

Particle physics is at a crossroads. Behind us is the standard model (SM), the remarkably successful theory of all known elementary particles and their interactions. Ahead of us is an equally remarkable array of possibilities for new phenomena at the weak scale. Never before has an energy scale been so widely anticipated to yield profound insights, and never before have there been so many ideas about exactly what these insights could be. In this article, we review the current state of experimental searches for supersymmetry, the most widely studied extension of the SM.

### 9.1.1. *Motivations for new phenomena*

There are at present many reasons to expect new physics at the weak scale $m_{\rm weak} \sim 100$ GeV $-$ 1 TeV. Chief among these is the Higgs boson, an essential component of the SM that has yet to be discovered. At the same time, there are also strong motivations for new phenomena beyond the Higgs boson. These motivations include naturalness, dark matter, and unification.

#### 9.1.1.1. *Naturalness*

The physical mass of the SM Higgs boson is given by

$$m_h^2 = m_h^{0\,2} + \Delta m_h^2 \ , \tag{9.1}$$

where $m_h^{0\,2}$ is the bare mass parameter present in the Lagrangian, and the quantum corrections are

$$\Delta m_h^2 \sim \frac{\lambda^2}{16\pi^2} \int^{\Lambda} \frac{d^4 p}{p^2} \sim \frac{\lambda^2}{16\pi^2} \Lambda^2 \ , \tag{9.2}$$

where $\lambda$ is a dimensionless gauge or Yukawa coupling, and $\Lambda$ is the energy scale at which the SM is no longer a valid description of nature. Because $\Delta m_h^2$ is proportional to $\Lambda^2$ ("quadratically divergent"), it is natural to expect the Higgs mass to be pulled up to within an order of magnitude of $\Lambda$ by quantum corrections.[1–4] Given that unitarity and precision constraints require $m_h$ to be at the weak scale,[5] this implies $\Lambda \lesssim 1$ TeV, and new physics should appear at the current energy frontier. Of course, the Higgs boson may not be a fundamental scalar, but in this case, too, its structure requires new physics at the weak scale.[6] For these reasons, naturalness is among the most robust motivations for new physics at an energy scale accessible to accelerator-based experiments.

#### 9.1.1.2. *Dark matter*

In the last decade, a wealth of cosmological observations have constrained the energy densities of baryons, non-baryonic dark matter, and dark energy, in units of the critical density, to be[7]

$$\begin{aligned} \Omega_{\rm B} &= 0.0462 \pm 0.0015 \\ \Omega_{\rm DM} &= 0.233 \pm 0.013 \\ \Omega_{\Lambda} &= 0.721 \pm 0.015 \ . \end{aligned} \tag{9.3}$$

The non-baryonic dark matter must be stable or very long-lived and dominantly cold or warm. None of the particles of the SM satisfies these conditions, and so cosmology requires new particles.

Perhaps the simplest production mechanism for dark matter is thermal freeze out.[8–11] In this scenario, a new particle is initially in thermal contact with the SM, but as the Universe cools and expands, this particle loses thermal contact and its energy density approaches a constant. Under very general assumptions, this relic energy density satisfies

$$\Omega_X \propto \frac{1}{\langle \sigma v \rangle} \,, \tag{9.4}$$

where $\langle \sigma v \rangle$ is the dark matter's thermally-averaged annihilation cross section. It is a tantalizing fact that, when this cross section is typical of weak-scale particles, that is, $\sigma v \sim \alpha^2 / m_{\text{weak}}^2$, where $m_{\text{weak}} \sim 100$ GeV, $\Omega_X$ is near the observed value of $\Omega_{\text{DM}}$ given in (9.3). If thermal freeze out is the mechanism by which dark matter is produced in the early Universe, then, cosmological data therefore also point to the weak scale as the natural scale for new physics.

#### 9.1.1.3. *Unification*

The SM is consistent with the observed properties of all known elementary particles. It also elegantly explains why some phenomena, such as proton decay and large flavor-changing neutral currents, are not observed. The latter fact is highly non-trivial, as evidenced by the intellectual contortions required of model builders who try to extend the SM.

At the same time, the SM contains many free parameters with values constrained by experiment, but not explained. The number of free parameters may be reduced in unified theories, in which the symmetries of the SM are extended to larger symmetries. In particular, grand unified theories, in which the SU(3) × SU(2) × U(1) gauge structure is extended to larger groups, are significantly motivated by the fact that the SM particle content fits perfectly into multiplets of SU(5) and larger groups,[12] potentially explaining the seemingly random assignment of quantum numbers, such as hypercharge.

One straightforward implication of the simplest ideas of grand unification is that the gauge couplings of the SM must unify when extrapolated to higher scales through renormalization group evolution. The gauge couplings do not unify at any scale given the particle content of the SM, but they do unify at the value $g_U \simeq 0.7$ at $M_{\text{GUT}} \simeq 2 \times 10^{16}$ GeV if the SM is minimally extended by supersymmetry (SUSY) *and* the supersymmetric particles are at the weak scale.[13–17] This unification is highly non-trivial, not only because the couplings are now so precisely measured, but also because $g_U$ is in the perturbative regime and $M_{\mathrm{GUT}}$ is in the narrow range that is both high enough to suppress proton decay and low enough to avoid quantum gravitational effects. This unification is only logarithmically sensitive to the superpartner mass scale, and the degree of its success is somewhat model-dependent; see, *e.g.*, the review by Raby in Ref. 18. In conjunction with the previous two motivations, however, it provides still more evidence for new physics at the weak scale, and selects supersymmetry as a particularly motivated possibility.

### 9.1.2. *Experimental context*

There are two main areas where new phenomena could appear in particle physics. Deviations from SM predictions could show up in measurements performed with increasing precision. Examples are the anomalies observed in the forward-backward asymmetry in the production of $b\bar{b}$ pairs in $e^+e^-$ collisions at the $Z$ peak (see Ref. 19, particularly section 7.3.5), or in the anomalous magnetic moment of the muon (see, *e.g.*, the review by Höcker and Marciano in Ref. 18). Even if such anomalies receive experimental confirmation at a sufficient significance level, their interpretation will however remain ambiguous, because it will involve virtual contributions to the relevant amplitudes of yet undiscovered, therefore most likely very massive, particles. The alternative approach is to try to observe directly the production of these new particles, which is among the goals of the experiments at colliders operating at the highest possible energies.

The Large Hadron Collider (LHC) at CERN will soon occupy the energy frontier. When it comes into operation, $pp$ collisions will take place at a center-of-mass energy of 10 TeV, and of 14 TeV later on. The instantaneous luminosity will be raised first to $10^{32}$ cm$^{-2}$ s$^{-1}$ and progressively to $10^{34}$ cm$^{-2}$ s$^{-1}$. With the enormous data samples accumulated, the two general purpose experiments at the LHC, ATLAS[20,21] and CMS,[22,23] will be in a position to explore in great detail the physics at the TeV scale. Since this is an entirely new domain, and since there are strong reasons to expect new phenomena at that scale, as advocated in the preceding section of this review, it may well be that ground breaking discoveries are made at the LHC, even after a short period of operation, once the detectors are properly aligned, calibrated, and well understood.

Until then, the most constraining results on searches for new phenomena at high energy have been or are still being obtained at LEP, HERA, and the Tevatron. Providing a comprehensive account of such searches for supersymmetry is the purpose of this review.

The large $e^+e^-$ collider (LEP) at CERN operated from 1989 to 2000. In a first phase (LEP1), the center-of-mass energy was set at or close to 91 GeV, the peak of the $Z$ boson resonance. Four experiments, ALEPH,[24,25] DELPHI,[26,27] L3,[28] and OPAL[29] studied millions of $Z$ decays that allowed them to perform stringent precision tests of the SM. From the end of 1995 on, the energy was progressively increased (LEP2) to reach 209 GeV in the center of mass during the last year of operation. Altogether, each of the experiments collected a total of $\sim$ 1 fb$^{-1}$ of data, of which $\sim$ 235 pb$^{-1}$ in 2000 at and above 204 GeV, the data set most relevant for new particle searches.

At DESY, the HERA collider operation was terminated in June 2007. There, $e^{\pm}p$ collisions were collected by two experiments, H1[30] and ZEUS,[31] at a center-of-mass energy of $\sim$ 300 GeV. This was an asymmetric collider, with $e^{\pm}$ and proton beam energies of 30 and 820 GeV, respectively. An upgrade took place in 2001 (HERA2), leading to higher luminosities than in the previous phase (HERA1), and allowing operation with polarized $e^{\pm}$ beams. The data sets collected at HERA1 and HERA2 with electron or positron beams altogether correspond to an integrated luminosity of $\sim$ 0.5 fb$^{-1}$ per experiment.

Until the LHC comes into operation, the highest energy collisions are provided by the Tevatron $p\bar{p}$ collider at Fermilab. During its first phase of operation (Run I), the center-of-mass energy was set to 1.8 TeV, and a data sample of $\sim$ 110 pb$^{-1}$ was collected by each of the two experiments, CDF[32] and DØ.[33] The highlight of that period was the discovery of the top quark in 1995. Major upgrades of the accelerator complex and of the two detectors took place for the second phase (Run II), which began in 2001. The center-of-mass energy was raised to 1.96 TeV, and the instantaneous luminosity was progressively increased to regularly approach or exceed $3 \times 10^{32}$ cm$^{-2}$ s$^{-1}$ in 2008. More than 5 fb$^{-1}$ of integrated luminosity had been delivered by the Tevatron by the end of fiscal year (FY) 2008, and it is expected that another $\sim$ 1.5 fb$^{-1}$ of luminosity will be provided per additional year of operation. At the time of writing, running in FY 2009 is underway, running in FY 2010 is increasingly likely, and running in FY 2011 is kept as an option.

All general purpose detectors at colliders share similar features. A cylindrical "barrel" structure parallel to the beam axis surrounds the interaction region, and is closed by "end caps" perpendicular to the beam. The first elements encountered beyond the beam pipe are charged-particle detectors, with those closest to the interaction point benefiting from the highest spatial precision. This tracking system is immersed in an axial magnetic field provided by a solenoidal magnet. Beyond the tracking system, electromagnetic calorimeters provide electron and photon identification and energy measurement. These are followed by hadron calorimeters for the measurement of jet energies. Finally, track detectors are used to identify and measure the muons which have penetrated through the calorimeters and possibly additional absorber material.

Non-interacting particles, such as neutrinos, are detected by an apparent non-conservation of energy and momentum. In $e^+e^-$ annihilation, the missing energy and momentum can be directly inferred from a measurement of the final state particles, by comparison with the center-of-mass energy of the collision. In hadronic or $ep$ collisions, the partons participating in the hard process carry only a fraction of the beam energy, and the beam remnants associated with the spectator partons largely escape undetected in the beam pipe. As a consequence, only conservation of the momentum in the direction transverse to the beams can be used, and the relevant quantity is the missing transverse energy $\not{E}_T$, rather than the total missing energy.

The mass reach in $p\bar{p}$ collisions at the Tevatron is expected to be substantially larger than at LEP because of the higher center-of-mass energy. However, since the initial partons participating in the hard process carry fractions $x_1$ and $x_2$ of the beam energy, the effective center-of-mass energy is only $\sqrt{\hat{s}} = \sqrt{x_1 x_2 s}$. Because of the rapidly falling parton distribution functions (PDFs) as a function of those energy fractions, increasingly large integrated luminosities are needed to probe larger and larger $\sqrt{\hat{s}}$ values. At HERA, furthermore, the center-of-mass energy in the $eq$ collision cannot be fully used for new particle production, except in some very specific instances. This is in contrast to $e^+e^-$ or $q\bar{q}$ annihilation, and to $gg$ fusion. As a consequence, the most constraining results on new particle searches typically come from LEP and from the Tevatron.

In the following, all limits quoted are given at a confidence level of 95%.

## 9.2. Supersymmetric Models and Particles

Supersymmetry (SUSY)[34–36] is an extension of Poincaré symmetry, which encompasses the known spacetime symmetries of translations, rotations, and boosts. As with the Poincaré and internal symmetries, SUSY transforms particle states to other particle states. In contrast to these other symmetries, however, SUSY relates states of different spin, transforming fermions into bosons and vice versa. None of the known particles can be supersymmetric partners of other known particles. As a result, SUSY predicts many new particle states. If SUSY were exact, these particles would be degenerate with known particles. Since this is experimentally excluded, if SUSY is a symmetry of nature, it must be broken.

SUSY is the most studied extension of the SM because it directly addresses several of the motivations for new physics discussed in Sec. 9.1. In supersymmetric theories, the quadratically divergent loop contributions to the Higgs boson mass from SM particles are canceled by similar contributions from superpartners, ameliorating the gauge hierarchy problem.[37–39] Supersymmetric theories also include excellent dark matter candidates, in the form of neutralinos[40,41] and gravitinos,[42,43] that may naturally have the desired relic density. Finally, SUSY is strongly motivated by the hope for unifying forces, as it makes gauge coupling unification possible in simple grand unified theories (GUTs).[13–17] It is important to note that *all* of these virtues are preserved only if the superpartner mass scale is around the weak scale. The existence of SUSY in nature, although not necessarily at the weak scale, is also motivated by string theory and the beautiful mathematical properties of SUSY that are beyond the scope of this review.

For these reasons, this review is devoted to searches for SUSY at colliders. In this Section, we present brief summaries of the supersymmetric spectrum, parameters, and unifying frameworks to establish our conventions and notation. More extensive phenomenological reviews of SUSY may be found in Refs. 44–47.

### 9.2.1. *Superpartners*

In this review, we focus our attention on the minimal supersymmetric extension of the standard model (MSSM), the supersymmetric model with minimal field content. Bosonic superpartners are given names with the prefix "s–," and fermionic superpartners are denoted by the suffix "–ino." Squarks and sleptons are collectively known as "sfermions," and the entire group of superpartner particles are often called "sparticles."

The particle content of the MSSM is in fact slightly more than a doubling of the SM particle content. This is because, in addition to introducing superpartners for all known particles, the MSSM requires two electroweak Higgs doublets. There are two reasons for this. First, in the SM, mass terms are generated for up- and down-type particles by Yukawa couplings to $\varphi^*$ and $\varphi$, respectively, where $\varphi$ is the SM Higgs field. In SUSY, Yukawa couplings are generalized to terms in a superpotential, a function of superfields that contain both SM particles and their superpartners, which generates the SM Yukawa couplings as well as all other terms related to these by SUSY. Complex-conjugated fields are not allowed in the superpotential, however. As a result, two separate Higgs fields, denoted $H_u$ and $H_d$, are required to generate masses through the superpotential terms

$$W = \lambda_u H_u Q\bar{U} + \lambda_d H_d Q\bar{D} + \lambda_e H_d L\bar{E} \ , \tag{9.5}$$

where $Q$, $U$, $D$, $L$, and $E$ are the SU(2) quark doublet, up-type quark singlet, down-type quark singlet, lepton doublet, and lepton singlet superfields, respectively, and the $\lambda$ couplings are Yukawa couplings. Second, SUSY requires that the SM Higgs field have fermion partners, the Higgsinos. The introduction of these additional fermions charged under SM gauge groups ruins anomaly cancellation, making this theory mathematically untenable. The introduction of an additional Higgs doublet, with its extra Higgsinos, restores anomaly cancellation.

The MSSM Higgs boson sector therefore consists of eight degrees of freedom. As in the SM, three of these are eaten to make massive $W$ and $Z$ bosons, but five remain, which form four physical particles:

$$\text{MSSM Higgs Bosons (Spin 0)}: \quad h, H, A, H^{\pm} \ , \tag{9.6}$$

where $h$ and $H$ are the CP-even neutral Higgs bosons, with $h$ lighter than $H$, $A$ is the CP-odd neutral Higgs boson, and $H^{\pm}$ is the charged Higgs boson.

The remaining supersymmetric particle content of the MSSM is straightforward to determine and consists of the following states:

$$\begin{aligned} \text{Neutralinos (Spin 1/2)}: &\ \tilde{B}, \tilde{W}^0, \tilde{H}_u^0, \tilde{H}_d^0 \\ \text{Charginos (Spin 1/2)}: &\ \tilde{W}^+, \tilde{H}_u^+ \\ &\ \tilde{W}^-, \tilde{H}_d^- \end{aligned}$$

$$\begin{aligned}
\text{Sleptons (Spin 0)}: &\ \tilde{e}_{L,R}, \tilde{\mu}_{L,R}, \tilde{\tau}_{L,R} \\
&\ \tilde{\nu}_e, \tilde{\nu}_\mu, \tilde{\nu}_\tau \\
\text{Squarks (Spin 0)}: &\ \tilde{u}_{L,R}, \tilde{c}_{L,R}, \tilde{t}_{L,R} \\
&\ \tilde{d}_{L,R}, \tilde{s}_{L,R}, \tilde{b}_{L,R} \\
\text{Gluinos (Spin 1/2)}: &\ \tilde{g}\ . \qquad (9.7)
\end{aligned}$$

Each SM chiral fermion has a (complex) scalar partner, denoted by the appropriate chirality subscript. The dimensionless couplings of all of these particles are fixed by SUSY to be identical to those of their SM partners. Note, however, that, as described in the appropriate sections below, the states in each line of (9.7) (except for the last one) may mix, and mass eigenstates are in general linear combinations of these gauge eigenstates.

Finally, most analyses of SUSY include the supersymmetric partner of the graviton:

$$\text{Gravitino (Spin 3/2)}: \quad \tilde{G}\ . \qquad (9.8)$$

Although not technically required as a part of the MSSM, when SUSY is promoted to a local symmetry, it necessarily includes gravity, and the resulting supergravity theories include both gravitons and gravitinos. The gravitino is therefore present if SUSY plays a role in unifying the SM with gravity, as in string theory.

If SUSY were exact, the gravitino's properties would be determined precisely by the graviton's, and it would be massless and have gravitational couplings suppressed by the reduced Planck mass $M_* \simeq 2.4 \times 10^{18}$ GeV. However, just as Goldstone bosons appear when conventional symmetries are spontaneously broken, a fermion, the Goldstino $\tilde{G}_{1/2}$, appears when SUSY is broken. The gravitino then becomes massive by eating the Goldstino. In terms of $F$, the mass dimension-2 order parameter of SUSY breaking, the gravitino mass becomes

$$m_{\tilde{G}} \sim \frac{F}{M_*}\ , \qquad (9.9)$$

and, very roughly, its interactions in processes probing energy scale $E$ may be characterized by a dimensionless coupling

$$g_{\tilde{G}} \sim \frac{E^2}{F} \sim \frac{E^2}{m_{\tilde{G}} M_*}\ . \qquad (9.10)$$

Light gravitinos couple more strongly. As we will see below, in well-motivated supersymmetric theories, these properties may take values in the range

$$\text{eV} \lesssim m_{\tilde{G}} \lesssim 10 \text{ TeV} \tag{9.11}$$

$$10^{-5} \gtrsim g_{\tilde{G}} \gtrsim 10^{-18} , \tag{9.12}$$

where we have assumed colliders probing $E \sim m_{\text{weak}}$.

### 9.2.2. *Supersymmetry parameters*

As noted above, if SUSY exists in nature, it must be broken. Although many different Lagrangian terms could be added to break SUSY, only some of these are allowed if SUSY is to stabilize the gauge hierarchy. These terms, known as "soft" SUSY-breaking terms, include most, but not all, Lagrangian terms with mass dimension 3 and below.[48] For the MSSM, they are

$$\begin{aligned} & m_{\tilde{Q}}^2|\tilde{Q}|^2+m_{\tilde{U}}^2|\tilde{U}|^2+m_{\tilde{D}}^2|\tilde{D}|^2+m_{\tilde{L}}^2|\tilde{L}|^2+m_{\tilde{E}}^2|\tilde{E}|^2 \\ & +\frac{1}{2}\left\{\left[M_1\tilde{B}\tilde{B}+M_2\tilde{W}^j\tilde{W}^j+M_3\,\tilde{g}^k\tilde{g}^k\right]+\text{h.c.}\right\} \\ & +\lambda_u A_U H_u\tilde{Q}\tilde{U}+\lambda_d A_D H_d\tilde{Q}\tilde{D}+\lambda_e A_E H_d\tilde{L}\tilde{E} \\ & +m_{H_u}^2|H_u|^2+m_{H_d}^2|H_d|^2+(BH_uH_d+\text{h.c.})\,. \end{aligned} \tag{9.13}$$

These lines are sfermion masses, gaugino masses, trilinear scalar couplings ("$A$-terms"), and Higgs boson couplings. In addition to the parameters above, there are two other key parameters: the $\mu$ parameter, which enters in the Higgsino mass terms $\mu\tilde{H}_u^i\tilde{H}_d^i$, and

$$\tan\beta \equiv \frac{\langle H_u^0\rangle}{\langle H_d^0\rangle} , \tag{9.14}$$

which parameterizes how the SM Higgs vacuum expectation value is divided between the two neutral Higgs scalars.

The interactions of (9.13) conserve $R$-parity.[49,50] With $R = (-1)^{3(B-L)+2S}$, where $B$ and $L$ are the baryon and lepton numbers, respectively, and $S$ the spin, all superpartners are odd and all SM particles are even under $R$-parity. This implies that all interactions involve an even number of superpartners, and so the lightest superpartner is stable, and a potential dark matter candidate. $R$-parity violation generically violates both baryon and lepton number, leading to too-rapid proton decay, which is why, for most of this review, we limit ourselves to the $R$-parity conserving case.

Even restricting ourselves to the $R$-parity-preserving terms of (9.13), however, we see that SUSY introduces many new parameters. Note that the terms involving sfermions need not be flavor-diagonal, and so the sfermion masses and $A$-terms are in fact matrices of parameters in the most general case. At the same time, fully general flavor mixing terms violate low energy constraints on flavor-changing neutral currents. In addition, arbitrary complex parameters also violate bounds on CP-violation from, for example, $\epsilon_K$ and the electric dipole moments of the electron and neutron. These considerations motivate unifying frameworks, to which we now turn.

### 9.2.3. *Unifying frameworks*

In collider searches, it is desirable to consider theories that are both viable and simple enough to be explored fully. For this reason, it is common to work in simple model frameworks that reduce the number of independent SUSY parameters. In some cases, these model frameworks also motivate particular collider signatures that might otherwise appear highly unlikely or fine-tuned.

In the most common unifying frameworks, SUSY is assumed to be broken in some other sector. SUSY breaking is then mediated to the MSSM through a mechanism that defines the framework. This sets SUSY-breaking parameters at some high energy scale. Renormalization group evolution to the weak scale then determines the physical soft SUSY-breaking parameters and the physical spectrum of the MSSM. A representative example of renormalization group evolution is shown in Fig. 9.1. In this evolution from the high scale to the weak scale, gauge couplings increase masses and Yukawa couplings decrease masses. This is central to understanding the sparticle spectrum of many models. In addition, it explains why $m^2_{H_u}$ becomes negative at the weak scale — it is the only particle to receive large negative contributions from Yukawa couplings without compensating large positive contributions from the strong coupling. When $H_u$ becomes tachyonic, it breaks electroweak symmetry, and this feature, known as "radiative electroweak symmetry breaking," is a virtue of many supersymmetric frameworks. Note, however, that radiative electroweak symmetry breaking makes essential use of the large top quark mass, and so shifts the burden of understanding why electroweak symmetry is broken to the question of why the top quark is heavy.

In this section, we discuss several common unifying frameworks that have been used in collider searches, namely, models with gravity-, gauge-,

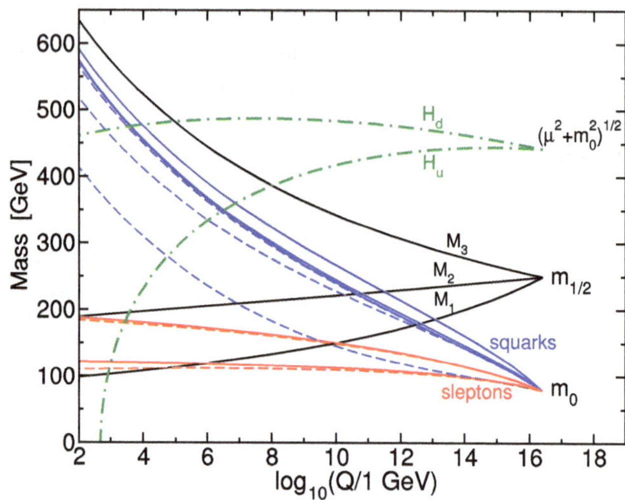


Fig. 9.1. Renormalization group evolution of scalar and gaugino mass parameters from the GUT scale $M_{\rm GUT} \simeq 2 \times 10^{16}$ GeV to the weak scale in a representative mSUGRA model. From Ref. 44.

and anomaly-mediated SUSY breaking. Each of these has its distinctive characteristics. As a rough guide, in Fig. 9.2 we show representative spectra resulting from each of these frameworks. These spectra may be generated using publicly available computer programs, including ISAJET,[51] SOFTSUSY,[52] SPHENO,[53] and SUSPECT.[54]

### 9.2.3.1. *Gravity mediation (SUGRA)*

In gravity-mediated SUSY-breaking models,[56–61] sometimes referred to as supergravity (SUGRA) models, SUSY breaking in a hidden sector is mediated to the MSSM through terms suppressed by the reduced Planck mass $M_*$. For example, sfermion masses are $m_{\tilde{f}} \sim F/M_*$. For these to be at the weak scale, $\sqrt{F}$ must be around $10^{11}$ GeV. Given Eqs. (9.9) and (9.10), the gravitino also has a weak scale mass and couples with gravitational strength in SUGRA models.

Without a quantum theory of gravity, the structure of gravity-mediated SUSY parameters is unconstrained and generically violates low-energy constraints. To make these theories viable, *ad hoc* unifying assumptions must be made. By far the most common assumptions are those of minimal supergravity (mSUGRA), which is specified by 4 continuous and 1 discrete

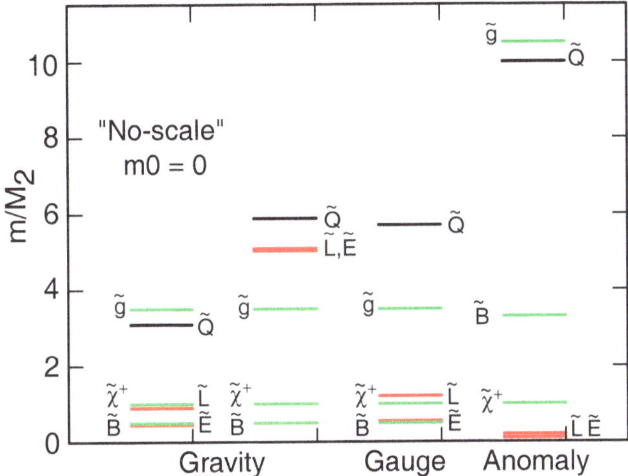


Fig. 9.2. Sparticle spectra for representative models with gravity-, gauge-, and anomaly-mediated SUSY breaking. The masses are normalized to $M_2$, the Wino mass parameter at the weak scale. In the gravity-mediated case, two example spectra are presented: one for "no-scale" models with $m_0 = 0$, and another for $m_0 = 5M_2$. In the anomaly-mediated case, the sleptons are tachyonic in the minimal case — additional effects are required to raise these to a viable range. From Ref. 55.

parameter choice:

$$\text{mSUGRA: } m_0, m_{1/2}, A_0, \tan\beta, \text{sign}(\mu) \ , \tag{9.15}$$

where the first three parameters are the universal scalar mass (including the two Higgs scalars), unified gaugino mass, and universal $A$-parameter, which are all specified at the grand unified theory (GUT) scale $M_{\text{GUT}} \simeq 2 \times 10^{16}$ GeV. The remaining SUSY parameters $|\mu|$ and the dimension-2 Higgs boson mass parameter $B$ are determined by requiring that the Higgs potential at the weak scale give correct electroweak symmetry breaking. At tree-level, this requires

$$\frac{1}{2}m_Z^2 = \frac{m_{H_d}^2 - m_{H_u}^2 \tan^2\beta}{\tan^2\beta - 1} - |\mu|^2 \tag{9.16}$$

$$\sin 2\beta = \frac{2B}{m_{H_d}^2 + m_{H_u}^2 + 2|\mu|^2} \ . \tag{9.17}$$

Gaugino mass unification is motivated by the unification of gauge couplings at $M_{\text{GUT}}$ in the MSSM. It leads to the prediction that the Bino, Wino, and gluino masses are in the ratio $M_1 : M_2 : M_3 \simeq 1 : 2 : 7$ at the weak scale, as evident in Fig. 9.2. Scalar mass universality is on much less solid ground. Even in GUTs, for example, the Higgs scalars are not necessarily in the same multiplet as the squarks and sleptons. This motivates a slightly less restrictive framework, the non-universal Higgs model (NUHM) in which $m_0$ is the universal sfermion mass, but $m_{H_u}$ and $m_{H_d}$ are treated as independent parameters. One may exchange these new degrees of freedom for the more phenomenological parameters $\mu$ and $m_A$ at the weak scale:

$$\text{NUHM: } m_0, m_{1/2}, A_0, \tan\beta, \mu, m_A \ , \tag{9.18}$$

The NUHM framework is employed in some MSSM Higgs boson studies discussed in Sec. 9.2.4.

#### 9.2.3.2. *GMSB*

In gauge-mediated SUSY-breaking (GMSB) models,[62–67] in addition to the gravity-mediated contributions to soft parameters discussed above, each sparticle receives contributions to its mass determined by its gauge quantum numbers. These new contributions to sfermion masses are $\sim F/M_{\text{mess}}$, where $M_{\text{mess}}$ is the mass scale of the messenger particles that transmit the SUSY breaking. The GMSB contributions are flavor-blind, and do not violate low energy bounds. For these to be dominant, one requires $M_{\text{mess}} \lesssim 10^{14}$ GeV, and so we find that $m_{\tilde{G}} \sim F/M_* \ll F/M_{\text{mess}} \sim m_{\text{weak}}$ in GMSB scenarios, and the lightest supersymmetric particle (LSP) is always the gravitino.

In GMSB models, the collider signatures are determined by the next-to-lightest supersymmetric particle (NLSP) and its lifetime, or equivalently, the gravitino's mass. If the NLSP is the lightest neutralino, the collider signature is either missing energy or prompt photons, $Z$ or Higgs bosons from $\tilde{\chi}^0 \to (\gamma, Z, h)\tilde{G}$;[68,69] if the NLSP is a slepton, the signature is typically either long-lived heavy charged particles or multi-lepton events.[70–72]

#### 9.2.3.3. *AMSB*

A third class of SUSY models are those with anomaly-mediated SUSY-breaking (AMSB).[73,74] These are extra dimensional scenarios in which SUSY is broken on another 3-dimensional subspace, and transmitted to our world through the conformal anomaly. As with all anomalies, this effect is one-loop suppressed. The fundamental scale of SUSY breaking as characterized by the gravitino mass is therefore $m_{\tilde{G}} \sim 10 - 100$ TeV, with MSSM sparticle masses one-loop suppressed and at the weak scale.

The AMSB contributions to sparticle masses are completely determined by the sparticle's gauge and Yukawa couplings. This leads to a highly predictive spectrum. Unfortunately, one of these predictions is $m_{\tilde{L}}^2, m_{\tilde{E}}^2 < 0$, but various mechanisms have been proposed to solve this tachyonic slepton problem; see, e.g., Refs. 75–77.

The gaugino masses are determined by the corresponding gauge group beta functions. In particular, AMSB predicts $M_1 : M_2 : M_3 \simeq 2.8 : 1 : 8$; because the SU(2) coupling is nearly scale-invariant in the MSSM, the Wino mass is the smallest. AMSB scenarios therefore motivate supersymmetric models with $\tilde{W}^0$ LSP and $\tilde{W}^\pm$ NLSP. This triplet may be extremely degenerate, with the chargino traveling macroscopic distances before decaying to soft and invisible decay products, which provides a distinctive and challenging signature for collider searches.[78]

### 9.2.4. *Supersymmetric Higgs bosons*

The MSSM Higgs potential is

$$
\begin{aligned}
V_H = {} & (m_{H_u}^2 + |\mu|^2)|H_u^0|^2 + (m_{H_d}^2 + |\mu|^2)|H_d^0|^2 \\
& - B(H_u^0 H_d^0 + \text{h.c.}) + \frac{1}{2} g^2 |H_u^{0*} H_d^0|^2 \\
& + \frac{1}{8}(g^2 + g'^2)(|H_u^0|^2 - |H_d^0|^2)^2 \,, \qquad (9.19)
\end{aligned}
$$

where the parameters $\mu$, $m_{H_u}^2$, and $m_{H_d}^2$ are as defined in Sec. 9.2.2, and SUSY implies that all the quartic couplings are determined by the SU(2) and U(1) hypercharge gauge couplings, denoted $g$ and $g'$, respectively. $V_H$ automatically conserves CP, since any phase in the $B$ parameter can be eliminated by a redefinition of the Higgs fields.

Assuming these parameters are such that the potential admits a stable, symmetry-breaking minimum, the three parameter combinations $m_{H_u}^2 + |\mu|^2$, $m_{H_d}^2 + |\mu|^2$, and $B$ may be exchanged for the two vacuum expectation values $v_u \equiv \sqrt{2}\langle H_u^0 \rangle$, $v_d \equiv \sqrt{2}\langle H_d^0 \rangle$, and one physical Higgs boson mass, conveniently taken to be $m_A$. The $W$ boson mass fixes $v_u^2 + v_d^2$, leaving one additional degree of freedom, usually taken to be $\tan\beta \equiv v_u/v_d$. Thus, at tree-level, the entire MSSM Higgs boson sector is determined by two parameters, $m_A$ and $\tan\beta$.

In terms of these parameters, the physical Higgs boson masses are

$$m^2_{\substack{H\\h}} = \frac{m_A^2 + m_Z^2 \pm \sqrt{(m_A^2+m_Z^2)^2 - 4m_A^2 m_Z^2 c_{2\beta}^2}}{2} \tag{9.20}$$

$$m^2_{H^\pm} = m_A^2 + m_W^2 \,, \tag{9.21}$$

where $c_{2\beta} \equiv \cos 2\beta$. The CP-even mass eigenstates are related to the gauge eigenstates through

$$\begin{pmatrix} H \\ h \end{pmatrix} = \begin{pmatrix} \cos\alpha & \sin\alpha \\ -\sin\alpha & \cos\alpha \end{pmatrix} \begin{pmatrix} \sqrt{2}\,\mathrm{Re}H_d^0 - v_d \\ \sqrt{2}\,\mathrm{Re}H_u^0 - v_u \end{pmatrix} , \tag{9.22}$$

where the rotation angle $\alpha$ satisfies

$$\cos 2\alpha = -\cos 2\beta \frac{m_A^2 - m_Z^2}{m_H^2 - m_h^2} \tag{9.23}$$

with $-\pi/2 < \alpha < 0$.

Equation (9.20) implies that $m_h < m_Z |\cos 2\beta|$, a rather disastrous relation, given that experimental bounds exclude $m_h < m_Z$. The results presented so far, however, are valid only at tree-level. Large radiative corrections from top squark/quark loops (see, *e.g.*, Ref. 79),

$$\Delta m_h^2 \sim \frac{1}{\sin^2\beta} \frac{3g^2 m_t^4}{8\pi^2 m_W^2} \log \frac{m_{\tilde{t}}^2}{m_t^2} \,, \tag{9.24}$$

can lift $m_h$ to values above the experimental bounds. Note, however, that for $\tan\beta = 1$, $m_h = 0$ at tree-level, and so large values of $m_h$ are not possible for $\tan\beta \approx 1$. From considerations of the Higgs mass alone, $\tan\beta < 1$ is possible. However, such values imply very large top Yukawa couplings, which become infinite well below the GUT or Planck scales. In addition, in simple frameworks, $\tan\beta < 1$ is incompatible with radiative electroweak symmetry breaking; for a review of bounds on $\tan\beta$, see Ref. 80.

### 9.2.5. *Neutralinos and charginos*

The neutralinos and charginos of the MSSM are the mass eigenstates that result from the mixing of the electroweak gauginos $\tilde{B}$ and $\tilde{W}^j$ with the Higgsinos.

The neutral mass terms are

$$\frac{1}{2}(\psi^0)^T M_N \psi^0 + \mathrm{h.c.}\,, \tag{9.25}$$

where $(\psi^0)^T = (-i\tilde{B}, -i\tilde{W}^3, \tilde{H}_d^0, \tilde{H}_u^0)$ and

$$M_N = \begin{pmatrix} M_1 & 0 & -\frac{1}{2}g'v_d & \frac{1}{2}g'v_u \\ 0 & M_2 & \frac{1}{2}gv_d & -\frac{1}{2}gv_u \\ -\frac{1}{2}g'v_d & \frac{1}{2}gv_d & 0 & -\mu \\ \frac{1}{2}g'v_u & -\frac{1}{2}gv_u & -\mu & 0 \end{pmatrix} . \tag{9.26}$$

The neutralino mass eigenstates are $\tilde{\chi}_i^0 = \mathbf{N}_{ij}\psi_j^0$, where $\mathbf{N}$ diagonalizes $M_N$. In order of increasing mass, the four neutralinos are labeled $\tilde{\chi}_1^0$, $\tilde{\chi}_2^0$, $\tilde{\chi}_3^0$, and $\tilde{\chi}_4^0$.

The charged mass terms are

$$(\psi^-)^T M_C \psi^+ + \text{h.c.} , \tag{9.27}$$

where $(\psi^\pm)^T = (-i\tilde{W}^\pm, \tilde{H}^\pm)$ and

$$M_C = \begin{pmatrix} M_2 & \frac{1}{\sqrt{2}}gv_u \\ \frac{1}{\sqrt{2}}gv_d & \mu \end{pmatrix} . \tag{9.28}$$

The chargino mass eigenstates are $\tilde{\chi}_i^+ = \mathbf{V}_{ij}\psi_j^+$ and $\tilde{\chi}_i^- = \mathbf{U}_{ij}\psi_j^-$, where the unitary matrices $\mathbf{U}$ and $\mathbf{V}$ are chosen to diagonalize $M_C$, and $\tilde{\chi}_1^\pm$ is lighter than $\tilde{\chi}_2^\pm$.

### 9.2.6. *Sleptons*

Sleptons are promising targets for colliders, as they are among the lightest sparticles in many models. As noted in Sec. 9.2.1, sleptons include both left- and right-handed charged sleptons and sneutrinos. The mass matrix for the charged sleptons is

$$\begin{pmatrix} m_{\tilde{L}}^2 + m_\tau^2 - m_Z^2(\frac{1}{2} - s_W^2)c_{2\beta} & m_\tau(A_\tau - \mu\tan\beta) \\ m_\tau(A_\tau - \mu\tan\beta) & m_{\tilde{E}}^2 + m_\tau^2 - m_Z^2 s_W^2 c_{2\beta} \end{pmatrix} \tag{9.29}$$

in the basis $(\tilde{\tau}_L, \tilde{\tau}_R)$, where $s_W \equiv \sin\theta_W$. The sneutrino has mass

$$m_{\tilde{\nu}}^2 = m_{\tilde{L}}^2 + \frac{1}{2}m_Z^2\cos 2\beta . \tag{9.30}$$

These masses are given in third-generation notation; in the presence of flavor mixing, these generalize to full six-by-six and three-by-three matrices.

The left-right mixing is proportional to lepton mass, and is therefore expected to be insignificant for selectrons and smuons, but may be important for staus, especially if $\tan\beta$ is large. Through level repulsion, this mixing lowers the lighter stau's mass. As noted in Sec. 9.2.3, Yukawa couplings also lower scalar masses through renormalization group evolution. Both of these effects imply that in many scenarios, the lighter stau is the lightest slepton, and often the lightest sfermion.

#### 9.2.7. *Squarks*

The mass matrix for up-type squarks is

$$\begin{pmatrix} m_{\tilde{Q}}^2 + m_t^2 + m_Z^2(\frac{1}{2} - \frac{2}{3}s_W^2)c_{2\beta} & m_t(A_t - \mu\cot\beta) \\ m_t(A_t - \mu\cot\beta) & m_{\tilde{U}}^2 + m_t^2 + m_Z^2 \frac{2}{3} s_W^2 c_{2\beta} \end{pmatrix} \tag{9.31}$$

in the basis $(\tilde{t}_L, \tilde{t}_R)$, and for down-type squarks is

$$\begin{pmatrix} m_{\tilde{Q}}^2 + m_b^2 - m_Z^2(\frac{1}{2} - \frac{1}{3}s_W^2)c_{2\beta} & m_b(A_b - \mu\tan\beta) \\ m_b(A_b - \mu\tan\beta) & m_{\tilde{D}}^2 + m_b^2 - m_Z^2 \frac{1}{3} s_W^2 c_{2\beta} \end{pmatrix} \tag{9.32}$$

in the basis $(\tilde{b}_L, \tilde{b}_R)$. Large mixing is expected in the stop sector, and possibly also in the sbottom sector if $\tan\beta$ is large. Because of these mixings and the impact of large Yukawa couplings in renormalization group evolution, the 3rd generation squarks are the lightest squarks in many models.[81]

### 9.3. Searches for MSSM Neutral Higgs Bosons

As already explained in Sec. 9.2, two Higgs doublets are needed in the MSSM to give mass to both up- and down-type quarks. Under the assumption that the Higgs sector is CP conserving, the physical states are two neutral CP-even Higgs bosons ($h$ and $H$, ordered by increasing mass), a neutral CP-odd Higgs boson ($A$), and a doublet of charged Higgs bosons ($H^{\pm}$). Further details on the Higgs sector of the MSSM have been given in Sec. 9.2. Here, we focus on searches for the neutral Higgs bosons of the MSSM, while searches for charged Higgs bosons will be discussed in Sec. 9.4.

#### 9.3.1. *MSSM benchmark scenarios*

It has been seen that two parameters are sufficient to fully determine the MSSM Higgs sector at tree level. These are commonly taken to be the $A$ boson mass $m_A$ and $\tan\beta$, the ratio of the vacuum expectation values of the Higgs fields giving mass to the up- and down-type quarks. This picture is modified significantly, however, by large radiative corrections, arising essentially from an incomplete cancellation of the top and stop loops. In particular, the important prediction $m_h < m_Z|\cos 2\beta|$ is invalidated.

Among the many parameters of the MSSM, a few have been identified as being most relevant for the determination of Higgs boson properties. In addition to $m_A$ and $\tan\beta$, an effective SUSY breaking scalar mass, $M_{SUSY}$, which sets the scale of all squark masses, and a term controlling the amount of mixing in the stop sector, $X_t$, play the leading role. (In (9.24), the stop mass is directly related to $M_{SUSY}$, and stop mixing is neglected.) The model is further specified by a weak gaugino mass, $M_2$, the gluino mass, $m_{\tilde{g}}$, and the SUSY Higgs mass term $\mu$. The relation $X_t = A - \mu\cot\beta$ then allows the trilinear Higgs-squark coupling $A$ (assumed to be universal) to be calculated. For large values of $\tan\beta$, mixing in the sbottom sector becomes relevant too; it is controlled by $X_b = A - \mu\tan\beta$. Finally, the top quark mass $m_t$ needs to be specified.

A few benchmark scenarios[82,83] were agreed upon to interpret the searches for MSSM Higgs bosons. The most widely considered are the so-called "$m_h$-max" and "no-mixing" ones, where $M_{SUSY} = 1$ TeV, $M_2 = 200$ GeV, $\mu = -200$ GeV and $m_{\tilde{g}} = 800$ GeV. In $m_h$-max, $X_t$ is set equal to $2M_{SUSY}$ (in the on-shell renormalization scheme), while it is set to 0 in the no-mixing scenario. The largest value of $m_h$ is obtained for large $m_A$ and $\tan\beta$, and is maximized (minimized) in the $m_h$-max (no-mixing) scenario. In the $m_h$-max scenario, the maximum value of $m_h$ is $\simeq 135$ GeV.

### 9.3.2. *Searches at LEP*

At LEP, the neutral Higgs bosons of the MSSM have been searched for in two production processes, the Higgsstrahlung process $e^+e^- \to hZ$[84–86] and the associated production $e^+e^- \to hA$.[87] Both processes are mediated by $s$-channel $Z$ boson exchange. With the notations of Sec. 9.2, the cross sections are

$$\sigma_{hZ} = \sin^2(\beta-\alpha)\sigma_{hZ}^{SM} \tag{9.33}$$

$$\sigma_{hA} = \cos^2(\beta-\alpha)\bar{\lambda}\sigma_{hZ}^{SM}\ , \tag{9.34}$$

where $\beta$ and $\alpha$ are defined in Eqs. (9.14) and (9.22),

$$\sigma_{hZ}^{SM} = \frac{G_F^2 m_Z^4}{96\pi s}(v_e^2+a_e^2)\lambda_{hZ}^{1/2}\frac{\lambda_{hZ}+12m_Z^2/s}{(1-m_Z^2/s)^2} \tag{9.35}$$

is the SM Higgs boson production cross section, $s$ is the square of the center-of-mass energy, and

$$\lambda_{ij} = [1-(m_i+m_j)^2/s][1-(m_i-m_j)^2/s]\ , \tag{9.36}$$

$$\bar{\lambda} = \lambda_{hA}^{3/2}[\lambda_{hZ}^{1/2}(\lambda_{hZ}+12m_Z^2/s)]\ . \tag{9.37}$$

It is apparent from the above formulae that the two processes are complementary. In practice, the Higgsstrahlung process dominates for values of $\tan\beta$ close to unity, while associated production dominates for large values of $\tan\beta$, if kinematically allowed. In large regions of the MSSM parameter space, the $h$ decay branching fractions are similar to those of the SM Higgs boson. For a mass of 115 GeV, these are 74% into $b\bar{b}$, 7% into both $\tau^+\tau^-$ and $gg$, 8% into $WW^*$, and 4% into $c\bar{c}$.[88] The $A$ boson couples only to fermions, so that its decay branching fraction into $b\bar{b}$ is always close to 90%, with most of the rest going into $\tau^+\tau^-$. These same branching fractions also hold for the $h$ boson for large values of $\tan\beta$.[88]

Searches for Higgs bosons were performed at LEP first in $Z$ boson decays during the LEP1 era, and subsequently at increasing center-of-mass energies at LEP2, up to 209 GeV in 2000. In the following, only the searches performed at the highest energies are described.

The four LEP experiments carried out searches for the SM Higgs boson produced via Higgsstrahlung, $e^+e^- \to HZ$, and the results were combined to maximize the sensitivity.[a] Four final state topologies were analyzed to cope with the various decay modes of the Higgs and $Z$ bosons: a four-jet topology with two $b$-tagged jets, for $(H \to b\bar{b})(Z \to q\bar{q})$; a two $b$-tagged jets and two-lepton topology, for $(H \to b\bar{b})(Z \to \ell^+\ell^-)$, with $\ell = e$ or $\mu$; a two $b$-tagged jets and missing energy topology, for $(H \to b\bar{b})(Z \to \nu\bar{\nu})$; and a two-jet and two-$\tau$ topology for $(H \to b\bar{b})(Z \to \tau^+\tau^-)$ and $(H \to \tau^+\tau^-)(Z \to q\bar{q})$. A few candidate events were observed at the edge of the sensitivity domain, but the overall significance was only at the level of $1.7\,\sigma$. A lower mass limit was therefore derived, excluding a SM Higgs boson with mass smaller than 114.4 GeV.[90]

The Higgs boson mass lower limit depends on the strength of the $HZZ$ coupling, and the LEP collaborations also provided, as a function of the mass of a SM-like Higgs boson, an upper limit on $\xi^2$, where $\xi$ is a multiplicative factor by which the SM $HZZ$ coupling is reduced.[90] By SM-like, it is meant that the decay branching fractions are similar to those expected from a SM Higgs boson. This result is shown in Fig. 9.3. Constraints on the MSSM parameter space can be deduced from this, since in that case $\xi = \sin(\beta - \alpha)$.

For Higgs boson masses accessible at LEP, the structure of the MSSM Higgs sector is such that the $h$ and $A$ masses are similar whenever as-

[a]Production by vector boson fusion, $e^+e^- \to He^+e^-$ or $H\nu\bar{\nu}$,[89] was also considered, but its contribution was found to be negligible in practice.

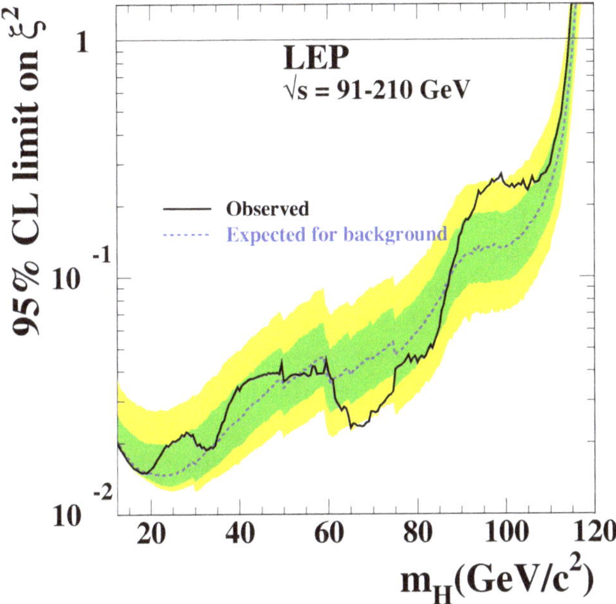


Fig. 9.3. The upper bound on the factor $\xi^2$ by which the square of the SM $HZZ$ coupling is multiplied, as provided by the LEP experiments.[90] The full curve is the observed limit, the dashed curve the median expected limit in the absence of signal, and the dark gray and light gray bands are the 68% and 95% probability regions around the expected limit.

sociated production is relevant, i.e., for large values of $\tan\beta$. Searches for $hA$ associated production were performed in the four $b$-jet final state for $(h \to b\bar{b})(A \to b\bar{b})$ and in the two $b$-jet and two-$\tau$ topology for $(h/A \to b\bar{b})(A/h \to \tau^+\tau^-)$. The constraint that the $h$ and $A$ boson candidate masses should be similar was imposed. The backgrounds from multijet and $WW$ production were largely reduced by the $b$-jet identification requirements, leaving $ZZ$ as an irreducible background.

No significant excess over the SM background expectation was observed, and production cross section upper limits were derived as a function of $m_h \simeq m_A$. For each benchmark scenario, a scan was performed as a function of $m_A$ and $\tan\beta$, and in each point of the scan the cross section upper limit was compared to the corresponding prediction, taking into account the slight modifications expected for the values of the $h$ and $A$ branching fractions into $b\bar{b}$ and $\tau^+\tau^-$, as well as the non-negligible difference between $m_h$ and $m_A$ which develops at lower values of $\tan\beta$. If the cross section upper limit was found to be smaller than the prediction, the $(m_A, \tan\beta)$ set was declared excluded. The result of the combination of the searches in the $hZ$ and $hA$ channels by the four LEP experiments[91] is shown in Fig. 9.4, projected onto the $(m_h, \tan\beta)$ plane in the $m_h$-max and no-mixing scenarios. In the derivation of those results, contributions of the $e^+e^- \to HZ$ and $HA$ processes were also taken into account whenever relevant, where $H$ is the heavier CP-even Higgs boson.

In the most conservative scenario, i.e., $m_h$-max, it can be seen in Fig. 9.4 that the lower limit on the mass of the SM Higgs boson holds also for $m_h$ as long as $\tan\beta$ is smaller than about 5, and that values of $\tan\beta$ between $\simeq 0.7$ and 2 are excluded for the current average value of the top quark mass, $173.1 \pm 1.3$ GeV.[92] A lower mass limit of 93 GeV is obtained for $m_h \simeq m_A$ for large values of $\tan\beta$.

The benchmark scenarios were chosen such that the Higgs bosons do not decay into SUSY particles. An interesting possibility is that the $h \to \tilde{\chi}_1^0\tilde{\chi}_1^0$ decay mode is kinematically allowed, where $\tilde{\chi}_1^0$ is the LSP. If $R$-parity is conserved, the LSP is stable and, since it is weakly interacting, the Higgs boson decay final state is invisible. Searches for such an "invisible" Higgs boson were performed by the LEP experiments, and the combination[93] yields a mass lower limit identical to that set on the SM Higgs boson if the production cross section is the SM one, as is the case for low values of $\tan\beta$.

To cope with fine-tuned choices of MSSM parameters, the LEP collaborations considered yet other possibilities, e.g., that the $h \to AA$ decay mode is kinematically allowed, or that the $h \to b\bar{b}$ decay is suppressed. For example, dedicated searches for $hA \to AAA \to b\bar{b}b\bar{b}b\bar{b}$ and for $hZ$, with $h \to q\bar{q}$ in a flavor-independent way, have been performed.[94] In the end, the sensitivity of the standard searches is only slightly reduced, except for rather extreme parameter choices leading, for instance, to $m_h \simeq 100$ GeV, while at the same time $m_A < 2m_b$. This last possibility is however less unnatural in extensions of the MSSM, such as the NMSSM where an additional Higgs singlet field is introduced.[95]

Finally, the possibility that CP is violated in the Higgs sector has also been considered. While CP is conserved at tree level, radiative corrections may introduce such a CP violation if the relative phase of $\mu$ and $A$ is not vanishing. In such a case, the three mass eigenstates all share properties of $h$, $H$ and $A$, so that the signatures of Higgs boson production are less distinct. The constraints are accordingly weaker. A dedicated "CPX" scenario[96,97] was set up to perform quantitative studies. As an example, a region around $m_h = 45$ GeV and $\tan\beta = 5$ is not excluded for $M_{SUSY} =$

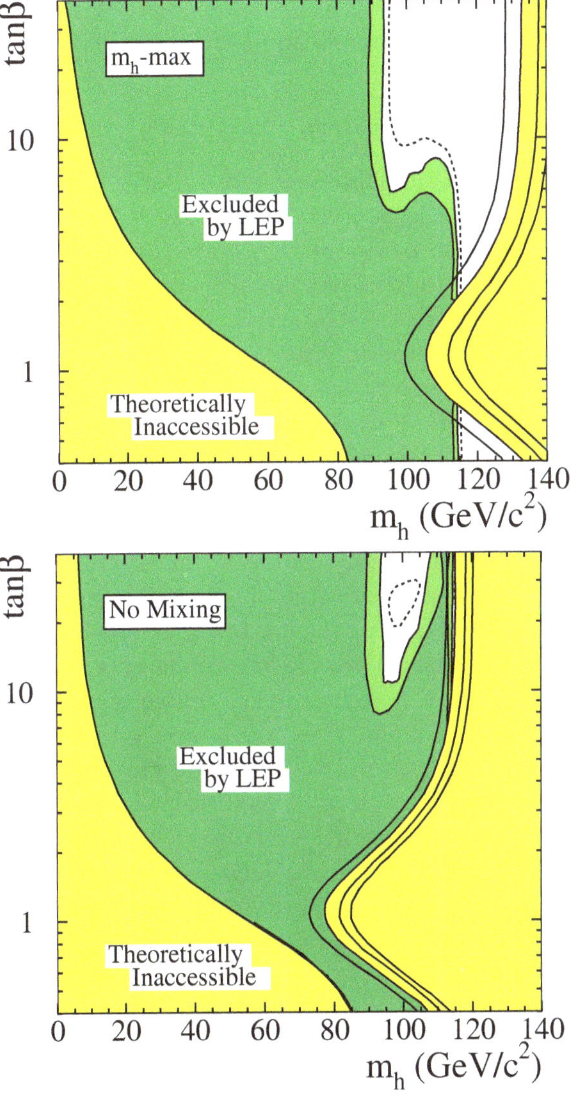


Fig. 9.4. Domains excluded at 95% CL (light green) and 99.7% CL (dark green) by the four LEP experiments[91] in the $(m_h, \tan\beta)$ plane in the $m_h$-max (top) and no-mixing (bottom) benchmark scenarios, with $m_t = 174.3$ GeV. The yellow regions are not accessible theoretically. The dashed lines represent the boundaries of the domains expected to be excluded at 95% CL in the absence of signal. The upper boundaries of the physical regions are indicated for four top quark masses: 169.3, 174.3, 179.3, and 183 GeV, from left to right.

500 GeV, $M_2 = 200$ GeV, $\mu = 2$ TeV, and $m_{\tilde{g}} = 1$ TeV, when $|A| = 1$ TeV and $\arg(A) = 90°$. Further details can be found in Ref. 91.

### 9.3.3. *Searches at the Tevatron*

At the Tevatron, i.e., in $p\bar{p}$ collisions at 1.96 TeV, the dominant production mechanism for the SM Higgs boson is via gluon fusion, $gg \to H$.[98,99] In the mass range that is of interest for a SM-like Higgs boson of the MSSM, namely $m_h < 135$ GeV, the dominant decay mode is $H \to b\bar{b}$. Such a two-jet final state is totally overwhelmed by standard jet production via the strong interaction, even after $b$-jet identification. This is why the SM Higgs boson searches at the Tevatron have been performed in the associated production processes $q\bar{q} \to (W/Z)H$,[100] which proceed via $s$-channel $W$ or $Z$ exchanges in a similar way to the Higgsstrahlung in $e^+e^-$ collisions. In spite of cross sections an order of magnitude smaller than that of gluon fusion, these processes offer better discrimination against the multijet background, by making use of the leptonic decays of the $W$ and $Z$ ($W \to \ell\nu$, $Z \to \ell^+\ell^-$ and $Z \to \nu\bar{\nu}$). These searches for the SM Higgs boson apply equally well for the $h$ boson of the MSSM in the low $\tan\beta$ regime. Their sensitivity is, however, still not sufficient to provide any significant constraint.

The situation is much more favorable for large values of $\tan\beta$. In this regime, the $A$ boson is almost mass degenerate with either the $h$ or $H$ boson, depending on whether $m_A$ is less than or greater than $m_h^{\max}$, where $m_h^{\max}$ is the maximum value that $m_h$ can take, e.g., 135 GeV in the $m_h$-max scenario. In the following, the two nearly degenerate Higgs bosons are collectively denoted $\phi$. Their couplings to $b$ quarks and $\tau$ leptons are enhanced by a factor $\tan\beta$ with respect to the SM couplings. As a result, the contribution of the $b$ quark loop to their production via gluon fusion is enhanced by a factor $2\tan^2\beta$. Although this is not sufficient to render feasible a detection in the $\phi \to b\bar{b}$ decay mode, this is not the case for the $\phi \to \tau^+\tau^-$ decay mode, which has a branching fraction of $\simeq 10\%$.

Both CDF and DØ required one of the two $\tau$ leptons to decay leptonically ($\tau \to (e/\mu)\nu\nu$) to ensure proper triggering. Three final state topologies were considered: $e\tau_{\rm had}$, $\mu\tau_{\rm had}$, and $e\mu$, all with missing transverse energy $\not\!\!E_T$ from the $\tau$ decay neutrinos. Here $\tau_{\rm had}$ denotes a $\tau$ lepton decaying into hadrons and a neutrino. The dominant, irreducible background comes from $Z$ production with $Z \to \tau^+\tau^-$, but there also remains a substantial component from $(W \to \ell\nu)$+jet, where the jet is misidentified as a $\tau$ lepton. This background was reduced, for instance, by requiring a low transverse mass of the lepton and the $\not{E}_T$. The final discriminating variable was chosen to be the visible mass $m_{\rm vis} = \sqrt{(P_{\tau_1} + P_{\tau_2} + \not{P}_T)^2}$, constructed from the $\tau$ visible products and from the $\not{E}_T$. The distribution of $m_{\rm vis}$ obtained by CDF[101] in a 1.8 fb$^{-1}$ data sample is shown in Fig. 9.5. From this distribution, as well as from a similar one in the $e\mu$ channel, a cross section upper limit on $\phi$ production was derived, which in turn was translated into exclusion domains in the $(m_A, \tan\beta)$ plane within benchmark scenarios. The result obtained in the $m_h$-max and no-mixing scenarios is shown in Fig. 9.6. Similar results have been obtained by DØ.[102] The calculations of Ref. 103 were used to derive these results as well as those reported in the rest of this section.

Because of the enhanced coupling of $\phi$ to $b$ quarks at high $\tan\beta$, the production of Higgs bosons radiated off a $b$ quark may be detectable in the $\phi \to b\bar{b}$ decay mode in spite of the large background from multijet events produced via the strong interaction ("QCD background"). This process can be described in the so-called four-flavor or five-flavor schemes, and it has been shown that the two approaches yield very similar results.[104] In the four-flavor scheme, the main contribution comes from gluon fusion, $gg \to b\bar{b}\phi$, while the main one in the five-flavor scheme comes from $gb \to b\phi$. Because one of the final state $b$ quarks (a spectator $b$ quark in the five-flavor scheme) tends to be emitted with a low transverse momentum, the searches required only three $b$ jets to be identified. The signal was searched for by inspecting the mass distribution of the two jets with highest transverse momenta in the sample of events with three $b$-tagged jets. Further discrimination against the QCD background was provided by the mass of the charged particles in the tagged jets (at CDF[105]) or by the inclusion of additional kinematic variables in a likelihood discriminant (at DØ[106]). The QCD background was modeled using a combination of information from control samples in the data, where one of the jets is not $b$-tagged, and from Monte Carlo simulations of the various processes contributing to the background ($bbb$, $bbc$, $bbq$, $ccc$, $ccq$, etc., where $q$ represents a light quark, $u$, $d$, $s$, or a gluon). The mass distribution obtained by CDF in a 1.9 fb$^{-1}$ data sample is shown in Fig. 9.7, with the individual background contributions displayed. No signal was observed, and production cross section upper limits were derived, from which exclusion domains in the $(m_A, \tan\beta)$ plane were determined in various benchmark scenarios. The DØ result obtained with 2.6 fb$^{-1}$ of data in the $m_h$-max scenario is shown in Fig.9.8. In the derivation of the cross section upper limits and exclusion domains, special attention was given to a proper handling of the Higgs boson width, which is

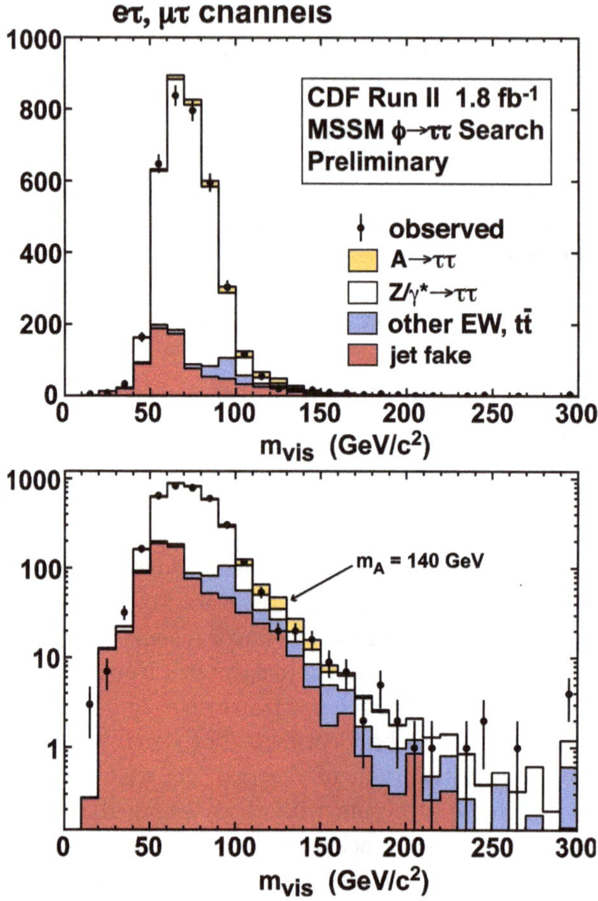


Fig. 9.5. Visible mass distribution in the $(e/\mu)\tau_{\text{had}}$ channels from the CDF search for $\phi \to \tau\tau$.[101] The signal contribution indicated corresponds to the cross section upper limit set with this data.

enhanced by a factor $\tan^2 \beta$ at tree level and therefore becomes large with respect to the mass resolution. (This is not the case for $\phi \to \tau^+\tau^-$ because of the degradation of the mass resolution due to the missing neutrinos.). It should also be noted that the exclusion domain is quite sensitive to the model parameters. It is smaller in the no-mixing scenario, and also if $\mu$ is positive. These effects due to potentially large SUSY loop corrections to the production cross sections and decay widths tend to cancel in the search for $\phi \to \tau^+\tau^-$ described above.[107,108]

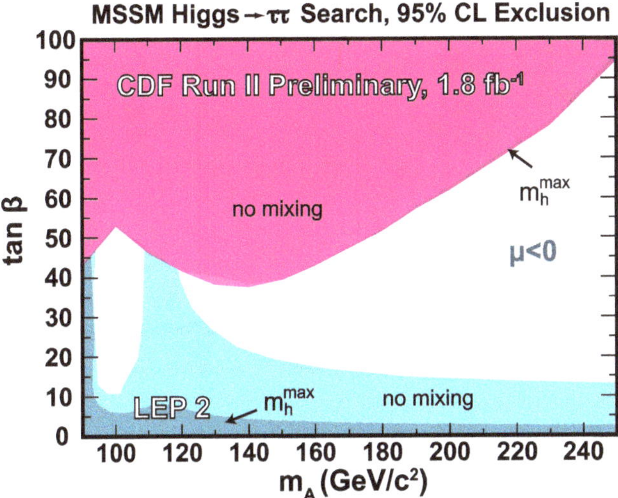


Fig. 9.6. Domains in the $(m_A, \tan\beta)$ plane excluded by the CDF search for $\phi \to \tau\tau$.[101] The domains excluded at LEP are also indicated.

Finally, Higgs bosons produced in association with $b$ quarks can also be searched for in the $\phi \to \tau^+\tau^-$ decay mode. Although the branching fraction is an order of magnitude smaller than the one of $\phi \to b\bar{b}$, the signal is much easier to disentangle from the background. A DØ analysis[109] was performed where one of the $\tau$ leptons decays into a muon and neutrinos, while the other decays into hadrons and a neutrino. Furthermore, a $b$-tagged jet was required, at which point the main background comes from top quark pair production, $t\bar{t} \to \mu\nu b\tau\nu\bar{b}$. A neural network was used to discriminate signal and $t\bar{t}$ background, taking advantage of the large differences in their kinematic properties. The result, based on 1.2 fb$^{-1}$ of data, is shown in Fig. 9.9. Given the limited amount of integrated luminosity used up to now, this channel appears to be quite promising.

## 9.4. Searches for Charged Higgs Bosons

Many extensions of the SM involve more than one complex doublet of Higgs fields. Two-Higgs doublet models (2HDMs) fall into three main categories. In Type I models, all quarks and leptons couple to the same Higgs doublet. In Type II models, down-type fermions couple to the first Higgs doublet, and up-type fermions couple to the second Higgs doublet. Flavor-changing

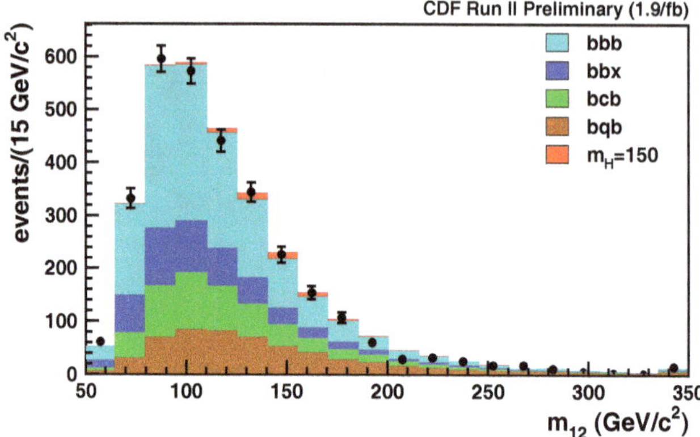


Fig. 9.7. Fit to the mass of the two jets with highest transverse momenta in the CDF sample of events with three $b$-tagged jets.[105] The contributions of the various multijet backgrounds and of a signal with a mass of 150 GeV are indicated.

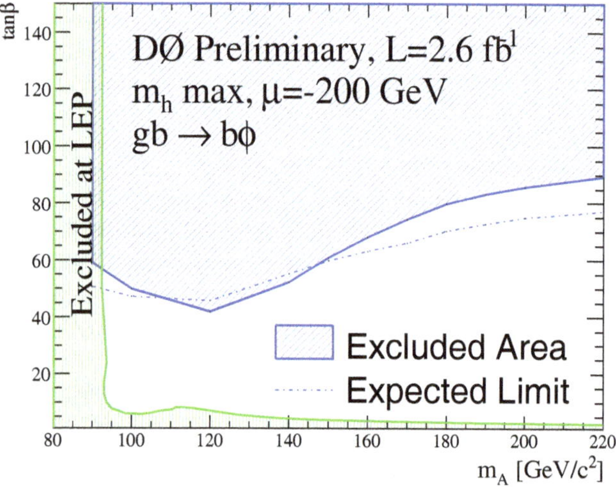


Fig. 9.8. Domain in the $(m_A, \tan\beta)$ plane excluded by the DØ search for $\phi \to b\bar{b}$ in events with three $b$-tagged jets in the $m_h$-max scenario.[106]

neutral currents are naturally avoided in Type I and Type II 2HDMs. In Type III models, fermions couple to both doublets, and flavor-changing neutral currents must be avoided using other strategies. In addition to the three neutral Higgs bosons discussed in the previous section, 2HDMs

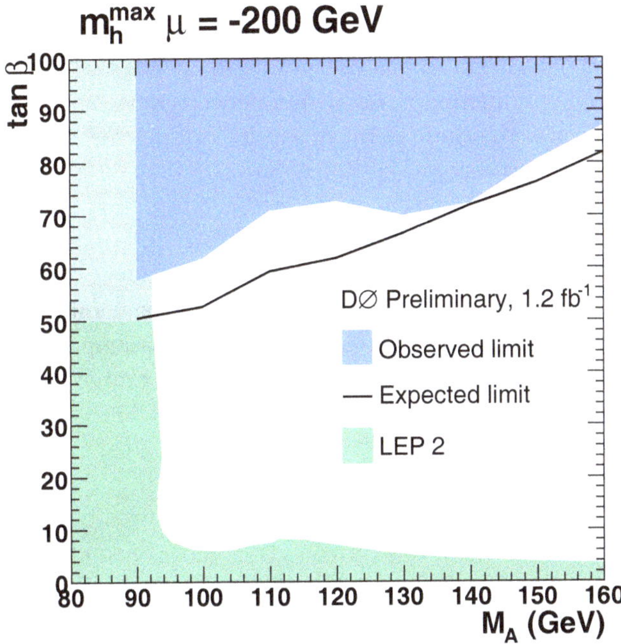


Fig. 9.9. Domain in the $(m_A, \tan\beta)$ plane excluded by the DØ search for $\phi \to \tau^+\tau^-$ produced in association with a $b$ quark in the $m_h$-max scenario.[109] The light green shaded region is the extension of the LEP exclusion region to $\tan\beta > 50$.

involve a pair of charged Higgs bosons, $H^\pm$. Most of the experimental results on charged Higgs bosons have been obtained within the context of Type II 2HDMs, of which the MSSM is a specific instance. Further details on extended Higgs boson sectors may be found in Ref. 110, 111.

In Type II 2HDMs, the charged Higgs boson decay width into a fermion pair $\bar{f}_u f_d$ is

$$\Gamma(H^- \to \bar{f}_u f_d) = \frac{N_c g^2 m_{H^\pm}}{32\pi m_W^2}\left(1 - \frac{m_{f_u}^2}{m_{H^\pm}^2}\right)^2 \times \left(m_{f_d}^2 \tan^2\beta + m_{f_u}^2 \cot^2\beta\right), \tag{9.38}$$

where $N_c$ is the number of colors, and we have approximated $m_{f_d} \ll m_{H^\pm}$ in the phase space factor. Charged Higgs bosons therefore decay into the heaviest kinematically-allowed fermions: $\tau^-\bar{\nu}_\tau$ at large $\tan\beta$ and $\bar{c}s$ at low $\tan\beta$ for charged Higgs boson masses to which current accelerators are sensitive.

### 9.4.1. *Searches at LEP*

At LEP, charged Higgs bosons are produced in pairs through $e^+e^- \to H^+H^-$.[112] The production cross section depends only on SM parameters and on the mass of the charged Higgs boson. The process $e^+e^- \to H^+W^-$ has a significantly lower cross section.

The charged Higgs boson can decay into $c\bar{s}$ or $\tau\nu_\tau$. In searches for Type I 2HDM Higgs bosons, the decay $H^\pm \to AW^{\pm *}$[113] was also considered, as in Refs. 114, 115. The interpretation of the search results generally assumed that $\text{Br}(H^\pm \to \tau\nu_\tau) + \text{Br}(H^\pm \to q\bar{q}') = 1$, where the dominant $q\bar{q}'$ flavors are $c\bar{s}$, due to the Cabbibo suppression of $c\bar{b}$. This assumption leads to the consideration of three topologies for pair-produced charged Higgs bosons: four jets from $H^+H^- \to c\bar{s}\bar{c}s$, two jets, a $\tau$ lepton and missing energy from $H^+H^- \to c\bar{s}\tau\bar{\nu}_\tau$ and two charge conjugate, acoplanar[b] $\tau$ leptons from $H^+H^- \to \tau^+\nu_\tau\tau^-\bar{\nu}_\tau$.

Direct searches for pair production of charged Higgs bosons have been published by all four LEP experiments.[114–117] Each topological analysis began with a general selection for the expected number of jets and $\tau$ leptons, followed by more sophisticated techniques. The main difficulty in these analyses was separating the signal from the nearly identical signature of $W^+W^-$ production; selection criteria usually included a mass-dependent optimization. Techniques such as linear discriminants, likelihood estimators, and jet-flavor tagging were used in these analyses. The $H^+H^- \to \tau^+\nu_\tau\tau^-\bar{\nu}_\tau$ channel had additional complexity due to the missing neutrinos, which removed the possibility of reconstructing the $H^\pm$ candidate masses and of improved discrimination from the equal-mass constraint. However, final states with $\tau$ leptons can benefit from extracting information about their polarization; the $\tau^+$ lepton from a $H^+$ boson (a scalar) is produced in a helicity state opposite to that of a $\tau^+$ lepton from $W^+$ decay.

The LEP experiments have combined the results of their searches for charged Higgs bosons into one result[118] based on common assumptions. The total dataset has an integrated luminosity of 2.5 fb$^{-1}$, collected at center-of-mass energies between 189 and 209 GeV. The possible decays were restricted to $H^+ \to c\bar{s}$ and $\tau^+\nu_\tau$ in a general 2HDM framework. The combined mass limit is shown in Fig. 9.10 as a function of $\text{Br}(H^+ \to \tau^+\nu_\tau)$. A lower bound of 78.6 GeV holds for any value of the branching ratio.

---

[b]The acoplanarity angle is the angle between the projections of the $\tau$ momenta on a plane transverse to the beam axis. If this angle is less than 180°, the $\tau$ leptons are said to be acoplanar.

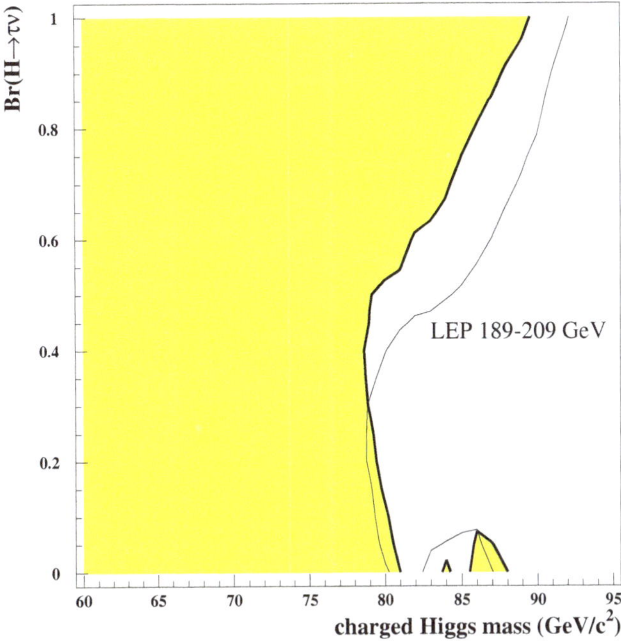


Fig. 9.10. Limit on the charged Higgs boson mass as a function of $\text{Br}(H^+ \to \tau^+\nu_\tau)$, from the combined data of the four LEP experiments at center-of-mass energies from 189 to 209 GeV. The expected exclusion limit is shown as a thin solid line and the observed limit as a thick solid line; the shaded region is excluded.[118]

### 9.4.2. *Searches at the Tevatron*

At the Tevatron, pair production of charged Higgs bosons is expected to occur at a very low rate. However, in contrast to searches at LEP, advantage can be taken of the large mass of the top quark, which opens new ways to search for evidence of charged Higgs bosons. Two approaches have been considered, depending on whether the charged Higgs boson is lighter or heavier than the top quark. In the first case, the top quark can decay into a $H^+$ boson and a $b$ quark.[112] For heavier charged Higgs bosons, resonant production of a single $H^+$ boson followed by the decay $H^+ \to t\bar{b}$ is the most promising process.[119]

In the SM, the top quark decays almost exclusively into a $W$ boson and a $b$ quark, and the possible signatures of $t\bar{t}$ pair production are associated with the various combinations of $W$-boson decay channels. If the charged Higgs boson is lighter than the top quark, the decay $t \to H^+b$ will compete

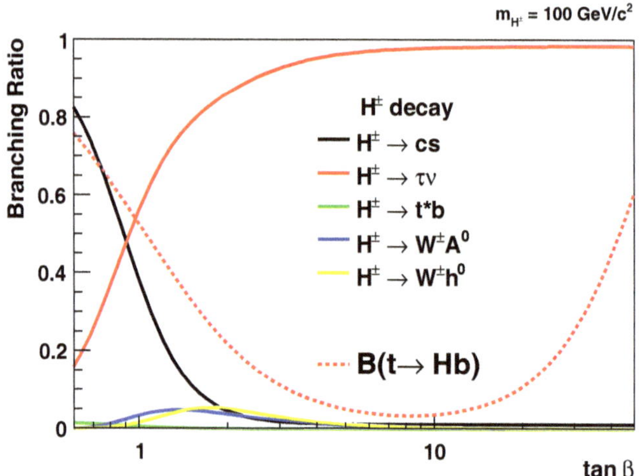


Fig. 9.11. For a charged Higgs boson mass of 100 GeV, the branching ratios for the top quark decay into $H^+b$, under the assumption that $\mathrm{Br}(t \to W^+b)+\mathrm{Br}(t \to H^+b)=1$, and for the various $H^+$ decay channels, as a function of $\tan\beta$. From Ref. 120.

with the standard $t \to W^+b$ mode. The decay of the charged Higgs boson, with branching ratios different from those of the $W$ boson, will modify the fractions of events observed in the various topologies, compared to the SM expectations.[c] The qualitative aspects and magnitude of these modifications depend on the model parameters. The dependence on $\tan\beta$ of the top quark decay ratio to $H^+b$ and of the various charged-Higgs boson decay channels is shown in Fig. 9.11 for $m_{H^+} = 100$ GeV and a typical set of MSSM parameters, with QCD, SUSY-QCD and electroweak radiative corrections to the top and bottom quark Yukawa couplings calculated with the CPsuperH code.[121] The dominant $H^+$ boson decay channels are $c\bar{s}$ at low values of $\tan\beta$ and $\tau^+\nu_\tau$ at high values; with this set of parameters, $H^+$ boson decays to $W^+A/h$ are also allowed, although always at a small rate. The $H^+ \to t^*\bar{b} \to W^+b\bar{b}$ decay mode becomes relevant for charged Higgs boson masses closer to the top quark mass. It can be seen that charged Higgs bosons will be most prominent at high and low values of $\tan\beta$. Two simplified models address each of these regions: the tauonic model, with

[c]The branching ratios for topologies arising from SM $t\bar{t}$ pair production are roughly 50% in six jets; 14% in each of $e$, $\mu$, and $\tau + 4$ jets $+\not{E}_T$; 1% in each of $ee$, $\mu\mu$, and $\tau\tau + 2$ jets $+\not{E}_T$; and 2% in each of $e\mu$, $e\tau$, and $\mu\tau + 2$ jets $+\not{E}_T$. In each of these channels, there are two $b$ jets.

$\mathrm{Br}(H^+ \to \tau^+\nu_\tau)$=1, and the leptophobic model, with $\mathrm{Br}(H^+ \to c\bar{s})$=1. The tauonic model is a very good approximation to the MSSM with $\tan\beta \gtrsim 15$, while purely leptophobic charged Higgs bosons can be found in some multi-Higgs-doublet models.[122]

Analyses based on measurements of $t\bar{t}$ final states include an earlier CDF search in 200 pb$^{-1}$ of data[123] and a recent DØ analysis of 1 fb$^{-1}$ of data.[120] The yields observed in the various topologies were compared to what would be expected in models with charged Higgs bosons, taking into account the $t \to H^+b$ and $H^\pm$ decay branching ratios predicted as a function of the Higgs boson mass and $\tan\beta$. In particular, no excess of final states involving $\tau$ leptons was observed, nor was any disappearance of final states with one or two leptons, jets and $\not\!\!E_T$, as would be expected at large and small $\tan\beta$, respectively. Figure 9.12 displays the exclusion domain in the plane of the charged Higgs boson mass and $\tan\beta$ from the DØ analysis,[120] for leptophobic and tauonic models. The CDF analysis excludes $\mathrm{Br}(t \to H^+b) > 0.4$ for a tauonic $H^\pm$ boson.[123]

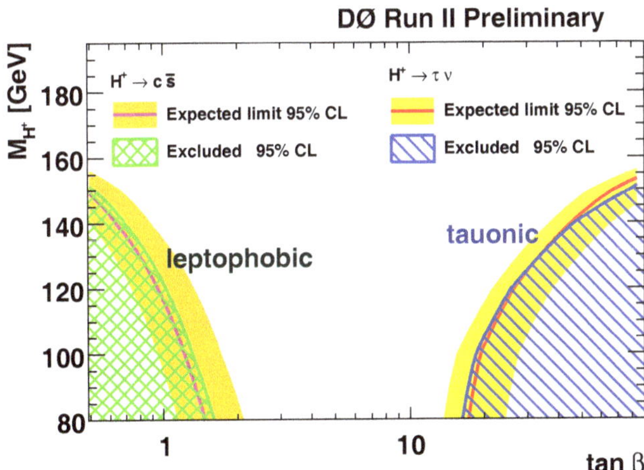


Fig. 9.12. Limit on the mass of the charged Higgs boson as a function of $\tan\beta$ from the DØ search in top quark decays.[120]

In a recent analysis based on a data sample of 2.2 fb$^{-1}$, the CDF collaboration used a different approach to search for a leptophobic charged Higgs boson in top quark decays.[124] The search was performed in the lepton + jets + $\not\!\!E_T$ final states with two $b$-tagged jets, where the lepton (electron or muon), the neutrino (responsible for the missing $E_T$), and a $b$ jet were the signature of a $t \to Wb \to \ell\nu b$ decay, while the other top quark of the $t\bar{t}$ pair was assumed to decay to either $Wb \to q\bar{q}'b$ or $Hb \to c\bar{s}b$. The $t\bar{t}$ events were fully reconstructed, taking the masses of the $W$ boson and of the top quark into account as constraints to assign correctly each of the $b$ jets to its parent $t$ or $\bar{t}$. Templates of the mass of the dijet system reconstructed from the non-$b$ jets were used to extract limits on the branching ratio of $t \to H^+b$, as shown in Fig. 9.13.

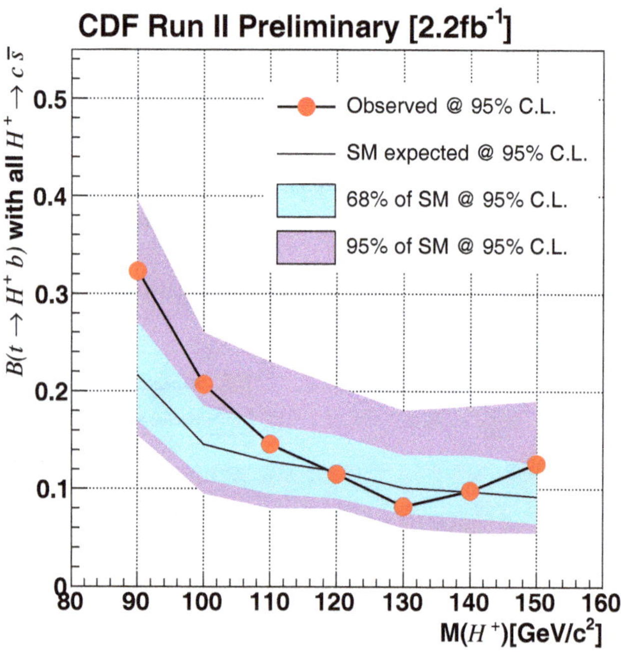


Fig. 9.13. For a leptophobic charged Higgs boson, upper limit on the branching ratio $\text{Br}(t \to H^+b)$ as a function of the Higgs boson mass from a CDF search in top quark decays.[124]

If the charged Higgs boson is heavier than the top quark, it will decay dominantly into $t\bar{b}$. The resonant production of such a charged Higgs boson leads to a final state similar to the one resulting from single top $s$-channel production, $q\bar{q} \to W^* \to t\bar{b}$. Therefore the analyses developed for the search for single top production can be applied to the search for a charged Higgs boson. Such an analysis was performed by the DØ collaboration,[125] in the topology arising from a subsequent $t \to Wb \to \ell\nu b$ decay. The large $H^\pm$ mass, reconstructed from the decay products imposing the $W$ boson and top quark mass constraints, was used as discriminating variable. No excess was observed over SM background predictions, and upper limits were set on the production of a charged Higgs boson. The results are, however, not sensitive to Type II 2HDMs, but provide some exclusion in Type I 2HDMs.

## 9.5. Searches for Supersymmetric Particles

### 9.5.1. *General features of SUSY models*

As explained in Sec. 9.2, the main features of SUSY models for phenomenology are related to the type of mediation mechanism for SUSY breaking, to the choice of soft breaking terms, and to whether or not $R$-parity is assumed to be conserved.

The most widely studied models involve gravity-mediation of SUSY breaking. In the minimal form of such models, mSUGRA, $R$-parity is conserved, and only five parameters are needed beyond those already present in the standard model: a universal gaugino mass $m_{1/2}$, a universal scalar mass $m_0$, and a universal trilinear coupling $A_0$, all defined at the scale of grand unification, and $\tan\beta$ and the sign of $\mu$. The low energy parameters, including $|\mu|$, are determined by the renormalization group equations and by the condition of electroweak symmetry breaking. In addition, it is commonly assumed that the LSP is the lightest neutralino $\tilde{\chi}_1^0$. A somewhat less constrained model keeps $\mu$ and $m_A$ as independent low energy parameters, which is in effect equivalent to decoupling the Higgs scalar masses from the masses of the other scalars. Such a model was largely used at LEP.

Many studies have been performed where the assumption of $R$-parity conservation is dropped, while keeping unchanged the other features of those mSUGRA inspired models. If $R$-parity is violated, the superpotential is allowed to contain lepton or baryon number-violating terms[126]

$$W_{R_p} = \lambda_{ijk} L_i L_j \bar{E}_k + \lambda'_{ijk} L_i Q_j \bar{D}_k + \lambda''_{ijk} \bar{U}_i \bar{D}_j \bar{D}_k \,, \tag{9.39}$$

where $L$ and $Q$ are lepton and quark doublet superfields, $E$ and $D$ are lepton and down-type quark singlet superfields, and $i$, $j$ and $k$ are generation indices. These terms are responsible for new couplings through which the LSP decays to SM particles. The simultaneous occurrence of different coupling types is however strongly constrained, e.g., by the bounds on the proton lifetime, which is why it is commonly assumed that only one of the $R$-parity violating terms is present in the superpotential.

In models with gauge-mediated SUSY breaking (GMSB), the LSP is a very light gravitino $\tilde{G}$, and the phenomenology is governed by the nature of the NLSP. In the minimal such model, mGMSB, all SUSY particle masses derive from a universal scale $\Lambda$, and in most of the parameter space the NLSP is either the lightest neutralino $\tilde{\chi}_1^0$ or the lighter stau $\tilde{\tau}_R$, the latter occurring preferentially at large $\tan\beta$. The couplings of the gravitino depend on yet another parameter, the SUSY-breaking scale $\sqrt{F}$, which can be traded for the lifetime of the NLSP.

Anomaly-mediation of SUSY breaking (AMSB) generically leads to a neutralino LSP which is almost a pure wino $\tilde{W}^0$, and has a small mass splitting with the lighter chargino. As a consequence, this chargino may acquire a phenomenologically relevant lifetime, possibly such that it behaves like a stable particle.

### 9.5.2. *Signatures and strategies*

Most of the searches for SUSY particles were performed within a "canonical scenario," the main features of which are borrowed from mSUGRA: $R$-parity conservation, universal gaugino mass terms, a universal sfermion mass term, and a neutralino LSP. Because of $R$-parity conservation, SUSY particles are produced in pairs, and each of the produced SUSY particles decays into SM particles accompanied by an LSP. Since the LSP is neutral and weakly interacting, it appears as missing energy, which is the celebrated signature of SUSY particle production.

Alternatively, if $R$-parity is not conserved, the LSP decays to SM particles, so that no missing energy is expected beyond that possibly arising from neutrinos. The signature of SUSY particle production is therefore to be sought in an anomalously large multiplicity of jets or leptons. The $R$-parity violating couplings can also make it possible that SUSY particles are produced singly, rather than in pairs.

In $R$-parity conserving scenarios other than the canonical one, additional or different features are expected. In GMSB, each of the pair-produced SUSY particles decays into SM particles and an NLSP. The NLSP further decays into its SM partner and a gravitino. With a neutralino NLSP in the mass range explored up to now, the dominant decay is $\tilde{\chi}_1^0 \to \gamma\tilde{G}$, so that the final state contains photons, with missing energy due to the escaping gravitinos. With a stau NLSP, the decay is $\tilde{\tau}_R \to \tau\tilde{G}$. If the stau lifetime is so long that it escapes the detector before decaying, the final state from stau pair production does not exhibit any missing energy, but rather appears as a pair of massive stable particles. A similar final state may also arise from chargino pair production in AMSB. Long-lived gluinos can lead to spectacular signatures if they are brought to rest by energy loss in the detector material.

Except for the gluino, all SUSY particles are produced in a democratic way in $e^+e^-$ collisions via electroweak interactions. It is therefore natural that the searches at LEP were targeted toward the lightest ones. The results of these searches could further be combined within a given model, thus providing constraints on the model parameters. In contrast, it is expected that the most copiously produced SUSY particles in hadron collisions, such as $p\bar{p}$ at the Tevatron, will be colored particles, namely squarks and gluinos. Their detailed signature however depends on the mass pattern of the other SUSY particles, which may be present in their decay chains. This is why a specific model, usually mSUGRA, is needed to express the search results in terms of mass constraints. Thanks to lower masses and more manageable backgrounds, the search for gauginos produced via electroweak interactions can be competitive at hadron colliders for model parameter configurations where their leptonic decays are enhanced.

In $e^+e^-$ collisions, the production cross sections of SUSY particles are similar to those of their SM partners, except for the phase space reduction due to their larger masses. The data collected at the highest LEP energies, up to 209 GeV, are therefore the most relevant for SUSY particle searches. Mixing effects may however reduce these cross sections, as is the case for instance for neutralinos with a small Higgsino component, in which case the large integrated luminosity accumulated by the LEP experiments at lower energies also contributes to the search sensitivity.

Although the center-of-mass energy of 1.96 TeV in $p\bar{p}$ collisions at the Tevatron allows higher new particle masses to be probed, large integrated luminosities are needed because of the rapid PDF fall off at high $x$, as explained in Sec. 9.1.2. The search for SUSY particles at the Tevatron is also rendered more challenging than at LEP because of the large cross sections of the background processes. In the searches for squarks and gluinos, signal production cross sections of the order of 0.1 pb at the edge of the sensitivity domain are to be compared to the total inelastic cross section of 80 mb. In the searches for gauginos, with similar signal production cross sections in the mass range probed, the main backgrounds are $W \to \ell\nu$ and $Z \to \ell\ell$, with cross sections at the 2.7 nb and 250 pb level per lepton flavor.

In $ep$ collisions at HERA, the most promising SUSY particle production process is single squark resonant production via an $R$-parity violating $\lambda'_{1j1}$

or $\lambda'_{11k}$ coupling, with a cross section depending not only on the squark mass, but also on the value of the coupling involved. The decay of the squark produced could be either direct, via the same $\lambda'$ coupling as for its production, or indirect through a cascade leading to the LSP, which in turn decays to two quarks and a neutrino or an electron. The mass reach at HERA is the full center-of-mass energy of 320 GeV, but the production of squarks with masses close to this bound involves quarks at large $x$ values, so that the effective reach is substantially smaller, even for large values of the $\lambda'$ coupling.

### 9.5.3. *Searches in the canonical scenario*

As mentioned above, the characteristic signature of SUSY particle production in the canonical scenario is missing energy carried away from the detector by the LSPs at the end of the decay chains.

#### 9.5.3.1. *Searches at LEP*

The main channels for SUSY particle searches in $e^+e^-$ collisions are slepton,[127,128] chargino[129,130] and neutralino[131–133] production. Squark pair production[134] can also be relevant in some specific cases.[135]

**Sleptons:** In $e^+e^-$ annihilation, the search for SUSY particles that involves the least set of hypotheses for its interpretation is the search for smuons. Pair production proceeds via $Z/\gamma^*$ exchange in the $s$-channel. Because of the small mass of the muon, the smuon mass eigenstates can be identified with the interaction eigenstates, of which $\tilde{\mu}_R$ is the lighter one in models with slepton and gaugino mass unification. The search results were interpreted under this assumption, which is furthermore conservative, as the coupling of the $\tilde{\mu}_R$ to the $Z$ boson is smaller than that of the $\tilde{\mu}_L$. Only one parameter is needed to calculate the smuon pair production cross section, the smuon mass $m_{\tilde{\mu}_R}$. The sole decay mode of a $\tilde{\mu}_R$ NLSP is $\tilde{\mu}_R \to \mu\tilde{\chi}_1^0$, so that smuon pair production leads to a final state consisting of two acoplanar muons with missing energy and momentum. The topology of this final state also depends on the mass of the LSP. If $m_{\tilde{\chi}_1^0}$ is small, the final state is very similar to that arising from $W$ pair production, with both $W$ bosons decaying to a muon and a neutrino. If the $\tilde{\mu}_R - \tilde{\chi}_1^0$ mass difference is small, the final state muons carry little momentum, so that the selection efficiency is reduced. In that configuration, the main background comes from "$\gamma\gamma$ interactions," $e^+e^- \to (e^+)\gamma^*\gamma^*(e^-) \to (e^+)\mu^+\mu^-(e^-)$, where the spectator electrons ($e^\pm$) escape undetected in the beam pipe. The LSP mass $m_{\tilde{\chi}_1^0}$ is therefore needed, in addition to the smuon mass, to interpret the search results. The constraints obtained in the $(m_{\tilde{\mu}_R}, m_{\tilde{\chi}_1^0})$ plane by the four LEP experiments[136] are shown in Fig. 9.14. If the assumption that the smuon is the NLSP is dropped, further specification of the model is needed to turn the search results into mass constraints. An example is shown in Fig. 9.14 in the case of gaugino mass unification, for the specified values of $\mu$ and $\tan\beta$. A slight reduction of the excluded domain is observed for low values of $m_{\tilde{\chi}_1^0}$, due to the competition of the $\tilde{\mu}_R \to \mu\tilde{\chi}_2^0$ decay mode, with $\tilde{\chi}_2^0 \to \gamma\tilde{\chi}_1^0$. Depending on $m_{\tilde{\chi}_1^0}$, smuon masses smaller than 95 to 99 GeV are excluded, except for $\tilde{\mu}_R - \tilde{\chi}_1^0$ mass differences below 5 GeV.

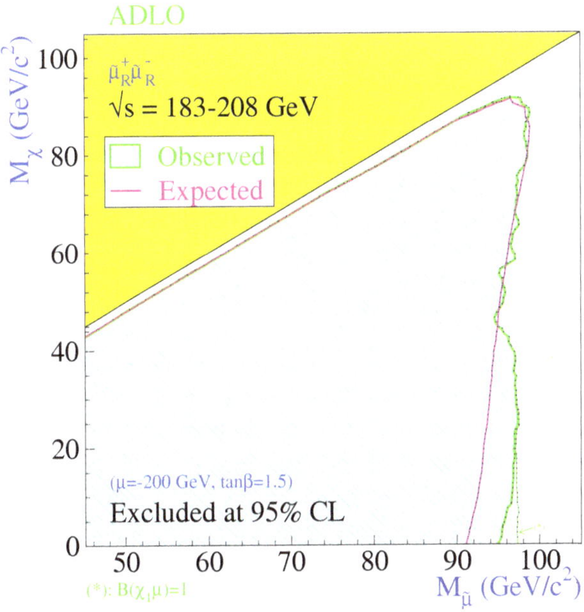


Fig. 9.14. Region in the $(m_{\tilde{\mu}_R}, m_{\tilde{\chi}_1^0})$ plane excluded by the searches for smuons at LEP.[136] The dotted contour is drawn under the assumption that the smuon decay branching ratio into $\mu\tilde{\chi}_1^0$ is 100%.

Because of the larger $\tau$ mass, compared to the muon mass, the hypothesis that the stau mass eigenstates can be identified with the interaction eigenstates may not hold, especially for large values of $\tan\beta$ that enhance the off-diagonal elements of the mass matrix in (9.29). The coupling to the $Z$ boson of the lighter stau mass eigenstate $\tilde{\tau}_1$ may therefore be reduced with respect to the smuon coupling, and even vanish. Moreover, because there is at least one neutrino in each $\tau$ decay, the visible energy of the final state arising from stau pair production is smaller than in the case of smuons, so that the selection efficiency is reduced. The mass lower limits obtained at LEP are therefore lower for staus than for smuons, from 86 to 95 GeV, depending on $m_{\tilde{\chi}_1^0}$, provided the $\tilde{\tau}_1 - \tilde{\chi}_1^0$ mass difference is larger than 7 GeV.[136]

As for smuons, the selectron mass eigenstates can be identified with the interaction eigenstates. But because of the contribution of $t$-channel neutralino exchange to selectron pair production, the gaugino sector of the model, mass spectrum and field contents, has to be specified to interpret the results of the searches for acoplanar electrons. With gaugino mass unification and for $\tan\beta = 1.5$ and $\mu = -200$ GeV, a selectron mass lower limit of 100 GeV was obtained for $m_{\tilde{\chi}_1^0} < 85$ GeV.[136] Neutralino $t$-channel exchange can furthermore mediate associated $\tilde{e}_L\tilde{e}_R$ production. This process is useful if the $\tilde{e}_R - \tilde{\chi}_1^0$ mass difference is small, because the electron from the $\tilde{e}_L \to e\tilde{\chi}_1^0$ decay can be energetic enough to lead to an apparent single electron final state. Both gaugino and slepton mass unifications have to be assumed for the masses of the two selectron species to be related. Under these assumptions, a lower limit of 73 GeV was set on $m_{\tilde{e}_R}$, independent of the $\tilde{e}_R - \tilde{\chi}_1^0$ mass difference.[137,138]

From the measurement of the invisible width of the $Z$ boson,[19] a general mass lower limit of 45 GeV can be deduced for a sneutrino LSP or NLSP.

**Charginos and neutralinos:** As evident from (9.28), three parameters are sufficient to fully specify the masses and field contents in the chargino sector. These may be taken to be $M_2$, $\mu$, and $\tan\beta$. The lighter of the two charginos will simply be denoted "chargino" in the following. To specify the neutralino mass matrix of (9.26), one more parameter, $M_1$, is needed. If gaugino mass unification is assumed, the two gaugino masses are related by $M_1 = (5/3)\tan^2\theta_W M_2 \simeq 0.5 M_2$. Unless otherwise specified, this relation is assumed to hold in the following. Charginos are pair produced via $s$-channel $Z/\gamma^*$ and $t$-channel $\tilde{\nu}_e$ exchanges, the two processes interfering destructively. The three-body final states $f\bar{f}'\tilde{\chi}_1^0$ are reached in chargino decays via virtual $W$ or sfermion exchange. If kinematically allowed, two-body decays such as $\tilde{\chi}^\pm \to \ell^\pm\tilde{\nu}$ are dominant. Similarly, neutralino pair or associated production proceed via $s$-channel $Z$ and $t$-channel selectron exchanges, and $\tilde{\chi}_2^0$ three-body decays to $f\bar{f}\tilde{\chi}_1^0$ via virtual $Z$ or sfermion exchange; whenever kinematically allowed, two-body decays such as $\tilde{\chi}_2^0 \to \nu\tilde{\nu}$ are dominant.

If sfermions are heavy, chargino decays are mediated by virtual $W$ exchange, so that the final states arising from chargino pair production are the same as for $W$ pairs, with additional missing energy from the two neutralino LSPs: all hadronic ($q\bar{q}'\tilde{\chi}_1^0 q\bar{q}'\tilde{\chi}_1^0$), mixed ($q\bar{q}'\tilde{\chi}_1^0 \ell\nu\tilde{\chi}_1^0$), and fully leptonic ($\ell\nu\tilde{\chi}_1^0\ell\nu\tilde{\chi}_1^0$). Selections were designed for these three topologies and for various $m_{\tilde{\chi}^\pm} - m_{\tilde{\chi}_1^0}$ regimes, with no excess observed over SM backgrounds. From a scan over $M_2$, $\mu$, and $\tan\beta$, a chargino mass lower limit of 103 GeV was derived for $m_{\tilde{\nu}} > 200$ GeV.[139] For smaller sneutrino masses, the limit is reduced by the destructive interference in the production. This limit holds for $M_2 <\simeq 1$ TeV. For larger $M_2$ values, the selection efficiency decreases rapidly as the $\tilde{\chi}^\pm - \tilde{\chi}_1^0$ mass difference becomes smaller. If this mass difference becomes so small that even the $\tilde{\chi}^\pm \to \pi^\pm\tilde{\chi}_1^0$ decay mode is closed, the chargino becomes long lived. Searches for charged massive stable particles, in which advantage is taken of their larger ionization power, were designed to cope with this configuration. For slightly larger mass differences, the visible final state is so soft that even triggering becomes problematic. Chargino pair production can however be "tagged" by an energetic photon from initial state radiation, $e^+e^- \to \gamma\tilde{\chi}^+\tilde{\chi}^-$, providing access to those almost invisible charginos, although at a reduced effective center of mass energy. The combination of these analysis techniques allowed chargino masses smaller than 92 GeV to be excluded, irrespective of the $\tilde{\chi}^\pm - \tilde{\chi}_1^0$ mass difference.[140]

For lower sfermion masses, the sensitivity of the former analyses is reduced first because of the destructive interference between the $s$-channel $Z/\gamma^*$ and $t$-channel sneutrino exchanges, and second because of the opening of two-body decays. The latter effect is specifically detrimental in the "corridor" of small $\tilde{\chi}^\pm - \tilde{\nu}$ mass differences, where the final state from the $\tilde{\chi}^\pm \to \ell\tilde{\nu}$ decays becomes invisible in practice. Gaugino mass unification allows indirect limits on charginos to be obtained, based on constraints on the parameter space resulting from searches for pair or associated neutralino production, e.g., $e^+e^- \to \tilde{\chi}_2^0\tilde{\chi}_2^0$ or $\tilde{\chi}_1^0\tilde{\chi}_2^0$. In order to relate all production cross sections and decay branching fractions, it is however necessary to fully specify the sfermion spectrum, which is done with the assumption of sfermion mass unification. The results of the chargino and neutralino searches are then expressed as exclusion domains in the $(\mu, M_2)$ plane for selected values of $\tan\beta$ and $m_0$. The invisible two-body decay $\tilde{\chi}_2^0 \to \nu\tilde{\nu}$ can however cause a large sensitivity reduction. Since this configuration occurs for low $m_0$ values, constraints arising from the slepton searches can be used to mitigate this effect. With gaugino and sfermion mass unification, the slepton masses are related to the model parameters by $m_{\tilde{\ell}_R}^2 \simeq m_0^2 + 0.22 M_2^2 - \sin^2\theta_W m_Z^2 \cos 2\beta$, so that a limit on $m_{\tilde{\ell}_R}$ can be turned into a limit on $M_2$ for given values of $\tan\beta$ and $m_0$. After a proper combination of the searches for charginos, neutralinos and sleptons, an example of which is shown in Fig. 9.15, it turns out that the chargino mass limit obtained in the case of heavy sfermions is only moderately degraded.

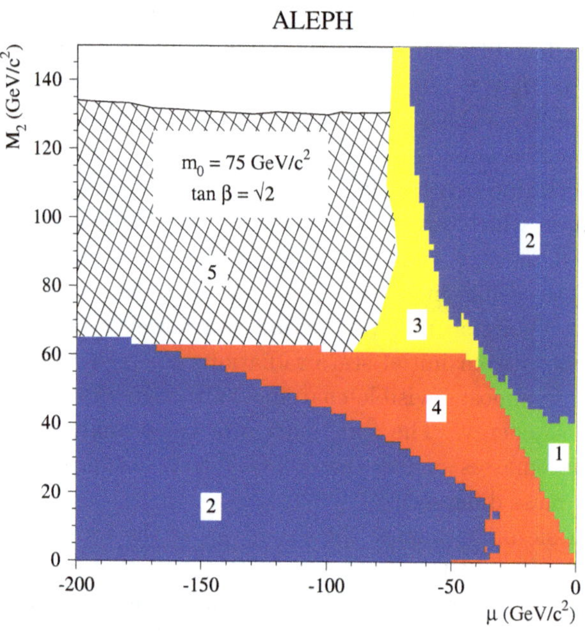


Fig. 9.15. Regions in the $(\mu, M_2)$ plane excluded by the LEP I constraints (1), and by the searches for charginos (2), neutralinos (3) and sleptons (4) at LEP II, for $\tan\beta = \sqrt{2}$ and $m_0 = 75$ GeV. The region (5) is excluded by the Higgs boson searches at LEP II. This figure is from Ref. 141.

Direct searches for the lightest neutralino had been performed at lower energy $e^+e^-$ colliders, PEP and PETRA, in the reaction $e^+e^- \to \gamma\tilde{\chi}_1^0\tilde{\chi}_1^0$, where the photon from initial state radiation is used to tag the production of an invisible final state. At LEP, at or above the $Z$ resonance, the irreducible background from $e^+e^- \to \gamma\nu\bar{\nu}$ is too large to obtain competitive results. Furthermore, production via $s$-channel $Z$ exchange may simply vanish, e.g., if the LSP is a pure photino, while production via $t$-channel selectron exchange can be made negligible if selectrons are sufficiently heavy.

Indirect limits on the mass of the LSP can however be obtained within constrained models. With gaugino mass unification, $m_{\tilde{\chi}_1^0}$ is typically half the chargino mass. As a result, the chargino mass limit translates into a $\tilde{\chi}_1^0$ mass lower limit of 52 GeV for heavy sfermions and large $\tan\beta$. If sfermion mass unification is used in addition, a limit of 47 GeV is obtained at large $\tan\beta$, independent of $m_0$. This limit is set by searches for sleptons in the corridor. For low values of $\tan\beta$, constraints from the Higgs boson searches can be used, as was shown in Sec. 9.3.2 for benchmark scenarios. A complete scan over $m_0$, $m_{1/2}$, $\mu$ and $\tan\beta$ was performed and, for each parameter set, the maximum $h$ mass predicted was compared to the experimental limit, and the constraints from chargino and slepton searches were included. The translation of the scan result in terms of excluded domain in the $(\tan\beta, m_{\tilde{\chi}_1^0})$ plane is shown in Fig. 9.16, from which a $\tilde{\chi}_1^0$ mass lower limit of 47 GeV is derived.[142] Within the more constrained mSUGRA scenario, wherein $\mu$ is calculated from the other parameters, this limit becomes 50 GeV.[143]

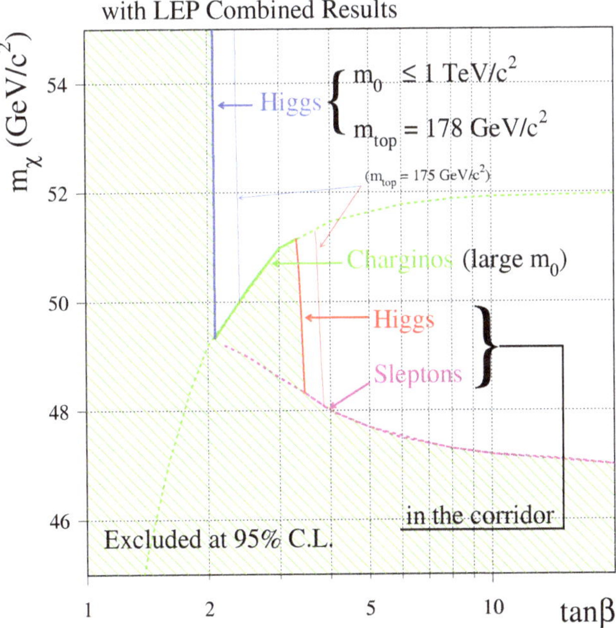


Fig. 9.16. Lower mass limit for the lightest neutralino as a function of $\tan\beta$, inferred in the conventional scenario from searches at LEP for charginos, sleptons, and neutral Higgs bosons.[142] The dashed contour is the limit obtained for large $m_0$.

**Squarks:** On general grounds, the mass reach for strongly interacting particles is expected to be substantially higher at the Tevatron than at LEP. For some specific configurations, however, the searches at the Tevatron become inefficient, in which cases the results obtained at LEP remain of interest. This is particularly relevant for third generation squarks which may be substantially lighter than the other squarks, as motivated in Sec. 9.2. The lighter third generation mass eigenstates are simply denoted stop and sbottom, $\tilde{t}$ and $\tilde{b}$, in the following.

In the mass range accessible at LEP, and given the chargino mass limit which effectively forbids $\tilde{t} \to b\tilde{\chi}^+$, the stop is expected to decay into a charm quark and a neutralino, $\tilde{t} \to c\tilde{\chi}_1^0$, as long as $m_{\tilde{t}} < m_W + m_b + m_{\tilde{\chi}_1^0}$.[144] Because this decay is a flavor-changing loop process, the stop lifetime can be large enough to compete with the hadronization time, and the simulation programs were adjusted to take this feature into account. The final state from stop pair production exhibits an acoplanar jet topology, for which no signal was observed above standard model backgrounds. As already explained for staus, the amount of mixing between the weak eigenstates can be such that the stop does not couple to the $Z$ boson. In this worst case scenario, stop mass lower limits ranging from 96 to 99 GeV were obtained, depending on the $\tilde{\chi}_1^0$ mass, as long as $m_{\tilde{t}} - m_{\tilde{\chi}_1^0} - m_c > 5$ GeV.[145] For smaller $\tilde{t} - \tilde{\chi}_1^0$ mass differences, long-lived $R$-hadrons may be produced in the stop hadronization process. The production of such $R$-hadrons and their interaction in the detector material were taken into account in a dedicated search, from which a stop mass lower limit of 63 GeV was derived, valid for any $m_{\tilde{t}} - m_{\tilde{\chi}_1^0}$.[146] For specific parameter choices, and in spite of the slepton mass limits, it can be that the $\tilde{t} \to b\ell\tilde{\nu}$ decay is kinematically allowed, in which case it is dominant. From a search for events exhibiting jets, leptons and missing energy, a stop mass lower limit of 96 GeV was obtained, valid for sneutrino masses smaller than 86 GeV.[145]

The case of a light sbottom is much simpler, as the tree-level $\tilde{b} \to b\tilde{\chi}_1^0$ decay mode is dominant. From searches for acoplanar $b$-flavored jets, a mass lower limit of about 95 GeV was obtained in the worst case scenario where the sbottom does not couple to the $Z$.[145]

### 9.5.3.2. *Searches at the Tevatron*

The program most widely used for the calculation of SUSY particle production cross sections at the Tevatron is PROSPINO,[147] which provides next-to-leading order accuracy. The results reported below were generally obtained with the CTEQ6.1M PDF set.[148,149] Various codes were used to calculate the low energy SUSY spectrum from initial parameters at the grand unification scale: SUSPECT,[54] SOFTSUSY,[52] and ISAJET.[51] This may introduce slight inconsistencies when comparing results in different channels or from different experiments in terms of parameters at the high scale. The production of SUSY particles was in general simulated with PYTHIA,[150] with decays modeled with SDECAY[151] or with ISASUGRA as implemented in PYTHIA. Typically, SM backgrounds were simulated with ALPGEN[152] for the production of $W$ and $Z$ bosons in association with jets, or with PYTHIA otherwise.

As already mentioned in Sec. 9.5.2, the channels most relevant for SUSY particle searches at hadron colliders are the production of squarks and gluinos on the one hand, of electroweak gauginos on the other. For squarks and gluino, the search is conducted in events exhibiting a jets+$\not\!\!E_T$ topology,[153–155] while for electroweak gauginos, it is conducted in the trilepton final state.[156–158]

**Generic squarks and gluinos:** Depending on the squark and gluino mass hierarchy, different pair production processes via the strong interaction are expected to contribute in $p\bar{p}$ collisions at the Tevatron: $\tilde{q}\bar{\tilde{q}}$ and, to a lesser extent, $\tilde{q}\tilde{q}$, if $m_{\tilde{q}} \ll m_{\tilde{g}}$; $\tilde{g}\tilde{g}$ if $m_{\tilde{g}} \ll m_{\tilde{q}}$; and all of these processes, as well as $\tilde{q}\tilde{g}$, if the squark and gluino masses are similar. If $m_{\tilde{q}} < m_{\tilde{g}}$, squarks are expected to decay directly into a quark and a gaugino, $\tilde{q} \to q\tilde{\chi}$, where $\tilde{\chi}$ is dominantly $\tilde{\chi}_1^0$ for $\tilde{q}_R$, and $\tilde{\chi}^\pm$ or $\tilde{\chi}_2^0$ for $\tilde{q}_L$. If $m_{\tilde{g}} < m_{\tilde{q}}$, gluinos are expected to decay via virtual squark exchange into a quark, an antiquark, and a gaugino, $\tilde{g} \to q\bar{q}\tilde{\chi}$, where $\tilde{\chi}$ is typically $\tilde{\chi}^\pm$ or $\tilde{\chi}_2^0$.[159,160] The heavier gauginos further decay into a fermion-antifermion pair and an LSP, $\tilde{\chi}_1^0$, so that there is always some missing $E_T$ in the final state. More detailed predictions can be made only within a specific model such as mSUGRA.

The aforementioned production processes have been searched for by CDF and DØ in topologies involving at least two jets, four jets and three jets, all with large $\not\!\!E_T$. Initial and final state radiation of soft jets can increase further those jet multiplicities. A first class of background to squark and gluino production arises from processes with intrinsic $\not\!\!E_T$, such as $(W \to \ell\nu)$+jets, where the lepton escapes detection, or $(Z \to \nu\nu)$+jets, which is irreducible. Monte Carlo simulations were used to estimate those backgrounds, after calibration on events where leptons from $W \to \ell\nu$ or $Z \to \ell\ell$ are detected. Another class of background is due to multijet production by strong interaction. Although there is no intrinsic $\not\!\!E_T$ in such events, fake $\not\!\!E_T$ can arise from jet energy mismeasurements (and also real

$\not\!\!E_T$ from semileptonic decays of heavy flavor hadrons). In such events, the $\not\!\!E_T$ distribution decreases quasi-exponentially, and the direction of the $\not\!\!E_T$ tends to be close to that of a mismeasured jet. Requiring sufficiently large $\not\!\!E_T$ and applying topological selection criteria allows this background to remain under control. While DØ applied criteria tight enough to reduce this background to a negligible level, CDF estimated its remaining contribution based on simulations calibrated on control samples.

No excesses of events were observed over SM backgrounds, which was translated into exclusion domains in the plane of squark and gluino masses. To this end, a specific SUSY model had to be chosen, so that the masses and decay modes of all the gauginos involved in the decay chains could be determined. The model used by both CDF and DØ was mSUGRA, with $A_0 = 0$, $\mu < 0$, and $\tan\beta = 5$ (CDF) or 3 (DØ). The production of all squark species was considered, except for the third generation (CDF) or for stops (DØ), and the squark mass quoted was the average of the masses of the squarks considered. Finally, the large theoretical uncertainties associated to the choices of PDFs and of the factorization and renormalization scales had to be taken into account when turning cross section upper limits into exclusion domains in terms of masses. Based on an integrated luminosity of 2.1 fb$^{-1}$, DØ excluded the domain shown in Fig. 9.17, from which lower limits of 379 and 308 GeV were derived for the squark and gluino masses, respectively, as well as a lower limit of 390 GeV if $m_{\tilde{q}} = m_{\tilde{g}}$.[161] Similar results were obtained by the CDF collaboration.[162]

**Third generation squarks:** As already mentioned, a stop NLSP decays into a charm quark and a neutralino as long as $m_{\tilde{t}} < m_W + m_b + m_{\tilde{\chi}_1^0}$. The final state from stop pair production therefore consists in acoplanar charm jets and $\not\!\!E_T$. Because only one of the squark species is now produced, the cross section is smaller than for generic squarks, and the mass reach is therefore lower. As a consequence, the jets are softer, and there is also less $\not\!\!E_T$. The corresponding loss of sensitivity was attenuated by making use of heavy-flavor tagging, which resulted in the exclusion domain shown in Fig. 9.18, obtained by DØ[163] from an analysis of 1 fb$^{-1}$ of data. It can be seen that a stop mass of 150 GeV is excluded for $m_{\tilde{\chi}_1^0} = 65$ GeV. In spite of the larger mass reach at the Tevatron, the LEP results remain the most constraining for $\tilde{t} - \tilde{\chi}_1^0$ mass differences smaller than $\simeq 40$ GeV. Similar searches were performed for a sbottom NLSP decaying into $b\tilde{\chi}_1^0$,[164,165] with better sensitivity due to a more efficient heavy-flavor tagging for $b$ than for $c$ quarks. A mass lower limit of 222 GeV was obtained by DØ for $m_{\tilde{\chi}_1^0} < 60$ GeV, based on 310 pb$^{-1}$ of data.

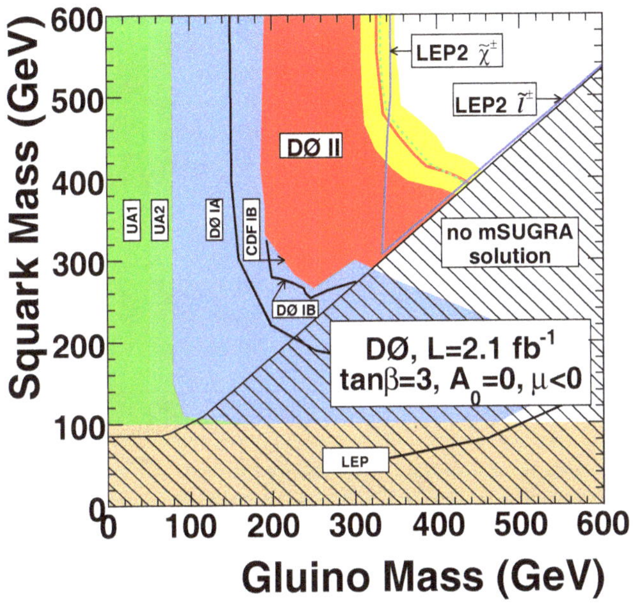


Fig. 9.17. Region in the $(m_{\tilde{g}}, m_{\tilde{q}})$ plane excluded by DØ[161] and by earlier experiments. The red curve corresponds to the nominal scale and PDF choices. The yellow band represents the uncertainty associated with these choices. The blue curves represent the indirect limits inferred from the LEP chargino and slepton searches.

Other mass hierarchies were considered, where the stop or sbottom is not the NLSP. Three-body stop decays, $\tilde{t} \to b\ell\tilde{\nu}$, are dominant if kinematically allowed and when $\tilde{t} \to \tilde{\chi}^+ b$ is not, which is possible for some model parameter choices in spite of the mass limits on charged sleptons available from LEP. The final states investigated by DØ comprised two muons or a muon and an electron, with $b$ jets and $\not\!\!E_T$. Based on an analysis of 400 pb$^{-1}$ of data, the largest stop mass excluded is 186 GeV, for $m_{\tilde{\nu}} = 71$ GeV.[166] If the chargino is lighter than the stop, the $\tilde{t} \to b\tilde{\chi}^+$ decay is dominant. A search was performed by CDF in the two lepton, two $b$ jets and $\not\!\!E_T$ final state, with a sensitivity depending on the branching fraction of the chargino leptonic decay, $\tilde{\chi}^{\pm} \to \ell\nu\tilde{\chi}_1^0$, which is enhanced for light sleptons. An example of an excluded domain in the $(m_{\tilde{t}}, m_{\tilde{\chi}_1^0})$ plane is shown in Fig. 9.19,[167] based on 2.7 fb$^{-1}$ of data. In both of those searches, the background from top quark pair production was a major challenge. Yet another mass hierarchy was considered by CDF, namely that where the sbottom is the only squark lighter than the gluino. In such a configuration, the $\tilde{g} \to b\tilde{b}$ decay is

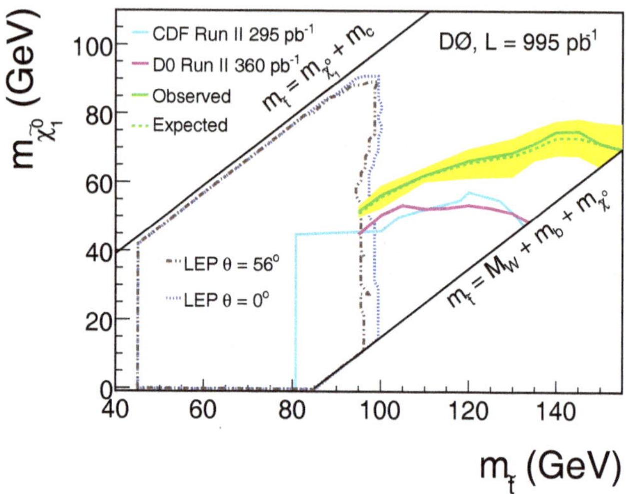


Fig. 9.18. Region in the $(m_{\tilde{t}}, m_{\tilde{\chi}_1^0})$ plane excluded by DØ[163] and by earlier experiments. The solid curve corresponds to the nominal scale and PDF choices. The gray band represents the uncertainty associated with these choices.

dominant, and gluino pair production then leads to a final state of four $b$ jets and $\not{E}_T$. This search was performed in a data sample of 2.5 fb$^{-1}$, and lead to excluded sbottom masses as large as 325 GeV for gluino and LSP masses of 340 and 60 GeV, respectively.[168]

**Charginos and neutralinos:** The associated production of charginos and neutralinos, $p\bar{p} \to \tilde{\chi}^{\pm}\tilde{\chi}_2^0$, is an electroweak process mediated by $s$-channel $W$ and $t$-channel squark exchanges. Leptonic decays, $\tilde{\chi}^{\pm} \to \ell^{\pm}\nu\tilde{\chi}_1^0$ and $\tilde{\chi}_2^0 \to \ell^+\ell^-\tilde{\chi}_1^0$, are mediated by $W$ and $Z$ exchange, respectively, and by slepton exchange. If sleptons are light, leptonic decays can be sufficiently enhanced for searches in final states consisting of three leptons and $\not{E}_T$ to become sensitive in spite of production cross sections of a fraction of a picobarn. An additional challenge is the rather small energy carried by the final state leptons in the chargino and neutralino mass domain to which the searches at the Tevatron are currently sensitive.

In both the CDF[169] and DØ[170] analyses, only two leptons were required to be positively identified as electrons or muons.[d] Allowing the third lepton to be detected as an isolated charged particle track provided sensitivity to

[d]The DØ analysis also considered final states with a muon and one or two $\tau$ leptons identified.

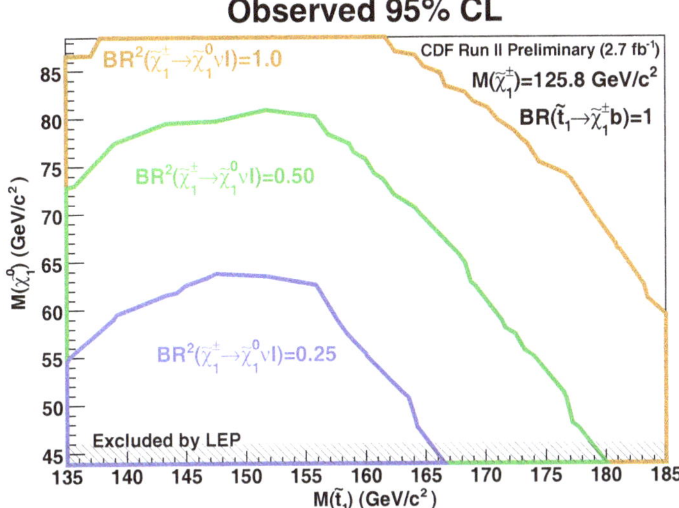


Fig. 9.19. Regions in the $(m_{\tilde{t}}, m_{\tilde{\chi}_1^0})$ plane excluded by CDF[167] for $m_{\tilde{\chi}^\pm} = 125.8$ GeV and for various values of the branching fraction for the $\tilde{\chi}^\pm \to \ell\nu\tilde{\chi}_1^0$ decay.

final states including a $\tau$ lepton that decays into hadrons. In the CDF analysis, the trilepton final state was split into topologies with different signal to background ratios, the purest being when the three leptons are positively identified as electrons or muons with tight criteria. In the DØ approach, different selections were optimized according to the amount of energy available to the lepton candidates. The ultimate background for these trilepton searches is associated $WZ$ production.

The DØ search, based on an integrated luminosity of 2.3 fb$^{-1}$, excludes regions in the mSUGRA parameter space as shown in Fig. 9.20 for $A_0 = 0$, $\tan\beta = 3$ and $\mu > 0$. It can be seen that the domain excluded at LEP is substantially extended by these trilepton searches. The interruption in the exclusion domain is due to configurations where the small $\tilde{\chi}_2^0 - \tilde{\ell}$ mass difference results in one of the final state leptons carrying too little energy, thus preventing efficient detection. Requiring only two leptons to be identified, but with same charge sign in order to reduce the otherwise overwhelming SM backgrounds, should provide sensitivity in that region, as was shown by DØ in an analysis based on a smaller data sample.[171] The same-sign dilepton signature was also considered in an earlier CDF analysis.[172]

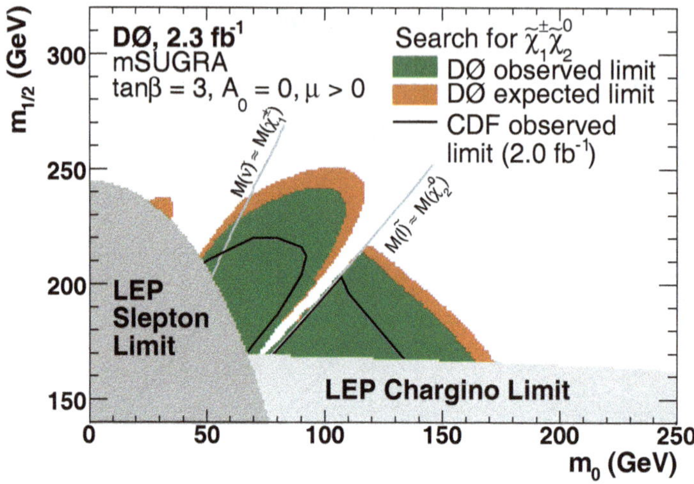


Fig. 9.20. Regions in the $(m_0, m_{1/2})$ plane excluded by the DØ search for trileptons.[170]

### 9.5.4. *Searches in non-canonical scenarios*

#### 9.5.4.1. *R-parity violation*

Searches for SUSY with $R$-parity violation were performed at LEP, the Tevatron and HERA. Both $R$-parity conserving pair production of SUSY particles and $R$-parity violating resonant single SUSY particle production were considered. The produced particles were subsequently subject to either direct or indirect (via a cascade to the LSP) $R$-parity violating decays. Unless otherwise specified, a single $R$-parity violating coupling is assumed to be non-vanishing in the following, large enough for the lifetime of the LSP to be safely assumed to be negligible.

**Searches at LEP:** Extensive searches for pair production were performed at LEP, involving all possible $R$-parity violating couplings. The possible final states are numerous, ranging from four leptons and missing energy for $\tilde{\chi}_1^0$ pair production, with decays mediated by a $\lambda$-type coupling, e.g., $\tilde{\chi}_1^0 \to e\mu\nu$, to ten hadronic jets and no missing energy for chargino pair production, with $\tilde{\chi}^\pm \to q\bar{q}'\tilde{\chi}_1^0$ followed by a $\tilde{\chi}_1^0$ decay into three quarks via a $\lambda''$-type coupling, e.g., $\tilde{\chi}_1^0 \to udd$. The results of these searches are at least as constraining as in the canonical scenario.[173–177]

The production of a sneutrino resonance via a $\lambda_{1j1}$ coupling was also investigated. No signal was observed, and mass lower limits almost up to the center-of-mass energy were set for sufficiently large values of the $R$-parity violating coupling involved.[178–182]

**Searches at HERA:** As explained in Sec. 9.5.2, the HERA $ep$ collider is most effective in the searches for $R$-parity violating resonant single squark production via a $\lambda'$-type coupling. Direct and indirect squark decays were investigated, and the search results were combined to lead to squark mass lower limits up to 275 GeV,[183–185] within mild model assumptions, for a $\lambda'$ coupling of 0.3, i.e., with electromagnetic strength.

**Searches at the Tevatron:** A fully general search for all $R$-parity violating couplings is not possible at the Tevatron, as it was at LEP. For instance, $\lambda''$ couplings lead to multijet final states with no or little missing energy, which cannot be distinguished from standard multijet production. Searches have therefore been designed for specific choices of couplings leading to distinct signatures.

Gaugino pair production followed by indirect decays has been extensively studied by both CDF[186] and DØ[187] in the case of a $\lambda$-type coupling. The final state is expected to contain four charged leptons, with flavors depending on the indices in the $\lambda_{ijk}$ coupling, and $\not\!\!E_T$ due to two neutrinos. For $m_0 = 1$ TeV, $\tan\beta = 5$, and $\mu > 0$, the chargino mass lower limits obtained by DØ from an analysis of 360 pb$^{-1}$ of data are 231, 229, and 166 GeV for the $\lambda_{121}$, $\lambda_{122}$, and $\lambda_{133}$ couplings, respectively, with reduced sensitivity in the last case due to the occurrence of $\tau$ leptons in the final state.

Stop pair production, with $\tilde{t} \to b\tau$ via a $\lambda'_{333}$ coupling has been searched by CDF[188] in the topology where one $\tau$ lepton decays into an electron or a muon, and the other into hadrons. From an analysis of 322 pb$^{-1}$ of data, a stop mass lower limit of 151 GeV was derived.

Resonant smuon or sneutrino production could be mediated by a $\lambda'_{211}$ coupling. With indirect decays, the final state would exhibit at least one muon and two jets. This topology was investigated by DØ,[189] and an excluded domain was set in the $(m_{\tilde{\mu}}, \lambda'_{211})$ plane, leading to a smuon mass lower limit of 363 GeV for $\lambda'_{211} = 0.1$, and for $A_0 = 0$, $\tan\beta = 5$, and $\mu < 0$.

Resonant sneutrino production mediated by a $\lambda'_{i11}$ coupling was also investigated by CDF and DØ,[190–193] now assuming that the sneutrino decays directly via a $\lambda$-type coupling. The final states considered were $ee$, $e\mu$, $\mu\mu$, and $\tau\tau$. The sneutrino mass limits obtained depend on the product of the two couplings involved.

9.5.4.2. *Gauge-mediated SUSY breaking*

As already explained, the LSP in GMSB is a very light gravitino, and the phenomenology depends essentially on the nature of the NLSP, a neutralino or a stau, possibly almost mass degenerate with $\tilde{e}_R$ and $\tilde{\mu}_R$, and on its lifetime.

**Neutralino NLSP:** In the mass range of current interest, a neutralino NLSP decays into a photon and a gravitino, $\tilde{\chi}_1^0 \to \gamma\tilde{G}$. Pair production of such a neutralino at LEP would therefore lead, assuming prompt decays, to a final state of two acoplanar photons and missing energy. As can be seen in Fig. 9.21, no excess was observed above the SM background from $e^+e^- \to (Z^{(*)} \to \nu\bar{\nu})\gamma\gamma$. In GMSB, $\tilde{\chi}_1^0$ has a large Bino component, so that pair production in $e^+e^-$ interactions proceeds via selectron $t$-channel exchange. An excluded domain in the $(m_{\tilde{e}_R}, m_{\tilde{\chi}_1^0})$ plane was therefore derived,[194] ruling out the GMSB interpretation (in terms of selectron pair production) of an anomalous $ee\gamma\gamma + \not{E}_T$ event that had been observed by CDF[195] during Run I of the Tevatron.

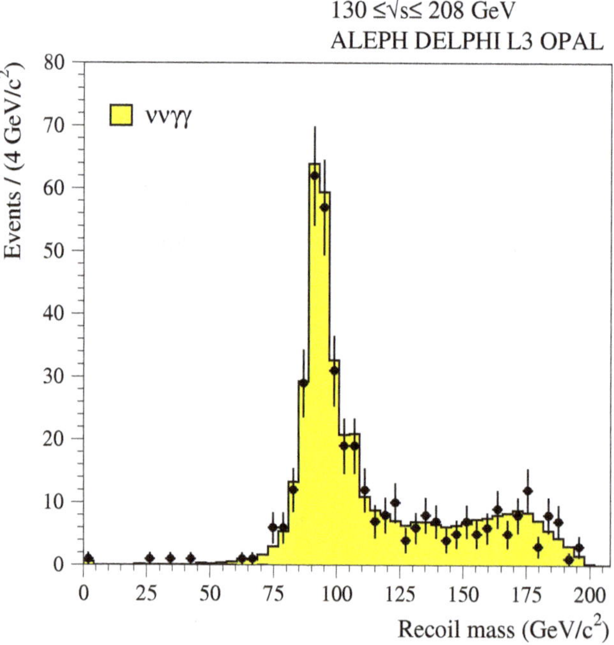


Fig. 9.21. Mass of the invisible system recoiling against pairs of photons at LEP.[194]

Searches were also performed at LEP for photons not pointing toward the interaction point, which could arise from non-prompt decays of a neutralino NLSP. For even longer lifetimes, the phenomenology becomes identical to that of the canonical scenario. The results of these various searches for a neutralino NLSP were combined with those in various topologies expected to arise from heavier SUSY particle production to lead to a robust neutralino mass lower limit of 54 GeV within the minimal GMSB framework.[196,197]

Searches for acoplanar photons with large $\not{E}_T$ were performed at the Tevatron by both CDF[198] and DØ.[199] This topology is expected to arise whenever SUSY particles are pair produced, which subsequently decay to a neutralino NLSP with negligible lifetime. No excess of events was observed over the backgrounds due to photon misidentification or from fake $\not{E}_T$, all determined from data. These results were interpreted within the "Snowmass slope SPS 8" benchmark GMSB model[200] where the only free parameter is the effective SUSY breaking scale $\Lambda$. The other parameters were fixed as follows: $N_5 = 1$ messenger, a messenger mass of $2\Lambda$, $\tan\beta = 15$, and $\mu > 0$. Neutralino NLSP masses smaller than 138 GeV are excluded by the CDF analysis, based on 2 fb$^{-1}$ of data.

The possibility of non-prompt neutralino NLSP decays was also investigated by CDF,[201] making use of the timing information of their calorimeter. No signal of delayed photons was observed in a data sample of 570 pb$^{-1}$, from which an excluded domain in the plane of the mass and lifetime of the NLSP was inferred, as shown in Fig. 9.22 together with the result of Ref. 198.

**Stau NLSP:** For prompt $\tilde{\tau} \to \tau\tilde{G}$ decays, the final state arising from stau pair production at LEP is the same as in the canonical scenario with a very light $\tilde{\chi}_1^0$. For very long lifetimes, the searches for long lived charginos already reported apply. Searches for in-flight decays along charged particle tracks were designed to address intermediate lifetimes. The combination of all these searches allowed a stau NLSP mass lower limit to be set from 87 to 97 GeV, depending on the stau lifetime, as shown in Fig. 9.23.[202]

#### 9.5.4.3. *Other non-canonical scenarios*

A number of searches were performed at LEP and at the Tevatron in other non-canonical scenarios.

**Stable charged particles:** In anomaly mediated SUSY breaking, the LSP is wino-like, and the $\tilde{\chi}^\pm - \tilde{\chi}_1^0$ mass difference is therefore small. As a

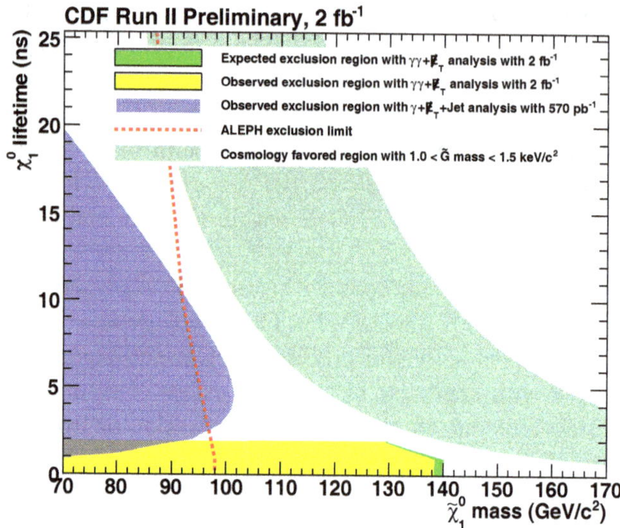


Fig. 9.22. Domain excluded by CDF in the plane of neutralino-NLSP mass and lifetime.[198,201]

result, stable charginos are not unlikely. The searches designed at LEP in the canonical scenario for large $M_2$ values apply here equally well. At the Tevatron, a search was performed by DØ for pairs of charged massive stable particles that could result from chargino pair production. Such particles would behave like slow moving muons that could be detected as delayed signals in the muon system. No significant excess of delayed muons was observed in 1.1 fb$^{-1}$ of data, and a mass lower limit of 206 GeV was set on long-lived wino-like charginos.[203]

A search for stable stops was performed by CDF in 1 fb$^{-1}$ of data, using a high $p_T$ muon trigger and their time-of-flight detector. Stable stops hadronize to form $R$-hadrons which behave like slow muons. A model for the interactions of those $R$-hadrons with the detector material was constructed, within which a stop mass lower limit of 249 GeV was derived.[204]

**Stable or long-lived gluinos:** Models have been built where the gluino could be the LSP and therefore stable, if $R$-parity is conserved.[205–207] Alternatively, gluinos may decay, but with long lifetimes. This occurs, for example, in models with "split SUSY," unnatural models in which all squarks and sleptons are very heavy, but the gauginos remain at the electroweak scale.[208–210] Since gluino decays are mediated by squark exchange, the gluino becomes long-lived.

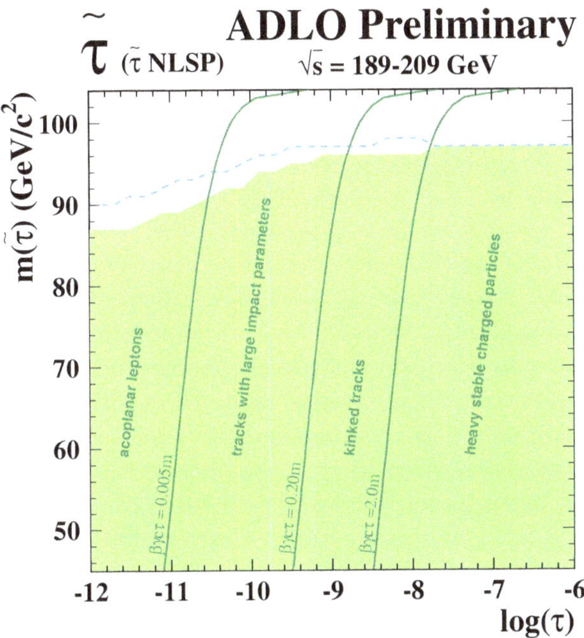


Fig. 9.23. Domain excluded at LEP[202] in the $(\log_{10}\tau, m_{\tilde{\tau}})$ plane, where $\tau$ is the lifetime in seconds and $m_{\tilde{\tau}}$ is the mass of a stau NLSP in GMSB. The shaded region is the excluded domain, and the dashed curve is the expected boundary of the exclusion region.

Although gluinos cannot be produced directly in $e^+e^-$ interactions, they could be produced via gluon splitting, e.g., $e^+e^- \to q\bar{q}g^* \to q\bar{q}\tilde{g}\tilde{g}$, and hadronize into metastable "$R$-hadrons." The QCD predictions for four-jet events would therefore be modified. Gluinos could also be produced in the decay of heavier squarks. Dedicated analyses were performed at LEP,[211,212] leading to a mass lower limit of 27 GeV for a stable gluino.

A search for long-lived gluinos was also performed by DØ with 410 pb$^{-1}$ of data.[213] After hadronization into an $R$-hadron, a long-lived gluino could come to rest in the calorimeter and decay later on, during a bunch crossing different from the one during which it was created.[214] The main decay mode expected is $\tilde{g} \to g\tilde{\chi}_1^0$, leading to an hadronic shower originating from within the calorimeter and not pointing toward the $p\bar{p}$ interaction region, in an otherwise empty event. No excess of this anomalous topology was observed over the background due to cosmic muons or to the beam halo. The gluino mass lower limits derived depend on the lifetime $\tau_{\tilde{g}}$, the branching fraction

$\mathcal{B}$ for the decay mode considered, the $\tilde{\chi}_1^0$ mass, and the cross section $\sigma_R$ for the conversion of a neutral $R$-hadron into a charged one in the calorimeter. As an example, a mass limit of 270 GeV was obtained for $\tau_{\tilde{g}} < 3$ hours, $\mathcal{B} = 1$, $m_{\tilde{\chi}_1^0} = 50$ GeV, and $\sigma_R = 3$ mb.

## 9.6. Summary

Supersymmetry is one of the most promising ideas for extending the standard model. When realized at the weak scale, many SUSY models provide natural, and even elegant, solutions to the most pressing problems in particle physics today by stabilizing the gauge hierarchy, providing dark matter candidates, and accommodating force unification, both with and without gravity. In addition, the general framework of weak-scale SUSY is flexible enough to encompass a wide variety of new phenomena, including extended Higgs sectors, missing energy, long-lived and metastable particles, and a host of other signatures of new physics. Searches for SUSY are therefore also searches for other forms of new physics which, even if less profoundly motivated, are, of course, also important to pursue.

In this review, we have comprehensively summarized the state of the art in searches for SUSY at the high energy frontier. Although this is a continuously evolving subject, this review provides a snapshot of the field at a particularly important time, when final results from LEP and HERA are in hand, the Tevatron experiments have reported deep probes of many supersymmetric models with several fb$^{-1}$ of data, and the LHC will soon begin operation.

This review has summarized searches for both supersymmetric Higgs bosons and standard model superpartners. In the Higgs sector, SUSY requires a light neutral Higgs boson. This Higgs boson could be standard-model like, but it could also have non-standard couplings. In addition, it is accompanied by other Higgs bosons, both neutral or charged. The most stringent constraints on a SM-like Higgs boson currently come from LEP, with a mass lower limit of 114.4 GeV that applies in the MSSM at low $\tan\beta$. Furthermore, the LEP experiments set a lower limit of 93 GeV on the lightest neutral Higgs boson of the MSSM, independent of $\tan\beta$. The MSSM parameter space has now been further restricted by the Tevatron experiments. For example, $\tan\beta$ values larger than 40 are excluded for $m_A = 140$ GeV. For charged Higgs bosons, LEP excludes masses below 78.6 GeV, and the Tevatron experiments have extended this mass limit to $\sim 150$ GeV for very large values of $\tan\beta$.

For superpartners, the bounds are, of course, model-dependent, but the main results may be summarized as follows. The searches at LEP have constrained the masses of all SUSY particles, except for the gluino and the LSP, to be larger than approximately 100 GeV in most SUSY scenarios. Furthermore, an indirect lower limit on the mass of a neutralino LSP has been set at 47 GeV in the MSSM with gaugino and sfermion mass unification. The higher center-of-mass energy at the Tevatron has allowed tighter mass limits to be obtained for strongly interacting SUSY particles: 379 and 308 GeV for squarks and gluinos, respectively, within the mSUGRA framework at low $\tan\beta$. In that same model, domains beyond the LEP reach were also probed by searches for associated chargino-neutralino production.

In the near future, the first indication for SUSY at high energy colliders could be the observation of a light neutral Higgs boson at the Tevatron. Of course, such a discovery is not proof of SUSY — only the discovery of superpartners would unambiguously establish SUSY as being realized in nature. Once collisions begin at the LHC and the detectors are sufficiently understood, it will not take more than $\sim 1$ fb$^{-1}$ to discover squarks and gluinos with masses less than $\sim 1.5$ TeV.[21,23] A new era will then begin during which the whole SUSY spectrum will have to be deciphered, and the properties of the SUSY model established. Many more fb$^{-1}$ will be needed for that purpose, and to unravel the spectrum of SUSY Higgs bosons.

## Acknowledgments

The work of JLF was supported in part by NSF grants PHY–0239817 and PHY–0653656, NASA grant NNG05GG44G, and the Alfred P. Sloan Foundation. JFG is supported by the CNRS/IN2P3 (France). The work of JN was supported by DOE grant DE–FG02–91ER40664.

## References

1. S. Weinberg, "Implications of Dynamical Symmetry Breaking," *Phys. Rev.* **D13** (1976) 974–996.
2. S. Weinberg, "Implications of Dynamical Symmetry Breaking: An Addendum," *Phys. Rev.* **D19** (1979) 1277–1280.
3. L. Susskind, "Dynamics of Spontaneous Symmetry Breaking in the Weinberg-Salam Theory," *Phys. Rev.* **D20** (1979) 2619–2625.
4. G. 't Hooft, in Recent Developments in Gauge Theories. Proceedings, NATO Advanced Study Institute, Cargese, France, August 26 - September 8, 1979,

New York, Plenum (1980) 438 p. (NATO Advanced Study Institutes Series: Series B, Physics, 59).

5. L. Reina, "TASI 2004 lecture notes on Higgs boson physics," arXiv:hep-ph/0512377.
6. C. T. Hill and E. H. Simmons, "Strong dynamics and electroweak symmetry breaking," *Phys. Rept.* **381** (2003) 235–402, arXiv:hep-ph/0203079.
7. **WMAP** Collaboration, E. Komatsu *et al.*, "Five-Year Wilkinson Microwave Anisotropy Probe (WMAP) Observations: Cosmological Interpretation," *Astrophys. J. Suppl.* **180** (2009) 330–376, arXiv:0803.0547 [astro-ph].
8. Y. B. Zeldovich *Adv. Astron. Astrophys.* **3** (1965) 241.
9. H.-Y. Chiu, "Symmetry between particle and anti-particle populations in the universe," *Phys. Rev. Lett.* **17** (1966) 712.
10. G. Steigman, "Cosmology Confronts Particle Physics," *Ann. Rev. Nucl. Part. Sci.* **29** (1979) 313–338.
11. R. J. Scherrer and M. S. Turner, "On the Relic, Cosmic Abundance of Stable Weakly Interacting Massive Particles," *Phys. Rev.* **D33** (1986) 1585.
12. H. Georgi and S. L. Glashow, "Unity of All Elementary Particle Forces," *Phys. Rev. Lett.* **32** (1974) 438–441.
13. S. Dimopoulos and H. Georgi, "Softly Broken Supersymmetry and SU(5)," *Nucl. Phys.* **B193** (1981) 150.
14. S. Dimopoulos, S. Raby, and F. Wilczek, "Supersymmetry and the Scale of Unification," *Phys. Rev.* **D24** (1981) 1681–1683.
15. N. Sakai, "Naturalness in Supersymmetric GUTs," *Zeit. Phys.* **C11** (1981) 153.
16. L. E. Ibanez and G. G. Ross, "Low-Energy Predictions in Supersymmetric Grand Unified Theories," *Phys. Lett.* **B105** (1981) 439.
17. M. B. Einhorn and D. R. T. Jones, "The Weak Mixing Angle and Unification Mass in Supersymmetric SU(5)," *Nucl. Phys.* **B196** (1982) 475.
18. **Particle Data Group** Collaboration, C. Amsler *et al.*, "Review of particle physics," *Phys. Lett.* **B667** (2008) 1.
19. **ALEPH, DELPHI, L3, OPAL, SLD** Collaborations, LEP Electroweak Working Group and SLD Electroweak Group and SLD Heavy Flavour Group, "Precision electroweak measurements on the $Z$ resonance," *Phys. Rept.* **427** (2006) 257, arXiv:hep-ex/0509008.
20. **ATLAS** Collaboration, G. Aad *et al.*, "The ATLAS Experiment at the CERN Large Hadron Collider," *J. Inst.* **3** (2008) S08003.
21. **ATLAS** Collaboration, G. Aad *et al.*, "Expected Performance of the ATLAS Experiment - Detector, Trigger and Physics," arXiv:0901.0512 [hep-ex].
22. CMS Collaboration, "CMS, the Compact Muon Solenoid: Technical proposal," CERN-LHCC-94-38.
23. **CMS** Collaboration, G. L. Bayatian *et al.*, "CMS technical design report, volume II: Physics performance," *J. Phys.* **G34** (2007) 995–1579.
24. **ALEPH** Collaboration, D. Decamp *et al.*, "ALEPH: a detector for electron-positron annihilations at LEP," *Nucl. Instrum. Meth.* **A294** (1990) 121–178.

25. **ALEPH** Collaboration, D. Buskulic *et al.*, "Performance of the ALEPH detector at LEP," *Nucl. Instrum. Meth.* **A360** (1995) 481–506.
26. **DELPHI** Collaboration, P. A. Aarnio *et al.*, "The DELPHI detector at LEP," *Nucl. Instrum. Meth.* **A303** (1991) 233–276.
27. **DELPHI** Collaboration, P. Abreu *et al.*, "Performance of the DELPHI detector," *Nucl. Instrum. Meth.* **A378** (1996) 57–100.
28. **L3** Collaboration, B. Adeva *et al.*, "The construction of the L3 experiment," *Nucl. Instrum. Meth.* **A289** (1990) 35–102.
29. **OPAL** Collaboration, K. Ahmet *et al.*, "The OPAL detector at LEP," *Nucl. Instrum. Meth.* **A305** (1991) 275–319.
30. **H1** Collaboration, I. Abt *et al.*, "The H1 detector at HERA," *Nucl. Instrum. Meth.* **A386** (1997) 310–347.
31. ZEUS Collaboration, "The ZEUS detector: Status report 1993," ZEUS-STATUS-REPT-1993. `http://www-zeus.desy.de/bluebook/bluebook.html`.
32. **CDF** Collaboration, D. E. Acosta *et al.*, "Measurement of the $J/\psi$ meson and $b-$hadron production cross sections in $p\bar{p}$ collisions at $\sqrt{s} = 1960$ GeV," *Phys. Rev.* **D71** (2005) 032001, `arXiv:hep-ex/0412071`.
33. **D0** Collaboration, V. M. Abazov *et al.*, "The Upgraded D0 Detector," *Nucl. Instrum. Meth.* **A565** (2006) 463–537, `arXiv:physics/0507191`.
34. Y. A. Golfand and E. P. Likhtman, "Extension of the Algebra of Poincare Group Generators and Violation of p Invariance," *JETP Lett.* **13** (1971) 323–326.
35. D. V. Volkov and V. P. Akulov, "Is the Neutrino a Goldstone Particle?," *Phys. Lett.* **B46** (1973) 109–110.
36. J. Wess and B. Zumino, "Supergauge Transformations in Four-Dimensions," *Nucl. Phys.* **B70** (1974) 39–50.
37. L. Maiani, in Vector bosons and Higgs bosons in the Weinberg-Salam theory of weak and electromagnetic interactions. Proceedings: Summer School on Particle Physics, Gif-sur- Yvette, 3-7 Sep 1979, Paris, France: Natl. Inst. Nucl. Phys. Part. Phys. (1979) 201 p.
38. M. J. G. Veltman, "The Infrared - Ultraviolet Connection," *Acta Phys. Polon.* **B12** (1981) 437.
39. E. Witten, "Dynamical Breaking of Supersymmetry," *Nucl. Phys.* **B188** (1981) 513.
40. H. Goldberg, "Constraint on the photino mass from cosmology," *Phys. Rev. Lett.* **50** (1983) 1419.
41. J. R. Ellis, J. S. Hagelin, D. V. Nanopoulos, K. A. Olive, and M. Srednicki, "Supersymmetric relics from the big bang," *Nucl. Phys.* **B238** (1984) 453–476.
42. H. Pagels and J. R. Primack, "Supersymmetry, Cosmology and New TeV Physics," *Phys. Rev. Lett.* **48** (1982) 223.
43. J. L. Feng, A. Rajaraman, and F. Takayama, "Superweakly-interacting massive particles," *Phys. Rev. Lett.* **91** (2003) 011302, `arXiv:hep-ph/0302215`.
44. S. P. Martin, "A Supersymmetry Primer," `arXiv:hep-ph/9709356`.

45. N. Polonsky, "Supersymmetry: Structure and phenomena. Extensions of the standard model," *Lect. Notes Phys.* **M68** (2001) 1–169, arXiv:hep-ph/0108236.
46. M. Drees, R. Godbole, and P. Roy, "Theory and phenomenology of sparticles: An account of four-dimensional N=1 supersymmetry in high energy physics," Hackensack, USA: World Scientific (2004) 555 p.
47. H. Baer and X. Tata, "Weak scale supersymmetry: From superfields to scattering events," Cambridge, UK: Univ. Pr. (2006) 537 p.
48. L. Girardello and M. T. Grisaru, "Soft Breaking of Supersymmetry," *Nucl. Phys.* **B194** (1982) 65.
49. P. Fayet, "Spontaneously Broken Supersymmetric Theories of Weak, Electromagnetic and Strong Interactions," *Phys. Lett.* **B69** (1977) 489.
50. G. R. Farrar and P. Fayet, "Phenomenology of the Production, Decay, and Detection of New Hadronic States Associated with Supersymmetry," *Phys. Lett.* **B76** (1978) 575–579.
51. F. E. Paige, S. D. Protopopescu, H. Baer, and X. Tata, "ISAJET 7.69: A Monte Carlo event generator for $pp$, $p\bar{p}$, and $e^+e^-$ reactions," arXiv:hep-ph/0312045.
52. B. C. Allanach, "SOFTSUSY: A C++ program for calculating supersymmetric spectra," *Comput. Phys. Commun.* **143** (2002) 305–331, arXiv:hep-ph/0104145.
53. W. Porod, "SPheno, a program for calculating supersymmetric spectra, SUSY particle decays and SUSY particle production at $e^+e^-$ colliders," *Comput. Phys. Commun.* **153** (2003) 275–315, arXiv:hep-ph/0301101.
54. A. Djouadi, J.-L. Kneur, and G. Moultaka, "SuSpect: A Fortran code for the supersymmetric and Higgs particle spectrum in the MSSM," *Comput. Phys. Commun.* **176** (2007) 426–455, arXiv:hep-ph/0211331.
55. M. E. Peskin, "Theoretical summary lecture for EPS HEP99," arXiv:hep-ph/0002041.
56. A. H. Chamseddine, R. L. Arnowitt, and P. Nath, "Locally Supersymmetric Grand Unification," *Phys. Rev. Lett.* **49** (1982) 970.
57. R. Barbieri, S. Ferrara, and C. A. Savoy, "Gauge Models with Spontaneously Broken Local Supersymmetry," *Phys. Lett.* **B119** (1982) 343.
58. N. Ohta, "Grand unified theories based on local supersymmetry," *Prog. Theor. Phys.* **70** (1983) 542.
59. L. J. Hall, J. D. Lykken, and S. Weinberg, "Supergravity as the Messenger of Supersymmetry Breaking," *Phys. Rev.* **D27** (1983) 2359–2378.
60. L. Alvarez-Gaume, J. Polchinski, and M. B. Wise, "Minimal Low-Energy Supergravity," *Nucl. Phys.* **B221** (1983) 495.
61. H. P. Nilles, "Supersymmetry, Supergravity and Particle Physics," *Phys. Rept.* **110** (1984) 1–162.
62. M. Dine, W. Fischler, and M. Srednicki, "Supersymmetric Technicolor," *Nucl. Phys.* **B189** (1981) 575–593.
63. S. Dimopoulos and S. Raby, "Supercolor," *Nucl. Phys.* **B192** (1981) 353.
64. C. R. Nappi and B. A. Ovrut, "Supersymmetric Extension of the SU(3) x SU(2) x U(1) Model," *Phys. Lett.* **B113** (1982) 175.

65. L. Alvarez-Gaume, M. Claudson, and M. B. Wise, "Low-Energy Supersymmetry," *Nucl. Phys.* **B207** (1982) 96.
66. M. Dine, A. E. Nelson, and Y. Shirman, "Low-energy dynamical supersymmetry breaking simplified," *Phys. Rev.* **D51** (1995) 1362–1370, arXiv:hep-ph/9408384.
67. M. Dine, A. E. Nelson, Y. Nir, and Y. Shirman, "New tools for low-energy dynamical supersymmetry breaking," *Phys. Rev.* **D53** (1996) 2658–2669, arXiv:hep-ph/9507378.
68. D. R. Stump, M. Wiest, and C. P. Yuan, "Detecting a Light Gravitino at Linear Collider to Probe the SUSY Breaking Scale," *Phys. Rev.* **D54** (1996) 1936–1943, arXiv:hep-ph/9601362.
69. S. Dimopoulos, M. Dine, S. Raby, and S. D. Thomas, "Experimental Signatures of Low Energy Gauge Mediated Supersymmetry Breaking," *Phys. Rev. Lett.* **76** (1996) 3494–3497, arXiv:hep-ph/9601367.
70. J. L. Feng and T. Moroi, "Tevatron signatures of longlived charged sleptons in gauge mediated supersymmetry breaking models," *Phys. Rev.* **D58** (1998) 035001, arXiv:hep-ph/9712499.
71. M. Drees and X. Tata, "Signals for heavy exotics at hadron colliders and supercolliders," *Phys. Lett.* **B252** (1990) 695–702.
72. J. L. Goity, W. J. Kossler, and M. Sher, "Production, collection and utilization of very longlived heavy charged leptons," *Phys. Rev.* **D48** (1993) 5437–5439, arXiv:hep-ph/9305244.
73. L. Randall and R. Sundrum, "Out of this world supersymmetry breaking," *Nucl. Phys.* **B557** (1999) 79–118, arXiv:hep-th/9810155.
74. G. F. Giudice, M. A. Luty, H. Murayama, and R. Rattazzi, "Gaugino Mass without Singlets," *JHEP* **12** (1998) 027, arXiv:hep-ph/9810442.
75. A. Pomarol and R. Rattazzi, "Sparticle masses from the superconformal anomaly," *JHEP* **05** (1999) 013, arXiv:hep-ph/9903448.
76. Z. Chacko, M. A. Luty, I. Maksymyk, and E. Ponton, "Realistic anomaly-mediated supersymmetry breaking," *JHEP* **04** (2000) 001, arXiv:hep-ph/9905390.
77. E. Katz, Y. Shadmi, and Y. Shirman, "Heavy thresholds, slepton masses and the mu term in anomaly mediated supersymmetry breaking," *JHEP* **08** (1999) 015, arXiv:hep-ph/9906296.
78. J. L. Feng, T. Moroi, L. Randall, M. Strassler, and S.-f. Su, "Discovering supersymmetry at the Tevatron in Wino LSP scenarios," *Phys. Rev. Lett.* **83** (1999) 1731–1734, arXiv:hep-ph/9904250.
79. R. Barbieri and M. Frigeni, "The Supersymmetric Higgs searches at LEP after radiative corrections," *Phys. Lett.* **B258** (1991) 395–398.
80. H. E. Haber, "Introductory low-energy supersymmetry," arXiv:hep-ph/9306207.
81. J. R. Ellis and S. Rudaz, "Search for Supersymmetry in Toponium Decays," *Phys. Lett.* **B128** (1983) 248.
82. M. S. Carena, S. Heinemeyer, C. E. M. Wagner, and G. Weiglein, "Suggestions for improved benchmark scenarios for Higgs boson searches at LEP2," arXiv:hep-ph/9912223.

83. M. S. Carena, S. Heinemeyer, C. E. M. Wagner, and G. Weiglein, "Suggestions for benchmark scenarios for MSSM Higgs boson searches at hadron colliders," *Eur. Phys. J.* **C26** (2003) 601–607, arXiv:hep-ph/0202167.
84. J. R. Ellis, M. K. Gaillard, and D. V. Nanopoulos, "A Phenomenological Profile of the Higgs Boson," *Nucl. Phys.* **B106** (1976) 292.
85. B. W. Lee, C. Quigg, and H. B. Thacker, "Weak Interactions at Very High-Energies: The Role of the Higgs Boson Mass," *Phys. Rev.* **D16** (1977) 1519.
86. B. L. Ioffe and V. A. Khoze, "What Can Be Expected from Experiments on Colliding $e^+e^-$ Beams with $E$ Approximately Equal to 100 GeV?," *Sov. J. Part. Nucl.* **9** (1978) 50.
87. G. Pocsik and G. Zsigmond, "On the production of neutral higgs bosons in the Weinberg-Salam model with two Higgs doublets," *Zeit. Phys.* **C10** (1981) 367.
88. A. Djouadi, J. Kalinowski, and M. Spira, "HDECAY: A program for Higgs boson decays in the standard model and its supersymmetric extension," *Comput. Phys. Commun.* **108** (1998) 56–74, arXiv:hep-ph/9704448.
89. D. R. T. Jones and S. T. Petcov, "Heavy Higgs Bosons at LEP," *Phys. Lett.* **B84** (1979) 440.
90. **LEP Higgs Working Group, ALEPH, DELPHI, L3, OPAL** Collaborations, R. Barate *et al.*, "Search for the standard model Higgs boson at LEP," *Phys. Lett.* **B565** (2003) 61–75, arXiv:hep-ex/0306033.
91. **LEP Higgs Working Group, ALEPH, DELPHI, L3, OPAL** Collaborations, S. Schael *et al.*, "Search for neutral MSSM Higgs bosons at LEP," *Eur. Phys. J.* **C47** (2006) 547–587, arXiv:hep-ex/0602042.
92. **CDF, D0** Collaborations, Tevatron Electroweak Working Group, "Combination of CDF and D0 Results on the Mass of the Top Quark," arXiv:0903.2503 [hep-ex].
93. **ALEPH, DELPHI, L3, OPAL** Collaborations, LEP Higgs Working Group, "Searches for invisible Higgs bosons: Preliminary combined results using LEP data collected at energies up to 209 GeV," arXiv:hep-ex/0107032.
94. **ALEPH, DELPHI, L3, OPAL** Collaborations, LEP Higgs Working Group, "Flavor independent search for hadronically decaying neutral Higgs bosons at LEP," arXiv:hep-ex/0107034.
95. R. Dermisek and J. F. Gunion, "Consistency of LEP event excesses with an $h \to aa$ decay scenario and low-fine-tuning NMSSM models," *Phys. Rev.* **D73** (2006) 111701, arXiv:hep-ph/0510322.
96. M. S. Carena, J. R. Ellis, A. Pilaftsis, and C. E. M. Wagner, "Renormalization group-improved effective potential for the MSSM Higgs sector with explicit CP violation," *Nucl. Phys.* **B586** (2000) 92–140, arXiv:hep-ph/0003180.
97. M. S. Carena, J. R. Ellis, A. Pilaftsis, and C. E. M. Wagner, "CP-violating MSSM Higgs bosons in the light of LEP 2," *Phys. Lett.* **B495** (2000) 155–163, arXiv:hep-ph/0009212.
98. F. Wilczek, "Decays of Heavy Vector Mesons Into Higgs Particles," *Phys. Rev. Lett.* **39** (1977) 1304.

99. H. M. Georgi, S. L. Glashow, M. E. Machacek, and D. V. Nanopoulos, "Higgs Bosons from Two Gluon Annihilation in Proton Proton Collisions," *Phys. Rev. Lett.* **40** (1978) 692.
100. S. L. Glashow, D. V. Nanopoulos, and A. Yildiz, "Associated Production of Higgs Bosons and $Z$ Particles," *Phys. Rev.* **D18** (1978) 1724–1727.
101. CDF Collaboration, "Search for neutral MSSM Higgs bosons decaying to tau pairs with 1.8 $\text{fb}^{-1}$ of data," CDF Note 9071.
102. D0 Collaboration, "Search for MSSM Higgs boson production in di-tau final states with $\mathcal{L} = 2.2$ $\text{fb}^{-1}$ at the DØ detector," D0 Note 5740-CONF.
103. T. Hahn, S. Heinemeyer, F. Maltoni, G. Weiglein, and S. Willenbrock, "SM and MSSM Higgs boson production cross sections at the Tevatron and the LHC," `arXiv:hep-ph/0607308`.
104. J. M. Campbell *et al.*, "Higgs boson production in association with bottom quarks," `arXiv:hep-ph/0405302`.
105. CDF Collaboration, "Search for Higgs bosons produced in association with $b$ quarks," CDF Note 9284.
106. D0 Collaboration, "Search for neutral Higgs bosons in multi-$b$-jet events in $p\bar{p}$ collisions at $\sqrt{s} = 1.96$ TeV," D0 Note 5726-CONF.
107. M. S. Carena, S. Mrenna, and C. E. M. Wagner, "MSSM Higgs boson phenomenology at the Tevatron collider," *Phys. Rev.* **D60** (1999) 075010, `arXiv:hep-ph/9808312`.
108. M. S. Carena, S. Mrenna, and C. E. M. Wagner, "The complementarity of LEP, the Tevatron and the LHC in the search for a light MSSM Higgs boson," *Phys. Rev.* **D62** (2000) 055008, `arXiv:hep-ph/9907422`.
109. D0 Collaboration, "A search for neutral Higgs bosons at high $\tan\beta$ in the mode $\phi b \to \tau_\mu \tau_h b$ in Run IIb data," D0 Note 5727-CONF.
110. J. F. Gunion, H. E. Haber, G. L. Kane, and S. Dawson, "The Higgs Hunter's Guide," Redwood City, USA: Addison-Wesley (1990) 425 p.
111. J. F. Gunion, H. E. Haber, G. L. Kane, and S. Dawson, "Errata for the Higgs hunter's guide," `arXiv:hep-ph/9302272`.
112. L. N. Chang and J. E. Kim, "Possible signature of charged Higgs particles in high-energy $e^+e^-$ annihilation," *Phys. Lett.* **B81** (1979) 233.
113. A. G. Akeroyd, "Three-body decays of Higgs bosons at LEP2 and application to a hidden fermiophobic Higgs," *Nucl. Phys.* **B544** (1999) 557–575, `arXiv:hep-ph/9806337`.
114. **DELPHI** Collaboration, J. Abdallah *et al.*, "Search for charged Higgs bosons at LEP in general two Higgs doublet models," *Eur. Phys. J.* **C34** (2004) 399–418, `arXiv:hep-ex/0404012`.
115. **OPAL** Collaboration, G. Abbiendi *et al.*, "Search for Charged Higgs Bosons in $e^+e^-$ Collisions at $\sqrt{s}$ = 189-209 GeV," `arXiv:0812.0267 [hep-ex]`.
116. **ALEPH** Collaboration, A. Heister *et al.*, "Search for charged Higgs bosons in $e^+e^-$ collisions at energies up to $\sqrt{s} = 209$ GeV," *Phys. Lett.* **B543** (2002) 1–13, `arXiv:hep-ex/0207054`.
117. **L3** Collaboration, P. Achard *et al.*, "Search for charged Higgs bosons at LEP," *Phys. Lett.* **B575** (2003) 208–220, `arXiv:hep-ex/0309056`.
118. **ALEPH, DELPHI, L3, OPAL** Collaborations, LEP Higgs Working

Group, "Search for charged Higgs bosons: Preliminary combined results using LEP data collected at energies up to 209 GeV," arXiv:hep-ex/0107031.
119. J. F. Gunion, H. E. Haber, F. E. Paige, W.-K. Tung, and S. S. D. Willenbrock, "Neutral and Charged Higgs Detection: Heavy Quark Fusion, Top Quark Mass Dependence and Rare Decays," *Nucl. Phys.* **B294** (1987) 621.
120. D0 Collaboration, "A search for charged Higgs bosons in $t\bar{t}$ events," D0 Note 5715-CONF.
121. J. S. Lee *et al.*, "CPsuperH: A computational tool for Higgs phenomenology in the minimal supersymmetric standard model with explicit CP violation," *Comput. Phys. Commun.* **156** (2004) 283–317, arXiv:hep-ph/0307377.
122. Y. Grossman, "Phenomenology of models with more than two Higgs doublets," *Nucl. Phys.* **B426** (1994) 355–384, arXiv:hep-ph/9401311.
123. **CDF** Collaboration, A. Abulencia *et al.*, "Search for charged Higgs bosons from top quark decays in $p\bar{p}$ collisions at $\sqrt{s} = 1.96$ TeV," *Phys. Rev. Lett.* **96** (2006) 042003, arXiv:hep-ex/0510065.
124. CDF Collaboration, "A search for charged Higgs in lepton + jets $t\bar{t}$ events using 2.2 fb$^{-1}$ of CDF data," CDF Note 9322.
125. **D0** Collaboration, V. M. Abazov *et al.*, "Search for charged Higgs bosons decaying to top and bottom quarks in $p\bar{p}$ collisions," *Phys. Rev. Lett.* **102** (2009) 191802, arXiv:0807.0859 [hep-ex].
126. V. D. Barger, G. F. Giudice, and T. Han, "Some New Aspects of Supersymmetry R-Parity Violating Interactions," *Phys. Rev.* **D40** (1989) 2987.
127. G. R. Farrar and P. Fayet, "Searching for the Spin 0 Leptons of Supersymmetry," *Phys. Lett.* **B89** (1980) 191.
128. A. Bartl, H. Fraas, and W. Majerotto, "Gaugino-Higgsino mixing in selectron and sneutrino pair production," *Z. Phys.* **C34** (1987) 411.
129. V. D. Barger, R. W. Robinett, W. Y. Keung, and R. J. N. Phillips, "Production of gauge fermions at colliders," *Phys. Lett.* **B131** (1983) 372.
130. A. Bartl, H. Fraas, and W. Majerotto, "Signatures of Chargino Production in $e^+e^-$ Collisions," *Z. Phys.* **C30** (1986) 441.
131. J. R. Ellis, J. M. Frere, J. S. Hagelin, G. L. Kane, and S. T. Petcov, "Search for neutral gauge fermions in $e^+e^-$ annihilation," *Phys. Lett.* **B132** (1983) 436.
132. D. A. Dicus, S. Nandi, W. W. Repko, and X. Tata, "Electron - positron annihilation cross-section in SU$(2)_L$ x U(1) supergravity," *Phys. Rev. Lett.* **51** (1983) 1030.
133. A. Bartl, H. Fraas, and W. Majerotto, "Production and Decay of Neutralinos in $e^+e^-$ Annihilation," *Nucl. Phys.* **B278** (1986) 1.
134. F. Almeida and D. H. Schiller, "Two jet distributions and energy-energy correlation for spin 0 quarks in $e^+e^-$ annihilation," *Z. Phys.* **C21** (1983) 103.
135. A. Bartl, H. Eberl, S. Kraml, W. Majerotto, and W. Porod, "Production of stop, sbottom, and stau at LEP2," *Z. Phys.* **C73** (1997) 469–476, arXiv:hep-ph/9603410.
136. **ALEPH, DELPHI, L3, OPAL** Collaborations, LEP SUSY Working Group, "Sleptons," Note LEPSUSYWG/04-01.1. http://lepsusy.web.cern.ch/lepsusy/Welcome.html.

137. **ALEPH** Collaboration, A. Heister *et al.*, "Absolute lower limits on the masses of selectrons and sneutrinos in the MSSM," *Phys. Lett.* **B544** (2002) 73–88, arXiv:hep-ex/0207056.
138. **L3** Collaboration, P. Achard *et al.*, "Search for scalar leptons and scalar quarks at LEP," *Phys. Lett.* **B580** (2004) 37–49, arXiv:hep-ex/0310007.
139. **ALEPH, DELPHI, L3, OPAL** Collaborations, LEP SUSY Working Group, "Charginos, Large $m_0$," Note LEPSUSYWG/01-03.1. http://lepsusy.web.cern.ch/lepsusy/Welcome.html.
140. **ALEPH, DELPHI, L3, OPAL** Collaborations, LEP SUSY Working Group, "Charginos, at small $\Delta m$," Note LEPSUSYWG/02-04.1. http://lepsusy.web.cern.ch/lepsusy/Welcome.html.
141. **ALEPH** Collaboration, R. Barate *et al.*, "Search for charginos and neutralinos in $e^+e^-$ collisions at center-of-mass energies near 183 GeV and constraints on the MSSM parameter space," *Eur. Phys. J.* **C11** (1999) 193–216.
142. **ALEPH, DELPHI, L3, OPAL** Collaborations, LEP SUSY Working Group, "Interpretation in Constrained MSSM and limit on the LSP mass," Note LEPSUSYWG/04-07.1. http://lepsusy.web.cern.ch/lepsusy/Welcome.html.
143. **ALEPH, DELPHI, L3, OPAL** Collaborations, LEP SUSY Working Group, "LSP mass limit in Minimal SUGRA," Note LEPSUSYWG/02-06.2. http://lepsusy.web.cern.ch/lepsusy/Welcome.html.
144. K.-i. Hikasa and M. Kobayashi, "Light Scalar Top at $e^+e^-$ Colliders," *Phys. Rev.* **D36** (1987) 724.
145. **ALEPH, DELPHI, L3, OPAL** Collaborations, LEP SUSY Working Group, "Stops and Sbottoms," Note LEPSUSYWG/04-02.1. http://lepsusy.web.cern.ch/lepsusy/Welcome.html.
146. **ALEPH** Collaboration, A. Heister *et al.*, "Search for scalar quarks in $e^+e^-$ collisions at $\sqrt{s}$ up to 209 GeV," *Phys. Lett.* **B537** (2002) 5–20, arXiv:hep-ex/0204036.
147. W. Beenakker, R. Hopker, M. Spira, and P. M. Zerwas, "Squark and gluino production at hadron colliders," *Nucl. Phys.* **B492** (1997) 51–103, arXiv:hep-ph/9610490.
148. J. Pumplin *et al.*, "New generation of parton distributions with uncertainties from global QCD analysis," *JHEP* **07** (2002) 012, arXiv:hep-ph/0201195.
149. D. Stump *et al.*, "Inclusive jet production, parton distributions, and the search for new physics," *JHEP* **10** (2003) 046, arXiv:hep-ph/0303013.
150. T. Sjostrand, S. Mrenna, and P. Skands, "PYTHIA 6.4 Physics and Manual," *JHEP* **05** (2006) 026, arXiv:hep-ph/0603175.
151. M. Muhlleitner, A. Djouadi, and Y. Mambrini, "SDECAY: A Fortran code for the decays of the supersymmetric particles in the MSSM," *Comput. Phys. Commun.* **168** (2005) 46–70, arXiv:hep-ph/0311167.
152. M. L. Mangano, M. Moretti, F. Piccinini, R. Pittau, and A. D. Polosa, "ALPGEN, a generator for hard multiparton processes in hadronic collisions," *JHEP* **07** (2003) 001, arXiv:hep-ph/0206293.

153. I. Hinchliffe and L. Littenberg, "Phenomenological Consequences of Supersymmetry," in Proc. of DPF Summer Study on Elementary Particle Physics and Future Facilities, Aspen, Colo., Jun 28 - Jul 16, 1982.
154. G. L. Kane and J. P. Leveille, "Experimental Constraints on Gluino Masses and Supersymmetric Theories," *Phys. Lett.* **B112** (1982) 227.
155. J. R. Ellis and H. Kowalski, "Supersymmetric particles at the cern $p\bar{p}$ collider," *Nucl. Phys.* **B246** (1984) 189.
156. H. Baer, K. Hagiwara, and X. Tata, "Gauginos as a Signal for Supersymmetry at $p\bar{p}$ Colliders," *Phys. Rev.* **D35** (1987) 1598.
157. P. Nath and R. L. Arnowitt, "Supersymmetric Signals at the Tevatron," *Mod. Phys. Lett.* **A2** (1987) 331–341.
158. R. Barbieri, F. Caravaglios, M. Frigeni, and M. L. Mangano, "Production and leptonic decays of charginos and neutralinos in hadronic collisions," *Nucl. Phys.* **B367** (1991) 28–59.
159. R. M. Barnett, J. F. Gunion, and H. E. Haber, "Gluino decay patterns and signatures," *Phys. Rev.* **D37** (1988) 1892.
160. H. Baer, X. Tata, and J. Woodside, "Gluino cascade decay signatures at the tevatron collider," *Phys. Rev.* **D41** (1990) 906.
161. **D0** Collaboration, V. M. Abazov *et al.*, "Search for squarks and gluinos in events with jets and missing transverse energy using 2.1 fb$^{-1}$ of $p\bar{p}$ collision data at $\sqrt{s}$ = 1.96 TeV," *Phys. Lett.* **B660** (2008) 449–457, `arXiv:0712.3805 [hep-ex]`.
162. **CDF** Collaboration, T. Aaltonen *et al.*, "Inclusive Search for Squark and Gluino Production in $p\bar{p}$ Collisions at $\sqrt{s}$ = 1.96 TeV," *Phys. Rev. Lett.* **102** (2009) 121801, `arXiv:0811.2512 [hep-ex]`.
163. **D0** Collaboration, V. M. Abazov *et al.*, "Search for scalar top quarks in the acoplanar charm jets and missing transverse energy final state in $p\bar{p}$ collisions at $\sqrt{s}$ = 1.96 TeV," *Phys. Lett.* **B665** (2008) 1–8, `arXiv:0803.2263 [hep-ex]`.
164. **D0** Collaboration, V. M. Abazov *et al.*, "Search for pair production of scalar bottom quarks in $p\bar{p}$ collisions at $\sqrt{s}$ = 1.96 TeV," *Phys. Rev. Lett.* **97** (2006) 171806, `arXiv:hep-ex/0608013`.
165. **CDF** Collaboration, T. Aaltonen *et al.*, "Search for Direct Pair Production of Supersymmetric Top and Supersymmetric Bottom Quarks in $p\bar{p}$ Collisions at $\sqrt{s}$ = 1.96 TeV," *Phys. Rev.* **D76** (2007) 072010, `arXiv:0707.2567 [hep-ex]`.
166. **D0** Collaboration, V. M. Abazov *et al.*, "Search for the lightest scalar top quark in events with two leptons in $p\bar{p}$ collisions at $\sqrt{s}$ = 1.96 TeV," *Phys. Lett.* **B659** (2008) 500–508, `arXiv:0707.2864 [hep-ex]`.
167. CDF Collaboration, "Search for pair production of stop quarks mimicking top events signatures," CDF Note 9439.
168. **CDF** Collaboration, T. Aaltonen *et al.*, "Search for Gluino-Mediated Sbottom Production in $p\bar{p}$ Collisions at $\sqrt{s}$ = 1.96 TeV," *Phys. Rev. Lett.* **102** (2009) 221801, `arXiv:0903.2618 [hep-ex]`.
169. **CDF** Collaboration, T. Aaltonen *et al.*, "Search for Supersymmetry in $p\bar{p}$ Collisions at $\sqrt{s}$ = 1.96 TeV Using the Trilepton Signature of

Chargino-Neutralino Production," *Phys. Rev. Lett.* **101** (2008) 251801, arXiv:0808.2446 [hep-ex].
170. **D0** Collaboration, V. M. Abazov *et al.*, "Search for associated production of charginos and neutralinos in the trilepton final state using 2.3 $\mathrm{fb}^{-1}$ of data," arXiv:0901.0646 [hep-ex].
171. **D0** Collaboration, V. M. Abazov *et al.*, "Search for supersymmetry via associated production of charginos and neutralinos in final states with three leptons," *Phys. Rev. Lett.* **95** (2005) 151805, arXiv:hep-ex/0504032.
172. **CDF** Collaboration, T. Aaltonen *et al.*, "Search for chargino-neutralino production in $p\bar{p}$ collisions at $\sqrt{s} = 1.96$ TeV," *Phys. Rev. Lett.* **99** (2007) 191806, arXiv:0707.2362 [hep-ex].
173. **ALEPH, DELPHI, L3, OPAL** Collaborations, LEP SUSY Working Group, "Rp Violation with LLE Couplings," Note LEPSUSYWG/02-10.1. http://lepsusy.web.cern.ch/lepsusy/Welcome.html.
174. **ALEPH** Collaboration, A. Heister *et al.*, "Search for supersymmetric particles with R parity violating decays in $e^+e^-$ collisions at $\sqrt{s}$ up to 209 GeV," *Eur. Phys. J.* **C31** (2003) 1–16, arXiv:hep-ex/0210014.
175. **DELPHI** Collaboration, J. Abdallah *et al.*, "Search for supersymmetric particles assuming R-parity non- conservation in $e^+e^-$ collisions at $\sqrt{s} = 192$ GeV to 208 GeV," *Eur. Phys. J.* **C36** (2004) 1–23, arXiv:hep-ex/0406009.
176. **L3** Collaboration, P. Achard *et al.*, "Search for R parity violating decays of supersymmetric particles in $e^+e^-$ collisions at LEP," *Phys. Lett.* **B524** (2002) 65–80, arXiv:hep-ex/0110057.
177. **OPAL** Collaboration, G. Abbiendi *et al.*, "Search for R parity violating decays of scalar fermions at LEP," *Eur. Phys. J.* **C33** (2004) 149–172, arXiv:hep-ex/0310054.
178. **ALEPH** Collaboration, R. Barate *et al.*, "Search for R-parity violating decays of supersymmetric particles in $e^+e^-$ collisions at center-of-mass energies from 189 GeV to 202 GeV," *Eur. Phys. J.* **C19** (2001) 415–428, arXiv:hep-ex/0011008.
179. **ALEPH** Collaboration, A. Heister *et al.*, "Search for R-parity violating production of single sneutrinos in $e^+e^-$ collisions at $\sqrt{s} = 189$ GeV to 209 GeV," *Eur. Phys. J.* **C25** (2002) 1–12, arXiv:hep-ex/0201013.
180. **DELPHI** Collaboration, J. Abdallah *et al.*, "Search for resonant sneutrino production at $\sqrt{s} = 183$ GeV to 208 GeV," *Eur. Phys. J.* **C28** (2003) 15–26, arXiv:hep-ex/0303033.
181. **L3** Collaboration, M. Acciarri *et al.*, "Search for R-parity breaking sneutrino exchange at LEP," *Phys. Lett.* **B414** (1997) 373–381.
182. **OPAL** Collaboration, G. Abbiendi *et al.*, "Tests of the standard model and constraints on new physics from measurements of fermion pair production at 189 GeV at LEP," *Eur. Phys. J.* **C13** (2000) 553–572, arXiv:hep-ex/9908008.
183. **H1** Collaboration, A. Aktas *et al.*, "Search for bosonic stop decays in R-parity violating supersymmetry in e+ p collisions at HERA," *Phys. Lett.* **B599** (2004) 159–172, arXiv:hep-ex/0405070.

184. **H1** Collaboration, A. Aktas *et al.*, "Search for squark production in R parity violating supersymmetry at HERA," *Eur. Phys. J.* **C36** (2004) 425–440, arXiv:hep-ex/0403027.
185. **ZEUS** Collaboration, S. Chekanov *et al.*, "Search for stop production in R-parity-violating supersymmetry at HERA," *Eur. Phys. J.* **C50** (2007) 269–281, arXiv:hep-ex/0611018.
186. **CDF** Collaboration, A. Abulencia *et al.*, "Search for anomalous production of multi-lepton events in $p\bar{p}$ collisions at $\sqrt{s} = 1.96$ TeV," *Phys. Rev. Lett.* **98** (2007) 131804, arXiv:0706.4448 [hep-ex].
187. **D0** Collaboration, V. M. Abazov *et al.*, "Search for R-parity violating supersymmetry via the LL anti-E couplings $\lambda_{121}$, $\lambda_{122}$ or $\lambda_{133}$ in $p\bar{p}$ collisions at $\sqrt{s} = 1.96$ TeV," *Phys. Lett.* **B638** (2006) 441–449, arXiv:hep-ex/0605005.
188. **CDF** Collaboration, T. Aaltonen *et al.*, "Search for Pair Production of Scalar Top Quarks Decaying to a $\tau$ Lepton and a $b$ Quark in $p\bar{p}$ Collisions at $\sqrt{s}$ =1.96 TeV," *Phys. Rev. Lett.* **101** (2008) 071802, arXiv:0802.3887 [hep-ex].
189. **D0** Collaboration, V. M. Abazov *et al.*, "Search for resonant second generation slepton production at the Tevatron," *Phys. Rev. Lett.* **97** (2006) 111801, arXiv:hep-ex/0605010.
190. **CDF** Collaboration, A. Abulencia *et al.*, "Search for new high mass particles decaying to lepton pairs in $p\bar{p}$ collisions at $\sqrt{s} = 1.96$ TeV," *Phys. Rev. Lett.* **95** (2005) 252001, arXiv:hep-ex/0507104.
191. **CDF** Collaboration, D. E. Acosta *et al.*, "Search for new physics using high mass tau pairs from 1.96 TeV $p\bar{p}$ collisions," *Phys. Rev. Lett.* **95** (2005) 131801, arXiv:hep-ex/0506034.
192. **CDF** Collaboration, A. Abulencia *et al.*, "Search for high-mass resonances decaying to e $\mu$ in $p\bar{p}$ collisions at $\sqrt{s} = 1.96$ TeV," *Phys. Rev. Lett.* **96** (2006) 211802, arXiv:hep-ex/0603006.
193. **D0** Collaboration, V. M. Abazov *et al.*, "Search for Scalar Neutrino Superpartners in $e + \mu$ Final States in $p\bar{p}$ Collisions at $\sqrt{s} = 1.96$ TeV," *Phys. Rev. Lett.* **100** (2008) 241803, arXiv:0711.3207 [hep-ex].
194. **ALEPH, DELPHI, L3, OPAL** Collaborations, LEP SUSY Working Group, "Acoplanar Two-Photon Events," Note LEPSUSYWG/04-09.1. http://lepsusy.web.cern.ch/lepsusy/Welcome.html.
195. **CDF** Collaboration, F. Abe *et al.*, "Searches for new physics in diphoton events in $p\bar{p}$ collisions at $\sqrt{s} = 1.8$ TeV," *Phys. Rev.* **D59** (1999) 092002, arXiv:hep-ex/9806034.
196. **ALEPH** Collaboration, A. Heister *et al.*, "Search for gauge mediated SUSY breaking topologies in $e^+e^-$ collisions at center-of-mass energies up to 209 GeV," *Eur. Phys. J.* **C25** (2002) 339–351, arXiv:hep-ex/0203024.
197. **OPAL** Collaboration, G. Abbiendi *et al.*, "Searches for gauge-mediated supersymmetry breaking topologies in $e^+e^-$ collisions at LEP2," *Eur. Phys. J.* **C46** (2006) 307–341, arXiv:hep-ex/0507048.
198. CDF Collaboration, "Limits on Gauge-Mediated Supersymmetry-Breaking Models using Diphoton Events with Missing Transverse Energy at CDF II," CDF Note 9625.

199. **D0** Collaboration, V. M. Abazov *et al.*, "Search for supersymmetry in diphoton final states at $\sqrt{s} = 1.96$ TeV," *Phys. Lett.* **B659** (2008) 856–863, arXiv:0710.3946 [hep-ex].
200. B. C. Allanach *et al.*, "The Snowmass points and slopes: Benchmarks for SUSY searches," *Eur. Phys. J.* **C25** (2002) 113–123, arXiv:hep-ph/0202233.
201. **CDF** Collaboration, A. Abulencia *et al.*, "Search for heavy, long-lived particles that decay to photons at CDF II," *Phys. Rev. Lett.* **99** (2007) 121801, arXiv:0704.0760 [hep-ex].
202. **ALEPH, DELPHI, L3, OPAL** Collaborations, LEP SUSY Working Group, "GMSB with Sleptons as NLSP," Note LEPSUSYWG/02-09.2. http://lepsusy.web.cern.ch/lepsusy/Welcome.html.
203. **D0** Collaboration, V. M. Abazov *et al.*, "Search for Long-Lived Charged Massive Particles with the D0 Detector," *Phys. Rev. Lett.* **102** (2009) 161802, arXiv:0809.4472 [hep-ex].
204. **CDF** Collaboration, T. Aaltonen *et al.*, "Search for Long-Lived Massive Charged Particles in 1.96 TeV $\bar{p}p$ Collisions," arXiv:0902.1266 [hep-ex].
205. S. Raby, "Gauge-mediated SUSY breaking with a gluino LSP," *Phys. Lett.* **B422** (1998) 158–162, arXiv:hep-ph/9712254.
206. H. Baer, K.-m. Cheung, and J. F. Gunion, "A Heavy gluino as the lightest supersymmetric particle," *Phys. Rev.* **D59** (1999) 075002, arXiv:hep-ph/9806361.
207. S. Raby and K. Tobe, "The phenomenology of SUSY models with a gluino LSP," *Nucl. Phys.* **B539** (1999) 3–22, arXiv:hep-ph/9807281.
208. N. Arkani-Hamed and S. Dimopoulos, "Supersymmetric unification without low energy supersymmetry and signatures for fine-tuning at the LHC," *JHEP* **06** (2005) 073, arXiv:hep-th/0405159.
209. G. F. Giudice and A. Romanino, "Split supersymmetry," *Nucl. Phys.* **B699** (2004) 65–89, arXiv:hep-ph/0406088.
210. G. F. Giudice and A. Romanino, "Split supersymmetry: Erratum," *Nucl. Phys.* **B706** (2005) 65.
211. **ALEPH** Collaboration, A. Heister *et al.*, "Search for stable hadronizing squarks and gluinos in $e^+e^-$ collisions up to $\sqrt{s} = 209$ GeV," *Eur. Phys. J.* **C31** (2003) 327–342, arXiv:hep-ex/0305071.
212. **DELPHI** Collaboration, J. Abdallah *et al.*, "Search for an LSP gluino at LEP with the DELPHI detector," *Eur. Phys. J.* **C26** (2003) 505–525, arXiv:hep-ex/0303024.
213. **D0** Collaboration, V. M. Abazov *et al.*, "Search for stopped gluinos from $p\bar{p}$ collisions at $\sqrt{s} = 1.96$ TeV," *Phys. Rev. Lett.* **99** (2007) 131801, arXiv:0705.0306 [hep-ex].
214. A. Arvanitaki, S. Dimopoulos, A. Pierce, S. Rajendran, and J. G. Wacker, "Stopping gluinos," *Phys. Rev.* **D76** (2007) 055007, arXiv:hep-ph/0506242.

# Low-Energy Supersymmetry at Future Colliders

John F. Gunion* and Howard E. Haber†

*Department of Physics,
University of California, Davis, CA 95616

†Santa Cruz Institute for Particle Physics,
University of California, Santa Cruz, CA 95064

We classify the variety of low-energy supersymmetric signatures that can be probed at future colliders. We focus on phenomena associated with the minimal supersymmetric extension of the Standard Model. The structure of the supersymmetry-breaking introduces additional model assumptions. The approaches considered here are supergravity-mediated and gauge-mediated supersymmetry-breaking. Alternative phenomenologies arising in non-minimal and/or R-parity-violating approaches are also briefly examined.

## 10.1. Introduction

In this chapter, we focus on the signatures for low-energy supersymmetry at future colliders in the context of the minimal supersymmetric extension of the Standard Model (MSSM).[1,2] In its most general form (with the assumption of R-parity conservation), the MSSM is a 124-parameter theory;[3,4] most of the parameter freedom is associated with the supersymmetry-breaking sector of the model.[a] This huge parameter space can be reduced by: (i) imposing phenomenological constraints, and (ii) imposing theoretical assumptions on the structure of supersymmetry-breaking. In addition, the scale of supersymmetry-breaking, $\sqrt{F}$, must be specified. It determines the properties of the gravitino, $\widetilde{g}_{3/2}$. In previous chapters, two broad model categories for supersymmetry-breaking were discussed, gravity-mediated

[a]The notation for the supersymmetric parameters used in this paper for the most part follows that of Ref. 2. The notation for supersymmetric particle names follows that of Ref. 1.

supersymmetry breaking (SUGRA) and gauge-mediated supersymmetry breaking (GMSB).

In most SUGRA models,[5] $\sqrt{F}$ is so large that the $\widetilde{g}_{3/2}$ interactions are too weak for it to play any role in collider phenomenology. In the *minimal* supergravity (mSUGRA) framework, the soft-supersymmetry-breaking parameters at the Planck scale take a particularly simple form and depend on essentially five new parameters. These include $m_0$ (a flavor universal soft supersymmetry-breaking scalar mass), $m_{1/2}$ (a universal gaugino mass), and $A_0$ (a flavor universal tri-linear scalar interaction). In particular, gaugino mass unification implies that at the unification scale ($M_X$), the U(1), SU(2) and SU(3) gaugino Majorana mass parameters are equal, *i.e.*, $M_1(M_X) = M_2(M_X) = M_3(M_X) = m_{1/2}$. This implies that the *low-energy* gaugino mass parameters satisfy:

$$M_3 = \frac{g_3^2}{g_2^2} M_2 \simeq 3.5 M_2 , \qquad M_1 = \tfrac{5}{3} \tan^2\theta_W M_2 \simeq 0.5 M_2 . \tag{10.1}$$

The other two mSUGRA parameters are the supersymmetric Higgs mass parameter $\mu$ and an off-diagonal soft Higgs squared-mass. After the imposition of electroweak symmetry breaking, these two parameters can be traded in for the two Higgs vacuum expectation values (modulo a sign ambiguity in $\mu$). By fixing the $Z$ mass, the remaining mSUGRA parameters are determined by the ratio of Higgs vacuum expectation values ($\tan\beta$) and the sign of $\mu$. The lightest supersymmetric particle (LSP) is nearly always the lightest neutralino (denoted in this chapter by $\widetilde{\chi}_1^0$). Non-minimal extensions of the mSUGRA model have also been considered in which some of the parameter universality assumptions have been relaxed.

In GMSB models,[6] $\sqrt{F}$ is sufficiently small that the $\widetilde{g}_{3/2}$ is almost always the lightest supersymmetric particle (LSP) and plays a prominent phenomenological role. Then, different choices for the next-to-lightest supersymmetric particle (NLSP) lead to different phenomenologies. In the simplest GMSB models, the gaugino and scalar soft-supersymmetry-breaking masses are given by SU(3), SU(2) and U(1) gauge group factors times an overall scale $\Lambda$, while the $A$ parameters are expected to be negligible. [The low-energy values of the gaugino mass parameters also satisfy Eq. (10.1).] The parameter set is then completed by $\tan\beta$ and sign($\mu$).

Finally, one can also consider alternative low-energy supersymmetric approaches. For example, if R-parity violation (RPV) is present additional supersymmetric parameters are introduced. These include parameters $\lambda_L$, $\lambda_L'$ and $\lambda_B$ which govern new lepton and baryon number violating scalar-fermion Yukawa couplings derived from the following supersymmetric interactions:

$$(\lambda_L)_{pmn}\widehat{L}_p\widehat{L}_m\widehat{E}^c_n + (\lambda'_L)_{pmn}\widehat{L}_p\widehat{Q}_m\widehat{D}^c_n + (\lambda_B)_{pmn}\widehat{U}^c_p\widehat{D}^c_m\widehat{D}^c_n\,, \tag{10.2}$$

where $p$, $m$, and $n$ are generation indices, and gauge group indices are suppressed. In the notation above, the "superfields" $\widehat{Q}$, $\widehat{U}^c$, $\widehat{D}^c$, $\widehat{L}$, and $\widehat{E}^c$ respectively represent $(u,d)_L$, $u^c_L$, $d^c_L$, $(\nu, e^-)_L$, and $e^c_L$ and the corresponding superpartners. The Yukawa interactions are obtained from Eq. (10.2) by taking all possible combinations involving two fermions and one scalar superpartner. A comprehensive review of R-parity-violating supersymmetry can be found in Refs. 7 and 8.

The Higgs sector of the MSSM is also a very rich source of collider phenomenology that is intimately connected with the supersymmetric structure of the theory. For example, in the MSSM, the tree-level mass of the lightest CP-even neutral Higgs boson ($h^0$) must be less than $m_Z$.[9] This upper bound is significantly modified by radiative corrections involving loops of virtual supersymmetric particles.[10] The most complete computations,[11] which incorporate renormalization group improvement and the leading two-loop contributions, yield $m_{h^0} \lesssim 135$ GeV (with an accuracy of a few GeV) for $m_t = 175$ GeV (assuming supersymmetric masses no larger than a few TeV). In this review, we focus on the phenomenological consequence of supersymmetric particle production. For a comprehensive treatment of the collider phenomenology of Higgs bosons in supersymmetric theories, see Refs. 12 and 13.

## 10.2. Classes of Supersymmetric Signals

The lack of knowledge of the origin and structure of the supersymmetry-breaking parameters implies that the predictions for low-energy supersymmetry and the consequent phenomenology depend on a plethora of unknown parameters. Nevertheless, we can broadly classify supersymmetric signals at future colliders by considering a variety of theoretical approaches. In this section, we delineate the possible supersymmetric signatures, and in the next section we explore their consequences for experimentation at future colliders.

### 10.2.1. *Missing energy signatures*

In R-parity-conserving low-energy supersymmetry, supersymmetric particles are produced in pairs. The subsequent decay of a heavy supersymmetric particle generally proceeds via a multistep decay chain,[14–16] ending in the production of at least one supersymmetric particle that (in conventional models) is weakly interacting and escapes the collider detector. Thus, supersymmetric particle production yields events that contain at least two escaping non-interacting particles, leading to a missing energy signature. At hadron colliders, it is only possible to detect missing transverse energy ($E_T^{\rm miss}$), since the center-of-mass energy of the hard collision is not known on an event-by-event basis.

In conventional SUGRA-based models, the weakly-interacting LSP's that escape the collider detector (which yields large missing transverse energy) are accompanied by energetic jets and/or leptons. This is the "smoking-gun" signature of low-energy supersymmetry. However, there are two unconventional approaches in which the smoking-gun signature is absent. First, consider a model in which the $\widetilde{\chi}_1^0$ is the LSP but the lightest neutralino and chargino are nearly degenerate in mass. If the mass difference is $\lesssim$ 100 MeV, then $\widetilde{\chi}_1^+$ is long-lived and decays outside the detector.[17,18] In this case, some supersymmetric events would yield *no* missing energy and two semi-stable charged particles that pass through the detector. Second, there are models in which a gluino (more precisely, the $R^0 = \widetilde{g}g$ bound state) is the LSP.[b] A massive $R^0$ is likely to simply pass through the detector without depositing significant energy. Even when it is light enough to be stopped, the hadronic calorimeter will measure only the kinetic energy of the $R^0$. In either case, there would be substantial missing energy.[21] However, there would be no jets arising from $\widetilde{g}$ decays in such models.

In conventional GMSB models with a gravitino-LSP,[c] all supersymmetric events contain at least two NLSP's, and the resulting signature depends on the NLSP properties. Four physically distinct possible scenarios emerge:

- The NLSP is electrically and color neutral and long-lived, and decays outside of the detector to its associated Standard Model partner and the gravitino.
- The NLSP is the sneutrino and decays invisibly into $\nu\widetilde{g}_{3/2}$ either inside or outside the detector.

[b]Farrar has advocated the existence of a very light gluino with a mass less than a few GeV.[19] Recent experimental data[20] show no evidence for a such a light gluino, although the assertion that light gluinos are definitively ruled out is still in dispute. The possibility of a more massive LSP gluino in SUGRA-based models has been considered in Ref. 18.

[c]It is also possible to construct a GMSB scenario in which the $\widetilde{g}$ is the LSP.[22] The resulting phenomenology corresponds to that of the massive gluino LSP discussed above.

In either of these two cases, the resulting missing-energy signal is similar to that of the SUGRA-based models where $\widetilde{\chi}_1^0$ or $\widetilde{\nu}$ is the LSP.

- The NLSP is the $\widetilde{\chi}_1^0$ and decays inside the detector to $N\widetilde{g}_{3/2}$, where $N = \gamma$, $Z$ or a neutral Higgs boson.

In this case, the gravitino-LSP behaves like the neutralino or sneutrino LSP of the SUGRA-based models. However, in contrast to SUGRA-based models, the missing energy events of the GMSB-based model are characterized by the associated production of (at least) two $N$'s, one for each NLSP.[d] Note that if $\widetilde{\chi}_1^0$ is lighter than the $Z$ and $h^0$ then $\mathrm{BR}(\widetilde{\chi}_1^0 \to \gamma\widetilde{g}_{3/2}) = 100\%$, and all supersymmetric production will result in missing energy events with at least two associated photons.

- The NLSP is a charged slepton (typically the $\widetilde{\tau}_R$ in GMSB models if $m_{\widetilde{\tau}_R} < m_{\widetilde{\chi}_1^0}$), which decays to the corresponding lepton and gravitino.

If the decay is prompt, then one finds missing energy events with associated leptons (taus). If the decay is not prompt, one observes a long-lived heavy semi-stable charged particle with *no* associated missing energy (prior to the decay of the NLSP).

There are also GMSB scenarios in which there are several nearly degenerate so-called co-NLSP's,[23] any one of which can be produced at the penultimate step of the supersymmetric decay chain.[e] The resulting supersymmetric signals would consist of events with two (or more) co-NLSP's, each one of which would decay according to one of the four scenarios delineated above. For additional details on the phenomenology of the co-NLSP's, see Ref. 23.

In R-parity violating SUGRA-based models the LSP is unstable. If the RPV-couplings are sufficiently weak, then the LSP will decay outside the detector, and the standard missing energy signal applies. If the LSP decays inside the detector, the phenomenology of RPV models depends on the identity of the LSP and the branching ratio of possible final state decay products. If the latter includes a neutrino, then the corresponding RPV supersymmetric events would result in missing energy (through neutrino

---

[d]If the decay of the NLSP is not prompt, it is possible to produce events in which one NLSP decays inside the detector and one NLSP decays outside of the detector.

[e]For example, if $\widetilde{\tau}_R^\pm$ and $\widetilde{\chi}_1^0$ are nearly degenerate in mass, then neither $\widetilde{\tau}_R^\pm \to \tau^\pm \widetilde{\chi}_1^0$ nor $\widetilde{\chi}_1^0 \to \widetilde{\tau}_R^\pm \tau^\mp$ are kinematically allowed decays. In this case, $\widetilde{\tau}_R^\pm$ and $\widetilde{\chi}_1^0$ are co-NLSP's, and each decays dominantly into its Standard Model superpartner plus a gravitino.

emission) in association with hadron jets and/or leptons. Other possibilities include decays into charged leptons in association with jets (with no neutrinos), and decays into purely hadronic final states. Clearly, these latter events would contain little missing energy. If R-parity violation is present in GMSB models, the RPV decays of the NLSP can easily dominate over the NLSP decay to the gravitino. In this case, the phenomenology of the NLSP resembles that of the LSP of SUGRA-based RPV models.

### 10.2.2. *Lepton ($e$, $\mu$ and $\tau$) signatures*

Once supersymmetric particles are produced at colliders, they do not necessarily decay to the LSP (or NLSP) in one step. The resulting decay chains can be complex, with a number of steps from the initial decay to the final state.[15] Along the way, decays can produce real or virtual $W$'s, $Z$'s, charginos, neutralinos and sleptons, which then can produce leptons in their subsequent decays. Thus, many models yield large numbers of supersymmetric events characterized by one or more leptons in association with missing energy, with or without hadronic jets.

One signature of particular note is events containing like-sign di-leptons.[24] The origin of such events is associated with the Majorana nature of the gaugino. For example, $\widetilde{g}\widetilde{g}$ production followed by $\widetilde{g} \to q\bar{q}\widetilde{\chi}_1^\pm \to q\bar{q}\ell^\pm\nu\widetilde{\chi}_1^0$ can result in like-sign leptons since the $\widetilde{g}$ decay leads with equal probability to either $\ell^+$ or $\ell^-$. If the masses and mass differences are both substantial (which is typical in mSUGRA models, for example), like-sign di-lepton events will be characterized by fairly energetic jets and isolated leptons and by large $E_T^{\rm miss}$ from the LSP's. Other like-sign di-lepton signatures can arise in a similar way from the decay chains initiated by the heavier neutralinos.

Distinctive tri-lepton signals[25] from $\widetilde{\chi}_1^\pm\widetilde{\chi}_2^0 \to (\ell^\pm\nu\widetilde{\chi}_1^0)(\ell^+\ell^-\widetilde{\chi}_1^0)$ can occur. Such events have little hadronic activity (apart from initial state radiation of jets off the annihilating quarks at hadron colliders). These events can have a variety of interesting characteristics depending on the fate of the final state neutralinos.

If the soft-supersymmetry breaking slepton masses are flavor universal at the high energy scale $M_X$ (as in mSUGRA models) and $\tan\beta \gg 1$, then the $\widetilde{\tau}_R$ will be significantly lighter than the other slepton states. As a result, supersymmetric decay chains involving (s)leptons will favor $\widetilde{\tau}_R$ production, leading to a predominance of events with multiple $\tau$-leptons in the final state.

In GMSB models with a charged slepton NLSP, the decay $\widetilde{\ell} \to \ell \widetilde{g}_{3/2}$ (if prompt) yields at least two leptons for every supersymmetric event in association with missing energy. In particular, in models with a $\widetilde{\tau}_R$ NLSP, supersymmetric events will characteristically contain at least two $\tau$'s.

In RPV models, decays of the LSP (in SUGRA models) or NLSP (in GMSB models) mediated by RPV-interactions proportional to $\lambda_L$ and $\lambda'_L$ will also yield supersymmetric events containing charged leptons. However, if the only significant RPV-interaction is the one proportional to $\lambda'_L$, then such events would *not* contain missing energy (in contrast to the GMSB signature described above).

### 10.2.3. *b-quark signatures*

The phenomenology of gluinos and squarks depends critically on their relative masses. If the gluino is heavier, it will decay dominantly into $q\widetilde{q}$,[f] while the squark can decay into quark plus chargino or neutralino. If the squark is heavier, it will decay dominantly into a quark plus gluino, while the gluino will decay into the three-body modes $q\bar{q}\widetilde{\chi}$ (where $\widetilde{\chi}$ can be either a neutralino or chargino, depending on the charge of the final state quarks). A number of special cases can arise when the possible mass splitting among squarks of different flavors is taken into account. For example, models of supersymmetric mass spectra have been considered where the third generation squarks are lighter than the squarks of the first two generations. If the gluino is lighter than the latter but heavier than the former, then the only open gluino two-body decay mode could be $b\widetilde{b}$.[g] In such a case, all $\widetilde{g}\widetilde{g}$ events will result in at least four $b$-quarks in the final state (in association with the usual missing energy signal, if appropriate). More generally, due to the flavor independence of the strong interactions, one expects three-body gluino decays into $b$-quarks in at least 20% of all gluino decays.[h] Additional $b$-quarks can arise from both top-quark and top-squark decays, and from neutral Higgs bosons produced somewhere in the chain decays.[26] Finally, at large $\tan\beta$, the enhanced Yukawa coupling to $b$-quarks can increase the

[f]In this section, we employ the notation $q\widetilde{q}$ to mean either $q\overline{\widetilde{q}}$ or $\overline{q}\widetilde{q}$.

[g]Although one top-squark mass-eigenstate ($\widetilde{t}_1$) is typically lighter than $\widetilde{b}$ in models, the heavy top-quark mass may result in a kinematically forbidden gluino decay mode into $t\widetilde{t}_1$.

[h]Here we assume the approximate degeneracy of the first two generations of squarks, as suggested from the absence of flavor-changing neutral-current decays. In many models, the $b$-squarks tend to be of similar mass or lighter than the squarks of the first two generations.

rate of $b$-quark production in neutralino and chargino decays occurring at some step in the gluino chain decay.

These observations suggest that many supersymmetric events at hadron colliders will be characterized by $b$-jets in association with missing energy.[16,27,28]

### 10.2.4. *Signatures involving photons*

In mSUGRA models, most supersymmetric events do not contain isolated energetic photons. However, some areas of low-energy supersymmetric parameter space do exist in which final state photons can arise in the decay chains of supersymmetric particles. If one relaxes the condition of gaugino mass unification, then the *low-energy* gaugino mass parameters no longer must satisfy Eq. (10.1). As a result, interesting alternative supersymmetric phenomenologies can arise. For example, if the low-energy mass parameters satisfy $M_1 \simeq M_2$, then the branching ratio for $\widetilde{\chi}_2^0 \to \widetilde{\chi}_1^0\gamma$ can be significant.[29] In the model of Ref. 30, the $\widetilde{\chi}_1^0$-LSP is dominantly higgsino, while $\widetilde{\chi}_2^0$ is dominantly gaugino. Thus, many supersymmetric decay chains end in the production of $\widetilde{\chi}_2^0$, which then decays to $\widetilde{\chi}_1^0\gamma$. In this picture, the pair production of supersymmetric particles often yields two photons plus associated missing energy. At LEP-2, one can also produce $\widetilde{\chi}_1^0\widetilde{\chi}_2^0$ which would then yield single photon events in association with large missing energy.

In GMSB models with a $\widetilde{\chi}_1^0$-NLSP, all supersymmetric decay chains would end up with the production of $\widetilde{\chi}_1^0$. Assuming that $\widetilde{\chi}_1^0$ decays inside the collider detector, one possible decay mode is $\widetilde{\chi}_1^0 \to \gamma\widetilde{g}_{3/2}$. In many models, the branching ratio for this radiative decay is significant (and could be as high as 100% if other possible two-body decay modes are not kinematically allowed). In the latter case, supersymmetric pair production would also yield events with two photons in associated with large missing energy. The characteristics of these events differ in detail from those of the corresponding events expected in the model of Ref. 30.

### 10.2.5. *Kinks and long-lived heavy particles*

In most SUGRA-based models, all supersymmetric particles in the decay chain decay promptly until the LSP is reached. The LSP is exactly stable and escapes the collider detector. However, exceptions are possible. In particular, if there is a supersymmetric particle that is just barely heavier than the LSP, then its (three-body) decay rate to the LSP will be significantly suppressed and it could be long lived. For example, in the models with

$|\mu| \gg M_1 > M_2$[17,18] implying $m_{\widetilde{\chi}_1^\pm} \simeq m_{\widetilde{\chi}_1^0}$, the $\widetilde{\chi}_1^\pm$ can be sufficiently long lived to yield a detectable vertex, or perhaps even exit the detector.

In GMSB models, the NLSP may be long-lived, depending on its mass and the scale of supersymmetry breaking, $\sqrt{F}$. The NLSP is unstable and eventually decays to the gravitino. For example, in the case of the $\widetilde{\chi}_1^0$-NLSP (which is dominated by its U(1)-gaugino component), one finds $\Gamma(\widetilde{\chi}_1^0 \to \gamma \widetilde{g}_{3/2}) = m_{\widetilde{\chi}_1^0}^5 \cos^2\theta_W / 16\pi F^2$. It then follows that

$$(c\tau)_{\widetilde{\chi}_1^0 \to \gamma \widetilde{g}_{3/2}} \simeq 130 \left(\frac{100 \text{ GeV}}{m_{\widetilde{\chi}_1^0}}\right)^5 \left(\frac{\sqrt{F}}{100 \text{ TeV}}\right)^4 \mu\text{m}\,. \tag{10.3}$$

For simplicity, assume that $\widetilde{\chi}_1^0 \to \gamma \widetilde{g}_{3/2}$ is the dominant NLSP decay mode. If $\sqrt{F} \sim 10^4$ TeV, then the decay length for the NLSP is $c\tau \sim 10$ km for $m_{\widetilde{\chi}_1^0} = 100$ GeV; while $\sqrt{F} \sim 100$ TeV implies a short but vertexable decay length. A similar result is obtained in the case of a charged NLSP. Thus, if $\sqrt{F}$ is sufficiently large, the charged NLSP will be semi-stable and may decay outside of the collider detector.

Finally, if R-parity violation is present, the decay rate of the LSP in SUGRA-based models (or the NLSP in R-parity-violating GMSB models) could be in the relevant range to yield visible secondary vertices.

## 10.3. Supersymmetry Searches at Future Colliders

In this section, we consider the potential for discovering low-energy supersymmetry at future colliders. A variety of supersymmetric signatures have been reviewed in Section 10.2, and we now apply these to supersymmetry searches at future colliders. Ideally, experimental studies of supersymmetry should be as model-independent as possible. Ultimately, the goal of experimental studies of supersymmetry is to measure as many of the 124 MSSM parameters (and any additional parameters that can arise in non-minimal extensions) as possible. In practice, a fully general analysis will be difficult, particularly during the initial supersymmetry discovery phase. Thus, we focus the discussion in this section on the expected phenomenology of supersymmetry at the various future facilities under a number of different model assumptions. Eventually, if candidates for supersymmetric phenomena are discovered, one would utilize precision experimental measurements to map out the supersymmetric parameter space and uncover the structure of the underlying supersymmetry-breaking.

### 10.3.1. *SUGRA-based models*

We begin with the phenomenology of mSUGRA. Of particular importance are the relative sizes of the different supersymmetric particle masses. Generic properties of the resulting superpartner mass spectrum are discussed in Ref. 2. An important consequence of the mSUGRA mass spectrum is that substantial phase space is available for most decays occurring at each step in a given chain decay of a heavy supersymmetric particle.

Extensive Monte Carlo studies have examined the region of mSUGRA parameter space for which direct discovery of supersymmetric particles at the Tevatron and the LHC will be possible.[31] At the hadron colliders, the ultimate supersymmetric mass reach is determined by the searches for both the strongly-interacting superpartners (squarks and gluinos) and the charginos/neutralinos. Cascade decays of the produced squarks and gluinos lead to events with jets, missing energy, and various numbers of leptons. Pair production of charginos and/or neutralinos can produce distinctive multi-lepton signatures. The chargino/neutralino searches primarily constrain the mSUGRA parameter $m_{1/2}$, which can be translated into an equivalent bound on the gluino mass. As a result, gluino and squark masses up to about 400 GeV can be probed at the upcoming Tevatron Run-II; further improvements are projected at the proposed TeV-33 upgrade,[32] where supersymmetric masses up to about 600 GeV can be reached. The maximum reach at the LHC is generally attained by searching for the $1\ell+\text{jets}+E_T^{\text{miss}}$ channel; one will be able to discover squarks and gluinos with masses up to several TeV.[26] Some particularly important classes of events include:

- $pp \to \widetilde{g}\widetilde{g} \to \text{jets}+E_T^{\text{miss}}$ and $pp \to \widetilde{g}\widetilde{g} \to \ell^\pm\ell^\pm+\text{jets}+E_T^{\text{miss}}$ (the like-sign dilepton signal[24]). The mass difference $m_{\widetilde{g}} - m_{\widetilde{\chi}_1^\pm}$ can be determined from jet spectra end points, while $m_{\widetilde{\chi}_1^\pm} - m_{\widetilde{\chi}_1^0}$ can be roughly determined by analyzing various distributions of kinematic observables in the like-sign channel.[24,27,33] An absolute scale for $m_{\widetilde{g}}$ can be estimated (within an accuracy of roughly $\pm 15\%$) by separating the like-sign events into two hemispheres corresponding to the two $\widetilde{g}$'s,[24] by a similar separation in the $\text{jets}+E_T^{\text{miss}}$ channel,[26] or variations thereof.[27,33]
- $pp \to \widetilde{\chi}_1^\pm\widetilde{\chi}_2^0 \to (\ell^\pm\nu\widetilde{\chi}_1^0)(\ell^+\ell^-\widetilde{\chi}_1^0)$, which yields a tri-lepton + $E_T^{\text{miss}}$ final state. The mass difference $m_{\widetilde{\chi}_2^0} - m_{\widetilde{\chi}_1^0}$ is easily determined[i] if enough events are available.[25]

---

[i]In some cases, $m_{\widetilde{\chi}_2^0} - m_{\widetilde{\chi}_1^0}$ can still be determined if $\widetilde{\chi}_2^0$ is produced at some step in a supersymmetric decay chain.

- $pp \to \widetilde{\ell\ell} \to 2\ell + E_T^{\rm miss}$, detectable at the LHC for $m_{\tilde{\ell}} \lesssim 300$ GeV.[26]
- Squarks will be pair produced and, for $m_0 \gg m_{1/2}$, would lead to $\widetilde{g}\widetilde{g}$ events with two extra jets emerging from the primary $\widetilde{q} \to q\widetilde{g}$ decays.

The LHC provides significant opportunities for precision measurements of the mSUGRA parameters.[27] In general, one expects large samples of supersymmetric events with distinguishing features that allow an efficient separation from Standard Model backgrounds. The biggest challenge in analyzing these events may be in distinguishing one set of supersymmetric signals from another. Within the mSUGRA framework, the parameter space is small enough to permit the untangling of the various signals and allows one to extract the mSUGRA parameters with some precision.

Important discovery modes at the ILC include the following:[34]

- $e^+e^- \to \widetilde{\chi}_1^+\widetilde{\chi}_1^- \to (q\bar{q}\widetilde{\chi}_1^0 \text{ or } \ell\nu\widetilde{\chi}_1^0) + (q\bar{q}\widetilde{\chi}_1^0 \text{ or } \ell\nu\widetilde{\chi}_1^0)$;
- $e^+e^- \to \widetilde{\ell}^+\widetilde{\ell}^- \to (\ell^+\widetilde{\chi}_1^0 \text{ or } \bar{\nu}\widetilde{\chi}_1^+) + (\ell^-\widetilde{\chi}_1^0 \text{ or } \nu\widetilde{\chi}_1^-)$.

In both cases, the masses of the initially produced supersymmetric particles as well as the final state neutralinos and charginos will be well-measured. Here, one is able to make use of the energy spectra end points and beam energy constraints to make precision measurements of masses and determine the underlying supersymmetric parameters. Polarization of the beams is an essential tool that can be used to enhance signals while suppressing Standard Model backgrounds. Moreover, polarization can be employed to separate out various supersymmetric contributions in order to explore the inherent chiral structure of the interactions. The supersymmetric mass reach is limited by the center-of-mass energy of the ILC. For example, if the scalar mass parameter $m_0$ is too large, squark and slepton pair production will be kinematically forbidden. To probe values of $m_0 \sim 1$—1.5 TeV requires a collider energy in the range of $\sqrt{s} \gtrsim 2$—3 TeV. It could be that such energies will be more easily achieved at a future $\mu^+\mu^-$ collider.

The strength of the lepton colliders lies in the ability to analyze supersymmetric signals and make precision measurements of observables. Ideally, one would like to measure the underlying supersymmetric parameters without prejudice. One could then test the model assumptions, and study possible deviations. The most efficient way to carry out such a program is to set the lepton collider center-of-mass energy to the appropriate value of $\sqrt{s}$ in order to first study the light supersymmetric spectrum (lightest charginos and neutralinos and sleptons). In this way, one limits the interference among competing supersymmetric signals. Experimentation at the lepton colliders then can provide model-independent measurements of the associated underlying supersymmetric parameters. Once these parameters are ascertained, one can analyze with more confidence events with heavy supersymmetric particles decaying via complex decay chains. Thus, the ILC and LHC supersymmetric searches are complementary.

Beyond mSUGRA, the MSSM parameter space becomes more complex. It is possible to perturb the mSUGRA model by adding some non-universality among the scalar mass parameters without generating phenomenologically unacceptable flavor changing neutral currents. There has been no systematic analysis of the resulting phenomenology at future colliders. (The implications of non-universal scalar masses for LHC phenomenology were briefly addressed in Ref. 28.) Nevertheless, the possible non-degeneracy of squarks could have a significant impact on the search for squarks at hadron colliders. In particular, in mSUGRA models one typically finds that four flavors of squarks (with two squark eigenstates per flavor) and $\widetilde{b}_R$ are nearly mass-degenerate, while the masses of $\widetilde{b}_L$ and the top-squark mass eigenstates could be significantly different.[j] This means that the *observed* cross-section for the production of squark pairs at hadron colliders would be enhanced by a multiplicity factor of eight or larger (depending on the number of approximately mass-degenerate squark species). Clearly, if some of the first and second generation squarks are split in mass, the relevant effective cross-sections are smaller. This could lead to more background contamination of squark signals at hadron colliders. The impact of squark non-degeneracy on the discovery mass reach for squarks at the Tevatron and LHC has not yet been analyzed.

It is also possible to introduce arbitrary non-universal gaugino mass parameters (at the high-energy scale). For example, suppose that the non-universal gaugino masses at the high-energy scale imply that the gaugino mass parameters at the low-energy scale satisfy $M_2 < M_1$, *i.e.*, the SU(2)-gaugino component is dominant in the lightest chargino and neutralino.[17,18] In this case, the $\widetilde{\chi}_1^0$ and $\widetilde{\chi}_1^{\pm}$ can be closely degenerate, in which case the visible decay products in $\widetilde{\chi}_1^+ \to \widetilde{\chi}_1^0 + X$ decays will be very soft and difficult to detect. Consequences for chargino and neutralino detection in $e^+e^-$ and $\mu^+\mu^-$ collisions, including the importance of the $e^+e^- \to \gamma\widetilde{\chi}_1^+\widetilde{\chi}_1^-$ production channel, are discussed in Refs. 17 and 18. There is also the possibility that $m_{\widetilde{g}} \sim m_{\widetilde{\chi}_1^{\pm}} \simeq m_{\widetilde{\chi}_1^0}$. The decay products in the $\widetilde{g}$ decay chain would

[j] If $\tan\beta \gg 1$, then $\widetilde{b}_L$–$\widetilde{b}_R$ mixing can be significant, in which case the two bottom-squark mass eigenstates could also be significantly split in mass from the first two generations of squarks.

then be very soft, and isolation of $\widetilde{g}\widetilde{g}$ events would be much more difficult at hadron colliders than in the usual mSUGRA case. In particular, hard jets in association with missing energy would be much rarer, since they would only arise from initial state radiation. The corresponding reduction in supersymmetric parameter space coverage at the Tevatron Main Injector is explored in Ref. 18.

As a second example, consider the case where the low-energy gaugino mass parameters satisfy $M_2 \sim M_1$.[k] If we also assume that $\tan\beta \sim 1$ and $|\mu| < M_1$, $M_2$,[l] then the lightest two neutralinos are nearly a pure photino and higgsino respectively, *i.e.*, $\widetilde{\chi}_2^0 \simeq \widetilde{\gamma}$ and $\widetilde{\chi}_1^0 \simeq \widetilde{H}$. For this choice of MSSM parameters, one finds that the rate for the one-loop decay $\widetilde{\chi}_2^0 \to \gamma\widetilde{\chi}_1^0$ dominates over all tree level decays of $\widetilde{\chi}_2^0$ and $\mathrm{BR}(\widetilde{\mathrm{e}} \to \mathrm{e}\widetilde{\chi}_2^0) \gg \mathrm{BR}(\widetilde{\mathrm{e}} \to \mathrm{e}\widetilde{\chi}_1^0)$. Clearly, the resulting phenomenology[30] differs substantially from mSUGRA expectations. This scenario was inspired by the CDF $ee\gamma\gamma$ event.[35] Suppose that the $ee\gamma\gamma$ event resulted from $\widetilde{e}\widetilde{e}$ production, where $\widetilde{e} \to e\widetilde{\chi}_2^0 \to e\gamma\widetilde{\chi}_1^0$. Then in the model of Ref. 30, one would expect a number of other distinctive supersymmetric signals to be observable at LEP-2 (running at its maximal energy) and at Run-II of the Tevatron. In particular, LEP-2 would expect events of the type: $\ell\ell + X + E_T^{\rm miss}$ and $\gamma\gamma + X + E_T^{\rm miss}$, while Tevatron would expect events of the type: $\ell\ell + X + E_T^{\rm miss}$, $\gamma\gamma + X + E_T^{\rm miss}$, $\ell\gamma + X + E_T^{\rm miss}$, $\ell\ell\gamma + X + E_T^{\rm miss}$, $\ell\gamma\gamma + X + E_T^{\rm miss}$, and $\ell\ell\ell + X + E_T^{\rm miss}$. In the above signatures, $X$ stands for additional leptons, photons, and/or jets. These signatures can also arise in GMSB models, although the kinematics of the various events can often be distinguished.

### 10.3.2. *GMSB-based models*

The collider signals for GMSB models depend critically on the NLSP identity and its lifetime (or equivalently, its decay length). Thus, we examine the phenomenology of both promptly-decaying and longer-lived NLSP's. In the latter case, the number of decays where one or both NLSP's decay within a radial distance $R$ is proportional to $[1 - \exp(-2R/c\tau)] \simeq 2R/(c\tau)$. For large $c\tau$, most decays would be non-prompt, with many occurring in the outer parts of the detector or completely outside the detector. To maximize sensitivity to GMSB models and fully cover the $(\sqrt{F}, \Lambda)$ parameter

[k] We remind the reader that gaugino mass unification at the high-energy scale would predict $M_2 \simeq 2M_1$.

[l] To achieve such a small $\mu$-parameter requires, *e.g.*, some non-universality among scalar masses of the form $m_{H_1}^2 \neq m_{\widetilde{q}}^2, m_{\widetilde{\ell}}^2$.

space, we must develop strategies to detect decays that are delayed, but not necessarily so delayed as to be beyond current detector coverage and/or specialized extensions of current detectors.

In the discussion below, we focus on various cases, where the NLSP is a neutralino dominated by its U(1)-gaugino ($\widetilde{B}$) or Higgsino ($\widetilde{H}$) components, and where the NLSP is the lightest charged slepton (usually the $\widetilde{\tau}_R$). We first address the case of prompt decays, and then indicate the appropriate strategies for the case of the longer-lived NLSP.

- **Promptly-decaying NLSP: $\widetilde{\chi}_1^0 \simeq \widetilde{B}$**

We focus on the production of the neutralinos, charginos, and sleptons since these are the lightest of the supersymmetric particles in the GMSB models. The possible decays of the NLSP in this case are: $\widetilde{B} \to \gamma \widetilde{g}_{3/2}$ or $\widetilde{B} \to Z\widetilde{g}_{3/2}$. The latter is only relevant for the case of a heavier NLSP (and moreover is suppressed by $\tan^2\theta_W$). It will be ignored in the following discussion.

At hadronic colliders, the $\widetilde{\chi}_1^0\widetilde{\chi}_1^0$ production rate is small, but rates for $\widetilde{\chi}_1^+\widetilde{\chi}_1^- \to W^{(\star)}W^{(\star)}\widetilde{\chi}_1^0\widetilde{\chi}_1^0 \to W^{(\star)}W^{(\star)}\gamma\gamma + E_T^{\rm miss}$, $\widetilde{\ell}_R\widetilde{\ell}_R \to \ell^+\ell^-\widetilde{\chi}_1^0\widetilde{\chi}_1^0 \to \ell^+\ell^-\gamma\gamma + E_T^{\rm miss}$, $\widetilde{\ell}_L\widetilde{\ell}_L \to \ell^+\ell^-\widetilde{\chi}_1^0\widetilde{\chi}_1^0 \to \ell^+\ell^-\gamma\gamma + E_T^{\rm miss}$, *etc.* will all be substantial. Implications for GMSB phenomenology at the Tevatron can be found in Refs. 36, 37 and 38. It is possible to envision GMSB parameters such that the $ee\gamma\gamma + E_T^{\rm miss}$ CDF event[35] corresponds to selectron pair production followed by $\widetilde{e} \to e\widetilde{\chi}_1^0$ with $\widetilde{\chi}_1^0 \to \gamma\widetilde{g}_{3/2}$.[30,36,38,39] However, in this region of GMSB parameter space, other supersymmetric signals should be prevalent, such as $\widetilde{\nu}_L\widetilde{\ell}_L \to \ell\gamma\gamma + E_T^{\rm miss}$ and $\widetilde{\nu}_L\widetilde{\nu}_L \to \gamma\gamma + E_T^{\rm miss}$. The $\widetilde{\chi}_2^0\widetilde{\chi}_1^\pm$ and $\widetilde{\chi}_1^+\widetilde{\chi}_1^-$ rates would also be significant and lead to $X\gamma\gamma + E_T^{\rm miss}$ with $X = \ell^\pm, \ell^+\ell'^-, \ell^+\ell^-\ell'^\pm$. Limits on these event rates from current CDF and D0 data already eliminate much, if not all, of the parameter space that could lead to the CDF $ee\gamma\gamma$ event.[40]

At LEP-2/ILC,[41] the rate for the simplest signal, $e^+e^- \to \widetilde{\chi}_1^0\widetilde{g}_{3/2} \to \gamma + E_T^{\rm miss}$, is expected to be very small. A more robust channel is $e^+e^- \to \widetilde{\chi}_1^0\widetilde{\chi}_1^0 \to \gamma\gamma + E_T^{\rm miss}$ with a (flat) spectrum of photon energies in the range $\frac{1}{4}\sqrt{s}(1-\beta) \leq E_\gamma \leq \frac{1}{4}\sqrt{s}(1+\beta)$.

- **Promptly decaying NLSP: $\widetilde{\chi}_1^0 \simeq \widetilde{H}$**

The possible decays of the NLSP in this case are: $\widetilde{H} \to \widetilde{g}_{3/2} + h^0, H^0, A^0$, depending on the Higgs masses. If the corresponding two-body decays are not kinematically possible, then three-body decays (where the corresponding Higgs state is virtual) may become relevant. However, in realistic cases, one expects $\widetilde{\chi}_1^0$ to contain small but non-negligible gaugino components, in which case the rate for $\widetilde{\chi}_1^0 \to \widetilde{g}_{3/2}\gamma$ would dominate all three-body decays. In what follows, we assume that the two-body decay $\widetilde{H} \to \widetilde{g}_{3/2}h^0$ is kinematically allowed and dominant. The supersymmetric signals that would emerge at both Tevatron/LHC and LEP-2/ILC would then be $4b+X+E_T^{\rm miss}$ final states, where $X$ represents the decay products emerging from the cascade chain decays of the more massive supersymmetric particles. Of course, at LEP-2/ILC direct production of higgsino pairs, $e^+e^- \to \widetilde{H}\widetilde{H}$ (via virtual $s$-channel $Z$-exchange) would be possible in general, leading to pure $4b+E_T^{\rm miss}$ final states.

- **Promptly decaying NLSP: $\widetilde{\ell}_R$**

The dominant slepton decay modes are: $\widetilde{\ell}_R^{\pm} \to \ell^{\pm}\widetilde{g}_{3/2}$ and $\widetilde{\ell}_L^{\pm} \to \ell^{\pm}\widetilde{\chi}_1^{0\star} \to \ell^{\pm}(\widetilde{\ell}_R^{\pm}\ell^{\mp})' \to \ell^{\pm}(\ell^{\pm}\ell^{\mp})'\widetilde{g}_{3/2}$. The $\widetilde{\chi}_1^0$ will first decay to $\ell\widetilde{\ell}_L$ and $\ell\widetilde{\ell}_R$, followed by the above decays.

At both the Tevatron/LHC and LEP-2/ILC, typical pair production events will end with $\widetilde{\ell}_R\widetilde{\ell}_R \to \ell^+\ell^- + E_T^{\rm miss}$, generally in association with a variety of cascade chain decay products. The lepton energy spectrum will be flat in the $\widetilde{\ell}_R\widetilde{\ell}_R$ center of mass. Of course, pure $\widetilde{\ell}_R\widetilde{\ell}_R$ production is possible at LEP/ILC and the $\widetilde{\ell}_R\widetilde{\ell}_R$ center of mass would be the same as the $e^+e^-$ center of mass. Other simple signals at LEP/ILC, would include $\widetilde{\ell}_L\widetilde{\ell}_L \to 6\ell + E_T^{\rm miss}$.

If a slepton is the NLSP, it is most likely to be the $\widetilde{\tau}_R$. If this state is sufficiently lighter than the $\widetilde{e}_R$ and $\widetilde{\mu}_R$, then $\widetilde{e}_R \to e\widetilde{\tau}_R\tau$ and $\widetilde{\mu}_R \to \mu\widetilde{\tau}_R\tau$ decays (via the $\widetilde{B}$ component of the mediating virtual neutralino) might dominate over the direct $\widetilde{e}_R \to e\widetilde{g}_{3/2}$ and $\widetilde{\mu}_R \to \mu\widetilde{g}_{3/2}$ decays, and all final states would cascade to $\tau$'s. The relative importance of these different possible decays has been examined in Ref. 23. A study of this scenario at LEP-2 has been performed in Ref. 42.

- **Longer-lived NLSP: $\widetilde{\ell}_R$**

If the $\widetilde{\ell}_R$ mainly decays before reaching the electromagnetic calorimeter, then one should look for a charged lepton that suddenly appears a finite distance from the interaction region, with non-zero impact parameter as measured by either the vertex detector or the electromagnetic calorimeter. Leading up to this decay would be a heavily ionizing track with $\beta < 1$ (as could be measured if a magnetic field is present).

If the $\widetilde{\ell}_R$ reaches the electromagnetic and hadronic calorimeters, then it behaves much like a heavy muon, presumably interacting in the muon chambers or exiting the detector if it does not decay first. Limits on such objects should be pursued. There will be many sources of $\widetilde{\ell}_R$ production, including direct slepton pair production, and cascade decays resulting from the production of gluinos, squarks, and charginos.[43] Based on Tevatron data, a charged pseudo-stable $\widetilde{\ell}_R$ can be ruled out with a mass up to about 80–100 GeV. Similar limits can probably be extracted from LEP-2 data.

- **Longer-lived NLSP: $\widetilde{\chi}_1^0$**

This is a much more difficult case. As before, we assume that the dominant decay of the NLSP in this case is $\widetilde{\chi}_1^0 \to \gamma \widetilde{g}_{3/2}$. Clearly, the sensitivity of detectors to delayed $\gamma$ appearance signals will be of great importance. If the $\widetilde{\chi}_1^0$ escapes the detector before decaying, then the corresponding missing energy signatures are the same as those occurring in SUGRA-based models.

At the Tevatron, standard supersymmetry signals (*e.g.*, jets or trileptons plus $E_T^{\rm miss}$) are viable if $\Lambda \lesssim 30$–$70$ TeV (given an integrated luminosity of $L = 0.1$–$30\,{\rm fb}^{-1}$) independent of the magnitude of $\sqrt{F}$.[44,45] Meanwhile, the prompt $\widetilde{\chi}_1^0 \to \gamma \widetilde{g}_{3/2}$ decay signals discussed earlier are viable only in a region defined by $\sqrt{F} \lesssim 500$ TeV at low $\Lambda$, rising to $\sqrt{F} \lesssim 1000$ TeV at $\Lambda \sim 120$ TeV.[44,45] This leaves a significant region of $(\sqrt{F}, \Lambda)$ parameter space that can only be probed by the delayed $\widetilde{\chi}_1^0 \to \gamma \widetilde{g}_{3/2}$ decays.[44,45]

The ability to search for delayed-decay signals is rather critically dependent upon the detector design. The possible signals include the following:[44,45] (i) looking for isolated energy deposits (due to the $\gamma$ from the $\widetilde{\chi}_1^0$ decay) in the outer hadronic calorimeter cells of the D0 detector; (ii) searching for events where the delayed-decay photon is identified by a large (transverse) impact parameter as it passes into the electromagnetic calorimeter; and (iii) looking for delayed decays where the photon first emerges outside the main detector and is instead observed in a scintillator array (or similar device) placed at a substantial distance from the detector. The observed signal will always contain missing energy from one or more emitted gravitinos and/or from $\widetilde{\chi}_1^0$'s that do not decay inside the detector. Thus, by requiring large missing energy, the backgrounds can be greatly reduced while maintaining good efficiency for the GMSB signal. In combination, the above techniques may[m] allow the detection of supersym-

[m]Event rates are significant even after very strong cuts on jets, photon energy and missing energy, but detailed background calculations remain to be done.

metric particle production at the Tevatron in the GMSB parameter region $\sqrt{F} \lesssim 3000$ TeV and $\Lambda \lesssim 150$ TeV.

### 10.3.3. *R-parity violating (RPV) models*

In R-parity violating models, the LSP is no longer stable.[n] The relevant signals depend upon the nature of the LSP decay. The phenomenology depends on which R-parity violating couplings [Eq. (10.2)] are present.

If the RPV coupling strengths are very small, then the RPV-violating decay of the LSP (*e.g.*, $\widetilde{\chi}_1^0$) could occur a substantial distance from the primary interaction point, but still within the detector (or at least not far outside the detector). The general techniques for detecting such delayed decays outlined at the end of Section 10.3.2 would again be relevant. It is particularly important to note that observation of the delayed decays would allow a determination of the absolute strengths of the RPV couplings.

One should not neglect the possibility that RPV couplings could be present in GMSB models. If the RPV couplings are substantial, then the RPV decays of the NLSP will dominate its R-parity-conserving decays into $\widetilde{g}_{3/2} + X$,[46] and all the RPV phenomenology described in this section will apply. For smaller RPV couplings, there could be competition between the RPV decays and the $\widetilde{g}_{3/2} + X$ decays of the NLSP.

Only a brief discussion of the phenomenology of RPV models at hadron and lepton colliders will be given here. For further details, see Refs. 7 and 8.

At the Tevatron and LHC,[47] consider $\widetilde{g}\widetilde{g}$ production followed by gluino decay via the usual set of possible decay chains ending up with the LSP plus Standard Model particles. Until this point, all decays have involved only R-parity conserving interactions.[o] The RPV-interactions now enter in the decay of the LSP. We shall assume in the following discussion that the $\widetilde{\chi}_1^0$ is the LSP, although other possible choices can also be considered.

If $\lambda_B \neq 0$, then the dominant decay of $\widetilde{\chi}_1^0$ would result in the production of a three-jet final state ($\widetilde{\chi}_1^0 \to jjj$). The large jet backgrounds imply that we would need to rely on the like-sign dilepton signal (which would still be viable despite the absence of missing energy in the events). In general, this signal is sufficient for supersymmetry discovery up to gluino masses somewhat above 1 TeV. However, if the leptons of the like-sign dilepton signal are very soft, then the discovery reach would be much reduced. For

[n] We assume that the gravitino is not relevant for RPV phenomenology, as in SUGRA-based models.

[o] By assumption, the strengths of the R-parity conserving interactions are significantly larger than the corresponding RPV-interaction strengths.

example, soft leptons would occur in models where $m_{\widetilde{\chi}_1^\pm} \sim m_{\widetilde{\chi}_1^0}$, which requires non-universal gaugino masses.[17,28] This is one of the few cases where one could miss discovering low-energy supersymmetry at the LHC. If $\lambda_L$ dominates $\widetilde{\chi}_1^0$ decays, $\widetilde{\chi}_1^0 \to \mu^\pm e^\mp \nu, e^\pm e^\mp \nu$, and there would be many very distinctive multi-lepton signals. If $\lambda'_L$ is dominant, then $\widetilde{\chi}_1^0 \to \ell jj$ and again there would be distinctive multi-lepton signals.

More generally, many normally invisible events become visible. An important example is sneutrino pair production. Even if the dominant decay of the sneutrino is $\widetilde{\nu} \to \nu \widetilde{\chi}_1^0$ (which is likely if $m_{\widetilde{\nu}} > m_{\widetilde{\chi}_1^0}$), a visible signal emerges from the $\widetilde{\chi}_1^0$ decay as sketched above. Of course, for large enough $\lambda_L$ or $\lambda'_L$ the $\widetilde{\nu}$'s would have significant branching ratio for decay to charged lepton pairs or jet pairs, respectively. Indeed, such decays might dominate if $m_{\widetilde{\nu}} < m_{\widetilde{\chi}_1^0}$.

At LEP-2, the ILC or the muon collider,[48–50] the simplest supersymmetric production process for observing RPV phenomena is

$$e^+e^- \to \widetilde{\chi}_1^0 \widetilde{\chi}_1^0 \to \underbrace{(jjj)(jjj)}_{\lambda_B}, \quad \underbrace{(\ell\ell\nu)(\ell\ell\nu)}_{\lambda_L}, \quad \underbrace{(\ell jj)(\ell jj)}_{\lambda'_L} \tag{10.4}$$

(or the $\mu^+\mu^-$ collision analogue), where the relevant RPV-coupling is indicated below the corresponding signal. Substantial rates for equally distinctive signals from production of more massive supersymmetric particles (including sneutrino pair production) would also be present. All these processes (if kinematically allowed) should yield observable supersymmetric signals. Some limits from LEP data already exist.[51] Of particular potential importance for non-zero $\lambda_L$ is $s$-channel resonant production of a sneutrino in $e^+e^-$[49] and $\mu^+\mu^-$[50] collisions. In particular, at $\mu^+\mu^-$ colliders this process is detectable down to quite small values of the appropriate $\lambda_L$, and could be of great importance as a means of actually determining the R-parity-violating couplings. Indeed, for small R-parity-violating couplings, absolute measurements of the couplings through other processes are extremely difficult. This is because such a measurement would typically require the R-parity-violating effects to be competitive with an R-parity-conserving process of known interaction strength. (For example, R-parity-violating neutralino branching ratios constrain only ratios of the R-parity-violating couplings.) Since sneutrino pair production would have been observed at the LHC, the ILC and/or the muon collider, it would be easy to center on the sneutrino resonance in order to perform the crucial sneutrino factory measurements.

## 10.4. Supersymmetry at Future Colliders: An Update

Since this chapter was first written in 1998, new phenomenological considerations have arisen motivated by a number of new theoretical developments concerning the mechanism of supersymmetry-breaking. In this brief update, we will provide a short survey of some of the more recent theoretical and phenomenological works relevant to the search for supersymmetry at colliders. Additional references can be found in Ref. 52.

With the overwhelming evidence for neutrino masses and mixing,[53] it is clear that any viable supersymmetric model of fundamental particles must incorporate some form of lepton-number violation in the low-energy theory.[54] This requires an extension of the MSSM, which (as in the case of the minimal Standard Model) contains three generations of massless neutrinos. RPV-supersymmetry provides one possible mechanism for neutrino masses.[55,56] An alternative approach extends the Standard Model by introducing three generations of singlet right-handed neutrinos ($\nu_R$) and superheavy Majorana masses for the $\nu_R$. Diagonalizing the neutrino mass matrix yields three superheavy neutrino states, and three very light neutrino states that are identified as the light neutrino states observed in nature. This is the seesaw mechanism.[57] The supersymmetric generalization of the seesaw model of neutrino masses is now easily constructed.[58,59] In particular, the supersymmetric seesaw model conserves R-parity. The supersymmetric analogue of the Majorana neutrino mass term in the sneutrino sector leads to sneutrino–antisneutrino mixing phenomena.[59,60]

In some supergravity models, tree-level masses for the gauginos are absent. The gaugino mass parameters arise at one-loop and do not satisfy Eq. (10.1). In this case, one finds a model-independent contribution to the gaugino mass whose origin can be traced to the super-conformal (super-Weyl) anomaly, which is common to all supergravity models.[61] This approach is called anomaly-mediated supersymmetry breaking (AMSB). Eq. (10.1) is then replaced (in the one-loop approximation) by:

$$M_i \simeq \frac{b_i g_i^2}{16\pi^2} m_{3/2}\,, \tag{10.5}$$

where $m_{3/2}$ is the gravitino mass (assumed to be on the order of 1 TeV), and the $b_i$ are the coefficients of the MSSM gauge beta-functions corresponding to the corresponding U(1), SU(2), and SU(3) gauge groups: $(b_1, b_2, b_3) = (\frac{33}{5}, 1, -3)$. Eq. (10.5) yields $M_1 \simeq 2.8M_2$ and $M_3 \simeq -8.3M_2$, which implies that the lightest chargino pair and neutralino comprise a nearly mass-degenerate triplet of winos, $\widetilde{W}^{\pm}$, $\widetilde{W}^0$, over most of the MSSM parameter space. For example, if $|\mu| \gg m_Z$, then Eq. (10.5) implies that $M_{\widetilde{\chi}_1^\pm} \simeq M_{\widetilde{\chi}_1^0} \simeq M_2$.[62] The corresponding supersymmetric phenomenology differs significantly from the standard phenomenology based on Eq. (10.1), and is explored in detail in Ref. 63.[p]

Approaches to supersymmetry breaking have also been developed in the context of theories in which the number of space dimensions is greater than three. In particular, a number of supersymmetry-breaking mechanisms have been proposed that are inherently extra-dimensional.[65] For example, one can consider a higher-dimensional theory that is compactified to four spacetime dimensions. In this approach, supersymmetry is broken by boundary conditions on the compactified space that distinguish between fermions and bosons. The phenomenology of such models can be strikingly different from that of the usual MSSM.[66]

If supersymmetry is not connected with the origin of the electroweak scale, a remnant of supersymmetry at a much higher energy scale could still be manifest at the TeV-scale. This is the idea of *split-supersymmetry*,[67] in which supersymmetric scalar partners of the quarks and leptons are significantly heavier (perhaps by many orders of magnitude) than 1 TeV, whereas the fermionic partners of the gauge and Higgs bosons have masses on the order of 1 TeV or below (presumably protected by some chiral symmetry). With the exception of a single light neutral scalar whose properties are indistinguishable from those of the Standard Model Higgs boson, all other Higgs bosons are also taken to be very heavy. Although this framework cannot provide a natural explanation for the existence of the light Standard-Model-like Higgs boson, whose mass lies orders of magnitude below the mass scale of the heavy scalars, models of split-supersymmetry can account for the dark matter (which is assumed to be the LSP) and gauge coupling unification. Thus, there is some motivation for pursuing the phenomenology of such approaches.[68] One notable difference from the usual MSSM phenomenology is the existence of a long-lived gluino.[69]

Extensions of the MSSM to include one or more additional singlet superfields maintain all the desirable features of the MSSM (gauge coupling unification, solution to hierarchy problem, etc.), while potentially addressing some of the theoretical shortcomings of the simplest supersymmetric models. In the next-to-minimal supersymmetric extension of the Standard

---

[p]In its simplest form, AMSB models possess negative squared-masses for the charged sleptons. It may be possible to cure this fatal flaw in approaches beyond the minimal supersymmetric model.[64] Alternatively, one can simply posit that anomaly-mediation is not the sole source of supersymmetry-breaking in the slepton sector.

Model (NMSSM), the addition of the superpotential term $\lambda\widehat{S}\widehat{H}_u\widehat{H}_d$, which couples the two doublet Higgs superfields to a new complex singlet superfield $\widehat{S}$, leads to substantial changes in supersymmetric and Higgs boson phenomenology at colliders. For example, the model includes an additional CP-even and CP-odd Higgs boson (which can significantly modify the interpretation of the LEP supersymmetric Higgs mass bounds[70]), and a fifth neutralino that is often dominantly singlet and could easily be the LSP. A comprehensive review of the theory and phenomenology of the NMSSM can be found in Ref. 71.

Of course, any of the theoretical assumptions employed in this chapter could be wrong and must eventually be tested experimentally. The measurements of low-energy supersymmetric parameters at future colliders may eventually provide sufficient information to determine the organizing principle governing supersymmetry breaking and yield significant constraints on the values of the fundamental supersymmetric parameters. A key component will be the determination of the masses of all the supersymmetric particles. This would appear to be very difficult for typical hadron collider events containing two chain decays, initiated by production of a pair of supersymmetric particles with each ending in an invisible LSP. However, recently a variety of techniques for accurately determining the absolute masses (as opposed to just the mass differences) of all the superparticles in the chain, including the mass of the LSP, have been developed (e.g., see Refs. 72 and 73). In addition, a number of sophisticated global fitting techniques have been recently established for analyzing experimental data to test the viability of a particular supersymmetric framework and for measuring the fundamental model parameters and their uncertainties.[74]

## 10.5. Summary and Conclusions

Much effort has been directed at trying to develop strategies for precision measurements to establish the underlying supersymmetric structure of the interactions and to distinguish among models. However, we are far from understanding all possible facets of the most general MSSM parameter space (even restricted to those regions that are phenomenologically viable). Moreover, the phenomenology of non-minimal and alternative low-energy supersymmetric models (such as models with R-parity violation) and the consequences for collider physics have only recently begun to attract significant attention. The variety of possible non-minimal models of low-energy supersymmetry presents an additional challenge to experimenters who plan on searching for supersymmetry at future colliders.

If supersymmetry is discovered, it will provide a plethora of experimental signals and theoretical analyses. The many phenomenological manifestations and parameters of supersymmetry suggest that many years of experimental work will be required before it will be possible to determine the precise nature of supersymmetry-breaking and its implications for a more fundamental theory of particle interactions.

## Acknowledgments

This work was supported in part by the U.S. Department of Energy.

## References

1. H.E. Haber and G.L. Kane, *Phys. Rept.* **117** (1985) 75.
2. S.P. Martin, "A Supersymmetry Primer," hep-ph/9709356. A shortened version of the Primer appears as a chapter in this volume.
3. S. Dimopoulos and D. Sutter, *Nucl. Phys.* **B452** (1995) 496; D.W. Sutter, Stanford Ph. D. thesis, hep-ph/9704390.
4. H.E. Haber, *Nucl. Phys. B (Proc. Suppl.)* **62A-C** (1998) 469.
5. H.P. Nilles, *Phys. Rept.* **110** (1984) 1; P. Nath, R. Arnowitt and A.H. Chamseddine, *Applied* $N=1$ *Supergravity* (World Scientific, Singapore, 1984).
6. For a review of gauge-mediated supersymmetry breaking, see G.F. Giudice and R. Rattazzi, *Phys. Rept.* **322** (1999) 419.
7. H. Dreiner, hep-ph/9707435, chapter in this volume.
8. M. Chemtob, *Prog. Part. Nucl. Phys.* **54** (2005) 71; R. Barbier *et al.*, *Phys. Rept.* **420** (2005) 1.
9. K. Inoue, A. Kakuto, H. Komatsu and S. Takeshita, *Prog. Theor. Phys.* **68** (1982) 927 [Erratum: **70** (1983) 330]; **71** (1984) 413; R. Flores and M. Sher, Ann. Phys. (NY) **148** (1983) 95; J.F. Gunion and H.E. Haber, *Nucl. Phys.* **B272** (1986) 1 [Erratum: **B402** (1993) 567].
10. H.E. Haber and R. Hempfling, *Phys. Rev. Lett.* **66** (1991) 1815; Y. Okada, M. Yamaguchi and T. Yanagida, *Prog. Theor. Phys.* **85** (1991) 1; J. Ellis, G. Ridolfi and F. Zwirner, *Phys. Lett.* **B257** (1991) 83.
11. See, *e.g.*, G. Degrassi, S. Heinemeyer, W. Hollik, P. Slavich and G. Weiglein, *Eur. Phys. J.* **C28** (2003) 133; B.C. Allanach, A. Djouadi, J.L. Kneur, W. Porod and P. Slavich, *JHEP* **0409** (2004) 044; S.P. Martin, *Phys. Rev.* **D75** (2007) 055005.
12. J.F. Gunion, H.E. Haber, G. Kane and S. Dawson, *The Higgs Hunter's Guide* (Westview Press, Boulder, CO, 2000).
13. M. Carena and H.E. Haber, *Prog. Part. Nucl. Phys.* **50** (2003) 63; A. Djouadi, *Phys. Rept.* **459** (2008) 1.
14. G.L. Kane and J.P. Leveille, *Phys. Lett.* **112B** (1982) 227.

15. H. Baer, J. Ellis, G. Gelmini, D. Nanopoulos and X. Tata, *Phys. Lett.* **161B** (1985) 175; G. Gamberini, *Z. Phys.* **C30** (1986) 605; H. Baer, V. Barger, D. Karatas and X. Tata, *Phys. Rev.* **D36** (1987) 96; R.M. Barnett, J.F. Gunion and H.E. Haber, *Phys. Rev. Lett.* **60** (1988) 401; *Phys. Rev.* **D37** (1988) 1892.
16. H. Baer, X. Tata and J. Woodside, *Phys. Rev.* **D41** (1990) 906; *Phys. Rev.* **D45** (1992) 142.
17. C.H. Chen, M. Drees and J.F. Gunion, *Phys. Rev. Lett.* **76** (1996) 2002.
18. C.H. Chen, M. Drees and J.F. Gunion, *Phys. Rev.* **D55** (1997) 330.
19. G.R. Farrar, *Phys. Rev. Lett.* **76** (1996) 4111, 4115; *Phys. Rev.* **D51** (1995) 3904. For a review, see G.R. Farrar, *Nucl. Phys. B (Proc. Suppl.)* **62A-C** (1998) 485.
20. F. Csikor and Z. Fodor, *Phys. Rev. Lett.* **78** (1997) 4335; hep-ph/9712269; Z. Nagy and Z. Trocsanyi, hep-ph/9708343; *Phys. Rev.* **D57** (1988) 5793; J. Adams *et al.* [KTeV Collaboration], *Phys. Rev. Lett.* **79** (1997) 4083; P. Abreu *et al.* [DELPHI Collaboration], *Phys. Lett.* **B414** (1997) 401; R. Barate *et al.* [ALEPH Collaboration], *Z. Phys.* **C96** (1997) 1.
21. J.F. Gunion, in the Proceedings of the *International Workshop on Quantum Effects in the MSSM*, UAB, Barcelona, 9–13 September 1997, edited by J. Sola (World Scientific, Singapore, 1997) pp. 30–86.
22. S. Raby, *Phys. Rev.* **D56** (1997) 2852; *Phys. Lett.* **B422** (1998) 158.
23. S. Ambrosanio, G.D. Kribs and S.P. Martin, *Nucl. Phys.* **B516** (1997) 55.
24. R.M. Barnett, J.F. Gunion and H.E. Haber, *Phys. Lett.* **B315** (1993) 349.
25. For an LHC study, see H. Baer, C.H. Chen, F. Paige and X. Tata, *Phys. Rev.* **D50** (1994) 4508. Tevatron studies by theorists include: H. Baer and X. Tata, *Phys. Rev.* **D47** (1993) 2739; H. Baer, C. Kao and X. Tata, *Phys. Rev.* **D48** (1993) 5175; H. Baer, C.-H. Chen, C. Kao and X. Tata, *Phys. Rev.* **D52** (1995) 1565; S. Ambrosanio, G.L. Kane, G.D. Kribs, S.P. Martin and S. Mrenna, *Phys. Rev.* **D54** (1996) 5395. The most recent supersymmetric limits based on Tevatron tri-lepton searches as of 1998 are given in B. Abbott *et al.* [D0 Collaboration], *Phys. Rev. Lett.* **80** (1998) 1591; and F. Abe *et al.* [CDF Collaboration], *Phys. Rev. Lett.* **76** (1996) 4307.
26. H. Baer, C.-H. Chen, F. Paige and X. Tata, *Phys. Rev.* **D52** (1995) 2746; *Phys. Rev.* **D53** (1996) 6241.
27. I. Hinchliffe, F.E. Paige, M.D. Shapiro, J. Soderqvist and W. Yao, *Phys. Rev.* **D55** (1997) 5520.
28. G. Anderson, C.H. Chen, J.F. Gunion, J. Lykken, T. Moroi, Y. Yamada, in *New Directions for High-Energy Physics*, Proceedings of the 1996 DPF/DPB Summer Study on High Energy Physics, Snowmass '96, edited by D.G. Cassel, L.T. Gennari and R.H. Siemann (Stanford Linear Accelerator Center, Stanford, CA, 1997) pp. 669–673.
29. H. Komatsu and J. Kubo, *Phys. Lett. 157B* (1985) 90; *Nucl. Phys.* **B263** (1986) 265; H.E. Haber, G.L. Kane and M. Quiros, *Phys. Lett.* **160B** (1985) 297; *Nucl. Phys.* **B273** (1986) 333; R. Barbieri, G. Gamberini, G.F. Giudice and G. Ridolfi, *Nucl. Phys.* **B296** (1988) 75; H.E. Haber and D. Wyler, *Nucl. Phys.* **B323** (1989) 267; S. Ambrosanio and B. Mele, *Phys. Rev.* **D53** (1996) 2541; **D55** (1997) 1399 [Erratum: **D56** (1997) 3157].

30. S. Ambrosanio, G.L. Kane, G.D. Kribs, S.P. Martin and S. Mrenna, *Phys. Rev.* **D55** (1997) 1372.
31. H. Baer *et al.*, in *Electroweak Symmetry Breaking and New Physics at the TeV Scale*, edited by T.L. Barklow, S. Dawson, H.E. Haber and J.L. Siegrist (World Scientific, Singapore, 1996) pp. 216–291.
32. See, *e.g.*, H. Baer. C.-H. Chen, F. Paige and X. Tata, *Phys. Rev.* **D54** (1996) 5866; D. Amidei *et al.* [TeV-2000 Study Group], FERMILAB-PUB-96-082; and references therein.
33. A. Bartl *et al.*, in *New Directions for High-Energy Physics*, Proceedings of the 1996 DPF/DPB Summer Study on High Energy Physics, Snowmass '96, edited by D.G. Cassel, L.T. Gennari and R.H. Siemann (Stanford Linear Accelerator Center, Stanford, CA, 1997) pp. 693–707.
34. T. Tsukamoto, K. Fujii, H. Murayama, M. Yamaguchi and Y. Okada, *Phys. Rev.* **D51** (1995) 3153; J.L. Feng, M.E. Peskin, H. Murayama and X. Tata, *Phys. Rev.* **D52** (1995) 1418; H. Baer, R. Munroe and X. Tata, *Phys. Rev.* **D54** (1996) 6735; J.L. Feng and M.J. Strassler, *Phys. Rev.* **D55** (1997) 1326; H. Murayama and M.E. Peskin, *Ann. Rev. Nucl. Part. Sci.* **46** (1996) 533.
35. F. Abe *et al.* [CDF Collaboration], *Phys. Rev. Lett.* **81** (1998) 1791.
36. S. Dimopoulos, S. Thomas and J.D. Wells, *Phys. Rev.* **D54** (1996) 3283; *Nucl. Phys.* **B488** (1997) 39.
37. H. Baer, M. Brhlik, C.-H. Chen and X. Tata, *Phys. Rev.* **D55** (1997) 4463.
38. S. Ambrosanio, G.L. Kane, G.D. Kribs, S.P. Martin, *Phys. Rev.* **D54** (1996) 5395.
39. S. Dimopoulos, M. Dine, S. Raby and S. Thomas, *Phys. Rev. Lett.* **76** (1996) 3494; S. Ambrosanio, G.L. Kane, G.D. Kribs, S.P. Martin and S. Mrenna, *Phys. Rev. Lett.* **76** (1996) 3498.
40. B. Abbott *et al.* [D0 Collaboration], *Phys. Rev. Lett.* **80** (1998) 442; E. Flattum [for the CDF and D0 Collaborations], FERMILAB-Conf-97/404-E.
41. J. Lopez, D. Nanopoulos and A. Zichichi, *Phys. Rev.* **D55** (1997) 5813; *Phys. Rev. Lett.* **77** (1996) 5168; S. Dimopoulos, M. Dine, S. Raby and S. Thomas, *Phys. Rev. Lett.* **76** (1996) 3494; S. Ambrosanio, G.D. Kribs and S.P. Martin, *Phys. Rev.* **D56** (1997) 1761; D.R. Stump, M. Wiest and C.-P. Yuan, *Phys. Rev.* **D54** (1996) 1936; J. Bagger, K. Matchev, D. Pierce and R.-J. Zhang, *Phys. Rev. Lett.* **78** (1997) 1002.
42. D.A. Dicus, B. Dutta and S. Nandi, *Phys. Rev. Lett.* **78** (1997) 3055.
43. J.L. Feng and T. Moroi, *Phys. Rev.* **D58** (1998) 035001.
44. C.-H. Chen and J.F. Gunion, *Phys. Lett.* **B420** (1998) 77.
45. C.-H. Chen and J.F. Gunion, *Phys. Rev.* **D58** (1998) 075005.
46. M. Carena, S. Pokorski and C.E.M. Wagner, *Phys. Lett.* **B430** (1998) 281.
47. P. Binetruy and J.F. Gunion, in *Heavy Flavors and High Energy Collisions in the 1—100 TeV Range*, Proceedings of the INFN Eloisatron Project Workshop, Erice, Italy, June 10–27, 1988, edited by A. Ali and L. Cifarelli (Plenum Press, New York, 1989) pp. 489–496; H. Dreiner and G.G. Ross, *Nucl. Phys.* **B365** (1991) 597; H. Dreiner, M. Guchait and D.P. Roy, *Phys. Rev.* **D49** (1994) 3270; V. Barger, M.S. Berger,

P. Ohmann, R.J.N. Phillips, *Phys. Rev.* **D50** (1994) 4299; H. Baer, C. Kao and X. Tata, *Phys. Rev.* **D51** (1995) 2180; H. Baer, C.-H. Chen and X. Tata, *Phys. Rev.* **D55** (1997) 1466; A. Bartl *et al.*, *Nucl. Phys.* **B502** (1997) 19.

48. G. Bhattacharyya, D. Choudhury and K. Sridhar, *Phys. Lett.* **B355** (1995) 193; G. Bhattacharyya, J. Ellis and K. Sridhar, *Mod. Phys. Lett.* **A10** (1995) 1583; J.C. Romao, F. de Campos, M.A. Garcia-Jareno, M.B. Magro and J.W.F. Valle, *Nucl. Phys.* **B482** (1996) 3; K. Huitu, J. Maalampi and K. Puolamaki, *Eur. Phys. J* **C6** (1999) 159; F. de Campos, O.J.P. Eboli, M.A. Garcia-Jareno and J.W.F. Valle, *Nucl. Phys.* **B546** (1999) 33.
49. S. Dimopoulos and L.J. Hall, *Phys. Lett.* **B207** (1988) 210; V. Barger, G.F. Giudice and T. Han, *Phys. Rev.* **D40** (1989) 2987; R.M. Godbole, P. Roy and X. Tata, *Nucl. Phys.* **B401** (1993) 67; J. Erler, J.L. Feng and N. Polonsky, *Phys. Rev. Lett.* **78** (1997) 3063; J. Kalinowski, R. Ruckl, H. Spiesberger and P.M. Zerwas, *Phys. Lett.* **B406** (1997) 314.
50. J.L. Feng, J.F. Gunion and T. Han, *Phys. Rev.* **D58** (1998) 071701.
51. D. Buskuli *et al.* [ALEPH Collaboration], *Phys. Lett.* **B384** (1996) 461.
52. H.E. Haber, *Supersymmetry, Part I (Theory)*, in C. Amsler *et al.* [Particle Data Group], Phys. Lett. **B667** (2008) 1.
53. For a review of the current status of neutrino masses and mixing, see: M.C. Gonzalez-Garcia and M. Maltoni, *Phys. Rept.* **B460** (2008) 1; A. Strumia and F. Vissani, hep-ph/0606054.
54. For a recent review of neutrino masses in supersymmetry, see B. Mukhopadhyaya, Proc. Indian National Science Academy **A70** (2004) 239.
55. See e.g., M. Hirsch, M.A. Diaz, W. Porod, J.C. Romao and J.W.F. Valle, *Phys. Rev.* **D62** (2000) 113008 [Erratum: **D65** (2002) 119901]; **D68** (2003) 013009; A. Abada, G. Bhattacharyya and M. Losada, *Phys. Rev.* **D66** (2002) 071701; M. Hirsch and J.W.F. Valle, *New J. Phys.* **6** (2004) 76.
56. For a review, see J.C. Romao, *Nucl. Phys. Proc. Suppl.* **81** (2000) 231.
57. P. Minkowski, *Phys. Lett.* **67B** (1977) 421; M. Gell-Mann, P. Ramond and R. Slansky, in *Supergravity*, edited by D. Freedman and P. van Nieuwenhuizen (North Holland, Amsterdam, 1979) p. 315; T. Yanagida, in *Proceedings of the Workshop on Unified Theory and Baryon Number in the Universe*, edited by O. Sawada and A. Sugamoto (KEK, Tsukuba, Japan, 1979).
58. J. Hisano, T. Moroi, K. Tobe, M. Yamaguchi and T. Yanagida, *Phys. Lett.* **B357** (1995) 579; J. Hisano, T. Moroi, K. Tobe and M. Yamaguchi, *Phys. Rev.* **D53** (1996) 2442; J.A. Casas and A. Ibarra, *Nucl. Phys.* **B618** (2001) 171; J. Ellis, J. Hisano, M. Raidal and Y. Shimizu, *Phys. Rev.* **D66** (2002) 115013; A. Masiero, S.K. Vempati and O. Vives, *New J. Phys.* **6** (2004) 202; E. Arganda, A.M. Curiel, M.J. Herrero and D. Temes, *Phys. Rev.* **D71** (2005) 035011; J.R. Ellis and O. Lebedev, *Phys. Lett.* **B653** (2007) 411.
59. Y. Grossman and H.E. Haber, *Phys. Rev. Lett.* **78** (1997) 3438; A. Dedes, H.E. Haber and J. Rosiek, *JHEP* **0711** (2007) 059.
60. M. Hirsch, H.V. Klapdor-Kleingrothaus and S.G. Kovalenko, *Phys. Lett.* **B398** (1997) 311; L.J. Hall, T. Moroi and H. Murayama, *Phys. Lett.* **B424** (1998) 305; K. Choi, K. Hwang and W.Y. Song, *Phys. Rev. Lett.* **88** (2002) 141801; T. Honkavaara, K. Huitu and S. Roy, *Phys. Rev.* **D73** (2006) 055011.

61. L. Randall and R. Sundrum, *Nucl. Phys.* **B557** (1999) 79.
62. J.F. Gunion and H.E. Haber, *Phys. Rev.* **D37** (1988) 2515; S.Y. Choi, M. Drees and B. Gaissmaier, *Phys. Rev.* **D70** (2004) 014010.
63. J.L. Feng, T. Moroi, L. Randall, M. Strassler and S.-F. Su, *Phys. Rev. Lett.* **83** (1999) 1731; T. Gherghetta, G.F. Giudice and J.D. Wells, *Nucl. Phys.* **B559** (1999) 27; J.F. Gunion and S. Mrenna, *Phys. Rev.* **D62** (2000) 015002.
64. Some attempts to resolve the tachyonic slepton problem can be found in: I. Jack, D.R.T. Jones and R. Wild, *Phys. Lett.* **B535** (2002) 193; B. Murakami and J.D. Wells, *Phys. Rev.* **D68** (2003) 035006; R. Kitano, G.D. Kribs and H. Murayama, *Phys. Rev.* **D70** (2004) 035001; R. Hodgson, I. Jack, D.R.T. Jones and G.G. Ross, *Nucl. Phys.* **B728** (2005) 192; D.R.T. Jones and G.G. Ross, *Phys. Lett.* **B642** (2006) 540.
65. For a review, see e.g., C.A. Scrucca, *Mod. Phys. Lett.* **A20** (2005) 297.
66. See, e.g., R. Barbieri, L.J. Hall and Y. Nomura, *Phys. Rev.* **D66** (2002) 045025; *Nucl. Phys.* **B624** (2002) 63.
67. N. Arkani-Hamed and S. Dimopoulos, *JHEP* **0506** (2005) 073; G.F. Giudice and A. Romanino, *Nucl. Phys.* **B699** (2004) 65 [Erratum: **B706** (2005) 487].
68. N. Arkani-Hamed, S. Dimopoulos, G.F. Giudice and A. Romanino, *Nucl. Phys.* **B709** (2005) 3; W. Kilian, T. Plehn, P. Richardson and E. Schmidt, *Eur. Phys. J.* **C39** (2005) 229.
69. K. Cheung and W. Y. Keung, *Phys. Rev.* **D71** (2005) 015015; P. Gambino, G.F. Giudice and P. Slavich, *Nucl. Phys.* **B726** (2005) 35.
70. R. Dermisek and J.F. Gunion, *Phys. Rev.* **D73** (2006) 111701; **D75** (2007) 075019; **D76** (2007) 051105, 095006; **D77** (2008) 015013; **D79** (2009) 055014.
71. U. Ellwanger, C. Hugonie and A.M. Teixeira, arXiv:0910.1785 [hep-ph].
72. C.G. Lester and D.J. Summers, *Phys. Lett.* **B463** (1999) 99; A.J. Barr, C.G. Lester, M.A. Parker, B.C. Allanach and P. Richardson, *JHEP* **0303** (2003) 045; A. Barr, C. Lester and P. Stephens, *J. Phys.* **G29** (2003) 2343.
73. H.C. Cheng, J.F. Gunion, Z. Han, G. Marandella and B. McElrath, *JHEP* **0712** (2007) 076. B. Gripaios, *JHEP* **0802** (2008) 053; A.J. Barr, B. Gripaios and C.G. Lester, *JHEP* **0802** (2008) 014; W.S. Cho, K. Choi, Y.G. Kim and C.B. Park, *Phys. Rev. Lett.* **100** (2008) 171801; M.M. Nojiri, G. Polesello and D.R. Tovey, *JHEP* **0805** (2008) 014; H.C. Cheng and Z. Han, *JHEP* **0812** (2008) 063; H.C. Cheng, J.F. Gunion, Z. Han and B. McElrath, *Phys. Rev.* **D80** (2009) 035020; M. Burns, K. Kong, K.T. Matchev and M. Park, *JHEP* **0903** (2009) 143; K.T. Matchev, F. Moortgat, L. Pape and M. Park, *JHEP* **0908** (2009) 104; W.S. Cho, J.E. Kim and J.-H. Kim, arXiv:0912.2354 [hep-ph].
74. B.C. Allanach, K. Cranmer, C.G. Lester and A.M. Weber, *JHEP* **0708** (2007) 023; R. Lafaye, T. Plehn, M. Rauch and D. Zerwas, *Eur. Phys. J.* **C54** (2008) 617; O. Buchmueller *et al.*, *JHEP* **0809** (2008) 117; O. Buchmueller *et al.*, *Eur. Phys. J.* **C64** (2009) 391; S.S. AbdusSalam, B.C. Allanach, M.J. Dolan, F. Feroz and M.P. Hobson, *Phys. Rev.* **D80** (2009) 035017; S.S. AbdusSalam, B.C. Allanach, F. Quevedo, F. Feroz and M. Hobson, arXiv:0904.2548 [hep-ph]; P. Bechtle, K. Desch, M. Uhlenbrock and P. Wienemann, arXiv:0907.2589 [hep-ph].

## Computational Tools for Supersymmetry Calculations

Howard Baer
*Homer L. Dodge Department of Physics and Astronomy,*
*University of Oklahoma, Norman, OK 73019 USA*
*baer@nhn.ou.edu*

I present a brief overview of a variety of computational tools for supersymmetry calculations, including: spectrum generators, cross section and branching fraction calculators, low energy constraints, general purpose event generators, matrix element event generators, SUSY dark matter codes, parameter extraction codes and Les Houches interface tools.

### 11.1. Introduction

The Standard Model (SM) of particle physics provides an excellent description of almost all physical processes as measured in terrestrial experiments, and is rightly regarded as the crowning achievement of many decades of experimental and theoretical work in elementary particle physics.[1]

As exciting as this is, it is also apparent that the SM cannot account for a wide assortment of astrophysical data, including neutrino oscillations, the matter-anti-matter content of the universe, the presence of dark energy and the presence of dark matter in the universe, and it doesn't include gravitation. Even before these astrophysical anomalies became evident, it was apparent on theoretical grounds, mainly associated with quadratic divergences in the scalar (Higgs) sector, that the SM was to be regarded as an effective theory valid only at the energy scale of $\sim 100$ GeV and below. At higher energies, it seemed likely that some new physics must arise, which would be associated with the mechanism for electroweak symmetry breaking

While a vast array of physics theories beyond the SM have been proposed, the general class of theories including *weak scale supersymmetry* seem to most naturally solve the theoretical ills of the SM, while at the same time they receive support from a variety of precision experimental measurements.[2] Most impressive is the measured values of three SM gauge couplings at the weak scale: when extrapolated to high energies using the renormalization group group equations, the gauge couplings very nearly meet at a point under supersymmetric standard model evolution, while they miss badly under SM evolution.[3] Gauge coupling unification suggests that physics at scales $M_{GUT} \sim 2 \times 10^{16}$ GeV is described by a supersymmetric grand unified theory, and that below $M_{GUT}$, the correct effective field theory is the Minimal Supersymmetric Standard Model (MSSM), or the MSSM plus gauge singlets (since gauge singlets don't affect the running of gauge couplings at one loop).

Supersymmetric models predict the existence of a whole new class of matter states at or around the weak scale: the so-called super-partners. Gluinos, charginos, neutralinos, squarks, sleptons plus additional Higgs scalars ($H$, $A$ and $H^{\pm}$) should all be present in addition to the usual states of matter present in the SM. In order to fully test the hypothesis of weak scale supersymmetry, it seems necessary to actually produce at least some of the superpartners at high energy collider experiments, and to measure many of their properties (mass, spin, coupling strengths and mixing), in order to verify that any new physics signal indeed comes from superpartner production. In addition, the properties of the superpartners will be key to understanding the next level of understanding in the laws of physics, perhaps opening windows to the physics of grand unification and even string theory.

The key link between theoretical musings about various theories of SUSY or other new physics, and the experimental observation of particle tracks and calorimeter depositions in collider detectors is the *event generator program*. Given some theory of new physics, which usually predicts the existence of new matter states or new interactions, the event generator program allows us to compute how such a theory would manifest itself at high energy colliding beam experiments. Thus, event generator programs function as a sort of beacon, showing the way to finding new physics in a vast assortment of collider data.

Searches for new matter states at the CERN LEP2 collider have found no firm new physics signals. We thus conclude that the SM Higgs boson must have mass $m_{H_{SM}} \gtrsim 114$ GeV, while the charginos of supersymmetry must have mass $m_{\tilde{\chi}_1} \gtrsim 103.5$ GeV. The Fermilab Tevatron is probing sparticle masses such as the gluino up to the 300 GeV level. The CERN LHC is a $pp$ collider which is just now beginning to explore the energy regime where the SM breaks down, and where new physics ought to lie. LHC is expected to operate at energy scales $\sqrt{s} = 7 - 14$ TeV; this ought to be sufficient to either produce some superpartners, or rule out most particle physics models which include weak scale supersymmetry.

As we enter the LHC era, it is important to review the available calculational tools which are available, that aid in connecting theory to experiment. In this chapter, I present a brief overview of some of the publicly available tools. In Section 11.2, I examine the various

- sparticle mass spectrum calculators,

and the *Les Houche Accord files* which provide a handy interface between these and event generator programs. Sec. 11.3 lists some

- codes which calculate sparticle production rates, decay widths and branching fractions.

Sec. 11.4 reviews

- event generators for SUSY processes, including
  - multi-purpose generators, complete with parton showers, hadronization and underlying events, and
  - more specialized matrix element generators, which tend to focus on specific reactions.

The *Les Houche Event files* allow parton level collider events to be easily passed into general purpose event generators so that showering, hadronization and underlying events can be included. Sec. 11.5 lists

- codes relevant to supersymmetric dark matter calculations,

while Sec. 11.6 examines

- codes designed to extract fundamental theory parameters from sets of experimental measurements.

The supersymmetry parameter analysis (SPA) project seeks to develop a uniform set of conventions which would allow unambiguous extraction of high energy model parameters from various collider measurements of supersymmetric production and decay reactions.

I note here that this Chapter is an updated version of the 1997 version by H. Baer and S. Mrenna which appeared in the volume *Perspectives on Supersymmetry*, edited by G. Kane (World Scientific).[4]

## 11.2. SUSY Spectrum Calculators

The first step in connecting supersymmetric theory to experiment is to begin with a supersymmetric model, and calculate the expected spectrum of superpartner and Higgs boson masses and couplings. The models we will be focusing on are 4-d supersymmetric quantum field theories with softly broken supersymmetry at the TeV scale. These models might be the low energy effective theory resulting from some even more encompassing theory, such as superstring theory, or a particular SUSY GUT model, or which may invoke some specific mechanism for supersymmetry breaking.

The effective field theory is specified[5] by adopting 1. the gauge symmetry, 2. the (super-) field content and 3. the Lagrangian. In the case of supersymmetric theories, the Lagrangian is derived from the more fundamental superpotential and Kähler potential, and for non-renormalizable models, the gauge kinetic function. The effects of SUSY breaking are encoded in the Lagrangian soft SUSY breaking (SSB) terms. One must also specify the energy scale at which the effective theory and Lagrangian parameters are valid. Since collider experiments will be testing physics at the weak scale $Q \sim 1$ TeV, while the Lagrangian parameters are frequently specified at much higher scales (*e.g.* $M_{GUT}$ or $M_P$), the *renormalization group equations* (RGEs) must be used to connect the disparate scales in the model.

Once the Lagrangian parameters are known at the weak scale, then the physical (s)particle masses must be identified, often by diagonalizing the relevant *mass matrices.* Higher order perturbative corrections to the mass eigenstates– at least at 1-loop– are nowadays necessary to gain sufficient accuracy in the predictions.[6]

Numerous researchers have developed private codes to calculate sparticle masses given high scale model inputs. Here, we will focus only on *publicly available codes*, since these are available to the general user, and are frequently kept up-to-date and user friendly. The first of the publicly available spectrum calculator codes to appear was the ISASUGRA[7] subprogram of the event generator ISAJET,[8] in 1994. This was followed by SUSPECT[9] (1997), SOFTSUSY[10] (2002) and SPHENO[11] (2003).

### 11.2.1. *Isasusy, Isasugra and Isajet*

ISASUSY is a subprogram of the ISAJET event generator which calculates sparticle mass spectra given a set of 24 SSB input parameters at the weak scale. The program includes full 1-loop corrections to all sparticle masses. For Higgs masses and couplings, the full 1-loop effective potential is minimized at an optimized scale choice which accounts for leading 2-loop effects.[12] Yukawa couplings which are necessary for the loop calculations are evaluated using simple SM running mass expressions.

The ISASUGRA[7] program starts with models defined at a much higher mass scale (*e.g.* $Q = M_{GUT}$), and calculates the weak scale SUSY parameters via the full set of 2-loop RGEs.[13] An iterative approach to solving the RGEs is employed, since weak scale threshold corrections which depend on the entire SUSY mass spectrum are included. Once convergence is achieved, then the complete set of 1-loop corrected sparticle and Higgs masses are computed as in ISASUSY. Since ISASUGRA employs full 2-loop running of gauge and Yukawa couplings including threshold corrections– while ISASUSY does not– the sparticle masses will differ between ISASUSY and ISASUGRA even for the same weak scale parameter inputs.

A listing of pre-programmed ISASUGRA models include the following:

- mSUGRA (or CMSSM) model: 4 parameters plus a sign ($m_0$, $m_{1/2}$, $A_0$, $\tan\beta$, $sign(\mu)$),
- minimal gauge-mediated SUSY breaking (mGMSB, 4 param's plus sign plus $C_{grav}$) and non-minimal GMSB,
- non-universal supergravity (19 param's plus sign)
  - SSB terms can be assigned at any intermediate scale $M_{weak} < Q < M_{GUT}$,
  - non-universal Higgs model with weak scale $\mu$ and $m_A$ inputs in lieu of $m_{H_u}^2$ and $m_{H_d}^2$,
- mSUGRA or NUSUGRA plus right-hand neutrino (RHN),
- minimal and non-minimal anomaly mediation (AMSB),
- mixed moduli-AMSB (mirage mediation),
- hypercharged AMSB.

The related program `RGEFLAV` computes the complete flavor matrix structure of the SSB terms and Yukawa couplings, including $CP$-violating effects.[14] The webpage is located at `http://www.nhn.ou.edu/~isajet/`.

### 11.2.2. *Suspect*

SUSPECT[9] runs the 2-loop MSSM RGEs to determine weak scale SUSY parameters in the mSUGRA, GMSB and AMSB models, and in the pMSSM (a more general MSSM model). One-loop sparticle mass corrections are included. Some two loop corrections to Higgs masses are included. The webpage is located at `http://www.lpta.univ-montp2.fr/users/kneur/Suspect/`.

### 11.2.3. *SoftSUSY*

SOFTSUSY[10] is a $C++$ code that calculates 2-loop MSSM RGEs to determine weak scale SUSY parameters in the mSUGRA, mGMSB and mAMSB models, and in the general MSSM. $R$-parity violating effects are possible. One-loop sparticle mass corrections are included. Some two loop corrections to Higgs masses are included. SOFTSUSY calculates the complete flavor matrix structure of the MSSM soft terms and Yukawa couplings. The webpage is located at `http://projects.hepforge.org/softsusy/`.

### 11.2.4. *Spheno*

SPHENO[11] is a Fortran 90 code that calculates 2-loop MSSM RGEs to determine weak scale SUSY parameters in the mSUGRA, mGMSB and mAMSB models, and in the general MSSM. One-loop sparticle mass corrections are included. Some two loop corrections to Higgs masses are included. The webpage is located at `http://www.physik.uni-wuerzburg.de/~porod/SPheno.html`.

### 11.2.5. *Les Houches Accord (LHA) files*

A standard input/output file under the name of Les Houches Accord (LHA) files has been created. All of the above codes can create LHA output files. The advantage of LHA output files is that various event generator and dark matter codes (see below) can use these as *inputs* for generating collider events or dark matter observables. The specific form of the LHA files is presented in Ref. 15.

In addition, the ISASUGRA and ISASUSY codes output to a special `IsaWIG` file, which is created expressly for input of sparticle mass spectra and decay branching fractions to the event generator HERWIG.

### 11.2.6. *Comparison of spectra generator codes*

Several papers have been written comparing the SUSY spectra codes,[16] although these tend to be all dated material, as the codes are continually being updated and debugged. While many features of these codes are similar, and so agreement between spectra tends to be good in generic parameter space regions, there are some differences as well. In particular, the codes SuSpect, Softsusy and SPheno all adopt a *sharp cut-off scale* between the MSSM and SM effective theories. Allowance for the sharp cut-off is compensated for by log terms in the 1-loop sparticle and Higgs boson mass corrections. The Isasugra code instead adopts a "tower of effective theories" approach, and incorporates threshold corrections in the 1-loop RGEs. Here, the beta-functions changes each time a sparticle mass threshold is passed over. One loop corrections to non-mixing sparticle masses are implemented at each sparticle's mass scale, so all logs are minimized. Sparticles that mix have all their SUSY parameters evaluated at the $M_{SUSY} \equiv \sqrt{m_{\tilde{t}_L} m_{\tilde{t}_R}}$ scale due to a need for consistency amongst the various soft terms that enter the mass matrices.[17] In this way, better accuracy is expected in cases where the sparticle mass spectra is spread across a large energy range, as happens– for instance– in focus point or split SUSY, where scalars are at multi-TeV values or beyond, whereas gauginos can be quite light.

## 11.3. Sparticle Production and Decay Codes

### 11.3.1. *Production cross sections*

The multi-purpose event generators Isajet, Pythia, Herwig, SUSYGen and Sherpa all have a complete set of tree-level SUSY particle production reactions encoded, and can be used to calculate tree-level sparticle production cross sections. In the case of Pythia or Herwig, the LHA files from spectra generators can be used as input to calculate these, or general SUSY parameter inputs are allowed. Isajet does not allow LHA input since it has its own spectra generator. The Isajet code also calculates all sparticle and Higgs boson production reactions for $e^+e^-$ colliders including variable beam polarization, and bremsstrahlung and beamstrahlung.[18] The SPheno code also calculates lowest order $e^+e^- \rightarrow SUSY$ cross sections.

The code Prospino[19] has been created to calculate all $2 \rightarrow 2$ supersymmetric production cross sections at hadron colliders at both leading order (LO) and next-to-leading order (NLO) in QCD. The current version of Prospino takes LHA files as its input format.

### 11.3.2. *Decay widths and branching fractions*

The programs ISASUSY and ISASUGRA also calculate all sparticle and Higgs boson $1 \to 2$-body and $1 \to 3$-body decay widths and branching fractions (BFs). These widths and BFs are output in ISAJET standard output files, and are used internally for event generation. The chargino and neutralino branching fractions are sensitive to the parameter $\tan\beta$ in that at large $\tan\beta$, decays to third generation quarks and leptons are enhanced relative to decays to first/second generation fermions.

The program SUSYHIT[20] is a relatively new release that combines SUSPECT with the branching fraction codes SDECAY and HDECAY to also generate a table of sparticle and Higgs boson decay widths and branching fractions. Some of the decay modes in SUSYHIT are calculated at NLO in QCD.

The program SPHENO also computes sparticle decay widths and branching fractions.

The branching fractions from ISAJET, SUSYHIT and SPHENO all seem to enjoy excellent agreement with each other. The branching fractions of all these codes can be input to event generators via the LHA input/output files. Early versions of HERWIG took branching fraction inputs from the `IsaWIG` output files.

Care must be taken in extracting branching fractions for neutralinos and charginos computed internally from PYTHIA in that they may not be valid at large $\tan\beta$ values $\gtrsim 10$, since Yukawa couplings, mixing effects, and decays through intermediate Higgs bosons are neglected.

Some specialized codes are available for calculating decays modes of the SUSY Higgs bosons. These include FEYNHIGGS,[21] which calculates MSSM Higgs boson masses at two-loop level, along with branching fractions, CPSUPERH[22] which calculates Higgs boson branching fractions including CP-violating parameters, and NHMDECAY,[23] which calculates Higgs boson masses and branching fractions in the next-to-minimal supersymmetric Standard Model (NMSSM).

## 11.4. Event Generators

Supersymmetric models can be used to calculate sparticle masses and mixings, which in turn allow for a prediction of various sparticle production rates and decay widths into final states containing quarks, leptons, photons, gluons (and LSPs in $R$-parity conserving models). However, quarks and gluons are never directly measured in any collider detector. Instead, detectors measure tracks of (quasi-)stable charged particles and their momenta as they bend in a magnetic field. They also measure energy deposited in calorimeter cells by hadrons, charged leptons and photons. There is thus a gap between the predictions of supersymmetric models in terms of final states involving quarks, gluons, leptons and photons, with what is actually detected in the experimental apparatus. This gap is bridged by event generator computer programs. Once a collider type and supersymmetric model is specified, the event generator program can produce a complete simulation of the sorts of scattering events that are to be expected. The final state of any simulated scattering event is composed of a listing of electrons, muons, photons and the long-lived hadrons (pions, kaons, nucleons etc.) and their associated 4-vectors that may be measured in a collider experiment.

The underlying idea of SUSY event generator programs is that for a specified collider type ($e^+e^-$, $pp$, $p\bar{p}$, $\cdots$) and center of mass energy, the event generator will, for any set of model parameters, generate various sparticle pair production events in the ratio of their production cross sections, and with distributions as given by their differential cross sections. Moreover, the produced sparticles will undergo a (possibly multi-step cascade) decay[24] into a partonic final state, according to branching ratios as fixed by the model. Finally, this partonic final state is converted to one that is comprised of particles that are detected in an experimental apparatus. By generating a large number of "SUSY events" using these computer codes, the user can statistically simulate the various final states that are expected to be produced within the framework of any particular model.

Several general purpose event generator programs that incorporate SUSY are currently available, including ISAJET,[8] PYTHIA,[25] HERWIG,[26] SUSYGEN[27] and SHERPA.[28] These include usually just the leading order $2 \to 2$ SUSY production processes.

In addition, specific $2 \to n$ ($n \leq 6$) SUSY reactions may be generated by such programs as COMPHEP, CALCHEP, MADGRAPH, SUSY-GRACE, AMEGIC++ and O'MEGA. The output of these latter programs can be interfaced with the PYTHIA or HERWIG programs to yield complete scattering event simulations by generating output in the *Les Houches Event file* (LHE) format. (Isajet generates LHE output, but does not accept LHE files as input, since it includes its own mass and branching fraction generator). Ideally, event generator programs should be flexible enough to enable simulation of SUSY events from a variety of models such as mSUGRA, GMSB, AMSB *etc.*. This is usually accomplished nowadays by reading in the LHA model files into the event generators PYTHIA or HERWIG. Since ISAJET does its own spectra calculation, it only outputs LHA files, but does not accept them as input.

The simulation of hadron collider scattering events may be broken up into several steps, as illustrated in Fig. 11.1. The steps include:

- the perturbative calculation of the hard scattering subprocess in the parton model, and convolution with parton distribution functions (PDFs),
- inclusion of sparticle cascade decays,
- implementation of perturbative parton showers for initial and final state colored particles, and for other colored particles which may be produced as decay products of heavier objects,
- implementation of a hadronization model which describes the formation of mesons and baryons from quarks and gluons. Also, unstable particles must be decayed to the (quasi-)stable daughters that are ultimately detected in the apparatus, with rates and distributions in accord with their measured or predicted values.
- Finally, the debris from the colored remnants of the initial beams must be modeled to obtain a valid description of physics in the forward regions of the collider detector.

For $e^+e^-$ collider simulations, in addition we have to allow for the possibility of polarized initial beams, and beam-strahlung effects.

### 11.4.1. *Hard scattering*

The hard scattering and convolution with PDFs forms the central calculation of event generator programs. The calculations are usually performed at lowest order in perturbation theory, so that the hard scattering is either a $2 \to 2$ or $2 \to 1$ scattering process. Matrix element generators are usually used for $2 \to n$ processes, with $n \geq 3$.

For supersymmetric particle production at a high energy hadron collider such as the LHC, a large number of hard scattering subprocesses are likely to be kinematically accessible. Each subprocess reaction must be convoluted with PDFs so that a total hadronic cross section for each reaction may be determined. The $Q^2$-dependent PDFs commonly used are constructed to be solutions of the Dokshitzer, Gribov, Lipatov, Altarelli, Parisi (DGLAP) QCD evolution equations, which account for multiple *collinear* emissions of quarks and gluons from the initial state in the leading log approximation.

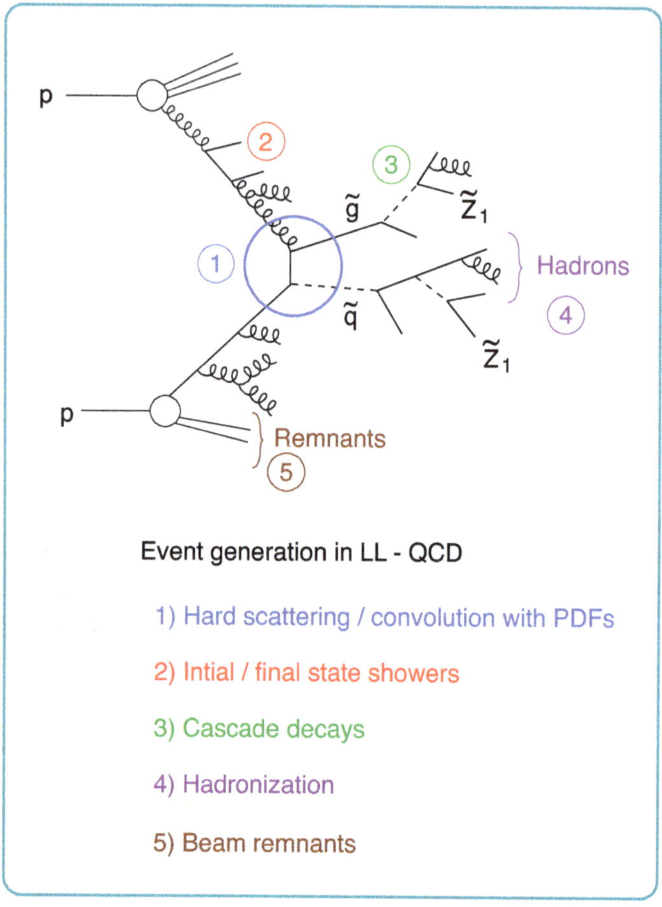


Fig. 11.1. Steps in any event generation procedure.

As $Q^2$ increases, more gluons are radiated, so that the distributions soften for large values of $x$, and correspondingly increase at small $x$ values. Use of a running QCD coupling constant makes the entire calculation valid at leading log level.

Once the total cross sections are evaluated for all the allowed subprocesses, then reactions may be selected probabilistically (with an assigned weight) using a random number generator. This will yield sparticle events in the ratio predicted by the particular model being simulated.

For sparticle production at $e^+e^-$ colliders, it may also be necessary to convolute with *electron and positron* PDFs to incorporate bremsstrahlung and beamstrahlung effects. In addition, if beam polarization is used, then each subprocess cross section will depend on beam polarization parameters as well.

### 11.4.2. *Parton showers*

For reactions occurring at both hadron and lepton colliders, to obtain a realistic portrait of supersymmetric (or Standard Model) events, it is necessary to account for multiple *non-collinear* QCD radiation effects. The evaluation of the cross section using matrix elements for multi-parton final states is prohibitively difficult. Instead, these multiple emissions are approximately included in an event simulation via a parton shower (PS) algorithm. They give rise to effects such as jet broadening, radiation in the forward regions and energy flow into detector regions that are not described by calculations with only a limited number of final state partons.

In leading log approximation (LLA), the cross section for *single* gluon emission from a quark line is given by

$$d\sigma = \sigma_0 \frac{\alpha_s}{2\pi}\frac{dt}{t}P_{qq}(z)dz, \tag{11.1}$$

where $\sigma_0$ is the overall hard scattering cross section, $t$ is the intermediate state virtual quark mass, and $P_{qq}(z) = \frac{4}{3}\left(\frac{1+z^2}{1-z}\right)$ coincides with the Altarelli-Parisi splitting function for $q' \to qg$ for the fractional momentum of the final quark $z \equiv \frac{|\vec{p}_q|}{|\vec{p}_{q'}|} < 1$. Interference between various *multiple* gluon emission Feynman graphs, where the gluons are ordered differently, is a subleading effect. Thus, Eq. (11.1) can be applied successively, and gives a factorized probability for each gluon emission. The idea behind the PS algorithm is then to use these approximate emission probabilities (which are exact in the collinear limit), along with exact (non-collinear) kinematics to construct a program which describes multiple non-collinear parton emissions. Notice, however, that the cross section (11.1) is singular as $t \to 0$ and as $z \to 1$, *i.e.* in the regime of collinear and also soft gluon emission. These singularities can be regulated by introducing physically appropriate cut-offs. A cutoff on the value of $|t|$ of order $|t_c| \sim 1$ GeV corresponds to the scale below which QCD perturbation theory is no longer valid. A cutoff on $z$ is also necessary, and physically corresponds to the limit beyond which the gluon is too soft to be resolved.

The PS algorithms available vary in their degree of sophistication. The simplest algorithm was created by Fox and Wolfram in 1979.[29] Their method was improved to account for interference effects in the angle-ordered algorithm of Marchesini and Webber.[30] In addition, parton emission from heavy particles results in a dead-cone effect, where emissions in the direction of the heavy particle are suppressed. Furthermore, it is possible to include spin correlations in the PS algorithm.

PS algorithms are also applied to the initial state partons. In this case, a backwards shower algorithm is most efficient, which develops the emissions from the hard scattering backwards in time towards the initial state. The backward shower algorithm developed by Sjöstrand[31] makes use of the PDFs evaluated at different energy scales to calculate the initial state parton emission probabilities.

### 11.4.3. *Cascade decays*

Not only are there many reactions available via which SUSY particles may be produced at colliders, but once produced, there exist many ways in which superparticles may decay. For the next-to-lightest SUSY particle (NLSP), there may be only one or at most a few ways to decay to the LSP. Thus, for a collider such as LEP or even the Fermilab Tevatron, where only the lightest sparticles will have significant production rates, we might expect that their associated decay patterns will be relatively simple. However, the number of possible final states increases rapidly if squarks and gluinos that can decay into the heavier charginos and neutralinos are accessible, and the book-keeping becomes correspondingly more complicated. Indeed, at the CERN LHC, where the massive strongly interacting sparticles such as gluinos and squarks are expected to be produced at large rates, sparticle cascade decay patterns can be very complex. As a rough estimate, of order $10^3$ subprocess cross sections may be active at LHC energies, with of order 10 decay modes for each sparticle. Naively, this would give of order $10^5$ $2 \to n$-body subprocesses that would need to be calculated.

Monte Carlo event generators immensely facilitate the analysis of signals from such complex cascade decays, especially in the case where no single decay chain dominates. An event generator can select different cascade decay branches by generating a random number which picks out a particular decay choice, with a weight proportional to the corresponding branching fraction, at each step of the cascade decay. Quarks and gluons produced as the end products of cascade decays will shower off still more quarks and gluons, with probabilities determined by the PS algorithm.

The procedure that we have just described is exact for cascade decays of spinless particles into two other spinless particles at each step in the cascade. This is because the squared matrix element is just a constant, and there are no spin correlations possible. This is not true in general and in many cases, it can be very important to include the decay matrix element and/or spin correlations in the calculation of cascade decays of sparticles. A general method for incorporating spin correlations based on density matrices has been put forth by Richardson,[32] and incorporated into HERWIG.

Spin correlation effects are especially important for precision measurements at $e^+e^-$ linear colliders. While retaining spin correlations may be less crucial in many situations at a hadron collider, this is not always the case. For instance, relativistic $\tau^-$ leptons produced from $W$ decay are always left-handed, while those produced from a charged Higgs decay are always right-handed. Likewise, the polarization of the taus from $\tilde{\tau}_1$ decays depends on the stau mixing angle. Since the undetectable energy carried off by $\nu_\tau$ from tau decay depends sensitively on the parent tau helicity, it is necessary to include effects of tau polarization in any consideration involving the energy of "tau jets". By evaluating the mean polarization of taus in any particular process, these effects can be incorporated, at least on average, into event generator programs such as ISAJET. Of course, such a procedure would not include correlations between decay products of two taus produced in the same reaction.

Another aspect is to include appropriately the complete 3-body decay matrix elements. While some programs merely use a flat phase space distribution, ISAJET and HERWIG include pre-programmed exact decay matrix elements.

### 11.4.4. *Models of hadronization*

Once sparticles have been produced and have decayed through their cascades, and parton showers have been evolved up to the point where the partons have virtuality smaller than $\sim 1$ GeV$^2$, the partons must be converted to hadrons. This is a non-perturbative process, and one must appeal to phenomenological models for its description. The independent hadronization (IH) model of Field and Feynman[33] is the simplest such model to implement. In this picture, a new quark anti-quark pair $q_1\bar{q}_1$ can be created in the color field of the parent quark $q_0$. Then the $q_0\bar{q}_1$ pair can turn into a meson with a longitudinal momentum fraction described by a phenomenological function, with the remainder of the longitudinal momentum carried by the quark $q_1$. This process is repeated by the creation of a $q_2\bar{q}_2$ pair in the color field of $q_1$, and so on down the line to $q_n\bar{q}_n$. A host of mesons are thus produced, and decayed to the quasi-stable $\pi$, $K, \cdots$ mesons according to their experimental properties. The final residual quark $q_n$ will have very little energy, and can be discarded without significantly affecting jet physics. Finally, a small transverse momentum can be added according to a pre-assigned Gaussian probability distribution to obtain a better description of the data. Quark fragmentation into baryons is also possible by creation of diquark pairs in its color field, and can be incorporated. The IH scheme, with many parameters tuned to fit the data, will thus describe the bulk features of hadronization needed for event simulation programs.

The string model of hadronization developed at Lund[34] is a more sophisticated model than IH, which treats hadron production as a universal process independent of the environment of the fragmenting quark. In the string model, a produced quark-antiquark pair is assumed to be connected by a color flux tube or string. As the quark-antiquark pair moves apart, more and more energy is stored in the string until it is energetically favorable for the string to break, creating a new quark-antiquark pair. Gluons are regarded as kinks in the string. The string model correctly accounts for color flow in the hadronization process, as opposed to the IH model. In $e^+e^- \to q\bar{q}g$ (3-jet) events, the string model predicts fewer produced hadrons in the regions between jets than the IH model, in accord with observation.

A third scheme for hadronization is known as the cluster hadronization model.[35] In this case, color flow is still accounted for, but quarks and antiquarks that are nearby in phase space will form a cluster, and will hadronize according to preassigned probabilities. This model avoids non-locality problems associated with the string hadronization model, where quarks and antiquarks separated by spacelike distances can affect the hadronization process.

### 11.4.5. *Beam remnants*

Finally, at a hadron collider the colored remnants of the nucleon that did not participate in the hard scattering must be accounted for. These beam remnant effects produce additional energy flow, especially in the far forward regions of the detector. A variety of approaches are available to describe these non-perturbative processes, including models involving Pomeron exchange and multiple scatterings. In addition, the beam remnants must be hadronized as well, and appear to require a different parametrization from "minimum bias" events where there are only beam jets but no hard scattering.

### 11.4.6. *Multi-purpose event generators*

Publicly available event generators for SUSY processes include,

- **ISAJET:** (H. Baer, F. Paige, S. Protopopescu and X. Tata), `http://www.nhn.ou.edu/~isajet/`
- **PYTHIA:** (T. Sjöstrand, L. Lönnblad and S. Mrenna), `http://www.thep.lu.se/~torbjorn/Pythia.html`
- **HERWIG:** (G. Corcella, I. G. Knowles, G. Marchesini, S. Moretti, K. Odagiri, P. Richardson, M. Seymour and B. R. Webber), `http://hepwww.rl.ac.uk/theory/seymour/herwig/`
- **SUSYGEN:** (N. Ghodbane, S. Katsanevas, P. Morawitz and E. Perez), `http://lyoinfo.in2p3.fr/susygen/susygen3.html`
- **SHERPA:** (T. Gleisberg, S. Höche, F. Krauss, M. Schönherr, S. Schumann, F. Siegert and J. Winter) `http://projects.hepforge.org/sherpa/dokuwiki/doku.php`

The event generator program ISAJET was originally developed in the late 1970's to describe scattering events at the ill-fated ISABELLE $pp$ collider at Brookhaven National Laboratory. It was developed by F. Paige and S. Protopopescu to generate SM and beyond scattering events at hadron and $e^+e^-$ colliders. H. Baer and X. Tata, in collaboration with Paige and Protopopescu, developed ISAJET to give a realistic portrayal of SUSY scattering events. ISAJET uses the IH model for hadronization, and the original Fox-Wolfram (Sjöstrand) PS shower algorithm for final state (initial state) parton showers. It includes an $n$-cut Pomeron model to describe beam-jet evolution.

The event generator PYTHIA was developed mainly by T. Sjöstrand in the early 1980s to implement the Lund string model for event generation. PYTHIA uses the FW virtuality-ordered shower model, but with an angle-ordered veto. S. Mrenna contributed the inclusion of SUSY processes in PYTHIA.

The event generator HERWIG was developed in the mid-1980s to describe scattering events with angle-ordered parton showers, which accounted for interference effects neglected in the FW shower approach. HERWIG implements a cluster hadronization model. HERWIG is notable in that it includes sparticle production and decay spin correlations using density matrix techniques.[32]

The program SUSYGEN was developed by S. Katsanevas and P. Morawitz to generate $e^+e^- \to SUSY$ events for the LEP experiments. SUSYGEN interfaces with PYTHIA for hadronization and showering. SUSYGEN has since been upgraded to also generate events for hadron colliders.

The program SHERPA was developed as a new generation event generator in the $C++$ language. It calculates subprocess reactions using AMEGIC++. It includes its own shower and cluster hadronization routines.

### 11.4.7. *Matrix element generators*

For generating various $2 \to n$ scattering reactions using complete matrix elements, a number of automated tree-level codes are available.

The code COMPHEP by E. Boos *et al.*[36] is designed to take one directly from a Lagrangian to distributions. Feynman rules can be calculated using the LANHEP code,[37] and then COMPHEP will generate the squared matrix element by constructing the squared amplitude, taking traces, and storing the output as subroutines. COMPHEP also includes code for doing the phase space integration, convolution with PDFs, and after integration, numerical output, or output in terms of histograms.

The code CALCHEP[38] by Pukhov, Belyaev and Christensen is very similar to COMPHEP, and was in fact created as a spin-off by some of the original authors of COMPHEP.

The code MADGRAPH /MADEVENT was developed by Stelzer and Long and others.[39] It allows the user to input initial and final state particles, and then generates all Feynman diagrams along with a subroutine which evaluates the scattering amplitude as a complex number using the HELAS helicity amplitude subroutines developed by Hagiwara and Murayama.[40] Since MADGRAPH directly evaluates the amplitude, and not amplitude squared, computational sampling of the squared matrix element should be faster than programs which evaluate traces over gamma matrices. The latest versions of MADGRAPH, updated to MADEVENT, will convolute with PDFs and perform phase space integration and evaluate distributions as well. A number of models for BSM physics, including the MSSM, are available in MADGRAPH /MADEVENT.

The program O'MEGA by Ohl, Reuter and Schwinn, also generates tree-level SM and MSSM amplitudes, and can work in concert with the WHIZARD program for event generation.[41,42]

The program SUSY-GRACE by Tanaka, Kuroda, Kaneko, Jimbo and Kon also generates SM and MSSM amplitudes, and generates scattering events in association with the GRAPPA program.[43,44]

### 11.4.8. *Les Houches Event (LHE) files*

A Les Houches Event (LHE) file format has been proposed[45] which allows for a simple communication between parton level event generators and all purpose generators such as PYTHIA and HERWIG. This is particularly useful when matrix element generators like CALCHEP or MADGRAPH are used, but the user needs a complete event output including parton showers, hadronization and underlying event simulation.

The LHE file is an `ascii` file which includes lines pertaining to the generator initialization. It then follows with a listing of partons (particle ID code), their associated 4-vectors and color flow information. The generators PYTHIA and HERWIG then can read in these files, to add on showering, hadronization and underlying event. A sample SUSY event in LHE format is listed below. It lists a reaction with $sg \to \tilde{s}\tilde{g}$, with $\tilde{s} \to s\tilde{\chi}_1^0$ and $\tilde{g} \to \tilde{d}\bar{d}$ and then $\tilde{d} \to d\tilde{\chi}_1^0$. After listing a line of event characteristics, the event listing follows. The first column corresponds to particle ID, 2nd column to stability of particle, 3rd and 4th columns list the source of the particle, 5th and 6th columns relate to color flow, and 7th column is the $x$-component of the energy-momentum four-vector. The four-vector listing has been truncated to fit on the page.

```
<event>
    10         2160         1.00000    0.768145E+06          ...
             3    -1     0      0  101      0  0.000000E+00  ...
            21    -1     0      0  102    101  0.000000E+00  ...
       2000003     2     1      2  103      0  0.220402E+03  ...
       1000021     2     1      2  102    103  -.220402E+03  ...
       1000022     1     3      0    0      0  0.778961E+02  ...
             3     1     3      0  103      0  0.142506E+03  ...
       2000002     2     4      0  102      0  -.185972E+03  ...
            -2     1     4      0    0    103  -.344299E+02  ...
       1000022     1     7      0    0      0  0.800434E+02  ...
             2     1     7      0  102      0  -.266016E+03  ...
</event>
```

## 11.5. Dark Matter Codes

In response to the increasing precision of data corresponding to the density of dark matter in the universe, several public codes have been developed which evaluate key astrophysical observables in supersymmetric (and other) models.

### 11.5.1. *DarkSUSY*

The DARKSUSY code, developed by Gondolo *et al.*,[46] evaluates the relic density of neutralino dark matter in SUSY models. DarkSUSY computes all relevant neutralino annihilation and co-annihilation processes, and solves the Boltzmann equation to output the current density of neutralino CDM. It accepts input files from Isajet/Isasugra or from LHA input files. DarkSUSY also calculates: spin-independent and spin-dependent neutralino-nucleon scattering rates (direct WIMP detection), and indirect neutralino detection rates, such as: muon flux from neutralino annihilation in the core of earth or sun, flux of $\gamma$ rays, $\bar{p}$s, $e^+$s and $\bar{d}$s from neutralino annihilation in the galactic core or halo. The halo annihilation rates all depend on an assumed form for the galactic dark matter density profile.

### 11.5.2. *Micromegas*

MICROMEGAS was developed by Belanger *et al.*,[47] and also evaluates the neutralino relic density due to all annihilation and co-annihilation processes. It also computes the WIMP relic density for a variety of other non-SUSY models. It also outputs neutralino direct and indirect detection rates, $b \to s\gamma$ branching fraction and $(g-2)_\mu$.

### 11.5.3. *Isatools*

ISATOOLS is part of the ISAJET package. It includes a subroutine ISARED[48] to evaluate the neutralino relic density, the direct neutralino detection rates via spin-independent and spin-dependent scattering, the $b \to s\gamma$ branching fraction, $(g-2)_\mu^{SUSY}$, $BF(B_s \to \mu^+\mu^-)$ and the thermally averaged neutralino annihilation cross section, which is key input to neutralino halo annihilation calculations.

### 11.6. Parameter Fitting Codes

If supersymmetry is indeed discovered at the Tevatron, LHC and/or a linear $e^+e^-$ collider, then an exciting task will be to make precision measurements of all sparticle masses, spins, couplings and mixings. Once these are known, then, if the MSSM is indeed the correct effective theory all the way from $M_{weak}$ to $M_{GUT}$, it is possible to map out the GUT scale values of the soft SUSY breaking parameters. Once these are known, important information will be gained which will allow for the construction of SUSY models at or beyond the GUT scale. Two such codes are available which accomplish this task: SFITTER[49] and FITTINO.[50]

### 11.7. SPA Convention

The supersymmetry parameter analysis (SPA) project[51] is an attempt to achieve co-ordination between the various sparticle mass generation codes, event generators, relic density codes, and parameter fitting codes, with a goal in mind to determine the fundamental SUSY Lagrangian. In the SPA convention, all programs should input/ouput SUSY parameters in the $\overline{DR}$ scheme at the $Q = 1$ TeV scale. Once this benchmark is set, then all remaining calculations may proceed from this common agreed upon point.

### 11.8. Summary

In the past decade, there has been an explosion of interest in supersymmetry phenomenology. This is exhibited in part by the corresponding development of numerous computational tools to aid in supersymmetry calculations for expected collider events and for dark matter observables. Supersymmetry has certainly been an enduring theme in high energy physics. Hopefully, at the dawn of the LHC era, we are on the verge of actual discovery of supersymmetry. In this case, many of these tools for SUSY will be put to good hard use, as the community analyzes the upcoming collider data.

We expect that new tools for SUSY will emerge, which will be more focused on the new matter states that might appear. As an example, if SUSY is discovered, then the MSSM (or perhaps NMSSM) may become the new SM, and radiative corrections will have to be calculated for any remaining production and decay reactions, and in a form suitable for embedding in event generator programs. The clues we find pertaining to dark matter will impact on all astrophysical codes. In addition, new tools should also emerge that facilitate model building, as the clues we expect to emerge from the data point the way to a new paradigm in physics beyond the Standard Model.

## Acknowledgments

I thank Xerxes Tata for his comments on the manuscript, and Gordy Kane for urging me to write this updated chapter.

## References

1. For a recent review, see *e.g.* G. Altarelli, arXiv:0902.2797 [hep-ph].
2. H. Baer and X. Tata, *Weak Scale Supersymmetry: From Superfields to Scattering Events*, (Cambridge University Press, 2006); M. Drees, R. Godbole and P. Roy, *Sparticles*, (World Scientific, 2004); S. P. Martin in *Perspectives on Supersymmetry*, edited by G. Kane (World Scientific, 1998); P. Nath, hep-ph/0307123.
3. S. Dimopoulos, S. Raby and F. Wilczek, Phys. Rev. D **24** (1981) 1681; U. Amaldi, W. de Boer and H. Furstenau, Phys. Lett. B **260** (1991) 447; J. R. Ellis, S. Kelley and D. V. Nanopoulos, Phys. Lett. B **260** (1991) 131; P. Langacker and M. x. Luo, Phys. Rev. D **44** (1991) 817.
4. H. Baer and S. Mrenna, in *Perspectives on Supersymmetry*, edited by G. Kane (World Scientific, 1998).
5. D. J. H. Chung, L. L. Everett, G. L. Kane, S. F. King, J. D. Lykken and L. T. Wang, Phys. Rept. **407** (2005) 1.
6. D. M. Pierce, J. A. Bagger, K. T. Matchev and R. j. Zhang, Nucl. Phys. B **491** (1997) 3.
7. ISASUGRA, by H. Baer, F. Paige, S. Protopopescu and X. Tata; see H. Baer, C. H. Chen, R. B. Munroe, F. E. Paige and X. Tata, Phys. Rev. D **51**, 1046 (1995).
8. ISAJET, by H. Baer, F. Paige, S. Protopopescu and X. Tata, hep-ph/0312045.
9. SuSPECT, by A. Djouadi, J. L. Kneur and G. Moultaka, Comput. Phys. Commun. **176**, 426 (2007).
10. SOFTSUSY, by B. C. Allanach, Comput. Phys. Commun. **143**, 305 (2002).
11. SPHENO, by W. Porod, *Comput. Phys. Commun.***153** (2003) 275.
12. H. E. Haber, R. Hempfling and A. Hoang, *Z. Physik* **C 75** (1996) 539.
13. S. P. Martin and M. T. Vaughn, Phys. Rev. D **50** (1994) 2282 [Erratum-ibid. D **78** (2008) 039903]; Y. Yamada, *Phys. Rev.* **D 50** (1994) 3537; I. Jack and D. R. T. Jones, *Phys. Lett.* **B 333** (1994) 372.
14. A. D. Box and X. Tata, Phys. Rev. D **77** (2008) 055007 and Phys. Rev. D **79** (2009) 035004.
15. P. Skands *et al.*, JHEP **0407** (2004) 036.
16. B.C. Allanach, S. Kraml and W. Porod, *J. High Energy Phys.* **03** (2003) 016; G. Belanger, S. Kraml and A. Pukhov, *Phys. Rev.* **D 72** (2005) 015003; S.

Kraml and S. Sekmen in: M.M. Nojiri *et al.*, *Physics at TeV Colliders 2007, BSM working group report*, arXiv:0802.3672 [hep-ph].
17. H. Baer, J. Ferrandis, S. Kraml and W. Porod, Phys. Rev. D **73** (2006) 015010.
18. H. Baer, R. B. Munroe and X. Tata, Phys. Rev. D **54** (1996) 6735 [Erratum-ibid. D **56** (1997) 4424].
19. PROSPINO, by W. Beenakker, R. Hopker, M. Spira, hep-ph/9611232 (1996).
20. SUSYHIT, by A. Djouadi, M. M. Muhlleitner and M. Spira, Acta Phys. Polon. B **38**, 635 (2007).
21. FEYNHIGGS, byS. Heinemeyer, W. Hollik and G. Weiglein, Comput. Phys. Commun. **124** (2000) 76.
22. CPSUPERH, by J. S. Lee, A. Pilaftsis, M. S. Carena, S. Y. Choi, M. Drees, J. R. Ellis and C. E. M. Wagner, Comput. Phys. Commun. **156** (2004) 283.
23. NHMDECAY, by U. Ellwanger, J. F. Gunion and C. Hugonie, JHEP **0502** (2005) 066.
24. H. Baer, J. Ellis, G. Gelmini, D. V. Nanopoulos and X. Tata, *Phys. Lett.* **B 161** (1985) 175; G. Gamberini, *Z. Physik* **C 30** (1986) 605; H. Baer, V. Barger, D. Karatas and X. Tata, *Phys. Rev.* **D 36** (1987) 96.
25. PYTHIA, by T. Sjostrand, S. Mrenna and P. Skands, *J. High Energy Phys.* **0605** (2006) 026.
26. HERWIG, by G. Corcella *et al.*, *J. High Energy Phys.* **0101** (2001) 010.
27. SUSYGEN, by S. Katsanevas and P. Morawitz, Comput. Phys. Commun. **112** (1998) 227.
28. SHERPA, by T. Gleisberg, S. Hoche, F. Krauss, M. Schonherr, S. Schumann, F. Siegert and J. Winter, JHEP **0902** (2009) 007.
29. G. C. Fox and S. Wolfram, Nucl. Phys. B **168** (1980) 285.
30. G. Marchesini and B. R. Webber, Nucl. Phys. B **238** (1984) 1.
31. T. Sjostrand, Phys. Lett. B **157** (1985) 321.
32. P. Richardson, JHEP **0111** (2001) 029.
33. R. D. Field and R. P. Feynman, Nucl. Phys. B **136** (1978) 1.
34. B. Andersson, G. Gustafson and B. Soderberg, Z. Phys. C **20** (1983) 317 and Nucl. Phys. B **264** (1986) 29; for a review, see B. Andersson, G. Gustafson, G. Ingelman and T. Sjostrand, Phys. Rept. **97** (1983) 31.
35. B. Webber, Nucl. Phys. B **238** (1984) 492.
36. COMPHEP, by E. Boos *et al.* [CompHEP Collaboration], Nucl. Instrum. Meth. A **534** (2004) 250.
37. LANHEP, by A. Semenov, Comput. Phys. Commun. **180** (2009) 431.
38. CALCHEP, by A. Pukhov, hep-ph/0412191.
39. MADGRAPH, by T. Stelzer and W. F. Long, Comput. Phys. Commun. **81** (1994) 357; MADEVENT, by F. Maltoni and T. Stelzer, *J. High Energy Phys.* **0302** (2003) 027; J. Alwall *et al.*, *J. High Energy Phys.* **0709** (2007) 028.
40. HELAS, by H. Murayama, I. Watanabe and K. Hagiwara, KEK-91-11.
41. O'MEGA, by M. Moretti, T. Ohl and J. Reuter, arXiv:hep-ph/0102195.
42. WHIZARD, by W. Kilian, T. Ohl and J. Reuter, arXiv:0708.4233 [hep-ph].
43. SUSY-GRACE, by M. Jimbo, H. Tanaka, T. Kaneko and T. Kon [Minami-Tateya Collaboration], arXiv:hep-ph/9503363.

44. GRAPPA, by S. Tsuno, hep-ph/0501174 (2005).
45. J. Alwall *et al.*, *Comput. Phys. Commun.***176** (2007) 300.
46. DARKSUSY, by P. Gondolo, J. Edsjo, P. Ullio, L. Bergstrom, M. Schelke and E. A. Baltz, JCAP **0407** (2004) 008.
47. MICROMEGAS, by G. Belanger, F. Boudjema, A. Pukhov and A. Semenov, *Comput. Phys. Commun.***174** (2006) 577; *Comput. Phys. Commun.***176** (2007) 367.
48. ISAREED, by H. Baer, C. Balazs and A. Belyaev, *J. High Energy Phys.* **0203** (2002) 042.
49. SFITTER, by R. Lafaye, T. Plehn and D. Zerwas, arXiv:hep-ph/0404282.
50. FITTINO, by P. Bechtle, K. Desch and P. Wienemann, Comput. Phys. Commun. **174** (2006) 47.
51. J. A. Aguilar-Saavedra *et al.*, Eur. Phys. J. C **46** (2006) 43.

# Charge and Color Breaking

J. Alberto Casas

*Instituto de Estructura de la Materia, Serrano 123 (CSIC), 28006 Madrid, Spain*

The presence of scalar fields with color and electric charge in supersymmetric theories makes feasible the existence of dangerous charge and color breaking (CCB) minima and unbounded from below directions (UFB) in the effective potential, which would make the standard vacuum unstable. The avoidance of these occurrences imposes severe constraints on the supersymmetric parameter space. We give here a comprehensive and updated account of this topic.

## 12.1. Introduction

Experimental observation tells us that color and electric charge are gauge quantum numbers preserved in nature. From the thoretical point of view, in the Standard Model they are certainly conserved in an automatical way since the only fundamental scalar field is the Higgs boson, a colorless electroweak doublet. The Higgs potential has a continuum of degenerate minima, but these are all physically equivalent and one can always define the unbroken $U(1)$ generator to be the electric charge. In supersymmetric (SUSY) extensions of the Standard Model things become more complicated. First, the Higgs sector must contain for consistency at least two Higgs doublets $H_1$, $H_2$ (plus perhaps some singlets or triplets). Hence, one has to check that the minimum of the Higgs potential $V(H_1, H_2)$ still occurs for values of $H_1$, $H_2$ which are apropriately aligned in order to preserve the electric charge; otherwise the whole electroweak symmetry becomes spontaneously broken. Second, the supersymmetric theory has a large number of additional charged and colored scalar fields, namely all the sleptons and squarks, say $\tilde{l}_i$, $\tilde{q}_i$. Consequently one has to verify that the minimum of the whole potential $V(H_1, H_2, \tilde{q}_i, \tilde{l}_i)$ still occurs at a point in the field space, which we will call "realistic minimum" in what follows, where $\tilde{q}_i, \tilde{l}_i = 0$, thus preserving color and electric charge.

The generic situation is that the scalar potential does not present just a single minimum, and, besides the realistic minimum, there is a number of additional charge and color breaking (CCB) minima. Then, a reasonable requirement is that the realistic minimum is the deepest one, i.e the global minimum of the theory. This is certainly the usual constraint imposed in the literature and represents the most conservative attitude in order to be safe. Nevertheless, a situation with CCB minima deeper than the realistic minimum could still be acceptable if the cosmology leads the universe to the latter and this is stable enough. This issue will be discussed in the last section of this chapter.

CCB minima are not the only disease that the supersymmetric scalar potential can present. It may also happen that the field space contains directions along which the potential becomes unbounded form below (UFB), which is obviously undesirable. Both issues, CCB and UFB, are closely related, as we will see throughout the chapter.

In order to introduce some notation and to illustrate some relevant aspects and warnings concerning CCB, let us briefly review the CCB condition which has been most extensively used in the literature, namely the "traditional" bound, first studied by Frere et al. and subsequently by others.[1,2] The *tree-level* scalar potential, $V_0$, in the minimal supersymmetric standard model (MSSM) is given by

$$V_0 = V_F + V_D + V_{\rm soft} \ , \tag{12.1}$$

with

$$V_F = \sum_\alpha \left| \frac{\partial W}{\partial \phi_\alpha} \right|^2 \ , \tag{12.2a}$$

$$V_D = \frac{1}{2} \sum_a g_a^2 \left( \sum_\alpha \phi_\alpha^\dagger T^a \phi_\alpha \right)^2 \ , \tag{12.2b}$$

$$\begin{aligned} V_{\rm soft} = \sum_\alpha m_{\phi_\alpha}^2 |\phi_\alpha|^2 + \sum_{i \equiv generations} \{ & A_{u_i} \lambda_{u_i} Q_i H_2 u_i + A_{d_i} \lambda_{d_i} Q_i H_1 d_i \\ & + A_{e_i} \lambda_{e_i} L_i H_1 e_i + \text{h.c.} \} + (B\mu H_1 H_2 + \text{h.c.}) \ , \end{aligned} \tag{12.2c}$$

where $W$ is the MSSM superpotential

$$W = \sum_{i \equiv generations} \{\lambda_{u_i} Q_i H_2 u_i + \lambda_{d_i} Q_i H_1 d_i + \lambda_{e_i} L_i H_1 e_i\} + \mu H_1 H_2 \ , \tag{12.3}$$

$\phi_\alpha$ runs over all the scalar components of the chiral superfields and $a, i$ are gauge group and generation indices respectively. $Q_i$ ($L_i$) are the scalar partners of the quark (lepton) $SU(2)_L$ doublets and $u_i, d_i$ ($e_i$) are the scalar partners of the quark (lepton) $SU(2)_L$ singlets. In our notation $Q_i \equiv (u_L,\ d_L)_i$, $L_i \equiv (\nu_L,\ e_L)_i$, $u_i \equiv u_{Ri}$, $d_i \equiv d_{Ri}$, $e_i \equiv e_{Ri}$. Finally, $H_{1,2}$ are the two SUSY Higgs doublets. The first observation is that the previous potential is extremely involved since it has a large number of independent fields. Furthermore, even assuming universality of the soft breaking terms at the unification scale, $M_X$, it contains a large number of independent parameters: $m$, $M$, $A$, $B$, $\mu$, i.e. the universal scalar and gaugino masses, the universal coefficients of the trilinear and bilinear scalar terms, and the Higgs mixing mass, respectively. In addition, there are the gauge ($g$) and Yukawa ($\lambda$) couplings which are constrained by the experimental data. Notice that $M$ does not appear explicitely in $V_0$, but it does through the renormalization group equations (RGEs) of all the remaining parameters.

The complexity of $V$ has made that until recently only particular directions in the field-space have been explored. The best-known example of this is the "traditional" bound, first studied by Frere et al. and subsequently by others.[1,2] These authors considered just the three fields present in a particular trilinear scalar coupling, e.g. $\lambda_u A_u Q_u H_2 u$, assuming equal vacuum expectation values (VEVs) for them:

$$|Q_u| = |H_2| = |u| \ , \tag{12.4}$$

where only the $u_L$-component of $Q_u$ takes a VEV in order to cancel the D–terms. The phases of the three fields are taken in such way that the trilinear scalar term in the potential gets negative sign. Then, the potential (12.1) gets extremely simplified and it is easy to show that a very deep CCB minimum appears *unless* the famous constraint

$$|A_u|^2 \le 3\left(m_{Q_u}^2 + m_u^2 + m_2^2\right) \tag{12.5}$$

is satisfied. In the previous equation $m_{Q_u}^2, m_u^2, m_2^2$ are the mass parameters of $Q_u$, $u$, $H_2$. Notice from eq.(12.1) that $m_2^2$ is the sum of the $H_2$ squared soft mass, $m_{H_2}^2$, plus $\mu^2$. Similar constraints for the other trilinear terms can straightforwardly be written. These "traditional" bounds have extensively been used in the literature. Notice that the trilinear coefficient, $A$, plays a crucial role for the appearance of a CCB minimum. This is logical since the scalar trilinear terms are essentially negative contributions to the scalar potential (they are negative for a certain combination of the phases of the fields). However, we will see in Sec. 12.4 that they are irrelevant for UFB directions.

From the previous bound we can extract two important lessons. First, many ordinary CCB bounds (as the one of eq.(12.5)) come from the analysis of particular directions in the field-space, thus corresponding to *necessary but not sufficient* conditions to avoid dangerous CCB minima. Consequently a complete analysis requires a more exhaustive exploration of the field space. Second, the bound of eq.(12.5) has been obtained from the analysis of the tree-level potential $V_0$. Hence, the *radiative corrections* should be incorporated in some way. With regard to this point a usual practice has been to consider the tree-level scalar potential improved by one-loop RGEs, so that all the parameters appearing in it (see eq.(12.1)) are running with the renormalization scale, $Q$. Then it is demanded that the previous CCB constraints, i.e. eq.(12.5) and others, are satisfied at any scale between $M_X$ and $M_Z$. As we will see in Sec. 12.2 this procedure is not correct and leads to an overestimate of the restrictive power of the bounds. Therefore a more careful treatment of the radiative corrections is necessary when analyzing CCB bounds.

The chapter is organized as follows. Section 12.2 is devoted to analyze and give prescriptions to handle the above-mentioned issue of the radiative corrections. Section 12.3 deals with the Higgs part of the potential, which is a requirement for subsequent analyses. In Sections 12.4 and 12.5 a complete analysis of the UFB and UFB directions of the MSSM field space is performed, giving a complete set of optimized bounds. Special attention will be paid to the most powerful one, the so-called UFB-3 bound. The effective restrictive power of these bounds is examined in Section 12.6. Section 12.7 is devoted to the bounds that CCB pose on flavour mixing couplings, which turn out to be surprisingly strong. Finally, the cosmological considerations are left for Section 12.8.

## 12.2. The Role of the Radiative Corrections

As has been mentioned in Sec. 12.1, in the CCB analysis the scalar potential is usually considered at tree-level, improved by one-loop RGEs, so that all the parameters appearing in it (see eq.(12.1)) are running with the renormalization scale, $Q$. The two questions that arise are:

- What is the appropriate scale, say $Q = \hat{Q}$, to evaluate $V_0$?
- How important are the radiative corrections that are being ignored?

These two questions are intimately related. To understand this it is important to recall that the *exact* effective potential

$$V(Q, \lambda_\alpha(Q),\ m_\beta(Q), \phi(Q)) \tag{12.6}$$

(in short $V(Q,\phi)$), where $\lambda_\alpha(Q), m_\beta(Q)$ are running parameters and masses and $\phi(Q)$ are the generic classical fields, is scale-independent, i.e.

$$\frac{dV}{dQ} = 0\ . \tag{12.7}$$

This property allows in principle any choice of $Q$, and in particular a different one for each value of the classical fields, i.e. $Q = f(\phi)$. When analyzing CCB bounds, one is interested in possible CCB minima , so one has to minimize the scalar potential. Denoting by $\langle\phi\rangle$ the VEVs of the $\phi$–fields obtained from the minimization of $V$, it is clear from (12.7) that the two following minimization conditions

$$\frac{\partial V(Q = f(\phi), \phi)}{\partial \phi} = 0 \tag{12.8}$$

$$\left.\frac{\partial V(Q,\phi)}{\partial \phi}\right|_{Q=f(\phi)} = 0 \tag{12.9}$$

yield equivalent results for $\langle\phi\rangle$ (for a more detailed discussion see refs. 3, 4).

The previous results apply exactly *only* to the exact effective potential. In practice, however, we can only know $V$ with a certain degree of accuracy in a perturbative expansion. In particular, at one-loop level

$$V_1 = V_0(Q,\phi) + \Delta V_1(Q,\phi) \tag{12.10}$$

where $V_0$ is the (one-loop improved) tree-level potential and $\Delta V_1$ is the one-loop radiative correction to the effective potential

$$\Delta V_1 = \sum_\alpha \frac{n_\alpha}{64\pi^2} M_\alpha^4 \left[\log\frac{M_\alpha^2}{Q^2} - \frac{3}{2}\right]\ . \tag{12.11}$$

Here $M_\alpha^2(Q)$ are the improved tree-level squared mass eigenstates and $n_\alpha = (-1)^{2s_\alpha}(2s_\alpha + 1)$, where $s_\alpha$ is the spin of the corresponding particle. It is important to notice that $M_\alpha^2(Q)$ are in general field–dependent quantities since they are the eigenvalues of the $(\partial^2 V_0/\partial\phi_i\partial\phi_j)$ matrix. Hence, the values of $M_\alpha^2(Q)$ depend on the values of the fields and thus on which direction in the field space is being analyzed. $V_1(Q,\phi)$ does *not* obey eq.(12.7) for all values of $Q$. However, in the region of $Q$ of the order of the most significant masses appearing in (12.11), the logarithms involved in the radiative corrections, and the radiative corrections themselves (i.e. $\Delta V_1$), are minimized, thus improving the perturbative expansion. So we expect $V$ to be well approximated by $V_1$ and it is not surprising that in that region of $Q$, $V_1$ is approximately scale-independent,[5,6] i.e. eq.(12.7) is nearly satisfied. On the other hand, due to the smallness of $\Delta V_1$, $V_1$ and $V_0$ are, in this region, very similar. Consequently (always in this region of $Q$) we can safely approximate $V$ by $V_1$ or even[a] $V_0$, and minimize by using either eq.(12.8) or eq.(12.9), although of course eq.(12.9) is much easier to handle. This statement can be numerically confirmed, see e.g. Refs. 4, 7.

In conclusion, the radiative corrections are reasonably well incorporated by using the tree-level potential $V_0(\phi,\hat{Q})$, where the renormalization scale $\hat{Q}$ is of the order of the most significant mass, normally $\hat{Q}\sim\phi$. The application of these recipes to our task of determining the CCB minima and extract the corresponding CCB bounds will be shown in sects.4,5.

## 12.3. The Higgs Potential and the Realistic Minimum

The Higgs part of the MSSM potential can be extracted (at tree level) from eq.(12.1). It reads

$$\begin{aligned} V_{\rm Higgs} &= m_1^2|H_1|^2+m_2^2|H_2|^2-m_3^2\left(\epsilon_{ij}H_1^iH_2^j+{\rm h.c.}\right)-\frac{1}{2}g_2^2\left|\epsilon_{ij}H_1^iH_2^j\right|^2 \\ &\quad+\frac{1}{8}(g_2^2+g'^2)(|H_2|^4+|H_1|^4)+\frac{1}{8}(g_2^2-g'^2)|H_2|^2|H_1|^2, \end{aligned} \tag{12.12}$$

where $H_1\equiv(H_1^0,\ H_1^-)$, $H_2\equiv(H_2^+,\ H_2^0)$, $m_1^2\equiv m_{H_1}^2+\mu^2$, $m_2^2\equiv m_{H_2}^2+\mu^2$, $m_3^2\equiv-\mu B$ and $g_2$, $g'$ are the gauge couplings of $SU(2)\times SU(1)_Y$. All these parameters are understood to be running parameters evaluated at some renormalization scale $Q$.

Our first interest in $V_{\rm Higgs}$ comes from the fact that $V_{\rm Higgs}$ depends not only on the neutral components of $H_1$, $H_2$, but also on the charged ones, i.e. $H_1^-$, $H_2^+$. Hence, one should check that $\langle H_1^-\rangle$, $\langle H_2^+\rangle$ remain vanishing when $V_{\rm Higgs}$ is minimized (one of them, say $\langle H_2^+\rangle$, can always be chosen as vanishing through an $SU(2)$ rotation). Fortunately, it is easy to show

[a]More precisely, for a choice of $Q$ such that $\partial\Delta V_1/\partial\phi=0$ the results from $V_0$ and $V_1$ are the same. In practice this precise condition is quite involved and such a degree of precision is not necessary.

from (12.12) that the minimum of $V_{\rm Higgs}$ always lies at $H_2^+ = H_1^- = 0$. So the MSSM is safe from this point of view. It is worth remarking that non-minimal supersymmetric extensions of the standard model do not have this nice property, at least in such an automatic way. (This is e.g. the case of the so-called next-to-minimal supersymmetric standard model (NMSSM), which contains an extra singlet in the Higgs sector.[8]) Therefore we can set $H_2^+ = H_1^- = 0$ and focuss our attention on the neutral part of $V_{\rm Higgs}$, which reads

$$V_{\rm Higgs} = m_1^2|H_1|^2 + m_2^2|H_2|^2 - 2|m_3^2||H_1||H_2| + \frac{1}{8}(g'^2 + g_2^2)(|H_2|^2 - |H_1|^2)^2. \tag{12.13}$$

Notice that, since we are interested in the minimization of the potential, we have implicitely chosen in (12.13) a phase of $H_1$, $H_2$ such that the mixing term $\propto (\epsilon_{ij} H_1^i H_2^j + \text{h.c.})$ gets negative.

The second aspect of $V_{\rm Higgs}$ which interests us is that $V_{\rm Higgs}$ should develope a minimum at $|H_1^0| = v_1$, $|H_2^0| = v_2$, such that $SU(2) \times U(1)_Y$ is broken in the correct way, i.e. $v_1^2 + v_2^2 = 2M_W^2/g_2^2 \simeq (175\ rmGeV)^2$. This is the realistic minimum that corresponds to the standard vacuum. This requirement fixes one of the five independent parameters $(m, M, A, B, \mu)$ of the MSSM, say $\mu$, in terms of the others. Actually, for some choices of the four remaining parameters $(m, M, A, B)$, there is no value of $\mu$ capable of producing the correct electroweak breaking. Therefore, this requirement restricts the parameter space further, as is illustrated in Fig. 12.1 (darked region) with a representative example (which will be discussed in detail in Sec. 12.6). In addition, the actual value of the potential at the realistic minimum, say $V_{\rm real\ min}$, is important for the CCB analysis since the possible CCB vacua are dangerous as long as they deeper than $V_{\rm real\ min}$. From (12.13) it is straightforward to get $V_{\rm real\ min}$

$$V_{\rm real\ min} = -\frac{1}{8}(g'^2 + g_2^2)(v_2^2 - v_1^2)^2 = -\frac{\{[(m_1^2 + m_2^2)^2 - 4|m_3|^4]^{1/2} - m_1^2 + m_2^2\}^2}{2(g'^2 + g_2^2)} \tag{12.14}$$

Note that this is the result obtained by minimizing just the tree-level part of (12.13). As explained in Sec. 12.2 this procedure is correct if the minimization is performed at some sensible scale $Q$, which should be of the order of the most relevant mass entering $\Delta V_1$, see eq.(12.11). Since we are dealing here with the Higgs-dependent part of the potential, that mass is necessarily of the order of the largest Higgs-dependent mass, namely the largest stop mass. From now on we will denote this scale by $M_S$[b].

Finally, to be considered as realistic, the previous minimum must be really a minimum in the *whole* field-space. This simply implies that all the scalar squared mass eigenvalues (squarks and sleptons) must be positive. Actually, we should go further and demand that all the not yet observed particles, i.e. charginos, squarks, etc., have masses compatible with the experimental bounds.

## 12.4. Unbounded from Below (UFB) Constraints

In this section we analyze the constraints that arise from directions in the field-space along which the (tree-level) potential can become unbounded from below (UFB). It is in fact possible to give a *complete* clasification of the potentially dangerous UFB directions and the corresponding constraints in the MSSM. In order to understand what are the dangerous directions and the form of the corresponding bounds it is useful to notice the following two general properties about UFB in the MSSM:

**1** Contrary to what happens to the CCB minima (see Sec. 12.1), the trilinear scalar terms cannot play a significant role along an UFB direction since for large enough values of the fields the corresponding quartic (and positive) F–terms become unavoidably larger.

**2** Since all the physical masses must be positive at $Q = M_S$, the only negative terms in the (tree-level) potential that can play a relevant role along an UFB direction are[c]

$$m_2^2|H_2|^2\,, \quad -2|m_3^2||H_1||H_2| \quad . \tag{12.15}$$

Therefore, any UFB direction must involve, $H_2$ and, *perhaps*, $H_1$. Furthermore, since the previous terms are cuadratic, all the quartic (positive) terms coming from F– and D–terms must be vanishing or kept under control along an UFB direction. This means that, in any case, besides $H_2$ some additional field(s) are required for that purpose. In all the instances, the preferred additional fields are $H_1$ and/or sleptons since they normally have smaller soft masses and therefore amount to a less positive contribution to the potential.

[b] A more precise estimate of $M_S$ was given in Ref. 7, but for our purposes this is accurate enough.

[c] The only possible exception are the stop soft mass terms $m_{Q_t}^2|Q_t|^2 + m_t^2|t|^2$ since the stop masses are given by $\sim (m_{Q_t,t}^2 + M_{top}^2 \pm$ mixing), but this possibility is barely consistent with the present bounds on squark masses.

Using the previous general properties we can completely clasify the possible UFB directions in the MSSM. Special attention should be paid to the UFB–3 bound, which is the strongest one:

**UFB-1**

The first possibility is to play just with $H_1$ and $H_2$. Then, the relevant terms of the potential are those written in eq.(12.13). Obviously, the only possible UFB direction corresponds to choose $H_1 = H_2$ (up to $O(m_i)$ differences which are negligible for large enough values of the fields), so that the quartic D–term is cancelled. Thus, the (tree-level) potential along the UFB-1 direction is

$$V_{\rm UFB-1} = (m_1^2 + m_2^2 - 2|m_3^2|)|H_2|^2 \ . \tag{12.16}$$

The constraint to be imposed is that, for *any* value of $|H_2| < M_X$,

$$V_{\rm UFB-1}(Q = \hat{Q}) > V_{\rm real\ min}(Q = M_S) \ , \tag{12.17}$$

where $V_{\rm real\ min}$ is the value of the realistic minimum, given by eq.(12.14), and $V_{\rm UFB-1}$ is evaluated at an appropriate scale $\hat{Q}$ (see Sec. 12.2). $\hat{Q}$ must be of the same order as the most significant mass along this UFB-1 direction, which is obviously of order $H_2$. More precisely $\hat{Q} \sim {\rm Max}(g_2|H_2|,\ \lambda_{top}|H_2|,\ M_S)$. Consequently, from (12.16) the bound (12.17) is accurately equivalent to the well-known condition

$$m_1^2 + m_2^2 \geq 2|m_3^2|. \tag{12.18}$$

From the previous discussion, it is clear that the bound (12.18) must be satisfied at any $Q > M_S$ and, in particular, at $Q = M_X$.

**UFB-2**

If, besides $H_2, H_1$, we consider additional fields in the game, it is easy to check by simple inspection (see property 2 above) that the best possible choice is a slepton $L_i$ (along the $\nu_L$ direction), since it has the lightest mass without contributing to further quartic terms in $V$. Consequently, from eq.(12.1), the relevant potential reads

$$\begin{aligned} V &= m_1^2|H_1|^2 + m_2^2|H_2|^2 - 2|m_3^2||H_1||H_2| + m_{L_i}^2|L_i|^2 \\ &+ \frac{1}{8}(g'^2 + g_2^2)(|H_2|^2 - |H_1|^2 - |L_i|^2)^2. \end{aligned} \tag{12.19}$$

By minimizing $V$ with respect to $H_1, L_i$, it is possible to write these two fields in terms of $H_2$. This step leads to non-trivial results provided that $|m_3^2| < \mu^2$, $|H_2|^2 > 4m_{L_i}^2/(g'^2 + g_2^2)\left[1 - \frac{|m_3|^4}{\mu^4}\right]$; otherwise the optimum value for $L_i$ is $L_i = 0$, and we come back to the direction UFB-1. Then, the potential along the UFB-2 direction reads[d]

$$V_{\rm UFB-2} = \left[m_2^2 + m_{L_i}^2 - \frac{|m_3|^4}{\mu^2}\right]|H_2|^2 - \frac{2m_{L_i}^4}{g'^2+g_2^2} \ . \qquad (12.20)$$

From (12.20) it might seem that the potential is unbounded from below unless $m_2^2 + m_{L_i}^2 - \frac{|m_3|^4}{\mu^2} \geq 0$. However, strictly, the UFB-2 constraint reads

$$V_{\rm UFB-2}(Q=\hat{Q}) > V_{\rm real\ min}(Q=M_S) \ , \qquad (12.21)$$

where $V_{\rm real\ min}$ is the value of the realistic minimum, given by eq.(12.14), and $V_{\rm UFB-2}$ is evaluated at an appropriate scale $\hat{Q}$. Again $\hat{Q} \sim {\rm Max}(g_2|H_2|,\ \lambda_{top}|H_2|,\ M_S)$.

**UFB-3**

The only remaining possibility is to take $H_1 = 0$. Then, the $H_1$ F–term can be cancelled with the help of the VEVs of sleptons of a particular generation, say $e_{L_j}, e_{R_j}$, without contributing to further quartic terms. More precisely

$$\left|\frac{\partial W}{\partial H_1}\right|^2 = \left|\mu H_2 + \lambda_{e_j} e_{L_j} e_{R_j}\right|^2 = 0 \ , \qquad (12.22)$$

where $\lambda_{e_j}$ is the corresponding Yukawa coupling. It is important to note that this trick is not useful if $H_1 \neq 0$, as it happens in the UFB–2 direction, since then the $e_{L_j}, e_{R_j}$ F–terms would eventually dominate. Now, in order to cancel (or keep under control) the $SU(2)_L$ and $U(1)_Y$ D–terms we need the VEV of some additional field, which cannot be $H_1$ for the above mentioned reason. Once again the optimum choice is a slepton $L_i$ (with $i \neq j$) along the $\nu_L$ direction, as in the UFB–2 case. Denoting $|e_{L_j}| = |e_{R_j}| \equiv |e| = \sqrt{\frac{|\mu|}{\lambda_{e_j}}}|H_2|$, the relevant potential reads

$$\begin{aligned} V &= (m_2^2 - \mu^2)|H_2|^2 + (m_{L_j}^2 + m_{e_j}^2)|e|^2 + m_{L_i}^2|L_i|^2 \\ &+ \frac{1}{8}(g'^2+g_2^2)(|H_2|^2 + |e|^2 - |L_i|^2)^2. \end{aligned} \qquad (12.23)$$

Now, the value of $L_i$ can be written, by simple minimization, in terms of $H_2$, namely $|L_i|^2 = -\frac{4m_{L_i}^2}{g'^2+g_2^2} + (|H_2|^2 + |e|^2)$. It turns out that for any value

[d]Eq.(12.20) relies on the equality $m_1^2 - m_{L_i}^2 = \mu^2$, which only holds under the assumption of degenerate soft scalar masses for $H_1$ and $L_i$ at $M_X$ and in the approximation of neglecting the bottom and tau Yukawa couplings in the RGEs. Otherwise, one simply must replace $\mu^2$ by $m_1^2 - m_{L_i}^2$ in eq.(12.20).

of $|H_2| < M_X$ satisfying

$$|H_2| > \sqrt{\frac{\mu^2}{4\lambda_{e_j}^2} + \frac{4m_{L_i}^2}{g'^2 + g_2^2}} - \frac{|\mu|}{2\lambda_{e_j}} , \tag{12.24}$$

the value of the potential along the UFB-3 direction is simply given by

$$V_{\text{UFB}-3} = [m_2^2 - \mu^2 + m_{L_i}^2]|H_2|^2 + \frac{|\mu|}{\lambda_{e_j}}[m_{L_j}^2 + m_{e_j}^2 + m_{L_i}^2]|H_2| - \frac{2m_{L_i}^4}{g'^2 + g_2^2} , \tag{12.25}$$

Otherwise

$$V_{\text{UFB}-3} = \left[m_2^2 - \mu^2\right]|H_2|^2 + \frac{|\mu|}{\lambda_{e_j}}\left[m_{L_j}^2 + m_{e_j}^2\right]|H_2| + \frac{1}{8}(g'^2 + g_2^2)\left[|H_2|^2 + \frac{|\mu|}{\lambda_{e_j}}|H_2|\right]^2 . \tag{12.26}$$

Then, the UFB-3 condition reads

$$V_{\text{UFB}-3}(Q = \hat{Q}) > V_{\text{real min}}(Q = M_S) , \tag{12.27}$$

where $V_{\text{real min}}$ is given by eq.(12.14), $\hat{Q} \sim \text{Max}(g_2|e|, \lambda_{top}|H_2|, g_2|L_i|, M_S)$.

It is interesting to mention that the previous constraint (12.27) with the replacements $e \to d$ , $\lambda_{e_j} \to \lambda_{d_j}$ , $L_j \to Q_j$ , must also be imposed. Now $i$ may be equal to $j$ (the optimum choice is $d_j$ = sbottom) and $\hat{Q} \sim \text{Max}\ (\lambda_{top}|H_2|,\ g_3|d|,\ \lambda_{u_j}|d|,\ g_2|L_i|,\ M_S)$.

Anyway, the optimum condition is the one with the sleptons (note e.g. that the second term in eqs.(12.25, 12.26) is proportional to the slepton masses and thus smaller) and indeed represents, as we will see in Sec. 12.6, the *strongest* one of *all* the UFB and CCB constraints in the parameter space of the MSSM.

## 12.5. Charge and Color Breaking (CCB) Constraints

These constraints arise from the existence of CCB minima in the potential deeper than the realistic minimum. We have already mentioned the "traditional" CCB constraint[1] of eq.(12.5). Other particular CCB constraints have been explored in the literature.[9–11] In this section we will perform a *complete* analysis of the CCB minima, obtaining a set of analytic constraints that represent the *necessary and sufficient* conditions to avoid the dangerous ones. As we will see, for certain values of the initial parameters, the

CCB constraints "degenerate" into the previously found UFB constraints since the minima become unbounded from below directions. In this sense, the following CCB constraints comprise the UFB bounds of the previous section, which can be considered as special (but extremely important) limits of the former.

In order to gain intuition about CCB, let us enumerate a number of general properties which are relevant when one is looking for CCB constraints in the MSSM. (Formal proofs of the following statements can be found in Ref. 7.)

**1** The most dangerous, i.e. the deepest, CCB directions in the MSSM potential involve only one particular trilinear soft term of one generation (see eq.(12.2c)). This can be either of the leptonic type (i.e. $A_{e_i}\lambda_{e_i}L_iH_1e_i$) or the hadronic type (i.e. $A_{u_i}\lambda_{u_i}Q_iH_2u_i$ or $A_{d_i}\lambda_{d_i}Q_iH_1d_i$). Along one of these CCB directions the remaining trilinear terms are vanishing or negligible. This is because the presence of a non-vanishing trilinear term in the potential gives a net negative contribution only in a region of the field space where the relevant fields are of order $A/\lambda$ with $\lambda$ and $A$ the corresponding Yukawa coupling and soft trilinear coefficient; otherwise either the (positive) mass terms or the (positive) quartic F–terms associated with these fields dominate the potential. In consequence two trilinear couplings with different values of $\lambda$ cannot efficiently "cooperate" in any region of the field space to deepen the potential. Accordingly, to any optimized CCB constraint there corresponds a unique relevant trilinear coupling, which makes the analysis much easier.

**2** If the trilinear term under consideration has a Yukawa coupling $\lambda^2 \ll g^2$, where $g$ represents a generic gauge coupling constant, then along the corresponding deepest CCB direction the D-term must be vanishing or negligible. This occurs essentially in all the cases except for the top, and simplifies enormously the analysis.

From the previous properties it can be checked that for a given trilinear coupling under consideration there are *two* different relevant directions to explore. Next, we illustrate them taking the trilinear coupling of the first generation, $A_u\lambda_uQ_uH_2u_R$, as a guiding example.

**Direction (a)**

It exploits the trick expounded in the direction UFB-3. Namely, if $H_1 = 0$, then one can take two $d$-type squarks $d_{L_j}, d_{R_j}$ (or sleptons $e_{L_j}, e_{R_j}$)

such that $\lambda_{d_j} \gg \lambda_u$ (or $\lambda_{e_j} \gg \lambda_u$), so that their VEVs cancel the $H_1$ F–term, i.e.

$$\left|\frac{\partial W}{\partial H_1}\right|^2 = \left|\mu H_2 + \lambda_{d_j} d_{L_j} d_{R_j}\right|^2 = 0 \ . \tag{12.28}$$

Notice that $H_1$ must be very small or vanishing, otherwise the (positive) $d_{L_j}$ and $d_{R_j}$ F–terms, $\lambda_{d_j}^2 \left\{|H_1 d_{R_j}|^2 + |d_{L_j} H_1|^2\right\}$, would clearly dominate the potential (this is also in agreement with the property 1 above). Since $|d_{L_j}|^2, |d_{R_j}|^2 \ll |H_2|^2, |Q_u|^2, |u_R|^2$, the $d_{L_j}, d_{R_j}$ mass terms are negligible and the net effect of eq.(12.28) is to decrease the $H_2$ squared mass from $m_2^2$ to[e] $m_2^2 - \mu^2$. Furthermore, in addition to $H_2, Q_u, u_R, d_{L_j}, d_{R_j}$, other fields could take extra non-vanishing VEVs. As in the above-explained UFB-2 direction (see Sec. 12.4) and for similar reasons, it turns out that the optimum choice is $L_i \neq 0$, with the VEV along the $\nu_L$ direction. Therefore, along the direction $(a)$

$$H_2, Q_u, u_R \neq 0 \ , \quad \text{Possibly} \quad L_i \neq 0 \ , \tag{12.29a}$$

$$|d_{L_j}|^2 = |d_{R_j}|^2 \ ; \ \ d_{L_j} d_{R_j} = -\frac{\mu}{\lambda_{d_j}} H_2 \tag{12.29b}$$

**Direction (b)**

If we allow for $H_1 \neq 0$, then we cannot play the trick of eq.(12.28) to cancel the $H_1$ F–term. Therefore, along this alternative direction

$$H_2, Q_u, u_R, H_1 \neq 0 \ , \quad \text{Possibly} \quad L_i \neq 0 \ . \tag{12.30}$$

Let us now write the potential along the directions $(a)$, $(b)$. It is useful for this task to express the various VEVs in terms of the $H_2$ one, using the following notation[10]

$$\begin{aligned} |Q_u| &= \alpha |H_2| \ , \quad & |u_R| &= \beta |H_2| \ , \\ |H_1| &= \gamma |H_2| \ , \quad & |L_i| &= \gamma_L |H_2| \ . \end{aligned} \tag{12.31}$$

E.g. the "traditional" direction, eq.(12.4), is recovered for the particular values $\alpha = \beta = 1$, $\gamma = \gamma_L = 0$. Now, the basic expression for the scalar potential (see eq.(12.1)) is

$$V = \lambda_u^2 F(\alpha, \beta, \gamma, \gamma_L) \alpha^2 \beta^2 |H_2|^4 - 2\lambda_u \hat{A}(\gamma) \alpha\beta |H_2|^3 + \hat{m}^2(\alpha, \beta, \gamma, \gamma_L) |H_2|^2 \ , \tag{12.32}$$

[e]Recall that $m_2^2 - \mu^2 = m_{H_2}^2$, i.e. the $H_2$ soft mass, see Sec. 12.3.

where

$$F(\alpha,\beta,\gamma,\gamma_L) = 1 + \frac{1}{\alpha^2} + \frac{1}{\beta^2} + \frac{f(\alpha,\beta,\gamma,\gamma_L)}{\alpha^2\beta^2} \ ,$$

$$f(\alpha,\beta,\gamma,\gamma_L) = \frac{1}{\lambda_u^2}\left\{\frac{1}{8}g_2^2\left(1-\alpha^2-\gamma^2-\gamma_L^2\right)^2 + \frac{1}{8}g'^2\left(1+\frac{1}{3}\alpha^2-\frac{4}{3}\beta^2-\gamma^2-\gamma_L^2\right)^2 + \frac{1}{6}g_3^2(\alpha^2-\beta^2)^2\right\} ,$$

$$\hat{A}(\gamma) = |A_u| + |\mu|\gamma \ ,$$

$$\hat{m}^2(\alpha,\beta,\gamma,\gamma_L) = m_2^2 + m_{Q_u}^2\alpha^2 + m_u^2\beta^2 + m_1^2\gamma^2 + m_{L_i}^2\gamma_L^2 - 2|m_3^2|\gamma \ . \quad (12.33)$$

For the ($a$)–direction eqs.(12.32,12.33) hold replacing $\gamma = 0$, $m_2^2 \to m_2^2 - \mu^2$ in eq.(12.33). For the ($b$)–direction, when sign($A_u$) = sign($B$), it is not possible to choose the phases of the fields in such a way that the trilineal scalar coupling ($\propto A_u\lambda_u Q_u H_2 u_R$), the cross term in the $H_2$ F–term ($\propto \mu\lambda_u Q_u H_1 u_R$) and the Higgs mixing term ($\propto \mu B H_1 H_2$) become negative at the same time.[7] Correspondingly, one (any) of the three terms $\{|A_u|, |\mu|\gamma, -2|m_3^2|\gamma\}$ in eq.(12.33) must flip the sign.

Minimizing $V$ with respect to $|H_2|$ for fixed values of $\alpha,\beta,\gamma,\gamma_L$, we find, besides the $|H_2| = 0$ extremal (all VEVs vanishing), the following CCB solution

$$|H_2|_{ext} = |H_2(\alpha,\beta,\gamma,\gamma_L)|_{ext} = \frac{3\hat{A}}{4\lambda_u\alpha\beta F}\left\{1 + \sqrt{1-\frac{8\hat{m}^2F}{9\hat{A}^2}}\right\} . \quad (12.34)$$

$$V_{\text{CCB min}} = -\frac{1}{2}\alpha\beta|H_2|_{ext}^2\left(\hat{A}\lambda_u|H_2|_{ext} - \frac{\hat{m}^2}{\alpha\beta}\right) . \quad (12.35)$$

Notice that, as was stated above (see property 1), the typical VEVs at a CCB minimum are indeed of order $A/\lambda$. The previous CCB minimum will be negative[f] (and much deeper than the realistic minimum) unless

$$\hat{A}^2 \leq F\hat{m}^2 \quad (12.36)$$

This is in fact the most general form of a CCB constraint.

The previous CCB bound takes a more handy form if we realize that since $\lambda_u^2 \ll 1$ (see property 3 above) the D–terms should vanish. This implies $\alpha^2 - \beta^2 = 0$, $1 - \alpha^2 - \gamma^2 - \gamma_L^2 = 0$. As a consequence $f(\alpha,\beta,\gamma,\gamma_L)$ becomes vanishing and $F = 1 + \frac{2}{\alpha^2}$.

[f]The mere existence of a CCB minimum is discarded by demanding $\hat{A}^2 < (8/9)F\hat{m}^2$, see eq.(12.34).

Now, we can write the explicite form of the bounds for the directions ($a$, $b$):

**CCB-1**

This bound arises by considering the direction ($a$) and thus the general condition (12.36) takes the form

$$|A_u|^2 \leq \left(1+\frac{2}{\alpha^2}\right)\left[m_2^2-\mu^2+(m_{Q_u}^2+m_u^2)\alpha^2+m_{L_i}^2\gamma_L^2\right] \ , \quad (12.37)$$

where $\alpha^2$ is arbitrary and $\gamma_L^2$ is given by $\gamma_L^2 = 1-\alpha^2$. More precisely

(1) If $m_2^2-\mu^2+m_{L_i}^2 > 0$ and $3m_{L_i}^2-(m_{Q_u}^2+m_u^2)+2(m_2^2-\mu^2) > 0$, then the optimized CCB-1 occurs for $\alpha = 1$, i.e.

$$|A_u|^2 \leq 3\left[m_2^2-\mu^2+m_{Q_u}^2+m_u^2\right] \quad (12.38)$$

(2) If $m_2^2-\mu^2+m_{L_i}^2 > 0$ and $3m_{L_i}^2-(m_{Q_u}^2+m_u^2)+2(m_2^2-\mu^2) < 0$, then the optimized CCB-1 bound is

$$|A_u|^2 \leq \left(1+\frac{2}{\alpha^2}\right)\left[m_2^2-\mu^2+(m_{Q_u}^2+m_u^2)\alpha^2+m_{L_i}^2(1-\alpha^2)\right] \quad (12.39)$$

with $\alpha^2 = \sqrt{\frac{2(m_{L_i}^2+m_2^2-\mu^2)}{m_{Q_u}^2+m_u^2-m_{L_i}^2}}$.

(3) If $m_2^2-\mu^2+m_{L_i}^2 < 0$, then the CCB-1 bound is automatically violated. In fact the minimization of the potential in this case gives $\alpha^2 \to 0$, and we are exactly led to the UFB-3 direction shown above, which represents the correct analysis in this instance.

**CCB-2**

This bound arises by considering the direction ($b$). Then the general condition (12.36) takes the form

$$\begin{aligned}(|A_u|+|\mu|\gamma)^2 \ \leq & \left(1+\frac{2}{\alpha^2}\right)\left[m_2^2+(m_{Q_u}^2+m_u^2)\alpha^2\right. \\ & \left.+\ m_1^2\gamma^2+m_{L_i}^2\gamma_L^2-2|m_3^2|\gamma\right] \quad (12.40)\end{aligned}$$

where $\alpha^2, \gamma^2$ are arbitrary and $\gamma_L^2$ is given by $\gamma_L^2 = 1-\alpha^2-\gamma^2$. Rules to handle this bound in an efficient way (i.e. to take the values of $\alpha^2, \gamma^2$ that make the bound as strong as possible) can be found in Ref. 7.

If $\text{sign}(A_u) = \text{sign}(B)$, the sign of one of the three terms $\{|A_u|, |\mu|\gamma, -2|m_3^2|\gamma\}$ in (12.40) must be flipped (see comments after eq.(12.33)). Notice that, due to the form of (12.40) flipping the sign of $|A_u|$ or the sign of $|\mu|\gamma$ leads to the same result. Therefore, there are only two choices to examine.

Concerning the renormalization scale at which the previous CCB-1, CCB-2 constraints must be evaluated, a sensible choice is $\hat{Q} \sim \text{Max}(M_S, g_3\frac{A_u}{4\lambda_u}, \lambda_t\frac{A_u}{4\lambda_u})$, since $H_2 \sim \frac{A_u}{4\lambda_u}$, see eq.(12.34).

The previous CCB-1, CCB-2 bounds are straightforwardly generalized to all the couplings with coupling constant $\lambda \ll 1$. This essentialy includes all the couplings apart from the top. The generalization to the top Yukawa coupling case is more involved since $\lambda_{top} = O(1)$, so the D-terms should not be assumed to vanish anymore. Furthermore, the CCB-1 bounds are not longer applicable due to the absence of $d$–type squarks such that $\lambda_{d_j} \gg \lambda_{top}$. Finally, the associated CCB minima have in many cases a similar size to the realistic one. So, it is important to examine explicitely the condition $V_{\text{CCB min}} > V_{\text{real min}}$.

For more details, the interested reader is referred to Ref. 7.

## 12.6. Constraints on the SUSY Parameter Space

In Sections 12.4–12.6 a complete analysis of all the potentially dangerous unbounded from below (UFB) and charge and color breaking (CCB) directions has been carried out. Now, we wish to show explicitely, through a numerical analysis, the restrictive power of the constraints on the MSSM parameter space. We will see that this is certainly remarkable.

We will consider the whole parameter space of the MSSM, $m$, $M$, $A$, $B$, $\mu$, with the only assumption of universality[g]. Actually, universality of the soft SUSY-breaking terms at $M_X$ is a desirable property not only to reduce the number of independent parameters, but also for phenomenological reasons, particularly to avoid flavour-changing neutral currents (see, e.g. Ref. 12). As discussed in Sec. 12.3, the requirement of correct electroweak breaking fixes one of the five independent parameters of the MSSM, say $\mu$, so we are left with only four parameters ($m$, $M$, $A$, $B$). In order to present the results in a clear way we will start by considering the particular case $m = 100$ GeV and $B = A - m$ (i.e. the well–known minimal SUGRA relation[13]), and later we will let $B$ to vary freely.

[g]Let us remark, however, that the constraints found in previous sections are general and they can also be applied to the non-universal case.

Figure 12.1(a) shows the region excluded by the "traditional" CCB bounds of the type of eq.(12.5), evaluated at an *appropriate* scale (see Sec. 12.2). Clearly, the "traditional" bounds, when correctly evaluated, turn out to be very weak. In fact, only the leptonic (circles) and the $d$–type (diamonds) terms do restrict, very modestly, the parameter space. Let us

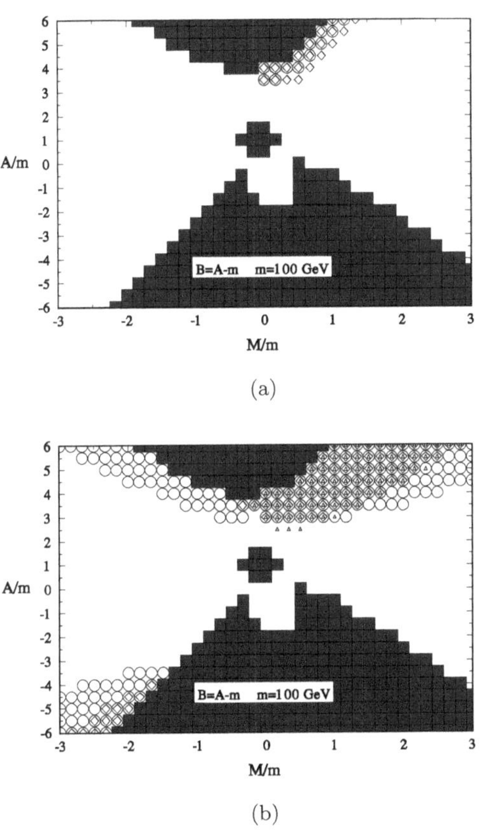


Fig. 12.1. Excluded regions in the parameter space of the MSSM, with $B = A - m$, $m = 100$ GeV and $M_{\text{top}}^{\text{phys}} = 174$ GeV. The darked region is excluded because there is no solution for $\mu$ capable of producing the correct electroweak breaking. (a) The circles and diamonds indicate regions excluded by the "traditional" CCB constraints associated with the $e$ and $d$-type trilinear terms respectively. (b) The same as (a) but using the "improved" CCB constraints. The triangles correspond to the $u$-type trilinear terms. (c) The crosses, squares and small filled squares indicate regions excluded by the UFB-1,2,3 constraints respectively. (d) The previous excluded regions together with the one arising from the experimental lower bounds on supersymmetric particle masses (filled diamonds).

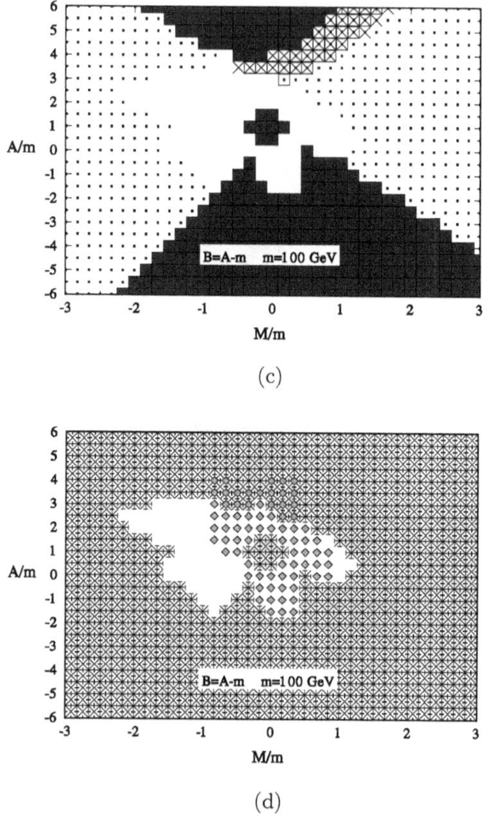


(c)

(d)

Fig. 12.1. – Continued.

recall here that it has been a common (incorrect) practice in the literature to evaluate these traditional bounds at all the scales between $M_X$ and $M_W$, thus obtaining very important (and of course overestimated) restrictions in the parameter space. Figure 12.1(b) shows the region excluded by the "improved" CCB constraints obtained in Sec. 12.5. Clearly, the excluded region becomes dramatically increased. Notice also that all the trilinear couplings (except the top one in this case) give restrictions, producing areas constrained by different types of bounds simultaneously. The restrictions coming from the UFB constraints, obtained in Sec. 12.4, are shown in Fig. 12.1(c). By far, the most restrictive bound is the UFB–3 one (small filled squares). Indeed, the UFB–3 constraint is the *strongest* one of *all* the UFB and CCB constraints, excluding extensive areas of the parameter space. This is a most remarkable result. Finally, in Fig. 12.1(d) we summarize all the constraints plotting also the excluded region due to the experimental bounds on SUSY particle masses (filled diamonds). The finally allowed region (white) is quite small.

How do these results evolve when we vary the values of $m$ and $B$? The results indicate that the smaller the value of $m$ the more restrictive the bounds become (an explanation of this behavior will be given below). More precisely for $m < 50$ GeV the whole parameter space becomes forbidden (for any value of the remaining parameters). So, from UFB and CCB constraints we can conclude

$$m \geq 50 \text{ GeV} \ . \tag{12.41}$$

Obviously, the limiting case $m = 0$ is excluded. This is very relevant for no-scale models, since $m = 0$ is a typical prediction in that kind of scenarios. Concerning the remaing parameter, $B$, the results indicate that the larger the value of $B$, the more restrictive the bounds. In general, for $m \lesssim 500$ GeV [h], $B$ has to satisfy the bound

$$|B| \lesssim 3.5\ m \ . \tag{12.42}$$

Figures illustrating eqs.(12.41,12.42) can be found in Refs. 7, 15.

So far, we have just presented the numerical results in the figs.1a–d and eqs.(12.41,12.42) with no attempt to explain the physical reasons underlying them. It is, however, very instructive to examine this question. The first thing to note is that, due to their structure, the CCB bounds on $A/m$ (see eqs.(12.37,12.40)) are essentially $m$–invariant and $B$–invariant. The numerical analysis confirms this fact.[7,15] On the other hand, the UFB–3, which is the strongest (CCB and UFB) bound, becomes more stringent as $m_{H_2}^2 = m_2^2 - \mu^2$ (i.e. the $H_2$ soft mass) becomes more negative. This is clear from eqs.(12.25–12.27). The precise value of $m_{H_2}^2$ at low energy depends on its initial value at $M_X$, i.e. $m$, and on the RG running that, due to the effect of $\lambda_{top}$, brings $m_{H_2}^2$ to negative values. Consequently, the smaller $m$ and the larger $\lambda_{top}$, the stronger the UFB–3 bound becomes. Concerning $m$ this result is certainly well reflected in eq.(12.41). Concerning $\lambda_{top}$, since $m_{top} \sim \lambda_{top}\langle H_2\rangle$, where $\langle H_2\rangle = 2M_W^2 \sin\beta/g_2^2$, it is clear that the smaller $\tan\beta$, the larger $\lambda_{top}$ and therefore the stronger the UFB–3 bound. But $\tan\beta$ decreases as $B$ increases, thus the form of eq.(12.42). On the other

[h] Larger values of $m$ start to conflict clearly the naturality bounds for electroweak breaking,[6,14] so they are not realistic.

hand, values of $\tan\beta$ too close to 1 demand a value of $\lambda_{top}$ at low energy higher than the infrared fixed point value, which is impossible to get from the running from high energies. This fact also contributes to the upper bound on $|B|$, eq.(12.42).

To summarize, the UFB and CCB bounds, specially the UFB-3 bound, put important constraints on the MSSM parameter space. Contrary to a common believe, the bounds affect not only the trilinear parameter, $A$, but also the values of the universal scalar mass, $m$, the bilinear term parameter, $B$, and the universal gaugino mass, $M$. This can be noted from the figures 1a–d and eqs.(12.41, 12.42). Also, the frequently used constraint $|A| \leq 3m$ is not in general a good approximation. The actual bounds on $A$ depend on the values of the other SUSY parameters $(m, M, B)$.

The application of the UFB and CCB bounds to particular SUSY scenarios has been considered in some works. It is worth–mentioning that the string-inspired dilaton-dominated scenario is completely excluded on these grounds[16] (as the above-mentioned no-scale scenarios). The infrared fixed point scenario is also severely constrained[15] (in particular it requires $M < 1.1\ m$).

## 12.7. CCB Constraints on Flavor-Mixing Couplings

Supersymmetry has sources of flavor violation which are not present in the Standard Model.[17] These arise from the possible presence of non-diagonal terms in the squark and slepton mass matrices, coming from the soft-breaking potential (see [i] eq.(12.2c))

$$\begin{aligned} V_{\rm soft} = {} & \left(m_L^2\right)_{ij} \bar{L}_{L_i} L_{L_j} \ + \ \left(m_{e_R}^2\right)_{ij} \bar{e}_{R_i} e_{R_j} \\ & + \ \left(m_Q^2\right)_{ij} \bar{Q}_{L_i} Q_{L_j} \ + \ \left(m_{u_R}^2\right)_{ij} \bar{u}_{R_i} u_{R_j} \ + \ \left(m_{d_R}^2\right)_{ij} \bar{d}_{R_i} d_{R_j} \\ & + \left[A_{ij}^l \bar{L}_{L_i} H_1 e_{R_j} + A_{ij}^u \bar{Q}_{L_i} H_2 u_{R_j} + A_{ij}^d Q_{L_i} H_1 d_{R_j} + \text{h.c.}\right] + \dots \end{aligned} \tag{12.43}$$

where $i, j = 1, 2, 3$ are generation indices. A usual simplifying assumption of the MSSM is that $m_{ij}^2$ is diagonal and universal and $A_{ij}$ is proportional to the corresponding Yukawa matrix. Actually, we have implicitely used this assumption in all the previous sections. However, there is no compelling theoretical argument for these hypotheses [j].

---

[i] We work in a basis for the superfields where the Yukawa coupling matrices are diagonal.

[j] This is why, contrary to eq.(12.1), we have not factorized the Yukawa couplings, $\lambda$, in the trilinear terms in eq.(12.43).

The size of the off-diagonal entries in $m^2_{ij}$ and $A_{ij}$ is strongly restricted by FCNC experimental data.[17–20] Here, we will focus our attention on the $A^{(f)}_{ij}$ terms; a summary of the corresponding FCNC bounds is given in the second column of Table 1.[18,20] The $\left(\delta^{(f)}_{LR}\right)_{ij}$ parameters used in the table are defined as

$$\left(\delta^{(f)}_{LR}\right)_{ij} \equiv \frac{\left(\Delta M^{2\ (f)}_{LR}\right)_{ij}}{M^{2\ (f)}_{av}} \ , \tag{12.44}$$

where $f = u, d, l$; $M^{2\ (f)}_{av}$ is the average of the squared sfermion ($\tilde{f}_L$ and $\tilde{f}_R$) masses and $\left(\Delta M^{2\ (f)}_{LR}\right)_{ij} = A^{(f)}_{ij}\langle H^0_f\rangle$, with $H^0_u \equiv H^0_2$, $H^0_{d,l} \equiv H^0_1$, are the off-diagonal entries in the sfermion mass matrices. It is remarkable that the $A^{(f)}_{ij}$ terms are also restricted on completely different grounds, namely from the requirement of the absence of dangerous charge and color breaking (CCB) minima or unbounded from below (UFB) directions. These bounds are in general *stronger than the FCNC ones.* Other properties of these bounds are the following:

*i)* Some of the bounds, particularly the UFB ones, are genuine effects of the non-diagonal $A^{(f)}_{ij}$ structure, i.e. they do not have a "diagonal counterpart".

*ii)* Contrary to the FCNC bounds, the strength of the CCB and UFB bounds does not decrease as the scale of supersymmetry breaking increases.

There is no room here to review in detail how these bounds arise, although the philosophy is similar to that explained in sects.4, 5 (for further details see Ref. 21). Let us write however the final form of the constraints

**CCB bounds**

$$\begin{aligned}
\left|A^{(u)}_{ij}\right|^2 &\le \lambda^2_{u_k}\left(m^2_{u_{L_i}} + m^2_{u_{R_j}} + m^2_2\right), \qquad k = \text{Max}\ (i,j)\\
\left|A^{(d)}_{ij}\right|^2 &\le \lambda^2_{d_k}\left(m^2_{d_{L_i}} + m^2_{d_{R_j}} + m^2_1\right), \qquad k = \text{Max}\ (i,j)\\
\left|A^{(l)}_{ij}\right|^2 &\le \lambda^2_{e_k}\left(m^2_{e_{L_i}} + m^2_{e_{R_j}} + m^2_1\right), \qquad k = \text{Max}\ (i,j)
\end{aligned} \tag{12.45}$$

**UFB bounds**

$$\left|A^{(u)}_{ij}\right|^2 \le \lambda^2_{u_k}\left(m^2_{u_{L_i}} + m^2_{u_{R_j}} + m^2_{e_{L_p}} + m^2_{e_{R_q}}\right), \qquad k = \text{Max}\ (i,j),\ \ p \neq q\ .$$

$$\left|A_{ij}^{(d)}\right|^2 \le \lambda_{d_k}^2\left(m_{d_{L_i}}^2 + m_{d_{R_j}}^2 + m_{\nu_m}^2\right), \qquad k = \text{Max}\,(i,j)$$

$$\left|A_{ij}^{(l)}\right|^2 \le \lambda_{e_k}^2\left(m_{e_{L_i}}^2 + m_{e_{R_j}}^2 + m_{\nu_m}^2\right), \qquad k = \text{Max}\,(i,j), \quad m \ne i,j. \tag{12.46}$$

The CCB bounds must be evaluated at a renormalization scale $Q \sim 2A_{ij}^{(f)}/\lambda_{f_k}^2$, while the UFB bounds must be imposed at any $Q^2 \gg (m/\lambda_{f_k})^2$. This can be relevant in many instances. For example, for universal gaugino and scalar masses ($M_{1/2}$ and $m$ respectively) satisfying $M_{1/2} \gtrsim m$, the UFB bounds are more restrictive at $M_X$ than at low energies (especially the hadronic ones). This trend gets stronger as the ratio $M_{1/2}/m$ increases.

The previous CCB and UFB bounds can be expressed in terms of the $\left(\delta_{LR}^{(f)}\right)_{ij}$ parameters defined in eq.(12.44) and compared with the corresponding FCNC bounds. It turns out that the former are almost always stronger. This is illustrated in Table 1 for the particular case $M_{av}^{(f)} = 500$ GeV. The only exception is $\left(\delta_{LR}^{(l)}\right)_{12}$, which is experimentally constrained by the $\mu \to e, \gamma$ process. As the scale of supersymmetry breaking increases the FCNC bounds are easily satisfied whereas the CCB and UFB bounds continue to strongly constrain the theory.

Another case in which the FCNC constraints are satisfied is when approximate "infrared universality" emerges from the RG equations.[18,19,22] Again, the CCB and UFB bounds continue to impose strong constraints on such theories. This is because, as argued before, these bounds have to be evaluated at different large scales and do not benefit from RG running.

## 12.8. Cosmological Considerations and Final Comments

As has been mentioned in Sec. 12.1, the CCB and UFB bounds presented here are conservative; they correspond to sufficient, but not necessary, conditions for the viability of the standard vacuum. It is possible that we live in a metastable vacuum,[2,23] whose lifetime is longer than the age of the universe. This certainly softens the constraints obtained here.

The first study on CCB-metastability bounds was performed by Claudson et al.[2] They showed that only the top-Yukawa CCB bounds are dangerous from this point of view. In other words, among the various CCB minima (see Sec. 12.5), the one associated with the top-Yukawa coupling

Table 12.1. FCNC bounds versus CCB and UFB bounds on $(\delta^{(f)}_{LR})_{ij}$ for $M^{(f)}_{av} = 500$ GeV. The bounds have been obtained from Ref. 20 taking $x = (m_{\rm gaugino}/M^{(f)}_{av})^2 = 1$.

| | FCNC | CCB and UFB |
|---|---|---|
| $\left(\delta^{(d)}_{LR}\right)_{12}$ | $4.4 \times 10^{-3}$ | $2.9 \times 10^{-4}$ |
| $\left(\delta^{(d)}_{LR}\right)_{13}$ | $3.3 \times 10^{-2}$ | $10^{-2}$ |
| $\left(\delta^{(d)}_{LR}\right)_{23}$ | $1.6 \times 10^{-2}$ | $10^{-2}$ |
| $\left(\delta^{(u)}_{LR}\right)_{12}$ | $3.1 \times 10^{-2}$ | $2.3 \times 10^{-3}$ |
| $\left(\delta^{(l)}_{LR}\right)_{12}$ | $8.5 \times 10^{-6}$ | $3.6 \times 10^{-4}$ |
| $\left(\delta^{(l)}_{LR}\right)_{13}$ | $5.5 \times 10^{-1}$ | $6.1 \times 10^{-3}$ |
| $\left(\delta^{(l)}_{LR}\right)_{23}$ | $10^{-1}$ | $6.1 \times 10^{-3}$ |

is the only one to which the realistic minimum has a substantial probability to decay during the universe life-time. The remaining CCB minima, although deeper, present too high barriers for an efficient tunnelling. That analysis has been re–done by Kusenko et al.,[23] taking into account some subtleties when analyzing the transition probabilities. Their results are qualitatively similar to those of Ref. 2. Quantitatively, they obtain a bound similar to the CCB-1 bound (see eq.(12.38)), empirically modified as $|A_t|^2 \lesssim 3\left[m_2^2 - \mu^2 + 2.5(m^2_{Q_t} + m^2_{\tilde{t}})\right]$, which of course is weaker than the pure stability bound. On the other hand, the UFB bounds (in particular the UFB-3 bound, which is the strongest of all the CCB and UFB bounds) have not been analyzed yet from this point of view.

It is important to keep in mind that the metastability bounds represent necessary but perhaps not sufficient conditions to be safe (in the same sense that the stability bounds presented in sects.4–5 represent sufficient but perhaps not necessary conditions). The reason is that for the applicability of the metastability bounds, the universe should be driven by some mechanism into the realistic (but local and metastable) minimum. This problem has been treated in several papers.[23,24] Of course, a definite answer (not based in an anthropic principle) requires the consideration of a particular cosmological scenario in order to determine the initial values of the relevant fields at early times. Apparently, the realistic minimum is indeed favoured in many cosmological scenarios. Namely, if the initial conditions are dicted by thermal effects, the universe tends to fall into the realistic minimum since it is the closest one to the origin. This can also be the case in some inflationary scenarios. However, a more systematic analysis of these issues would be welcome.

Finally, from a more philosophic point of view, it is conceptually difficult to understand how the cosmological constant is vanishing precisely in a local "interim" vacuum (especially from an inflationary point of view). It is also interesting that many of the (tree-level) UFB directions presented here, particularly the ones associated with flavour violating couplings, are really unbounded from below (after radiative corrections) and, if present, make the theory ill-defined, at least until Planckean physics comes to the rescue. These issues, however, enter the realm of still unknown pieces of fundamental physics.

## References


1. J.M. Frere, D.R.T. Jones and S. Raby, *Nucl. Phys.* **B222** (1983) 11; L. Alvarez-Gaumé, J. Polchinski and M. Wise, *Nucl. Phys.* **B221** (1983) 495; J.P. Derendinger and C.A. Savoy, *Nucl. Phys.* **B237** (1984) 307; C. Kounnas, A.B. Lahanas, D.V. Nanopoulos and M. Quirós, *Nucl. Phys.* **B236** (1984) 438.
2. M. Claudson, L.J. Hall and I.Hinchliffe, *Nucl. Phys.* **B228** (1983) 501.
3. C. Ford, D.R.T. Jones, P.W. Stephenson and M.B. Einhorn, *Nucl. Phys.* **B395** (1993) 17.
4. J.A. Casas, J.R. Espinosa, M. Quirós and A. Riotto, *Nucl. Phys.* **B436** (1995) 3.
5. G. Gamberini, G. Ridolfi and F. Zwirner, *Nucl. Phys.* **B331** (1990) 331.
6. B. de Carlos and J.A. Casas, *Phys. Lett.* **B309** (1993) 320.
7. J.A. Casas, A. Lleyda and C. Muñoz, *Nucl. Phys.* **B471** (1996) 3.
8. J. Ellis, J.F. Gunion, H.E. Haber, L. Roszkowski and F. Zwirner, *Phys. Rev.* **D39** (1989) 844.
9. M. Drees, M. Glück and K. Grassie, *Phys. Lett.* **B157** (1985) 164.
10. J.F. Gunion, H.E. Haber and M. Sher, *Nucl. Phys.* **B306** (1988) 1.
11. H. Komatsu, *Phys. Lett.* **B215** (1988) 323.
12. G.G. Ross, "Grand Unified Theories", Benjamin Publishing Co. (1985).
13. R. Barbieri, S. Ferrara and C.A. Savoy, *Phys. Lett.* **119B** (1982) 343; L. Hall, J. Lykken and S. Weinberg, *Phys. Rev.* **D27** (1983) 2359.
14. R. Barbieri and G.F. Giudice, *Nucl. Phys.* **B306** (1988) 63.
15. J.A. Casas, A. Lleyda and C. Muñoz, *Phys. Lett.* **B389** (1996) 305.
16. J.A. Casas, A. Lleyda and C. Muñoz, *Phys. Lett.* **B380** (1996) 59.
17. S. Dimopoulos and H. Georgi, *Nucl. Phys.* **B193** (1981) 150; F. Gabbiani and A. Masiero *Nucl. Phys.* **B322** (1989) 235; J.S. Hagelin, S. Kelley and T. Tanaka *Nucl. Phys.* **B415** (1994) 293; R. Barbieri and L.J. Hall,

*Phys. Lett.* **B338** (1994) 212; S. Dimopoulos and D. Sutter *Nucl. Phys.* **B452** (1995) 49.
18. D. Choudhury, F. Eberlein, A. Konig, J. Louis and S. Pokorski, *Phys. Lett.* **B342** (1995) 180
19. B. de Carlos, J.A. Casas and J.M. Moreno, *Phys. Rev.* **D53** (1996) 6398.
20. F. Gabbiani, E. Gabrielli, A. Masiero and L. Silvestrini, *Nucl. Phys.* **B477** (1996) 321
21. J.A. Casas and S. Dimopoulos, *Phys. Lett.* **B387** (1996) 107.
22. M. Dine, A. Kagan and S. Samuel, *Phys. Lett.* **B243** (1990) 250
23. A. Kusenko, P. Langacker and G. Segre, *Phys. Rev.* **D54** (1996) 5824.
24. T. Falk, K. Olive, L. Roszkowski and M. Srednicki, *Phys. Lett.* **B367** (1996) 183; A. Riotto and E. Roulet, *Phys. Lett.* **B377** (1996) 60; A. Strumia, *Nucl. Phys.* **B482** (1996) 24.

# Regularisation of Supersymmetric Theories

I. Jack[*] and D. R. T. Jones[†]
*Department of Mathematical Sciences, University of Liverpool,*
*Peach Street, Liverpool L69 3BX, UK*
[*] *dij@liv.ac.uk*
[†] *drtj@liv.ac.uk*

We discuss issues that arise in the regularisation of supersymmetric theories.

## 13.1. Beyond the Tree Approximation

In this chapter we consider issues of both practice and principle that arise when we take supersymmetric theories and calculate radiative corrections. It is usually the case that a symmetry of the Lagrangian is still a symmetry of the full quantum effective action; in which case we say that radiative corrections preserve the symmetry. There are important exceptions to this rule, however: for example conformal invariance is in general violated by radiative corrections, and massless quantum electrodynamics has a global $U_1$ axial symmetry which is violated at the one-loop level (in accordance with the famous Adler-Bardeen theorem). It is not a priori obvious, therefore, that supersymmetry is a symmetry of the full quantum theory in any particular case. Indeed it has been occasionally claimed that there exists a supersymmetry anomaly. In some cases these claims have been erroneous, and have occurred because it is difficult to distinguish between a genuine anomaly and an apparent violation of a supersymmetric Ward identity due to use of a regularisation method that itself violates supersymmetry. Contrariwise, a detailed formal renormalisation program has been pursued in a series of papers by Piguet and collaborators[1] including one[2] where a proof that supersymmetry is *not* anomalous was presented. There is no real reason to doubt this conclusion (although the treatment of infra-red singularities in the program is a possible weakness), for the class of theories considered; the evidence adduced so far points to supersymmetry being a symmetry of the full quantum theory.

The existence of an anomaly is intimately related to the question of regularisation. Beyond the tree level, certain amplitudes in any given quantum field theory are not defined, due to divergences caused by the need to integrate over all momenta for particles in intermediate states. Regularisation is the process whereby the result of an ill-defined correction to a given amplitude is separated into a finite part (which is retained) and an "infinite" part (or more precisely, a part which tends to infinity in the limit that a certain parameter, or parameters, specific to the regularisation method is removed) which is removed from the theory ("subtracted") by introducing a counter-term which precisely cancels it. If the regularised theory fails to respect any given symmetry then the finite amplitudes will fail to satisfy the Ward identities of the symmetry, giving rise to an apparent anomaly and the confusion alluded to above. When the anomaly really is specious, it is possible to restore invariance by modifying the counter-terms by finite amounts. (Obviously the counter-terms are ambiguous, in that their defining role is to cancel something which becomes infinite as the regulator is removed, so that adding a finite quantity to any counter–term leaves its *raison d'être* intact. If no modification of the counter-terms will restore the Ward identity then there is an anomaly.[a]

From a formal point of view, the choice of regularisation scheme made in the implementation of a renormalisation is not of great significance; it is important only that it corresponds to addition of *local* counter-terms. For the extraction of physical predictions, however, the choice becomes a matter of considerable practical significance. It is convenient, for example, to use a regularisation method that preserves symmetries. It should be clear, in fact, from the above discussion that the existence of a regulator consistent with a given symmetry suffices to prove that symmetry to be anomaly–free. In this context the approach of West,[3] consisting of higher derivative regularisation supplemented by Pauli-Villars [b] at one loop, is worthy of consideration; but as the author himself remarks, the issue of a possible anomaly is not thereby fully resolved because of the infra-red difficulties already alluded to.

---

[a]Sometimes an anomaly can be apparently removed only to reappear in another guise. This is the case with the Adler–Bardeen anomaly, which is a property of the fermion triangle with two vector and one axial–vector vertices. The anomaly may affect the axial current or the vector current, depending on how the theory is regularised.

[b]Use of Pauli-Villars at one loop in supersymmetry was also advocated by Gaillard[4] in 1995.

Dimensional regularisation (DREG) is an elegant and convenient way of dealing with the infinities that arise in quantum field theory beyond the tree approximation.[5] It is well adapted to gauge theories because it preserves gauge invariance; it is less well-suited, however, for supersymmetry because invariance of an action with respect to supersymmetric transformations only holds in general for specific values of the space-time dimension $d$. This is essentially due to the fact that a necessary condition for supersymmetry is equality of Bose and Fermi degrees of freedom. In non-gauge theories it is relatively easy to circumvent this problem, and DREG as usually employed is, in fact, a supersymmetric procedure. Gauge theories are a different matter, however, and the question as to whether there exists a completely satisfactory supersymmetric regulator for gauge theories remains controversial.

An elegant attempt to modify DREG so as to render it compatible with supersymmetry was made by Siegel.[6] The essential difference between Siegel's method (DRED) and DREG is that the continuation from 4 to $d$ dimensions is made by *compactification*, or *dimensional reduction.* Thus while the momentum (or space-time) integrals are $d$-dimensional in the usual way, the number of field components remains unchanged and consequently supersymmetry is undisturbed. (A pedagogical introduction to DRED was given by Capper et al.[7])

As pointed out by Siegel himself,[8] there remain potential ambiguities with DRED associated with treatment of the Levi-Civita symbol, $\epsilon^{\mu\nu\rho\sigma}$. We will address this difficulty and the related one involving $\gamma^5$ in Section 13.3.

We must also address problems which arise only when DRED is applied to non-supersymmetric theories. That DRED is a viable alternative to DREG in the non–supersymmetric case was claimed early on.[7] Subsequently it has been so used, occasionally, motivated usually by the fact that Dirac matrix algebra is easier in four dimensions–and in particular by the desire to use Fierz identities.[9,10] One must, however, be very careful in applying DRED to non-supersymmetric theories because of the existence of *evanescent couplings.* These were first described[11] in 1979. Their existence was used by van Damme and 't Hooft[12] to argue that while DRED is a satisfactory procedure for supersymmetric theories[13,14] (modulo the subtleties alluded to above), it leads to a catastrophic loss of unitarity in the non-supersymmetric case. Evidently there is an important issue to be resolved here–is use of DRED in fact forbidden (except in the supersymmetric case) in spite of its apparent convenience? It has, in fact, been conclusively demonstrated[15,16] that if DRED is employed in the manner envisaged by

Capper et al,[7] (which as we shall see differs in an important way from the definition of DRED primarily used by 't Hooft and van Damme) then there is no problem with unitarity. There exist a set of transformations whereby the $\beta$–functions of a particular theory (calculated using DRED) may be related to the $\beta$–functions of the same theory (calculated using DREG) by means of coupling constant reparametrisation. The key is that a correct description of any non-supersymmetric theory impels us to a recognition of the fundamental fact that in general the evanescent couplings renormalise in a manner different from the "real" couplings with which we may be tempted to associate them. This means that care must be taken as we go beyond one loop; nevertheless it is still possible to exploit the simplifications in the Dirac algebra which have motivated the use of DRED. We will return to this point later.

At this point the reader may wonder why, in a book about supersymmetry, we should worry about renormalising non–supersymmetric theories at all. The main practical reason is that the supersymmetric standard model is an effective theory in which supersymmetry is *explicitly* broken, albeit by terms with non–zero dimension of mass.

The reader may also feel that, given the problems with DRED, we should explore other regulators. One example is differential regularisation.[17] The fact is, however, that the convenience of DREG for calculations beyond one loop makes the use of some variant of it very desirable. Use of other proposed regulators is rarely pursued beyond verification of some already known (and usually one-loop) results.

## 13.2. Introduction to DRED

As a concrete example, let us consider a non-abelian gauge theory with fermions but no elementary scalars. The theory to be studied consists of a Yang-Mills multiplet $W^a_\mu(x)$ with a multiplet of spin $\frac{1}{2}$ fields [c] $\psi^\alpha(x)$ transforming according to an irreducible representation $R$ of the gauge group $G$. Of course if $\psi$ *is* Majorana, then $R$ must be a real representation, since the Majorana condition is not preserved by a unitary transformation.

The Lagrangian density (in terms of bare fields) is

$$L_B = -\frac{1}{4}G^2_{\mu\nu} - \frac{1}{2\alpha}(\partial^\mu W_\mu)^2 + C^{a*}\partial^\mu D^{ab}_\mu C^b + i\overline{\psi}^\alpha\gamma^\mu D^{\alpha\beta}_\mu\psi^\beta \tag{13.1}$$

[c]Which may be Dirac or Majorana at this stage.

where

$$G^a_{\mu\nu} = \partial_\mu W^a_\nu - \partial_\nu W^a_\mu + g f^{abc} W^b_\mu W^c_\nu \tag{13.2}$$

and

$$D^{\alpha\beta}_\mu = \delta^{\alpha\beta}\partial_\mu - ig(R^a)^{\alpha\beta} W^a_\mu \tag{13.3}$$

and the usual covariant gauge fixing and ghost terms have been introduced.

The process of dimensional reduction consists of imposing that all field variables depend only on a subset of the total number of space-time dimensions– in this case $d$ out of 4 where $d = 4 - \epsilon$. We will use $\mu, \nu \cdots$ to denote 4-dimensional indices and $i, j$ to denote $d$-dimensional ones, with corresponding metric tensors $g_{\mu\nu}$ and $g_{ij}$. It is also convenient to introduce "hatted" quantities (such as $\hat{g}_{\mu\nu}$ and $\hat{\gamma}^\mu$) which are identical to the corresponding $d$-dimensional quantities $(g_{ij}, \gamma^i \cdots)$ within the $d$-dimensional subspace, but whose remaining components are zero. Momenta $p_\mu$ exist only in the $d$ dimensional subspace so we do not bother to "hat" them. Thus we have for example

$$\not{p} = p_\mu \gamma^\mu = p_\mu \hat{\gamma}^\mu \tag{13.4}$$

and

$$g^{\mu\nu} g_{\mu\nu} = 4, \quad \hat{g}^{\mu\nu}\hat{g}_{\mu\nu} = g^{ij} g_{ij} = d. \tag{13.5}$$

In particular, we have that

$$\hat{g}^{\mu\nu} g_\nu{}^\lambda = \hat{g}^{\mu\lambda} \qquad \text{and} \qquad \hat{g}^{\mu\nu}\gamma_\nu = \hat{\gamma}^\mu. \tag{13.6}$$

These apparently innocuous relations will cause us trouble in the next section.

In order to fully appreciate the consequences of DRED for $L_B$ we must make the decomposition

$$W^a_\mu(x^j) = \{W^a_i(x^j), W^a_\sigma(x^j)\} \tag{13.7}$$

where

$$\delta^i_i = \delta^j_j = d \qquad \text{and} \qquad \delta_{\sigma\sigma} = \epsilon. \tag{13.8}$$

It is then easy to show that

$$L_B = L^d_B + L^\epsilon_B \tag{13.9}$$

where

$$L^d_B = -\frac{1}{4}G^2_{ij} - \frac{1}{2\alpha}(\partial^i W_i)^2 + C^{a*}\partial^i D^{ab}_i C^b + i\overline{\psi}^\alpha \gamma^i D^{\alpha\beta}_i \psi^\beta \tag{13.10}$$

and

$$L_B^\epsilon = \frac{1}{2}(D_i^{ab}W_\sigma^b)^2 - g\overline{\psi}\gamma_\sigma R^a\psi W_\sigma^a - \frac{1}{4}g^2 f^{abc}f^{ade}W_\sigma^b W_{\sigma'}^c W_\sigma^d W_{\sigma'}^e. \quad (13.11)$$

Conventional dimensional regularisation (DREG) amounts to using Eq. 13.10 and discarding Eq. 13.11. For DRED , on the other hand we include both. [d] In simple applications it is in general more convenient to eschew the separation performed above and calculate with 4-dimensional and $d$-dimensional indices rather than $d$-dimensional and $\epsilon$-dimensional ones. As a simple illustration, consider the following typical calculation:

$$\gamma^\mu \not{p}\gamma_\mu = p_\nu\gamma^\mu\gamma^\nu\gamma_\mu = -2p_\nu\gamma^\nu = -2\not{p} \quad (13.12)$$

or equivalently

$$\gamma^\mu \not{p}\gamma_\mu = \gamma^i \not{p}\gamma_i + \gamma^\sigma \not{p}\gamma_\sigma = (2-d)\not{p} + (d-4)\not{p} = -2\not{p}. \quad (13.13)$$

From the dimensionally reduced form of the gauge transformations:

$$\begin{aligned} \delta W_i^a &= \partial_i\Lambda^a + gf^{abc}W_i^b\Lambda^c \\ \delta W_\sigma^a &= gf^{abc}W_\sigma^b\Lambda^c \\ \delta\psi^\alpha &= ig(R^a)^{\alpha\beta}\psi^\beta\Lambda^a \end{aligned} \quad (13.14)$$

we see that each term in Eq. 13.11 is separately invariant under gauge transformations. The $W_\sigma$-fields behave exactly like scalar fields, and are hence known as $\epsilon$-scalars. The significance of this is that gauge invariance *per se* provides no reason to expect the $\overline{\psi}\psi W_\sigma$ vertex to renormalise in the same way as the $\overline{\psi}\psi W_i$ vertex. In the case of the quartic $\epsilon$-scalar coupling the situation is more complex since in general of course more than one such coupling is permitted by Eq. 13.14. In other words, we cannot in general expect the $f-f$ tensor structure present in Eq. 13.11 to be preserved under renormalisation: and indeed it is not. This is clear from the abelian case, where there is no such quartic interaction in $L_B^\epsilon$ but there is a divergent contribution at one loop from a fermion loop. The significance of all this and its impact on the 3-loop DRED calculation of the QCD gauge $\beta$-function has recent been explored in some detail.[18]

In the case of supersymmetric theories, however, these difficulties do not arise. If $\psi$ above is a Majorana fermion in the adjoint representation, then $L_B$ is supersymmetric. This links $W_i$ and $W_\sigma$ in a way that is not severed

[d]The additional contributions from $L_B^\epsilon$ are precisely what is required to restore the supersymmetric Ward identities at one loop in supersymmetric theories, as described in section 13.4.

by the dimensional reduction. Thus the $\overline{\psi}\psi W_\sigma$ and $\overline{\psi}\psi W_i$ vertices (both equal to $g$ at the tree level) remain equal under renormalisation. We will return in section 13.7 to the application of DRED to non-supersymmetric theories.

## 13.3. DRED Ambiguities

With DRED it would seem that necessarily $d < 4$, since the regulated action is, after all, defined by dimensional *reduction*. Then, given $d < 4$, one can define an object $\hat{\epsilon}^{\mu\nu\rho\sigma}$ as follows:

$$\hat{\epsilon}^{\mu\nu\rho\sigma} = \hat{g}^{\mu\alpha}\hat{g}^{\nu\beta}\hat{g}^{\rho\gamma}\hat{g}^{\sigma\delta}\epsilon_{\alpha\beta\gamma\delta} \tag{13.15}$$

where $\epsilon_{\alpha\beta\gamma\delta}$ is the usual 4-dimensional tensor. Unfortunately it is now possible to show that algebraic inconsistencies result[8] unless $d = 4$. Let us illustrate these problems in the two dimensional case. The alternating tensor $\epsilon^{\mu\nu}$ satisfies (in two Euclidean dimensions) the relation

$$\epsilon^{\mu\nu}\epsilon^{\rho\sigma} = g^{\mu\rho}g^{\nu\sigma} - g^{\mu\sigma}g^{\nu\rho}. \tag{13.16}$$

Using Eq. 13.6 it is easy to show that

$$\hat{\epsilon}^{\mu\nu}\hat{\epsilon}^{\rho\sigma} = \hat{g}^{\mu\rho}\hat{g}^{\nu\sigma} - \hat{g}^{\mu\sigma}\hat{g}^{\nu\rho} \tag{13.17}$$

where $\hat{\epsilon}^{\mu\nu}$ is defined similarly to Eq. 13.15.

However it is trivial to demonstrate that the result of applying Eq. 13.17 to the tensor

$$A^{\mu\nu} = \hat{\epsilon}^{\mu\nu}\hat{\epsilon}^{\rho\sigma}\hat{\epsilon}_{\rho\sigma} \tag{13.18}$$

is ambiguous inasmuch that it differs according to which pair of $\hat{\epsilon}$-tensors are selected: the result is the identity

$$(d+1)(d-2)\hat{\epsilon}^{\mu\nu} = 0. \tag{13.19}$$

A related problem (of course) is the fact that the only mathematically consistent treatment of $\gamma^5$ within DREG is predicated[5] on having $d > 4$. Given Eq. 13.6 and the usual relation

$$\{\gamma_\mu, \gamma^5\} = 0, \tag{13.20}$$

it follows that

$$\{\hat{\gamma}_\mu, \gamma^5\} = 0, \tag{13.21}$$

and hence that

$$(d-4)\mathrm{Tr}\left[\gamma^5\hat{\gamma}^\mu\hat{\gamma}^\nu\hat{\gamma}^\rho\hat{\gamma}^\sigma\right] = 0. \tag{13.22}$$

This is unfortunate since it renders problematic the discussion of the axial anomaly.

For $d > 4$, however, Eq. 13.6 does not hold and so Eq. 13.21 no longer follows. Instead we impose

$$\left[\gamma_\sigma, \gamma^5\right] = 0, \quad \text{for} \quad 4 < \sigma < d\,, \tag{13.23}$$

and this leads to a straightforward and unambiguous derivation of the axial anomaly. It has been verified[19] that this prescription correctly reproduces the Adler-Bardeen theorem at the two-loop level using DREG.

Returning to the DRED prescription, there are a number of possible "fixes" at one loop;[20] at two loops, it was shown[19] that the Adler-Bardeen theorem could indeed still be satisfied if relations like

$$\gamma^i \gamma^5 \gamma_i = (d-8)\gamma^5 \tag{13.24}$$

which follow in the $d > 4$ case, are used in conjunction with DRED .

A possible point of view concerning all this[21] is that DRED is terminally inconsistent and should not be used. We believe, however, that the difficulties are essentially technical and can be evaded. For example, one well-defined procedure would be to write

$$\gamma^5 = \frac{1}{4!}\epsilon^{\mu\nu\rho\sigma}\gamma_\mu\gamma_\nu\gamma_\rho\gamma_\sigma \tag{13.25}$$

and factor all out $\epsilon$-tensors. Renormalised amplitudes may then be calculated, which, being finite as $d \to 4$, are unambiguous when the $\epsilon$-tensors are contracted in. [e] We would claim also that other modes of procedure which would give different answers because of the ambiguities detailed above, correspond nevertheless to *the same physical results.* This assertion has in fact been verified in one particular case,[22] where a prescription first suggested by Hull and Townsend[23] was used. This amounted to employing as $\epsilon^{\mu\nu}$ not the usual alternating tensor but instead a structure satisfying

$$\hat{\epsilon}^{\mu}{}_{\nu}\hat{\epsilon}^{\nu\rho} = (1 + c\epsilon)\hat{g}^{\mu\rho} \tag{13.26}$$

(where here $\epsilon = 2 - d$). It turns out[22] that the dependence of the results on the parameter $c$ can be absorbed into redefinitions of the renormalised metric and torsion tensors. In the special case $c = 0$, $\hat{\epsilon}^{\mu\nu}$ is an almost complex structure.[24]

[e] This somewhat cumbersome procedure can usually be finessed. For instance, if only even-parity fermion loops are present, then there is no problem with a fully anticommuting $\gamma^5$.

There have been a considerable number of papers discussing the interpretation of $\gamma^5$ in both DREG and DRED , and the reader may consult them for further enlightenment.[25] We turn in the next section to another (but again related) problem with DRED , arising from the fact that in spite of the correct counting of degrees of freedom, there are still ambiguities associated with establishing invariance of the action: we will look at this in more detail below in the context of the supersymmetry Ward identity.

### 13.4. The Supersymmetry Ward Identity

The first concrete illustration of the different results provided by DRED and DREG for a supersymmetric theory was as follows. Consider the basic supersymmetric gauge theory in the Wess-Zumino gauge as defined by the Lagrangian $L_S$ where

$$L_S = -\frac{1}{4}G^{\mu\nu}G_{\mu\nu} + i\frac{1}{2}\overline{\lambda}^{\alpha}\gamma^{\mu}D^{\alpha\beta}_{\mu}\lambda^{\beta} + \frac{1}{2}D^2. \tag{13.27}$$

In $d = 4$, $L_S$ is invariant (up to a total derivative) under the transformations

$$\begin{aligned} \delta W^a_\mu &= i\overline{\epsilon}\gamma_\mu\lambda^a, \qquad \delta\lambda^a = \frac{1}{2}G^a_{\mu\nu}\gamma^\mu\gamma^\nu\epsilon - iD^a\gamma^5\epsilon \\ \delta D^a &= -\overline{\epsilon}\gamma_\mu\gamma^5(D^\mu\lambda)^a. \end{aligned} \tag{13.28}$$

It is an excellent exercise in spinor algebra to verify this invariance. Note the presence in Eq. 13.28 of $\gamma^5$-terms; to obtain invariance one must assume that $\gamma^5$ is totally anti-commuting. Of course in this particular case we could set $D^a = 0$, and still have an invariance (not involving $\gamma^5$). However this does not escape the Siegel ambiguity, as we shall now show. With due care, one obtains (up to total derivatives)

$$\delta L_S = g\frac{1}{2}f^{abc}\overline{\epsilon}\gamma^\mu\lambda^a\overline{\lambda}^b\gamma_\mu\lambda^c. \tag{13.29}$$

This is identically zero for $d = 4$, though this is not obvious even if we rewrite in two–component formalism; a Fierz re-ordering is required. For $d \neq 4$, $\delta L_S$ is not zero; and the key to the distinction between DRED and DREG lies in the $\gamma^\mu \otimes \gamma_\mu$ contraction, which is $d$-dimensional for DREG and four-dimensional for DRED . There are important consequences for the regularisation of supersymmetric theories, as we shall now see. Let us add to $L_S$ gauge fixing and ghost terms:

$$L_S \to L_T = L_S + L_G \tag{13.30}$$

where

$$L_G = -\frac{1}{2\alpha}(\partial^\mu W_\mu)^2 + C^{a*}\partial^\mu D_\mu^{ab} C^b. \tag{13.31}$$

Then we introduce the functional $Z(J, j, j_D)$ where

$$Z = \int d\{W_\mu\}d\{\lambda\}d\{D\}e^{i\int d^dx\left[L_T + J^\mu W_\mu + \bar{j}\lambda + j_D D\right]} \tag{13.32}$$

and use of Eq. 13.28 leads to the following Ward identity:

$$0 = < \int d^dx \left[J^\mu \delta W_\mu + \bar{j}\delta\lambda + j_D \delta D + \delta L_S + \delta L_G\right] > \tag{13.33}$$

where

$$\delta L_G = -\frac{1}{\alpha}\partial^\mu W_\mu \partial^\mu \delta W_\mu + g f^{abc} C^{a*}\partial^\mu \delta W_\mu^c C^b \tag{13.34}$$

and

$$< X > = \int d\{W_\mu\}d\{\lambda\}d\{D\}X e^{i\int d^dx\left[L_T + J^\mu W_\mu + \bar{j}\lambda + j_D D\right]}. \tag{13.35}$$

Notice we have included the term $\delta L_S$ from Eq. 13.29 to allow for the fact that this may not be zero away from $d = 4$. When this Ward identity was investigated at one loop,[7] it was found to be true with DRED and false with DREG; which conclusion was arrived at because the contribution from $\delta L_S$ was ignored. It is easy to show that this contribution is zero for DRED , and in the case of DREG serves precisely to restore Eq. 13.34. The distinction between DRED and DREG is manifest in the fact that the contribution of $\delta L_S$ is zero in the former case. It is in this sense that DRED is more consistent with supersymmetry. In terms of this Ward identity the difference may not seem crucial, but the fact that $\delta L_S$ is effectively non-zero (with DREG) means that if we employ DREG in a supersymmetric theory then great care must be taken with the formulation of physical predictions. This becomes particularly clear when we generalise to include matter fields; in supersymmetric QCD, for example, the fact that the gluino-quark-squark coupling is equal to the gauge coupling $g$ is a consequence of supersymmetry and will not be preserved under renormalisation if DREG is employed. In fact, this is very similar to the problems that occur when we want to apply DRED to non–supersymmetric theories; once again there are coupling constant relations that are not preserved by radiative corrections.

While DRED was successful in the above application, it does not follow that an insertion of $\delta L_S$ in a diagram of arbitrary complexity gives zero. It can be shown[26] that such an insertion depends on the quantity $\Delta$ where

$$\Delta = \mathrm{Tr}(A\gamma^\mu B\gamma_\mu) + \mathrm{Tr}(A\gamma^\mu)\mathrm{Tr}(B\gamma_\mu) - (-1)^k \mathrm{Tr}(A\gamma^\mu B^R \gamma_\mu). \tag{13.36}$$

Here $A$ and $B$ are products of Dirac $\gamma$-matrices; $k$ is the number of such matrices in $B$, and $B^R$ consists of the the same set of matrices as $B$ but written down in reverse order. In strict $d = 4$, $\Delta$ is zero; but because $A$, $B$ may contain $d$-dimensional $\gamma$-matrices (due to contraction with momenta) $\Delta$ is non-zero in general. If we set

$$A = \gamma^{\mu_1}\gamma^{\mu_2}\cdots\gamma^{\mu_5} \quad \text{and} \quad B = \gamma_{\nu_1}\gamma_{\nu_2}\cdots\gamma_{\nu_5} \tag{13.37}$$

then

$$\Delta = 48\delta^{[\mu_1}_{\nu_1}\delta^{\mu_2}_{\nu_2}\cdots\delta^{\mu_5]}_{\nu_5} \tag{13.38}$$

where the brackets $[\cdots]$ denote antisymmetrisation. This is clearly zero for integer $d \leq 4$ but if the various indices are $d$-dimensional then it is not. For instance if we calculate the trace by contracting $\Delta$ with $\delta^{\nu_1}_{\mu_1}\cdots\delta^{\nu_5}_{\mu_5}$ then we obtain $\mathrm{Tr}\Delta = 48d(d-1)(d-2)(d-3)(d-4)$. A diagram with at least four loops is required[26] so that we get enough $\gamma$-matrices to activate this problem in the propagator Ward identity.[7] This is clearly the same ambiguity at bottom addressed by Siegel.[8]

One might hope that this problem is somehow resolved by use of superfields. Using superfield perturbation theory Feynman rules maintains supersymmetry in a manifest way; but a crucial part of the calculational procedure relies on the reduction of products of supercovariant $D_\alpha$ and $\overline{D}_{\dot\alpha}$ derivatives to products of four or less, and this is possible only when the fact that the $\alpha, \dot\alpha$ indices are two-valued is used. Thus the same ambiguity must be present, albeit in a somewhat different form. However as we argued in the previous section, the ambiguity will not affect physical results since it is equivalent to the freedom available in choice of regularisation scheme, as long as a systematic procedure is adopted. Stockinger has discussed[27] such a systematic procedure; note however that in his approach use of Fierz identities is disallowed, which removes one of the main motivations for using DRED in non-supersymmetric theories.

Despite all difficulties, DRED remains the regulator of choice for supersymmetric theories, and has survived many practical tests.

## 13.5. $N = 2$ and $N = 4$ Supersymmetry

$N = 2$ supersymmetry corresponds, in the language of $N = 1$ superfields, to the special case of a superpotential taking the form:

$$W = \sqrt{2}g\phi^a\xi^T S_a\chi, \tag{13.39}$$

where $\xi, \chi, \phi$ are multiplets transforming under the $S^*$, $S$ and adjoint representations of the gauge group $\mathcal{G}$ respectively. In the special case that $S$ is the adjoint representation we have $N = 4$ supersymmetry. $N = 2$ theories are extraordinary in that they have only one-loop divergences.[28] This means that for $N = 2$, $\beta_g$ vanishes beyond one-loop if computed using DRED; crucial here is the fact that DRED incorporates *minimal subtraction*. In the $N = 4$ case the one-loop contribution also vanishes, so $N = 4$ theories are ultra-violet finite to all orders of perturbation theory.

$N = 2$ and $N = 4$ theories, although obviously of great interest, possess a property that is unfortunate if we want to try and incorporate them into a realistic theory. This property is that since the chiral superfields are either adjoint or in $S, S^*$ pairs, gauge invariant mass terms are possible for the fermionic components of the multiplets. It is difficult, therefore, to arrange for fermion masses (such as the electron mass) to be much less than the scale of supersymmetry–breaking, at least. Nevertheless there have been occasional attempts to construct phenomenologically viable models,[29] and explore their consequences.[30]

## 13.6. The Supersymmetric $\beta$-Functions

In this section we examine the $\beta$-functions for an $N = 1$ supersymmetric theory defined by the superpotential

$$W = \frac{1}{6}Y^{ijk}\Phi_i\Phi_j\Phi_k + \frac{1}{2}\mu^{ij}\Phi_i\Phi_j. \tag{13.40}$$

The multiplet of chiral superfields $\Phi_i$ transforms as a representation $R$ of the gauge group $\mathcal{G}$, which has structure constants $f_{abc}$. In accordance with the non-renormalisation theorem,[31] the $\beta$-functions for the Yukawa couplings $\beta_Y^{ijk}$ are given by

$$\beta_Y^{ijk} = Y^{p(ij}\gamma^{k)}{}_p = Y^{ijp}\gamma^k{}_p + (k \leftrightarrow i) + (k \leftrightarrow j), \tag{13.41}$$

where $\gamma(g, Y)$ is the anomalous dimension for $\Phi$. There exists an all-orders relation between the gauge $\beta$-function $\beta_g(g, Y)$ and $\gamma$ which was first derived (in the special case of a theory without a chiral multiplet) using an argument based on the Adler-Bardeen theorem,[32] and subsequently for the general case using instanton calculus:[33]

$$\beta_g^{\mathrm{NSVZ}} = \frac{g^3}{16\pi^2}\left[\frac{Q - 2r^{-1}\mathrm{Tr}\left[\gamma^{\mathrm{NSVZ}}C(R)\right]}{1 - 2C(G)g^2(16\pi^2)^{-1}}\right]. \tag{13.42}$$

Here $Q = T(R) - 3C(G)$, $T(R)\delta_{ab} = \mathrm{Tr}(R_a R_b)$, $C(G)\delta_{ab} = f_{acd}f_{bcd}$, $r = \delta_{aa}$ and $C(R)^i{}_j = (R_a R_a)^i{}_j$.

We have added a "NSVZ" label to both $\beta_g$ and $\gamma$ in Eq. 13.42 because of scheme dependence issues which we will discuss shortly. In the special case $Y = 0$, the fixed point $g^* = 0$, defined by

$$Q = \frac{2}{r}\mathrm{Tr}\left[\gamma^{\mathrm{NSVZ}} C(R)\right] \tag{13.43}$$

is important for duality, in the context of the conformal window identified by Seiberg.[34] We will return to this fixed point in the context of large-$N$ expansions.

It turns out that if $\beta_g$ and $\gamma$ are calculated using DRED, then they begin to deviate from Eq. 13.42 at three loops. The relationship between $\beta_g^{\mathrm{NSVZ}}$ and $\beta_g^{\mathrm{DRED}}$ has been explored in detail,[35–37] with the conclusion that there exists an analytic redefinition of $g$, $g \to g'(g, Y)$ which connects them. We emphasise that it is quite non-trivial that the redefinition exists at all; in the abelian case for example, the redefinition consists of a single term, but it affects four distinct terms (with different tensor structure) in the $\beta$-functions. By exploiting the fact that $N = 2$ theories are finite beyond one loop[28] it was possible to determine $\beta_g^{\mathrm{DRED}}$ at three loops by a comparatively simple calculation. and at four loops in the general case except for one undetermined parameter, determined subsequently by an ingenious method, paradoxically involving the softly-broken theory.[38]

Use of what regularisation scheme would lead to the NSVZ result, which is associated with the holomorphic nature of the Wilsonian action? Presumably, for example, a combination of Pauli-Villars and higher derivatives.[3] Notwithstanding the existence of the exact NSVZ result, however, it is still important to have $\beta_g^{\mathrm{DRED}}$ as accurately as possible, because in calculating physical predictions DRED (or, as we shall see, more accurately DRED$'$) is the scheme most often used. The three loop results for $\beta_g^{\mathrm{DRED}}$ and $\gamma^{\mathrm{DRED}}$ were first given by Jack et al.[35,36] (the $\beta_g$ result was subsequently verified by an explicit calculation[39]). These results have found phenomenological applications.[40,41] The results for $\beta_g$ in supersymmetric QCD (SQCD) with $N_f$ flavours and $N_c$ colours are:

$$16\pi^2 \beta_g^{(1)} = (N_f - 3N_c)\, g^3,$$

$$(16\pi^2)^2 \beta_g^{(2)} = \left(\left[4N_c - \frac{2}{N_c}\right] N_f - 6N_c^2\right) g^5,$$

$$(16\pi^2)^3\beta_g^{(3)} = \left(\left[\frac{3}{N_c} - 4N_c\right]N_f^2 + \left[21N_c^2 - \frac{2}{N_c^2} - 9\right]N_f - 21N_c^3\right)g^7. \tag{13.44}$$

For $\beta_g^{(4)}$ we have:[38]

$$(16\pi^2)^4\beta_g^{(4)} = \left(-\tfrac{2}{3N_c}N_f^3 + \left[44 + \tfrac{36\zeta(3)-20}{3N_c^2} - (42+12\zeta(3))N_c^2\right]N_f^2 \right.$$
$$\left. + \left[132N_c^3 - 66N_c - \tfrac{8}{N_c} - \tfrac{4}{N_c^3}\right]N_f - 102N_c^4\right)g^9. \tag{13.45}$$

It is very interesting that the higher order group theory invariants found by van Ritbergen et al[42] in the corresponding calculation for QCD do not appear here. Of course the QCD calculation was done with DREG rather than DRED; but since these group structures first appear at four loops we would expect, for these particular terms, that DRED and DREG should give the same result at this order. It is an excellent check on both calculations, therefore, that when in the QCD case we go to the special case of $N = 1$ supersymmetry (by setting $N_f = \frac{1}{2}$ and putting the fermions in the adjoint representation) the new invariants cancel. It is also interesting to note that in the pure gauge theory, these invariants signalled the first contribution from non-planar structures to $\beta_g$ in QCD; for SQCD, it remains possible that $\beta_g^{\mathrm{DRED}}$ is free of such structures to all orders (this is manifestly so for $\beta_g^{\mathrm{NSVZ}}$ in the absence of chiral superfields, of course).

## 13.7. Soft supersymmetry Breaking

We saw in section 13.2 that under renormalisation the $\epsilon$-scalars behave differently from the gauge fields, except in supersymmetric theories. On might be tempted to assert that it doesn't matter if Green's functions with external $\epsilon$-scalars are divergent (since they are anyway unphysical) and introduce a common wave function subtraction for $W_i$ and $W_\sigma$, a wave function subtraction for $\psi$ and a coupling constant subtraction for $g$, these being determined (as usual) by the requirement that Green's functions with real particles be rendered finite. This was the procedure adopted in the main by van Damme and 't Hooft.[12] On the other hand we could insist on all Green's functions (including those with external $\epsilon$-scalars) being finite, leading to the introduction of a plethora of new subtractions or equivalently coupling constants. We have shown[15,16] that it is only the latter procedure which leads to a consistent theory; the former manifestly breaks unitarity.

Now in a supersymmetric theory the complications described above can be safely ignored. The wave function renormalisations of $W_\sigma$ and $W_i$ are equal because of supersymmetry, and the evanescent couplings remain equal to their "natural" values after renormalisation. At first sight, this conclusion also appears to obtain when supersymmetry is softly broken, since the dimensionless couplings renormalise exactly as in the fully supersymmetric theory. This is not quite true, however, since there is nothing to protect the $\epsilon$–scalars acquiring a mass through interacting with the genuine fields; and, indeed, precisely this happens.[43] In other words, the $\beta$–function for the $\epsilon$-scalar mass $\tilde{m}$ is inhomogeneous with respect to $\tilde{m}$:

$$\beta_{\tilde{m}^2} = A(g, Y)\tilde{m}^2 + \sum_i B_i(g, Y)m_i^2 + \cdots, \tag{13.46}$$

where the $m_i^2$ are the genuine scalar masses, $Y$ represents the Yukawa couplings and the $+\cdots$ denotes terms involving the gaugino mass(es) and the $A$-parameter(s). Moreover, the two–loop $\beta$–functions for the genuine scalar masses depend explicitly on the $\epsilon$–scalar masses, when calculated using DRED. This fact would annoyingly complicate an extension to two loops of the standard running analysis relating the low energy values of the various soft parameters to the corresponding values at gauge unification. Fortunately, however, there exists a hybrid scheme[44] which decouples this $\epsilon$–scalar dependence both from the $\beta$–functions and from the threshold corrections to the physical masses. At leading order, this scheme is arrived at from DRED by redefining the masses $m_i^2$ as follows:

$$m_i^2|_{\text{DRED}'} = m_i^2|_{\text{DRED}} - C_i(g)\tilde{m}^2 \tag{13.47}$$

where $C_i(g)$ is easily calculated.[44] The resulting $\beta_{m^2}$ is independent of $\tilde{m}^2$ though two loops; and conveniently the same transformation removes the $\tilde{m}^2$ term from the one-loop relationship between the renormalised and physical scalar masses $m^2$. So in the DRED$'$ scheme, although $\tilde{m}$ evolves under the renormalisation group, it is decoupled from physical quantities and so can be safely ignored.

Remarkably, we shall see now that the DRED$'$ scheme exists to all orders in perturbation theory, because it is possible to express the $\beta$-functions for the soft supersymmetry-breaking masses and couplings in terms of the $\beta$-functions of the underlying supersymmetric theory by expressions that are *exact*. The key to this revelation is the fact that soft breaking terms may be accommodated within the superfield formalism by the introduction of an external "spurion" field.[45] For a $N = 1$ supersymmetric gauge theory with superpotential as in Eq. 13.40 we take the soft breaking Lagrangian $L_{SB}$ as follows:

$$L_{SB}(\Phi, W_A) = -\left\{\int d^2\theta\eta\Big(\frac{1}{6}h^{ijk}\Phi_i\Phi_j\Phi_k + \frac{1}{2}b^{ij}\Phi_i\Phi_j + \frac{1}{2}MW_A{}^{\alpha}W_{A\alpha}\Big) + \text{h.c.}\right\}$$
$$- \int d^4\theta\eta\eta^{\dagger}\Phi^j(m^2)^i{}_j(e^{2gV})_i{}^k\Phi_k. \qquad (13.48)$$

Here $\eta = \theta^2$ is the spurion external field and $M$ is the gaugino mass. Jack et al[46–48] now showed that all the soft $\beta$-functions can be related to the underlying supersymmetric theory. The $\phi^3$, $\phi^2$ and gaugino mass $\beta$-functions, $\beta_h$, $\beta_b$ and $\beta_M$, are given by the following simple expressions:

$$\begin{aligned}
\beta_h^{ijk} &= \gamma^i{}_l h^{ljk} + \gamma^j{}_l h^{ilk} + \gamma^k{}_l h^{ijl} - 2\gamma^i_{1}{}_l Y^{ljk} - 2\gamma^j_{1}{}_l Y^{ilk} - 2\gamma^k_{1}{}_l Y^{ijl} \\
\beta_b^{ij} &= \gamma^i{}_l b^{lj} + \gamma^j{}_l b^{il} - 2\gamma^i_{1}{}_l \mu^{lj} - 2\gamma^j_{1}{}_l \mu^{il} \\
\beta_M &= \mathcal{O}\left(\frac{\beta_\alpha}{\alpha}\right) \qquad (13.49)
\end{aligned}$$

where we have written $\alpha = g^2$,

$$\mathcal{O} = \left(M\alpha\frac{\partial}{\partial\alpha} - h^{lmn}\frac{\partial}{\partial Y^{lmn}}\right), \qquad (13.50)$$

and

$$(\gamma_1)^i{}_j = \mathcal{O}\gamma^i{}_j. \qquad (13.51)$$

The result for $\beta_{m^2}$ is[48]

$$(\beta_{m^2})^i{}_j = \left[\Delta + \tilde{X}(\alpha, Y, Y^*, h, h^*, m, M)\frac{\partial}{\partial\alpha}\right]\gamma^i{}_j. \qquad (13.52)$$

Here

$$\Delta = 2\mathcal{O}\mathcal{O}^* + 2MM^*\alpha\frac{\partial}{\partial\alpha} + \tilde{Y}_{lmn}\frac{\partial}{\partial Y_{lmn}} + \tilde{Y}^{lmn}\frac{\partial}{\partial Y^{lmn}}, \qquad (13.53)$$

$Y_{lmn} = (Y^{lmn})^*$, and

$$\tilde{Y}^{ijk} = (m^2)^i{}_l Y^{ljk} + (m^2)^j{}_l Y^{ilk} + (m^2)^k{}_l Y^{ijl}. \qquad (13.54)$$

The term in $\tilde{X}$ does not appear in a naive application of the spurion formalism, because (when using DRED) it fails to allow for the fact that, as we have discussed, the $\epsilon$-scalars associated with DRED acquire a mass through radiative corrections.[43] Indeed, in DRED, $\beta_{m^2}$ will actually depend on the $\epsilon$-scalar mass. In DRED$'$ , on the other hand, $\beta_{m^2}$ is given by Eq. 13.52 with the leading contribution to $\tilde{X}$ given by[46]

$$\tilde{X} = -4S\alpha^2(16\pi^2)^{-1} \qquad (13.55)$$

where

$$S = r^{-1}\mathrm{Tr}[m^2 C(R)] - MM^* C(G). \tag{13.56}$$

An interesting feature of the above results for the soft $\beta$-functions is that armed with them we are able to construct the following solutions[49] for the soft parameters, which are *renormalisation group invariant* to all orders of perturbation theory:

$$\begin{aligned} M &= m_0 \beta_g / g \\ h &= -m_0 \beta_Y \\ m^2 &= \frac{1}{2} m_0^2 \mu \frac{d}{d\mu} \gamma \\ b &= \kappa m_0 \mu - m_0 \beta_\mu \end{aligned} \tag{13.57}$$

where $m_0$ and $\kappa$ are constant. Remarkably, the above results for $M, h$ and $m^2$ are those associated with Anomaly Mediated Supersymmetry Breaking.[50–52]

As a result of the exact $\beta$-function results described above it is straightforward to perform the standard running analysis (at two or even three[53] loops) for the MSSM from the gauge unification scale, with whatever boundary conditions there on the masses and couplings seem appropriate. To extract physical predictions for masses or cross-sections requires further calculation of course, and there is a large and growing literature on the subject (see for example Ref. 54). One well known example is that the relationship between the running top quark mass and its physical (pole) mass using DRED′ and retaining only the gluon contribution is given at one loop by

$$m_t^{\mathrm{pole}} = m_t(\mu) \left[ 1 + \frac{\alpha_3(\mu)}{3\pi} \left( 5 - 3 \ln \frac{m_t^2}{\mu^2} \right) \right] \tag{13.58}$$

whereas in DREG one has

$$m_t^{\mathrm{pole}} = m_t(\mu) \left[ 1 + \frac{\alpha_3(\mu)}{3\pi} \left( 4 - 3 \ln \frac{m_t^2}{\mu^2} \right) \right]. \tag{13.59}$$

## 13.8. Large-$N_f$ Supersymmetric Gauge Theories

The large-$N$ expansion is an alternative to conventional perturbation theory that requires calculation (and regularisation) of diagrams with arbitrarily high numbers of loops. In both QCD and SQCD, the large $N_c$ expansion is of particular interest;[55] more tractable, however, is the large $N_f$ expansion.

Recently the leading and $O(1/N_f)$ terms in $\beta_g$ and $\gamma$ have been calculated (using DRED) for a number of supersymmetric theories.[56] We give below the results for SQCD (noting that we have rescaled the gauge coupling, $g \to g/\sqrt{N_f}$):

$$\gamma = -\frac{(N_c^2 - 1)}{N_f N_c} \hat{K} G(\hat{K}), \tag{13.60}$$

and

$$\begin{aligned} \beta_g = g\hat{K} - \tfrac{3N_c}{N_f} g\hat{K} + 4g\hat{K}\tfrac{N_c}{N_f} \int_0^{\hat{K}} (1-x)G(x)\,dx \\ - \tfrac{2g\hat{K}}{N_c N_f} \int_0^{\hat{K}} (1-2x)G(x)\,dx, \end{aligned} \tag{13.61}$$

where $\hat{K} = g^2/(16\pi^2)$, and

$$G(x) = \frac{\Gamma(2-2x)}{\Gamma(2-x)\Gamma(1-x)^2\Gamma(1+x)}. \tag{13.62}$$

These results do not satisfy Eq. 13.42, because they were calculated using DRED. It is quite remarkable that the $O(1/N_f)$ corrections to the SQCD $\beta$-function depend only on simple integrals involving $G(x)$. $G$ has a simple pole at $x = 3/2$ and consequently $\beta_g$ has a logarithmic singularity at $g^2 = 24\pi^2$ and a finite radius of convergence in $g$. Using Eq. 13.61 for $N_f = 6$ and $N_c = 3$, values which lie in the conformal window[34] $3N_c/2 < N_f < 3N_c$, we indeed find an infra-red fixed point in the gauge coupling evolution, corresponding to $g^* \approx 8$. It is interesting that the range of $N_f$ such that $g^* < 24\pi^2$ is given by

$$aN_c < N_f < 3N_c,$$

where $a \approx 1.7$ depends weakly on $N_c$. This is remarkably close to the exact conformal window.

Of course the result for $g^*$ is scheme dependent. In the NSVZ scheme it is possible to show that the $O(1/N_f)$ contribution to $\gamma$ is in fact the same as in DRED, with the corresponding result for $\beta_g$ being easily calculated from Eq. 13.42. One then finds that the fixed point (for $N_c = 3$) corresponds to $g^* \approx 7$.

Of course it is not clear that this regime is within the region of validity of our approximation: do we believe that the appropriate expansion parameter is $N_c/N_f$ or $3N_c/N_f$? It would obviously be interesting if we could calculate more terms in the $1/N_f$ expansion. Even the $O(1/N_f^2)$ contribution presents considerable technical problems, however.

## References

1. O. Piguet, Lectures at the 2nd International Winter School on High-Energy Physics (WHEP 96), Brazil, 1996, hep-th/9611003
2. O. Piguet, M. Schweda and K. Sibold, *Nucl. Phys.* B **174**, 183 (1980)
3. P. West, *Nucl. Phys.* B **268**, 113 (1986)
4. M.K. Gaillard, *Phys. Lett.* B **342**, 125 (1995); *Phys. Lett.* B **347**, 284 (1995)
5. G. 't Hooft and M. Veltman, *Nucl. Phys.* B **44**, 189 (1972)
6. W. Siegel, *Phys. Lett.* B **84**, 193 (1979)
7. D.M. Capper, D.R.T. Jones and P. van Nieuwenhuizen, *Nucl. Phys.* B **167**, 479 (1980)
8. W. Siegel, *Phys. Lett.* B **94**, 37 (1980)
9. J.G. Körner and M.M. Tung, *Z. Phys.* C **64**, 255 (1994)
10. M. Misiak, *Phys. Lett.* B **321**, 113 (1994)
11. D.R.T. Jones (unpublished) 1979;
see also W. Siegel, P.K. Townsend and P van Nieuwenhuizen, Proc. 1980 Cambridge meeting on supergravity, ITP-SB-80-65
12. R. van Damme and G. 't Hooft, *Phys. Lett.* B **150**, 133 (1985)
13. I. Jack, *Phys. Lett.* B **147**, 405 (1984);
G. Curci and G. Paffuti, *Phys. Lett.* B **148**, 78 (1984);
D. Maison, *Phys. Lett.* B **150**, 139 (1985)
14. I. Jack and H. Osborn, *Nucl. Phys.* B **249**, 472 (1985)
15. I. Jack, D.R.T. Jones and K.L. Roberts, *Z. Phys.* C **62**, 161 (1994)
16. I. Jack, D.R.T. Jones and K.L. Roberts, *Z. Phys.* C **63**, 151 (1994)
17. D.Z. Freedman, K. Johnson and J.I. Latorre, *Nucl. Phys.* B **371**, 353 (1992)
18. R. V. Harlander, D.R.T. Jones, P. Kant, L. Mihaila and M. Steinhauser, *JHEP* **0612**, 024 (2006)
19. D.R.T. Jones and J.P. Leveille, *Nucl. Phys.* B **206**, 473 (1982)
20. H. Nicolai and P.K. Townsend, *Phys. Lett.* B **93**, 111 (1980)
21. G. Bonneau, *Int. J. Mod. Phys.* A **5**, 3831 (1990)
22. R.W. Allen and D.R.T. Jones, *Nucl. Phys.* B **303**, 271 (1988)
23. C.M. Hull and P.K. Townsend, *Phys. Lett.* B **191**, 115 (1987)
24. H. Osborn, *Annals Phys.* **200**, 1 (1990)
25. S.A. Larin, *Phys. Lett.* B **303**, 113 (1993);
C. Schubert, *Nucl. Phys.* B **323**, 478 (1989);
J.G. Körner, D. Kreimer and K. Schilcher, *Z. Phys.* C **54**, 503 (1992)
26. L.V. Avdeev, G.A. Chochia and A.A. Vladimirov, *Phys. Lett.* B **105**, 272 (1981)
27. D. Stockinger, JHEP **0503**, 076 (2005)
28. P.S. Howe, K.S. Stelle and P. West, *Phys. Lett.* B **124**, 55 (1983);
P.S. Howe, K.S. Stelle and P.K. Townsend, *Nucl. Phys.* B **236**, 125 (1984)
29. F. del Aguila et al, *Nucl. Phys.* B **250**, 225 (1985)
30. I. Antoniadis, J. Ellis and G.K. Leontaris, *Phys. Lett.* B **399**, 92 (1997)
31. M.T. Grisaru, W. Siegel and M. Roček, *Nucl. Phys.* B **159**, 429 (1979)
32. D.R.T. Jones, *Phys. Lett.* B **123**, 45 (1983)
33. V. Novikov et al, *Nucl. Phys.* B **229**, 381 (1983);

V. Novikov et al, *Phys. Lett.* B **166**, 329 (1986);
M. Shifman and A. Vainstein, *Nucl. Phys.* B **277**, 456 (1986)
34. N. Seiberg, *Nucl. Phys.* B **435**, 129 (1995)
35. I. Jack, D.R.T. Jones and C.G. North, *Phys. Lett.* B **386**, 138 (1996)
36. I. Jack, D.R.T. Jones and C.G. North, *Nucl. Phys.* B **473**, 308 (1996)
37. I. Jack, D.R.T. Jones and C.G. North, *Nucl. Phys.* B **486**, 479 (1997)
38. I. Jack, D.R.T. Jones and A. Pickering, *Phys. Lett.* B **435**, 61 (1997)
39. R. V. Harlander, L. Mihaila and M. Steinhauser, *Eur. Phys. J.* C **63**, 383 (2009)
40. P.M. Ferreira, I. Jack and D.R.T. Jones, *Phys. Lett.* B **387**, 80 (1996)
41. C. Kolda and J. March-Russell, *Phys. Rev.* D **55**, 4252 (1997)
42. T. van Ritbergen, J.A.M. Vermaseren and S.A. Larin, *Phys. Lett.* B **400**, 379 (1997)
43. I. Jack and D.R.T. Jones, *Phys. Lett.* B **333**, 372 (1994)
44. I. Jack, D.R.T. Jones, S.P. Martin, M.T. Vaughn and Y. Yamada, *Phys. Rev.* D **50**, R5481 (1994)
45. L. Girardello and M.T. Grisaru, *Nucl. Phys.* B **194**, 65 (1982)
46. I. Jack and D.R.T. Jones, *Phys. Lett.* B **415**, 383 (1997)
47. I. Jack, D.R.T. Jones and A. Pickering, *Phys. Lett.* B **426**, 73 (1998)
48. I. Jack, D.R.T. Jones and A. Pickering, *Phys. Lett.* B **432**, 114 (1998)
49. I. Jack and D.R.T. Jones, *Phys. Lett.* B **465**, 148 (1999)
50. L. Randall and R. Sundrum, *Nucl. Phys.* B **557**, 79 (1999)
51. G.F. Giudice, M.A. Luty, H. Murayama and R. Rattazzi, JHEP **9812**, 27 (1998)
52. A. Pomarol and R. Rattazzi, JHEP **9905**, 013 (1999)
53. I. Jack, D. R. T. Jones and A. F. Kord, *Annals Phys.* **316**, 213 (2005)
54. J. A. Aguilar-Saavedra *et al.*, *Eur. Phys. J.* C **46**, 43 (2006)
55. E. Witten, *Annals Phys.* **128**, 363 (1980)
56. P.M. Ferreira, I. Jack and D.R.T. Jones, *Phys. Lett.* B **399**, 258 (1997);
P.M. Ferreira, I. Jack and D.R.T. Jones, *Nucl. Phys.* B **504**, 108 (1997)

# Probing Physics at Short Distances with Supersymmetry

Hitoshi Murayama

*Department of Physics, University of California, Berkeley, CA 94720*

We discuss the prospect of studying physics at short distances, such as Planck length or GUT scale, using supersymmetry as a probe. Supersymmetry breaking parameters contain information on all physics below the scale where they are induced. We will gain insights into grand unification (or in some cases string theory) and its symmetry breaking pattern combining measurements of gauge coupling constants, gaugino masses and scalar masses. Once the superparticle masses are known, it removes the main uncertainty in the analysis of proton decay, flavor violation and electric dipole moments. We will be able to discuss the consequence of flavor physics at short distances quantitatively.

## 14.1. Introduction

The aim of particle physics is very simple: to understand the structure of matter and their interactions at as short distant scale as possible. This is the ultimate form of the reductionist approach of physics. This approach has revealed several layers of distance scales in nature, bulk (1 cm), atomic ($10^{-8}$ cm), nuclear ($10^{-13}$ cm). Understanding physics at shorter distance scales always gave us better understanding of physics at a previously known longer distance scale. Knowing the structure of atoms, we can deduce the chemical properties of atoms and molecules. Knowing the statistics of nuclei, we understand the levels of molecular excitations. The quest continues to understand the origin of the known distance scales, such as the electroweak scale of $10^{-16}$ cm, and to discover new layer of physics at shorter distance scales.

The main motivation of supersymmetry is to stabilize the electroweak scale against radiative corrections, which tend to make it much shorter (as short as the Planck length) or much longer (no electroweak symmetry breaking). Whenever we speculate about physics at shorter distance scales, we cannot go around the problem of the stability of the electroweak scale. However, once we accept supersymmetry as the stabilization mechanism, we are allowed to speculate physics at much shorter distances, and ask questions about the origin of gauge forces, fermion masses, and even cosmological issues such as baryon asymmetry.

The aim of this short article is to further elaborate on this point. Not only that supersymmetry allows us to speculate physics as the shortest distance scales, it actually provides probes of it. One can even dream about exploring physics at the GUT scale ($10^{-30}$ cm) or the Planck scale ($10^{-33}$ cm) once we see superparticles. We will present various possibilities how we may be able to probe physics at such short distance scales using supersymmetry as a probe (see, *e.g.*, a review on this point[1]). We therefore assume that we will find and can study superparticles at collider experiments.[a]

Of course, such a dream scenario cannot be discussed without certain assumptions. For each examples of such probes in the following sections, we try to make explicit what the underlying assumptions are. One of the main assumptions in any of these discussions is that the layers in distance scales are exponentially apart from each other. This is not an unreasonable assumption from the historic perspective. All the layers of physics came at very disparate distance scales. There appears to be nothing new between the characteristics scales: *deserts*. We do not know if this is the way nature is organized; we can only assume that the shorter distance scales also come with exponential hierarchy and discuss its consequences in our further exploration of physics at yet shorter distance scales.

## 14.2. Grand Unification

The simplest example in our approach is the grand unification. Needless to explain, a grand unified theory intends to explain the rather baroque pattern of quark, lepton quantum numbers in the Standard Model by embedding its gauge groups into a simple gauge group. Such a theory would resolve bizarre puzzles in the Standard Model. The fact that the matter is neutral (at the level of at least $10^{-21}$) requires a cancellation of electric charges between and electron, two up-quarks and one down-quark. The cancellation of anomalies in the Standard Model appears miraculous. And

[a]Needless to say, it is important to confirm experimentally that the discovered new particles have properties consistent with supersymmetry.[2,3]

probably most importantly, the grand unification explains why the strong interaction is stronger than the electromagnetism as a simple consequence of the difference in the size of the gauge groups.

### 14.2.1. *Gauge coupling constants*

The supersymmetric grand unification received a strong attention in the past seven years after the precise measurement of the weak mixing angle at LEP in 1991. Many took the agreement of the observed and predicted value of the weak mixing angle as an experimental support for supersymmetry because the prediction was quite off if the minimal non-supersymmetric Standard Model was used to predict the weak mixing angle. The situation has not changed qualitatively since. The detailed discussions on the dependence on superparticle spectrum, GUT-scale threshold, and its correlation to the proton decay are all important issues, and we refer to another chapter on gauge unification.

Since our question is what we will learn by using supersymmetry as a probe, let us suppose that we already have found the superparticles. Then the question on grand unification changes dramatically. First of all, we do not need to motivate supersymmetry *assuming* grand unification. The question goes the other way around. Since we know that the supersymmetry is there, we will rather ask if the gauge coupling constants unify given the particle content seen at the electroweak scale. More importantly, we measure all the masses of superparticles, which give us quantitative inputs on the supersymmetry thresholds in the renormalization group analysis. Knowing the particle content at the TeV scale and their masses will completely change the rule of the research. Note, however, that this analysis assumes a desert between the electroweak scale under experimental study and the GUT scale.

If they do unify within a certain accuracy, say within a few percents, we will begin asking what the origin of the small mismatch is (if any). For each of the GUT models we construct, we calculate the GUT-scale threshold corrections and compare them to the data. Such an analysis would certainly exclude parts of the parameter space in each model, and in some cases, the model itself. Especially the correlation to the proton decay becomes important, since the GUT-scale threshold corrections contain information about the mass of the color-triplet Higgsino which mediates the dimension-five proton decay such as $p \to e^+ K^0$ and the mass of the GUT gauge bosons which mediate the dimension-six proton decay such as

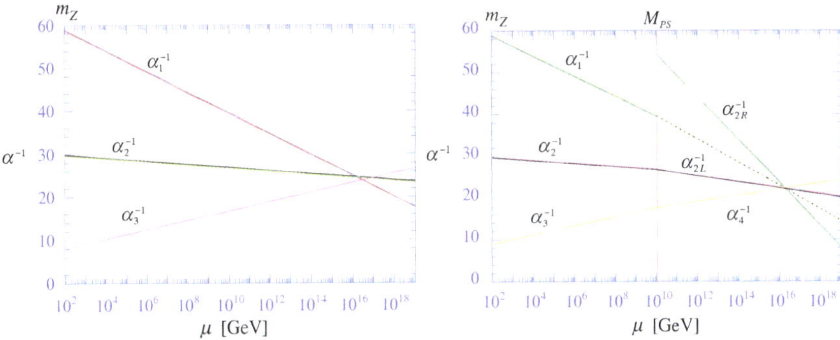


Fig. 14.1. The renormalization group evolution of the gauge coupling constants in two models, $SU(5)$ GUT with grand desert, and $SO(10)$ GUT with an intermediate Pati–Salam symmetry and a particular particle content above the Pati–Salam scale.[8]

$p \to e^+\pi^0$.[4] We will come back to this point in the section on proton decay. Another origin of a small mismatch may be a higher dimension operator in the gauge kinetic function which depends on the GUT-Higgs field, such as $\int d^2\theta \mathrm{Tr}(\frac{\Sigma}{M} W_\alpha W^\alpha)$, where $\Sigma$ is the adjoint Higgs in $SU(5)$ GUT.[5,6] If we assume that $M$ is the reduced Planck scale and $\Sigma$ has a VEV at the conventional GUT-scale of order $10^{16}$ GeV, such an operator gives an order percent correction to the gauge coupling unification. This possibility may unfortunately contaminate the information on GUT-scale threshold. In any case, however, it is clear that the rule of the game changes from motivating supersymmetry using GUT to making selection of GUT models from observed supersymmetry spectrum.

Unfortunately, the fact that the observed gauge couplings appear consistent with the $SU(5)$ unification does not rule out other possibilities, such as intermediate gauge groups with certain matter content.[7,8] Fig. 14.1 shows two patterns of gauge coupling unifications, one with grand-desert $SU(5)$ and the other with intermediate Pati–Salam symmetry. The latter model is intended to be a comparison toy model to make the points clear for the later discussions, how the study of superparticles would help us to sort out the physics at shorter distance scales.

### 14.2.2. *Gaugino masses*

Another aspect of the grand unification is the unification of superparticle masses. This discussion assumes that the supersymmetry breaking parameters are generated at a scale higher than the GUT-scale, such as the string or Planck scales, and hence respect grand-unified symmetry. Under this assumption, we will see if the superparticle masses unify *at the same scale* as where the gauge coupling constants unify. In fact, the gaugino mass unification,

$$\frac{M_1}{\alpha_1} = \frac{M_2}{\alpha_2} = \frac{M_3}{\alpha_3} \tag{14.1}$$

holds even when the GUT-group breaks to the Standard Model gauge group in several steps, *i.e.*, with intermediate gauge groups such as Pati–Salam $SU(4) \times SU(2) \times SU(2)$ or its subgroup starting from $SO(10)$ or $E_6$, as long as the Standard Model gauge groups are embedded in a simple group with a single gaugino mass.[8] At low-energy, the gaugino masses run in the exactly the same way as the gauge coupling constants squared do, which can be read off from Fig. 14.1 in these two examples. Therefore, the gaugino mass unification tests the idea of grand unification in a highly model-independent manner.

Experimental strategies have been discussed how to disentangle the mixing in the neutralino-chargino sector to measure the supersymmetry breaking masses for $SU(2)$ and $U(1)$ gauginos, $M_2$ for wino and $M_1$ for bino, at future $e^+e^-$ linear colliders.[9,10] At the LHC, mass differences can be measured well by identifying the end points in the decay distributions. Upon the assumption that the first and second neutralinos are close to pure bino and wino eigenstates (which may be cross-checked by other analyses of the data), the mass differences also test the gaugino unification.[11] Putting information from both types of experiments, we will have three new numbers to deal with. This will provide us two more independent test if they unify at the scale determined from the gauge coupling unification.

It is an important question if the gaugino mass unification can be spoiled even within GUT models. One possible effect is the threshold correction at the GUT scale. Fortunately, there is no logarithmic threshold corrections unlike to the gauge coupling constants[12] and hence the gaugino mass unification is a better prediction of GUT than the gauge coupling unification. The only way to spoil the gaugino unification from the threshold corrections is to have extremely large representations under the grand unified group with highly non-universal trilinear and bilinear supersymmetry breaking parameters. Another possible effect is a higher dimension operator in the gauge kinetic function which depends on the GUT-Higgs field, such as $\int d^2\theta \mathrm{Tr}(\frac{\Sigma}{M} W_\alpha W^\alpha)$ we discussed before, with an $F$-component VEV of the $\Sigma$ field.[9] If we take $M$ at the reduced Planck scale and $F_\Sigma \simeq m_{SUSY} M_{GUT}$,

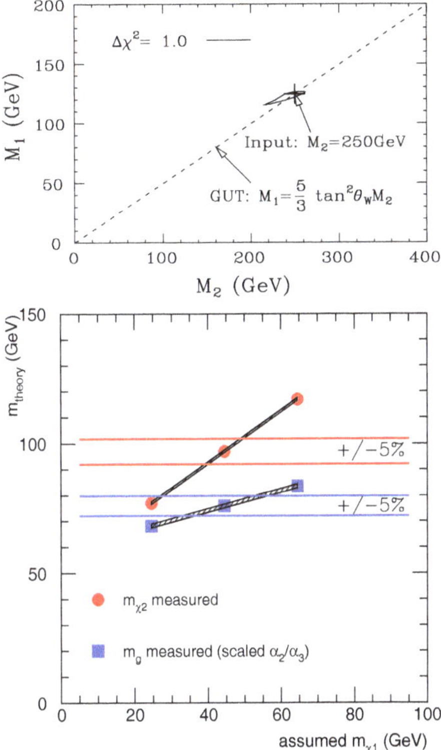


Fig. 14.2. Experimental tests of gaugino mass unification at a future $e^+e^-$ collider[9] and the LHC.[11]

this operator generates an order percent correction to the gaugino mass unification. Note, however, that the size of the $F$-component VEV tends to be only of $m^2_{SUSY}$ in a wide class of supersymmetry breaking.[13,14]

On the other hand, there is a case where we may be fooled by the apparent gaugino mass unification. In the models of gauge mediated supersymmetry breaking,[15–17] the gaugino masses may satisfy the same relation as the case with grand unification even though the supersymmetry breaking gaugino masses have nothing to do with the physics of grand unification. However, this happens only when the messenger fields fall into full $SU(5)$ multiplets and when they acquire masses from the same field which has both $A$- and $F$-component VEVs. This is naturally expected in the GUT models, even when the supersymmetry breaking is induced well below the GUT-scale. On the other hand, there is no reason for the messengers to fall into full $SU(5)$ multiplets and acquire masses from the same field if the theory is not grand unified. Even though this remains as a logically possibility that the data can fool us, the apparent gaugino mass unification still strongly suggests grand unification. One case which probably cannot be distinguished on the bases of gaugino masses is the dilaton-dominant supersymmetry breaking in superstring models.[21] In this case, however, there is a specific prediction on the ratio of the scalar masses (universal to all scalars) and gaugino masses (universal to all gauginos) and can be confronted to the data.

There are GUT-like models which do not lead to unified gaugino masses, when the unified group is not simple. One example is flipped $SU(5)$.[18] In this case, we expect to see the unification of $M_2$ and $M_3$ at the scale where the gauge coupling constants $\alpha_2$ and $\alpha_3$ meet, while $M_1$ may not. This is an interesting discriminator. The other is the model of dynamical GUT-breaking based on $SU(5) \times SU(3) \times U(1)$.[19,20] Here the gaugino masses do not appear unified at all.

### 14.2.3. *Scalar masses*

Under $SU(5)$ grand unified group, quarks and leptons belong to either **10** or **5***  multiplets. Under the same assumption that the supersymemtry breaking masses respect grand-unified symmetry, we can extrapolate the observed scalar masses to higher energies and see if they unify at the same scale where the gauge couplings and gaugino masses unify (if they do).

The scalar mass unification will be an independent useful piece of information beyond that from gauge couplings and gaugino masses. One probably very convincing case for grand-desert $SU(5)$ unification is when both the gauge couplings and gaugino masses all unify at the same scale, and also scalars in **10** and **5*** unify there but with different masses. On the other hand, the dilaton-dominated supersymmetry breaking predicts the universal scalar mass, not separate for **10** and **5***, and a definite ratio of the scalar mass to the gaugino mass ratio.

There is a possibility that we get fooled by an apparent unification of gauge couplings and gaugino masses. This happens, for instance, when the supersymmetry breaking is induced by gauge mediation with messengers in full $SU(5)$ multiplets which acquire both supersymmetric and supersymmetry-breaking masses from a single field. We argued in the previous section that such a case already strongly suggests grand unification, but there remains a possibility that it is not. In this case, the scalar masses do not appear grand unified, and provide a way to differentiate the conventional grand-desert $SU(5)$ GUT from gauge-mediated supersymmetry breaking.

Unlike the gaugino masses, the scalar mass spectrum is sensitive to the pattern of GUT symmetry breaking.[8] Many different patterns of scalar masses were discussed from GUT models.[22,23] An important effect is that the scalars which originally resided in the same GUT multiplet may acquire different contributions from the $D$-term VEV when the rank of the gauge group is reduced.[24] For instance, all quarks and lepton fields live in the same **16** multiplet under $SO(10)$; but the breaking of $SO(10)$ to $SU(5)$ generally splits the **10** and $\mathbf{5}^*$ masses because of the $D$-term. The $D$-term contributions are determined solely by the gauge quantum numbers under the broken gauge group and hence generation blind, and are safe from the point of view of flavor-changing effects. A complicated superpotential interactions may modify the scalar masses as well.[13,14,25] An extreme case is when the quarks and leptons in the Standard Model come from different GUT multiplets; then their scalar masses can be totally unrelated.[26] However, the constraint from the flavor-changing neutral currents and smallness of Yukawa couplings for the first, second generation give us a prejudice that the superpotential couplings, which can potentially split the mass of first and second generation scalars in a non-universal manner, are small. Therefore, it is likely that the scalar masses unify according to the patterns of the GUT symmetry breaking, at least for the first and second generations.

By allowing $D$-term contributions but not $F$-term contributions motivated by the above argument, one can try to fit the observed scalar mass spectrum as a function of the symmetry breaking scales and original supersymmetry breaking parameters. For more complicated symmetry breaking patterns, there are less relations and hence the model is harder to test. But still in many interesting symmetry breaking patterns, there remain non-trivial relations among scalar masses which can be confronted to data.[8,22,23] The Fig. 14.3 shows how the scalar masses acquire different patterns at the electroweak scale between the grand-desert $SU(5)$ and the toy Pati–Salam model, which could not be distinguished based on the gauge coupling constants and the gaugino masses. Therefore the role of gaugino mass unification and scalar mass unification are complimentary; the former gives a model-independent test of the grand unification, while the latter selects out particular symmetry breaking patterns and their energy scales.[8]

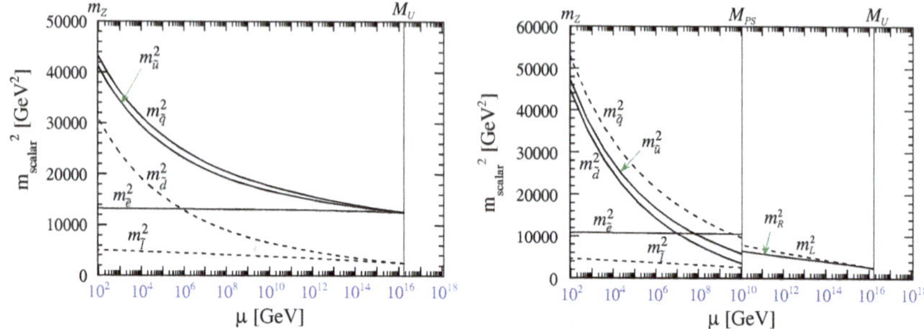


Fig. 14.3. The renormalization group evolution of the scalar masses in two models, $SU(5)$ GUT with grand desert and $SO(10)$ GUT with an intermediate Pati–Salam symmetry used in Fig. 14.1, which cannot be discriminated based on gauge coupling constants and gaugino masses.

Even in the case of non-simple GUT groups, the scalar masses still give us useful information. In flipped $SU(5)$, different sets of fields $(Q, d^c, \nu)$ belong to **10** and $(L, u^c)$ to $\mathbf{5}^*$, and $e^c$ is a singlet by itself. Therefore, there is still the sclar-mass unification of $Q$ and $d^c$, and $L$ and $u^c$ separately. In the model of dynamical GUT-breaking based on $SU(5) \times SU(3) \times U(1)$,,[19] still all matter fields belong to the ordinary $\mathbf{10} + \mathbf{5}^*$ multiplets and the pattern of scalar masses is the same as in the grand-desert $SU(5)$.

Experimentally, measurement of scalar masses is also feasible. At an $e^+e^-$ collider, a well-defined kinematics allows a simple kinematic fit to the decay distributions to extract the mass of the parent scalar particle. This comment applies both to the sleptons[9,10] and squarks[27] using beam polarizations, as long as they are within the kinematic reach. At the LHC, the mass differences can be measured well as before; especially when the second neutralino decays into on-shell sleptons, one has a high rate and the mass difference between the slepton and the lightest neutralino is measured very well. Many other mass patterns also allow certain mass differences to be measured accurately at a few percent level.[11] It is quite imaginable that the spectroscopy of superparticles will be the main experimental project of the next decade.

### 14.3. Proton Decay

Proton decay has been virtually the only direct probe of the physics at the GUT-scale and discussed extensively in the literature. The original idea is that the gauge bosons in $SU(5)$ GUT cause transitions between quarks and leptons in the same $SU(5)$ multiplets and hence allow proton to decay. Assuming that the quarks and leptons of the first generation belong to the same $SU(5)$ multiplets, the exchange of the heavy $SU(5)$ gauge boson generates an operator

$$\mathcal{L} = \frac{1}{M_V^2} uude, \tag{14.2}$$

which gives rise to a decay $p \to e^+\pi^0$. The current experimental bound excludes the process for heavy gauge bosons approximately up to $1.5 \times 10^{15}$ GeV,[29] where we estimated the bound conservatively.[4] Because the operator has a suppression by two powers of a high mass scale, the proton decay rate is suppressed by the fourth power in the mass scale $\Gamma_p \propto m_p^5/M^4$. It is not easy to extend the experimental reach on $M$. SuperKamiokande will probably extend the limit on the lifetime by a factor of 30 beyond the current one, which translates to a modest improvement by a factor of 2.3 on the GUT-scale. ICARUS may reach the mass scale of the supersymmetric GUT or $10^{16}$ GeV.

### 14.3.1. $D = 5$ operators

The important and novel feature in supersymmetric models is that there are operators of dimension-five which violate baryon and lepton numbers and hence can cause proton decay.[30b] For instance, the following operator is possible in the superpotential:

$$W = \frac{\lambda}{M}(Q_1 Q_1)(Q_2 L_i), \tag{14.3}$$

where the subscript refers to the generation, and $\lambda$ is a coupling constant. The operator involves squarks and sleptons, which need to be converted to quarks or leptons by a loop diagram. The proton decay rate therefore scales as $\Gamma_p \propto m_p^5 \lambda^2/(16\pi^2)^2 M^2 m_{SUSY}^2$ where $m_{SUSY}$ is the mass scale of superparticles. As a result, the reach in the energy scale is drastically improved. The current experimental limit does not allow $M$ below $10^{24}$ GeV if $\lambda \sim 1$. Therefore, we are sensitive to even Planck-scale suppressed operators which, actually, are excluded with $O(1)$ couplings.

It is interesting that the dimension-five operators necessarily involve quark superfields of different generations (at least two). This is a simple

[b]In this discussion, we assume that there is no dimension-*four* operators which violate baryon and lepton numbers. Such operators are conveniently forbidden by imposing the $R$-parity.

consequence of the Bose symmetry among superfields and the Standard Model gauge invariance. The interesting consequence of this fact is that the proton (or neutron) decay modes preferentially involve kaons in the final state, such as $p \to K^+ \bar{\nu}_\mu$ as predicted to be dominant in the minimal $SU(5)$ GUT. If the dominant proton decay mode will be seen to involve kaons, it is likely to be a consequence of dimension-five operators possible only in supersymmetric theories.

### 14.3.2. *Minimal SUSY SU(5)*

In the minimal SUSY $SU(5)$ GUT,[31] the dimension-five operators are generated by the exchange of color-triplet Higgs supermultiplet. The Yukawa couplings of quarks to the Higgs doublets are known from the quark masses, and the couplings to the color-triplet Higgs ($SU(5)$ partner of the doublets) are the those to the Higgs doublets at the GUT-scale because of the $SU(5)$ invariance. Therefore, there is little freedom in this model and the size of the dimension-five operators is completely fixed except possible relative phases which become unobservable below the GUT-scale.[32] The mass of the color-triplet Higgs at first appears to be a free parameter. However, the gauge unification constrains its mass through its threshold correction.[4] At the one-loop level of the renormalization group equations, one obtains

$$(3\alpha_2^{-1} - 2\alpha_3^{-1} - \alpha_1^{-1})(m_Z) = \frac{1}{2\pi}\left\{\frac{12}{5}\ln\frac{M_H}{m_Z} - 2\frac{m_{SUSY}}{m_Z}\right\}, \tag{14.4}$$

$$(5\alpha_1^{-1} - 3\alpha_2^{-1} - 2\alpha_3^{-1})(m_Z) = \frac{1}{2\pi}\left\{12\ln\frac{M_V^2 M_\Sigma}{m_Z^3} + 8\ln\frac{m_{SUSY}}{m_Z}\right\}. \tag{14.5}$$

Here, $m_{SUSY}$ stands for some weighted average of the superparticle masses. Once the mass spectrum of the superparticle is measured, one can determine $m_{SUSY}$ in the above formulae, and then extract the mass of the colored Higgs $M_{H_C}$, and a combination of $M_V$ and $M_\Sigma$ from the renormalization group equations. In fact, the measured $\alpha_s$ is smaller than the preferred value from the GUT and as a result prefers a low-value of color-triplet Higgs mass. Since various $\alpha_s$ measurements basically converged recently to $\alpha_s(m_Z) = 0.118 \pm 0.003$, the minimal $SU(5)$ model is almost excluded[28,29] unless extreme parameters are chosen for gauginos (preferentially light) and squarks (preferentially heavy). Given the uncertainties in the superparticle spectrum, it is hard to announce the definite exclusion of the model. Once the superparticles are found, however, we will be able to make a

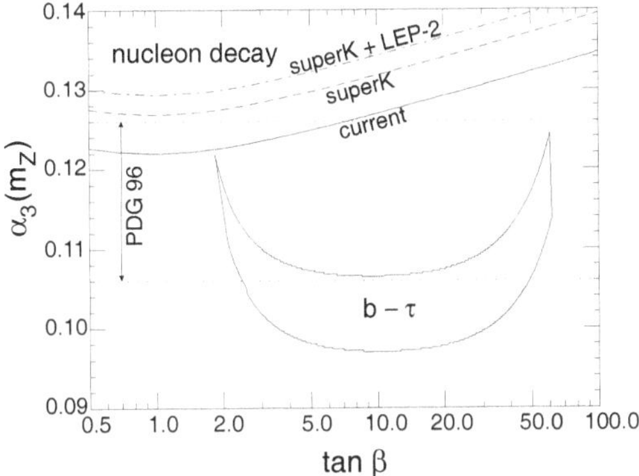


Fig. 14.4. Excluded region on $(\tan\beta, \alpha_s(m_Z))$ plane from nucleon decay based on very conservative assumptions using constraints on superparticle spectrum from LEP-1 only.[29] Expected improvements from superKamiokande and LEP-2 are also shown. The range shown for $\alpha_s(m_Z)$ from PDG96 is two-sigma range. The preferred region from $b-\tau$ unification is also shown for $m_t = 176$ GeV as a crescent-shaped region.

final word on the model, assuming the current value of $\alpha_s$ persists and the superKamiokande will not find proton decay.

### 14.3.3. *Non-minimal SUSY-GUT*

There are many good reasons to discuss extensions of the minimal SUSY SU(5) GUT. Among them, there are two points directly relevant to the nucleon decay. (1) The triplet-doublet splitting problem. In minimal SUSY SU(5) GUT, one needs to fine-tune independent parameters at the level of $10^{-14}$ to keep Higgs doublets light while making the color-triplet Higgs heavy. (2) The wrong fermion mass relations. It predicts $m_s = m_\mu$ and $m_d = m_e$ at the GUT-scale, which are off from the phenomenologically preferred Georgi–Jarlskog relations $m_s = m_\mu/3$, $m_d = 3m_e$.

Solutions to the above-mentioned problems modify the predicted rate and branching ratios of the nucleon decay. One possible attempt to obtain Georgi–Jarlskog relations is to use the SU(5)-adjoint Higgs to construct an effective **45** Higgs doublets as composites of ordinary Higgs doublets in **5** and the adjoint. This modification leads to a factor-of-two enhancement in the amplitude; a factor of four in the rate.[36] The relative branching ratios can be also different. It remains true that the $K^{+,0}\bar{\nu}_\mu$ modes are the dominant ones, while the $K^0\mu^+$ mode may be much less suppressed than in the minimal SU(5).[37,38] the proton decay in $SO(10)$ models with realistic fermion mass texture has been also discussed extensively.[39,40]

There are various proposals to solve the triplet-doublet splitting problem, which lead to different nucleon decay phenomenology. I discuss three of them here. (1) The missing partner model,[41] (2) Dimopoulos–Wilczek–Srednicki mechanism,[42] and (3) flipped SU(5) model.[18]

In the missing partner model, one employs **75** representation to break SU(5) instead of the adjoint **24**, and further introduces **50** and **50***  representations which mix with the color-triplet Higgs to make them massive. Since the model involves such large representations, the size of the GUT-scale threshold corrections are significantly larger than that in the minimal model. And the correction changes the determination of the color-triplet Higgs mass as done in Eq. (14.4), and the measured values of the gauge coupling constants prefer larger $M_{H_C}$ than in the minimal model.[43] In this case the proton decay rates are much more suppressed, by a few orders of magnitudes. One drawback of the model is that it becomes non-perturbative well below the Planck scale due to large representations and one needs to complicate the model further to keep it perturbative.[44] It is worth to recall that the minimal SU(5) model is marginally allowed only with very conservative assumptions. Even though there is an additional suppression to the proton decay rate in this class of models, the decay rate may still well be within the reach of superKamiokande experiment.

The mechanism proposed by Dimopoulos, Wilczek and further by Srednicki employs SO(10) unification with Higgs fields in adjoint and symmetric tensor representations which naturally keep Higgs doublets light. However, their model breaks SO(10) only to $SU(3) \times SU(2)_L \times SU(2)_R$ and has to be extended to achieve the desired symmetry breaking down to the standard model gauge group. One of such extensions by Babu and Barr[45] eliminates $D = 5$ entirely; but it involves rather complicated Higgs sector, and one needs to forbid some allowed interactions in the superpotential arbitrarily. A later attempt to guarantee the special form of the superpotential by symmetries did not eliminate the $D = 5$ operators entirely, but resulted in a weak suppression of the operators. Again in view of the very marginal situation in the minimal model, the decay rate could be within the reach of the superKamiokande.

The flipped SU(5) model solves the triplet-doublet splitting problem in a way that it also eliminates the $D = 5$ operators entirely. A possible problem with this model is that the gauge unification becomes more or less an accident rather than a prediction. On the other hand, the elimination of the $D = 5$ operator is a natural consequence of the structure of the Higgs sector, and is rather a robust prediction of the model except the Planck-scale effects which will be discussed below. An interesting feature of the model is that the GUT-scale is determined by $\alpha_2$ and $\alpha_3$ and hence can be *lower* than the scale in the minimal SU(5) which is determined by $\alpha_2$ and $\alpha_1$. Since the model does not predict the relation between $\alpha_1$ and $\alpha_{SU(5)}$, $\alpha_1$ does not need to meet with the other coupling constants at the same scale. Therefore, the GUT-scale can be as low as $M_{GUT}^{\text{flipped}} = 4\text{–}20 \times 10^{15}$ GeV. If the $M_{GUT}$ is at the low side within this range, the $D = 6$ operator may be observable in the $\pi^0 e^+$ mode,[46] since the superKamiokande is expected to extend the reach by a factor of 30.

Certain models of direct gauge mediation[33] also have $SU(5)$ group broken below the typical GUT-scale and can lead to dimension-six proton decay at a rate observable at superKamiokande. There is also a variant of missing-partner model with dimension-six proton decay within the reach of superKamiokande.[34]

### 14.3.4. *Planck-scale operators*

Planck-scale physics may generate $D = 5$ operators suppressed by the reduced Planck scale $M_* = 2 \times 10^{18}$ GeV. Even when there is no color-triplet Higgs, such as in string compactifications which breaks the gauge group down to the standard model (with possible U(1) factors) directly, the higher string excitations may give rise to effective non-renormalizable $D = 5$ operators which break baryon- and/or lepton-number symmetries. For $D = 5$ operators which involve first- and second-generation fields, $1/M_*$ suppression is far from enough: one needs a coupling constant of order $10^{-7}$ to keep the nucleons stable enough as required by experiments.

It is a serious question in supersymmetry phenomenology why the Planck-scale $D = 5$ operators are so much suppressed. One possibility is to forbid them by employing discrete gauge symmetries[47] which are believed to be respected by quantum gravitational effects unlike global symmetries. In this case, there is no baryon-number-violating $D = 5$ operator from Planck-scale physics and we do not have any handle on it. A different type of solution is probably more interesting: the $D = 5$ operators are suppressed because of the same reason why the Yukawa couplings of light generations are suppressed.[35] One way to understand why the Yukawa couplings are so small, such as $10^{-6}$ for the case of the electron, may be a natural consequence of an approximate flavor symmetry. If a flavor symmetry exists and is only weakly broken to explain smallness of the Yukawa couplings, the same flavor symmetry can well suppress the $D = 5$ operators at the

Planck-scale. The $D = 5$ operators with such a flavor origin may have very different flavor structure from those in the GUT models, and may lead to quite different decay modes like $p \to K^0 e^+$.

Suppression of Planck-scale $D = 5$ operators based on certain flavor symmetries were discussed.[48,49] For instance the $S_3^3$ model[49] explains the hierarchical Yukawa matrices as a consequence of sequential breaking of the flavor symmetry while the symmetry preserves sufficient degeneracy among the squarks and sleptons to suppress flavor-changing neutral currents. It happens that the flavor symmetry in this model also suppresses $D = 5$ operators to the level of about 1/9 of the minimal $SU(5)$ model, so that it can well be within the reach of superKamiokande.[50] What is particularly interesting in this model is that it predicts $p \to K^0 e^+$ as the *dominant* mode over the $K^+ \bar{\nu}$, while $n \to K^0 \bar{\nu}_e$ is the dominant mode in neutron decay with a comparable rate. In general, decay modes of proton, if observed, will provide interesting information on the flavor physics.[37,38,40]

## 14.4. Flavor Physics

Another interesting topic is how well we will be able to understand the origin of flavor, fermion masses and mixing based on the study of supersymmetry possible at the electroweak scale. Unlike the case of grand unification and proton decay, the answer to this question depends heavily on what the true story is.

### 14.4.1. *Neutrino physics*

An analogue of proton decay discussed in the previous section is a consequence of flavor physics suppressed by powers of the mass scale, such as the neutrino mass via the seesaw mechanism.[51] The neutrino masses are generated from their Dirac masses $m_D$ with right-handed neutrinos neutral under the Standard Model gauge groups and their Majorana masses $M$ which violate the lepton numbers by two units. The one-generation case is given by a two-by-two mass matrix

$$\mathcal{L}_{mass} = \frac{1}{2}(\nu, N^c) \begin{pmatrix} 0 & m_D \\ m_D & M \end{pmatrix} \begin{pmatrix} \nu \\ N^c \end{pmatrix}, \tag{14.6}$$

where $\nu$ is the Weyl field of the left-handed neutrino, and $N$ the right-handed neutrino. The Lagrangian is written in terms of the charge-conjugated Weyl spinor $N^c$ with the same chirality as the left-handed field $\nu$. After diagonalization of the mass matrix, one obtains a mass for the left-handed neutrino of $m_D^2/M$, which is power suppressed. This mechanism naturally explains why the neutrino masses are so small, if finite, and leaves imprint of short-distance physics in the pattern of neutrino masses and mixings.

Let us emphasize that this is the area where a dramatic progress is likely to be made in the next few years, from superKamikande (together with neutrino beam from KEK), CHORUS, NOMAD, KARMEN, SNO, BOREXINO, MINOS, and more. Even though supersymmetry does not necessarily help to study the physics at the scale of right-handed neutrinos, many flavor models in supersymmetry predict interesting patterns on neutrino masses.[52] We will certainly be making selections on different flavor models based on neutrino physics if finite neutrino masses and their mixings will be established.

### 14.4.2. *Flavor-changing neutral currents*

As discussed in other chapters, there are severe constraints on the superparticle masses and mixings from the flavor-changing neutral currents (FCNC). There are broadly three categories of models which naturally suppress the FCNC. (1) Flavor symmetry enforces the squarks, sleptons to be degenerate,[53] or aligns their mass basis to that of the down-quark, charged-lepton mass basis.[54] (2) The string theory generates universal scalar mass.[21] (3) The supersymmetry breaking is generated in a flavor-blind fashion below the scale of flavor physics.[15–17]

In cases (1) and (2), there may be interesting imprints of flavor physics in the small mixing between squarks and sleptons. The case of GUT also belongs to the category: the large top Yukawa coupling above the GUT-scale may affect the slepton masses-squared with small mixing between, for instance, selectron and smuon.[55,56] The search for rare decays such as $\mu \to e\gamma$,[56,57] $K$-physics,[58] or electric dipole moments of electron or neutron[59] may reveal the imprints of flavor physics in scalar masses. At present, the main uncertainty in the quantitative analysis is the mass of superparticles. Once they are measured, however, we can try to extract the mixing effects in the scalar mass matrices from the FCNC data.

An interesting case where the flavor-mixing effects in scalar masses can be probed at colliders was discussed.[60] Analogous to neutrino oscillation, a selectron produced from an $e^+e^-$ or $e^-e^-$ collider can oscillate to a

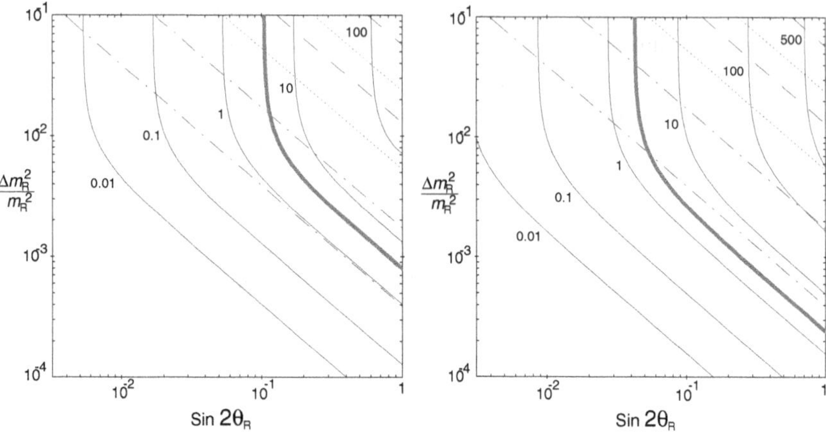


Fig. 14.5. Contours of constant $\sigma(e^+e_R^- \to e^\pm\mu^\mp\tilde{\chi}^0\tilde{\chi}^0)$ (solid) and $\sigma(e_R^-e_R^- \to e^-\mu^-\tilde{\chi}^0\tilde{\chi}^0)$ (solid) in fb for $e^+e^-$ or $e^-e^-$ linear colliders, with $\sqrt{s} = 500$ GeV, $m_{\tilde{e}_R}, m_{\tilde{\mu}_R} \approx 200$ GeV, and $M_1 = 100$ GeV (solid). The thick gray contour represents the experimental reach in one year. Constant contours of $B(\mu \to e\gamma) = 4.9 \times 10^{-11}$ and $2.5 \times 10^{-12}$ are also plotted for degenerate left-handed sleptons with mass 120 GeV and $\tilde{t} \equiv -(A + \mu \tan\beta)/\bar{m}_R = 0$ (dotted), 2 (dashed), and 50 (dot-dashed), with left-handed sleptons degenerate at 350 GeV.

smuon as a result of the flavor mixing and is detected as $e\mu$ final state (see Fig. 14.5).

In the case (3) where the scalar masses are generated in a flavor-blind fashion below the flavor physics scale, such as in the models of low-energy gauge mediation, we unfortunately may not learn about the origin of flavor from the study of flavor signatures at the electroweak scale.

## 14.5. Conclusion

Experiments at the electroweak scale will remove the cloud which masks the physics at yet shorter distance scales. If supersymmetry turns out to be the mechanism of stabilizing the electroweak scale, we will have a wealth of new data on superparticle spectroscopy. Combined with data on proton decay, neutrino physics, and FCNC, we will obtain useful information on physics such as grand unification, string, flavor physics. At this point it is just a dream; but we may be able to glimpse the physics at the shortest possible distance scales by this program, which is nothing but the goal of particle physics after all.

## Acknowledgments

This work was supported in part by the U.S. Department of Energy under Contracts DE-AC03-76SF00098, in part by the National Science Foundation under grant PHY-95-14797, and also by Alfred P. Sloan Foundation.

## References

1. H. Murayama, Invited Talk at the ICEPP Symposium "From LEP to the Planck World," University of Tokyo, Dec 17–18, 1992. In Proceedings of the ICEPP Symposium "From LEP to the Planck World," eds. K. Kawagoe and T. Kobayashi, UT-ICEPP 93-12, TU-451, 11pp.
2. J.L. Feng, M.E. Peskin, H. Murayama, and X. Tata, *Phys. Rev.* D **52**, 1418 (1995).
3. M. M. Nojiri, K. Fujii, and T.Tsukamoto, *Phys. Rev.* D **54**, 6756 (1996).
4. J. Hisano, H. Murayama, and T. Yanagida, *Phys. Rev. Lett.* **69**, 1014, (1992); *Nucl. Phys.* **B402**, 46 (1993).
5. L. J. Hall and U. Sarid, *Phys. Rev. Lett.* **70**, 2673 (1993).
6. T. Dasgupta, P. Mamales, and P. Nath, *Phys. Rev.* D **52**, 5366 (1995).
7. N.G. Deshpande, E. Keith, and T.G. Rizzo, *Phys. Rev. Lett.* **70**, 3189 (1993).
8. Y. Kawamura, H. Murayama, and M. Yamaguchi, *Phys. Lett.* **B324**, 52 (1994).
9. T. Tsukamoto, K. Fujii, H. Murayama, M. Yamaguchi, and Y. Okada, *Phys. Rev.* D **51**, 3153 (1995).
10. J. L. Feng and M. J. Strassler, *Phys. Rev.* D **51**, 4661 (1995); *Phys. Rev.* D **55**, 1326 (1997); H. Baer, R. Munroe, and X. Tata, *Phys. Rev.* D **54**, 6735 (1996); Erratum-ibid **56**, 4424 (1997).
11. I. Hinchliffe, F.E. Paige, M.D. Shapiro, J. Soderqvist, and W. Yao, *Phys. Rev.* D **55**, 5520 (1997).
12. J. Hisano, H. Murayama, and T. Goto, *Phys. Rev.* **D49**, 1446 (1994).
13. Y. Kawamura, H. Murayama, and M. Yamaguchi, *Phys. Rev.* **D51**, 1337 (1995).
14. A. Pomarol and S. Dimopoulos, *Nucl. Phys.* B **453**, 83 (1995); R. Rattazzi, *Phys. Lett.* B **375**, 181 (1996).
15. M. Dine and A.E. Nelson, *Phys. Rev.* **D48**, 1277 (1993).
16. M. Dine, A.E. Nelson and Y. Shirman, *Phys. Rev.* **D51**, 1362 (1995).
17. M. Dine, A.E. Nelson, Y. Nir and Y. Shirman, *Phys. Rev.* **D53**, 2658 (1996).
18. S. M. Barr, *Phys. Lett.* B **112**, 219 (1982); J.P. Derendinger, J. E. Kim, and D.V. Nanopoulos, *Phys. Lett.* B **139**, 170 (1984); I. Antoniadis, J. Ellis, J.S. Hagelin, and D.V. Nanopoulos, *Phys. Lett.* B **194**, 231 (1987).
19. T. Yanagida, *Phys. Lett.* B **344**, 211 (1995);
20. N. Arkani-Hamed, H.-C. Cheng, and T. Moroi, *Phys. Lett.* B **387**, 529 (1996).
21. V. S. Kaplunovsky and J. Louis, *Phys. Lett.* B **306**, 269 (1993); A. Brignole, L.E. Ibanez, and C. Munoz, *Nucl. Phys.* B **422**, 125 (1994), Erratum-ibid.

**436**, 747 (1995); A. Brignole, L.E. Ibanez, C. Munoz, and C. Scheich, *Z. Phys.* C **74**, 157 (1997).
22. C. Kolda, and S. P. Martin, *Phys. Rev.* D **53**, 3871 (1996).
23. H.C. Cheng, and L.J. Hall, *Phys. Rev.* D **51**, 5289 (1995).
24. M. Drees, *Phys. Lett.* **181B**, 279 (1986); J.S. Hagelin and S. Kelley, *Nucl. Phys.* **B342**, 95 (1990); A.E. Faraggi, J.S. Hagelin, S. Kelley, and D.V. Nanopoulos, *Phys. Rev.* **D45** 3272 (1992).
25. H. Murayama, Invited plenary talk given at 4th International Conference on "Physics Beyond the Standard Model," Lake Tahoe, CA, 13-18 Dec 1994. Proceedings, eds. by J. Gunion, T. Han, J. Ohnemus, World Scientific, 1995.
26. S. Dimopoulos and A. Pomarol, *Phys. Lett.* B **353**, 222 (1995).
27. J. L. Feng and D. E. Finnell, *Phys. Rev.* D **49**, 2369 (1994).
28. J. Hisano, T. Moroi, K. Tobe, and T. Yanagida, *Mod. Phys. Lett.* A **10**, 2267 (1995); J. Bagger, K. Matchev, and D. Pierce, *Phys. Lett.* B **348**, 443 (1995).
29. H. Murayama, Invited talk presented at 28th International Conference on High-energy Physics (ICHEP 96), Warsaw, Poland, 25-31 Jul 1996. Published in the Proceedings of the 28th International Conference on High Energy Physics, *eds.*, Z. Ajduk and A. K. Wroblewski, World Scientific, 1997, pp. 1377–1382.
30. N. Sakai and Tsutomu Yanagida, *Nucl. Phys.* B **197**, 533 (1982); S. Weinberg, *Phys. Rev.* D **26**, 287 (1982).
31. S. Dimopoulos and H. Georgi, *Nucl. Phys.* B **193**, 150 (1981); N. Sakai, *Z. Phys.* C **11**, 153 (1981).
32. P. Nath and R. Arnowitt, *Phys. Rev.* D **38**, 1479 (1988).
33. H. Murayama, *Phys. Rev. Lett.* **79**, 18 (1997); S. Dimopoulos, G. Dvali, R. Rattazzi, and G.F. Giudice, CERN-TH/97-98, hep-ph/9705307.
34. J. Hisano, Y. Nomura, and T. Yanagida, KEK-TH-547, hep-ph/9710279.
35. H. Murayama and D. B. Kaplan, *Phys. Lett.* **B336**, 221 (1994).
36. H. Murayama, Invited talk presented at the 22nd INS International Symposium on Physics with High Energy Colliders, Tokyo, Japan, March 8–10, 1994, published in Proceedings of INS Symposium, World Scientific, 1994.
37. A. Antaramian, LBL-36819, Feb 1995, Ph.D. Thesis.
38. K.S. Babu and S.M. Barr, *Phys. Lett.* B **381**, 137 (1996).
39. V. Lucas and S. Raby, *Phys. Rev.* D **55**, 6986 (1997).
40. K.S. Babu, J. C. Pati, and F. Wilczek, IASSNS-HEP-97-136, hep-ph/9712307.
41. A. Masiero, D.V. Nanopoulos, K. Tamvakis, and T. Yanagida, *Phys. Lett.* B **115**, 380 (1982); B. Grinstein, *Nucl. Phys.* B **206**, 387 (1982).
42. S. Dimopoulos and F. Wilczek, NSF-ITP-82-07 (unpublished); M. Srednicki, *Nucl. Phys.* B **202**, 327 (1982).
43. K. Hagiwara and Y. Yamada, *Phys. Rev. Lett.* **70**, 709 (1993); Y. Yamada, *Z. Phys.* C **60**, 83 (1993).
44. J. Hisano, T. Moroi, K. Tobe, and T. Yanagida, *Phys. Lett.* B **342**, 138 (1995).
45. K.S. Babu and S.M. Barr, *Phys. Rev.* D **48**, 5354 (1993).
46. J. Ellis, J. L. Lopez, and D.V. Nanopoulos, *Phys. Lett.* B **371**, 65 (1996).

47. L. E. Ibanez and G. G. Ross, *Nucl. Phys.* **B368**, 3 (1992).
48. V. Ben-Hamo and Y. Nir, *Phys. Lett.* B **339**, 77 (1994).
49. L. J. Hall and H. Murayama, *Phys. Rev. Lett.* **75**, 3985 (1995).
50. C. D. Carone, L. J. Hall, and H. Murayama, *Phys. Rev.* **D53**, 6282 (1996).
51. T. Yanagida, in *Proceedings of Workshop on the Unified Theory and the Baryon Number in the Universe*, Tsukuba, Japan, 1979, edited by A. Sawada and A. Sugamoto (KEK, Tsukuba, 1979), p. 95; M. Gell-Mann, P. Ramond and R. Slansky, in *Supergravity*, proceedings of the Workshop, Stony Brook, New York, 1979, edited by P. Van Nieuwenhuizen and D.Z. Freedman (North-Holland, Amsterdam, 1979), p. 315.
52. Y. Grossman and Y. Nir, *Nucl. Phys.* B **448**, 30 (1995); M. Schmaltz, *Phys. Rev.* D **52**, 1643 (1995); C. D. Carone and L. J. Hall, *Phys. Rev.* D **56**, 4198 (1997); P. Binetruy, S. Lavignac, S. Petcov, and P. Ramond, *Nucl. Phys.* B **496**, 3 (1997).
53. M. Dine, A. Kagan, and R. Leigh, *Phys. Rev.* D **48**, 4269 (1993) ; P. Pouliot and N. Seiberg, *Phys. Lett.* B **318**, 169 (1993); D.B. Kaplan and M. Schmaltz, *Phys. Rev.* D **49**, 3741 (1994); A. Pomarol and D. Tommasini, *Nucl. Phys.* B **466**, 3 (1996); R. Barbieri, G. Dvali, and L. J. Hall, *Phys. Lett.* B **377**, 76 (1996); R. Barbieri, L. J. Hall, S. Raby, and A. Romanino, *Nucl. Phys.* B **493**, 3 (1997).
54. Y. Nir and N. Seiberg, *Phys. Lett.* B **309**, 337 (1993).
55. L. J. Hall, V. A. Kostelecky, and S. Raby, *Nucl. Phys.* B **267**, 415 (1986).
56. R. Barbieri and L.J. Hall, *Phys. Lett.* B **338**, 212 (1994).
57. R. Barbieri, L.J. Hall, and A. Strumia, *Nucl. Phys.* B **445**, 219 (1995); J. Hisano, T. Moroi, K. Tobe, M. Yamaguchi, and T. Yanagida, *Phys. Lett.* B **357**, 579 (1995).
58. R. Barbieri, L.J. Hall, and A. Strumia, *Nucl. Phys.* B **449**, 437 (1995).
59. S. Dimopoulos and L.J. Hall, *Phys. Lett.* B **344**, 185 (1995).
60. N. Arkani-Hamed, H.-C. Cheng, J. L. Feng, and L. J. Hall, *Phys. Rev. Lett.* **77**, 1937 (1996).

# Supersymmetry and String Theory

Michael Dine

*Santa Cruz Institute for Particle Physics,*
*University of California, Santa Cruz, CA 95064*

Supersymmetry was first discovered in string theory. This essay describes how supersymmetry is crucial to our present understanding of string theory, and why if string theory is correct, low energy supersymmetry is a likely consequence.

## 15.1. Introduction

Many particle theorists are convinced that nature is supersymmetric. This despite the fact that, as of this writing, there is at best only one compelling piece of experimental evidence (the unification of couplings) to support this view, and that evidence is very indirect. We have gotten to this situation because there are at least three reasons to hypothesize a role for supersymmetry in low energy physics:

- The hierarchy problem
- The unification of couplings
- Low energy supersymmetry emerges naturally from string theory, the only known theory which unifies gravity and other interactions in a quantum-mechanically consistent way.

The hierarchy problem and the unification of couplings are discussed thoroughly in many reviews (including those in this volume), and I will not review them here. My focus will be principally on the role of supersymmetry in string theory, and on the role that string theory suggests for supersymmetry.

A lawyer might say that each of these three points above helps build a circumstantial case for supersymmetry. However, each can also be called into question. The hierarchy problem is often described as a problem of fine tuning. At a more primitive level, it is simply a problem of dimensional analysis: why isn't the Higgs mass as large as the largest mass scales in nature? When simple-minded dimensional analysis fails, it is often because there is some underlying symmetry principle. Supersymmetry is the only symmetry we know which could explain the lightness of the Higgs field. While this argument seems compelling, it is perhaps troubling that there is a problem which on its face looks rather similar: the cosmological constant problem. This is another striking failure of dimensional analysis. For this failure of dimensional analysis, we know of no symmetry which can provide an explanation. We can hope that we will yet find such a symmetry, or that alternatively the explanation may be dynamical, either involving a light field, or caught up in the subtleties of quantum gravity (perhaps in string theory). Still, we are left with the uneasy feeling that some fine tuning problems may be solved by mysterious cancellations between high and low energy phenomena. The unification of couplings, while reasonably impressive, involves only one number, and might turn out to be a coincidence. And while many feel that string theory is so extraordinary it could not fail to be correct, we seem to be a long way from understanding how the theory might make contact with nature, and what role supersymmetry might play.

This essay will not provide the reader with any pat answers to these questions. It will be largely descriptive, explaining the role of supersymmetry in our current understanding of string theory. It turns out that even at weak coupling it is difficult to describe string configurations which are not at least approximately supersymmetric, and supersymmetry has been crucial to all of the recent progress in understanding strongly coupled strings. Supersymmetry also figures crucially in all present ideas about string phenomenology. At this point it is clear that the phrase "superstring theory" is really just a code for some larger structure. This structure looks in some regimes like one or another type of string theory, in others like a field theory. It has regimes in which we have no idea how to describe at all. But, within our current understanding, supersymmetry is crucial to understanding the behavior in any of these regimes. The role of supersymmetry in string theory will be the subject of the second section. In the third, we will turn to the issue of relating string theory to nature. Given our limited understanding of the theory, this is clearly a risky endeavor. Still, there seem to be a set of fundamental questions which must be faced in understanding how string theory might describe the universe we observe. We will end with some timid speculations about how supersymmetry and string theory might resolve some of the deep questions of particle physics.

## 15.2. Supersymmetry in String Theory

Among the glaring limitations of the standard model is that it does not incorporate general relativity. Einstein's theory of gravitation does not seem to make sense when viewed as a quantum theory. It is non-renormalizable, which means that it almost certainly has to be incorporated in some larger structure. It suffers, as well, from Hawking's black hole information paradox, which seems to have no resolution within a field theory picture.[3]

Until 1984, it was perhaps appropriate to write off these problems as issues of ultra high energy physics, which would not be accessible to experimental study on any forseeable time scale. But in superstring theory,[1] we seem to have stumbled on a framework which does incorporate both general relativity and the gauge interactions of the standard model, and obey the rules of quantum mechanics. String theory has none of the ultraviolet divergences of conventional quantum field theories; it is completely finite. For example, in string theory, it is possible to calculate graviton scattering amplitudes without encountering divergences. Further theoretical support for the theory comes from the realization in recent years that there is only one such theory; all of the known consistent string theories are particular limiting cases of this larger, "M" theory[2] Moreover, in the last few years, a great deal of evidence has accumulated that string theory resolves one of the great puzzles of general relativity, the black hole information pardox.[3]

Remarkably, string theory is a structure which we have stumbled on by accident. String theory was discovered as an outgrowth of attempts to understand certain properties of strong interaction amplitudes. These could be explained if one supposed that hadrons were well described as excitations of strings. It is long since clear that the basic degrees of freedom strong interactions are quarks and gluons. The question of why QCD produces stringlike excitations remains an active area of research. But suppose one forgets this motivation, and simply hypothesizes that the fundamental objects in nature are strings, rather than point particles. Then one has a textbook problem, of quantizing a string no different than that described in standard undergraduate classical mechanics texts.[4] Such strings are characterized by a tension, and by a dimensionless number which describes the coupling of three strings. There are an infinite number of vibrational modes of the string, as well as translatonal modes (momenta). In a quantum theory, these states each have a definite energy and momentum, and thus a definite mass. The only subtlety is that one must make sure that the string interacts in a Lorentz invariant fashion (e.g. the states must fall in suitable representations of the Lorentz group), while at the same time the quantum mechanics of the string is unitary. When one imposes these requirements, one finds that the resulting theories possess:

- General coordinate invariance, i.e. general relativity.
- Yang-Mills Gauge Fields
- Supersymmetry – Supersymmetry appears in these theories as a local symmetry, analogous to general coordinate invariance or Yang-Mills gauge invariance, and presumably equally fundamental.

Again, it must be stressed that these features are not imposed on the theories, but emerge automatically, simply from the hypothesis that the fundamental entities are strings.

The most simple-minded treatment yields theories which are ten-dimensional. Five consistent theories emerge in this way: the Type I theory, with open and closed strings, the Type IIA and IIB theories, and two"heterotic" string theories, all theories of closed strings only. These theories are known to have classical solutions corresponding to compactification of some of the "extra" dimensions, and a huge number of four dimensional solutions are known. These include states with:

- General relativity
- Yang-Mills gauge groups, including the standard model gauge group.
- Three generations of quarks and leptons.
- Light Higgs particles
- Gauge coupling unification
- Calculable Yukawa couplings
- N=1 Supersymmetry.

It is exciting that supersymmetry appears in this list. When supersymmetry is introduced simply to solve the hierarchy problem, one sometimes has the uneasy feeling that it is simply a trick, concocted to arrange cancellation between various Feynman diagrams. When the symmetry is made local (supergravity), one obtains non-renormalizable theories, and it is not clear whether they make sense. But in superstring theory, supersymmetry is essential to the entire structure. Supersymmetry was actually *discovered* in the framework of string theory. String theory, then, provides a certain rationale for the existence of supersymmetry.

## 15.3. String Phenomenology

Given all of these remarkable features, what stands in the way of developing a string phenomenology, determining, for example, the soft breaking parameters, and subjecting the theory to experimental tests? In order to calculate physical quantities, we need a small parameter, and yet string theory is asserted to have no small parameters. What the theories do have is a huge ground state degeneracy. In ten dimensions, for example, in all of these theories, there are several massless states. In addition to the graviton, the gravitinos, gauge bosons and others, there is a scalar field known as the dilaton. There is no potential for this field, so the field can have *any* vacuum expectation value. This is something unfamiliar in conventional, non-supersymmetric field theories, but which makes perfectly good sense in the framework of supersymmetry, where non-renormalization theorems can forbid or suppress a potential. Such fields are called moduli. Now it turns out that for some values of the vev, it is possible to make a perturbation expansion, i.e. the theory is weakly coupled. For others, the theory is strongly coupled. So in some sense there is a free parameter, but, since it is the expectation value of a dynamical field, one can hope that it will be determined by the dynamics of the theory. For example, quantum effects might give rise to a potential for $\phi$.

When one considers the compactification of the theory to fewer dimensions (say $3 + 1$), there are typically many more moduli. One can think of these as describing the size and shape of the compact space, as well as the expectation values of gauge fields with indices on the compact space, and others. For some values of these moduli (some "regions of the moduli space") the theory is weakly coupled and one can perform a perturbative expansion. For others it is strongly coupled. What is often called the "second superstring revolution" is really an exploration of this moduli space. Exploiting supersymmetry, one can show that the strong coupling limit of various string theories are the weak coupling limits of others. For example, the strong coupling limit of the $O(32)$ heterotic string in ten dimensions is the weakly coupled Type I theory; the strong coupling limit of the IIB theory is the IIB theory; the strong coupling limit of the $IIA$ theory is a theory, not yet fully understood, whose low energy limit is *eleven dimensionl supergravity.*[2]

It should be stressed that apart from these continuous choices, there are many discrete choices. For example, while there are ground states with three generations of quarks and leptons, there are also ground states with four, five, 100, and many with no chiral generations at all. There are also potential ground states in more, and less, than four dimensions.

### 15.3.1. *Vacuum selection and supersymmetry breaking*

The great question in string theory, then, is what selects among these many possible vacua. In a non-supersymmetric theory, the existence of moduli at the classical level would be merely a curiousity; quantum effects would eliminate any such vacuum degeneracy in low orders of perturbation theory. In supersymmetric theories, these degeneracies often persist even *beyond* perturbation theory. For example, in ten dimensions the dilaton is part of the supergravity multiplet, and it turns out that all of the terms in the effective action with up to two derivatives are completely fixed by supersymmetry.[5] It is simply not possible to write a potential for $\phi$ consistent with the local supersymmetry. So any value of $\phi$ seems to be as good as any other. Similar arguments apply to compactifications of the theory with more than $N = 1$ supersymmetry. There are, for example, a large set of compactifications to four dimensions with $N = 4$ supersymmetry, and supersymmetry implies that these are all exact, non-perturbative ground states of the theory.

In four dimensional compactifications with only $N = 1$ supersymmetry, one can develop a potential for the moduli. This is the case in models of "gluino condensation," for example.[6] However, such a potential will always tend to zero as the coupling tends to zero. So it is difficult to see how there can be a minimum of the potential in any region of the moduli space where the coupling is small.[7]

The main hope that this problem has a solution is the fact that the couplings we observe in nature are all weak. The question, then, is how can the moduli be stabilized at such a point. The strong coupling problem described above means that, whatever the numerical values of the coupling, weak coupling string theory *cannot* be valid in any ground state which is to describe the real world. We have no tools at the present time with which we could explore such a state.[a] Here, again, supersymmetry can play a significant role, however. In Ref. 9, it was argued that if the moduli

---

[a]One might hope that using dualities, the relevant strong coupling region could be related to the weak coupling region of some other string theory. But in this other region, because the coupling is weak, it will not be possible to stabilize the moduli.[8]

are stabilized for small numerical values of the coupling, it may be possible to understand why there is a large hierarchy, why the low energy (i.e. multi-TeV) theory is approximately supersymmetric, why the couplings are unified. Even better, one can make precise, numerical predictions, in principle, of some quantities. The point is that in four dimensional string theories, the gauge couplings and superpotential are holomorphic functions of the various moduli. Holomorphy and discrete symmetries greatly constrain the form of these functions. If $e^{-8\pi^2/g^2}$ is numerically small (as it seems to be in nature), then these quantities are hardly corrected from their weak coupling values. Stabilization of the moduli occurs because the Kahler potential of these fields is not restricted by these symmetries. With the limitations of of our current understanding, one can only speculate that such phenomena occur. However, given that the couplings *are* weak, and that string theory has so many impressive features, it seems a good working hypothesis. From this point, one can try and build a phenomenology, by studying particular string compactifications, and calculating at low orders those quantities which receive only small corrections as one passes to the weak coupling limit.

This problem of the vacuum degeneracy is perhaps the most serious difficulty in relating string theory to nature. Related to this problem is the problem of the cosmological constant: in string theories in which the vacuum energy can be reliably calculated, it is generally of order the scale of supersymmetry breaking. If string theory does solve this problem, some entirely new ideas are probably required. There are at least two serious (promising is perhaps two strong a word at this point) suggestions for how the cosmological constant problem might be solved. The first is the observation of Susskind that string theory is "holographic," and that, in particular, there are not nearly as many degrees of freedom as one might naively expect.[10] The second, is due to Witten,[11] who notes that in three dimensions, it may make sense to have broken supersymmetry with vanishing cosmological constant. No one has yet developed either of these suggestions into a precise explanation of why the cosmological constant would vanish in superstring theory with broken supersymmetry.

## 15.4. Recent Progress and Insights

String theory is a mysterious structure. As we have described it, there are five string theories, described by considering the actual motion of strings. There are also numerous solutions, characterized by discrete choices as well as by continuous parameters called moduli. We don't possess (at least in any truly useful sense) a field theory of strings, one whose symmetries would be manifest, and which might permit some direct, non-perturbative attack on questions such as the moduli potential. As a result, we have to use more indirect methods to probe the theory.

One important tool in any such analysis is the low energy effective action for the light fields. This is a particularly powerful tool when the low energy theory possesses a certain amount of supersymmetry, as well perhaps as other symmetries. The light fields might include the graviton, the gravitino and other states related to them by supersymmetry, gauge bosons, fermions, moduli and others. We have already given an example of this sort of reasoning: in ten dimensions, with $N = 1$ supersymmetry, and in four dimensions with $N = 4$ supersymmetry, supersymmetry is enough to prevent the appearance of any potential for the moduli; vacuua which are classically allowed at tree level are therefore exact quantum mechanical ground states of the theory. For $N = 1$ (fortunately) this result does not hold, but it is still possible to derive general results by rather simple reasoning. For example, one can prove that no potential for the moduli is generated in perturbation theory. To do this, one notes, first, that the dilaton, which is the coupling constant of the theory, is in a supermultiplet with an axion. A simple argument[1] shows that string theory always has a "Peccei-Quinn" symmetry under which this axion field shifts by a constant. Because the superpotential must be a holomorphic function of the coupling and must respect this symmetry, it must be independent of the dilaton (coupling), at least in perturbation theory. $N = 1$ symmetry allows one to prove numerous other results. One can show that even if supersymmetry is broken non-perturbatively at weak coupling, the cosmological constant term in the lagrangian cannot vanish.[12] Numerous other results have been proven in the past by these methods.

### 15.4.1. *Supersymmetry as a tool for understanding string theory*

The past three years have seen a great deal of progress in string theory. In particular, under the heading of "duality," many aspects of non-perturbative string theory have been understood. A review of these developments would fill volumes, but supersymmetry has been crucial to all of them. Among the most striking discoveries is that there is really only one string theory. Starting with, say, the weakly coupled heterotic string theory, by varying the moduli one comes to the Type I theory. Similarly, if the heterotic theory is compactified, by varying the moduli one can find the various type II theories. Perhaps most suprising of all, if one takes the strongly coupled limit of the Type IIA theory, one finds a theory whose low energy limit is eleven-dimensional supergravity.

All of these strange connections have been found by exploiting supersymmetry. In particular, in most of the well-understood examples, one has more than $N = 1$ supersymmetry, in terms of four dimensional counting. In such cases, the supersymmetry algebra admits central charges. This, in turn, often means that one can prove exact mass formulas for certain states, known as "BPS" states. Because they are exact, these formulas are true for all values of the moduli. This means that one can determine the spectrum of such states, and their masses, at weak coupling, and then extrapolate reliably to strong coupling. One can then compare, for example, the weak coupling spectrum of one string theory with the strong coupling spectrum of another. Similarly, in many cases, the supersymmetry completely determines the form of the terms with two derivatives in the effective action, so one knows the action even at strong coupling. This, again, has been used to provide evidence for duality.

### 15.4.2. *Supersymmetry and the structure of space-time*

Perhaps the most remarkable recent development is a proposal for a non-perturbative formulation of string theory.[13] The conjecture, in its most basic form, is very simple. Supersymmetry is a crucial ingredient – supersymmetry is essential for the very existance of space-time!

It is easiest to describe how this proposal looks in the eleven dimensional limit. There, the theory is supposed to be described an $SU(N)$ matrix quantum mechanics. This quantum mechanics is supersymmetric; it is the reduction to $0+1$ dimensions of 10 dimensional supersymmetric Yang Mills theory. As such, it has 16 supersymmetries in ten dimensions. Without going into great detail, it is not difficult to explain some of the remarkable features of this theory (which has been dubbed "M(atrix) Theory"). The matrices are just the reduction of the ten dimensional gauge fields, i.e.

$$X_i(t) = A_i(t). \tag{15.1}$$

The potential for the $X$'s is just

$$V(X) = -\frac{-1}{4}\mathrm{Tr}([X_i, X_j])^2. \tag{15.2}$$

This potential has the feature that it vanishes if the $X_i$'s are diagonal, i.e. if

$$\vec{X} = \mathrm{diag}(\vec{x}_1 \ldots \vec{x}_N). \tag{15.3}$$

The $\vec{x}_i$'s are naturally interpreted as the location of particles. When the model is examined in more detail, it turns out that these particles have the correct properties to be identified as the graviton, gravitino, and anytisymmetric tensor field of eleven dimensional supergravity. At low energies, they scatter just as do gravitons in eleven dimensions.

There is significant evidence that this conjecture for a non-perturbative formulation of the theory is correct. Many of the intricate properties of string theory are reproduced, and there is hope of computing things which are new. One of the striking features of M(atrix) theory is the crucial role which supersymmetry seems to play. In a conventional quantum mechanical system, the vanishing of the potential described above would be a tree level accident; loop corrections would generate a potential, and the space-time interpretation would collapse. In this example, if there were only bosons, then this would already occur for the leading quantum correction. This is because, for large $x$, the off-diagonal modes of the matrices $\vec{X}$ have frequencies proportional to $|\vec{x}|$. Their zero point energies, then, grown with $x$, which is to say that there is a potential for $\vec{x}$. In the supersymmetric theory, however, for each bosonic oscillator of a given frequency, there is a fermionic mode of the opposite frequency. The ground state for this system has a filled, negative energy state which cancels the bosonic term. This can be shown to persist as an exact property of the quantum theory. The supersymmetry of the matrix model is also the space-time supersymmetry. So in these theories, the very existence of space-time (the ability to have large $\vec{x}$'s, i.e. to separate gravitons and other particles) is a consequence of supersymmetry.

### 15.4.3. *Outlook*

Supersymmetry plays a crucial role in all present thinking about superstring theory, to the point that it is hard to imagine that if superstring theory is correct, we will not see evidence for supersymmetry at weak-scale energies.[b] Using supersymmetry as a tool, many remarkable properties of

[b]It can hardly be said that we have an understanding of how supersymmetry breaking in string theory induces the breaking of electroweak symmetry; in making this identification I am relying on the standard ideas about supersymmetry breaking and the hierarchy problem.

string theory have been elucidated. Perhaps the most remarkable is that all of the known consistent theories of gravity are limits of just one theory. Many find these facts persuasive – they are certainly extremely impressive. This provides strong support for the view that nature is supersymmetric, and that supersymmetry should appear "soon." A skeptic might remind us that ten years ago two-dimensional conformal invariance was thought to be the deep, underlying principal of string theory, one of the interesting outcomes of recent developments has been the overthrow of this view. Supersymmetry, similarly, may turn out to be only a convenient crutch in the study of the theory. Unlike two dimensional conformal invariance, though, supersymmetry is a fundamental new symmetry of space and time, and I think it reasonable to bet that it has a much deeper significance.

## Acknowledgments

This work was supported in part by the U.S. Department of Energy.

## References


1. M.B. Green, J.H. Schwarz and E. Witten, *Superstring Theory*, Cambridge University Press, Cambridge (1987).
2. For a recent review, with extensive references, see J. Schwarz, "The Status of String Theory," hep-th/9711029.
3. For an up to date review, see J. Maldacena, "Black Holes in String Theory," hep-th/9705078; "Black Hoes and D-Branes," hep-th/9705078; D. Youm, "Black Holes and Solitons in String Theory," hep-tg.8619946.
4. J.B. Marion and S.T. Thornton, *Classical Dynamics of Particles and Systems*, Saunders, Fort Worth (1995).
5. E. Bergshoeff, M. De Roo, B. de Wit and P. Van Nieuwenhuizen, Nucl. Phys. **B195** (1982) 97.
6. J.P. Derendinger, L.E. Ibanez and H.P. Nilles, Phys. Lett. **155B** (1985) 65; M. Dine, R. Rohm, N. Seiberg and E. Witten, Phys. Lett. 156B (1985) 55.
7. M. Dine and N. Seiberg, Phys. Lett. **162B** (299) 1985 and in *Unified String Theories*, M. Green and D. Gross, Eds. World Scientific, Singapore (1986).
8. M. Dine and Y. Shirman, Phys. Lett. **B377** (1996) 36, hep-th/9601175.
9. T. Banks and M. Dine, Phys. Rev. **D50** (1994) 7454, hep-th/9406132.
10. L. Susskind, J. Math Phys. **36** (1995) 6377, hep-th/940989.
11. E. Witten, Mod. Phys. Lett. **A10** (1995) 2153, hep-th/9506101.
12. M. Dine and N. Seiberg, Nucl. Phys. **B301** (1988) 357.
13. T. Banks, W. Fischler, S.H. Shenker, and L. Susskind, Phys. Rev. **D55** (1997) 5112, hep-th/9610043.

# Supersymmetry and Inflation

L. Randall
*Massachusetts Institute of Technology, Center for Theoretical Physics, 77 Massachusetts Avenue, Cambridge, MA 02139*

Inflation is a promising solution to many problems of the standard Big-Bang cosmology. Nevertheless, inflationary models have proved less compelling. In this chapter, we discuss why supersymmetry has led to more natural models of inflation. We pay particular attention to multifield models, both with a high and a low Hubble parameter.

## 16.1. Introduction

Supersymmetric cosmology is necessarily a speculative subject, since the evolution of the universe is sensitive not only to the observed light degrees of freedom and their superpartners, but also to the as yet undetected heavy particle spectrum. The heavy degrees of freedom only decouple at low temperatures; in the early universe they can be very relevant. In fact, degrees of freedom which are heavy could have been light in the early universe, and vice versa. Furthermore, the vacuum structure of supersymmetric and superstring theories can be very rich and complex; we do not yet know how our vacuum is determined. Nevertheless, despite our ignorance of many aspects of high-energy particle physics, there are certain features of supersymmetric theories which have been shown in recent years to be relevant to cosmology. If the world is supersymmetric, it is clearly important for cosmology, both because of the many new particles which would be present and because of the many flat direction (moduli) fields. These are fields which have no potential in the supersymmetric limit. They can however get a small potential due to supersymmetry breaking, higher dimension operators, or interactions with other fields. These flat directions which only occur naturally in supersymmetric theories can provide large amounts of energy as they will almost certainly not start their evolution from their minimum. Many recent models of inflation are based on this observation, though often in different contexts. In this chapter, we will see how supersymmetric theories might provide more compelling models of inflation. We will consider some examples which demonstrate that supersymmetric theories might provide viable inflaton candidates. Even without knowing the correct particle physics model at high energy, we can identify what might be desirable features of this model if they are to simultaneously account for an earlier epoch of inflation.

In this chapter we will briefly review the motivation for inflation and the requirements for a successful inflationary cosmology.[3,4] We will then discuss in some detail the multifield models of inflation which potentially succeed in meeting the requirements of inflation with little or no fine-tuning. We will discuss several particular models, but cannot attempt a complete enumeration of all models to date-this list is changing very rapidly! We will instead focus on what we think are the additional requirements of the supersymmetric inflation models, their possible predictions, and important questions which remain and attempts to address them.

The standard Big Bang cosmology has many important successes. Most notable are the measured Hubble expansion of the universe, the predictions of the light element abundances from nucleosynthesis in the early universe, and the prediction of the 2.7° microwave background radiation spectrum. More recently, the measurment of the anisotropy in the cosmic microwave background is an indication that theories of structure formation are on the right track. The standard cosmology is simple and successful, but very likely incomplete. As with the standard model of particle physics, the major reason this is believed is that the model as it stands is unnatural, in that it requires very fine-tuned initial conditions.

The shortcomings of the standard cosmology are the problems of the large-scale smoothness of the universe, the spatial flatness problems, the origins of small inhomogeneities, and the potential presence of unwanted relics. Inflationary cosmology successfully resolves the first two problems for a sufficiently long-lived inflationary phase. If inflation involves the correct mass scales and/or parameters, inflation can also lead to the observed density perturbations. For relics which do not get produced late in the universe, inflation can solve the problem of unwanted relics, although it should be noted that the problem of unwanted relics is a serious consideration for most supersymmetric theories, even with an early inflationary epoch.

Most successful inflationary models are based on slow-roll inflation.[1,2] In its earliest implementation it is phenomenologically successful as a model of inflation but requires fine-tuning, either of the potential or of the initial conditions.[a] The requirements on the potential for the inflaton field $\phi$ for slow roll to be valid are $|V''(\phi)| < 9H^2$ and $|V'M_P/V| < \sqrt{48\pi}$. While these constraints are met, the potential is approximately constant as is the Hubble parameter, $H = \sqrt{8\pi V/3M_P^2}$. During the period of slow-roll, the universe can expand by an exponential factor, the value of which is determined by the time for which the slow-roll conditions are valid. Inflation ends when these conditions cease to apply.

So far, we see that the important condition for inflation is to have an approximately constant energy density for a finite interval. This in and of itself is not a serious constraint, particularly in a supersymmetric theory for which many light or massless scalars might be present. What makes the construction of inflationary models tricky (or fine-tuned) is that inflation needs to end, so that reheating can produce the known matter content of the universe. With the further requirement that inflation accounts for the density fluctuations in the microwave background (here given in the slow-roll approximation), $V^{3/2}/V'M_P^3 = 5.4 \cdot 10^{-4}$, one is led to the introduction of small parameters. These problems have been reviewed elsewhere.[3,4] In general,

$$\frac{\delta\rho}{\rho} \sim \delta N \sim \frac{H^2}{\dot{\phi}} \sim \frac{H^3}{V'} \sim \frac{H^3}{m^2\phi} \tag{16.1}$$

where use has been made of the slow-roll equation of motion $3H\dot{\phi} = -V'(\phi)$ and $N$ is the number of e-folds. It is important to notice that the Hubble parameter which will give the correct magnitude of density perturbations is determined not only by $m$, the mass of the inflaton, but also by $\phi$, which in this case means the magnitude of the field $\phi$ during the time density fluctuations are formed. Therefore, although it is conventionally assumed that $H$ during inflation is determined to get the density fluctuations right, this is not necessarily the case. By constructing an inflationary model with a different value of $\phi$, one can obtain more than one scale for $H$ which can yield sensible density perturbations.

We note that it is not an essential requirement that density fluctuations formed during inflation account for the observed structure. Other suggestions for producing density perturbations have been given.[6] However, it would certainly be more economical and greatly desirable to have density

[a]There is debate over whether an initial condition should be considered fine-tuned, particularly in an eternal inflationary scenario. We will not discuss this here but refer the reader to Ref. 4.

perturbations taken care of during inflation, since inflation automatically produces fluctuations. This is the assumption which we make here. Moreover, recent papers indicate a substantial vector and tensor contribution to the CMBR implying too low anisotropies on small angular scales in cosmic defect models.[7] Ultimately, measurements should conclusively distinguish inflationary perturbations from others.[8] Current evidence seems to favor inflation.[7]

In this chapter, we will give an overview of some recent ideas for implementing inflationary models in the context of supersymmetry. Most of them are based on "hybrid"[9,14] inflationary models, although there have been a few recent suggestions which try to implement slow-roll in the context of a single inflaton field. We will first review the motivation behind multifield inflation models, and discuss some examples. We will then briefly discuss recently suggested single field models.

## 16.2. Hybrid Inflation and Supersymmetry

Before introducing supersymmetry into our discussion, let us first consider the potential advantage to multifield inflation models. To do so, we consider a toy model [b] with potential

$$V = (\phi^2 - M^2)^2 + \lambda\phi^2\psi^2 + m^2\psi^2 \tag{16.2}$$

Notice that the $\psi$ potential is minimized when $\psi = 0$, at which point the $\phi$ potential is minimized with $\phi = M$, where the potential energy $V = 0$. On the other hand, it is unlikely that $\psi$ starts at its vacuum value. In fact, when the temperature exceeds the $\psi$ mass, the potential is negligible and $\psi$ evolves extremely slowly towards its minimum. Therefore in the early universe, one can reasonably expect large values of $\psi$. If $\psi$ is greater than $\psi_c = \sqrt{2}M/\sqrt{\lambda}$, and $\phi$ is sufficiently small, $\phi$ will rapidly move towards the origin where it will sit leading to a vacuum energy of approximately $V = M^4$ (assuming this dominates the $\psi$ contribution to the energy). This nonzero vacuum energy permits an inflationary stage.

In this model, inflation will end at around the time when the $\phi$ potential turns over, when $\psi = \psi_c$. In other implementations of "hybrid" inflation,[4] it could be that inflation ends when the $\psi$ potential ceases to correspond to a slow-roll situation. We will see an example of this shortly.

---

[b] I will generally use the GRS[31] conventions for the slow-rolling inflaton field $\psi$ and the field which controls the energy density will be denoted $\phi$. The reader should be aware of other conventions existing.

The above toy model is fine as a model of inflation. In fact, multifield models seem to resolve very nicely one of the major problems with the standard slow-roll potentials; how can the potential be very flat and then give rise to a rapid end and reheat? Having two fields to control inflation solves this problem beautifully. One field controls the vacuum energy, whereas the other field essentially acts as a "switch" for inflation.

However, there are several obvious questions. First, why should the mass parameters be small? And what sets the mass scales in the first place? Nonsupersymmetric field theory cannot in general address this question. Only for a Goldstone boson is there a reason to believe the mass of a scalar is small; in general we would not expect a slow-roll potential for the $\psi$ field.

Why can supersymmetry change this picture? First of all, flat directions are natural in supersymmetric theories. Not only does supersymmetry protect against radiative corrections; supersymmetric theories in general have a large moduli space of flat directions which need not be put in by hand. We should qualify what we mean by flat directions. In general, we are referring to fields with no potential in the supersymmetric limit, with no other fields away from their vacuum expectation value, and with nonrenormalizable terms neglected. The presence of any of these terms will in fact generate a potential, but one which can in general be consistent with the requirements of inflation if the curvature of the potential is sufficiently small (when compared to the scale set by the vacuum energy).

The other interesting aspect of supersymmetric theories is that in general they require at least one mass scale which is distinct from the Planck scale. This is necessary to account for the supersymmetry breaking scale, which is lower than the Planck scale if the standard low-energy picture of supersymmetry breaking as accounting for stabilization of the electrweak scale is correct. The precise value of this scale is model dependent. In hidden sector models, the new scale will be of order $10^{11}$GeV, while in models of supersymmetry breaking based on more direct communication, the supersymmetry breaking scale will be lower. The supersymmetry breaking scale, and in particular, the intermediate scale,seems to be an obvious candidate for application to inflationary models since it is associated with nonvanishing vacuum energy density. In some supersymmetric models, other scales can appear. Notable among these is the Grand Unification (GUT) scale. Many models try to associate directly particle physics models which incorporate a grand unified gauge theory to inflation.

To account for density perturbations, it is clear that there needs to be some small number in the particle physics theory, which could be a ratio of masses. Much of the work on supersymmetric inflation has been focussed on trying to exploit these mass scales to realize the necessary requirements of inflation.

It is useful to divide these efforts into two categories. In one class of theories, the density fluctuations are roughly of order $(M_G/M_p)^2$, whereas the second class of theories only exploits the intermediate scale, and obtains density fluctuations either as $(M_I/M_P)$ or as a result of various parameters which might appear. Here $M_G \approx 10^{16}$GeV is the GUT scale and $M_I \approx \sqrt{M_W M_{Pl}}$ is the intermediate scale of order $10^{11}$GeV which determines the soft supersymmetry breaking parameters in a hidden sector scenario for the communication of supersymmetry breaking. Notice that the first class of models involves the scale $M_G$, which is the VEV of some field and may or may not set the magnitude of the potential energy density. In the second case, $M_I$, we know it is associated with a vacuum energy density. We will discuss each of these models in turn.

Many of the models we discuss are given in the context of global supersymmetry, although some models are incorporated into a supergravity theory. Before proceeding, we mention two potential problems with supergravity inflaton models; only some of the models presented below address these issues. The first issue is that if $W$ is the superpotential and $K$ is the Kahler potential, the potential takes the form

$$e^{|K|}\left(\left(W_i + K_i W\right) K_{i\bar{j}}^{-1}\left(\bar{W}_{\bar{j}} + K_{\bar{j}}\bar{W}\right) - 3|W|^2\right) + D - \text{terms} \qquad (16.3)$$

Since during inflation, the term in parentheses is nonzero if inflation is due to nonvanishing $F$ terms, one will in general find a large potential for any field appearing in the Kahler potential, which will of course include the inflaton. This can destroy the flatness of the inflaton potential and thereby destroy inflation.[10–14]

The other potential problem is that in the supergravity theory where the cosmological contant at the desired minimum is cancelled by a constant, one generically finds a deeper minimum out at values of the field larger than $M_{Pl}$[c]. However, if the superpotential is purely cubic in the fields this problem will not arise[d]. In general though, it is difficult to know how to take this problem, as one is generally treating the potential as a Taylor expansion, and at field values beyond $M_p$ the theory is presumably no longer valid. Furthermore, it is not clear that our world is in the global minimum of the potential.

[c]I thank Paul Langacker for stressing this problem.
[d]I thank Gia Dvali for this comment.

## 16.3. Hybrid Inflation and High Scale Models

The idea of hybrid inflation in supersymmetric theories was studied in a seminal paper by Liddle, Lyth, Stewart, and Wands.[14] Dvali, Shaeffer, and Shafi [DSS][15] pointed out the importance of considering quantum supersymmetry breaking effects during inflation[e], which they then exploited to introduce an interesting model of inflation. However, their model still required arbitrary mass scales. There were subsequent models in which the authors tried to identify an appropriate mass scale. None of these models are perfect, but might nonetheless have the germ of truth.

We first discuss the DSS model. They have the superpotential

$$W = \kappa S\bar{\phi}\phi - \mu^2 S \tag{16.4}$$

where $S$ is a singlet and $\phi$ and $\bar{\phi}$ transform under a GUT group. Notice that when $\phi$ and $\bar{\phi}$ vanish, the $S$ field is a flat direction. The model also contains an $R$-symmetry under which $S$ transforms which forbids an $S^3$ term in the superpotential. This is important as it is essential that $S$ is a flat direction.

Let us now consider the potential for this model. We have

$$V(S,\phi,\bar{\phi}) = \kappa^2|S|^2(|\bar{\phi}|^2 + |\bar{\phi}|^2) + |\kappa\bar{\phi}\phi - \mu^2|^2 + D - terms \tag{16.5}$$

where the $D$-terms depend on the gauge representation of $\phi$ and $\bar{\phi}$.

Now at the supersymmetry preserving minimum, the $D$-term requirement imposes $\phi = \bar{\phi}$, whereas the superpotential imposes $\phi = \bar{\phi} = \mu/\sqrt{\kappa}$ and $S = 0$. However, in the early universe it is very likely that not all fields were at their supersymmetry-preserving minimum. In fact, $S$ might have started off at a value $S > S_c = \mu/\sqrt{\kappa}$, in which case the $\phi$ potential is minimized at vanishing $\phi$ and $\bar{\phi}$, where $V = \mu^4$. In other words, this is looking precisely like a hybrid inflation model, where $S$ plays the role of $\psi$ and $\phi$ and $\bar{\phi}$ play the role of $\phi$ in our toy model.

Now naively it looks like $S$ is exactly flat, which would be bad, since there would be no potential driving $\phi$ and inflation would never end. However, this neglects the fact that supersymmetry is broken during inflation! Here the nonzero breaking is due to the nonvanishing $F$ term; however we know this is generally true since inflation relies on nonvanishing vacuum energy. In fact, in general this can be a problem in models with more than one mass scale. Since supersymmetry is broken during inflation, this can be

[e]Classical supersymmetry breaking effects during inflation had been pointed out in Refs. 14 and 16.

unnatural as in a nonsupersymmetric model. However, in this model, the quantum corrections introduce a potential for the $S$ field which is desired.

The consequence of the nonvanishing $F$ term and the breaking of supersymmetry is that a potential for the $S$ field will be generated through radiative corrections. The one-loop effective potential for $S$ is

$$\Delta V(S) = \Sigma \frac{(-1)^F}{64\pi^2} M_i(S)^4 \log\left(\frac{M_i(S)}{\Lambda}\right)^2 \tag{16.6}$$

Here $M_i(S)$ are the $S$-dependent masses of the fields. This effective interaction introduces a slope to the $S$ potential. In fact, in this type of model, inflation generally ends when slow-roll ceases to apply, rather than when the "$\phi$" potential turns over.

Let us consider this in more detail. Because of the supersymmetry breaking $F_S$, the $\phi$ and $\bar{\phi}$ spectrum do not respect supersymmetry. The scalars have mass $\kappa^2 S^2 \pm \kappa\mu^2$, whereas the fermion has mass $\kappa S$. Substituting these $S$-dependent masses into the effective potential, one derives the $S$ potential at one-loop to be

$$V_{eff}(S) = \mu^4 + \frac{\kappa^2\mu^4}{32\pi^2}\left(\log\frac{\kappa^2 S^2}{\Lambda^2} + \frac{3}{2}\right) \tag{16.7}$$

This model succeeds as a hybrid inflation model. However, there are some important open questions. First, what is the origin of the scale $\mu$. To get the correct magnitude of density fluctuations, it turns out that $\mu$ is of order the GUT scale. This means one might want to tie $\phi$ to the field with GUT mass and VEV. However, if we take $\phi$ to be an adjoint, there is too much global symmetry and one obtains too many Goldstone bosons. Interactions which violate this symmetry can also destroy inflation. Alternatively, one can take a GUT group like SU(6) and let $\phi$ and $\bar{\phi}$ be Higgs fields in the 6 and $\bar{6}$. However, the VEV of this field is likely to be too low for a successful GUT model. So in summary, although the fact that the scale $\mu$ is of order the GUT scale is intriguing, in this basic model it is tricky to realize the connection.

Another potential problem when the scale $\mu$ is high is that the reheat temperature is likely to be too high, and can cause problems with the gravitino constraint.[27] This is readily seen by a simple estimate assuming instantaneous reheat. Reheat occurs when the Hubble parameter $H$ is of order of the inflaton width $\Gamma$. Since $H^2 \sim \rho/M_p^2 \sim T_R^4/M_p^2 \sim \Gamma^2$, we find $T_R \sim \sqrt{\Gamma M_p}$. The bound on the reheat temperature depends on the mass of the gravitino, but is generally of order $10^{10}$GeV. If the inflaton decays perturbatively, one expects $\Gamma \sim \frac{\alpha}{4\pi} M_{inf}$. If $M_{inf} \sim M_G$, this is clearly too big. Even if the reheat occurs through higher dimension Planck-suppressed operators, the reheat temperature is probably too high, since it is of order $\sqrt{M_G^3/M_P}$. This is not necessarily an insuperable problem, but it generally requires a more complicated model. In the context of reheat, it should be mentioned that there is still debate over the role of parametric resonance in the decay of the inflaton; however this would generally only increase the reheat temperature. A further point is that the reheat bound assumes only gravitino couplings suppressed by $M_{Pl}$. If the inflaton decays to particles in the sector in which supersymmetry is broken, the rate for gravitino production can be even larger and the reheat bound even stricter. One can also estimate a reheat bound in gauge-mediated models of supersymmetry breaking.[18] One generally finds even more stringent bounds in this case, since the gravitino is more strongly coupled.

A third problem is that our discussion so far has been in the context of global supersymmetry. Planck-suppressed operators however cannot be neglected, since the Hubble parameter itself is Planck-suppressed. Without some tuning, supergravity corrections can invalidate the conditions for slow-roll.

Subsequent models have tried to address the first type of problem, namely the origin of the scale $\mu$ and some have also addressed the third problem. Various authors[17] (see also Ref. 19) suggested $D$-term inflation. The idea is to generate the scale "$\mu$" though a Fayet-Iliopoulos $D$-term. They envision a model with an anomalous field content in the low-energy theory, with $n_+$ fields of charge 1 and $n_-$ fields of charge -1. The model contains a superpotential

$$W = \lambda_A X \phi_+^A \phi_-^A \tag{16.8}$$

The potential for this model is then

$$V = \lambda^2 |X|^2 \left(|\phi_-|^2 + |\phi_+|^2\right) + \lambda_A^2 |\phi_+ \phi_-|^2 + \frac{g^2}{2}\left(|\phi_+^i|^2 + |\phi_+^A|^2 - |\phi_-^A|^2 + \xi\right)^2 \tag{16.9}$$

If one looks at the potential along $\phi_+ = 0$, one sees that this potential takes precisely the form required for a successful hybrid inflation model. Here, $X$ plays the role of the $\psi$ field, and $\phi_-$ (or some linear combination) the role of the $\phi$ field. One can work out the requirements for sufficiently long slow-roll and for sufficient density fluctuations.

In this model, the vacuum energy density during inflation is given by

$$V = \frac{g^2}{2}\xi^2 \tag{16.10}$$

This model has the nice feature that because it is $D$-term inflation, one can control supergravity corrections which would destroy slow-roll. Recall that there is no symmetry to prevent the quadratic terms in the Kahler potential, which, in the presence of a nonzero $F$-term, will generate an inflaton mass. The situation is better with $D$-term inflation; however Lyth[20] has pointed out that even in this case, there can be large corrections is there is no symmetry preventing quadratic corrections to the holomorphic function of fields appearing in the gauge kinetic term.

One problem with $D$-term inflation is that it is difficult to get the scale right. If the parameter $\xi$ arises due to the Green-Schwartz mechanism, it is about $g^2 \mathrm{Tr} Q M_{Pl}^2/192\pi^2$, and is probably too big for the scale set by density fluctations. One would need a small parameter to get the correct mass scale.

Some authors have tried to address the question of getting the correct size of the $D$-term. One possible solution is that the $D$-term is generated at a scale below the Planck scale. However, it is difficult to see how this can be done without large $F$-term contributions to the energy as well, destroying the initial motivation for these models.

Matsuda[21] pointed out that the strength of the gauge coupling determines the magnitude of the $D$-term, and if this coupling is dynamically determined, the $D$-term at the time of inflation might be of a different size. He shows various ansatzes for the dependence of the coupling on mass scale; unfortunately however these are not motivated by any underlying physics. However, a realistic model of the scale dependence of the coupling would require a solution to the problem of dilaton stabilization.

March-Russell[22] suggests that in a model in which the string scale is reconciled with the GUT scale, one could obtain better numbers for the size of the $D$-term.

Lyth and Riotto[23] also tried to address the discrepancy of scales required for $D$-term inflation. They point out that the normalization of the magnitude of the $D$-term which is required to agree with density perturbations depends on the slope of the potential at the end of inflation. In some cases, this slope is given by the one-loop effective potential, so by altering the number of fields coupled to the inflaton, the slope can be increased. However adjusting the slope by this (or any other mechanism) will only buy you at most an order of magnitude (once consistency with the observations on the spectral index $n$ are imposed) if one does not take the coupling $g$ to be small.

Another interesting attempt to tie the $\mu$ scale to a physical scale (here a GUT scale) in the problem was made by Dimopoulos, Dvali, and Rattazzi.[24] Their model is based on a quantum corrected moduli space, where the strong interaction scale $\Lambda$ provides the scale for the overall energy density. The model they give has a gauged SU(2) group with four flavors, so there is a quantum modified moduli space. The superpotential, including the constraint, is

$$W_{eff} = A(DetM - \bar{B}B - \Lambda^4) + S(\mathrm{Tr}M + \frac{g'}{2}\mathrm{Tr}\Sigma^2) + \frac{h}{3}\mathrm{Tr}\Sigma^3 \quad (16.11)$$

where the last terms arise due to a tree-level superpotential, and $Q\bar{Q}$ has been replaced by the confined meson field $M$. If the field $\Sigma$ is the adjoint field which breaks $SU(5)$ down to the standard model, the scale $\Lambda$ should be of order the GUT scale, which works well for producing density fluctuations.

Inflation does not involve the $\Sigma$ field until the end. Initially, the model works as with other hybrid inflation models. $S$ is a flat direction. When $S$ is big, there is a mass, and $M$ sits at zero, so there is nonzero vacuum energy. In this model, inflation ends at a nonzero value of $S$ and $\Sigma$ and SU(5) is broken. In principle, inflation could also end with $\Sigma$ zero although the authors of Ref. 24 argue that this is not the case.

There are other models which try to incorporate hybrid inflation into a GUT model. For example, Covi, Mangano, Masiero, and Miele[25] implement hybrid inflation in an SU(5) model with an additional singlet, and an arbitrary parameter $\mu$ which they need to take of order the GUT scale. Another model is given by Lazarides, Panagiotakopoulos, and Vlachos,[26] who use a nonrenormalizable potential to introduce a slope to the inflaton field (which was given by supersymmetry breaking parameters in other models).

One potential worry with any model based on SU(5) is that most such models do not solve the doublet-triplet splitting problem, and are therefore unrealistic unless a severe fine-tuning is imposed. Since the point of inflationary model building is to eliminate small parameters, this is a less than satisfactory situation.

An even more severe problem in models which really try to tie inflation to an SU(5) GUT of the real world was pointed out by Dvali, Krauss, and Liu.[28] They point out that in SU(5) models with an adjoint which gets a nonzero VEV, there will be two choices of vacua, one in which $SU(4)\times U(1)$

is preserved, and one in which $SU(3) \times SU(2)$ is preserved They parameterize the vacua with three parameters, the overall scale of the symmetry breaking, and two angles (or orbit parameters). When the inflaton field (here we mean the field whose potential generates the vacuum energy density) begins to evolve away from its inflationary value, only one potential minimum is present, namely that corresponding to the bad $SU(4) \times U(1)$ vacuum. They argue further that the field will never make its way to the desired $SU(3) \times SU(2)$ vacuum. Furthermore, whatever vacuum is chosen, the transition happens *after* inflation, so the monopole problem is not solved. So without embellishment, the simplest hybrid inflationary models based on SU(5) GUTS are not successful. These authors suggest possible resolutions which involve somewhat more complicated theories. It is also possible that in a model such as that of Ref. 24 that a noncanonical Kahler potential invalidates the energy argument and that the suitable vacuum is obtained.

Most authors do not address the question of reheat. Lazarides[29] suggests a decay to a second generation neutrino to avoid too big renormalizable couplings. It is hard to think of a natural decay mode without small coupling which can avoid a high reheat temperature and overproduction of gravitinos. An alternative proposal of Dimopoulos and Dvali is that reheat is delayed by rolling along a flat direction;[30] however it is necessary to ensure that no other dangerous perturbations will be produced. It could be that there is some late entropy release which invalidates the gravitino bound; this might be required to solve the Polonyi problem[32] in any case. It is clear that the question of the high reheat temperature should be addressed in these high scale models.

So to summarize, models of hybrid inflation based on a high $H$ scale seem close to working. $D$-term inflation doesn't quite get the scale right, but is close. Models based on SU(5) generically suffer from the problem outlined in Ref. 28 which is unfortunate since it makes the very nice coincidence of scales less useful. However, it is not impossible to make these models work, and further advances might be forthcoming.

## 16.4. Low-Scale Models

We now go on to discuss another very promising class of models, which do not introduce a high scale Hubble parameter, but try to implement successful inflation only by assuming the existence of soft-supersymmetry breaking in a hidden sector. These models are intended as illustrations of how the moduli fields can be employed in a hybrid inflation scenario without strong or unnatural assumptions on the particle physics model. With the specific cases that were discussed, one can identify distinguishing characteristics of this class of model, which should be testable.

The goal of Randall, Soljačić, and Guth (RSG) was to construct models of inflation using *only* the intermediate mass scale $M_I$ and the Planck scale $M_p$. The aim was to see whether a natural model could be simply constructed which employed moduli fields with soft supersymmetry breaking masses. The essential observation is that the temperature fluctuations observed by COBE do not necessarily require high scale inflation, since the formula for density fluctuations actually depends both on the magnitude of the potential and its slope. Taking the potential quadratic at the time density fluctuations relevant to physical scales are formed, the formula for density fluctuations is

$$\frac{\delta\rho}{\rho} \approx \frac{H^2}{\dot{\psi}} \approx \frac{H^3}{m^2\psi} \tag{16.12}$$

where $\psi$ is the slow rolling field. Now if the energy density during inflation is $M^4$, the Hubble parameter is of order $M^2/M_p$. On the other hand, if the $\psi$ mass arises from hidden sector supersymmetry breaking, it is $m \approx M_I^2/M_{Pl}$. The assumption in this class of models is that $M \sim M_I$, in which case

$$\frac{\delta\rho}{\rho} \approx \frac{H}{\psi} \tag{16.13}$$

From this equation, it is clear that the magnitude of density fluctuations depends on the value of $\psi$ when inflation ends, which for hybrid inflation models is essentially $\psi_c$.

In Ref. 31, two types of potentials were considered. The first class of models assumed the fields were coupled through a higher dimension operator derived from the superpotential

$$W = \frac{\phi^2\psi^2}{2M'} \tag{16.14}$$

where $M'$ is a relevant physical mass scale. Perhaps the most natural possibility is $M_p$. However, it is conceivable there are heavy particle exchanges so that $M'$ can be identified with the GUT scale $M_G$ or $M_I$.

The other class of model assumed there was a potential coupling the two fields involved in hybrid inflation of the form

$$V = \frac{\lambda}{4}\psi^2\phi^2 \tag{16.15}$$

Such a coupling could arise for example if the superpotential coupled together three fields $W = \chi\phi\psi$, where $\chi = 0$ during inflation. This is in fact something which happens quite naturally, even in the context of the MSSM. For example, if $\psi = \bar{u}\bar{d}\bar{d}$ and $\phi = H_u H_d$, $W = \lambda_u Q_u H_u \bar{u}$, one realizes this situation. In fact, nonstandard GUT models[31,33] can realize this potential in a way consistent with the inflationary constraints. In this model, the magnitude of density fluctuations will be set by a Yukawa coupling; it is important to recognize that this might well be significantly less than unity.

The complete specification of the model requires the potential for the $\phi$ and $\psi$ fields (apart from their mutual coupling). Note that these potentials arise due to soft supersymmetry breaking and therefore should be characterized by potentials of the form $M_I^4 g(\phi/M_p)$ where $g$ is a function with a Taylor expansion with coefficients of order unity. To realize the hybrid inflationary scenario, the potential for $\phi$ is taken as $V = M^4 \cos^2(\phi/\sqrt{2}f)$ and the potential for $\psi$ is taken as $\frac{1}{2}m_\psi^2\psi^2$. To be consistent with the requirement that these are moduli fields with a potential generated by soft supersymmetry breaking, we would want to find $M \sim M_I$ and $m_\psi \sim M_I^2/M_p$. There has been much confusion over the very specific-looking form taken for the $\phi$ potential. Indeed, all that is relevant to inflation are the first two terms in the Taylor expansion! Only when inflation ends and $\phi$ moves from zero are the other terms relevant. This is simply a compact way of writing a function which has the correct negative curvature at the origin and zero vacuum energy for the true vacuum. It is interesting however that exactly such a potential could be produced by a nontrivially coupled pseudo-Goldstone mode [f]. However, this precise form of the potential is not at all essential, so the field $\phi$ can be any moduli field with negative mass squared and a supersymmetry-breaking source for its potential.

This model realizes very nicely the hybrid inflation scenario. Depending on the form of the soft-supersymmetry breaking potential and the couplings between moduli fields, it is very likely one can find suitable candidates for inflation. The major distinguishing characteristic of this type of model is that the $\phi$ field is light. Therefore, the dynamics controlling the end of inflation is very different. In many other models, the $\phi$ mass is large, so it very quickly rolls to its true minimum once inflation has stopped. In the RSG models, the field $\phi$ spends more time moving primarily due to de Sitter fluctuations, subsequent to which it rolls classically. Because of the initial motion when the field is moving relatively slowly, there is a

[f] I thank Gia Dvali for sharing this observation of Lawrence Krauss and himself.

spike in the density fluctuation spectrum. This spike can be interesting (or dangerous) in that it will lead to more structure on small scales. However, it is too large to be present in observed density fluctuations on scales from about 1 Mpc to $10^4$ Mpc, which gives a constraint on how quickly inflation must end. Including this constraint as well as the constraint from density fluctuations, one finds that the parameters of this model are such that it works rather well, with mild tuning depending on the particular model (numbers as small as 0.01 might be required; see Ref. 31 for details). That is, the original goal, to motivate the parameters by supersymmetry breaking scales, can be reasonably well accomodated.

It should be noted that the derivation of the parameters of the spike in the density perturbation spectrum is subtle. In Ref. 31, the calculation was based on using the Fokker-Planck equation to establish the $\phi$ mean $\phi$ distribution and the time delay given by a fluctuation in the $\phi$ field. Garcia-Bellido, Linde, and Wands[34] objected to the calculational method of Ref. 31 and instead calculated the fluctuations associated with each element of the ensemble assuming it was classical. However, it can be shown[35] that the method used is not valid, though the original[31] calculation needed to be improved to account for deviation from slow-roll and for a more exact calculation of the fluctuations for a massive field.

Stewart[36,37] has also constructed low-scale models for which the small tuning required to get a sufficiently flat potential is not required and for which the spike will have different properties. He points out that once quantum corrections are incorporated, there can be a special point (or more than one) where the potential is particularly flat. Initial conditions are probably different for this class of model; one relies on entering a phase of eternal inflation from which one will enter the desired hybrid inflationary phase.

In summary, the low-scale inflation models can have quite distinctive features, and do not have the problems associated with introducing a high scale in a particle physics context. However, they might involve a small amount of fine-tuning; on the other hand they might also involve a small parameter which is present. The spike is generically a test of the models; however if the field is rolling quickly at the phase transition, as is true for Ref. 37, this might be lessened; futher work is needed for these models to establish the detailed form of the spike. In general, the low-scale models are well motivated and worthy of further investigation.

Before discussing further models, it is worth noting a distinguishing feature of many hybrid inflation models, those with a mass for the inflaton (like the GRS models), namely the fact that the index $n$ is generally bigger than 1. Furthermore, by measuring the ratio of tensor to scalar perturbations, one can in principle (because it is a difficult measurement) distinguish high and low scale models. The deviation of the index $n$ from 1 measures the scale dependence of density fluctuatons. It can be determined from the potential at the time the relevant perturbation leaves the horizon from the formula

$$n = 1 - 3\left(\frac{V'}{V}\right)^2 + 2\frac{V''}{V} \tag{16.16}$$

whereas $R$, the ratio of tensor to scalar perturbations, is given by

$$R \approx 6\left(\frac{V'}{V}\right)^2 \tag{16.17}$$

where in these equations we have set $M_p$ to unity. One can obtain interesting qualitative information from these formulae. First consider the quantity $R$. If we can approximate $V'$ by a mass term near the end of inflaton, we have $V'/V \sim m^2\psi M_p/H^2M_p^2$. So if $H \sim m$, which is often the case, we find that $R$ is negligible unless $\psi \sim M_p$. As we have argued, models with low $H$ can achieve adequate density perturbations if $\psi$ is small (much less than $M_p$) at the end of inflation. We conclude that these type of models will always have neglibile $R$. For other models, with $\psi$ closer to $M_p$, it is a detailed question whether $R$ can be measurable.

Notice also that when $V'/V$ is neglibible, the sign of the mass squared term at the end of inflation determines whether $n$ is bigger, or less than unity. So models for which the inflaton field rolling towards the origin will have $n$ bigger than 1. The RSG models are of this type, as are hybrid inflation models with a mass term determining the evolution of the inflaton.

It should be noted that there are many models where the potential for the inflaton is not determined by a mass term. An interesting example of hybrid inflation which he dubbed "Mutated Hybrid Inflation" for which the index $n$ is less than 1 was given by Stewart.[38] He considers a toy model where inflation occurs along a nontrivial trajectory in field space. The net result is that along this trajectory, the potential can be written as a polynomial function of the inverse field, and the index $n$ can be shown to be less than 1. Generalizations of this idea and other mechanisms for producing an index less than 1, again in toy models, was given in Ref. 39.

## 16.5. Single Field Models

Aside from the many models of hybrid inflation based on supersymmetry, there are a couple of single field inflationary models worthy of note. By single field, we do not mean there is only one field in the potential, but that the inflationary dynamics can be viewed in terms of a single field (as opposed to hybrid inflation models). Garcia-Bellido[40] observed that the potential given by an $N = 2$ SU(2) gauge theory with supersymmetry breaking[41] incorporated takes a form which looks remarkably like a slow-roll potential along a particular trajectory in field space. However, the scale of supersymmetry breaking which is required has no particle physics motivation, so it is not yet clear if this can be tied to a particle physics theory of our world.

Another single field model is that of Adams, Ross, and Sarkar.[42] They are interested in the problem of the large quadratic terms which can be present in supergravity theories. They argue that there can be special points where the quadratic terms vanish, and these can be quasi-fixed points in the evolution of the field.

Another paper which addresses the issue of large supergravity corrections is by Gaillard, Murayama, and Olive.[43] They observe that at tree-level, the mass term for the inflaton which occurs generically in supergravity theories is absent if there is a Heisenberg symmetry. Although there is no symmetry reason, it is claimed that gravitational interactions preserve the symmetry (based on a one-loop calculation), so that the potential can be calculated from gauge and superpotential interactions. A more complete realization of this scenario could be interesting.

Stewart[16] also addressed the issue of large supergravity corrections to the inflaton mass. He identified conditions, which when imposed on the superpotential, guarantee the absence of such corrections. These conditions are $W = W_{\psi} = \phi = 0$ (in the GRS naming convention) during inflation, as well as some conditions on the Kahler potential. He argues that such potentials might arise naturally in superstring theories.

In fact, Copeland, Liddle, Lyth, Stewart, and Wands[14] had initially pointed out that supergravity corrections to the mass can cancel if there is a minimal Kahler term. Linde and Riotto[44] make the assumption that nonminimal terms which would destroy this cancellation are small, and then consider the model with both one-loop and gravitational effects taken into account.

In summary, there are currently many ideas on how to use the naturalness property of supersymmetry to provide candidates for inflaton fields. There are some clever ideas involved in high scale models; however the $D$-term models generally give too large density fluctuations while the GUT models often lead to the wrong vacuum following inflation. These models might however be incorporated into more complete and realistic models in the future. The low scale inflation models have the advantage that they require no new mass scales aside from that which was already required to give supersymmetry breaking parameters of order the weak scale in a hidden sector scenario. They provide an interesting signature of a spike in the spectrum so they should be subject to experimental verification in the future. Other objections might include the fact that this inflation is relatively late, so this might mean some prior nonstandard evolution (like a previous inflationary phase) is required. These models require mild tunings; presumably even this is not necessary if one will compromise with more complicated scenarios. It is intriguing that new models of particle physics might also lead to new potentials that could provide slow-roll. Exact superpotentials in the strong interaction regime often take nonpolynomial forms which would not have been anticipated on the basis of weakly coupled renormalizable field theories. Other models which fall outside the range of the supersymmetry-motivated field theories we have considered here include dilaton-based inflation[45] and string-theory-motivated-domain walls as the seed for inflation.[46]

Although there are many ideas, it should be remembered that there are many requirements for a good inflationary model. Given the indirectness of many cosmological constraints, it is remarkable how constrained models are. Requirements include a sufficiently long period of inflation, a mechanism for ending inflation sufficiently quickly to reheat to temperatures higher than the weak scale (this might be too stringent but in alternative scenarios one needs a mechanism for baryogenesis), a reheat temperature sufficiently low not to overproduce gravitinos, consistency with a spectral index which does not deviate by more than 20% from 1 (the exact constraint is subject to interpretation), and hopefully no ad hoc scales or small numbers. There are only a few models which meet all these criteria. And we have not even addressed the many issues of how inflation fits into a more complete picture of the cosmology of the early universe which includes baryogenesis[47] and a solution to the Polonyi problem.[32] So despite the many recent advances, the field still remains fertile, and it would not be surprising to see new and more compelling models of inflation in the future.

## Acknowledgments

I am very grateful to Ewan Stewart and Chris Kolda for discussions and their comments on the manuscript. I also thank Gia Dvali, Marc Kamionkowski, John March-Russell, and Riccardo Rattazzi for their comments. I also thank Princeton University and the Institute for Advanced Study for their hospitality while this work was completed. This work is supported in part by funds provided by the U.S. Department of Energy (D.O.E.) under cooperative agreement #DF-FC02-94ER40818.

## References

1. A. Albrecht and P. J. Steinhardt, *Phys. Rev. Lett.* **48** (1982) 1220.
2. A. D. Linde, *Phys. Lett.* **B108** (1982) 389.
3. M. S. Turner, Inflation after COBE: Lectures on Inflationary Cosmology, astro-ph/9304012, Boulder TASI 92: 165-234, Cargese Summer School 1992:23 1-398, and references therein.
4. A. D. Linde, Particle physics and inflationary cosmology (harwood Academic, Basel, 1990).
5. M. Sasaki and E. D. Stewart, *Prog. Theor. Phys.* 95, 71 (1996).
6. R. Brandenberger, Topological Defects and the Formation of Structure in the Universe, astro/ph-9604033, Talk given at Pacific Conference on Gravitation and Cosmology, Seoul, Korea, Feb, 1996.
7. B. Allen, R. R. Caldwell, S. Dodelson, L. Knox, E.P.S. Shellard, and A. Stebbins, astro-ph/9704160, U. Pen, U. Seljak, and N. Turok, astro-ph/9704165, N. Turok, U. Pen, and U. Seljak, astro-ph/9706250.
8. A. Albrecht, How to Falsify Scenarios with Primordial Fluctuations from Inflation, astro-ph/9612017, Presented at Conference on Critical Dialogs in Cosmology, Princeton, NJ, 24, June, 1996.
9. A. D. Linde, *Phys. Lett.* **B259** (1991) 38, A. R. Liddle and D. H. Lyth, *Phys. Rep. D* 52 (1995) 6789, A. D. Linde, *Phys. Rev.* **D49** (1994) 748;
10. M. Dine, W. Fischler, and D. Nemeschansky, *Phys. Lett.* **B136** (1984) 169;
11. G. Coughlan, R. Holman, P. Ramond, and G. Ross, *Phys. Lett.* **B140** (1984) 44.
12. M. Dine, L. Randall, and S. Thomas, hep-ph/9503303, *Phys. Rev. Lett.* **75** (1995) 398.
13. R. Barbieri, S. Ferrara, and C. A. Savoy, *Phys. Lett.* **B119** (1982) 343.
14. E. J. Copeland, A.R. Liddle, D.H. Lyth, E.D. Stewart, and D. Wands, *Phys. Rev.* **D49** (1994) 6410,E. Stewart, *Phys. Lett.* **B345** (1995) 414.
15. G. Dvali, Q. Shafi, and R. Schaefer, hep-ph/9406319*Phys. Rev. Lett.* **73** (1994) 1886.
16. E. Stewart, *Phys. Rev.* **D51** (1995) 6847.
17. E. Halyo, *Phys. Lett.* **B387** (1996) 43; P. Binetruy and G. Dvali, *Phys. Lett.* **B388** (1996) 241.

18. A. de Gouvea, Takeo Moroi, H. Murayama, *Phys. Rev.* **D56** (1997) 1281.
19. J. A. Casas and C. Munoz, *Phys. Lett.* **B216** (1989) 37; J. A. Casas, J. M. Moreno, C. Munoz, and M. Quiros, *Nucl. Phys.* **B328** (1989) 272; E. D. Stewart, *Phys. Rev.* **D51** (1995) 6847; M. Dine, L. Randall, and S. Thomas, *Nucl. Phys.* **B458** (1996) 291.
20. D. H. Lyth, hep-ph/9710347.
21. T. Matsuda, hep-ph/9705448, May, 1997.
22. John March-Russell, private communication.
23. D. H. Lyth and A. Riotto, hep-ph/9707273, June, 1997.
24. S. Dimopoulos, G. Dvali, and R. Rattazzi, hep-ph/9705348, May, 1997.
25. L. Covi, G. Mangano, A. Masiero, and G. Miele, hep-ph/9707405, July, 1997.
26. G. Lazarides, C. Panagiotakopoulos, and N. D. Vlachos, *Phys. Rev.* **D54** (1996) 54.
27. T. Moroi, Effects of the Gravitino on the Inflationary Universe, Ph. D. thesis, and refs. therein.
28. G. Dvali, L. Krauss, and H. Liu, hep-ph/9707456, July, 97.
29. G. Lazarides, R. K. Schaefer, and Q. Shafi, *Phys. Rev.* **D56** (1997) 1324.
30. Gia Dvali, private communication.
31. L. Randall, M. Soljačić, A. Guth, *Nucl. Phys.* **B472** (1996) 377.
32. T. Banks, D. Kaplan, and A. Nelson, *Phys. Rev.* **D49** (1994) 779; L. Randall, S. Thomas, *Nucl. Phys.* **B449** (1995) 229; D. H. Lyth and E. D. Stewart, *Phys. Rev. Lett.* **75** (1995) 201.
33. Z. Berezhiani, C. Csáki, and L. Randall, *Nucl. Phys.* **B444** (1995) 444.
34. J. Garcia-Bellido, A. Linde, and D. Wands *Phys. Rev.* **D54** (1996) 6040.
35. A. Guth, L. Randall, S. Su, in progress.
36. E. Stewart, *Phys. Lett.* **B391** (1997) 34.
37. E. Stewart, *Phys. Rev.* **D56** (1997) 2019.
38. E. Stewart, *Phys. Lett.* **B345** (1995) 414.
39. D. Lyth and E. Stewart, *Phys. Rev.* **D54** (1996) 7186.
40. J. Garcia-Bellido, hep-th/9707059.
41. L. Alvarez-Gaume, J. Distler, C. Kounnas, M. Marino, *Int. J. Mod. Phys. A* 11 (1996) 4745, hep-th/9604004.
42. J. Adams, G. G. Ross, S. Sarkar, *Phys. Lett.* **B391** (1997) 271.
43. M. Gaillard, H. Murayama, and K. Olive, *Phys. Lett.* **B355** (1995) 71.
44. A. Linde and A. Riotto, *Phys. Rev.* **D56** (1997) 1841.
45. R. Brustein, hep-th/9506045, Contribution to International Conference on Unified Symmetry in the Small and in the Large, Coral Gables, FL, Feb, 95 and references therein.
46. T. Banks, M. Berkooz, S. Shenker, G. Moore, P. Steinhardt, *Phys. Rev.* **D52** (1995) 3548.
47. A. Affleck and M. Dine *Nucl. Phys.* **B249** (1985) 361, M. Dine, L. Randall, and S. Thomas *Nucl. Phys.* **B456** (1996) 291 for example.

# An Introduction to Explicit R-Parity Violation

Herbi Dreiner
*Physikalisches Institut and BCTP, Universität Bonn, Nussallee 12, D53115 Bonn, Germany**

I discuss the theoretical motivations for R-parity violation, review the experimental bounds and outline the main changes in collider phenomenology compared to conserved R-parity. I briefly comment on the effects of R-parity violation on cosmology.

## 17.1. Introduction

Until recently, R-parity violation ($\not{R}_p$) has been considered an unlikely component of the supersymmetric extension of the Standard Model (SM). In the past two years, it has motivated potentially favoured solutions to experimentally observed discrepancies (*e.g.* $R_b$, $R_c$, ALEPH four-jet events, HERA high $Q^2$ excess). It is the purpose of this chapter to present $\not{R}_p$ as an equally well motivated supersymmetric extension of the SM and provide an introductory guide. I start out with the definition of $R_p$ and the most serious problem of proton decay. Then I discuss the various motivations for $\not{R}_p$, contrasting them with the $R_p$-conserving MSSM. Afterwards, I give an overview of the phenomenology of $\not{R}_p$. I finish with a discussion on cosmological effects.

## 17.2. What is R-Parity?

R-parity ($R_p$) is a discrete multiplicative symmetry. It can be written as[1]

$$R_p = (-\mathbf{1})^{3B+L+2S}. \tag{17.1}$$

---

*Originally published while at Rutherford Appleton Laboratory, Chilton, Didcot, Oxon OX11 0QX, UK.

Here $B$ denotes the baryon number, $L$ the lepton number and $S$ the spin of a particle. The electron has $R_p = +1$ and the selectron has $R_p = -1$. In fact, for all superfields of the supersymmetric SM, the SM field has $R_p = +1$ and its superpartner has [a] $R_p = -1$. $R_p$ is conserved in the MSSM, superpartners can only be produced in pairs (all initial states at colliders are $R_p$ even) and the LSP is stable. When extending the SM with supersymmetry one doubles the particle content to accomodate the superpartners and adds an additional Higgs doublet superfield. The minimal symmetries required to construct the Lagrangian are the gauge symmetry of the SM: $G_{SM} = SU(3)_c \times SU(2)_L \times U(1)_Y$ and supersymmetry (including Lorentz invariance). The most general superpotential with these symmetries and this particle content (*cf.* Ch. 1) is[4]

$$W = W_{MSSM} + W_{\not{R}_p}, \tag{17.2}$$

$$W_{MSSM} = h^e_{ij} L_i H_1 \bar{E}_j + h^d_{ij} Q_i H_1 \bar{D}_j + h^u_{ij} Q_i H_2 \bar{U}_j + \mu H_1 H_2, \tag{17.3}$$

$$W_{\not{R}_p} = \frac{1}{2}\lambda_{ijk} L_i L_j \bar{E}_k + \lambda'_{ijk} L_i Q_j \bar{D}_k + \frac{1}{2}\lambda''_{ijk} \bar{U}_i \bar{D}_j \bar{D}_k + \kappa_i L_i H_2. \tag{17.4}$$

$i, j = 1, 2, 3$ are generation indices and a summation is implied. $L_i$ ($Q_i$) are the lepton (quark) $SU(2)_L$ doublet superfields. $\bar{E}_j$ ($\bar{D}_j, \bar{U}_j$) are the electron (down- and up-quark) $SU(2)_L$ singlet superfields. $\lambda$, $\lambda'$, and $\lambda''$ are Yukawa couplings. The $\kappa_i$ are dimensionful mass parameters. The $SU(2)_L$ and $SU(3)_C$ indices have been suppressed. When including them we see that the first term in $W_{\not{R}_p}$ is anti-symmetric in $\{i, j\}$ and the third term is anti-symmetric in $\{j, k\}$. Therefore $i \neq j$ in $L_i L_j \bar{E}_k$ and $j \neq k$ in $\bar{U}_i \bar{D}_j \bar{D}_k$. Eq.(17.4) thus contains $9 + 27 + 9 + 3 = 48$ new terms beyond those of the MSSM.

The last term in Eq.(17.4), $L_i H_2$, mixes the lepton and the Higgs superfields. In supersymmetry $L_i$ and $H_1$ have the same gauge and Lorentz quantum numbers and we can redefine them by a rotation in $(H_1, L_i)$. The terms $\kappa_i L_i H_2$ can then be rotated to zero in the superpotential.[5] If the corresponding soft supersymmetry breaking parameters $B_i$ are aligned with the $\kappa_i$ they are simultaneously rotated away.[5,6] However, the alignment of the superpotential terms with the soft breaking terms is not stable under the renormalization group equations.[7] Assuming an alignment at the unification scale, the resulting effects are small[7] except for neutrino

[a] In general symmetries for which the anticommuting parameters, $\theta$, transform non-trivially (and thus superpartners differently) are denoted R-symmetries. They can be discrete ($R_p$), global continuous, or even gauged.[2,3] R-symmetries can be broken without supersymmetry being broken.

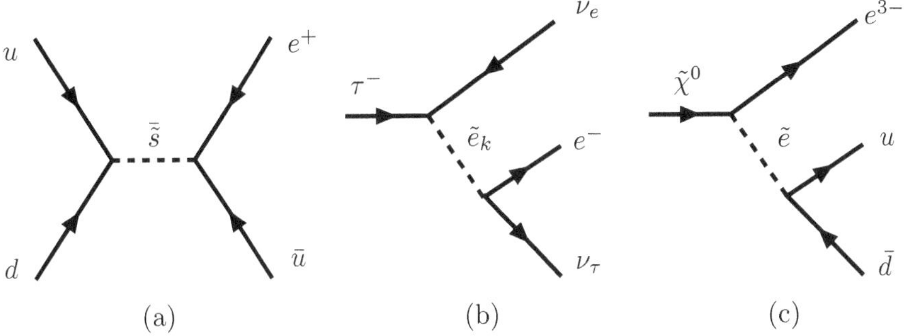


Fig. 17.1. (a) Proton decay via $\bar{U}_1\bar{D}_1\bar{D}_2$ and $L_1Q_1\bar{D}_2$, (b) Tau decay via two $L_1L_3\bar{E}_k$ insertions, (c) Neutralino decay via $L_1Q_1\bar{D}_1$.

masses.[7,8] The effects can be further suppressed by a horizontal symmetry. Throughout the rest of this Chapter, I will assume the $L_iH_2$ terms have been rotated away [b]

$$W_{\not R_p} = \lambda_{ijk}L_iL_j\bar{E}_k + \lambda'_{ijk}L_iQ_j\bar{D}_k + \lambda''_{ijk}\bar{U}_i\bar{D}_j\bar{D}_k. \tag{17.5}$$

Expanding for example the $LL\bar{E}$ term into the Yukawa couplings yields

$$\mathcal{L}_{LL\bar{E}} = \lambda_{ijk}\left[\tilde{\nu}^i_L\bar{e}^k_Re^j_L + \tilde{e}^j_L\bar{e}^k_R\nu^i_L + (\tilde{e}^k_R)^*(\bar{\nu}^i_L)^ce^j_L - (i\leftrightarrow j)\right] + h.c. \tag{17.6}$$

The tilde denotes the scalar fermion superpartners. These terms thus violate lepton-number. The $LQ\bar{D}$ terms also violate lepton number and the $\bar{U}\bar{D}\bar{D}$ terms violate baryon number. The entire superpotential (17.5) violates $R_p$.

## 17.3. Proton Decay and Discrete Symmetries

The combination of lepton- and baryon-number violating operators in the Lagrangian can possibly lead to rapid proton decay. For example the two operators $L_1Q_1\bar{D}_k$ and $\bar{U}_1\bar{D}_1\bar{D}_k$ ($k\neq 1$) can contribute to proton decay via the interaction shown in Figure 17.1(a). On dimensional grounds we estimate

$$\Gamma(P\to e^+\pi^0) \approx \frac{\alpha(\lambda'_{11k})\alpha(\lambda''_{11k})}{\tilde{m}^4_{dk}}M^5_{proton}. \tag{17.7}$$

---

[b]The ambiguity on bounds due to rotations in $(L_i, H_1)$ space has been discussed in Ref. 9.

Here $\alpha(\lambda) = \lambda^2/(4\pi)$. Given that[10] $\tau(P \to e\pi) > 10^{32}\, yr$, we obtain

$$\lambda'_{11k} \cdot \lambda''_{11k} \lesssim 2 \cdot 10^{-27} \left( \frac{\tilde{m}_{dk}}{100\, GeV} \right)^2 . \tag{17.8}$$

For a more detailed calculation see Ref. 11. This bound is so strict that the only natural explanation is for at least one of the couplings to be zero. Thus the simplest supersymmetric extension of the SM is excluded: an extra symmetry is required to protect the proton.

In the MSSM, $R_p$ is imposed by hand. This forbids all the terms in $W_{\not{R}_p}$ and thus protects the proton. An alternative discrete symmetry with the same physical result is matter parity

$$(L_i, \bar{E}_i, Q_i, \bar{U}_i, \bar{D}_i) \to -(L_i, \bar{E}_i, Q_i, \bar{U}_i, \bar{D}_i), \quad (H_1, H_2) \to (H_1, H_2). \tag{17.9}$$

This forbids all terms with an odd power of matter fields and thus forbids all the terms in $W_{\not{R}_p}$. However, there are other solutions, which protect the proton equally well. If baryon number is conserved the proton can not decay. Thus forbidding just the interactions $\bar{U}_i\bar{D}_j\bar{D}_k$ is sufficient. This can be achieved by baryon-parity

$$(Q_i, \bar{U}_i, \bar{D}_i) \to -(Q_i, \bar{U}_i, \bar{D}_i), \quad (L_i, \bar{E}_i, H_1, H_2) \to (L_i, \bar{E}_i, H_1, H_2). \tag{17.10}$$

This symmetry thus protects the proton but allows for $\not{R}_p$ via the $L_iL_j\bar{E}_k$ and $L_iQ_j\bar{D}_k$ operators. If only the the interactions $\bar{U}_i\bar{D}_j\bar{D}_k$ are allowed *and* the proton is lighter than the LSP the proton is stable as well. This can be achieved by lepton parity

$$(L_i, \bar{E}_i) \to -(L_i, \bar{E}_i), \quad (Q_i, \bar{U}_i, \bar{D}_i, H_1, H_2) \to (Q_i, \bar{U}_i, \bar{D}_i, H_1, H_2). \tag{17.11}$$

Baryon-parity and lepton parity are two possible solutions to maintain a stable proton *and* allow for $\not{R}_p$. There is a large number of discrete symmetries which can achieve this.[12]

### 17.4. Motivation

The symmetries discussed in the previous section were all imposed *ad hoc* with no deeper motivation than to ensure the stability of the proton. On this purely phenomenological level there is no reason to prefer the models with conserved $R_p$ versus those with $\not{R}_p$. However, this is not a satisfactory view of the weak-scale picture. Hopefully, the correct structure will emerge from a simpler theory at a higher energy predicting either $R_p$-conservation or $\not{R}_p$.

*Grand Unified Theories:* In GUTs quarks and leptons are typically in common multiplets and thus have the same quantum numbers. The discrete symmetries protecting the proton and resulting in $\not{R}_p$ typically treat quarks and leptons differently and thus seem incompatible with a GUT. All the same, several GUT models have been constructed[5,13–15] which have low-energy $\not{R}_p$. This is typically achieved by non-renormalisable GUT scale operators involving Higgs fields. These operators become renormalisable $\not{R}_p$-operators after the GUT symmetry has been broken. Such models have been constructed for the GUT gauge groups $SU(5)$,[5,14] $SO(10)$,[14] and $SU(5) \times U(1)$.[13,14] They have been constructed such that the only set of low-energy operators is $LL\bar{E}$ or $LQ\bar{D}$, or $\bar{U}\bar{D}\bar{D}$, respectively. There is thus no problem with proton decay. In order to ensure that only the required set of non-renormalisable operators are allowed, additional symmetries are required beyond the GUT gauge group. This is true for both $R_p$-conservation and $\not{R}_p$. Thus from a grand unified point of view there is no preference for either $R_p$-conservation or $\not{R}_p$.

*String Theory:* In string theories unification can be achieved without a simple gauge group. There is thus no difficulty in having distinct quantum numbers for quarks and lepton superfields. Indeed $R_p$-conserving and $\not{R}_p$ string theories have been constructed.[16] At present, there does not seem to be a preference at the string level for either of the two.

In both string theory and in GUTs, there is no generic prediction for the size of the $\not{R}_p$-Yukawa couplings. This is analogous to the fermion mass problem.

*Discrete Gauge Symmetries:* There has been a further attack on this problem from a slightly different angle. If a discrete symmetry is a remnant of a broken gauge symmetry it is called a discrete gauge symmetry. It has been argued that quantum gravity effects maximally violate all discrete symmetries unless they are discrete gauge symmetries.[17] The condition that the underlying gauge symmetry be anomaly-free can be translated into conditions on the discrete symmetry. A systematic analysis of all $\mathcal{Z}_N$ symmetries[18] has been performed. The result was that only two symmetries were discrete gauge anomaly-free: $R_p$ and baryon-parity (17.10). Baryon-parity was slightly favoured since in addition it prohibited dimension-5 proton-decay operators. It has since been shown[19] that the non-linear constraints in Ref. 18 are model dependent thus possibly allowing an even larger set of discrete symmetries.

Given the quantum gravity argument it is more appealing to determine the low-energy structure directly from gauge symmetries instead of discrete

Table 17.1. Strictest bounds on $R\!\!\!/_p$ Yukawa couplings for $\tilde{m} = 100\,GeV$. The physical processes from which they are obtained are summarized in the main text.

| $ijk$ | $\lambda_{ijk}$ | $ijk$ | $\lambda'_{ijk}$ | $ijk$ | $\lambda'_{ijk}$ | $ijk$ | $\lambda'_{ijk}$ | $ijk$ | $\lambda''_{ijk}$ |
|---|---|---|---|---|---|---|---|---|---|
| 121 | $0.05^{a\dagger}$ | 111 | $0.001^{d}$ | 211 | $0.09^{h}$ | 311 | $0.16^{k}$ | 112 | $10^{-6,\ell}$ |
| 122 | $0.05^{a\dagger}$ | 112 | $0.02^{a\dagger}$ | 212 | $0.09^{h}$ | 312 | $0.16^{k}$ | 113 | $10^{-5,m}$ |
| 123 | $0.05^{a\dagger}$ | 113 | $0.02^{a\dagger}$ | 213 | $0.09^{h}$ | 313 | $0.16^{k}$ | 123 | $1.25^{**}$ |
| 131 | $0.06^{b}$ | 121 | $0.035^{e\dagger}$ | 221 | $0.18^{i}$ | 321 | $0.20^{f*}$ | 212 | $1.25^{**}$ |
| 132 | $0.06^{b}$ | 122 | $0.06^{c}$ | 222 | $0.18^{i}$ | 322 | $0.20^{f*}$ | 213 | $1.25^{**}$ |
| 133 | $0.004^{c}$ | 123 | $0.20^{f*}$ | 223 | $0.18^{i}$ | 323 | $0.20^{f*}$ | 223 | $1.25^{**}$ |
| 231 | $0.06^{b}$ | 131 | $0.035^{e\dagger}$ | 231 | $0.22^{j\dagger}$ | 331 | $0.26^{g}$ | 312 | $0.43^{g}$ |
| 232 | $0.06^{b}$ | 132 | $0.33^{g}$ | 232 | $0.39^{g}$ | 332 | $0.26^{g}$ | 313 | $0.43^{g}$ |
| 233 | $0.06^{b}$ | 133 | $0.002^{c}$ | 233 | $0.39^{g}$ | 333 | $0.26^{g}$ | 323 | $0.43^{g}$ |

symmetries. This can possibly even be connected with the fermion-mass or flavour problem. This is an on-going field of research and it is too early to draw any conclusions. I just point out that gauged models with $R\!\!\!/_p$ have been constructed.[3,6]

In conclusion, from the theoretical understanding of unification, there is no clear preference between $R_p$ and $R\!\!\!/_p$. In light of the very distinct phenomenology which we discuss below, it is thus mandatory to experimentally search for both possibilities. $R_p$-conservation and $R\!\!\!/_p$ have the same minimal particle content. They also in principle have the same kind of symmetries, as we have just argued: $G_{SM}$ plus an additional symmetry to protect the proton. They should thus both be considered as different versions of the MSSM. We shall denote the $R_p$ conserving version of the MSSM as $R_p$-MSSM and the $R_p$-violating version as $R\!\!\!/_p$-MSSM.

## 17.5. Indirect Bounds

The $R\!\!\!/_p$ interactions can contribute to various (low-energy) processes through the virtual exchange of supersymmetric particles.[20] To date, all data are in good agreement with the SM. This leads directly to bounds on the $R\!\!\!/_p$ operators. When determining such limits one must make some simplifying assumptions due to the large number of operators. In the following, we shall assume that one $R\!\!\!/_p$ operator at a time is dominant while the others are negligible. We thus do not include the sometimes very strict bounds on products of operators, for example from $\mu \to e\gamma$.[21] This is an important assumption but not unreasonable. It holds for the SM for example, where the top quark Yukawa coupling is almost a factor 40 larger than the bottom Yukawa coupling. Since we do not know the origins of Yukawa couplings, we do not know whether this is a generic feature.

Before presenting the complete bounds, I shall discuss one example[22] to show how such bounds can be obtained. The operator $L_1 L_3 \bar{E}_k$ can contribute to the decay $\tau \to e\nu\bar{\nu}$ via the diagram in Figure 17.1(b). For large slepton masses, $\tilde{m}(\tilde{e}_R^k)$, this interaction is described by an effective 4-fermion Lagrangian (after Fierz re-ordering)[22]

$$\mathcal{L}_{eff} = \frac{|\lambda_{13k}|^2}{2\tilde{m}^2}(\bar{e}_L \gamma^\mu \nu_{eL})(\bar{\nu}_{\tau L}\gamma_\mu \tau_L). \tag{17.12}$$

This has the same structure as the term in the effective SM Lagrangian and thus leads to an apparent shift in the Fermi constant for tau decays. Considering the ratio $R_\tau \equiv \Gamma(\tau \to e\nu\bar{\nu})/\Gamma(\tau \to \mu\nu\bar{\nu})$, the contribution from $R\!\!\!/_p$ relative to the SM contribution is[22]

$$R_\tau = R_\tau(SM)\left[1 + 2\frac{M_W^2}{g^2}\left(\frac{|\lambda_{13k}|^2}{\tilde{m}^2(\tilde{e}_R^k)}\right)\right]. \tag{17.13}$$

Using the experimental value[23] $R_\tau/R_\tau(SM) = 1.0006 \pm 0.0103$ we obtain the bounds

$$|\lambda_{13k}| < 0.06\left(\frac{\tilde{m}(\tilde{e}_R^k)}{100\,GeV}\right), \quad k = 1, 2, 3, \tag{17.14}$$

which are given in Table 17.1. The strictest bounds on the remaining operators are also summarized in Table 17.1.

The bounds in Table 17.1 are obtained from the following physical processes: $^{(a)}$ charged current universality,[10,22] $^{(b)}$ $\Gamma(\tau \to e\nu\bar{\nu})/\Gamma(\tau \to \mu\nu\bar{\nu})$,[10,22] $^{(c)}$ bound on the mass of $\nu_e$,[5,24,25] $^{(d)}$ neutrinoless double-beta decay,[26,27] $^{(e)}$ atomic parity violation,[28–30] $^{(f)}$ $D^0 - \bar{D}^0$ mixing,[31–33] $^{(g)}$ $R_\ell = \Gamma_{had}(Z^0)/\Gamma_\ell(Z^0)$,[20,34] $^{(h)}$ $\Gamma(\pi \to e\bar{\nu})/\Gamma(\pi \to \mu\bar{\nu})$,[22] $^{(i)}$ $BR(D^+ \to \bar{K}^{0*}\mu^+\nu_\mu)/BR(D^+ \to \bar{K}^{0*}e^+\nu_\mu)$,[20,36] $^{(j)}$ $\nu_\mu$ deep-inelastic scattering,[22] $^{(k)}$ $BR(\tau \to \pi\nu_\tau)$,[20,36] $^{(\ell)}$ heavy nucleon decay,[11] and $^{(m)}$ $n-\bar{n}$ oscillations.[11,35]

The bounds are all given for $\tilde{m} = 100\,GeV$ and they become weaker with increasing $\tilde{m}$. They each depend on a specific scalar mass and have various functional dependences on this mass.

| | | | |
|---|---|---|---|
| $\lambda_{ijk}$ | $m_{\tilde{e}_{Rk}}/100\,GeV$ | $\lambda'_{111}$ | $(m_{\tilde{q}}/100\,GeV)^2(m_{\tilde{g}}/1\,TeV)^{1/2}$ |
| $\lambda'_{11k}, \lambda'_{21k}$ | $m_{\tilde{d}_{Rk}}/100\,GeV$ | $\lambda'_{1j1}$ | $m_{\tilde{q}_{Lj}}/100\,GeV$ |
| $\lambda_{133}, \lambda'_{1jj}$ | $\sqrt{m_{\tilde{\tau},\tilde{d}_j}/100\,GeV}$ | $\lambda'_{123}$ | $\sqrt{m_{\tilde{b}_R}/100\,GeV}$ |
| $\lambda'_{231}$ | $m_{\tilde{\nu}_{\tau L}}/100\,GeV$ | $\lambda'_{32k}$ | $\sqrt{m_{\tilde{d}_{Rk}}/100\,GeV}$ |

(17.15)

For $\lambda'_{111}$ the dependence can be on either $m_{\tilde{d}_{Rk}}$, or $m_{\tilde{u}_{Lk}}$. The bound in Table 17.1 is given for $\tilde{m}(\tilde{g}) = 1\,TeV$. For $\lambda'_{132}$, $\lambda'_{22k}$, $\lambda'_{23k}$, $\lambda'_{31k}$, and $\lambda'_{32k}$ one must consult the appropriate references since the dependence is only given numerically. The bounds on $\lambda''_{112,113}$ from heavy nucleon decay and $n - \bar{n}$ oscillations have very strong mass dependences.[11]

I have updated the previous bounds from charged current universality,[22] from lepton-universality[22] and from [c] $R_\ell$ using more recent data.[10,23] For the bound from the electron neutrino mass I have used the upper bound[25] $m_{\nu_e} < 5\,,eV$. The PDG number is $10 - 15\,eV$[10] and is very conservative.[37] The bound on $\lambda$ from $m_{\nu_e}$ scales with the square root of the upper bound on $m_{\nu_e}$. For the bound from atomic parity violation I have used the theory value:[30] $Q_W^{th} = -73.17 \pm 0.13$. The error includes the variations due to the unknown Higgs mass. I have also used the recent new experimental number.[29] For the bound from $D^0 - \bar{D}^0$ mixing, I have updated the bound from Ref. 33 to include a lattice calculation of[32] $B_D$ and a more updated value of $f_D$.[31] I have also included a 10% error to account for the quenched approximation. The bounds denoted by † are $2\sigma$ bounds, the other bounds are at the 1 sigma level. The bounds denoted by (**) are not direct experimental bounds. They are obtained[11,38] from the requirement that the $\not{R}_p$-coupling remains within the unitarity bound up to the grand unified scale of $10^{16}\,GeV$. This need not be the case.

The bounds denoted by * are based on a further *assumption* about the absolute mixing in the (SM) quark sector. As stated before, we do not know the physical origin of Yukawa couplings or superpotential terms. It is a reasonable (but not necessary) assumption that their structure is determined by some symmetry at an energy scale well above the electroweak scale, *e.g.* the GUT or the Planck scale. We would expect this symmetry to be in terms of the weak *current* eigenstates. Such a symmetry could then give us a single dominant operator, for example

$$L_1 Q_1 \bar{D}_1 = \lambda'_{111}(-\tilde{e}_L u_L \bar{d}_R + \tilde{\nu}_{eL} d_L \bar{d}_R \ldots). \qquad (17.16)$$

Below the electroweak scale the quarks become massive and we must rotate them to their *mass* eigenstate basis. (The squarks must separately also be rotated by a different rotation but that is not relevant to these bounds.) In (17.16) there are then separate rotations: $d_L \to \mathcal{D}_{1j} d'_{jL}$ and $u_L \to \mathcal{U}_{1j} u'_{jL}$ which generate extra $\not{R}_p$ terms suppressed by mixing angles.[39]

---

[c]I thank Gautam Bhattacharyya for providing me with updates on the bounds resulting from $R_\ell$.

For the quarks we do not know the *absolute* mixing of the down-quark sector, $\mathcal{D}_{ij}$, or of the up-quark sector $\mathcal{U}_{ij}$ and thus do not know by how much to rotate the up- and down-quark current eigenstates. The *relative* mixing of these two sectors is given by the CKM-matrix[10] of the SM. If we assume the relative rotation is solely due to an absolute mixing in the up-quark sector ($\mathcal{D}_{ij} = \mathbf{1}$) the best bounds are those given in Table 17.1. Those denoted by * are specifically based on this mixing assumption. If however the relative mixing is solely due to absolute mixing in the down-quark sector ($\mathcal{U}_{ij} = \mathbf{1}$) the $D^0$-$\bar{D}^0$ mixing bounds no longer apply. There are then significantly stricter bounds on many couplings from measurements of $K^+ \to \pi^+ \nu\nu$ decays[33]

$$\lambda'_{ijk} < 0.012,\ (90\%CL),\quad j \neq 3. \tag{17.17}$$

For Table 17.1 we have adopted the conservative estimate that the mixing is solely due to the up-quark sector since we do not know the absolute mixing. We therefore did not include the bounds (17.17).[d]

## 17.6. Changes to $R_p$-MSSM

On the Lagrangian level the only change to the $R_p$-MSSM is the inclusion of the operators in $W_{\not R_p}$ which give new lepton- and baryon number violating Yukawa couplings. There are several changes in the phenomenology of supersymmetry due to these couplings.[39]

(1) Lepton- or baryon number is violated as discussed in Sec. 17.5.
(2) The LSP is not stable and can decay in the detector. It is no longer a dark matter candidate.
(3) The neutralino is not necessarily the LSP.
(4) The single production of supersymmetric particles is possible.

**2.** If for example the neutralino is the LSP and the dominant $\not R_p$ operator is $L_1 Q_2 \bar{D}_1$ it can decay as shown in Figure 17.1(c). For LSP$=\tilde{\gamma}$ the decay rate is[40,41]

$$\Gamma_{\tilde{\gamma}} = \frac{3\alpha\lambda'^2_{121}}{128\pi^2}\frac{M^5_{\chi^0_1}}{\tilde{m}^4}. \tag{17.18}$$

[d]There is a possible loop-hole. The symmetry at the high energy scale could just produce such a combination of couplings that is rotated to one single dominant coupling at low energy. After all, it is possibly the same symmetry which produces the single dominant quark Yukawa coupling in the SM. However, I do not adopt this philosophy here.

The decay occurs in the detector if $c\gamma_L\tau(\tilde{\gamma}) \lesssim 1\,m$, or

$$\lambda'_{121} > 1.4\cdot 10^{-6}\sqrt{\gamma_L}\left(\frac{\tilde{m}}{200\,GeV}\right)^2\left(\frac{100\,GeV}{M_{\tilde{\gamma}}}\right)^{5/2}. \tag{17.19}$$

where $\gamma_L$ is the Lorentz boost factor. This is well below the bound of Table 17.1. Recall also for comparison, that in the SM Yukawa couplings can be very small: for the electron $h^e = 3\cdot 10^{-6}$. We have presented these numerical results for a photino for simplicity and clarity. The full analysis with a neutralino LSP has been performed in Refs. 42, 43. It involves several subtleties due to the $R_p$-MSSM parameter space which can have significant effects on the lifetime. Due to the LSP decay, supersymmetry with broken $R_p$ has no natural dark matter candidate.

**3.** In the $R_p$-MSSM the stable LSP must be charge and colour neutral for cosmological reasons (*cf.* Ref. 44). In the $\not{R}_p$-MSSM there is no preference for the nature of the unstable LSP. It can be any of the following[e]

$$LSP \;\;\epsilon\;\; \{\chi_1^0, \chi_1^{\pm}, \tilde{g}, \tilde{q}, \tilde{t}, \tilde{\ell}, \tilde{\nu}\}. \tag{17.20}$$

In each case the collider phenomenology can be quite distinct.

**4.** In the $\not{R}_p$-MSSM there are resonant and non-resonant single particle production mechanisms. The resonant production mechanisms are

$$e^+ + e^- \rightarrow \tilde{\nu}_{Lj}, \quad L_1L_j\bar{E}_1, \tag{17.21}$$
$$e^- + u_j \rightarrow \tilde{d}_{Rk}, \quad L_1Q_j\bar{D}_k, \tag{17.22}$$
$$e^- + \bar{d}_k \rightarrow \tilde{\bar{u}}_{Lj}, \quad L_1Q_j\bar{D}_k, \tag{17.23}$$
$$\bar{u}_j + d_k \rightarrow \tilde{e}^-_{Li}, \quad L_iQ_j\bar{D}_k, \tag{17.24}$$
$$d_j + \bar{d}_k \rightarrow \tilde{\bar{\nu}}_{Li}, \quad L_1Q_j\bar{D}_k, \tag{17.25}$$
$$\bar{u}_i + \bar{d}_j \rightarrow \tilde{d}_{Rk}, \quad \bar{U}_i\bar{D}_j\bar{D}_k, \tag{17.26}$$
$$d_j + d_k \rightarrow \tilde{\bar{u}}_{Ri}, \quad \bar{U}_i\bar{D}_j\bar{D}_k \tag{17.27}$$

These processes can be realized at $e^+e^-$-colliders, at HERA, and at hadron colliders, respectively. There are many further t-channel single sparticle production processes. For example at an $e^+e^-$-collider, we can have $e^+ + e^- \rightarrow \tilde{\chi}_1^0 + \nu_j$ via t-channel selectron exchange. The t-channel exchange of squarks (sleptons) can also contribute to $q\bar{q}$ ($\ell\tilde{\ell}$) pair production, leading to indirect bounds.[46]

[e]The stop is listed separately since it has a special theoretical motivation[45] and leads to quite distinct phenomenology given that the top quark is so heavy.

Table 17.2. Squark mass bounds from the Tevatron for various dominant $\not{R}_p$-operators.[45]

| | $L_{1,2}Q_{1,2}\bar{D}_k$, | $L_{1,2}L_3\bar{E}_3$ | $L_1L_2\bar{E}_3$ | $L_{1,2}L_3\bar{E}_{1,2}$ | $L_1L_2\bar{E}_{1,2}$ |
|---|---|---|---|---|---|
| $m_{\tilde{q}}$ | $100\,GeV$ | $100\,GeV$ | $140\,GeV$ | $160\,GeV$ | $175\,GeV$ |

## 17.7. Collider Phenomenology

The supersymmetric signals for $\not{R}_p$ will be a combination of supersymmetric production and decay to $R_p$ even final states. Supersymmetric particles can be produced in pairs via MSSM gauge couplings or singly as in (17.21)-(17.27). The former benefits from large couplings while being kinematically restricted to masses $< \sqrt{s}/2$. The latter case has double the kinematic reach but suffers from typically small Yukawa couplings. Combining the various production modes with the decays and the different dominant operators leads to a wide range of potential signals to search for. Instead of systematically listing them I shall focus on two examples. Throughout we shall assume a neutralino LSP.

### 17.7.1. *Squark pair production at the Tevatron*

Squark pair production at the Tevatron proceeds via the known gauge couplings of the $R_p$-MSSM

$$q\bar{q}, gg \to \tilde{q} + \bar{\tilde{q}}. \tag{17.28}$$

In $\not{R}_p$, once produced the squarks decay to an $R_p$ even final state. Let us consider a dominant $L_iL_j\bar{E}_k$ operator. The couplings $\lambda_{ijk}$ are bounded to be smaller than gauge couplings. Thus we expect the squarks to cascade decay to LSPs as in the MSSM. The LSPs in turn will then decay via the operator $L_iL_j\bar{E}_k$ to two charged leptons and a neutrino each (*cf.* Figure 17.1(c)). If each squark decays directly to the LSP (assuming it is the second lightest)

$$q'\bar{q}', gg \to \tilde{q} + \bar{\tilde{q}} \to q\bar{q} + \tilde{\chi}_1^0\tilde{\chi}_1^0 \to q\bar{q} + l^+l^-l^+l^-\nu\nu. \tag{17.29}$$

We therefore have a multi-lepton signal which is detectable.[47] To date it has not been searched for with $\not{R}_p$ in mind. However, before the top quark discovery there was a bound from CDF on a di-lepton production cross section. Making corresponding cuts and with some simple assumptions this can be translated into a bound on the rate of the process (17.29) and thus a lower bound on the squark mass.[47] The assumptions are: (i) $BR(\tilde{q} \to \tilde{\gamma}q) = 100\%$, $(m_{\tilde{q}} < m_{\tilde{g}})$, (ii) LSP$= \tilde{\gamma}$ with $M_{\tilde{\gamma}} = 30\,GeV$, (iii)

$\lambda, \lambda'$ satisfy the bound (17.19). For various dominant operators the bounds are given in Table 17.2. No attempt was made to consider final state $\tau$'s due to lack of data. These bounds are comparable to the $R_p$-MSSM squark mass bounds. Since, the theoretical analysis has been improved to allow for neutralino LSPs, more involved cascade decays and the operator $\bar{U}\bar{D}\bar{D}$.[39,48] However, to date no experimental analysis has been performed.

### 17.7.2. *Resonant squark production at HERA*

HERA offers the possibility to test the operators $L_1 Q_j \bar{D}_k$ via resonant squark production[49] [f]

$$e^+ + d_k \to \tilde{u}_j \to (e^+ + d_k,\ \tilde{\chi}_1^0 + u_j,\ \tilde{\chi}_1^+ + d_j), \tag{17.30}$$

$$e^+ + \bar{u}_j \to \bar{\tilde{d}}_k \to (e^+ + \bar{u}_j,\ \bar{\nu}_e + \bar{d}_j,\ \tilde{\chi}_1^0 + \bar{d}_k). \tag{17.31}$$

We have included what are most likely the dominant decay modes. The neutralino and chargino will decay as in Figure 17.1(c)

$$\tilde{\chi}_1^0 \to (e^\pm,\ \nu) + 2\,\text{jets}, \quad \tilde{\chi}_1^+ \to (e^+,\ \nu) + 2\,\text{jets}. \tag{17.32}$$

The neutralino can decay to the electron or positron since it is a Majorana fermion. We are thus left with several distinct decay topologies. *(i)* If the squark is the LSP it will decay to $e^+ + q$ or $\bar{\nu}_e + q$ ($\bar{\tilde{d}}_k$). The first looks just like neutral current DIS, except that for $x_{Bj} \approx \tilde{m}^2(\tilde{q})/s$ it results in a flat distribution in $y_e$ whereas NC-DIS gives a $1/y_e^2$ distribution. The latter looks just like CC-DIS. *(ii)* If the gauginos are lighter than the squark the gaugino decay will dominate[43] [g]. The clearest signal is a high $p_T$ electron which is essentially background free. The high $p_T$ positron or the missing $p_T$ of the neutrino can also be searched for.

All five signals have been searched for by the H1 collaboration[51] in the 1994 $e^+$ data ($\mathcal{L} = 2.83\,pb^{-1}$). The observations were in excellent agreement with the SM. The resulting bounds on the couplings are summarized in Table 17.3. After rescaling the bounds of Table 17.1 we see that the direct search is an improvement for $\lambda'_{121}$, $\lambda'_{131}$, and $\lambda'_{132}$. In the more recent data, an excess has been observed in high $Q^2$ NC-DIS.[52] If this persists it can possibly be interpreted as the resonant production of a squark via an $L_1 Q_j \bar{D}_k$ operator.[50,53]

---

[f] HERA has accumulated most of its data as a positron proton collider.

[g] The gaugino decays could be suppressed by phase space or by partial cancellations of the neutralino couplings.[42,50]

Table 17.3. Exclusion upper limits at 95% CL on $\lambda'_{1jk}$ for $\tilde{m}(\tilde{q}) = 150\,GeV$ and $\tilde{m}(\tilde{\chi}_1^0) = 40\,GeV$ for two different dominant admixtures of the neutralino.

| | $\lambda'_{111}$ | $\lambda'_{112}$ | $\lambda'_{113}$ | $\lambda'_{121}$ | $\lambda'_{122}$ | $\lambda'_{123}$ | $\lambda'_{131}$ | $\lambda'_{132}$ | $\lambda'_{133}$ |
|---|---|---|---|---|---|---|---|---|---|
| $\tilde{\gamma}$-like | 0.056 | 0.14 | 0.18 | 0.058 | 0.19 | 0.30 | 0.06 | 0.22 | 0.55 |
| $\tilde{Z}^0$-like | 0.048 | 0.12 | 0.15 | 0.048 | 0.16 | 0.26 | 0.05 | 0.19 | 0.48 |

## 17.8. Cosmology

### 17.8.1. *Bounds from GUT-scale baryogenesis*

There is a very strict bound on all $R\!\!\!/_p$ Yukawa couplings assuming the presently observed matter-asymmetry was created above the electroweak scale, *e.g.* at the GUT-scale[54] which I briefly recount. Assume that at the GUT scale a baryon- and lepton-asymmetry was created with possibly both $B + L \neq 0$ and $B - L \neq 0$. The electroweak sector of the SM (and MSSM) has baryon-number and lepton-number violating "sphaleron" interactions which conserve $B - L$ but violate $B + L$. These are in equilibrium above the electroweak phase transition and they thus erase the $B + L \neq 0$ component of the matter asymmetry.

Consider now adding one additional $R\!\!\!/_p$ operator, *e.g.* $\bar{U}\bar{D}\bar{D}$ which violates baryon number. If it is in thermal equilibrium during an epoch after the GUT epoch and together with the sphaleron-interactions then together they will erase the entire matter asymmetry. In order to avoid this scenario the $R\!\!\!/_p$ interactions should not be in thermal equilibrium above the electroweak scale resulting in the bounds[54–56]

$$\lambda, \lambda', \lambda'' < 5 \cdot 10^{-7} \left( \frac{\tilde{m}}{1\,TeV} \right)^{1/2} . \tag{17.33}$$

It should be clear that the argument holds for $LL\bar{E}$ or $LQ\bar{D}$ operators as well if lepton flavour is universal.[54] This is an extremely strict bound on *all* the couplings. If it is valid then $R\!\!\!/_p$ is irrelevant for collider physics and can only have cosmological effects.

There are two important loop-holes in this argument. The first and most obvious one is that the matter genesis occurred at the electroweak scale or below.[57] The second loop-hole has to do with the inclusion of all the symmetries and conserved quantum numbers.[56,58,59] The electroweak sphaleron interactions do not just conserve $B - L$. They conserve the three quantum numbers $B/3 - L_i$, one for each lepton flavour. These can also be written as $B - L$ and two independent combinations of $L_i - L_j$. First, again consider an additional $\bar{U}\bar{D}\bar{D}$ operator. If the matter genesis at the GUT scale is asymmetric in the lepton flavours, $(L_i - L_j)|_{M_{GUT}} \neq 0$, then this lepton-asymmetry is untouched by the sphalerons and by the $\Delta B \neq 0$ operators $\bar{U}\bar{D}\bar{D}$ operators. The baryon asymmetry is however erased. Below the electroweak scale, the lepton-asymmetry is partially converted into a baryon-asymmetry via (SM) leptonic *and* supersymmetric mass effects.[56] If now instead, we add a lepton-number violating operator, we will retain a matter asymmetry as long as one lepton flavour remains conserved. In order for $L_\tau$ for example to be conserved, all $L_\tau$ violating operators must remain out of thermal equilibrium above the elctroweak scale, *i.e.* satisfy the bound (17.33). From the low-energy point of view this is completely consistent with our Ansatz of considering only one large dominant coupling at a time. Thus in these simple scenarios the bounds (17.33) are evaded.

### 17.8.2. *Long-lived LSP*

One can consider three distinct ranges for the lifetime of the LSP

$$(i)\ \tau_{LSP} \lesssim 10^{-8}s, \quad (ii)\ 10^{-8}s \lesssim \tau_{LSP} \lesssim 10^{7}\tau_u, \quad (iii)\ \tau_{LSP} > 10^{7}\tau_u, \tag{17.34}$$

where $\tau_u \sim 10^{10}yr$ is the present age of the universe. We have discussed the first case in detail in the previous chapters. The third case is indistinguishable from the $R_p$-MSSM with the LSP being a good dark matter candidate. In the second case, the LSP can provide a long-lived relic whose decays can potentially lead to observable effects in the universe. There are bounds excluding any such relic with lifetimes[60]

$$1s < \tau_{LSP} < 10^{17}yr. \tag{17.35}$$

The lower end of the excluded region is due to the effects of hadron showers from LSP decays on the primordial abundances of light nuclei.[61] The upper bound is from searches for upward going muons in underground detectors which can result from $\nu_\mu$'s in LSP decays.[62] Note that even if $\tau_{LSP} > \tau_u$ the relic abundance is so large [h] that the decay of only a small fraction can lead to observable effects.

The above restrictions on decay lifetimes can be immediately applied to the case of $\not{R}_p$-MSSM. If we include LSP decays in collider experiments we are left with a gap of eight orders of magnitude in lifetimes $10^{-8}s < \tau_{LSP} < 1s$ where no observational tests are presently known. It is very important to

[h]This is for most values of the MSSM parameters, *cf.* Ref. 44.

find physical effects which could help to close this gap. Since the lifetime depends on the square of the $\not{R}_p$ Yukawa coupling this corresponds to a gap of four orders of magnitude in the coupling. For a photino LSP we can translate the above bounds into bounds on the $\not{R}_p$ Yukawa couplings.[55] Using Eq.(17.18) we obtain the excluded region for the couplings

$$10^{-22} < (\lambda, \lambda', \lambda'') \cdot \left(\frac{200\, GeV}{\tilde{m}}\right)^2 \left(\frac{M_{\tilde{\gamma}}}{100\, GeV}\right)^{5/2} < 10^{-10}. \qquad (17.36)$$

Note that the lower range of these bounds extends well beyond the already strict bound from proton decay (17.8). For a generic neutralino LSP the lifetime depends strongly on the MSSM parameters[42,43] and the bounds can only be transferred with caution.

## 17.9. Outlook

Once we include the $\not{R}_p$ terms in the superpotential we are left with a bewildering set of possibilities. We have 45 new Yukawa couplings of which any could be dominant and we have a set of seven different potential LSPs, each possibly leading to quite different phenomenology. This situation requires a systematic approach.

I would here like to suggest a two-fold approach. The theoretically best motivated model is one based on universal soft breaking terms at the unification scale $\sim 10^{16}\, GeV$, completely analogous to the MSSM. To obtain the low-energy spectrum one then employs the renormalisation group equations *including* the $\not{R}_p$-Yukawa couplings and all the soft breaking terms. This program has yet to be completed.[21,63] However, since most of the $\not{R}_p$-Yukawa couplings are bounded to be relatively small we expect for large regions in parameter space the spectrum of the $\not{R}_p$-MSSM to look just like that of the $R_p$-MSSM . The only difference will be a decaying neutralino LSP. To this extent the program has been implemented in SUSYGEN,[64] the supersymmetry Monte Carlo generator for $e^+e^-$-colliders. $\not{R}_p$ has only been implemented partially in ISAJET, a generator for hadron colliders.[65]

As a second step, I suggest a systematic listing of potential signal topologies which can arise for spectra not obtained in the simple unification approach. Any exotic topologies can easily be searched for on a qualitative level. These two approaches combined should ensure that we do not miss any signal for supersymmetry and also do not end up searching vigorously in the $\not{R}_p$-hat every time an experimental anomaly appears.

## Acknowledgments

I would like to thank the many people I have collaborated with on R-parity violation: Ben Allanach, Jon Butterworth, A. Chamseddine, Manoranjan Guchait, Smaragda Lola, Peter Morawitz, Felicitas Pauss, Emmanuelle Perez, Roger Phillips, Heath Pois, D.P. Roy and Yves Sirois. In particular I would like to thank Graham Ross who got me started on the subject and continuously encouraged me. I would furthermore like to thank Sacha Davidson and Gautam Bhattacharyya for discussions on updating the indirect bounds. I would like to thank Subir Sarkar for very helpful discussions on the long-lived LSP. I thank Ben Allanach, Gautam Bhattacharyya, Sacha Davidson, Gian Giudice, and Subir Sarkar for reading the manuscript and their helpful comments.

## References

1. G. Farrar, P. Fayet, Phys. Lett. B 76 (1978) 575.
2. D. Z. Freedman, Phys. Rev. D 15 (1977) 1173; S. Ferrara, L. Girardello, T. Kugo, A. van Proeyen, Nucl. Phys. B 223 (1983) 191.
3. A.H. Chamseddine, H. Dreiner, Nucl. Phys. B 458 (1996) 65, hep-ph/9504337.
4. S. Weinberg Phys. Rev. D 26 (1982) 287, N. Sakai, T. Yanagida, Nucl. Phys. B 197 (1982) 133.
5. L.J. Hall, M. Suzuki, Nucl. Phys. B 231 (1984) 419.
6. T. Banks, Y. Grossman, E. Nardi, Y. Nir, Phys. Rev. D 52 (1995) 5319, hep-ph/9505248.
7. B. de Carlos, P.L. White, Phys. Rev. D 54 (1996) 3427, hep-ph/9602381; E. Nardi, Phys. Rev. D 55 (1997) 5772, hep-ph/9610540.
8. R. Hempfling, Nucl. Phys. B 478 (1996) 3, hep-ph/9511288; H.-P. Nilles, N. Polonsky, Nucl. Phys. B 484 (1997) 33, hep-ph/9606388.
9. S. Davidson, J. Ellis, Phys. Lett. B 390 (1997) 210, hep-ph/9609451; and CERN-TH-97-14, hep-ph/9702247.
10. Particle Data Group, *Phys. Rev.* D **54**, 1 (1996).
11. J.L. Goity, Marc Sher, Phys. Lett. B 346 (1995) 69, erratum-ibid. B 385 (1996) 500, hep-ph/9412208.
12. A. Yu. Smirnov, F. Vissani, Phys. Lett. B 380 (1996) 317, hep-ph/9601387.
13. D. Brahm, L. Hall, Phys. Rev. D40 (1989) 2449; K. Tamvakis, Phys. Lett. B 382 (1996) 251, hep-ph/9604343.
14. G.F. Giudice, R. Rattazzi, CERN-TH-97-076, hep-ph/9704339.
15. R. Barbieri, A. Strumia, Z. Berezhiani, hep-ph/9704275; K. Tamvakis, Phys. Lett. B 383 (1996) 307, hep-ph/9602389; R. Hempfling, Nucl. Phys. B 478 (1996) 3, hep-ph/9511288, A. Yu. Smirnov, F. Vissani, Nucl. Phys. B460 (1996) 37, hep-ph/9506416.

16. M.C. Bento, L. Hall, G.G. Ross, Nucl. Phys. B 292 (1987) 400; N. Ganoulis, G. Lazarides, Q. Shafi, Nucl. Phys. B 323 (1989) 374.
17. L. M. Krauss, F. Wilczek, Phys. Rev. Lett. 62 (1989) 1221, T. Banks Nucl. Phys. B 323 (1989) 90.
18. L.E. Ibanez, G.G. Ross, Phys. Lett. B 260 (1991) 291; Nucl. Phys. B 368 (1992) 3.
19. T. Banks, M. Dine, Phys. Rev. D 45 (1992) 1424.
20. G. Bhattacharyya, Nucl. Phys. Proc. Suppl. 52A (1997) 83, hep-ph/9608415.
21. B. de Carlos, P.L. White, Phys. Rev. D 54 (1996) 3427, hep-ph/9602381.
22. V. Barger, G.F. Giudice, T. Han, Phys. Rev. D 40 (1989) 2987.
23. ALEPH Collaboration (D. Buskulic et al.). Z. Phys. C 70 (1996) 561.
24. R. M. Godbole, P. Roy, X. Tata, Nucl. Phys. B 401 (1993) 67, hep-ph/9209251.
25. A.I. Belesev et al, Phys. Lett. B 350 (1995) 263; C. Weinheimer, et al, Phys. Lett. B 300 (1993) 210 and update in http:// www.na.infn.it/ win97/second.html.
26. R. Mohapatra, Phys. Rev. D 34 (1986) 3457; J.D. Vergados Phys. Lett. B 184 (1987) 55; M. Hirsch, H.V. Klapdor-Kleingrothaus, S.G. Kovalenko, Phys. Lett. B 352 (1995) 1.
27. M. Hirsch, H.V. Klapdor-Kleingrothaus, S.G. Kovalenko, Phys. Rev. Lett. 75 (1995) 17; Phys. Rev. D 53 (1996) 1329, hep-ph/9502385.
28. S. Davidson, D. Bailey, B. A. Campbell, Z. Phys. C 61 (1994) 613, hep-ph/9309310.
29. C.S. Wood *et al.* , Science 275 (1997) 1759.
30. W. J. Marciano, J. L. Rosner, Phys. Rev. Lett. 65 (1990) 2963, ERRATUM-ibid. 68 (1992) 898; see also an article by W.J. Marciano in "Precision Tests of the Standard Electroweak Model", ed. P. Langacker, World Scientific, 1995. The number I present is an update of the last number by W. J. Marciano, talk given in the 1997 INT Summer Workshop.
31. I have updated the bound from Ref. 33 using better lattice data: H. Wittig, Oxford preprint, OUTP-97-20P, to be published in Int. J. Mod. Phys. A; hep-lat/9705034.
32. R. Gupta, T. Bhattacharya, S. Sharpe, Phys. Rev. D 55 (1996) 4036, hep-lat/9611023.
33. K. Agashe, M. Graeser, Phys. Rev. D 54 (1996) 4445.
34. G. Bhattacharyya, J. Ellis, K. Sridhar, Mod. Phys. Lett. A10 (1995) 1583, hep-ph/9503264; G. Bhattacharyya, D. Choudhury, K. Sridhar, Phys. Lett. B 355 (1995) 193, hep-ph/9504314; J. Ellis, S. Lola, K. Sridhar, e-Print Archive: hep-ph/9705416.
35. F. Zwirner, Phys. Lett. B 132 (1983) 103.
36. G. Bhattacharyya, D. Choudhury, Mod. Phys. Lett. A10 (1995) 1699.
37. V.M. Lobashev, talk at the "XVI edition of the International Workshop on Weak Interactions and Neutrinos," Capri, Italy, June 1997. The transparencies can be found at http://www.na.infn.it/win97/second.html. See in particular page 20.

38. B. Brahmachari, Probir Roy, Phys. Rev. D 50 (1994) 39, erratum-ibid. D 51 (1995) 3974, hep-ph/9403350.
39. H. Dreiner, G.G. Ross, Nucl. Phys. B 365 (1991) 597.
40. S. Dawson, Nucl. Phys. B 261 (1985) 297.
41. The more general formula for a neutralino LSP can be found in Ref. 42.
42. H. Dreiner, P. Morawitz, Nucl. Phys. B 428 (1994) 31, hep-ph/9405253.
43. E. Perez, Y. Sirois, H. Dreiner, Published in "Workshop on Future Physics at HERA", DESY, May 1996, hep-ph/9703444.
44. See the Chapter by J. Wells.
45. J. Ellis, S. Rudaz, Phys. Lett. B 128 (1983) 248.
46. S. Komamiya, CERN seminar, Feb. 25, 1997, OPAL, internal report PN280.
47. D.P. Roy, Phys. Lett. B 283 (1992) 270.
48. S. Dimopoulos, R. Esmailzadeh, L. Hall, G. Starkman, Phys. Rev. D 41 (1990) 2099, V. Barger, M.S. Berger, P. Ohmann, Phys. Rev. D 50 (1994) 4299, H. Baer, C. Kao, X. Tata, Phys. Rev. D 51 (1995) 2180, hep-ph/9410283, M. Guchait, D.P. Roy Phys. Rev. D 54 (1996) 3276, hep-ph/9603219.
49. J. Hewett, Proceedings of the 1990 Study on High Energy Physics, Snowmass; J. Butterworth, H. Dreiner, proceedings of the 2nd HERA Workshop, DESY, March - Oct. 1991; J. Butterworth, H. Dreiner, Nucl. Phys. B 397 (1993) 3; T. Kon and T. Kobayashi, Phys. Lett. B 270 (1991) 81.
50. G. Altarelli, J. Ellis, G.F. Giudice, S. Lola, M. L. Mangano, CERN-TH-97-040, hep-ph/9703276.
51. H1 Collaboration (S. Aid et al.), Z. Phys. C 71 (1996) 211, hep-ex/9604006.
52. H1 Collaboration (C. Adloff et al.), DESY-97-024, hep-ex/9702012; ZEUS Collaboration (J. Breitweg et al.), DESY-97-025, hep-ex/9702015.
53. D. Choudhury, S. Raychaudhuri, CERN-TH-97-026, hep-ph/9702392; H. Dreiner, P. Morawitz, hep-ph/9703279; J. Kalinowski, R. Ruckl, H. Spiesberger, P.M. Zerwas, DESY-97-038, hep-ph/9703288.
54. B. A. Campbell, S. Davidson, J. Ellis, K. A. Olive, Phys. Lett. B 256 (1991) 457; W. Fischler, G.F. Giudice, R.G. Leigh, S. Paban, Phys. Lett. B 258 (1991) 45.
55. B. A. Campbell, S. Davidson, J. Ellis, K. A. Olive, Astropart. Phys. 1 (1992) 77.
56. H. Dreiner, published in Marseille EPS HEP (1993) 424, (QCD 161: I48: 1993), hep-ph/9311286; H. Dreiner, G.G. Ross, *Nucl. Phys.* B **410**, 183 (1993), hep-ph/9207221.
57. For a recent review of baryogenesis at the electroweak scale within supersymmetry see: M. Carena, C.E.M. Wagner, to appear in 'Perspectives on Higgs Physics II', ed. G.L. Kane, World Scientific, Singapore, hep-ph/9704347.
58. For a related argument on majorana neutrino masses see A. E. Nelson, S. M. Barr, Phys. Lett. B 246 (1990) 141.
59. B. A. Campbell, S. Davidson, J. Ellis, K. A. Olive Phys. Lett. B 297 (1992) 118, hep-ph/9302221.
60. John Ellis, G.B. Gelmini, Jorge L. Lopez, D.V. Nanopoulos, S. Sarkar, Nucl. Phys. B 373 (1992) 399.

61. M. H. Reno, D. Seckel, Phys. Rev. D 37 (1988) 3441.
62. P. Gondolo, G.B. Gelmini, S. Sarkar, Nucl. Phys. B 392 (1993) 111.
63. H. Dreiner, H. Pois, ETH-TH-95-30, hep-ph/9511444; V. Barger, M.S. Berger, R.J.N. Phillips, T. Wohrmann, Phys. Rev. D 53 (1996) 6407, hep-ph/9511473
64. S.Katsanevas, P.Morawitz, "SUSYGEN 2.0 - A Monte Carlo Event Generator for MSSM Sparticle Production at e+ e- Colliders", http:// lyohp5.in2p3.fr/ delphi/katsan/susygen.html.
65. H. Baer, F. Paige, S. Protopopescu, X. Tata in proceedings of the Workshop on Physics at Current Accelerators and Supercolliders, Eds. J. Hewett, A. White and D. Zeppenfeld, hep-ph/9305342. Single LSP decay modes can be implemented via the FORCE command. If there are several LSP decays the ISAJET decay table must be modified, *e.g.* H. Baer, M. Brhlik, Chih-hao Chen, X. Tata, Phys. Rev. D 55 (1997) 4463, hep-ph/9610358.

www.ingramcontent.com/pod-product-compliance
Lightning Source LLC
Chambersburg PA
CBHW070712220326
41598CB00026B/3696

*9789814307499*